Chesapeake Sailing Craft

RECOLLECTIONS OF ROBERT H. BURGESS

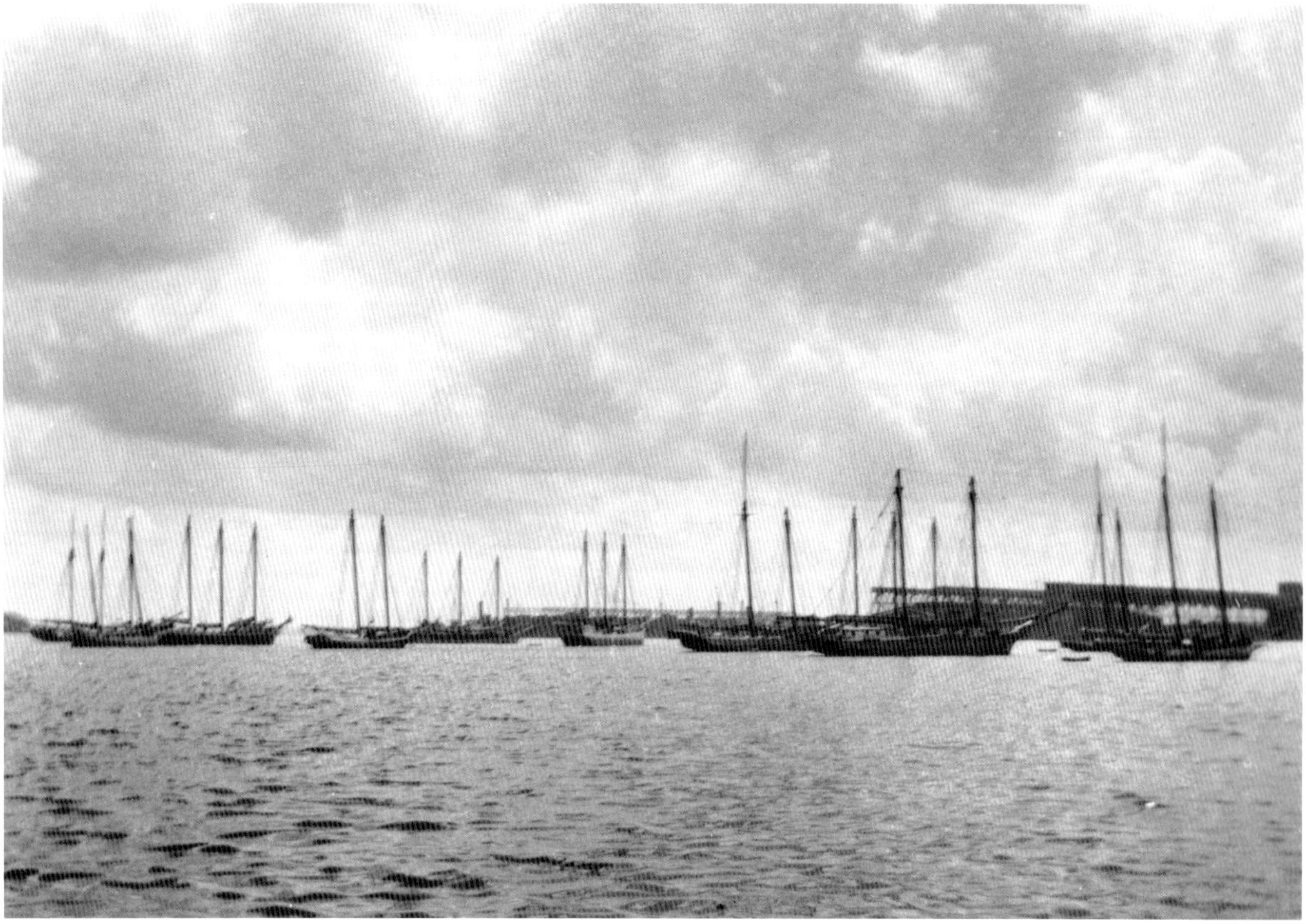

Canton Hollow, Baltimore harbor, was a favorite anchorage for Chesapeake sailing craft arriving from down the Bay or, after discharging their cargoes, preparing for a return trip down the Bay. Old-timers talked of the days, around the turn of the century, when sailing craft were so thick in this anchorage that one could virtually walk across the harbor on their decks. When this photograph was made, on September 3, 1933, the vessels had dwindled in number so that these eleven—four rams, five two-masted schooners, one three-masted schooner, and a bugeye—were a good representation of types of sail on these waters at that time. Of course, there were others at local piers, in other Bay ports in Maryland and Virginia, and sailing on the Chesapeake. But never again did the author see as many sailing craft in this anchorage at one time. Among the vessels in this scene are the rams *Edward R. Baird, Jr.* and *Mabel & Ruth,* schooners *Lucy May, Effie A. Chase,* and *Josephine Wimsatt,* and bugeye *Russell De Hughlett.* In the background are the Locust Point piers of the Baltimore and Ohio Railroad.

Chesapeake Sailing Craft

RECOLLECTIONS OF ROBERT H. BURGESS

Robert H. Burgess

Edited by William A. Fox

EXPANDED EDITION

Tidewater Publishers
Centreville, Maryland

Library of Congress Cataloging in Publication Data

Burgess, Robert H.
Chesapeake sailing craft: recollections of Robert H. Burgess / Robert H. Burgess; edited by William A. Fox.—Expanded ed.
p. cm.
Includes index.
ISBN-13: 978-0-87033-572-3
1. Boatbuilding—Chesapeake Bay (Md. and Va.) 2. Shipbuilding—Chesapeake Bay (Md. and Va.) 3. Sailing ships—Chesapeake Bay (Md. and Va.)—History. 4. Chesapeake Bay (Md. and Va.)—Navigation—History. 5. Shipping—Chesapeake Bay (Md. and Va.)—History. 6. Burgess, Robert H.—Travel. I. Fox, William A., 1943- II. Title.
VM24.M37B87 2005
623.822'609163470904—dc22
2005025547

Manufactured in China

To my father, William Kirk Burgess, who sailed on the Bay craft (several illustrated here) in their various trades on the Chesapeake and down to the Bahamas before going into steam. And to all those sailing men, shipyard operators, chandlers, ship wreckers, owners, brokers, and lighthouse keepers who tolerated my prying camera and inquisitive conversations during the past half century.

Robert H. Burgess

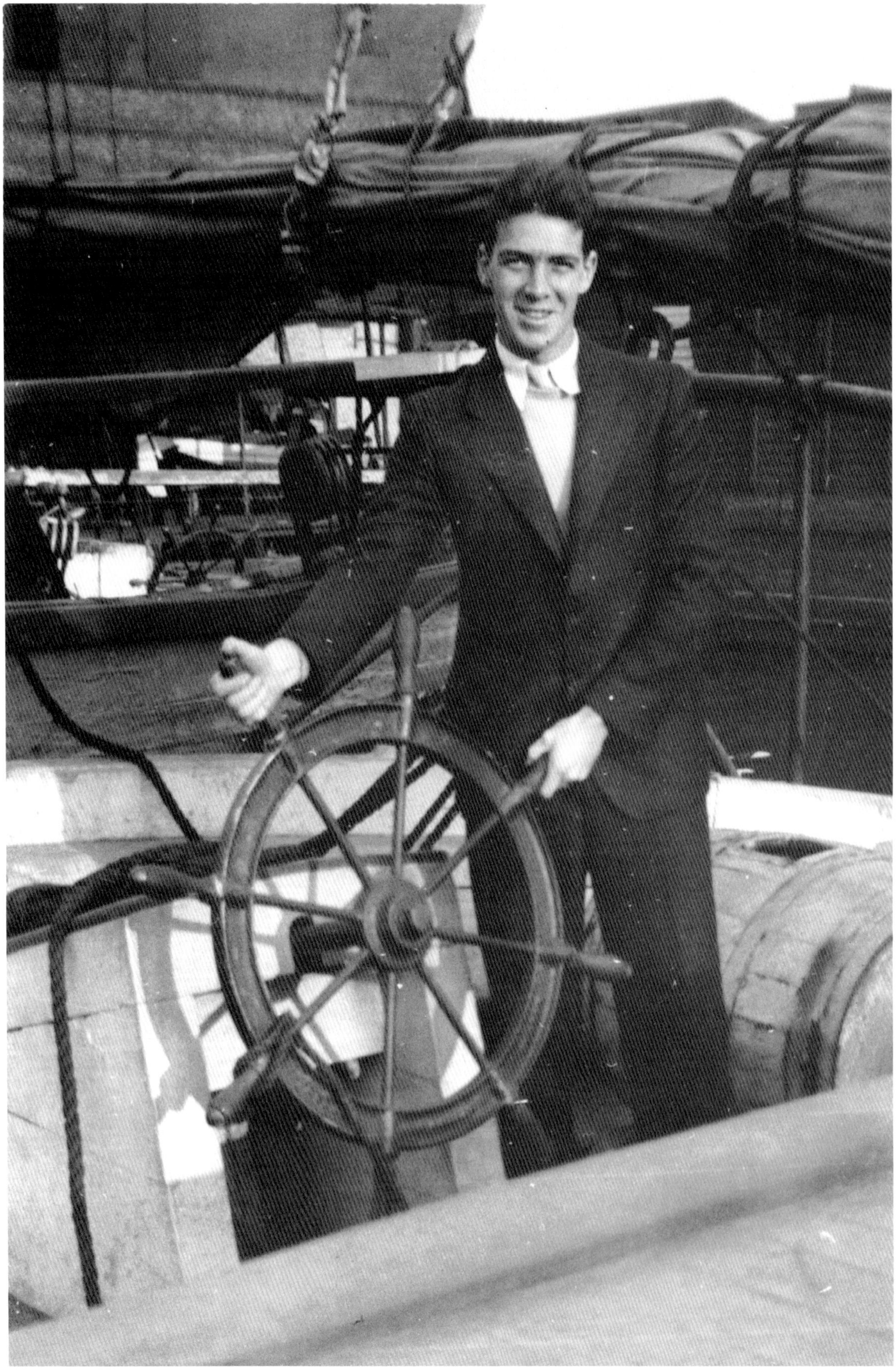

✪ Self portrait of a young Robert H. Burgess, aboard *Minnie May Kirwan* at Baltimore on November 30, 1933. He spent innumerable hours on the waterfront, visiting and photographing sailing vessels. The *Kirwan* may have been engaged in the lumber trade on this date, or she may have transitioned to oyster dredging for the season.

Contents

Foreword

TO THE EXPANDED EDITION

Thirty years have passed since the publication in 1975 of Robert H. Burgess's classic *Chesapeake Sailing Craft*. Mr. Burgess, like most of the people and vessels that he so enthusiastically documented in his photographs and writings, has passed on.

The original edition of this book has been out of print for many years. It was labeled Part I because the author had intended to write a second volume. Over the next 20 years or so, and with regular encouragement from the editors at Cornell Maritime Press, Burgess began to select photos for Part II. But as time passed and age began to take its toll, he was unable to summon the energy to complete the project.

In late 2002 his son R. Bruce Burgess and Bruce's wife Mary Alice Rath organized the photos that Robert Burgess had selected for Part II. Then I began to interview him to draw out his recollections of the vessels, people, and places depicted in the photos. Alas, we had barely started this work when he became ill early in 2003 and died on March 7 of that year.

At that time, Bruce, Mary Alice, Mr. and Mrs. Burgess's daughter Janet Burgess Loyd, and I decided to continue working on the book and to publish it as a memorial to Robert Herrmann Burgess and to his loving and patient wife, Adele Plitt Burgess, who had died in 1998. We had been working with about 150 photos that he had selected for Part II when, during an inventory of his estate, we uncovered more than 250 additional photos with which he had been working. Both groups of new photos depict many of the same vessels in the original edition, as well as many more vessels that were not. So the decision was made to complete the new edition using the original text, photos, and captions and to add new photos and captions only of those subjects that were already in the book.

The result is *Chesapeake Sailing Craft: Recollections of Robert H. Burgess,* a new and expanded edition of the original volume for Bay enthusiasts to enjoy. The original text remains unchanged from the 1975 edition. The reader is cautioned that when the author says, "today" or "40 years ago," he is writing in 1975. This foreword and captions for the new photos were written 30 years later. To avoid confusion, the 152 new photos and captions are identified by the symbol "✪". As in the original edition, all the photos in this book were taken by Robert Burgess. They appear as he saw them through the viewfinder of his camera and as he printed them in the darkroom, uncropped and unretouched.

Robert H. Burgess was my lifelong friend and mentor. Editing his book has enabled me to honor and thank him for his advice and encouragement. I wish to thank Bruce and Mary Alice, and Janet and her son Mark S. Hoffman for their work and help in completing this project. Thanks also to Peter Lesher, curator at the Chesapeake Bay Maritime Museum, for reading and commenting on the new captions and photos. My friend and co-worker Connie Jarvis helped considerably with the typing and editing. Thanks to Charlotte Kurst, former managing editor at Cornell Maritime Press, for her agreement to go forward with this project after a 30-year wait, and for her encouragement. We also want to thank Michelle Slavin, the current managing editor at Cornell, for her extreme patience and willingness to work with us, and Sandy Gallagher, associate editor, for her work during the final editing process. Finally, I would like to thank my wife, Mary, for reading, commenting on, and supporting me with this book over the past three years. With fond appreciation and enduring admiration, we dedicate this new edition of *Chesapeake Sailing Craft* to the memory of Robert and Adele Burgess.

William A. Fox
Williamsburg, Virginia
August 2005

Foreword

The 1925 issue of *Merchant Vessels of the United States* listed at least 720 sailing vessels registered out of the Chesapeake Bay ports of Baltimore, Annapolis, Crisfield, Cambridge, Washington, D.C., Alexandria, Reedville, Cape Charles, Newport News, and Norfolk. These were the places where the permanent documents of the vessels were issued and some were owned. Others of the craft would have been owned in the vicinity of their registered home port, but seldom if ever visited there. Some vessels that were employed commercially were too small to be registered in *Merchant Vessels of the United States* (hereafter referred to as *MVUS*): these were sloops, schooners with two to four masts, pungies, bugeyes, rams, and skipjacks. Most were the smaller vessels confining their activities to the Chesapeake Bay and its tributaries and nearby waters. The larger ones ranged coastwise and offshore.

The majority of these pure wooden sailing craft had been built in the latter 1800s, but in 1925 they were still earning a living freighting or in the fisheries. One can easily understand that in that era it was impossible to make a trip on the Bay without seeing craft under sail or encountering them in the harbors of the larger ports or bayside towns.

More commercial sailing craft plied Chesapeake waters than any other United States waterway at that time; the Bay was a natural, relatively sheltered body of water dividing the states of Maryland and Virginia. The major rivers of those states created peninsulas that further separated the people from the markets upon which they depended for supplies and outlets for their farm and seafood products. At that time, the Bay and its principal tributaries had not been bridged. Large ferries had been transporting motor vehicles for only a few years and many roads in the tidewater country were still of dirt or crushed shell, making truck, bus, or motor car transportation rather undesirable. Steamboats still operated and carried passengers and freight; sailing vessels hauled bulk cargo like lumber, fish scrap, oysters and shells, coal, fertilizer, grain, canned goods, and road-building materials. By way of the numerous rivers and creeks they could penetrate the rich farmland and forests to pick up and deliver their goods, none of which required fast transport. Then, as today, the state of Maryland permitted only sailing craft to dredge oysters from its waters; as a result the smaller vessels of the sailing fleet had their work cut out for them—in the winter months, at least. In warm-weather months many of those same craft turned to the freighting of grain, hay, baskets of tomatoes, fertilizer, watermelons, and other local commodities.

In the late 1920s and early '30s, age began to take its toll markedly among the Chesapeake's sailing vessels. Still in service at that time were schooners dating from the 1840s and '50s and pungies built in the 1860s. No doubt they had been rebuilt, but some eventually became hogged and structurally unsound. Repairs proved to be too costly as the volume of freight fell off and their hulls were shifted to remote creeks and marshes, pulled close to shore, and left to die. In many instances, after all other evidence of a vessel's existence had disappeared, there still remained the centerboard trunk to mark its grave. That part of a wooden vessel seemed to outlast all others.

But as commercial sail rapidly vanished from other American waters, on the Chesapeake it endured in a comparatively healthy manner well into the 1930s, in trades other than oystering. Lumber was the chief cargo of the schooners and rams; even several of the last surviving bugeyes and pungies could be seen around the Baltimore waterfront discharging lumber during that decade. Two- and three-masted schooners and rams picked up their cargoes at Virginia and North Carolina ports, while four-masted schooners brought in lumber from Georgetown, South Carolina, logwood from the West Indies, sheep and goat manure from Venezuela, and phosphate from Tampa.

Some of the two- and three-masted schooners, rams, pungies, and smaller craft were owned by their masters or by businessmen in Baltimore or the tidewater areas. Other larger schooners and rams were owned and/or managed by C.C. Paul & Company, Maryland Navigation Company, Charles M. Struven, W.B. Vane, Tom Webster, R.B. White, R. Duke Wathen, Mary Lee Carter, and Johnson & Wimsatt (the last two were from Washington, D.C.). Due to good management in providing cargoes, these firms and individuals kept their vessels active. During the 1930s Depression, the demand for lumber decreased and charters became scarce. Then World War II gave a spurt to the Bay's sailing vessel trade as lumber was needed for new construction. That conflict also resulted in some of the craft being sold away from the Bay to new owners in Florida and Caribbean ports—and they never came back. After the war only several schooners and rams remained in the Chesapeake trade, and a few bugeyes and skipjacks still found employment in Maryland's oyster fishery. Eventually, all but the skipjacks disappeared.

So this last 50-year period on Chesapeake Bay saw the procession of varied types of sailing craft that had been pursuing their trades just as thousands of wind-driven vessels had been doing on those waters since the tidewater region was first settled. The last cargo carried under sail on the Chesapeake in a large vessel was probably in the ram *Jennie D. Bell,* in 1954. At that time she was freighting fertilizer materials from Baltimore to Salisbury and soybeans from Kinsale, Virginia, to Baltimore. When she lowered her sails for the last time, only the oyster dredgers were left to uphold the tradition of sail on these waters. How much longer they will be able to remain active depends upon the legislators and scientists. Some want to change the system of oystering in Maryland, that is, permit the leasing of oyster grounds to large cooperatives that would plant and harvest their own oysters. That would be the end of dredging under sail, and those skipjacks not suitable for conversion into pleasure craft will be beached and abandoned.

Robert H. Burgess

Acknowledgments

Those I would most like to thank for having made this book a reality are unnamed and have passed on. They are the men of the Chesapeake's commercial sailing fleet, from captain to deckhand, not introduced to me by name but with whom I talked onboard their vessels as I took these photographs. Some became good friends, like Captain George H. Hopkins of the four-masted schooner *Doris Hamlin*. Others were cooperative but distant, as one might expect, when a stranger came aboard to photograph a battered old schooner or ram. I could never fully explain to them why I was doing it so that they would understand. They had been associated with the vessels all their lives and probably believed the ships would last forever; but I felt theirs was a way of life that was passing on. In some instances the vessels did outlast them.

Shipyards have been a good source of photographs, and especially so was that of A. Smith & Sons, Curtis Bay, Baltimore, where the pungies, bugeyes, schooners, and rams seemed to congregate. There the brothers Willis and Buck Smith, doubtful of my mission but never restrictive, saw me prowl their yard with a camera through the 1930s. There was real atmosphere in that yard during the period when sail was still prominent on the Bay. I thank them for the liberties granted me.

The Redman and Vane Shipyard on Key Highway, Baltimore, was another scene of Chesapeake sail activity until World War II, when the yard was swallowed by larger firms. The Krentzes, father and son, who operate the shipyard at Harryhogan, Virginia, I got to know in the 1950s when the Maryland skipjacks called there for repairs, and they were helpful in my efforts. During that period, also, I met J.P. Muir of Humphreys Railways, Weems, Virginia, who granted me access to his yard to photograph those former sailing craft converted to power. In addition, there were shipyards at Perrin and Deltaville in Virginia, and Cambridge and Crisfield in Maryland.

Among periodicals, the *Baltimore Sunpapers* led the list as a source of written information. I have been collecting clippings from those newspapers and checking microfilm of back issues for many years. The Mariners' Museum's excellent collection of *Merchant Vessels of the United States, Annual Reports of the United States Life-Saving Service, Nautical Gazette,* and *Shipping Illustrated* proved to be invaluable for research. The museum's library also contains a copy of the Henry Hall journal describing log canoe building in Poquoson, Virginia. Marion V. Brewington's *Chesapeake Bay Log Canoes and Bugeyes* and Captain T.C. Conwell's *Ships and Men of the Broad Kill* supplied necessary data. Another valuable reference was H.I. Chapelle's article on skipjacks in *The American Neptune,* Volume 4, 1944. John Lyman's *Log Chips* helped to clarify a few points, and individuals like W. Norris Parks, Spencer Wise, and Albert D. Gosper passed along bits of helpful information.

John C. North II, of Easton, Maryland, owner and sailer of log canoes, brought me up-to-date on those colorful craft. Mrs. W. Emmett Andrews, daughter of the late well-known Cambridge schooner operator, Captain Harvey H. Conway, furnished me with some facts about the man and his fleet. I thank them all.

Robert H. Burgess

Introduction

As I look back over a half century's association with Chesapeake Bay, it is difficult to realize that I have witnessed the panorama of its sailing craft, observed them in port and under sail, in shipyards and as abandoned hulks, and have seen them disappear in that time. My enthusiasm was born early, when as a boy traveling on the steamboats I viewed the sailing craft in Baltimore harbor and in Bay ports. The graceful hulls and slight rigging of the pungies and bugeyes, and sturdier hulls and more seagoing appearance of the schooners attracted my juvenile but inquisitive eye as nothing else did. I pursued the craft afloat and along the docks of the city and at down-the-Bay landings, admiring them at first, then photographing, researching, and writing about them.

Whenever I found myself in the vicinity of the Baltimore waterfront as a youngster I always searched for the tall, slender topmasts of schooners. When riding streetcars or buses on streets paralleling Pratt Street in the mid-1920s, I'd crane out the windows hoping to see a sailing craft. Trips down the harbor were extra bonuses, for then I'd see the many sails that stirred my adventurous mind.

At first I was satisfied just to look at them, but later I had to have a more lasting impression of the vessels, and photography seemed to be the answer. All that was available to me was a Brownie box camera, generally used to record outings and family pictures. With the Brownie, in 1924, I made my first photograph of a Chesapeake Bay sailing craft.

Fifty years later, that event is still clear in my mind. I was aboard a steamboat, with my mother, making a trip down the Bay to spend a vacation on my grandmother's farm in Northumberland County, Virginia. The steamer was the *Dorchester*. I'm certain of that, because before we left Light Street, Baltimore, my sister photographed my mother and me standing on the upper deck beside a lifeboat bearing that steamer's name. The *Dorchester* was bound down the Bay and up the Potomac and would stop at close to 25 landings before it reached its destination, Washington: I would get off the next morning at Mundy Point, on the Yeocomico River.

I remember, before taking that first picture in August 1924, the steamer was about abreast the Magothy River and off to port I saw two sailboats. That's what they were to me then—not types. They were both heading up the Bay, barely moving because of lack of wind. To me then, they were big vessels under sail, one a two-master and the other with three masts. As I see it now, they were too far away for a good picture, but these were vessels under sail and I had to photograph them. That distance also made it impossible to identify them, but at that time I wasn't that conscious of history. After a few years had passed, I realized I had photographed a bugeye and a ram, two types of Chesapeake sailing craft with which I was to become more familiar later. To serve as an illustration of my first attempt at marine photography, I enlarged that negative along with its accompanying defects, stains, scratches, and graininess. That scene marked the beginning of a long association with cameras and sailing craft. In the ensuing years, I photographed other sailing craft from Bay steamboats but was hardly more successful than on my first attempt, due to movement of the vessels and distance from the subject.

In August 1928, I had more success with my photographic efforts. I was on the side-wheel steamboat *Talbot,* then on the Baltimore-Washington route, and we were approaching Cowart's Wharf on the Coan River, Virginia. As we eased up to the wharf I saw a two-masted schooner tied up to a section of the pier adjoining a tomato cannery. The schooner and picturesque background composed of wharf, country store, cannery, and tall poplar trees created a pretty picture in my mind. Framing the scene in the frosted-glass finder, I snapped the shutter at 1/25th of a second, a sound which has been resounding in my ears ever since as I released other camera shutters on thousands of occasions.

I could see the name *Federal Hill* carved on the schooner's port quarterboard. What an appropriate name with which to initiate my hobby-career of photographing Chesapeake craft! Federal Hill, a promontory overlooking Baltimore harbor, had gazed down on Bay sailing craft since the beginning of trade on Chesapeake waters. That schooner had been built in Baltimore in 1856 and is one of the oldest Bay craft pictured in my photographic collection.

By coincidence, another of my early photographs of Chesapeake sailing craft was also that of the *Federal Hill.* In the summer of 1929 I was returning to Baltimore on the sidewheeler *Emma Giles* after a day's excursion to the West River, Maryland. In the Bay channel near Seven Foot Knoll light, the

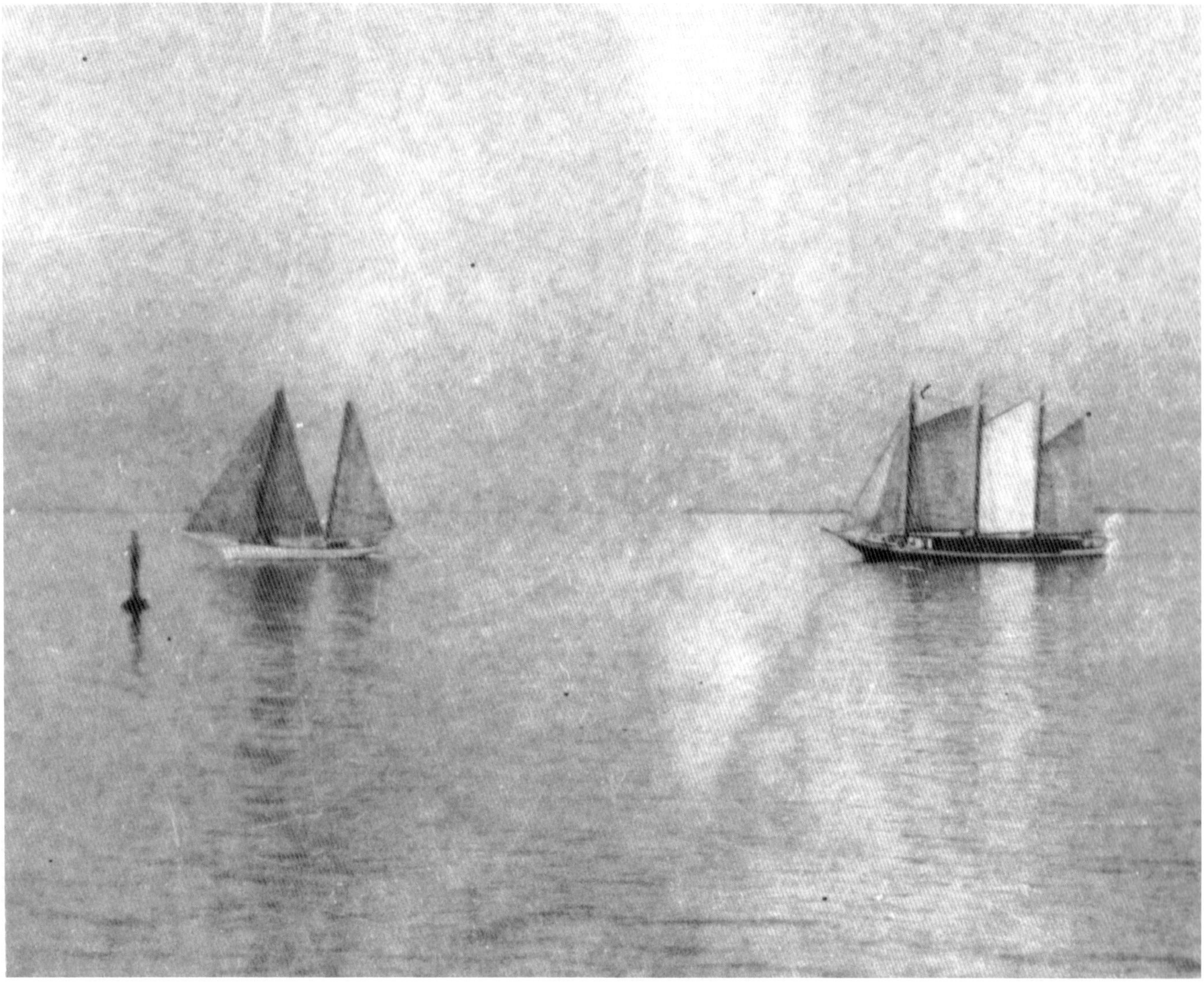

The author's first photograph of Chesapeake sailing craft, made in August 1924. An unidentified bugeye and ram move up the Bay with the help of their yawl boats astern. Love Point and the mouth of Chester River form the horizon in the background.

steamboat overtook a lumber-laden two-masted schooner. As the steamer passed, I aimed the box camera and clicked the shutter. Then, lifting my eyes from the finder, I was able to identify the vessel from the name painted on her starboard bow and, what a surprise—the *Federal Hill.* I was to encounter her many more times around Baltimore harbor in later years.

Needless to say, those early Brownies left something to be desired. It was difficult to stop action, and the image would be sharp only in the center of the negative. Upon graduation from high school, I was given a Kodak 116 folding, fixed-focus camera, a slight advance over the box camera. The new camera served well until I lost it overboard from the end of a pier several years later.

By then I had read books on photography and knew something about lenses and in 1936 invested in a Zeiss Ikon 116 Ikomat folding camera with f4.5 Tessar lens and speeds up to 250th of a second. That gave me razor-sharp negatives and now, 38 years later, is the camera I still use, although it is limited to a single type of film. No filters, range finders, or exposure meters did I use. Experience was my guide in camera settings, and it all worked out well.

The illustrations in this book were selected from thousands of photographs I have made of Chesapeake sailing craft during the past 50 years. Each scene represented is still fresh in my mind, and I must hold some kind of record for having boarded so many sailing craft, from sloops to five-masted schooners, and clambered over them from the hold to the crosstrees, from the ends of the jib booms of some to under the bilges of others as they lay hauled out on marine railways. As I look over this accumulation of photographs, it appears that I spent all my time around the waterfront. I will admit that most of my spare time was devoted to pursuing the Bay craft, and the sum of all those chases now gives the impression it was a full-time assignment.

Most of my visits to the craft were at dockside in Baltimore or other Bay ports. Some of the vessels I caught under sail or at anchor. Their shipyard visits afforded me the opportunity to view and photograph their underbodies and study the contrasts in hull design and structure. I followed many to their graveyard sites after they were abandoned.

Just within my time, as sail was passing on, I knew personally more than 200 individual Chesapeake sailing vessels of various rigs, the pungies, bugeyes, schooners, sloops, skipjacks, and rams. Most of these were gainfully employed. A few had sailed their last passage and were wasting away their days at a pier before final abandonment. Today, their number has dwindled to about 30 skipjacks, the only sailing type still in commercial use on the Bay.

How was it that these craft could manage to make a living within the Chesapeake Bay region when, except for a few schooners plodding along the New England coast, commercial sail had virtually disappeared from American waters? A look at a chart or map of the Chesapeake region reveals the extensive watershed, many creeks and rivers flowing into the Bay and linking the major ports of Maryland and Virginia with the smaller towns and country landings. At both ends of the Bay there were canals enabling the vessels to go to North Carolina for lumber, then back up the Chesapeake to make their way into the Delaware River. That is a large segment of the eastern seaboard, all in relatively sheltered waters.

Lumber was the cargo carried most frequently by the larger two-masted schooners and rams. It was easily stowed in the hold and on deck, required no speedy delivery, and could be left exposed to the weather. Many of the sources of lumber were along rivers not convenient to vehicular and rail transportation and could be loaded at the mill sites. And the sailing craft, with unobstructed holds, could carry sizable loads.

As long as the craft lasted there seemed to be cargoes for them. World War II reduced their number as some of the schooners were sold away from the Bay or converted to powerboats. Barges and trucks with pallet loading took over to transport lumber to city markets. Pungies disappeared with the war years. Bugeyes, sloops, and skipjacks continued to find employment on Maryland's oyster dredging grounds, since that state permitted only sailing craft to participate in this phase of the oyster industry. The law has been modified to permit dredging under power on certain days of the week; this accounts for the existence of the skipjacks.

No attempt has been made here to create a definitive study of the origins and development of Chesapeake commercial sailing craft. I merely touch on the origins in the introductory chapter of each type to give those unfamiliar

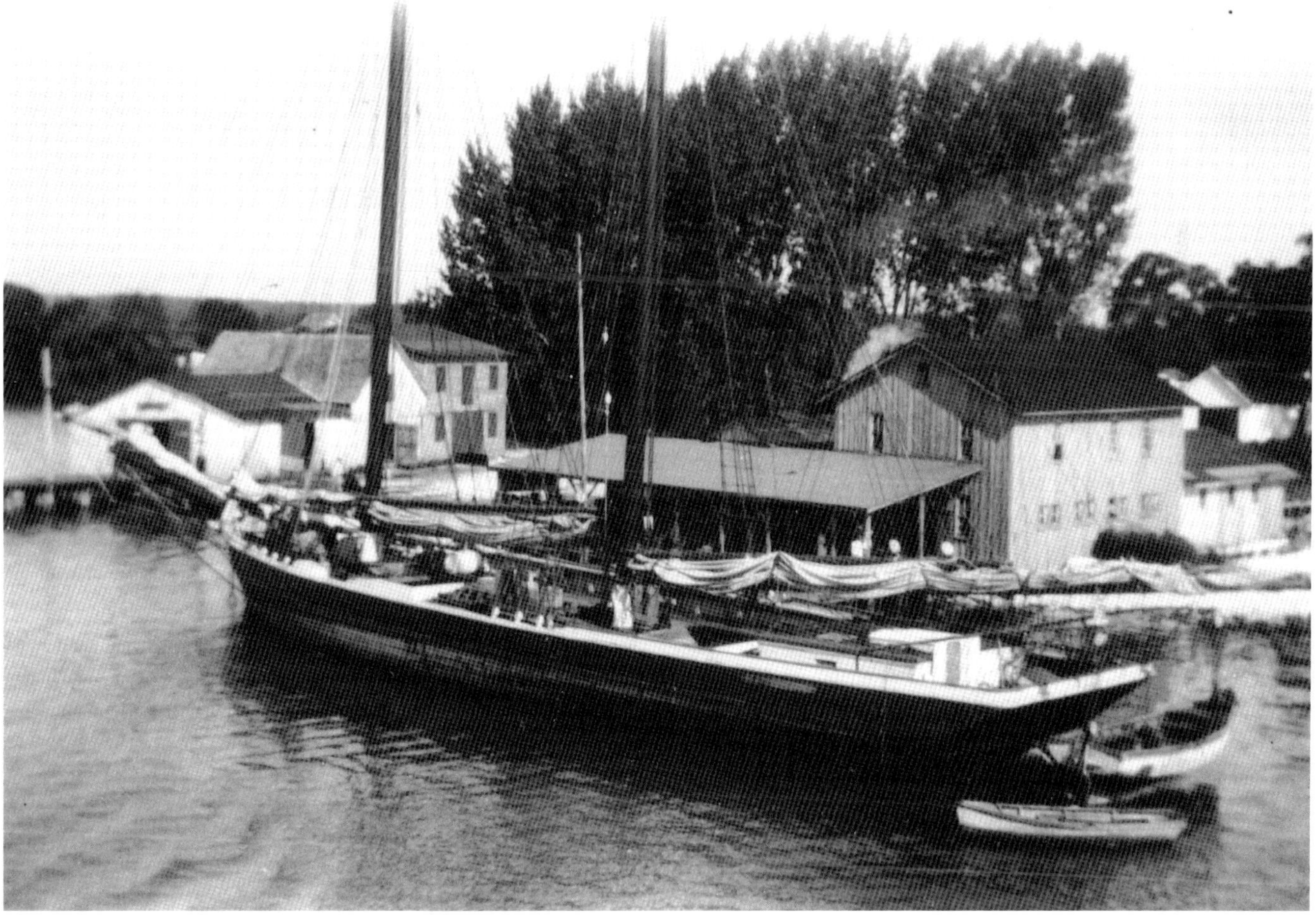

An early photograph of a Chesapeake Bay sailing craft taken by the author, from the side-wheel steamboat *Talbot,* in August 1928. The schooner *Federal Hill,* built in Baltimore in 1856, lies at Cowart's Wharf, Coan River, Virginia. The vessel has probably brought empty cans from Baltimore for the tomato cannery at right. Cowart's was a steamboat landing on the Baltimore-Washington-Potomac River route. The wharf and warehouse are at left. The general store and post office are nestled under the canopy of tall poplar trees in the background.

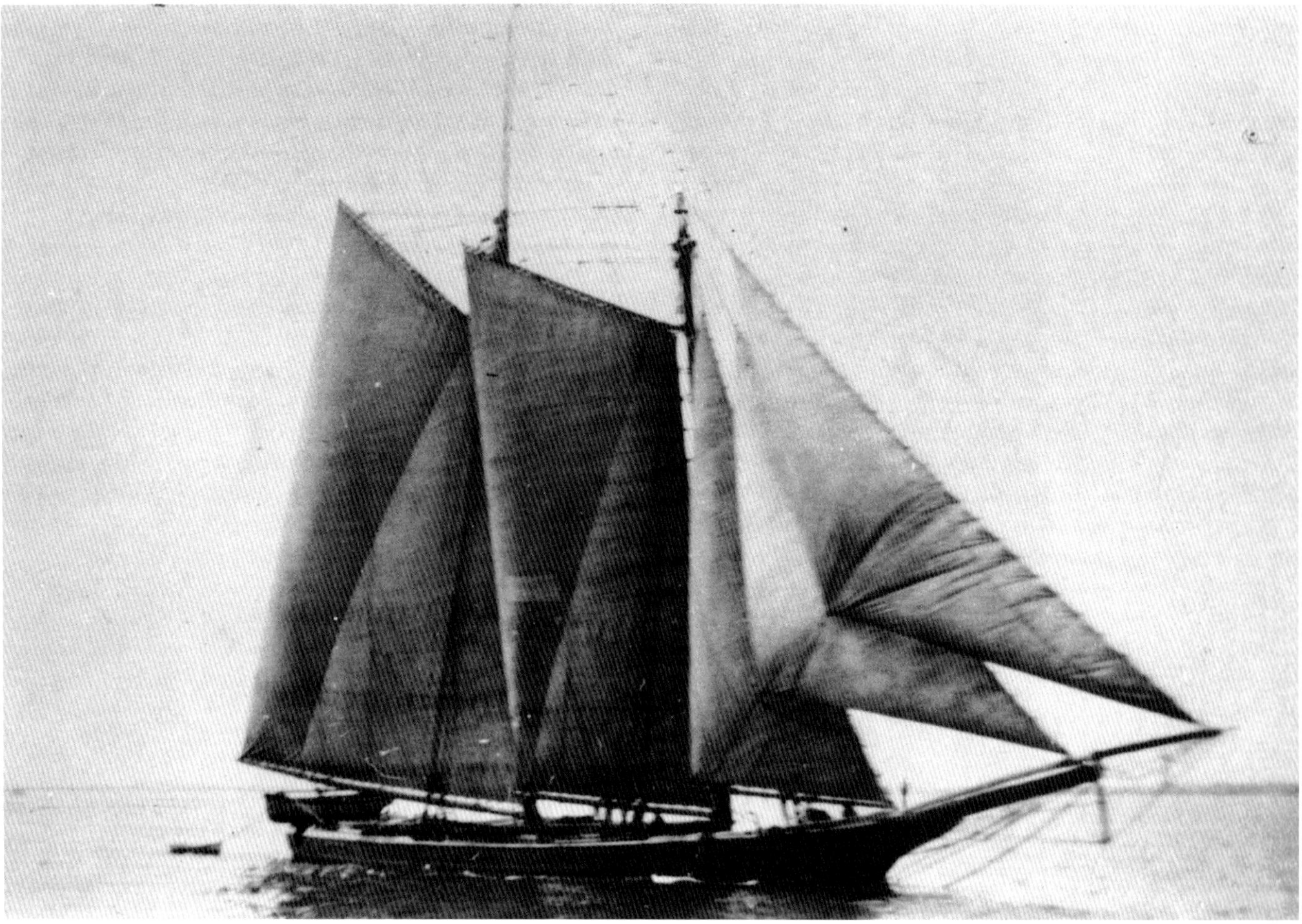

In July 1929, the author encounters the *Federal Hill* again as she sails up the Patapsco River with a cargo of lumber for Baltimore.

with the craft some brief information about them. My aim was solely to make a pictorial presentation of the vessels and reflect upon the period during which I knew and photographed them.

For those interested in pursuing further study of Chesapeake sail (the types stressed herein), I can recommend the works of other writers. In the large volume *Sailing Craft* by Edwin J. Schoettle and published in 1928, Peter C. Chambliss, of the *Baltimore Sun,* contributed an interesting chapter on log canoes and bugeyes. *Chesapeake Bay Log Canoes and Bugeyes,** by M.V. Brewington, published in 1941 originally by The Mariners' Museum, is the standard work on those types.

Howard I. Chapelle has written well on the skipjack in his *American Sailing Craft,* published in 1936. He also had a lengthy, informative article, with drawings, entitled "Notes on Chesapeake Bay Skipjacks," in Volume 4, 1944, of *The American Neptune,* a quarterly of maritime history. Dr. James E. Marvil published his study, *Sailing Rams,* in 1961, acknowledging the contributions made by his family and neighbors to the origin and development of that unique type.

The development of the pungy has been touched upon by Chapelle in his *History of American Sailing Ships,* published in 1935, and in *The Baltimore Clipper,* 1930. The Chesapeake sloop and schooner have yet to be researched in depth. A general article, "Three Hundred Years of Sail on Maryland's Chesapeake Bay Waters," by Ralph J. Robinson, appeared in the March, 1941, issue of *Baltimore* magazine.

I have boarded and photographed many other sailing vessels on Chesapeake waters that were owned elsewhere and came here only for an occasional visit. They are not covered in this volume, which is devoted solely to those vessels built or owned in the Bay country.

Some may wonder how I could have had such free access to so many ships, apparently coming and going as I pleased. In many instances there was no one onboard, so I just jumped on deck after getting no response to a call, walked around, studied details, and recorded on film what I thought was worth preserving. If there was anyone onboard, usually a deckhand or cook, he granted me permission to come on but expressed no interest in what I was doing. In fact, they thought it was odd that someone wanted to take a picture of "dat old sailboat."

I was never denied admission to any of the Bay vessels. At times I had problems at shipyards or docks where there were watchmen, but they got to know me. Some of the workers looked askance at me as I moved about photographing the old ships, failing to understand why I thought the craft worthy of

* Reprinted in 1963 by Cornell Maritime Press, Inc., Cambridge, Maryland.

such attention. Now, decades later, these same people revel in the scenes and enjoy talking about those days at great length.

I found the camera a wonderful device in helping me become accepted by the watermen. I became friendly with a number of them by giving them portraits of their vessels. In return I had the use of their skiffs to get me out to anchorages to photograph other craft. When the four-masted schooner *Doris Hamlin* was tied up for weeks at a time between charters at Young's dock in Baltimore, I became acquainted with the Norwegian mate who lived onboard and would visit on weekends to talk about big sailing vessels, climb the rigging, and photograph fittings. I got to know the watchman and had access through the locked gate. Eventually, I met the schooner's captain and within several years I had signed on her for a voyage to the West Indies. I later learned that the owner of the *Hamlin,* W.B. Vane, felt he would save money by allowing me to make a voyage instead of taking up so much of the crew's time standing around talking in port.

My companion on walks around the Baltimore waterfront and trips down the Bay on excursion steamers was the artist Louis J. Feuchter. He sketched as I photographed and he appears in some of my photos. He had little regard for photography at that time, although earlier he had photographed Bay craft. He claimed that the camera caused distortion and false perspective but he learned to rely on my photographs in later years as he was called upon to fulfill commissions of paintings of Chesapeake sailing craft. The details he wanted were recorded on my film.

I never met a Bay waterman who kept a camera aboard his vessel. What opportunities they would have had to record their craft under sail, others they passed or met in port, and at country landings! But it meant nothing to them at the time. Severely lacking, too, are scenes aboard the Bay craft as they sailed.

The illustrations in this book are straight prints from the original negatives, that is, enlarged to 5 × 7 size. Not a single scene has been retouched, printed from more than one negative, or specially posed. What I saw through the camera's finder, the reader sees in these pages. Perhaps I could apologize for bare skies in some views, but I couldn't select my days to make photographs. I had to take the weather as it came, on weekends, holidays, and vacations. My work in Baltimore wasn't near the waterfront, nor did I live in that vicinity, but I enjoyed long walks down and around the piers and to shipyards and lumber mills to seek out subject matter. Daily, there were new vessels in port as some departed after unloading and others arrived from down the Bay. Some of the craft evaded me completely; I just could not connect with them. When I saw them they were anchored out and I had no way to get to them, or they had sailed before I could get down to the piers. Though some would have Annapolis or Washington as ports of call and seldom came to Baltimore, the latter port was the focal point for most of the vessels.

The Baltimore Sun carried an excellent daily maritime column that recorded the arrival and departure of all ships—Bay and ocean steamers, powerboats, tugs and barges, and sailing craft. Names of masters, cargoes, and the ports they came from were reported. Then there were brief paragraphs of personal notes such as transfers of captains, sales of vessels, mishaps, and shipyard news. Reference to those columns alerted me as to vessels in port, especially the larger coasting schooners.

To those who never witnessed these sailing craft, or those who knew them and were unaware that a photographic record of them existed, I hope this book will prove to be enlightening. To model makers and artists who come along too late to see the real thing, the illustrations can serve as source material. Many times I have wished that someone had made an effort to preserve for posterity a record of Chesapeake shipping before my time. If the photographs prove of value to historians and researchers, then the project has been worthwhile. In any event, my association with the Bay craft has been most rewarding. To me this book is further recognition of my "old friends" that afforded me so much pleasure through the years as I sought them out around the Chesapeake and trod the decks of every vessel pictured herein.

The illustrations are arranged in this volume according to types. That is, all the bugeyes are together; the same with the pungies, schooners, etc. And in each group, after the log canoes, the vessels appear, for the most part, in alphabetical order. The sailing craft represented here are but a few active on the Bay during the past 50 years, but they are of the types and serve as a cross section of Chesapeake sail of that period.

Come aboard with me decades later and examine the decks and rigging of the vessels, the primitive steering arrangements and bulky windlasses of some. Board the largest two-masted schooner then along the coast, the *Harriet C. Whitehead,* and the smallest three-master here, the *William J. Stanford.* Visit Bay shipyards with me and get under the vessels. Gaze up at the last of the Bay's stern decorations on the pungy *Amanda F. Lewis.* Reach out and all but feel the finely carved trailboards. Look down the wide deck of the four-masted schooner *G.A. Kohler,* photographed shortly before leaving port on a voyage that ended with her wrecked on the North Carolina coast. View the intimate details of the little pungy *Wave,* built in 1862 while the Civil War was raging. See the veteran schooner *John W. Bell,* long on the Chesapeake and dating back to preclipper days. I've been aboard them all and knew them well, and now you can meet them. Their like will never be recorded on film again.

Robert H. Burgess

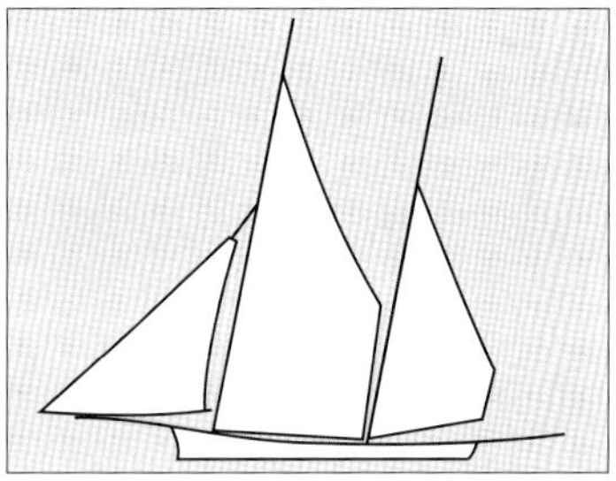

Log Canoes

Any book dealing with Chesapeake sail should start out with the log canoe. It was probably on a log, if not a hollowed-out one, that man first traversed the Bay's waters. It is quite likely that a log canoe will be the last pure Chesapeake sailing type to remain, long after the framed boats have disappeared from these waterways. The framed types have been reduced to the skipjack after running the gamut of the other types while the log canoes, now being sailed for pleasure, are being added to.

My introduction to craft native to the Chesapeake was by way of the log canoe. In 1923, while spending the summer at my grandmother Harriet Dixon's farm in the Northern Neck of Virginia, the outstanding event of that visit was a fishing trip on the Potomac River out of the Yeocomico River. My uncle knew a waterman, Anderson Kent, who lived and kept his boat near Mundy Point. This boat was a powered log canoe, or "cunner" as he called it. My association with that canoe left a deep impression on me. I recall today the sharp stern, clean entrance and graceful curves of her hull. I was too young to be conscious of the craft's lineage but I was aware that it was a "pretty" boat, and it kindled an interest in log canoes that hasn't waned for half a century. From that small dugout my interests in the larger Chesapeake sailing craft were generated.

On June 16, 1881, Henry Hall, who had taken leave of absence from the *New York Tribune* to visit shipyards in the eastern United States, stopped at the lower Virginia Peninsula to call at boatyards in that area. To reach the somewhat isolated region of Poquoson, Virginia's chief log canoe building center, he departed from Old Point Comfort by water and later entered the following in his journal:

> . . . At Poquosin (*sic*) about 21 miles by water north of here, there are several Moores who build canoes and buckeyes of logs. I visited the place, going thither in a canoe, leaving Old Point at 8 a.m. and reaching the Point again at 5 p.m. having been on the water all this time except for about half an hour. . . . One of the Moores, a large, tall man, was just completing two canoes, with the aid of a crowd of tall, sunburned boys, his sons probably. He thinks 50 canoes and buckeyes are built every year in this region. Many more up near York and Gloucester. The two canoes which Moore was engaged on were about 20 feet long. . . . These canoes were made of three logs. No model is used. The joining edges are gotten by means of a batten. The central log gives the stem, garboards, keel, and stern of the boat. The other two logs the bilges and wales. The three logs are almost as if hewn to two bow and buttock lines, one each side of the keel. Moore pays \$3 and \$3.50 apiece for the yellow pine logs, standing in the woods. He has to go back in the county 9 or 10 miles to get them, they being from 3 feet to 3½ feet in diameter. The hauling costs more than the tree. The canoes are worth \$100 each.

Three hundred years earlier (in 1585, to be more accurate), other observers were watching log canoes being constructed near present-day Roanoke Island, North Carolina, not too far from Virginia's Poquoson by today's standards. This account and drawings of the building process appear in DeBry's *Grandes Voyages*, published in 1590. Captain John Smith witnessed Indians building canoes in the Chesapeake region in a similar manner. He recorded, "These they make of one tree, by burning and scratching away the coles (*sic*) with stones and shells till they have made it in the form of a trough."

DeBry's description of canoe construction provides detail: "First they choose a tall, thick tree of the size required for the boat's frame. Then they light a fire close to its roots, feeding it bit by bit with dry moss and small chips of wood, keeping the flames from mounting too high. When the tree is almost burnt through, they make a good fire to cause it to fall. Then they burn off the top and boughs, taking care that the trunk should not be shortened.

"The tree is raised upon a platform built on forked posts at a height convenient for working. The bark is stripped off with sharp shells; the inner length of the trunk is kept for the bottom of the boat. A fire is made all along the length of the trunk, and when it has burned sufficiently it is quenched and the charred wood scraped away with shells. Then they build a new fire, burn out another piece, and so on, sometimes burning, sometimes scraping, until the boat has a good bottom. Thus God has endowed these savages with enough reason to make the things they need."

In those intervening 300 years the methods of building the canoes, the designs, and the tools used, naturally changed and improved. However, when the first settlers came to the Virginia shores, there were few skilled boat builders, if any, among them and they adopted the simple Indian dugout for their own use. Eventually, the water-trough appearance of the dugout was modified by the white man as he gave some shape to the bow and stern of the craft.

The use of the dugout canoe spread quickly through the Chesapeake Bay colonies. Then, in the late 1600s, the colonists found that their edged tools made it easier to build a canoe from more than a single log. Two similar logs were squared and placed side by side. The joint became the centerline and each log was shaped into the longitudinal half of the dugout. The halves were then separated and hollowed out with an adze. After the canoe had been shaped the logs were fastened with wooden dowels or with tenons fitted into mortises.

Most waterside residents of the Bay country had log canoes, which were about as valuable to them as horses would be to those who lived inland. Virginia law made it a felony to take a canoe without the owner's permission.

Log canoes gained in popularity as the Chesapeake region became more thickly settled. The raw material, pine, was readily available and many settlers possessed or acquired sufficient skill to produce shapely, seaworthy craft for fishing, crabbing, oystering, or for general transportation.

Several areas within the Chesapeake specialized in building canoes. These were the Tilghman Island and Pocomoke regions of the Eastern Shore of Maryland, and the Poquoson River, in York County, Virginia.

Construction techniques in these Chesapeake regions varied, and M.V. Brewington described them excellently in his book *Chesapeake Bay Log Canoes and Bugeyes.* Since we started out with Henry Hall's visit to the Poquoson region of Virginia, we can describe how log canoes were constructed there, particularly one of five logs.

Five pine logs, suitable for the size of canoe the builder plans to make, are selected. The keel, or center, log is laid out and cut to length. Its sides and bottom are squared and a centerline marked around the log. The width of the keel, stem, and stern is laid out and the ends cut to an angle. The deadrise is hewn into this log. Two other timbers, known as garboard logs, are selected to lie parallel to the keel log. These logs are cut to a length equal to the keel log between the stem and stern angles, and squared. The last two logs, unlike the other three which are selected because they are straight, are known as wing logs and

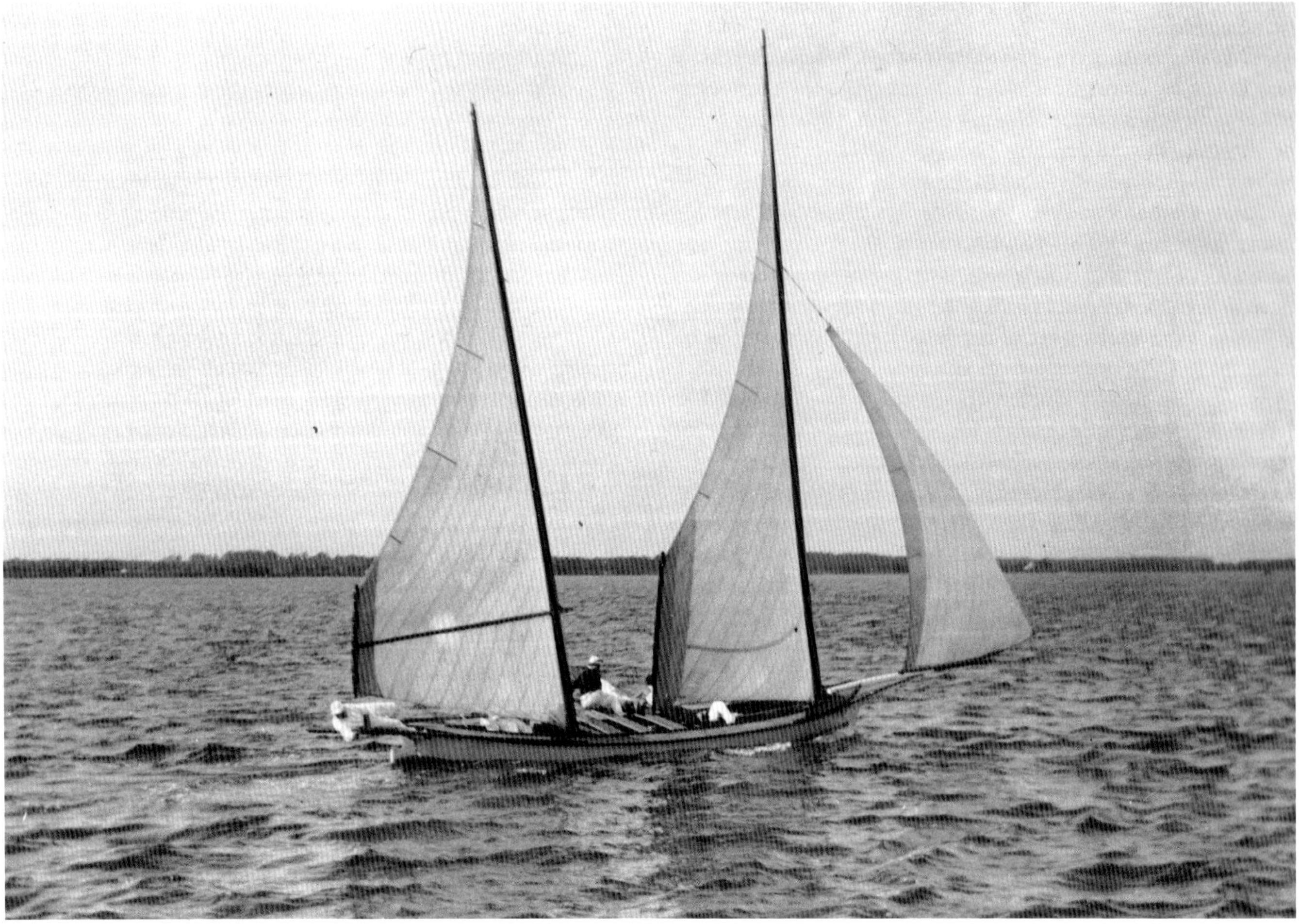

Log canoe *Mayflower,* owned by Howard B. Strong of Chestertown, Maryland, sails in the Tred Avon River off Oxford, Maryland, on Labor Day, 1932, as she wins second place in the log canoe event of the Tred Avon Yacht Club Regatta. *Mayflower* was built in 1880 by Wesley Stevens at Rock Hall, Maryland, and last raced about 1948 when owned by William Grieb of Chestertown. She won the highly cherished Governor's Cup in 1942.

are chosen for their curvature. They are sized, squared, and cut roughly to shape. These rough-hewn logs are then set up on blocks and raised to maintain the proper line of deadrise. The builder then begins to hollow the pieces with a gouge adze.

After the sides are shaped, the logs are separated, the keel log is turned on its side, and one garboard log is spiked temporarily in its place. A wing log is then spiked atop the garboard log. A broadax is then used to hew the bottom of the logs to the proper deadrise and to work off the hard chine into a round bilge. This operation completed, the logs are separated, the keel log is turned over, and the process is repeated on the other side. The next step is to assemble the logs on the blocks, bottom up. The builder can now see the shapely result of his hewing. At this point, any irregular spots are smoothed down.

The logs are separated again and the interior of the craft is roughed out. The keel log remains the thickest and the adjoining logs become thinner toward the sides. The shape of the canoe is now evident. The centerboard slot is cut through the keel log. To ensure a close fit after the roughing process, the logs are jammed together and a saw is run down their seams to ensure a perfectly matched joint. Holes are drilled along the inner seams and the garboard logs are secured to the keel log with long iron driftbolts. The wing logs are likewise fastened to the garboard logs.

By this time the canoe may be under a shelter so that the more detailed work can be carried on in bad weather without interruption. The canoe is placed upright on blocks for finishing. Filler pieces are placed at the ends of the logs and bolted in place. Other fillers are then added to achieve the proper depth and sheer. This requires some expert fitting and is usually done so well that the seams are not apparent. With the hull complete, all the holes are plugged with wooden dowels and the entire canoe is rubbed with a smoothing plane. The mast seat and mast's heel block are put in place. Then the centerboard well and seat locker are installed. The knees, natural crotches from the trees used in the canoe, are placed along the interior sides to support the washboards and coamings. The stempost, sternpost, and rubbing strakes complete the hull. After the sailing rig is installed on the canoe, the craft is taken down to the beach and she is ready for a long life of usefulness.

The commercial sailing canoe has disappeared from Chesapeake waters but many power-driven log canoes, some former sailing craft, still serve in the oyster and crab fisheries. Their thick bottoms seem almost indestructible. When hull repairs are required, it is necessary to cut out only the bad wood, insert a plug with fastenings, and finish it off. Even after the canoes are abandoned they seem to cling to their existence. Their metal fastenings may rust and the logs fall apart, but their heart pine logs seem to resist total decay. Quite often the remains of log canoes are found embedded in mud banks of tidewater streams and rivers, or in marshes, arousing speculation about their origin but usually being linked with Indian dugouts.

Quite a fleet of racing, sailing log canoes still assembles every summer in the waters around St. Michaels, Maryland, for a series of races. These are the *Island Bird, Island Blossom, Jay Dee, Magic, Mystery, Silver Heel, Daisy Belle, Billie P. Hall, Oliver's Gift, Noddy, Sandy, S.C. Dobson, Persistence,* and *Rover.* Others are the *Island Image, Eagle, Mary Ann,* and *Island Lark.* A new canoe, to be named *Spirit of Wye Town,* is being constructed by Sidney H. Dickson, of St. Michaels.

Each of these racing canoes carries a tremendous press of sail. Agile crewmen perch on springboards extended over the weather side to prevent the canoe from capsizing. Still, capsizes are frequent.

The major log canoe racing event is the Governor's Cup held each year, the first weekend in August, at the Miles River Yacht Club, St. Michaels. This has been staged since 1927 in Maryland waters. No regattas were held from 1942 to 1945 because of World War II, but in 1946 they were resumed.

A most fitting tribute was paid to the Chesapeake Bay log canoe by Sidney A. Vincent, eminent naval architect at the vast Newport News Shipbuilding and Dry Dock Company, Newport News, Virginia, in a letter to this author in August 1958. It read:

> In 1925, when I arrived in Newport News, some of those in the hull design division of the shipyard suggested that I might like to see a log canoe at the boat harbor. I expected to see a log burned out by fire and crudely chopped into shape by Indians. Instead I found what appeared to be an excellent hull form and I spent the weekend lifting the lines. The drawing was destroyed long ago but, and this is my point in writing, I did analyze the lines plotting the sectional area curve and entrance load waterline. To my surprise these basic hull form factors checked almost exactly with the latest (1925) naval destroyer hull forms. I assume that as one log canoe after another was built, the faster ones gave guidance for those to follow. This is exactly what is done in modern model basins where merchant and naval vessel lines are checked and developed. To me it is remarkable that non-technical oystermen of years before scientists with 1925 model basin facilities came to practically the same answer for relatively high speed craft.

The lower Chesapeake, in relatively recent years, was never too active in log canoe racing. The Hampton Yacht Club, in the late 1920s and early '30s, sponsored events for these colorful craft. World War II halted the races and they never were resumed. Still carrying on the tradition of the Poquoson-built canoe, however, is the *Kaynoo,* which participates in handicap class events at various yacht clubs.

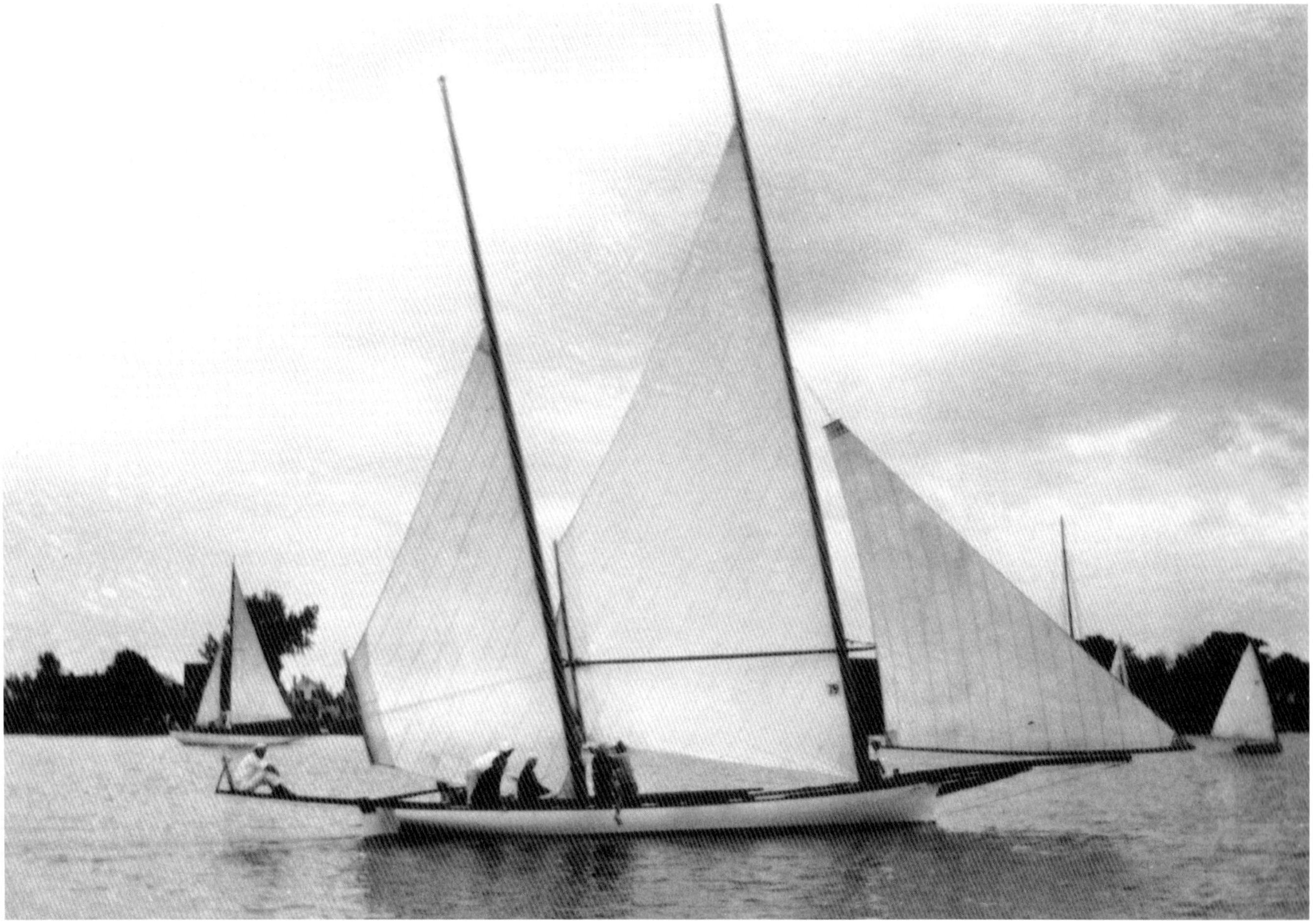

Flying Cloud, with a square stern and under a reduced storm rig, at the Tred Avon Yacht Club race off Oxford on Labor Day, 1933. Normally, she carried 11 men in her crew when competing in races, many of them acting as live ballast.

When the log canoe *Flying Cloud* was built in 1932 by John B. Harrison of Tilghman Island, Maryland, for A. Johnson Grymes of Easton, she had a "square stern" as opposed to the sharp stern of the traditional log canoe. The previous year, the *Jay Dee*, also built by Harrison and square-sterned, became waterborne. These canoes came upon the revival of log canoe racing at St. Michaels and Oxford. In 1933 the Chesapeake Bay Log Canoe Association was organized.

Flying Cloud, under Captain Buck Richardson, and *Jay Dee*, sailed by Captain Clarence Dobson, were pitched against each other for the first time in the Chesapeake Bay Championship (Free-For-All Sailing Race) of seven and a half miles over the Miles River, the feature event of the opening day's races of the eleventh annual Miles River Yacht Club Regatta in August 1932. The race drew every conceivable type of entry from small scows to representatives of the various purely racing classes, including Star boats. Fourteen boats started in the event and after 2 hours, 39 minutes, and 25 seconds *Flying Cloud* crossed the finish line with *Jay Dee* four minutes behind her. In third place was the *Mayflower*.

The next day the Governor's Cup race was held in the Miles River but was limited to sharp-sterned canoes, those with the square or "yawl-boat" stern being ineligible. In that event *Magic*, Captain Bob Wilson, triumphed. That afternoon *Magic* raced *Flying Cloud* in a free-for-all and the latter won. In 1933, in the second day's free-for-all event at the Miles River Yacht Club, *Flying Cloud* again romped around the course to lead a fleet of more than a dozen starters.

When the Governor's Cup race came around again on August 4, 1934, *Flying Cloud* was among those at the starting line. During the intervening year she had her square stern remodeled to the traditional sharp stern. Eight canoes faced the starter in the Cup race. *Flying Cloud* was the scratch boat; she was first over the line and was never headed. She rounded the eight-mile course in 1 hour, 51 minutes, and 36 seconds, with *Magic* second.

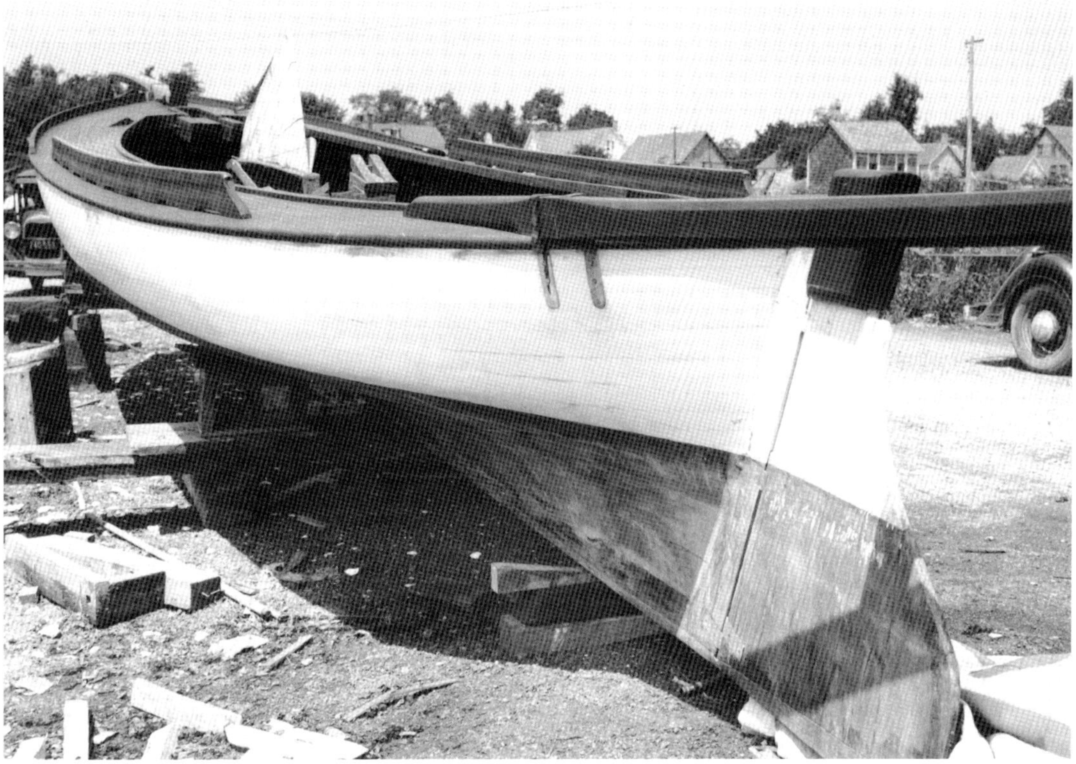

The clean lines of *Flying Cloud*, with her sharp stern, can be seen here as she rests on the blocks at Oxford in June 1936, preparing for the coming season.

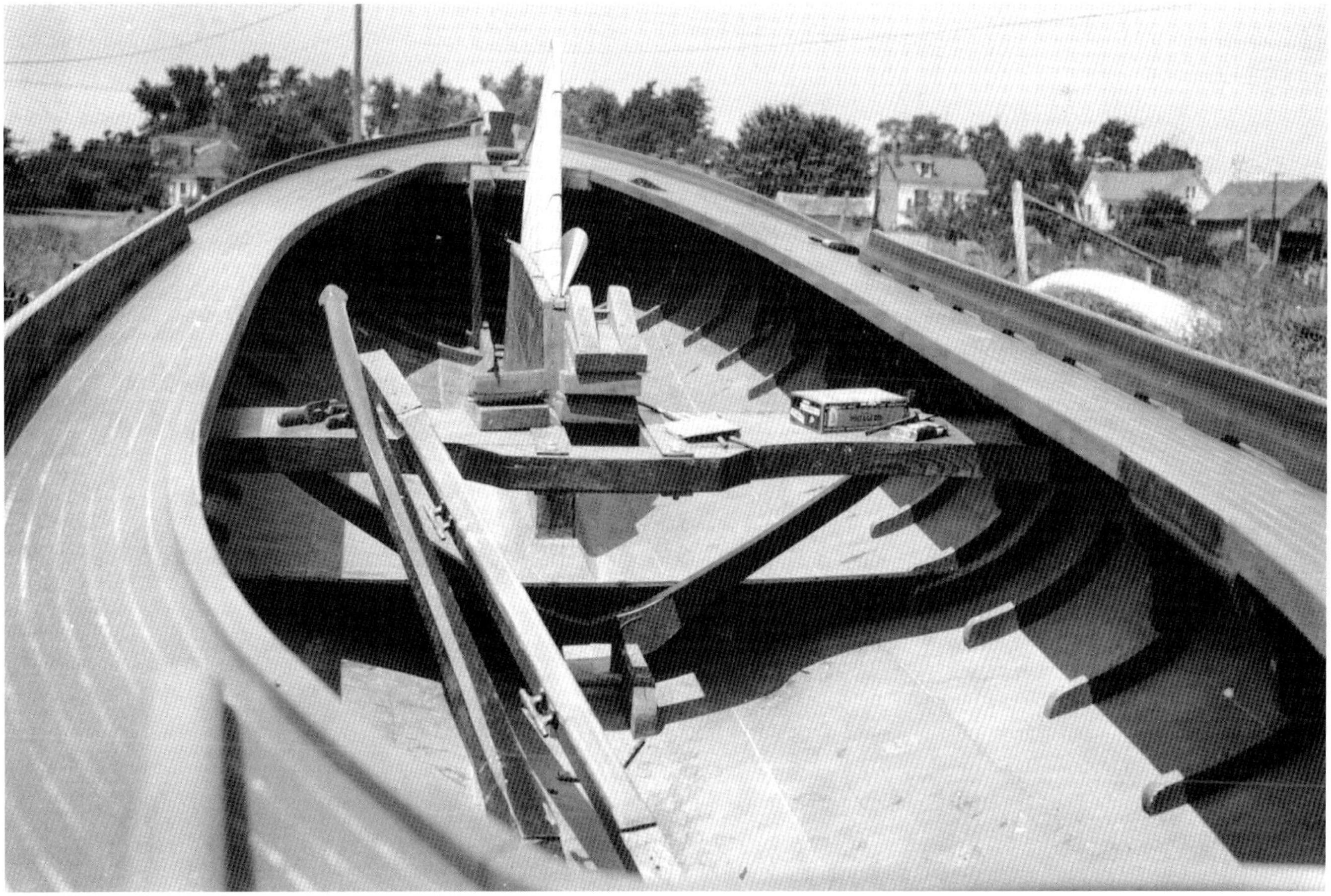

Looking forward on *Flying Cloud*. The bracing of the step for her mainmast shows the size of the material needed to withstand the strain of that tall-unstayed spar under a press of sail. This canoe is built of five logs, as are the majority of the racing canoes.

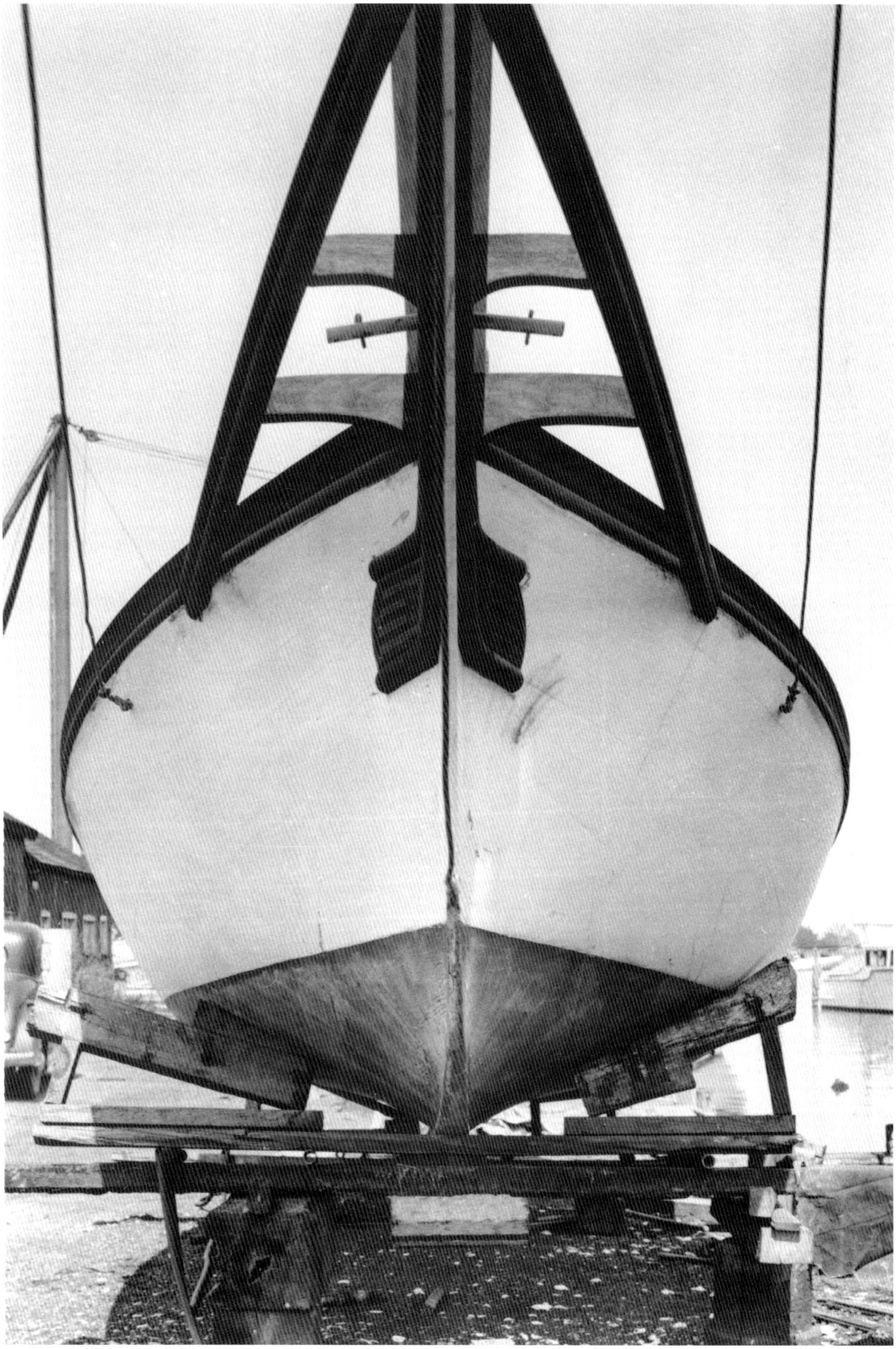

Fish-eye view of the *Flying Cloud* reveals a beautiful hull with scarcely a straight line to be seen.

By 1950 *Flying Cloud* had ended her racing days and had her sailing rig cut down and a portable cuddy cabin built over the forward section for use as a day cruiser sailing on the Choptank River. In 1952 she was acquired by Fred Kaiser of Hampton, Virginia, and for the next two years sailed on Hampton Roads. In 1954 the marine artist John A. Noble purchased the canoe and had her taken to the Staten Island, New York, area where she can be found today.

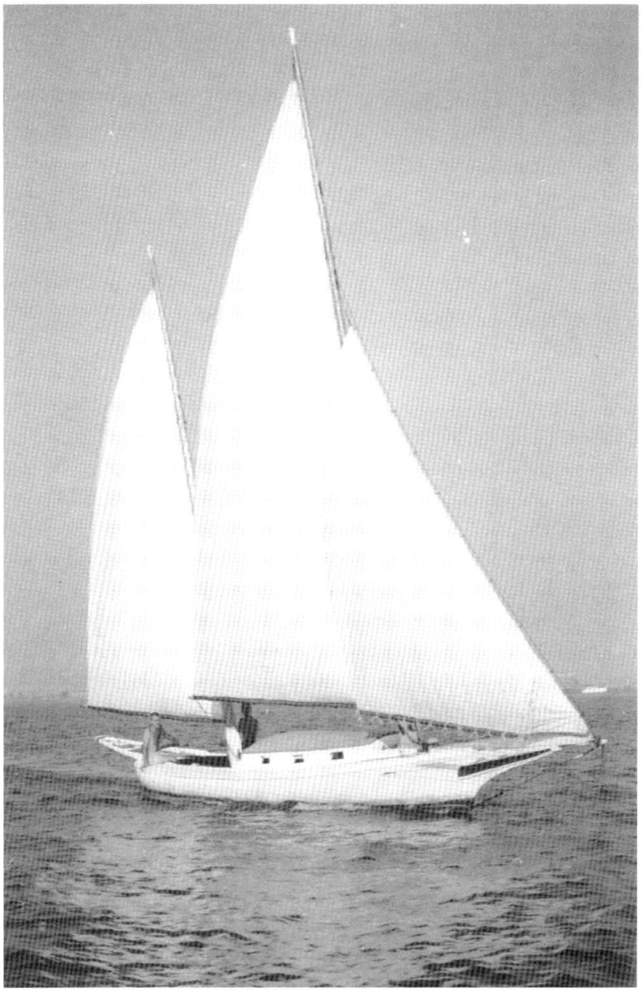

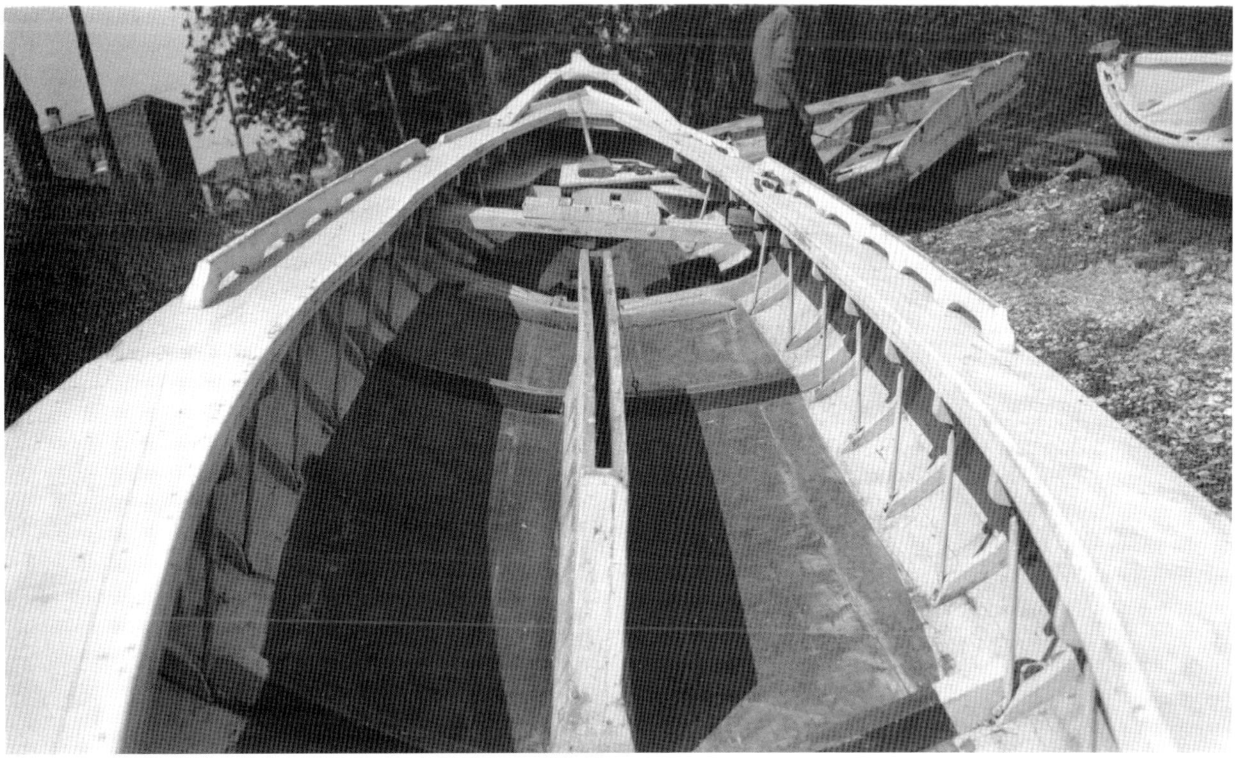

Looking aft in the five-log canoe *Belle M. Crane*, built by Greenbury Coffin at St. Michaels, Maryland, in 1897. Shown at Oxford, Maryland, June 17, 1936.

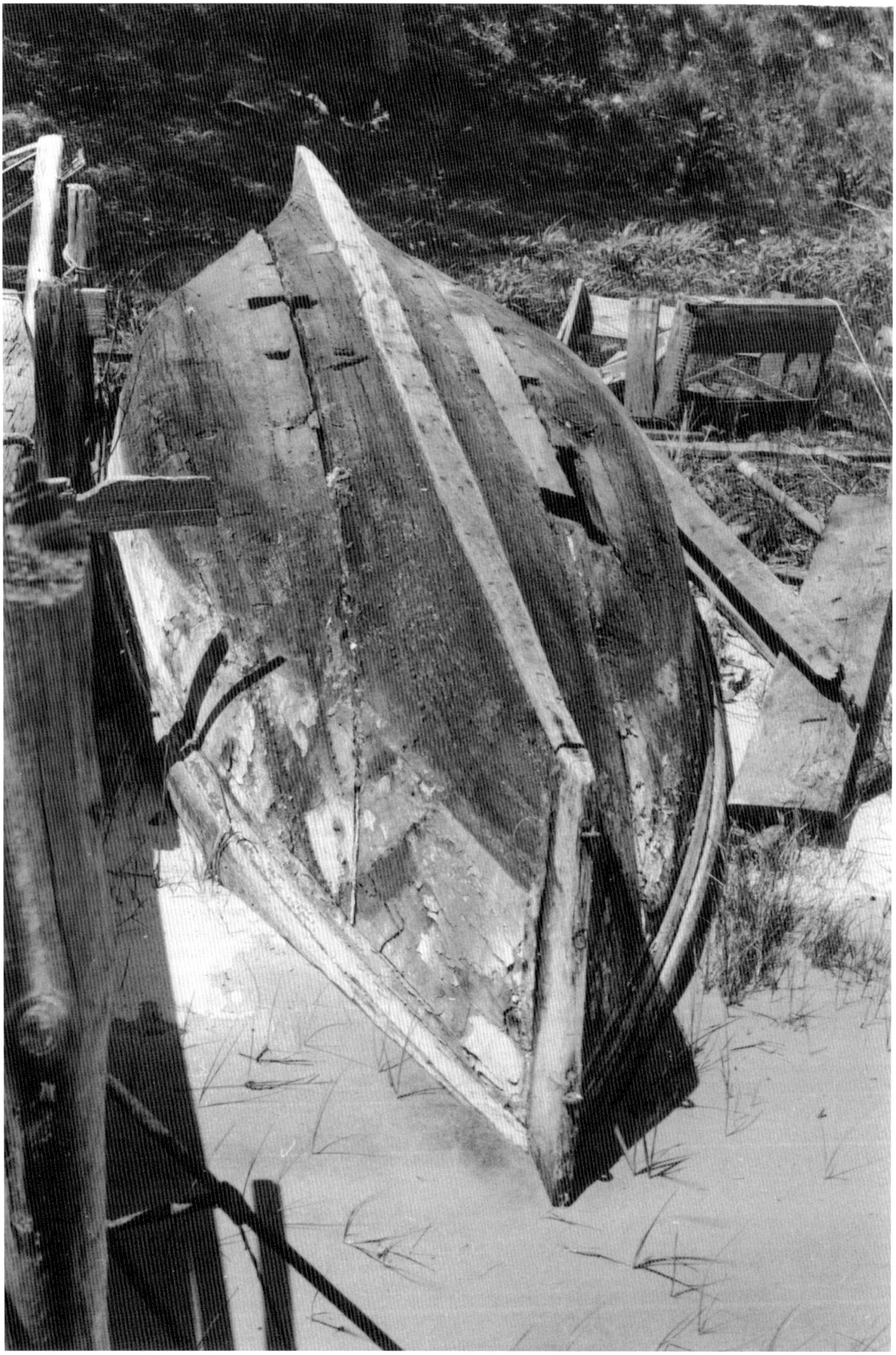

A three-log canoe; the seams of the logs are plainly visible as they have become separated. Rectangular holes in her bottom indicate where plugs had been placed at one time in making repairs.

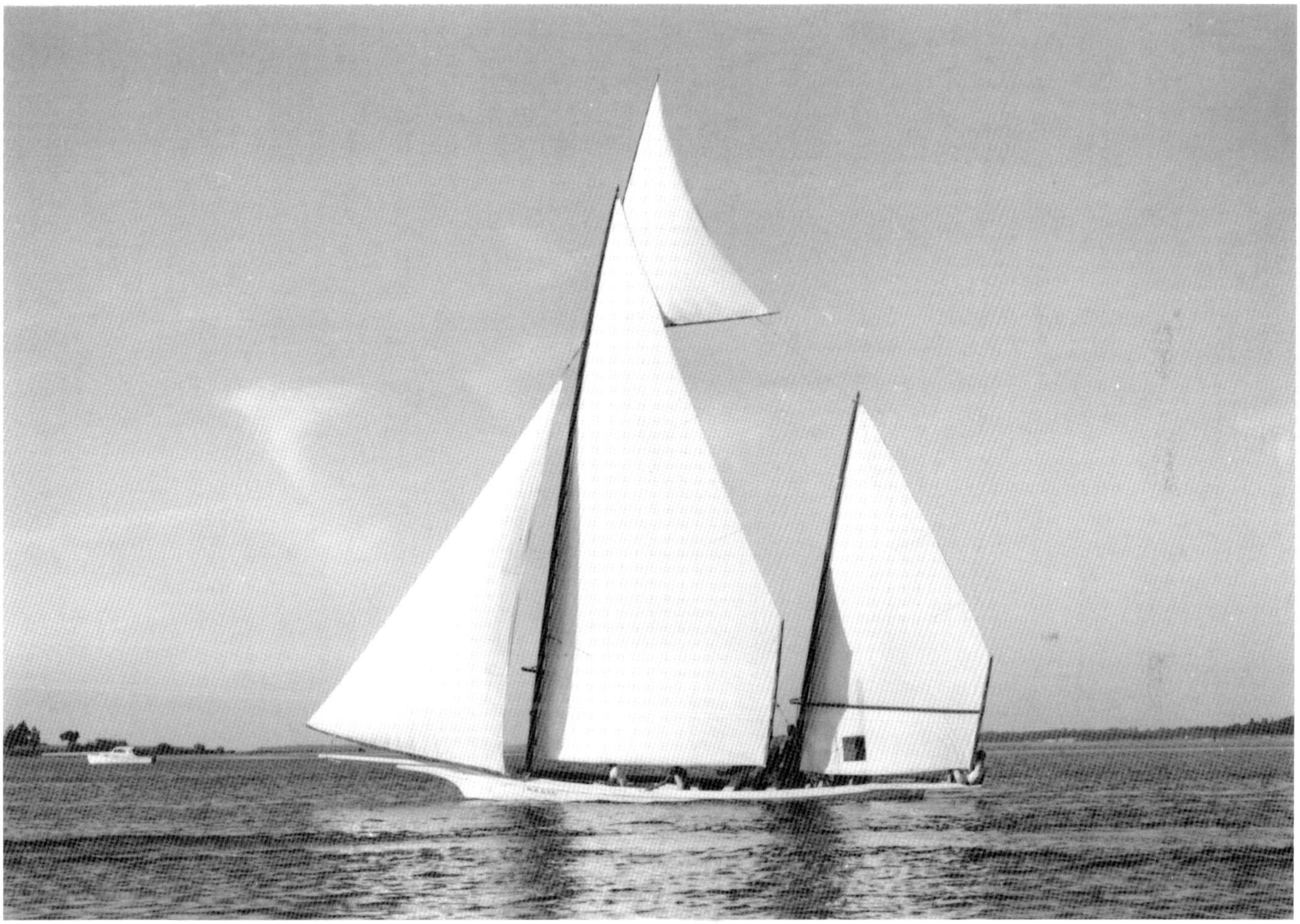

Magic, winner of numerous log canoe events in Maryland, including ten Governor's Cup races. She was built in 1894 by Charles Tarr at St. Michaels, and still sails. Captain George H. Wilson owned her from 1927 until his death about 1968. His youngest son, James, now owns the craft. Here she is shown in the Miles River Yacht Club races near St. Michaels in September 1964, sporting a kite atop her foremast and a plastic window in her mainsail so that her helmsman can have clear vision.

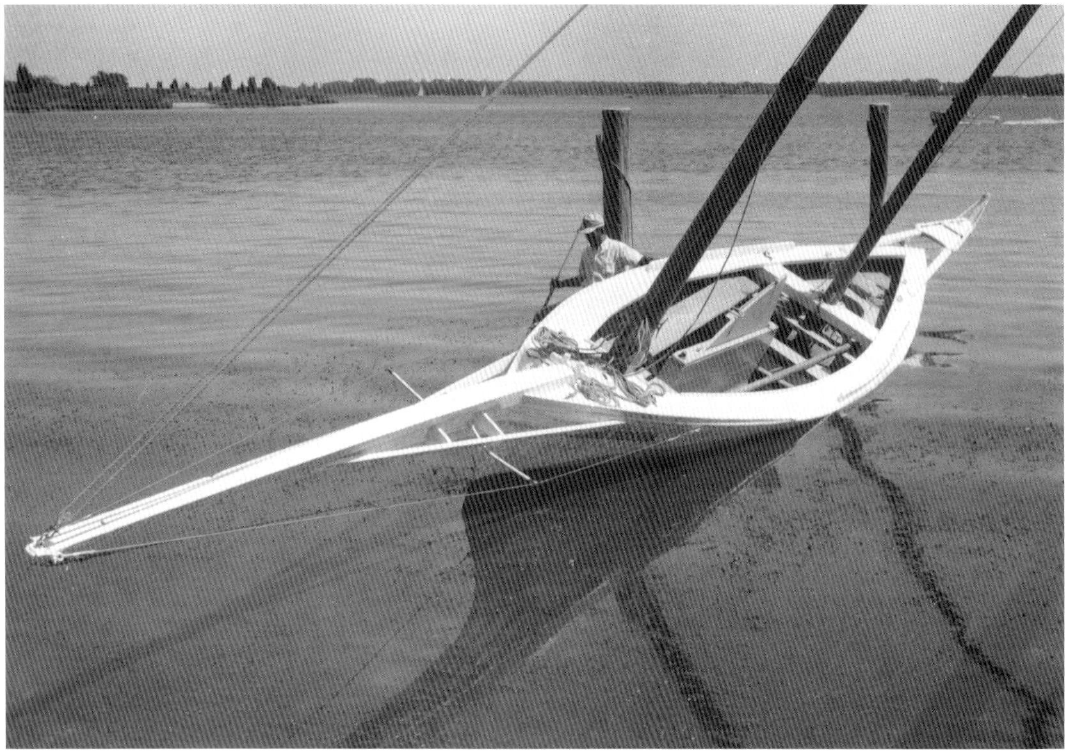

Log canoe *Island Blossom* resting on the sandy bottom of the Miles River while being scrubbed down prior to a race located by the Miles River Yacht Club, September 1964. This angle imparts a splinterlike effect to the canoe's hull and also reveals why these long narrow craft, with so little hull in the water, are so swift.

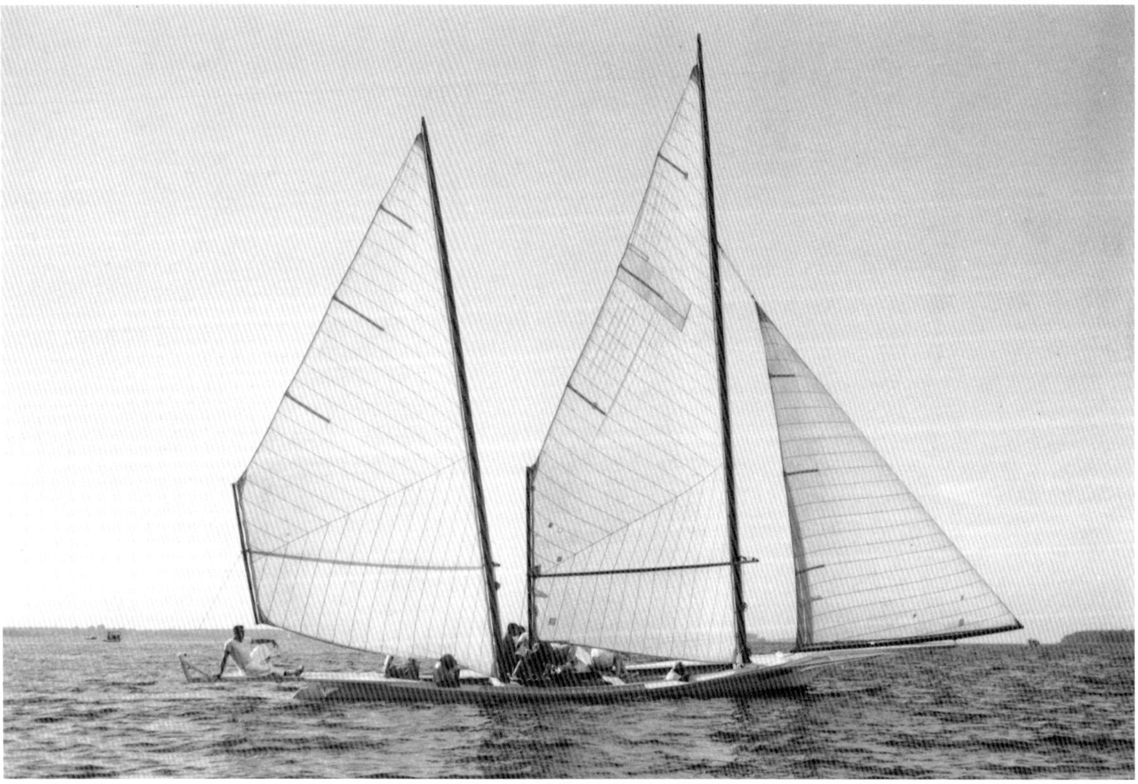

One of the older log canoes still sailing is the 32-foot-long *Island Blossom,* built by Captain Sid Covington at Tilghman Island and now owned by John C. North of Easton. His son, John C. North II, owns the 1888-built *Island Bird,* also constructed by Captain Covington, the owner's great-grandfather. *Blossom* has won the Governor's Cup seven times and *Bird* has won six times. The Norths also own *Jay Dee* and plan to retain her square stern. *Island Blossom* is shown here with eight of her crew visible during the race of the Miles River Yacht Club in September 1964.

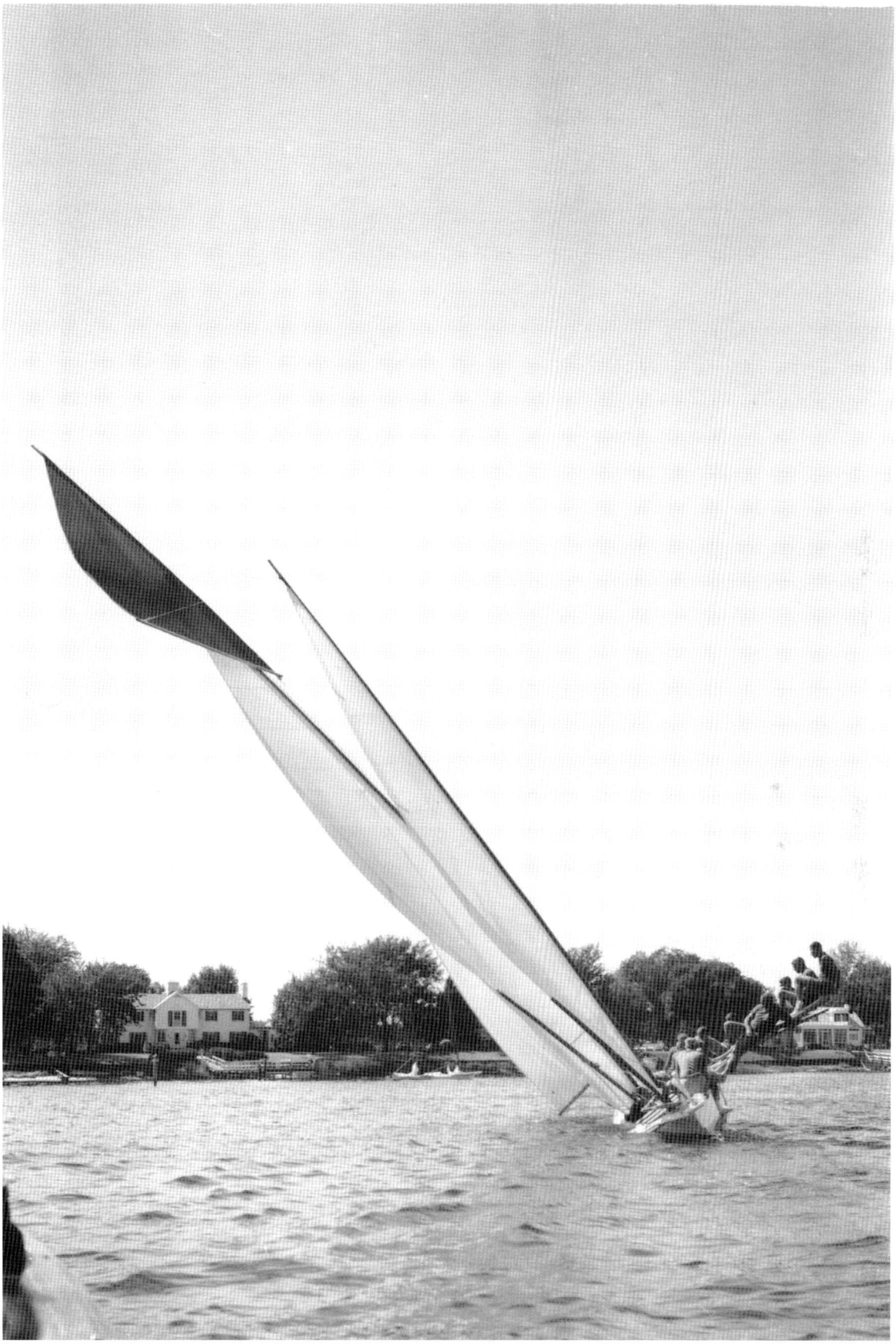

Log canoe *Silver Heel*, with her port washboard under water and her crew clambering out on the springboards to keep her upright, sails toward the finish line in the Miles River Yacht Club regatta for log canoes in September 1964. Her red kite atop her foremast matched the red shirts of her crew. This canoe was built about 1900 on Kent Island and at this time was owned by Robert Hewes of Chestertown.

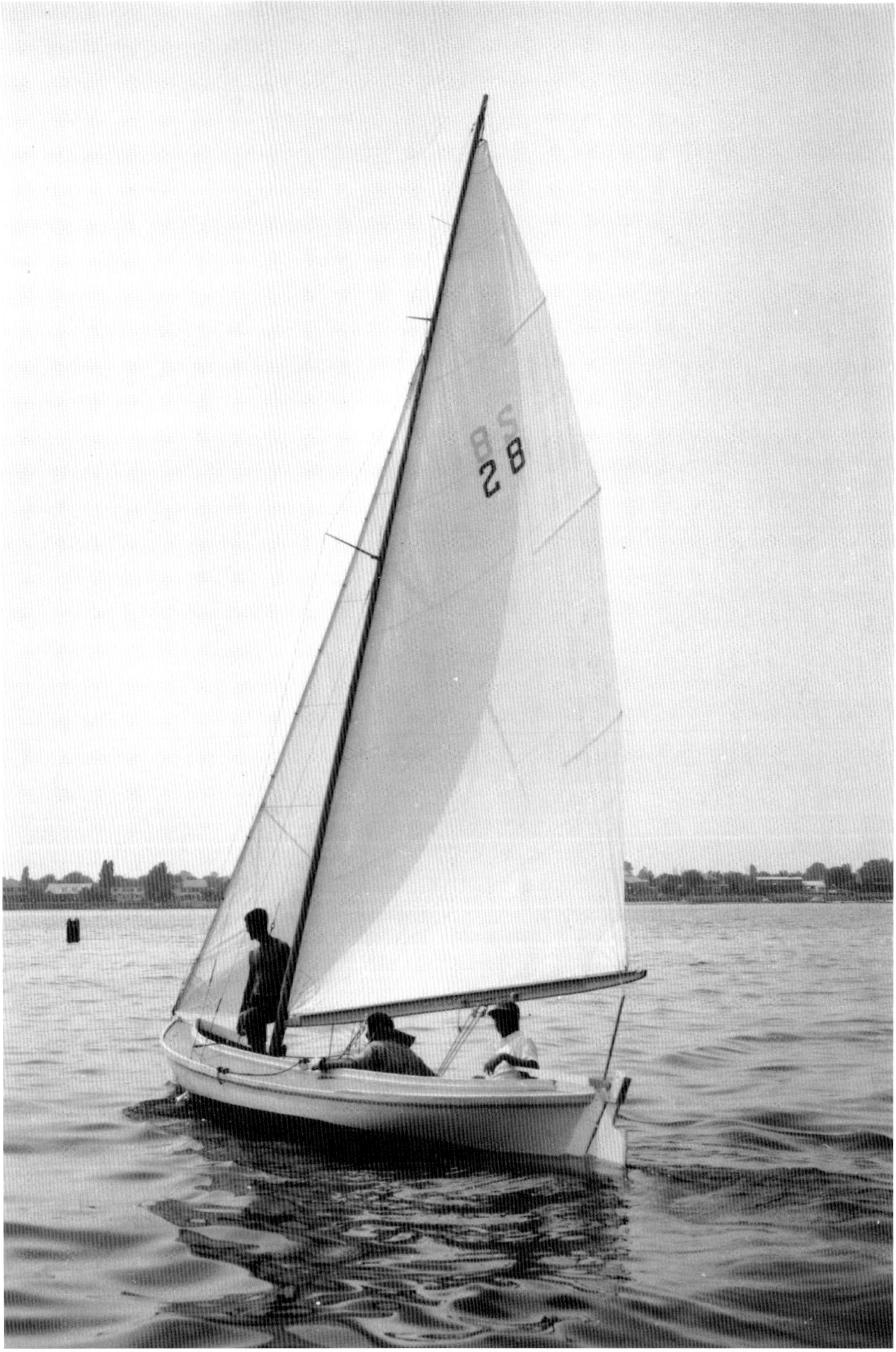

Kaynoo, 22-foot-long, two-log canoe built at Poquoson, Virginia, in 1876 by Harry Freeman and now owned by Milford V. Rollins of Hampton, Virginia. Rigged with a 30-foot mast bearing 260 square feet of sail, this craft has taken numerous trophies in the handicap class racing events in the lower Chesapeake Bay.

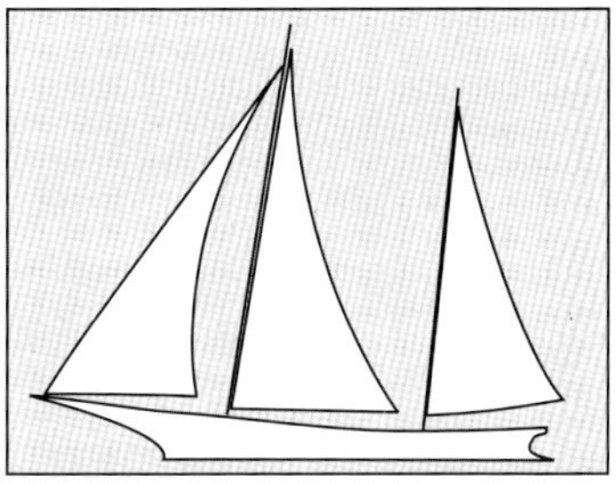

Bugeyes

In 1925 about 125 bugeyes still plied Chesapeake waters under sail alone. Other bugeyes had been converted to power but retained full sailing rig, using engines much as they would have used yawl boats—in calms and to assist maneuvering in harbor. A number were strictly powerboats. From that year on, the roster of those using sail diminished but up into the mid-1930s enough of them could still be found to convey the feeling that the Bay had not really changed very much. They were as symbolic of the Chesapeake as anything that ever floated there, and when one considers their origin it is easy to see why.

The bugeye came from genuine Chesapeake lineage and stock—the log canoe. The first ones were merely enlarged canoes, a result of the growth of the dugout into a type referred to as a "coasting canoe," which came into being during the first quarter of the 19th century. It measured 35 to 40 feet in length, had a fixed foremast, and removable mainmast. A cuddy cabin forward with a couple of bunks and wood stove made these small craft livable during cold weather and enabled them to be sailed to more distant city markets with an oyster catch rather than sell the shellfish to a buy boat at less profit.

The bugeye came into being because a lusty growing America demanded more and more of the succulent oysters that lived in coastal waters, most abundantly in the Bay and its tributaries. Oyster beds worked by tongers in the more shallow waters were becoming depleted. More productive beds were found, but these were in deeper waters beyond the reach of tongs. In the early 1800s a form of dredge, introduced by New England fishermen, had been used on the Bay, but its destructive effect on oyster beds caused it to be outlawed by the governments of Maryland and Virginia. During the mid-19th century Maryland, however, relaxed its ban on the dredge and permitted the use of a scrape in Somerset County waters. Coasting canoes were employed in dragging this device over the bottom, but it was soon realized that a larger, heavier craft could do a better job. The simple progressive step was to make the coasting canoe larger. This resulted in the log-built type, the brogan, measuring about 40 to 45 feet in length, and differing from her predecessor in that she had a deck, fixed masts with shrouds on the foremast, hatches, and a forward cabin. Because dredging had been permitted in other Maryland waters, the use of the log boats spread and toward the end of the Civil War, Maryland permitted the use of the dredge in Chesapeake waters over 15 feet deep. The sloop, pungy, and schooner were active on the Bay but certain inherent disadvantages made them unsuitable for dredging. For example, the bulwarks of the sloop and schooner were too high and the pungy's draft was too deep. Corrective measures were taken to make these types suitable for use but the larger log boats were ideal if they could be further increased in size. This was done through the use of additional logs. These materials were close at hand around the Bay country and there was a good supply of men who could fashion a strong, well-designed hull that could sail smartly and be sturdy enough to pull the heavy dredges over the bottom. Log boats were less expensive to build than the framed sloops and schooners of the period.

The name "bugeye" was not associated with the enlarged log boat until sometime in the late 1870s and early '80s, and even today there is no definite explanation as to the origin of the term. It is linked with the buckeye since the hull resembled the outer half shell of the horse chestnut. Then there was the bucklar, a type of Scotch boat, and the Scotch "buckie," meaning oyster shell. The craft's hawse holes looked like bugs' eyes when observed head-on. There are several other explanations, all having equal reliability.

The Eastern Shore of Maryland became the center of bugeye building and Bay watermen were quick to respond to the advantages of the type. When the supply of suitable logs became exhausted, bugeyes were framed and planked as the other Chesapeake sailing craft but still adhered to the hull design and other features characteristic of the bugeye. It was not until the introduction of the deadrise-hulled skipjack—smaller, cheaper to build, and more ideal for the smaller catches of oysters being made in the late 1890s—that the bugeye started to lose its place. Bugeyes were constructed in limited numbers up into the second decade of the 20th century and several held on to sail, working on Maryland's oyster beds, into the 1960s. Others, converted to yachts and powerboats, still bear the name on the Chesapeake and other waters.

In the 1920s and '30s the bugeyes were carrying on, in limited numbers, as they did when they were first introduced on the Bay. Two of the larger, later-built vessels were still under sail: the *Gladys L.* (ex *Miriam*) and the *Mollie V. Leonard*. The *Gladys L.*, however, was stripped about 1932 and converted into an outright powerboat. The *Mollie V. Leonard* had an engine but held on to her sailing rig until after World War II. They traded on the Bay just as the schooners did, transporting whatever cargoes they could get although oysters were to play a large part in their activities. They did not dredge but served as buyboats.

Few bugeyes could be found on the dredging grounds in the latter days of sail, compared to the numbers of skipjacks employed in that fishery. Most of them freighted, taking on wheat, tomatoes, and watermelons at the country landings for city markets in Baltimore, Norfolk, Washington, and Richmond. The boats would try to get a paying return cargo and, if successful, it might be coal, fertilizer, or empty cans. Even lumber from down the Bay filled the holds and overhung the decks of some of the bugeyes, like the *Maggie E. Maguire, M.A. Sommers, Majestic, John Branford,* and *Florence Northam.*

At least one, the round-sterned *Norma,* converted to power about 1922, engaged in bootlegging. In the summer of 1931 she was intercepted by the Coast Guard near Salisbury and was found to have a cargo of contraband whiskeys and Scotch malt valued at $100,000. She was taken to Baltimore and tied up at Pratt Street for removal of her valuable goods.

In this period there were bugeyes with square rig, that is, the gaff rig of the schooner, such as the *R.B. Haynie, Catherine, G.W. Glenn, Thomas Blades, Avalon,* and *A.J. Lawson.* The *Haynie*, *Blades*, and *Avalon* even had main topmasts. The *Haynie, Catherine, Glenn,* and *Blades* had round sterns—as did the "one-masted bugeyes with skipjack rig," the *Sallie L. Bramble, Agnes* and *Hattie Lloyd!*

There were fast bugeyes, vessels that were to leave their mark on the roster of bugeyes whenever the subject of speed was discussed in association with Chesapeake sailing craft. The races staged during the '20s and '30s contributed to their fame. There may have been swifter craft in the earlier years, their prowess known among the Bay watermen, but when official races were held, they were covered by newspapers and boating magazines causing their fame to become more widespread and permanent.

The remark, "I've got the fastest bugeye on the Bay," made by Henry du Pont Baldwin at a banquet held by the Gibson Island Yacht Squadron in 1935, led to one of the last free-for-all races between two Chesapeake bugeyes. Baldwin owned the *Bee*, and his claim was challenged by J. Linton Rigg, co-owner of the bugeye *Brown Smith Jones*, which resulted in a race on October 26, 1935. The *Bee* was originally the workboat *Carrie Moore* and the *Brown Smith Jones* was a former Maryland state conservation boat long noted for her speed on the Chesapeake—never a workboat but fitted out as a yacht and built to maintain peace on the oyster beds.

At 2 p.m. on the designated date the two bugeyes crossed the starting line off Gibson Island, Maryland, to establish once and for all, in the eyes of the owners, the fastest bugeye. The course was set down the Bay to Sharps Island Light and return to Gibson Island, a distance of approximately 60 miles. The boom of the starting gun found the *Brown Smith Jones* bearing down on the line with a spanking west wind on her starboard beam, leading the *Bee* by half a dozen lengths. Not long after she crossed the line, Rigg broke out a ballooner, increasing the speed of his boat considerably. The *Bee* followed suit, but as her huge sail filled a line parted and sail dragged in the water. After a second try the *Bee* got her big headsail to draw and she picked up speed toward the *Brown Smith Jones,* in the lead by 200 yards.

Misfortune stalked the *Bee,* for after a few minutes her ballooner blew out. This was replaced by a smaller sail. There were no sails barred in this race, unlike the official workboat races that permitted only sails used as a bugeye's working rig. The only requirement was to round the marker at Sharps Island Light and cross the finish line. The *Brown Smith Jones* had added an extension to her bowsprit to carry extra sail and had rigged a temporary fore topmast. The *Bee* carried two large square sails formerly used on the racing log canoes *Flying Cloud* and *Island Bird,* the latter owned by a member of the *Bee's* crew.

By the time the halfway mark had been reached, the *Bee* had crept up on her opponent so that the crews of the two boats could converse across the water. Right after making the turn, the *Brown Smith Jones* was maneuvered to cover the *Bee* and in an increasing wind she was never headed. At the finish, the *Bee* was almost an hour behind the lead boat with the wind failing rapidly. When the former Maryland police boat crossed the line she had earned the title, "fastest bugeye on the Bay"—that is, for those used as yachts.

The true working bugeye fleet had other claimants to that title. There were the fast bugeyes *Isaac H. Keeler* and *George Todd,* owned by the Todd family of Crisfield, the former winning once and the latter four times in the annual Chesapeake Bay Workboat Regatta between 1921 and 1931. In addition, there were the *Nettie B. Greenwell, Florence Northam, Agnes C. Leonard,* and *Emma A. Faulkner,* all having won the bugeye event in the official workboat regattas one or more times. A match between the *Brown Smith Jones* and one of the fast working bugeyes would have been interesting but that never came about. Yachtsmen appreciated the qualities of the swift workboats, for the *Agnes C. Leonard* and *Florence Northam* eventually became pleasure craft. Other bugeyes to take up a life of ease as yachts in their latter days were the *Little Jennie, Colonel R. Johnson Colton, Ethel Vail, Dorothy A. Parsons, Maggie E. Smith, Sanxton Hubbard,* and *Richard J. Vetra.* The *George Todd, Isaac H. Keeler,* and *Nettie B. Greenwell* were converted to working powerboats as engines outmoded their sails.

It seemed a sacrilege to convert a bugeye into a powerboat. Its graceful longhead and tapering bowsprit were cut off, sometimes leaving the heel of the latter in place as if a telephone pole had been lopped off. The mainmast was removed and a high pilothouse was erected over the after cabin. A powerful engine was installed deep in its bowels. The foremast was made more erect and shortened, at first carrying a sail to assist the engine in a fair wind and later abandoning that rig to rely solely on power. A boom would be fitted and rigged with block and fall to hoist oysters out of the hold.

Bugeyes relied upon wheat as a cargo to keep them busy during the summer months. In 1936 the superintendent of the Western Maryland Railway grain elevator in Baltimore stated that about 1,000,000 bushels of Eastern Shore wheat would be brought to Baltimore, principally by water. The first vessel to bring in a cargo of that year's crop was the *Mollie V. Leonard,* 3,700 bushels, on July 1—early for the season. Following in her wake were the bugeyes *Preston Lewis, Emma A. Faulkner,* and *Vaughan,* all contributing their small share to the total million bushels.

But the days of freighting bugeyes under sail were drawing to a close even then. Complete power propulsion was to become the fate of most of these colorful craft in the late 1930s and early '40s. If these large bugeyes not employed as oyster dredgers were to earn any kind of a living, the engine was the answer. Those not suitable for the transition, due to their condition, were abandoned and even now old age has caught up with most of the motorized ones and they, too, have been consigned to the shallows to await gradual disintegration. However, occasionally one can see the unmistakable hull of a former bugeye carrying out her role under power, much the same as she did under sail.

Some of the bugeyes struggled along under sail into the 1950s and 1960s, as part of Maryland's oyster dredging fleet, like the *J. C. Armiger, Richard Smith, Thomas M. Freeman, Hugh S. Orem, R.H. Daugherty, Clarence and Eva, Catherine,* and *Edna E. Lockwood.* With the exception of the *Freeman* and *Lockwood,* all have been abandoned and there are no more bugeyes working under sail, that is, unless one considers the role of the *Lockwood* that "works" during the summer months to teach youngsters the elements of sailing as part of the program of the Chesapeake Bay Maritime Museum, St. Michaels, Maryland, where she is an exhibition vessel.

The bugeye was on the Chesapeake just about a century; a small period in the matter of time, but it will be remembered as a most important part of the Bay scene as long as man sails those waters.

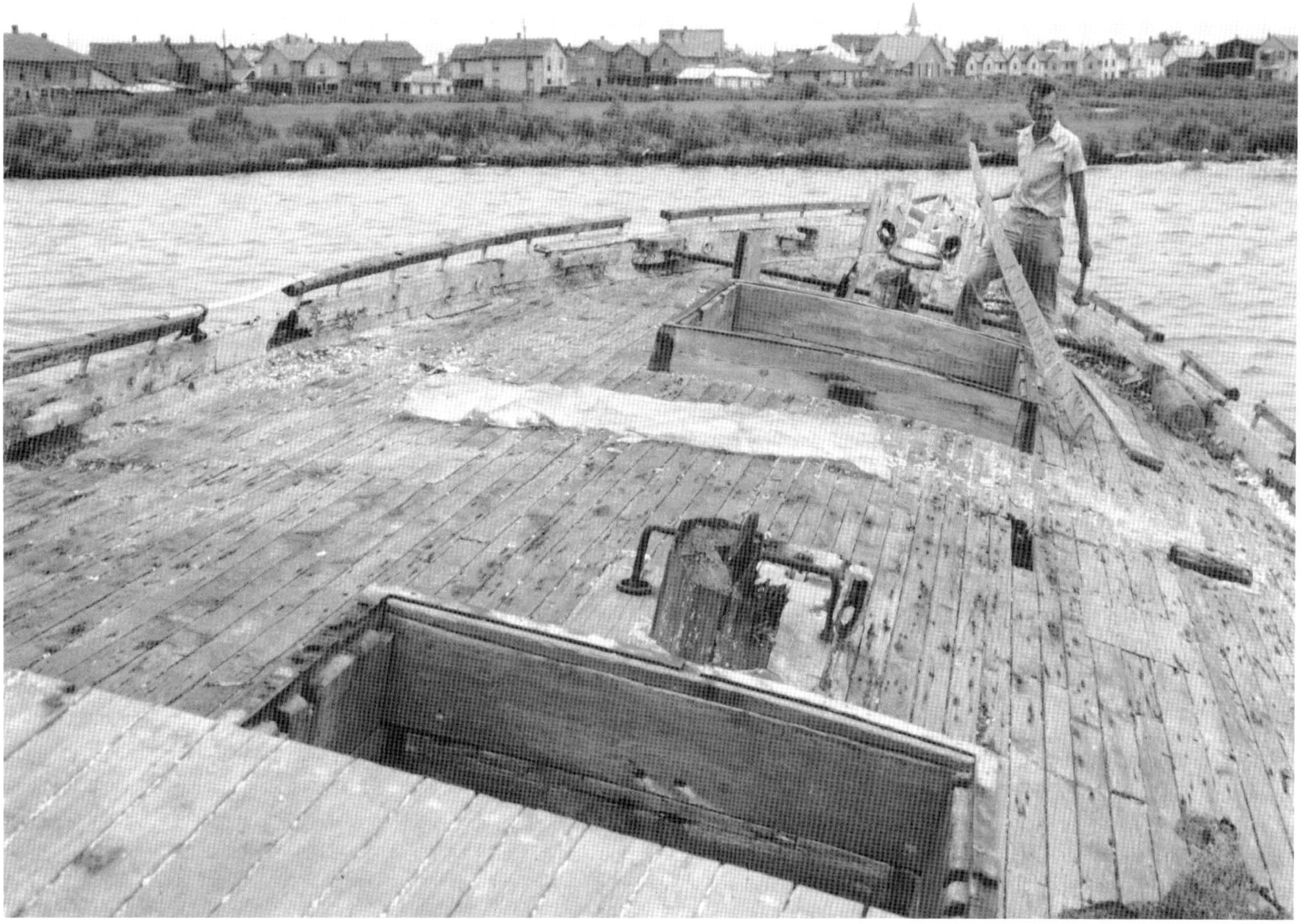

Hulks could be found around the Chesapeake region even up into the 1950s, if you knew where to look for them. Earlier, many of the sailing craft, old and worn out, were abandoned in shallow areas close to where they were owned. Here the bugeye *A.J. Lawson* lies in the backwaters of a cove at Crisfield, Maryland, on July 11, 1950. Built at Pocomoke City, Maryland, in 1883, this bugeye was unusual in that she had a gaff rig. Discarded about 1936, the *Lawson* shows utter decay about the deck and had been stripped of everything usable. These waters were eventually dredged, the *Lawson's* timbers removed, and a modern marina built along the marshy areas in the background. ✪ Hammer in hand, Fred Kaiser, owner of the log canoe *Flying Cloud* discussed in a previous chapter, prevents the loss of the *A.J. Lawson's* port trailboard.

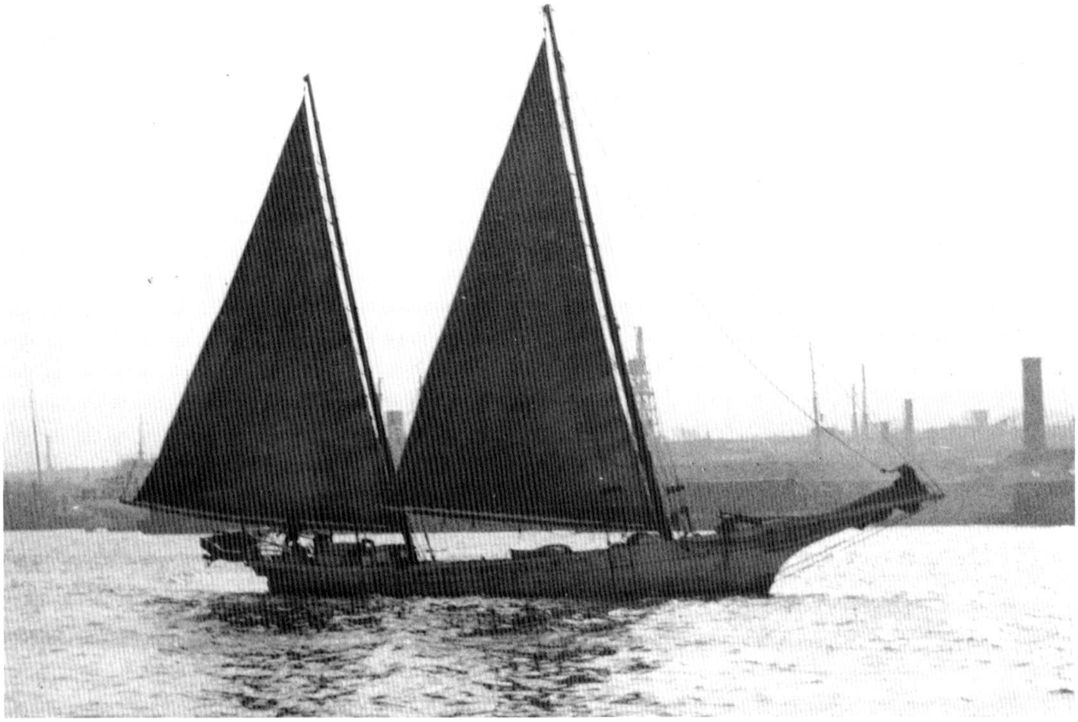

Agnes C. Leonard, built at Solomons, Maryland, as the *Willie H. White* by M.M. Davis in 1883, has beat her way up Baltimore harbor and is shown approaching Long Dock on December 21, 1933, with a load of oysters for Christmas tables. About 1940 she was converted into a yacht and renamed *North State.* In 1946 she went back to freighting and was still in the 1972 issue of *Merchant Vessels of the United States (MVUS).*

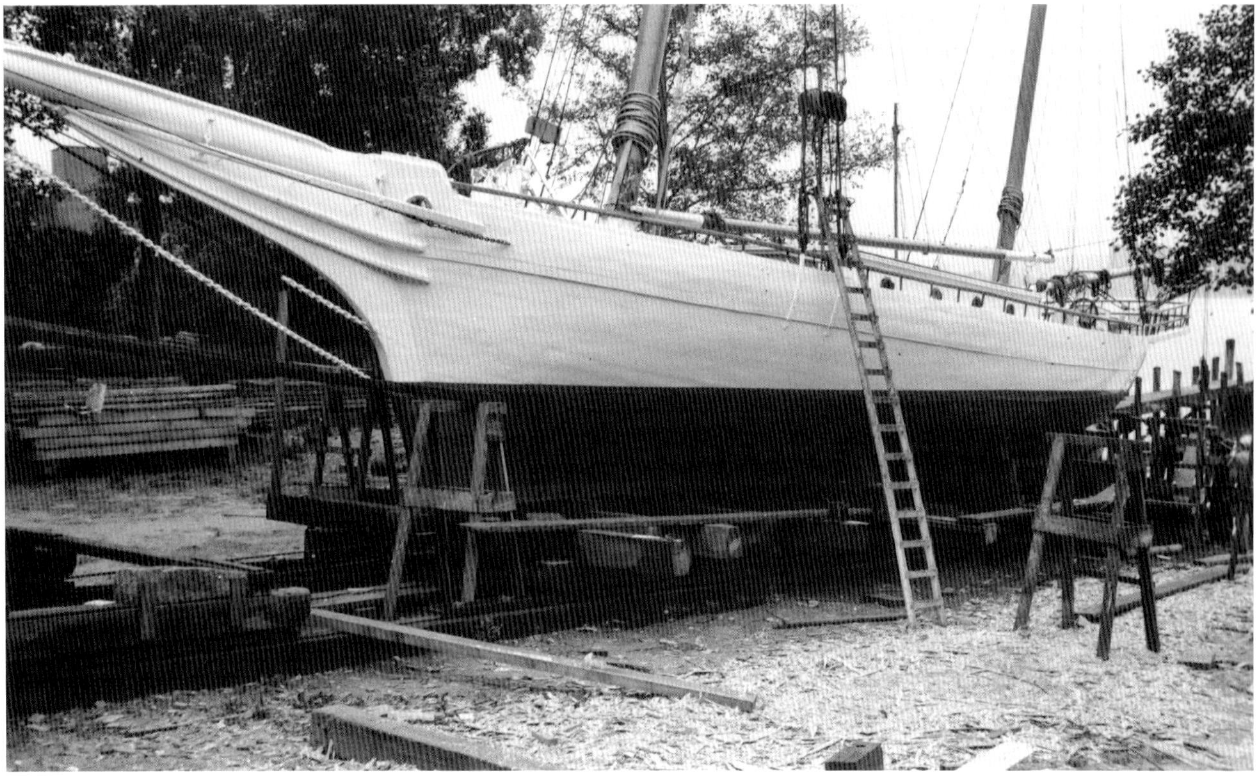

The *Brown Smith Jones* was built at Cambridge, Maryland, in 1894 by G.T. Johnson, for the Maryland State Navy to help enforce the state oyster laws and was named for Governor Frank Brown, Comptroller Marion de K. Smith, and Treasurer Spencer C. Jones. With yacht-like accommodations, she was eventually converted to that use, as she appears on the marine railway of A. Smith & Sons Shipyard, Curtis Bay, Baltimore, in June 1940.

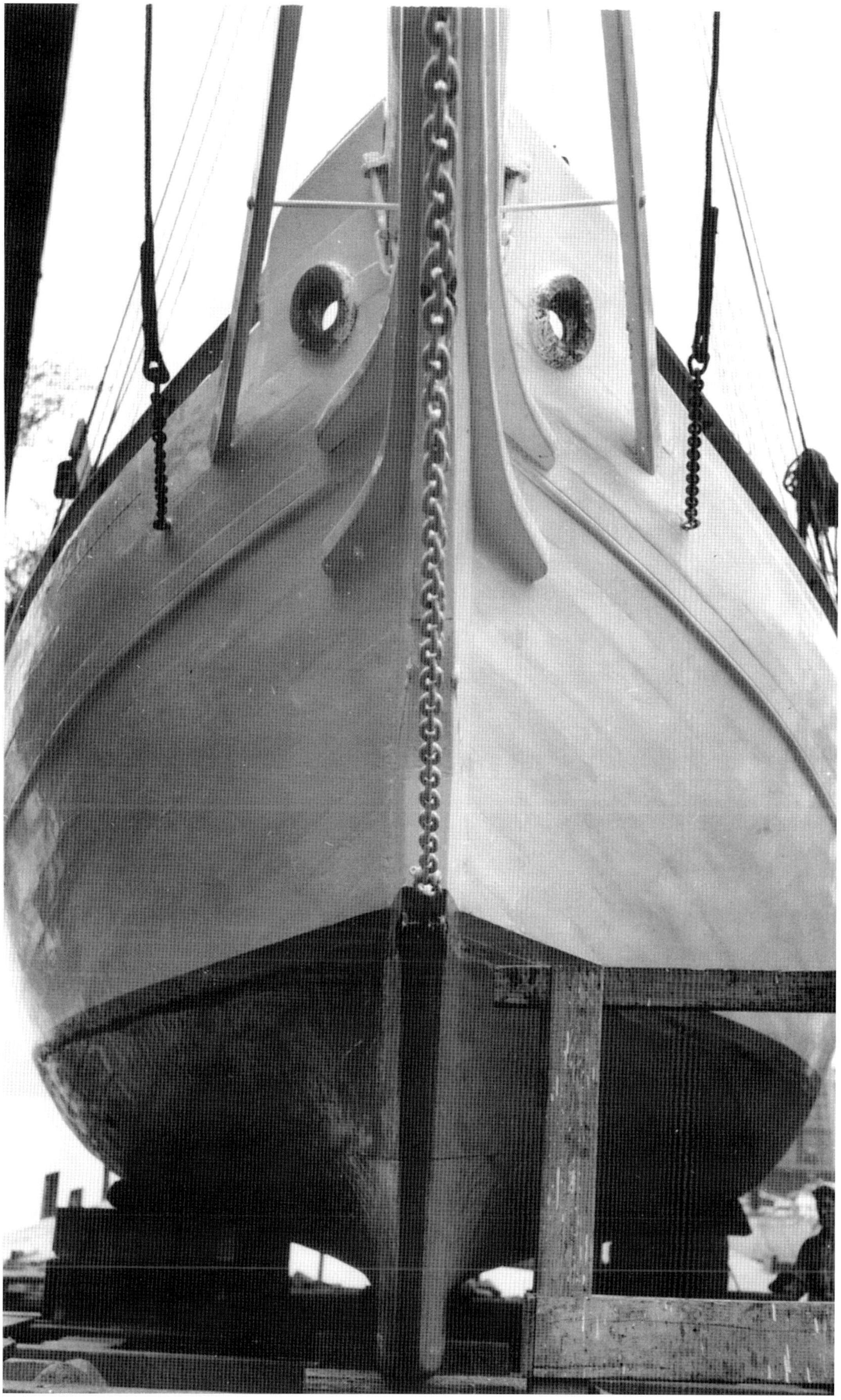

This bow view of the *Brown Smith Jones* supports the belief of some that bugeyes were given that name because the craft's hawsepipes resembled bug eyes.

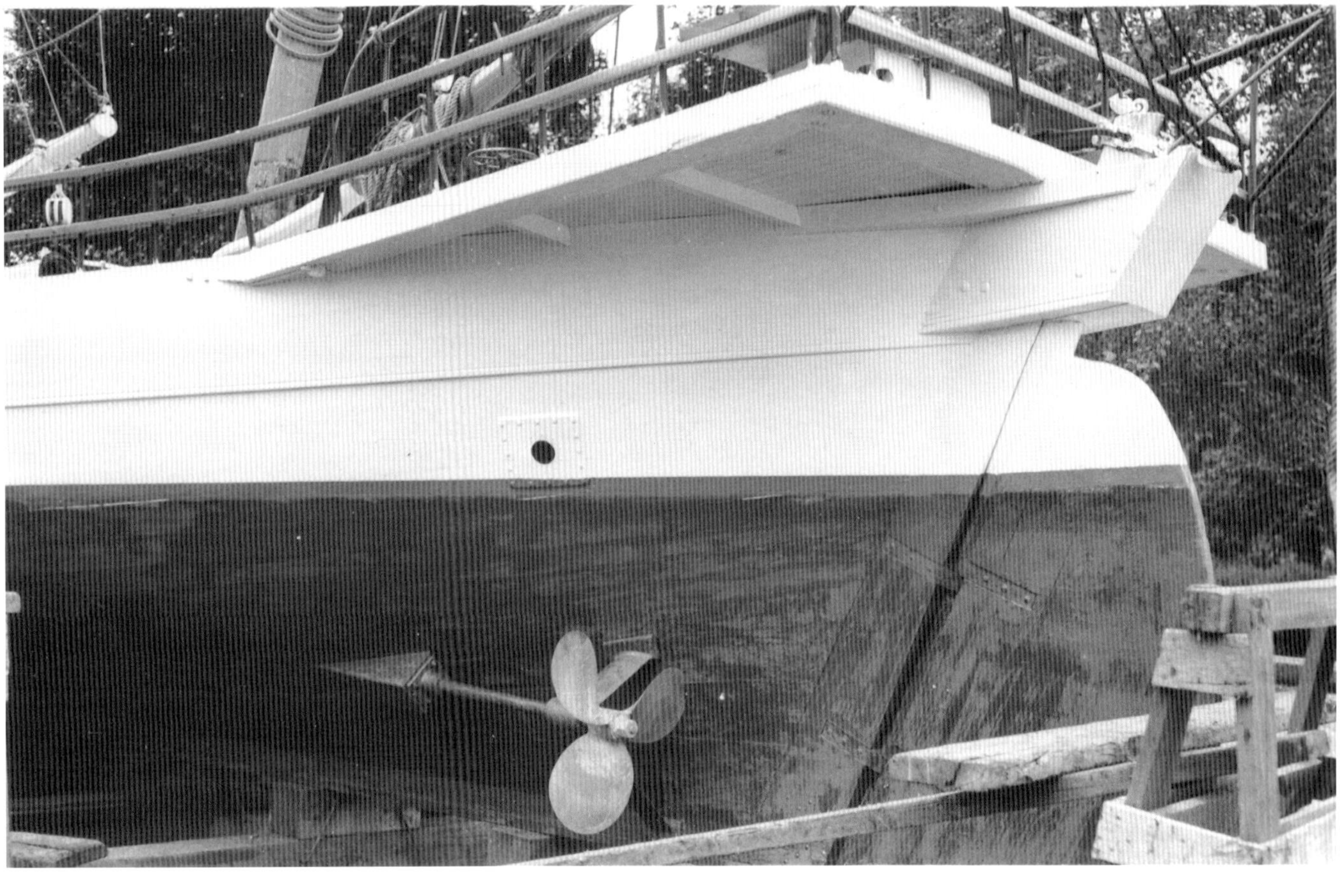

The drake tail and patent stern of the *Brown Smith Jones.* The patent stern was invented in 1908 by Captain Joseph E. Robbins of Cambridge, Maryland, to provide a means for supporting the pipe davits on the bugeye's sharp stern. Originally it was just a framework, but later grating or planking was placed over the beams to create the needed extra deck space. Captain Robbins patented his idea and all who used it paid him a small royalty. The off-center propeller was added when the bugeye became a yacht. About 1954 *Brown Smith Jones* was sold to Massachusetts and was dismantled about 1965 after being abandoned at Woods Hole.

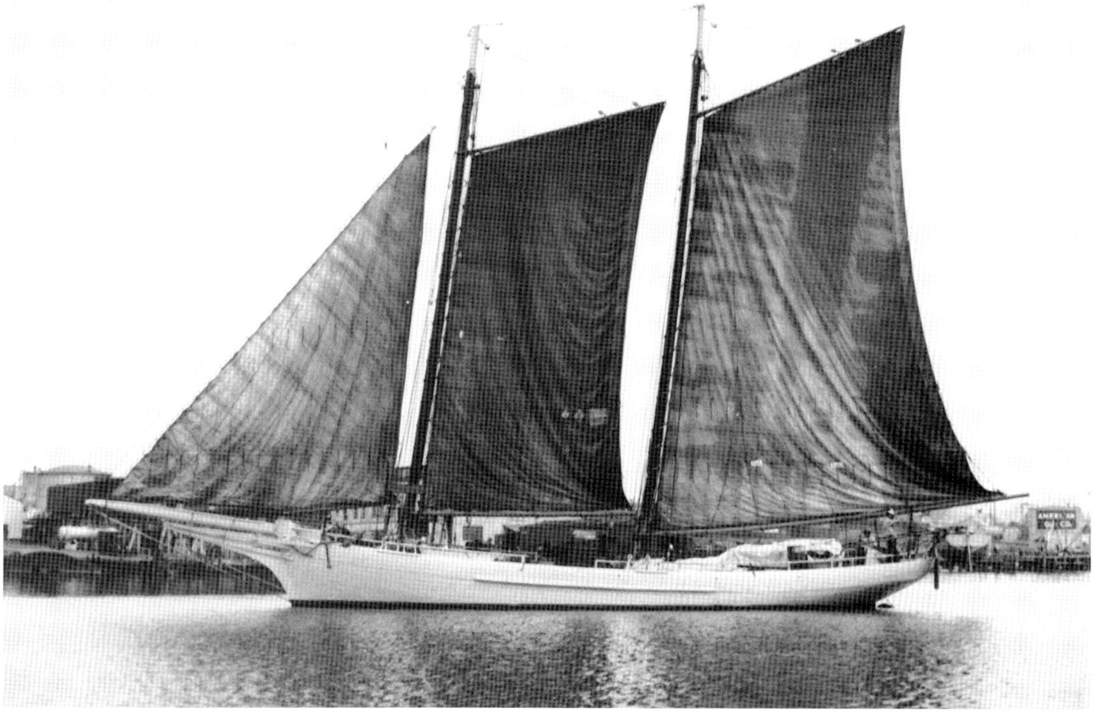

The round-sterned, gaff-rigged bugeye *Catherine* was one of several of that type to depart from the sharp stern and sharp rig. She had been built in 1911 at Solomons, Maryland, by M.M. Davis. When photographed at anchor in Curtis Creek on August 31, 1938, she had just come off the A. Smith & Sons marine railway. In the winter months she served as an oyster dredger.

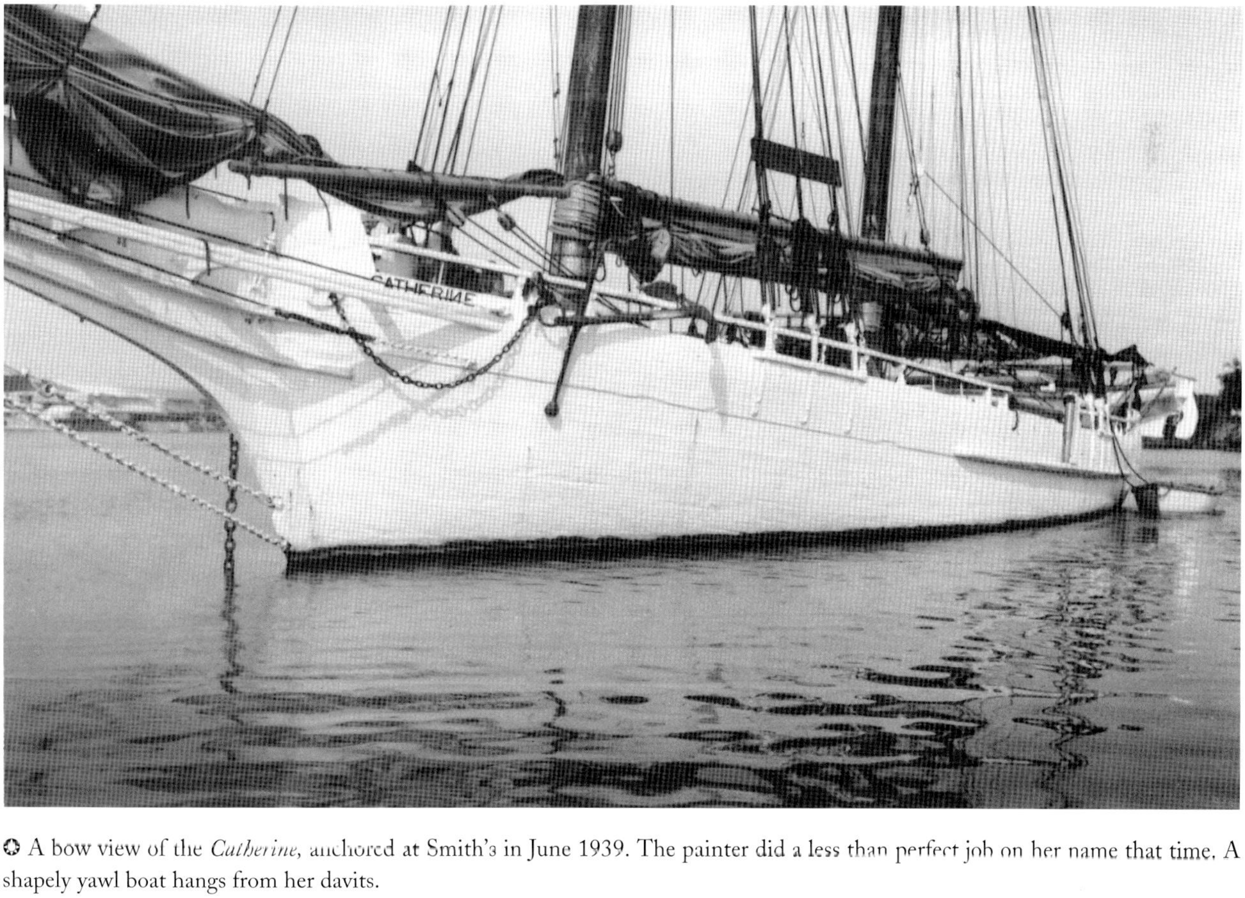

✪ A bow view of the *Catherine*, anchored at Smith's in June 1939. The painter did a less than perfect job on her name that time. A shapely yawl boat hangs from her davits.

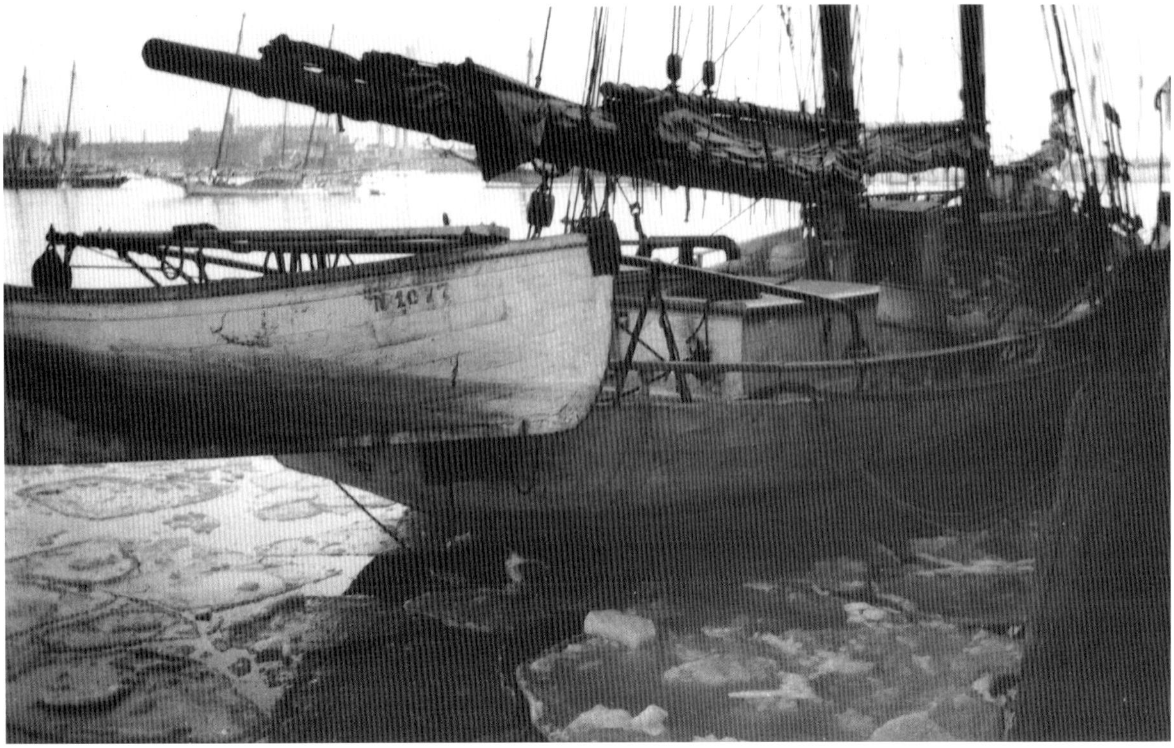

✪ A stern view of the *Catherine* in ice at the Lord-Mott Company, Fells Point packers, on February 18, 1934.

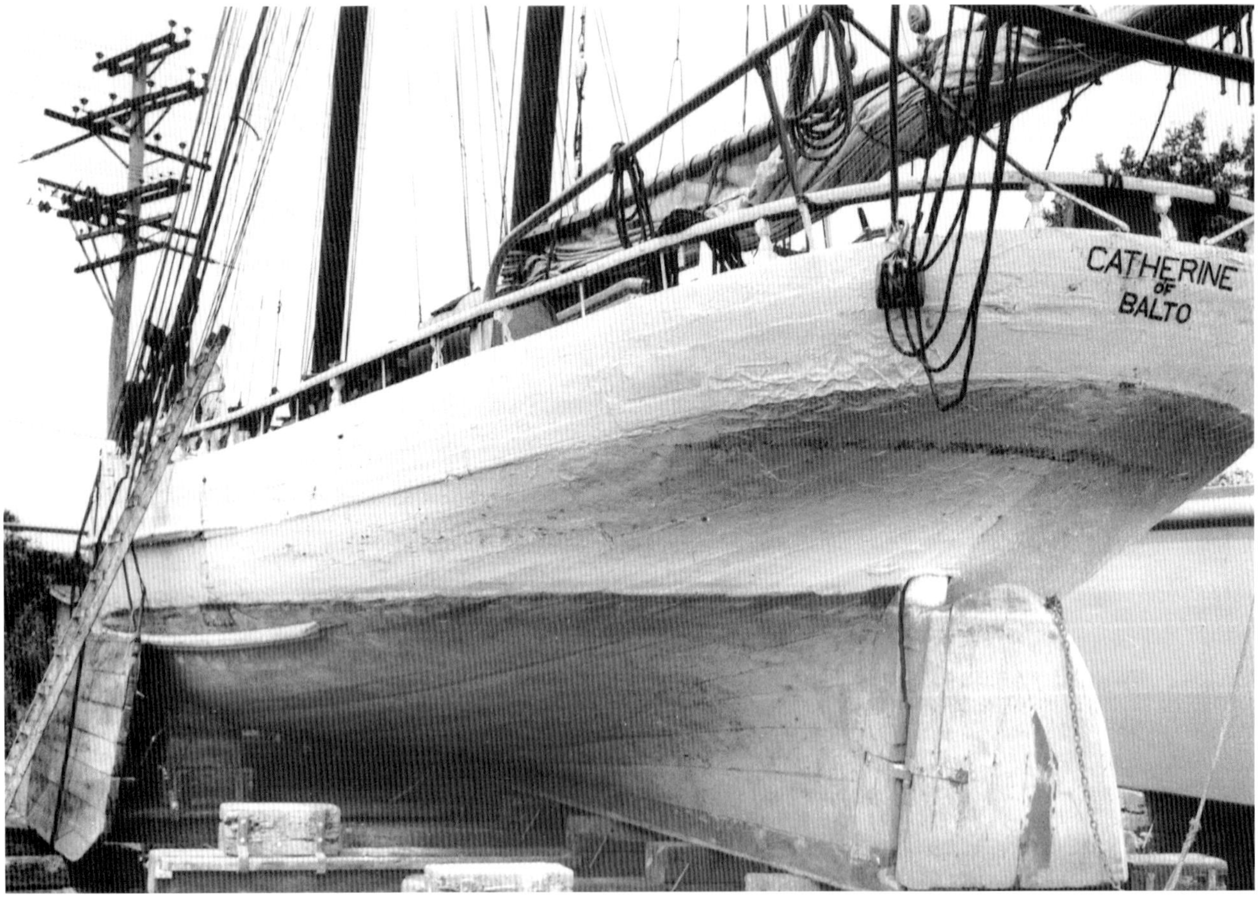

The *Catherine* exposes her underbody and shows her round stern to advantage on the marine railway of A. Smith & Sons, June 28, 1940. The advantages of the round stern were to provide more deck space aft, more room to handle the main sheet, support the stern davits to carry the heavy yawl boat, and protect the rudder. The centerboard lies along the port side, removed so that it and the interior of its trunk can be painted. The *Catherine* dredged oysters in the Chesapeake just about up to the time she was abandoned near Annapolis in 1956.

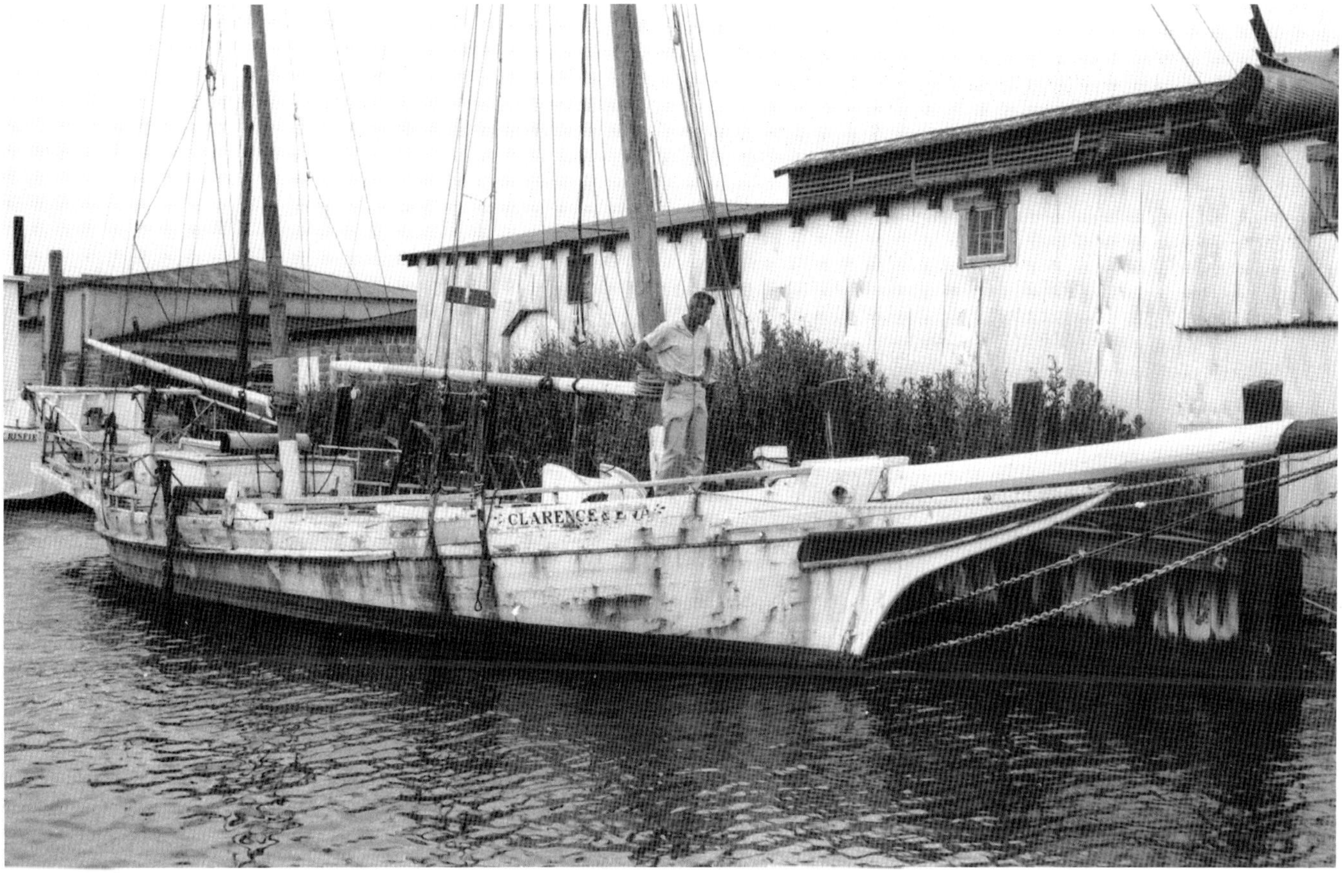

The 45-foot bugeye *Clarence & Eva*, shown here in Crisfield on July 10, 1950, as she was nearing the end of her career as an oyster dredger. She had been built at Cambridge in 1889 by R. Spear. In the 1950s she was abandoned near Crisfield but was still listed in the 1972 *MVUS*. ✪ Hampton yachtsman Fred Kaiser examines her deck.

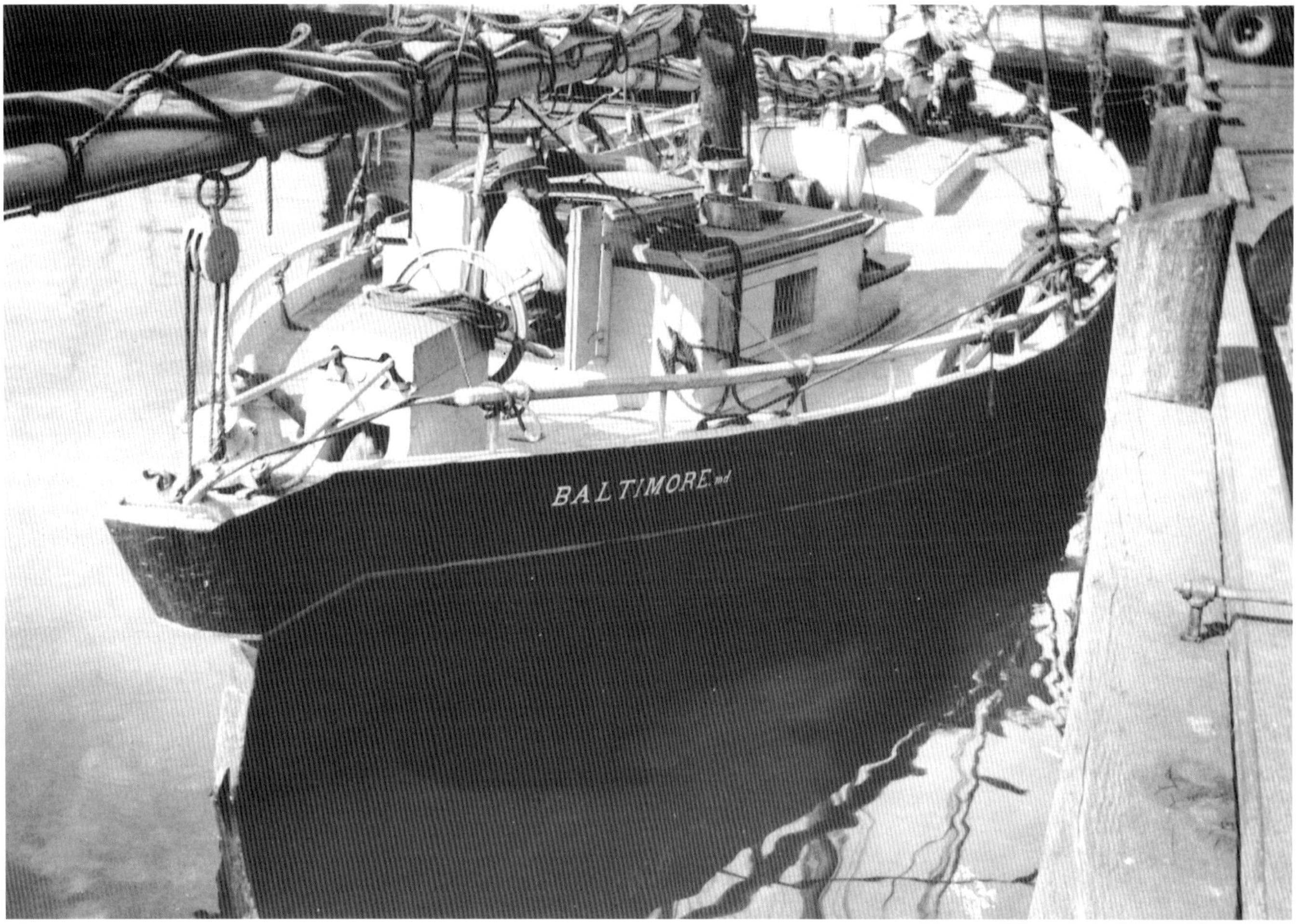

The drake-tail stern, without the accompanying patent stern, was unusual on the Chesapeake when the bugeye *Dan* was photographed at Pier 4, Pratt Street, Baltimore, October 12, 1934. She had been built by T. Kirby, at St. Michaels, Maryland, in 1891. When this photograph was taken she had been motorized, as shown by the exhaust pipe leading over the stern to the port of the wheelbox. As a result she did not require a yawl boat to push her nor a patent stern to support davits. She carried a skiff on deck to the port of her water barrel amidships. Her elderly captain sits in the companionway ready to converse with anyone who pauses on the wharf. Of interest in this view is the steering mechanism, undoubtedly the Lake steering gear, which suited the limited spaces of a sharp-sterned bugeye. Emerging from the wheelbox can be seen two large rods, or drag links, that are attached to a gear and then extend to a yoke on the rudder head.

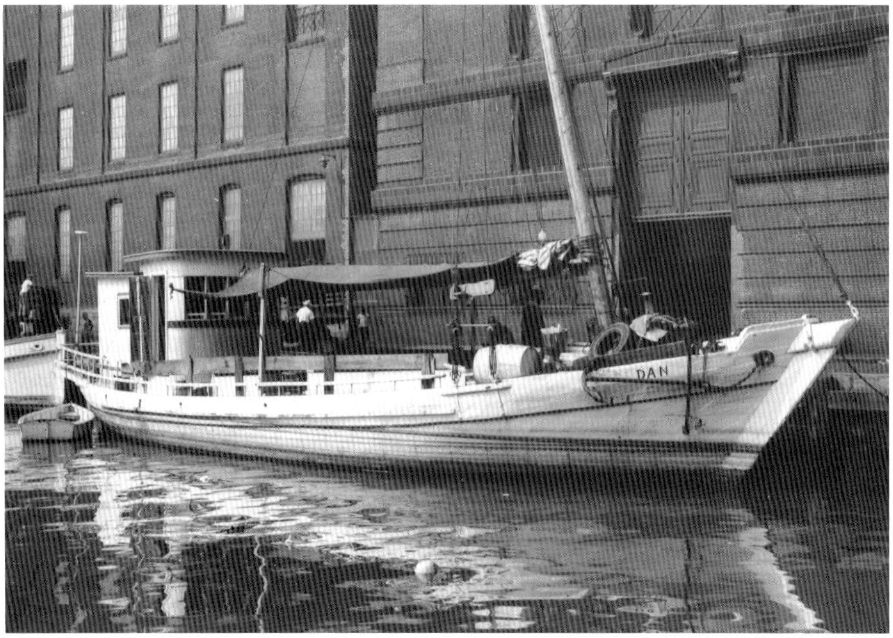

About 1936 *Dan* was converted into a full powerboat. With her graceful longhead and bowsprit cut off and mainmast removed, an engine was installed in her former cabin space and a pilothouse erected atop that. Then a patent stern was added to increase deck space aft. She retained her foremast and foresail to assist her engine in a favorable wind. In September 1948, she visits Long Dock, Baltimore, with a cargo of watermelons. ✪ She was reported as "Dismantled" in the 1978 *MVUS.*

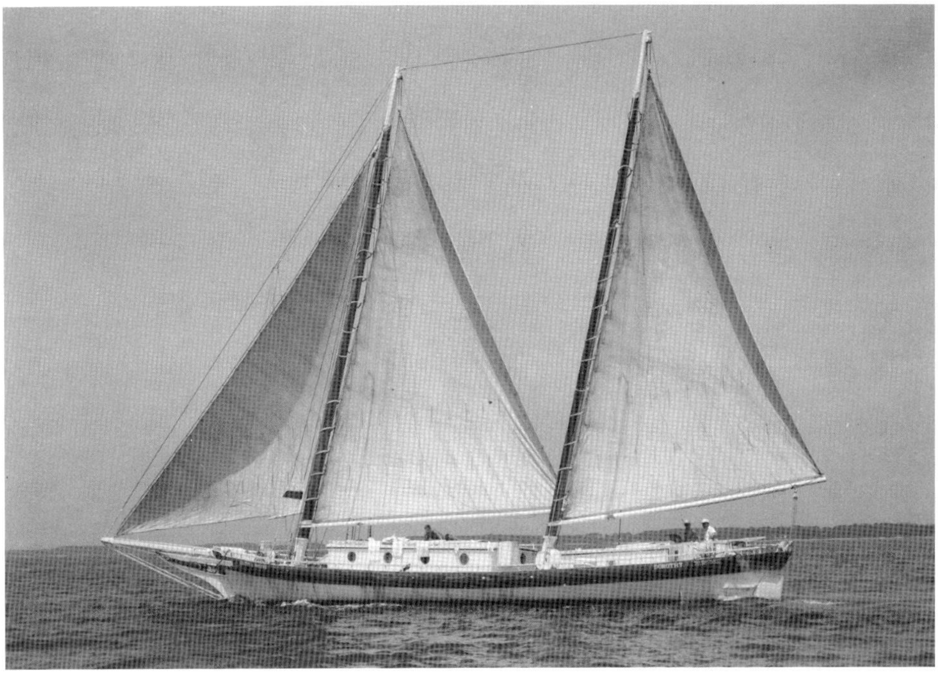

Dorothy A. Parsons sails and motors up the Rappahannock River as a yacht in July 1951. Built by B.P. Miles at Oriole, Maryland, in 1901, she had been powered about 1914 as a workboat and converted to a yacht about 1940. She was presented to Mystic Seaport as an exhibit and rebuilt to her original appearance. Later, the Seaport gave the bugeye to the Harry Lundeberg School, Piney Point, Maryland, to be preserved as an example of her type. A number of the Chesapeake sailing craft were converted into yachts, but the large deckhouse, to allow suitable headroom, usually ruined their appearance.

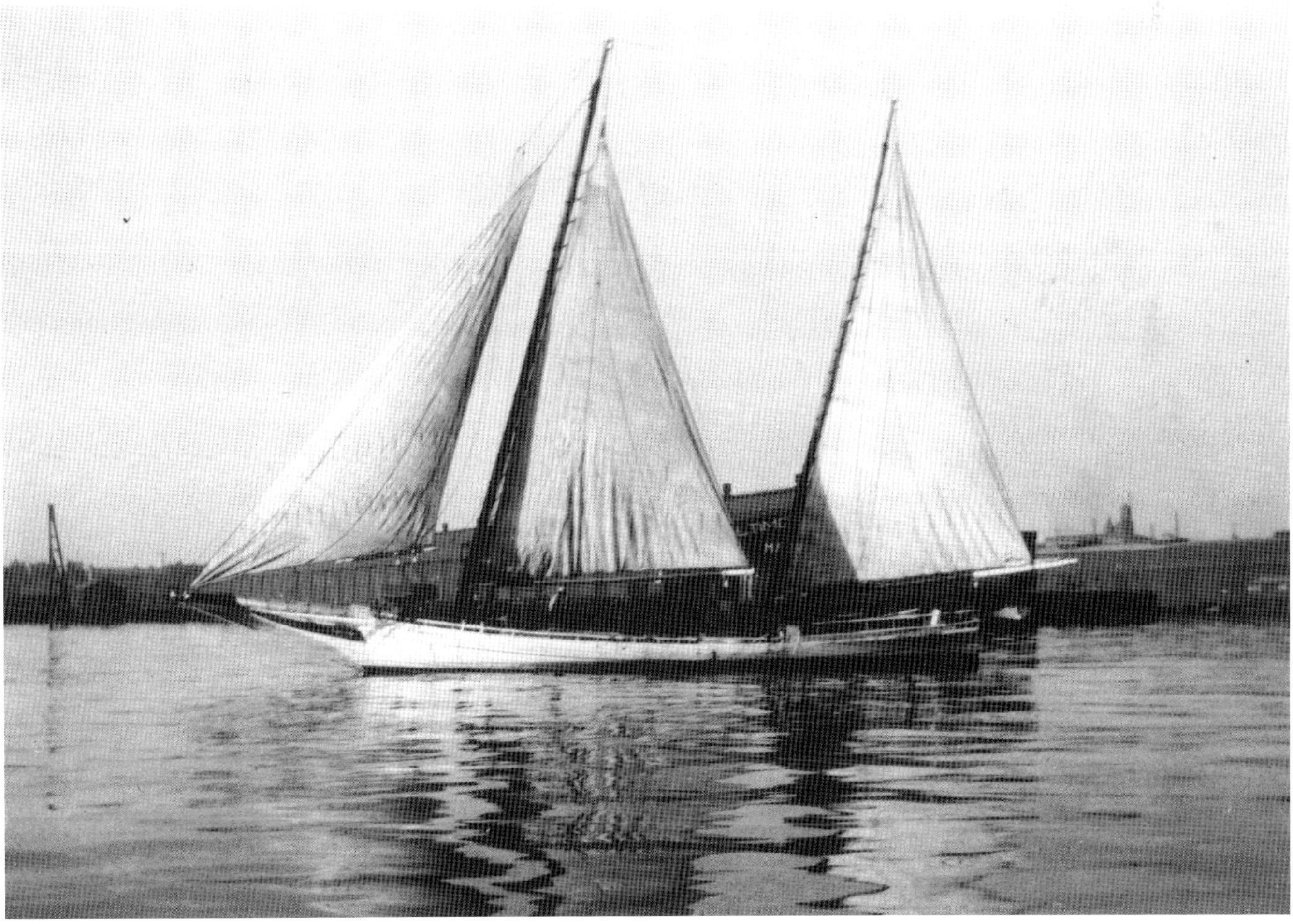

Virtually drifting out of Baltimore harbor, with the Locust Point docks in the background, is the *Edith F. Todd,* on September 13, 1933. She had brought in a cargo of watermelons to Long Dock and tied up there until she had sold them over the side. She has single reefs in her foresail and mainsail and probably was expecting heavier winds when she reached the open Chesapeake. The *Todd* was built in 1901 at Oriole, Maryland, and was listed as "Abandoned" in the 1947 issue of *MVUS.*

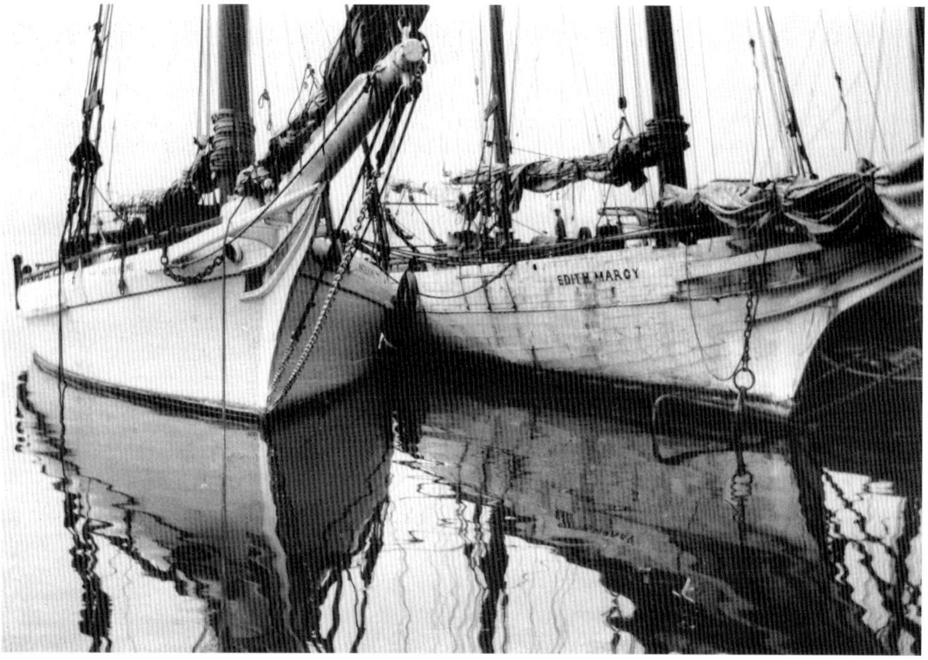

Bugeye *Edith Marcy,* at right, tied up at Montford Yard, Baltimore, with the skipjack *Katie Rowe,* October 27, 1935. The photograph was made a few days before the oyster season opened, which accounts for the *Katie Rowe* being freshly painted after a summer of idleness except, perhaps, for a trip or two to Baltimore with watermelons. The *Edith Marcy* was serving as a buyboat, or market boat, and did not take time out for refurbishing unless it was absolutely necessary. She had been built in 1895 at Solomons, Maryland, by M.M. Davis. Completely motorized about 1938, she freighted lumber from Harryhogan, Virginia, to Baltimore for many years. About 1967 she was sold to Philadelphia owners, and disappeared from *MVUS* in 1971.

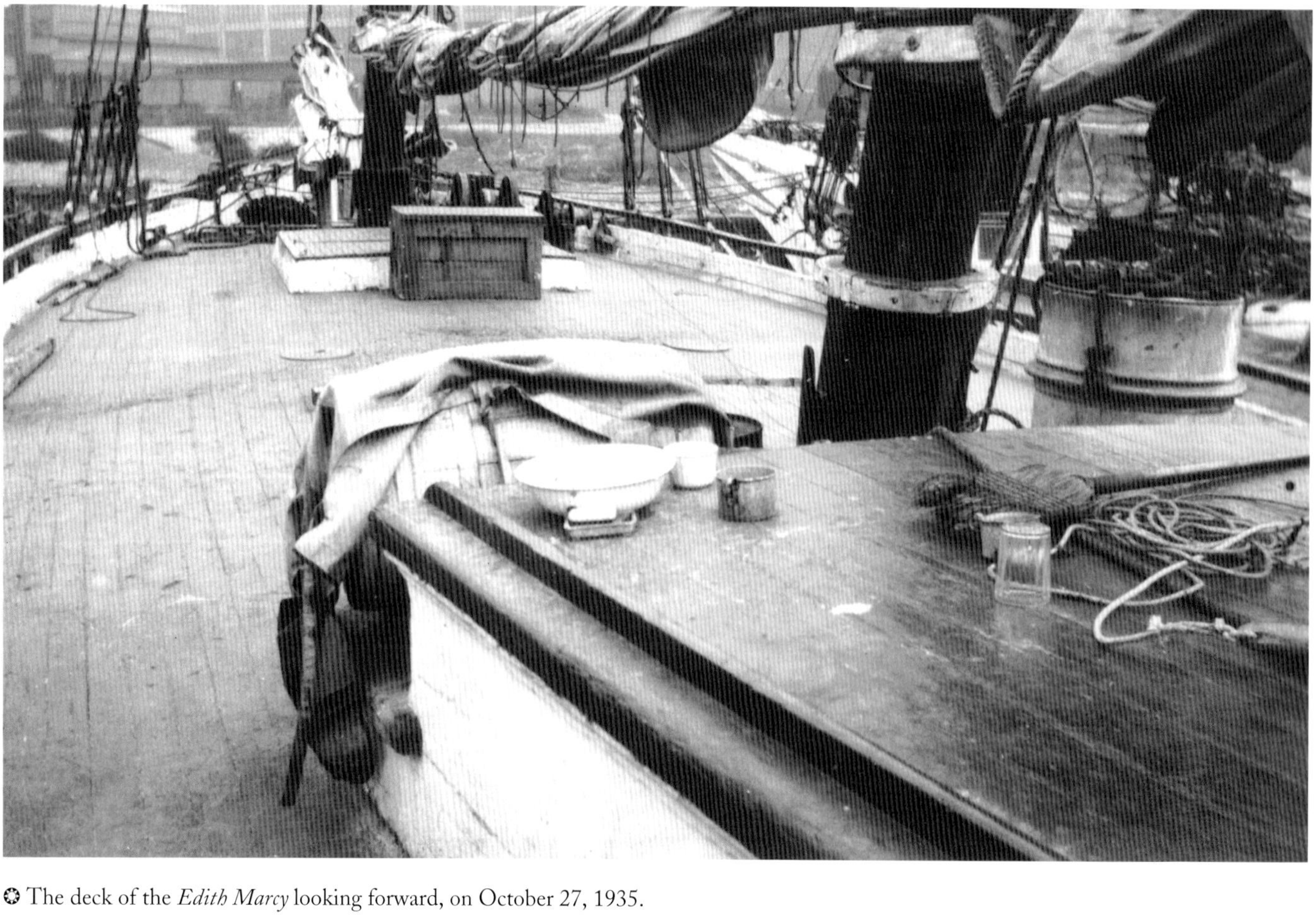

✪ The deck of the *Edith Marcy* looking forward, on October 27, 1935.

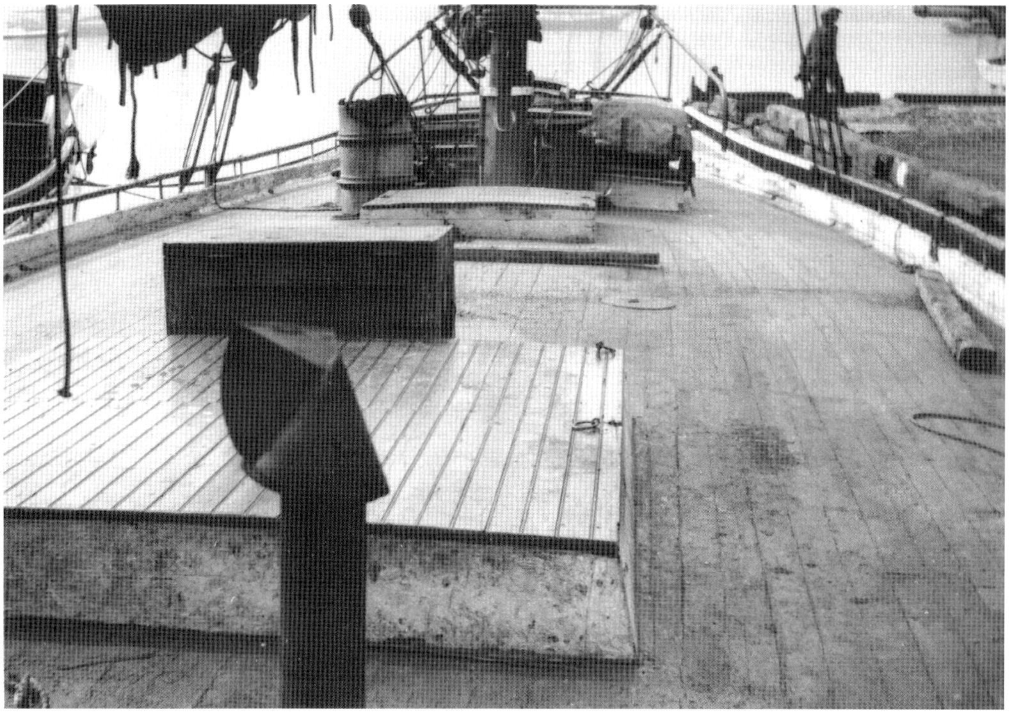

✪ The deck of the *Edith Marcy* looking aft, on the same day, with her Charley noble (galley stove-pipe) in the foreground.

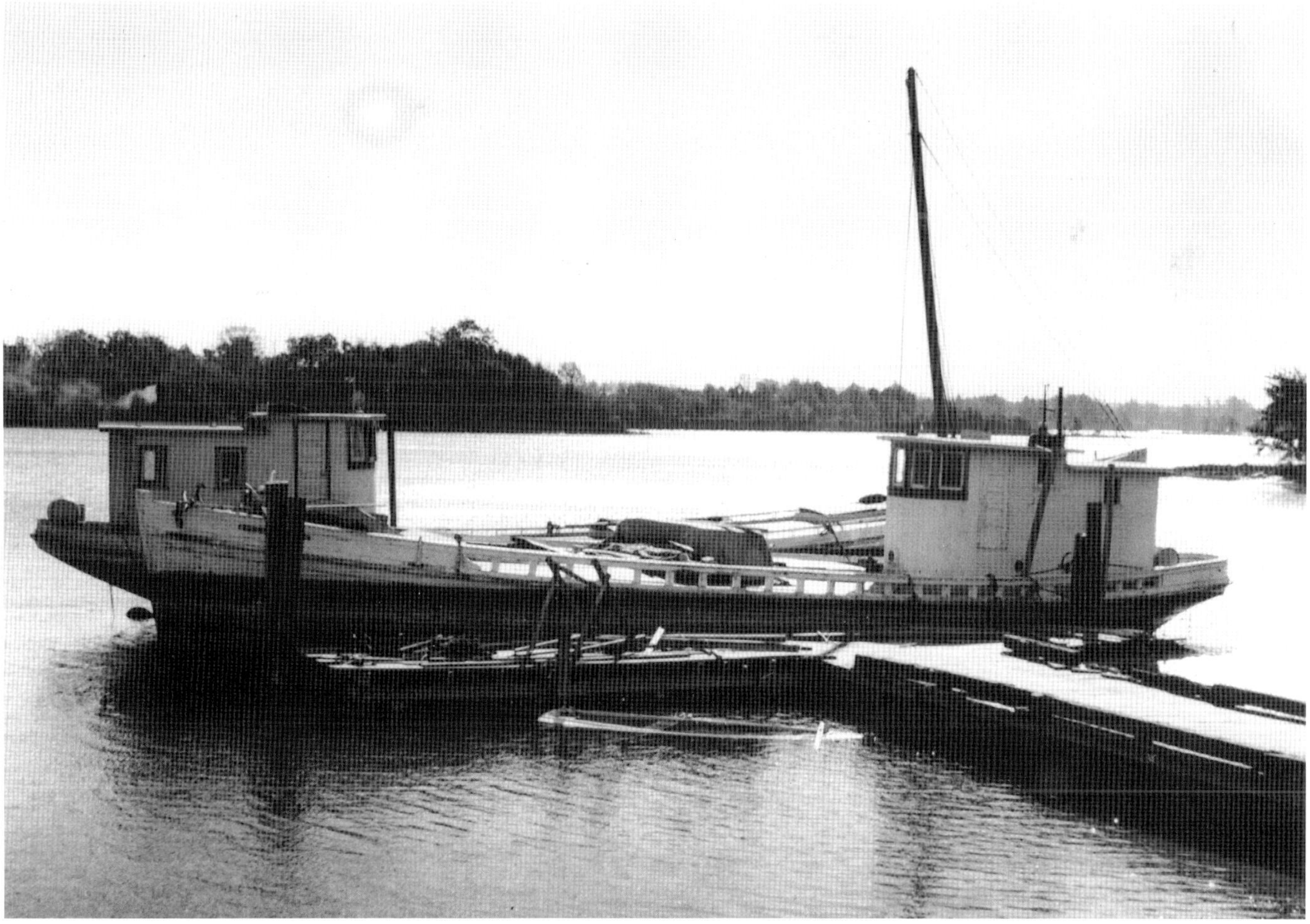

✪ The bugeye *Edith Marcy* after her conversion to diesel power, at Harryhogan, Virginia, on November 9, 1947. The former *Amanda C.*, also converted to power, is moored outboard of her.

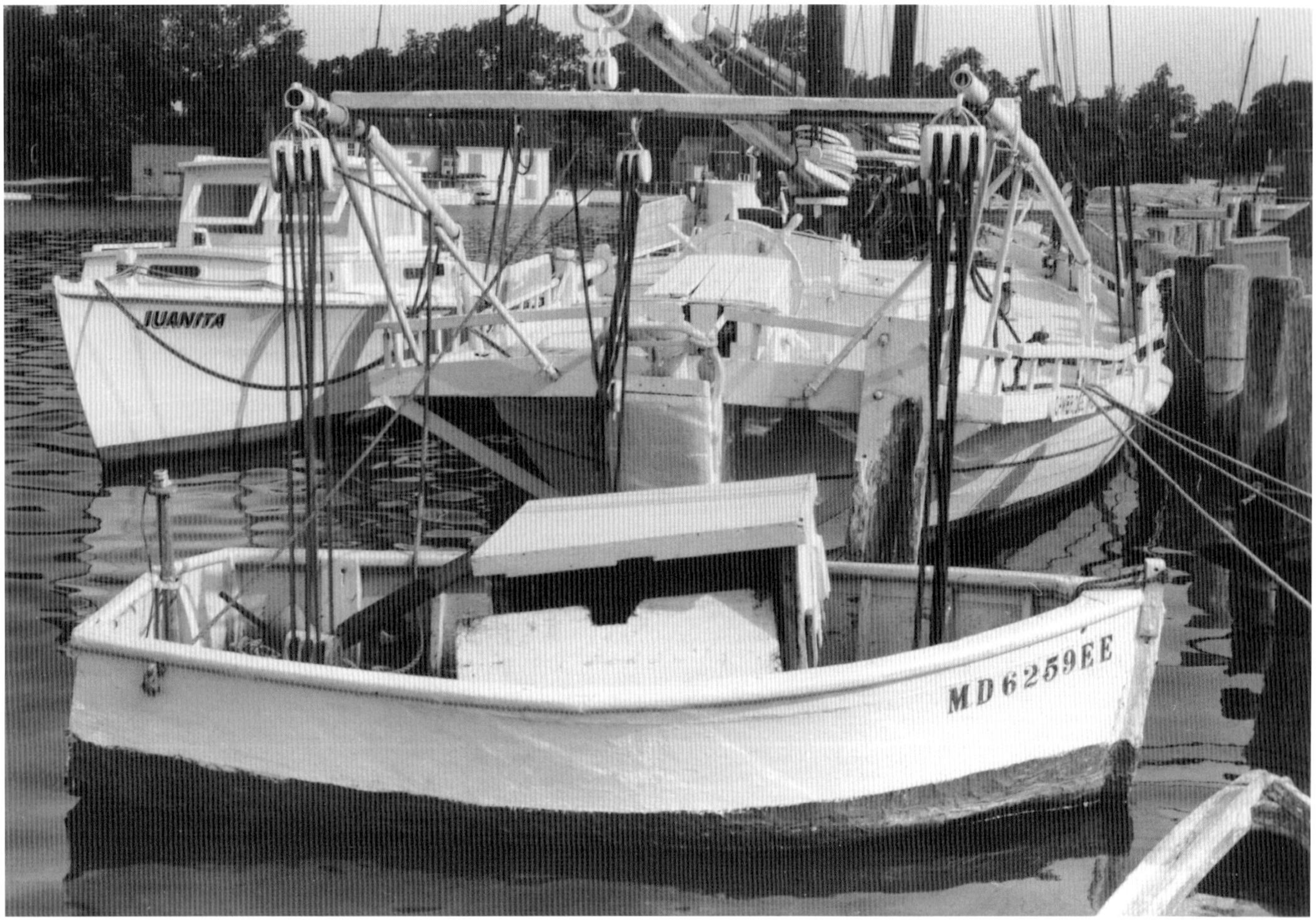

The last bugeye in the Chesapeake's oyster dredging fleet, and the last bugeye in commercial use on the Bay, was the *Edna E. Lockwood* shown here in Cambridge, Maryland, in October 1959, waiting for the oyster season to open. This log-bottomed craft was built in 1889 at Tilghman, Maryland, by D. Haddaway and after a hard career as a dredger was purchased by John R. Kimberly, of Queenstown, Maryland, about 1968. She was rebuilt but not altered from her working appearance and used solely for pleasure in the summer and tied up at the Chesapeake Bay Maritime Museum, St. Michaels, as an exhibit in the winter. She was eventually donated to that museum for preservation. This stern view shows the yawl boat and how it was rigged onto the davits. Hanging over the starboard portion of the patent stern is a board into which the bow of the yawl boat is raised when it is to push the bugeye.

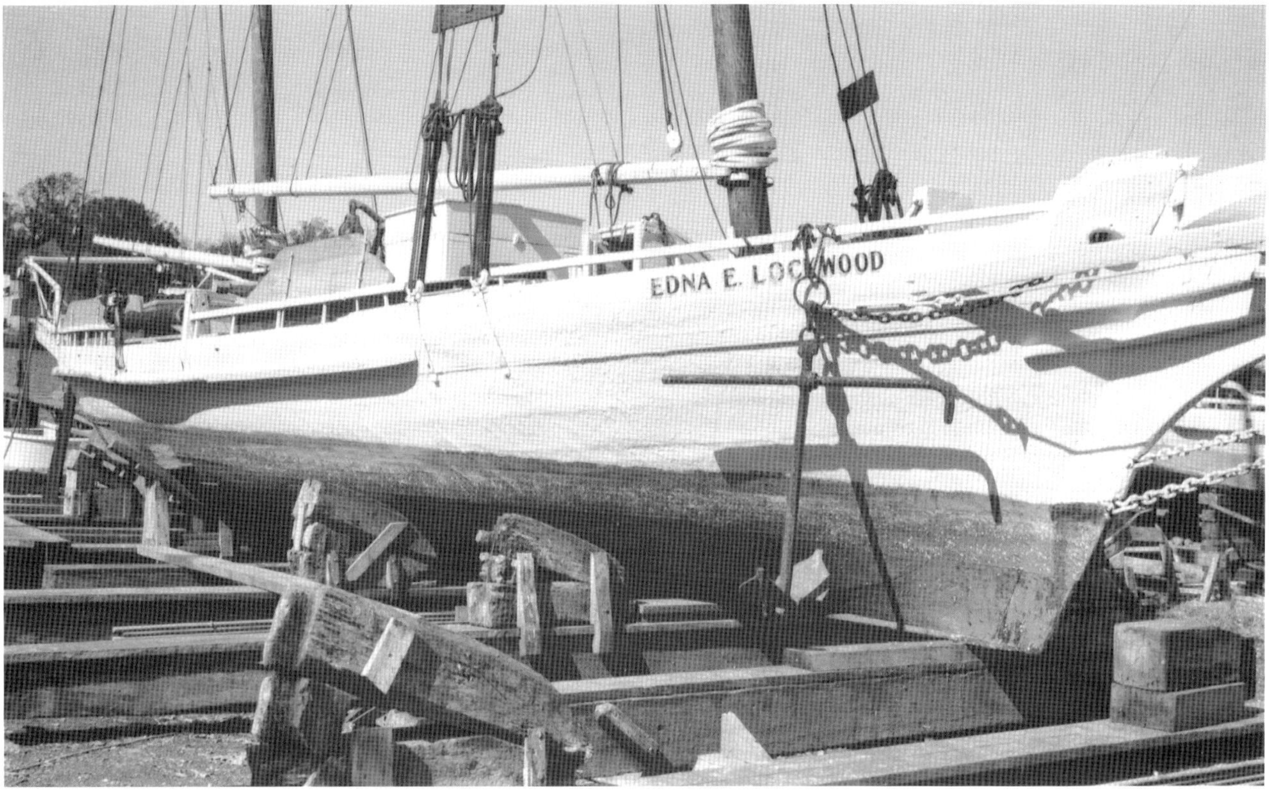

✪ The bugeye *Edna E. Lockwood* on the railway at Cambridge, Maryland, on October 17, 1954.

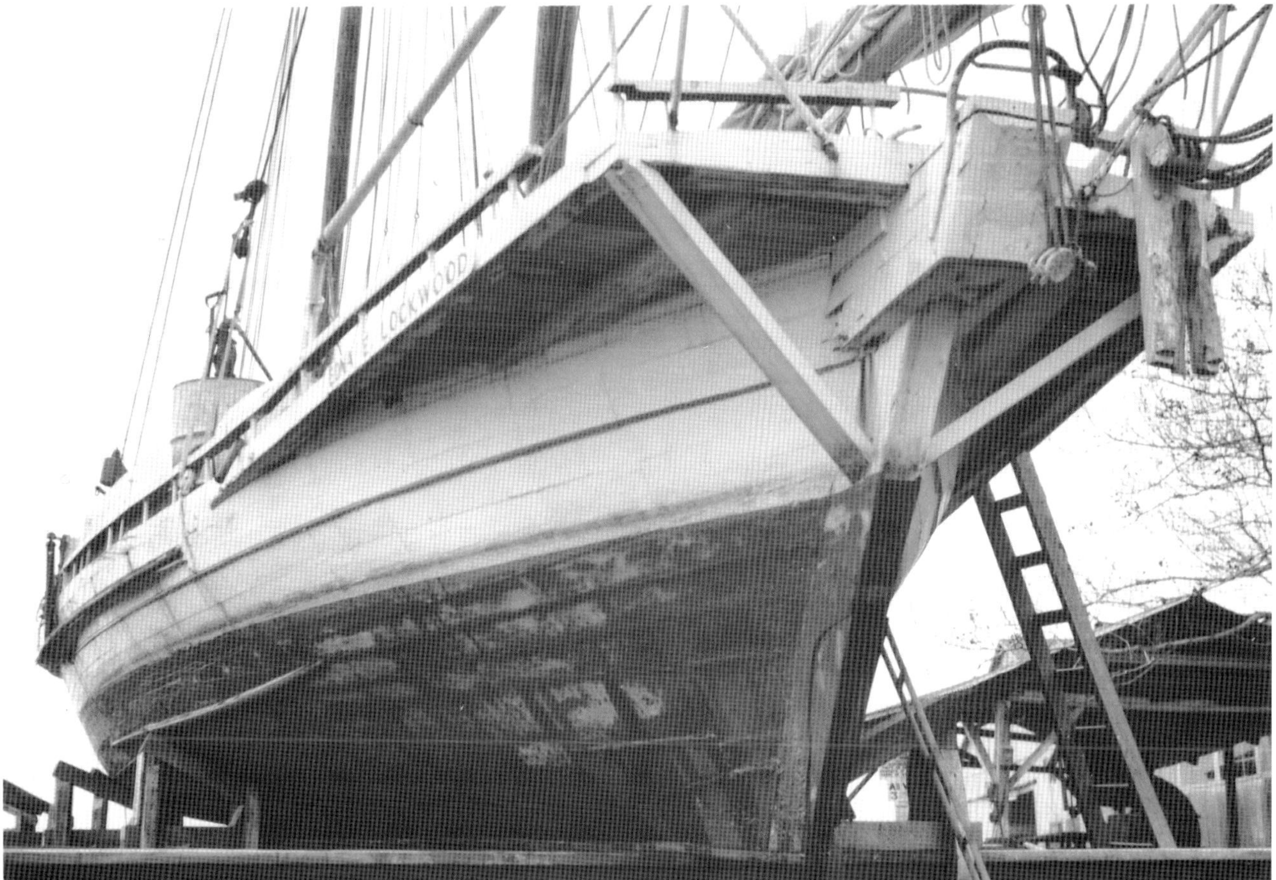

✪ The stern of the *Lockwood* on March 10, 1962, with her rudder removed. She survives today, at the Chesapeake Bay Maritime Museum. Note the worn chock for her push boat.

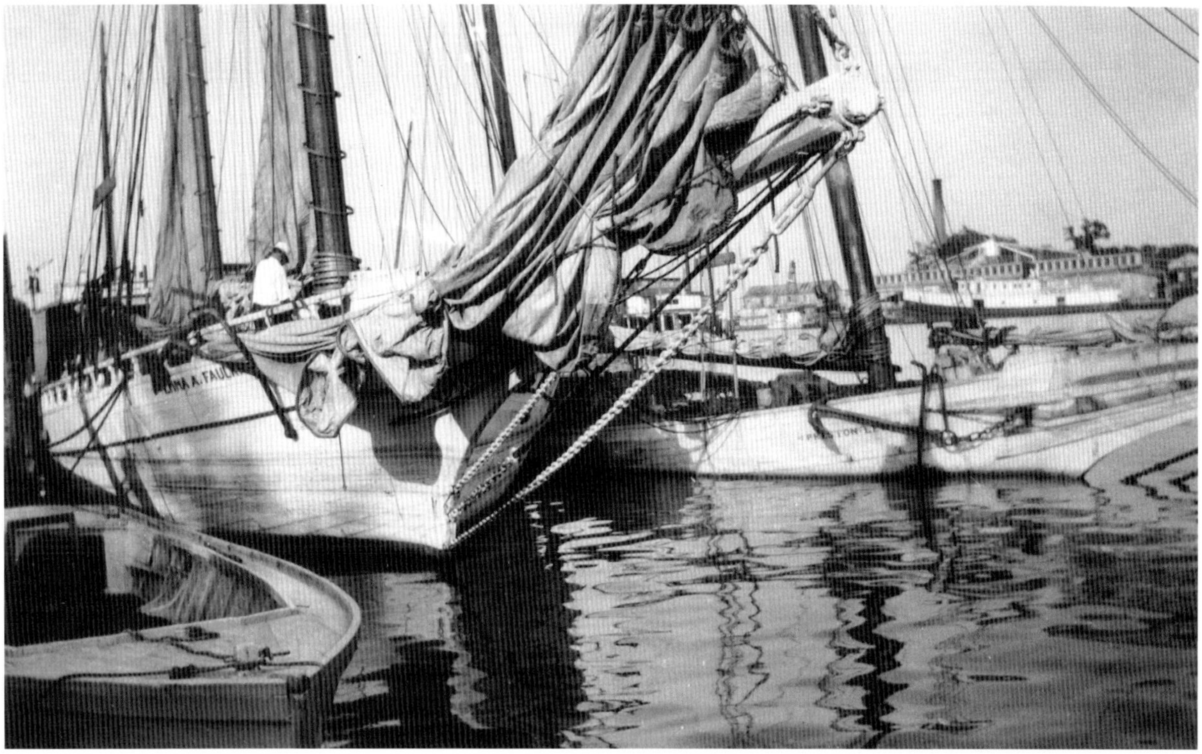

Emma A. Faulkner was unusual among latter-day bugeyes with her round stern and turned stanchion railing running full length of her deck. Here she is moored at her home port of Cambridge on August 23, 1935, with the bugeye *Preston Lewis* to her port. The Victor Lynn Lines diesel *Clio,* which ran between Cambridge and Baltimore, is in the right background. About 1949 the *Faulkner* lost her mainmast in a collision. She was abandoned in the shallows near Shad Point, Wicomico River, Maryland. *Preston Lewis* was motorized about 1942, sold to Washington, North Carolina, about 1953, and listed as "Abandoned" in the 1955 *MVUS.* The *Faulkner* was built at Tilghman, Maryland, by J.B. Harrison in 1901, and the *Lewis* the next year at Inverness, Maryland.

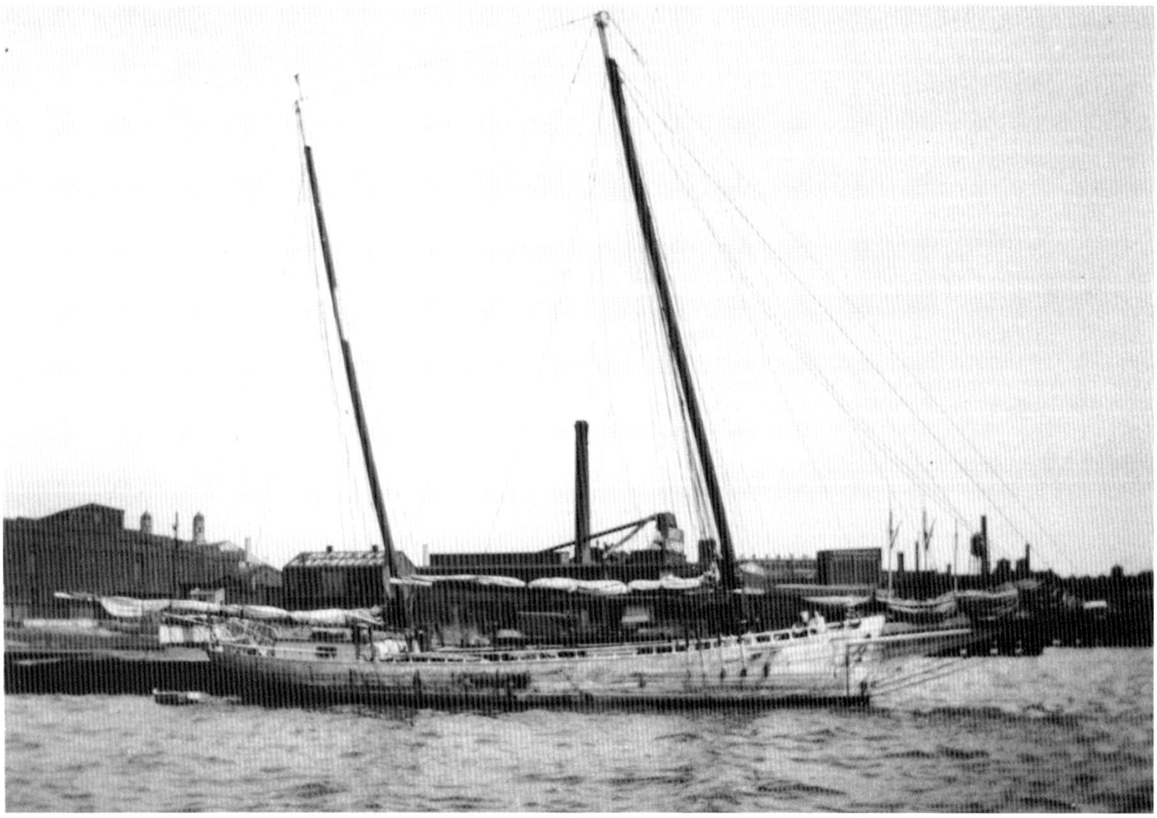

✪ A fine profile of the *Emma A. Faulkner* near the F. Bowie Smith Lumber Co. of Baltimore, sometime in the 1930s.

Florence Northam unloading a cargo of lumber at the foot of Philpot Street, Baltimore, May 19, 1934, proves she worked hard for a living. But she was a swift bugeye and also found time to play and won the bugeye event in the Chesapeake Bay Workboat Regatta, sponsored by *The Baltimore Sun,* in 1928 and 1929. Her builder was M.M. Davis at Solomons, Maryland, who launched the bugeye in 1890.

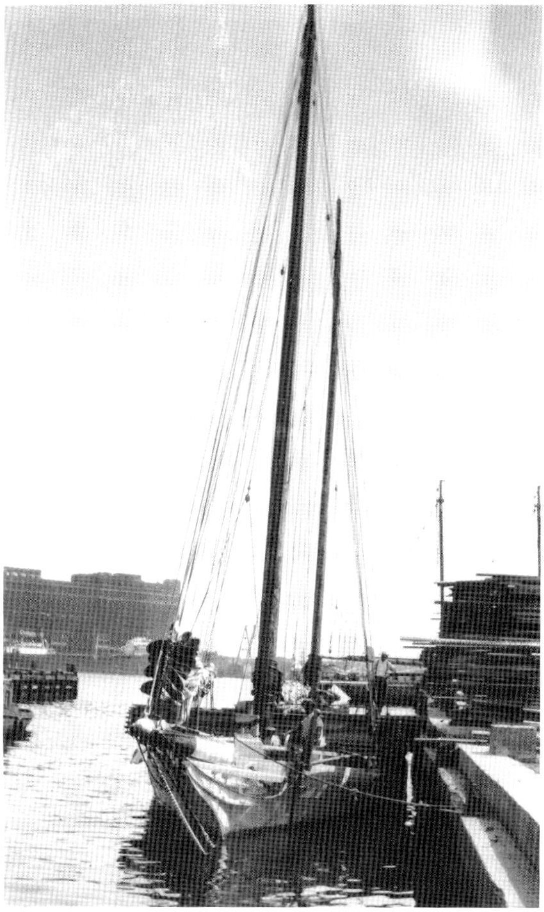

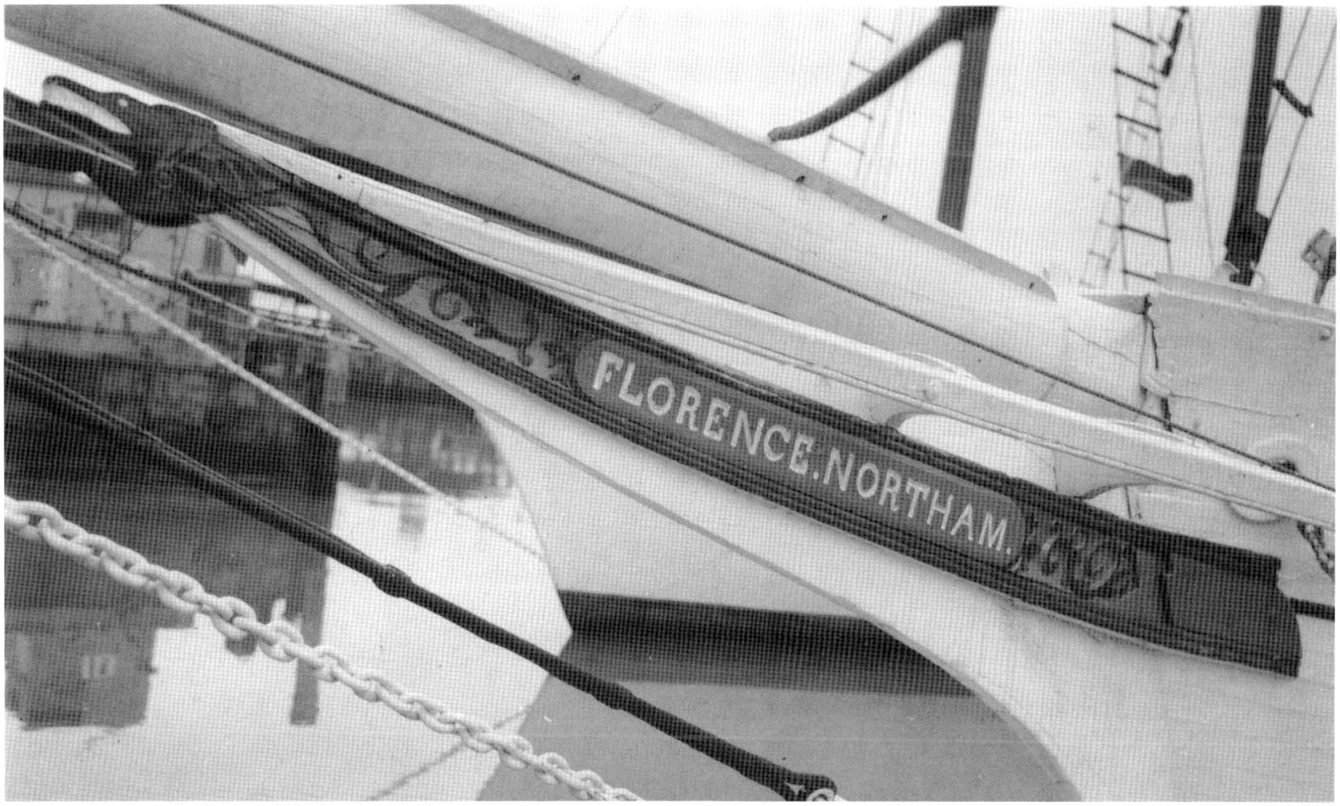

About 1945 the *Florence Northam* was purchased by Judge J. Abner Sayler, of Baltimore, who converted her into a yacht, sailing out of Gibson Island, Maryland. This photograph of her trailboard and sea monster figurehead was made at the A. Smith & Sons Shipyard, Curtis Bay, in August 1952. She is listed as "Dismantled" in the 1959 *MVUS.*

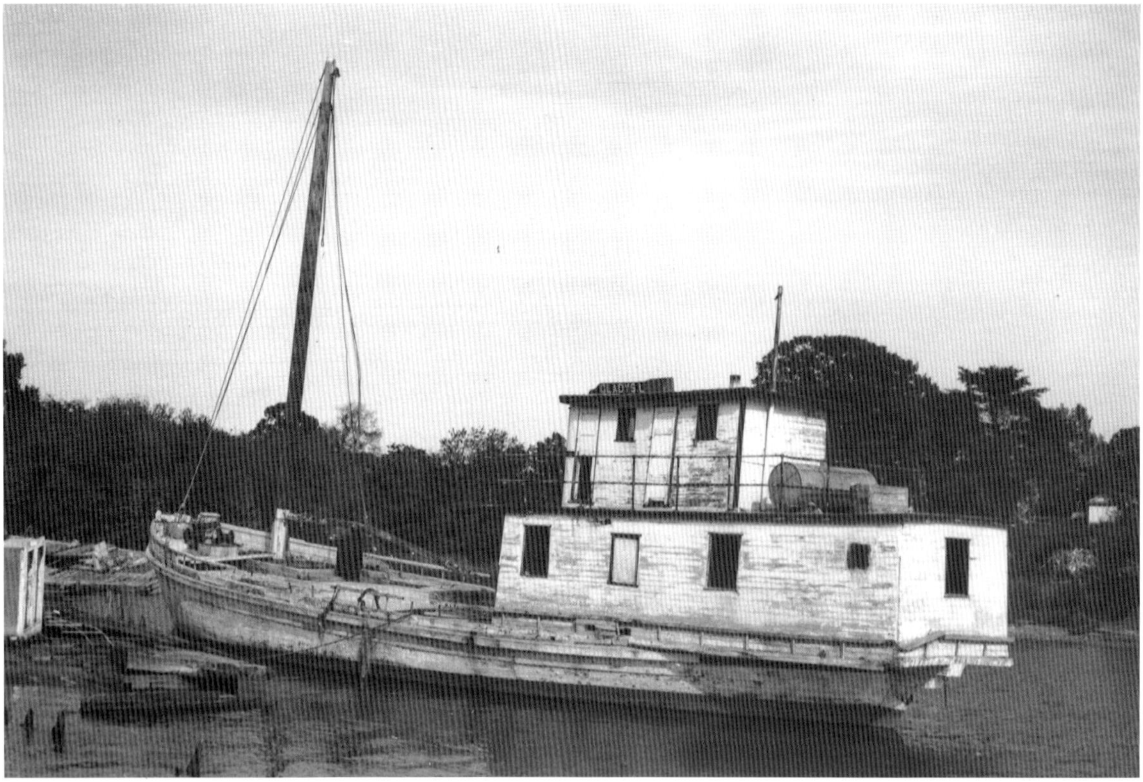

If the well-known Chesapeake shipbuilder Joseph W. Brooks, of Madison, Maryland, could have seen the largest bugeye he ever built, the *Miriam,* in her latter days as the powerboat *Gladys L.*, he would have been most disappointed. Launched in 1903 and the biggest bugeye built up to that time, the *Miriam* was a handsome vessel. She came under the ownership of Andrew J. Lewis of Walnut Point, Virginia, about 1920. When the Baltimore and Virginia Steamboat Company ceased operations in 1932, Mr. Lewis had converted the *Gladys L.* into a powerboat to operate with his other freighters picking up the steamers' business as the Potomac River Line. What he did to that bugeye is evident in this view showing a monstrosity of a structure built out over the patent stern. When photographed at Humphreys Shipyard, Weems, Virginia, on October 24, 1947, the *Gladys L.* had been idle for some time. Later she was renovated, but burned in the Rappahannock River, and was listed as "Abandoned" in the 1959 MVUS.

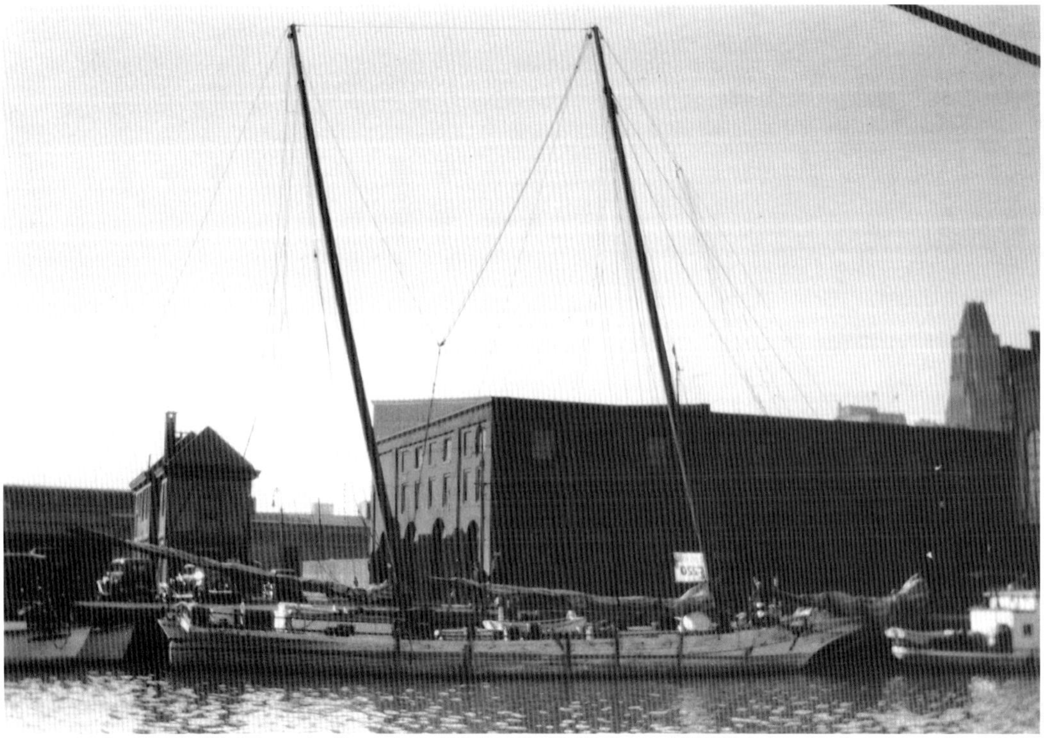

❂ The bugeye *Goldie C.* presents a sleek profile at Baltimore on December 1, 1933.

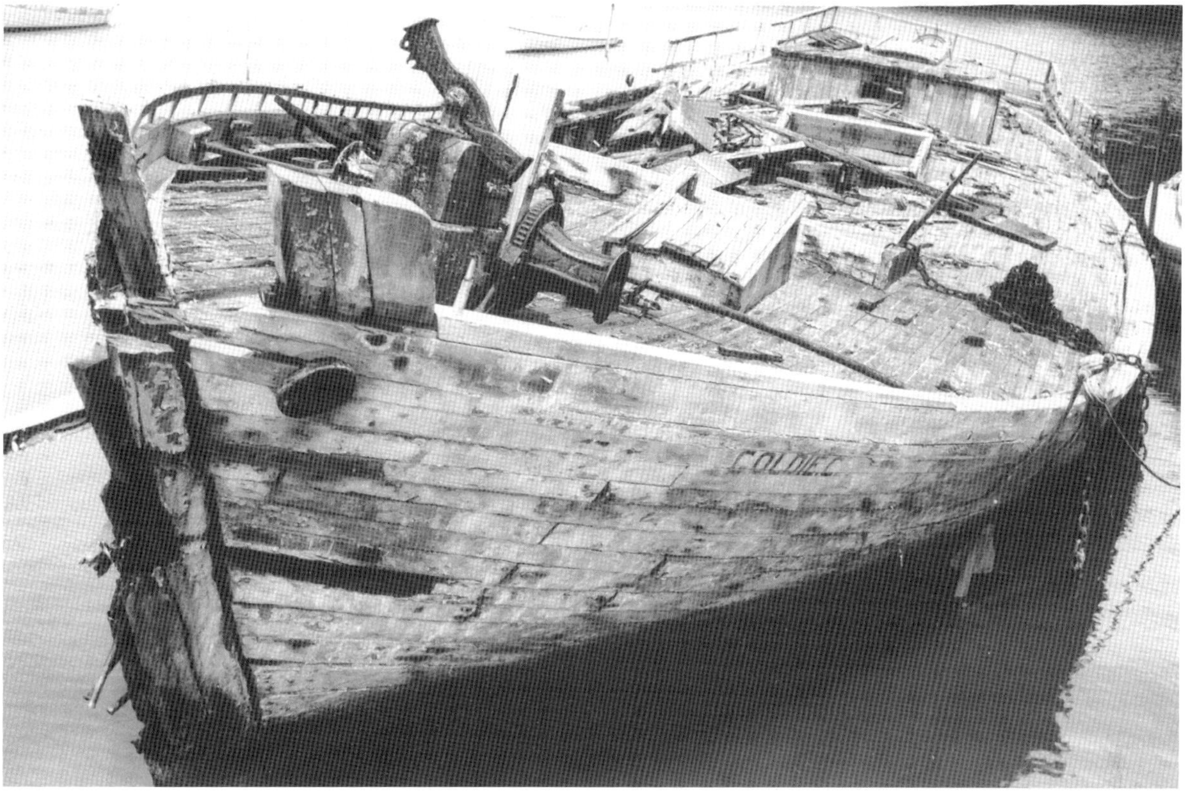

Goldie C. was one of 25 bugeyes built by the well-known John Branford, of Fairmount, Maryland, between 1883 and 1911. She came off the ways at Fishing Island in 1900. An engine was installed about 1929, but she retained her sailing rig. In the 1944 *MVUS Goldie C.* is "Out of Documentation," and on July 7, 1947, she appeared like this at Urbanna, Virginia. Although a total wreck, she shows on her deck a wealth of maritime relics—windlass, steering wheel and gear, hawsepipes, deadeyes, and chain and anchor, its fluke just jutting above the surface at the port waterline.

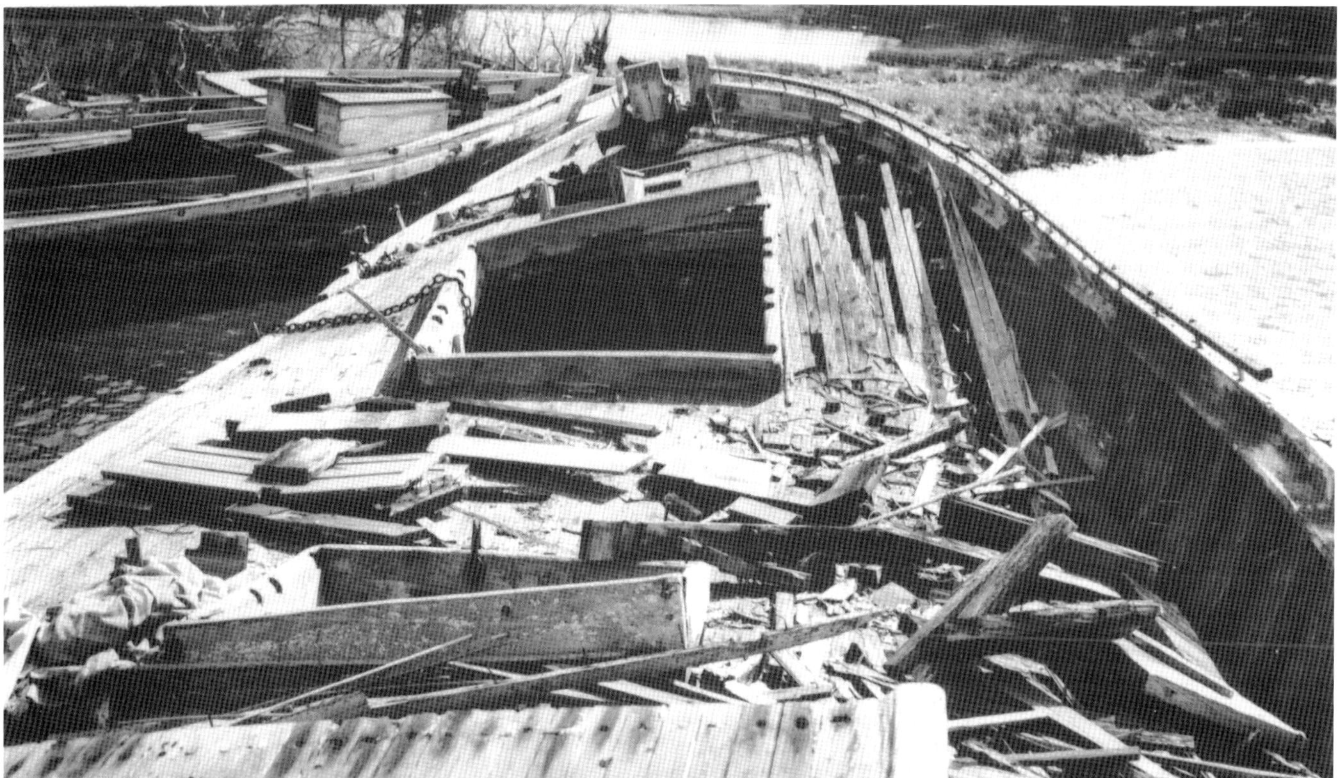

✪ The deck of the *Goldie C.* looking forward.

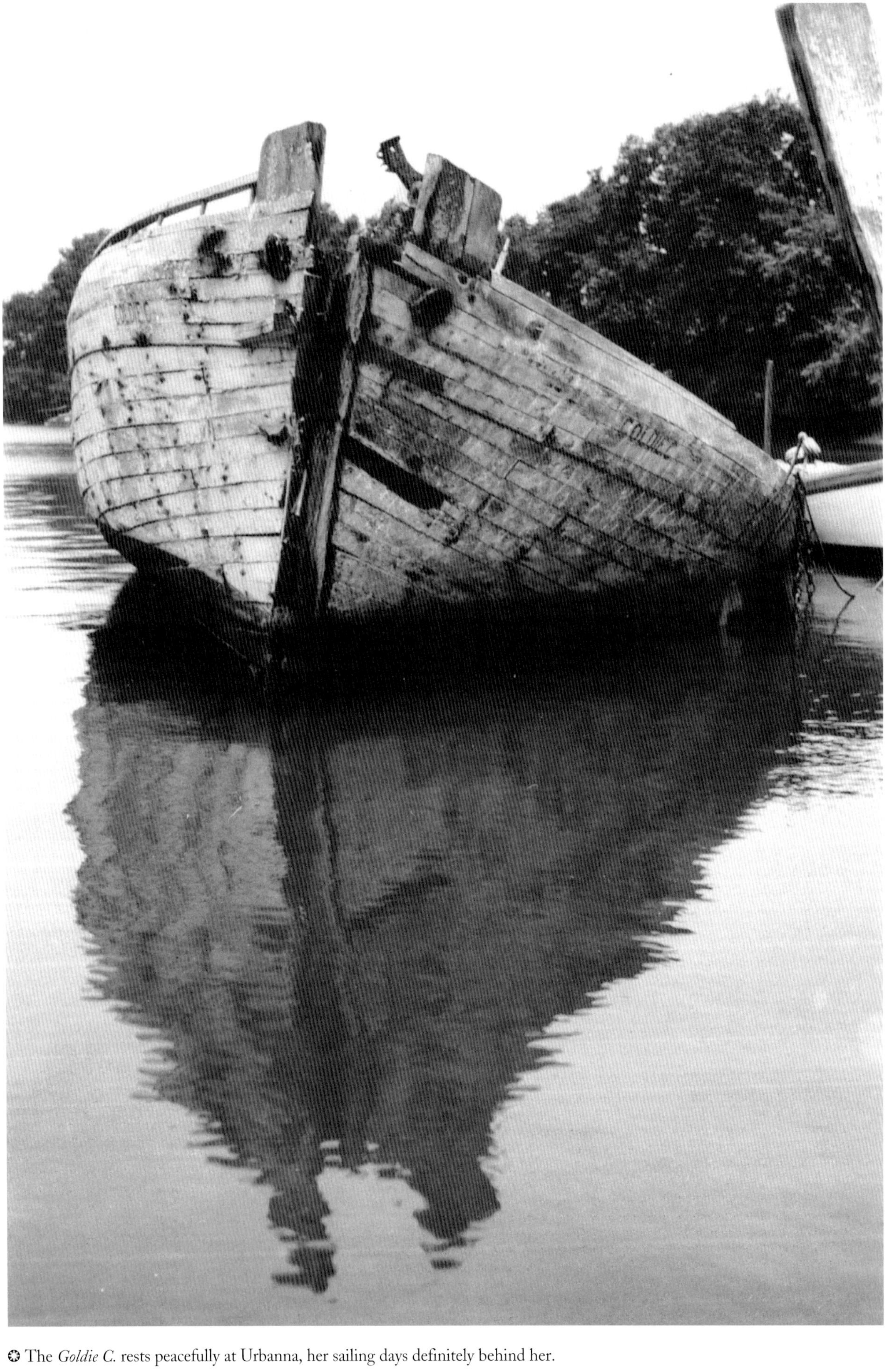

✪ The *Goldie C.* rests peacefully at Urbanna, her sailing days definitely behind her.

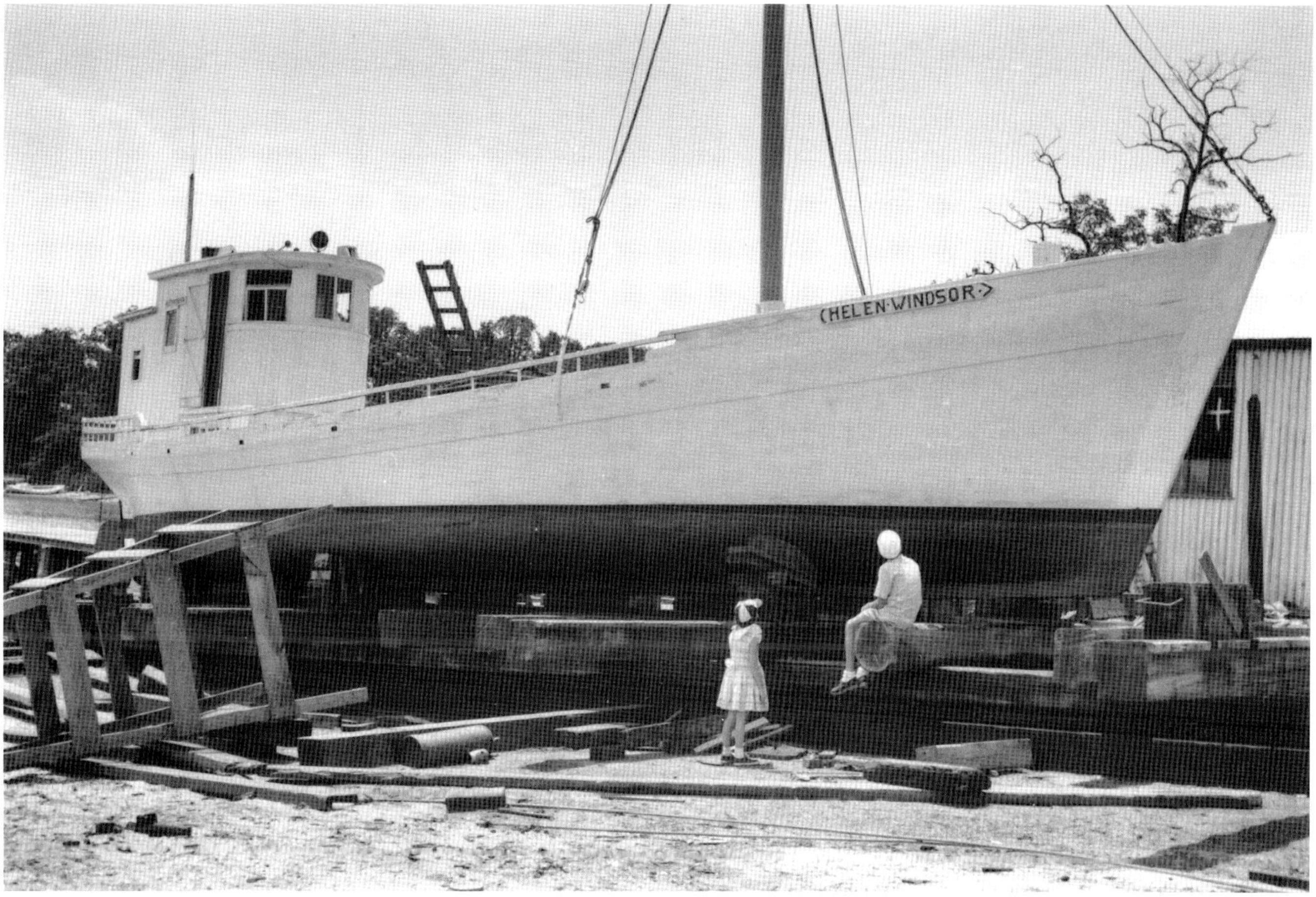

This is just about the typical appearance of a bugeye after it had been converted into a powerboat. In this instance, however, the sharp stern has been filled in through rebuilding. *Helen Windsor*, built in 1900 by B.P. Miles at Oriole, Maryland, was powered about 1919 but no doubt kept her sailing rig for some time thereafter before acquiring this appearance. In August 1955, she is on the marine railway at Humphreys Shipyard, Weems, Virginia, for an overhaul and repainting. She is listed as "Abandoned" in the 1968 *MVUS.* ✪ The author's daughter, Janet, and son, Bruce, dutifully pose in the foreground.

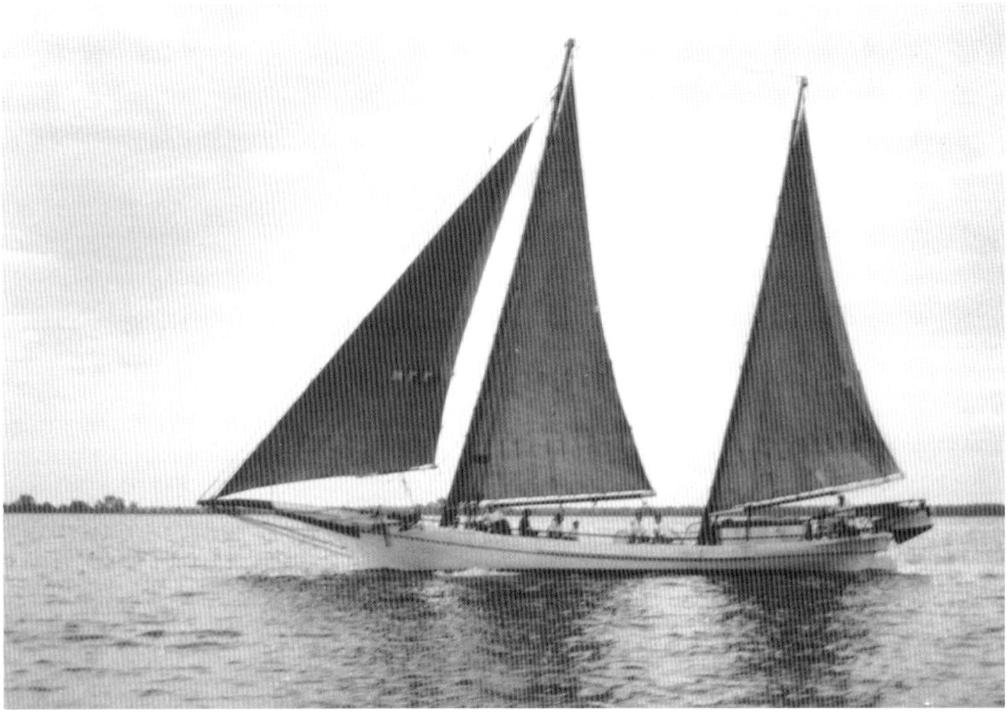

On Labor Day 1933, during the annual regatta of the Tred Avon Yacht Club, Oxford, Maryland, four of the local bugeyes were scheduled to race for the championship of that little Eastern Shore port. Only two of the bugeyes showed up, however—the *Susie E. Parriss* and *Hugh S. Orem*, the latter shown as she rounded the course that day. The other two bugeyes, *Little Jennie* and *Col. R. Johnson Colton*, were in Baltimore with watermelons. But the race went ahead between the *Parris*, *Orem*, the cutter yacht *Windrush*, and Pete Culler's replica of the yawl *Spray*. About 1950 the *Orem* was sold from the Chesapeake to Delaware Bay for oystering. Built at Pocomoke in 1892, she was listed as "Abandoned" in *MVUS* for 1964.

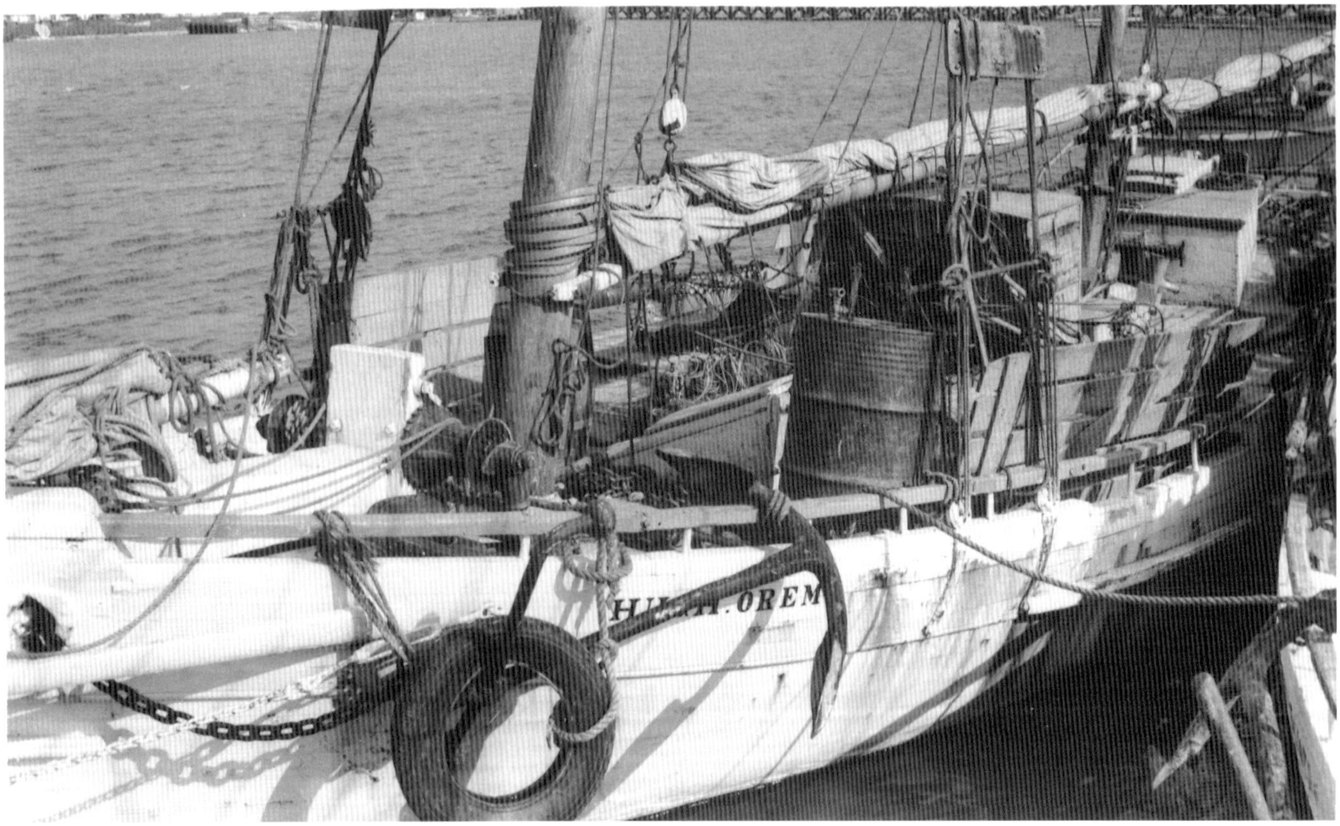

✪ A good view of the cluttered deck of the *Hugh S. Orem*, at Deal Island, Maryland, on January 28, 1949.

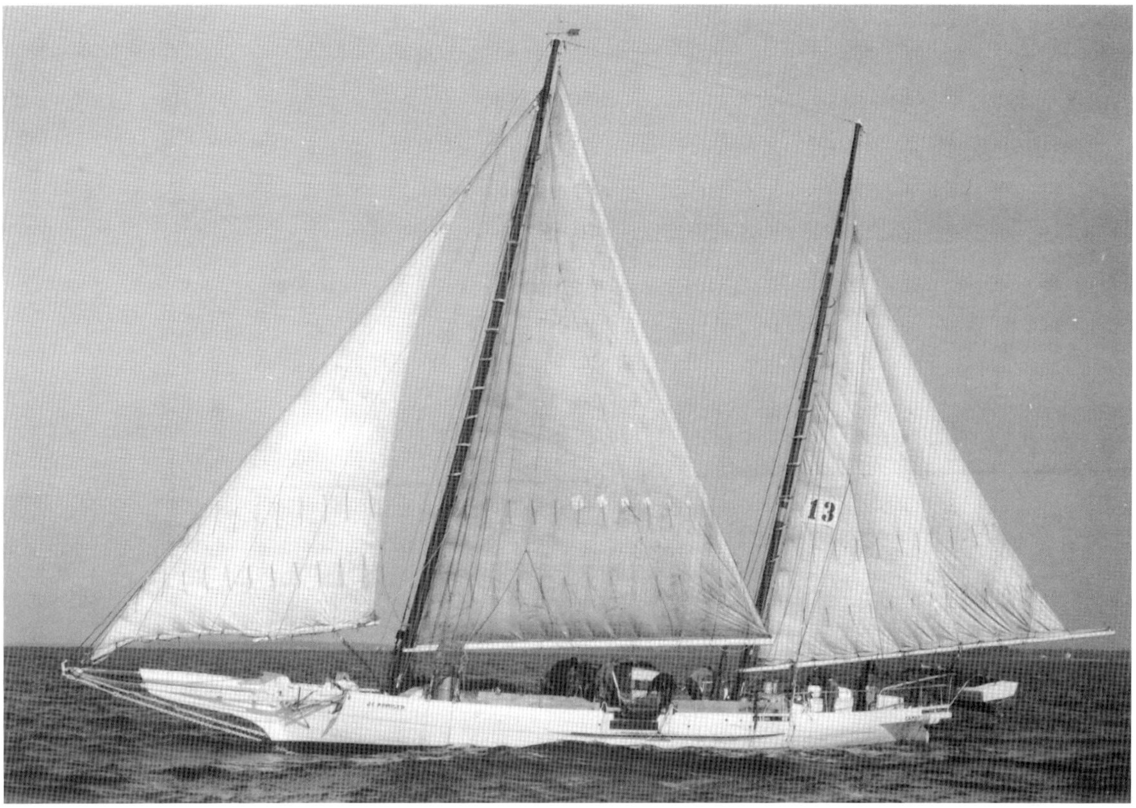

J.C. Armiger was one of the last surviving bugeyes on the Bay, operating out of Cambridge to the Choptank River oyster beds where she is pictured on November 5, 1957. It is the beginning of the oyster season, indicated by her freshly painted hull. The number 13 on her mainsail is her Maryland dredging license and is in large numerals so as to be easily identified by state police boats or aircraft if observed violating a regulation. Another way to identify the craft locally was by the arrow atop her foremast.

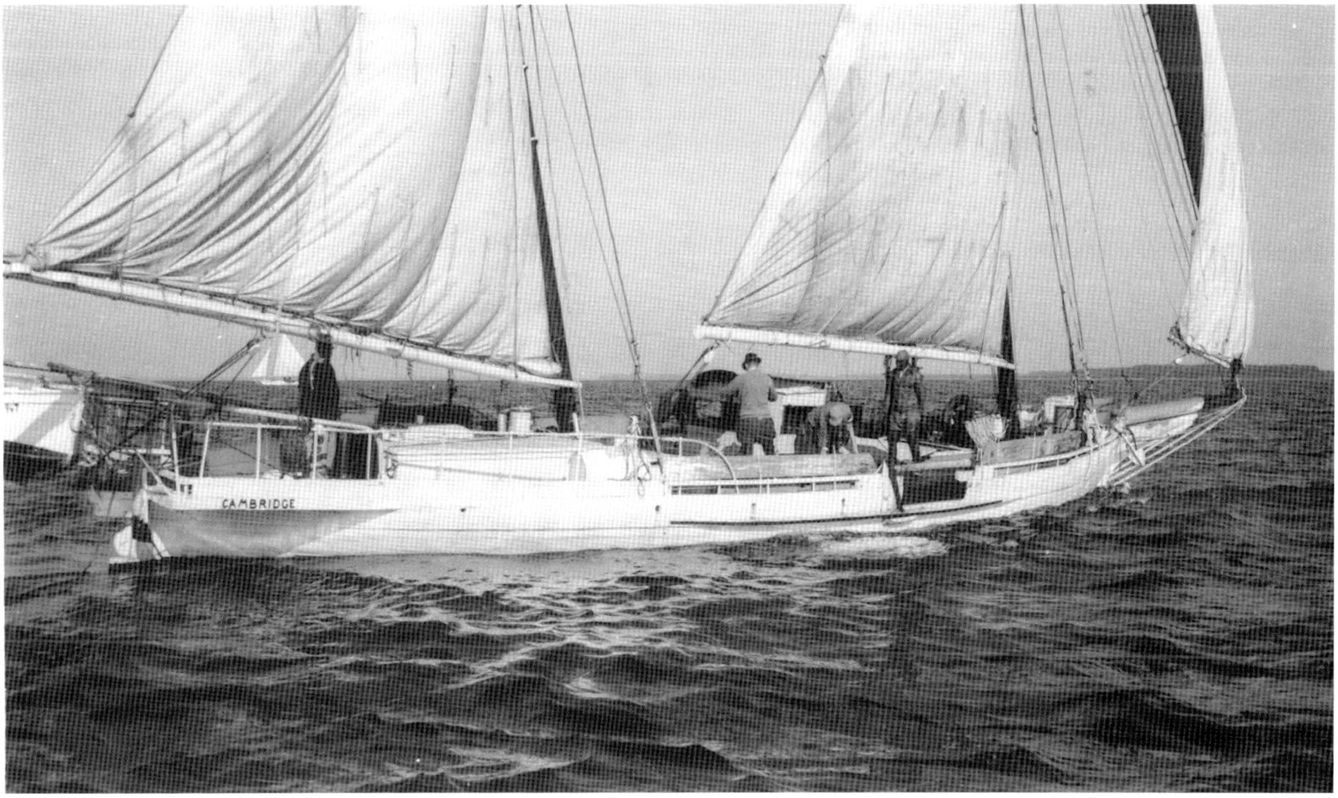

J.C. Armiger starts to make another "lick" over the Choptank River oyster beds as the dredge is thrown over the starboard side. Built at Solomons, Maryland, by M.M. Davis in 1890, the *J.C. Armiger* last appears in the 1964 *MVUS.*

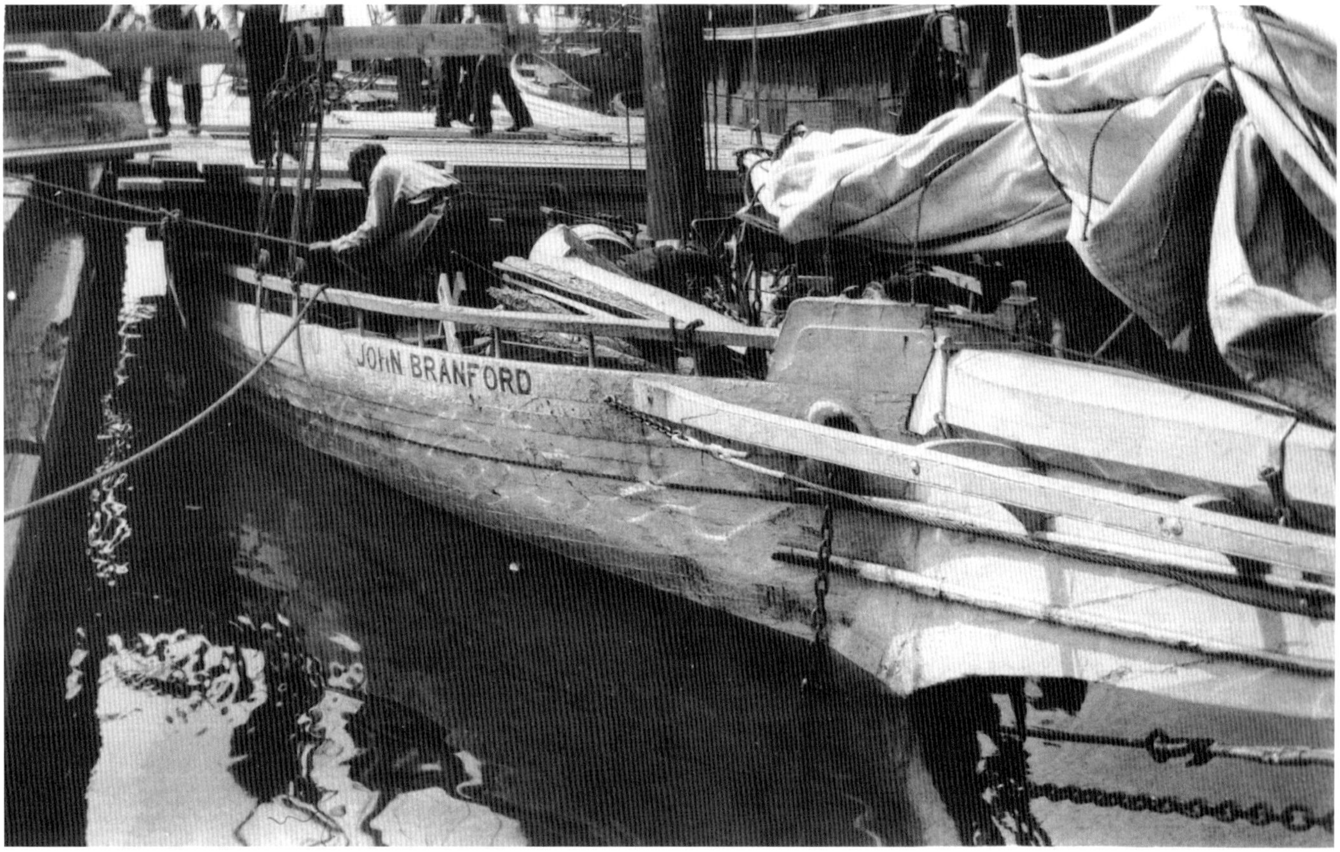

Named for the man who built her at Fairmount, Maryland, in 1900, the *John Branford* discharges a cargo of lumber at the foot of Philpot Street, Baltimore, on February 5, 1934. She was motorized about 1937 and stripped of her sailing rig but was still in the 1972 *MVUS*.

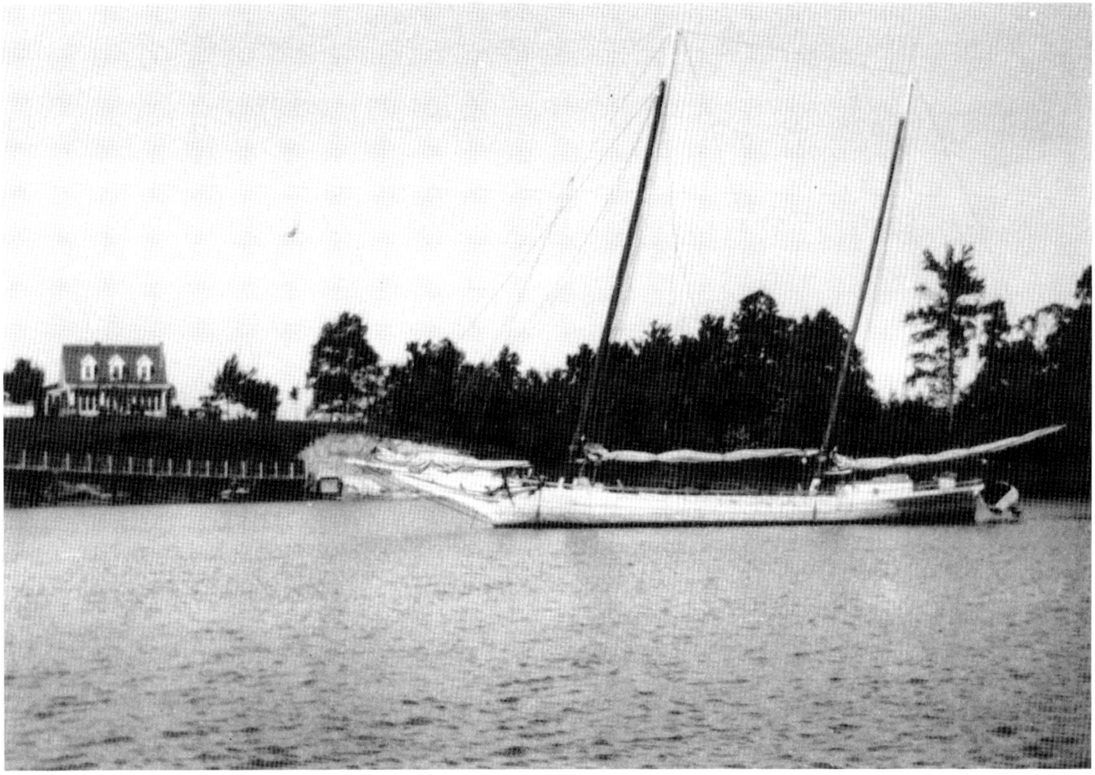

✪ The bugeye *John Branford,* at anchor at Kinsale, Virginia, on October 7, 1934.

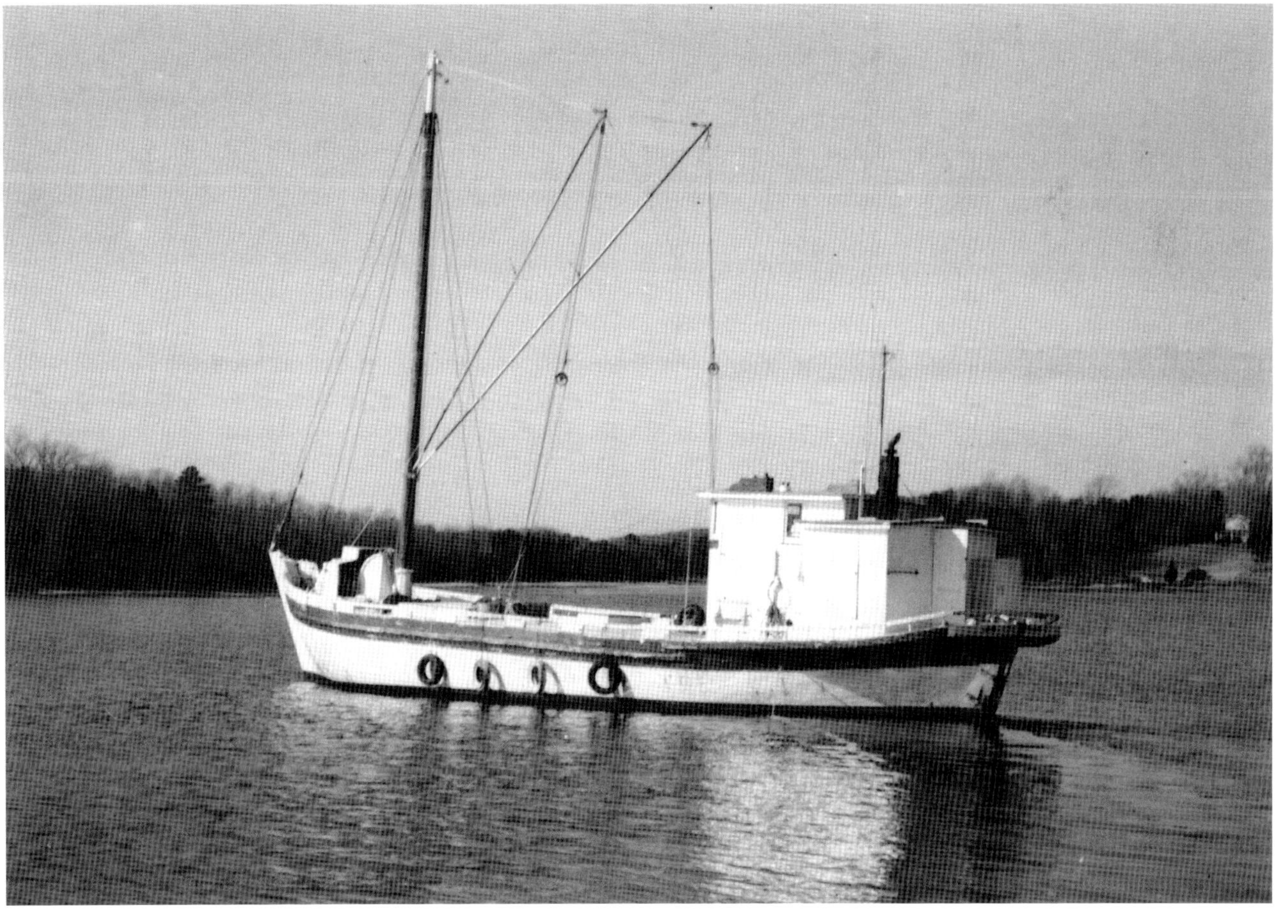

✪ A motorized *John Branford* leaves the wharf at A. Jackson Booth Oyster Co., Glebe Point, Virginia, on January 23, 1976. The *Branford,* later renamed *Anna Gloria,* was still afloat in 2004.

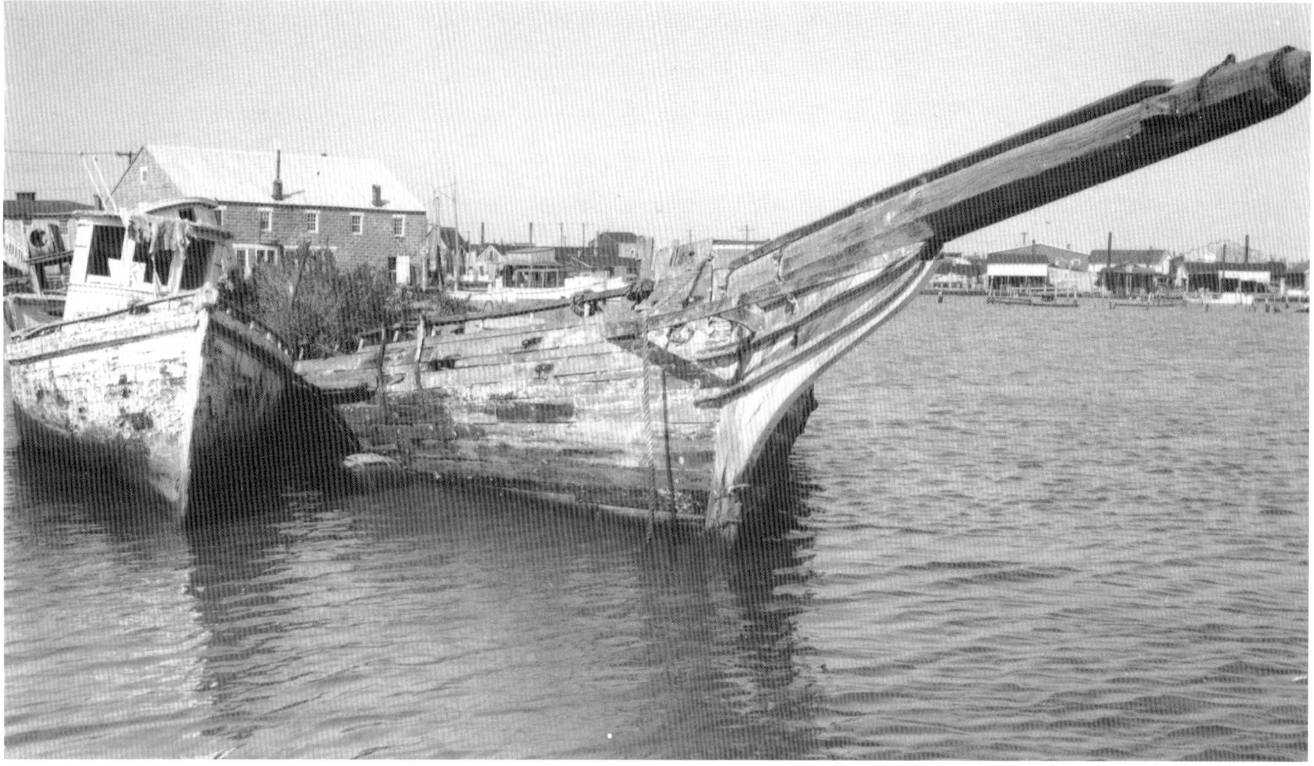

The ship "graveyard" at Crisfield, Maryland, held the remains of the *L.E. Coulbourn,* built in Somerset County, Maryland, in 1882 and abandoned about 1944. Although her deck had collapsed, her bow section, probably a result of a sturdy longhead that seemed to have been of newer construction, permitted her bowsprit to point skyward defiantly. To her starboard was the log-built *Evelyn,* a powerboat when abandoned but assuredly a sailing craft originally. Several years after this photograph was made on October 11, 1952, this section of the waterfront was cleaned up.

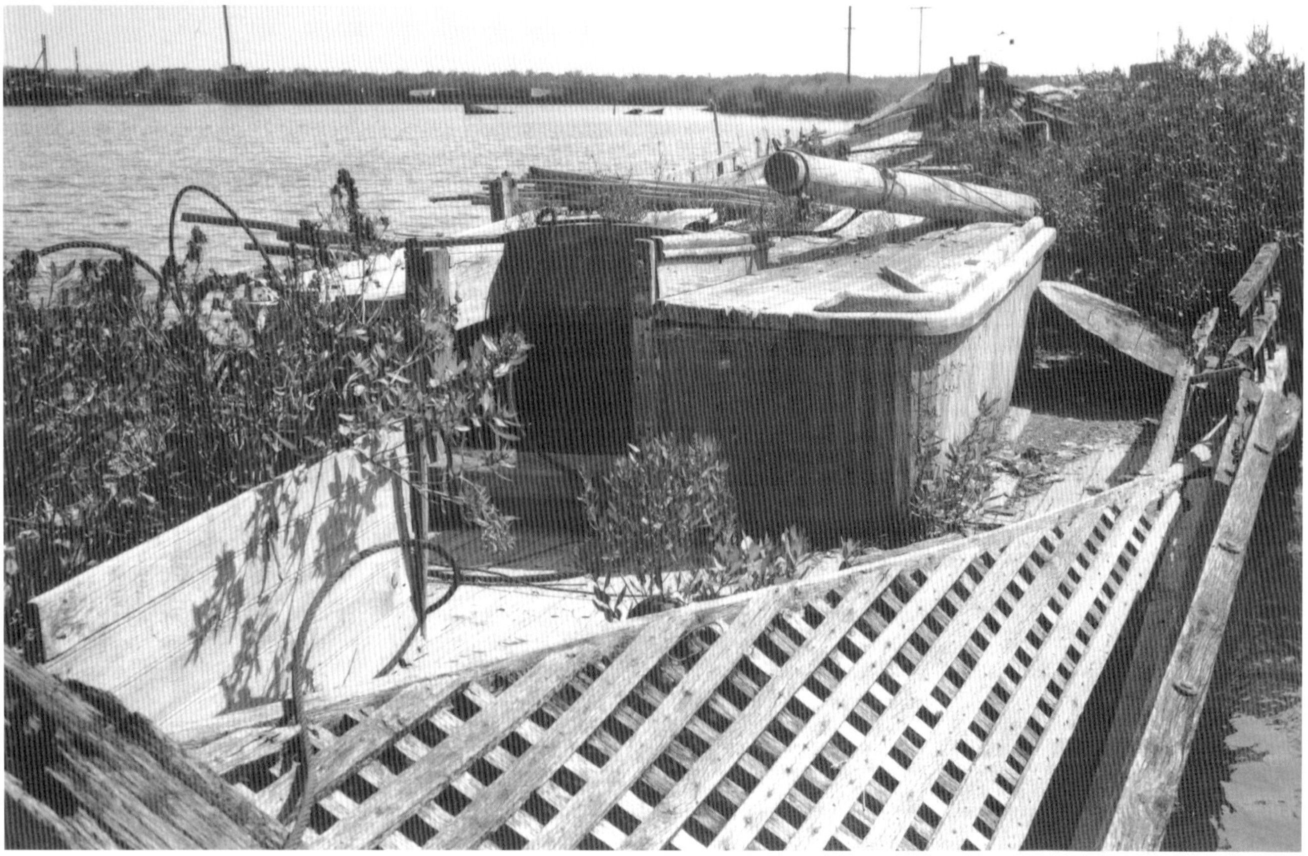

✪ The deck of the ruined bugeye *L.E. Coulbourn,* looking forward, at Crisfield in 1952.

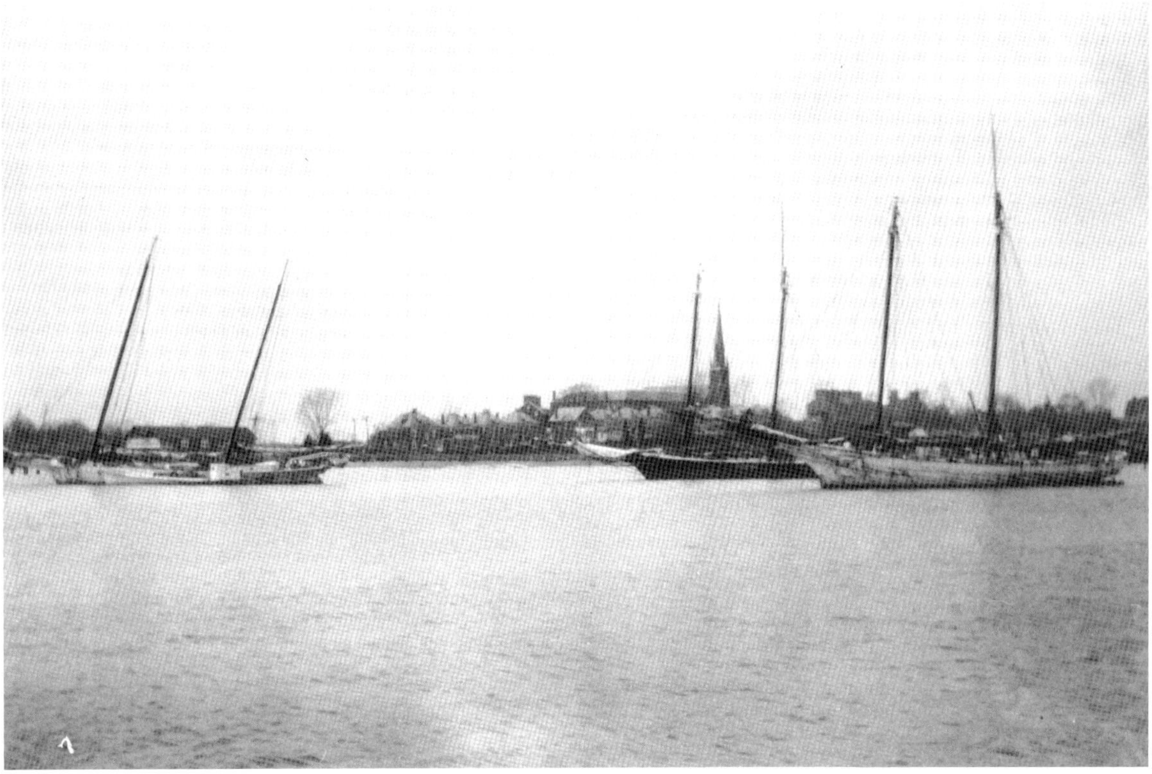

The *Lemuel Kirwan* and the schooners *Augusta* and *Smith K. Martin* anchored at Annapolis in the 1930s.

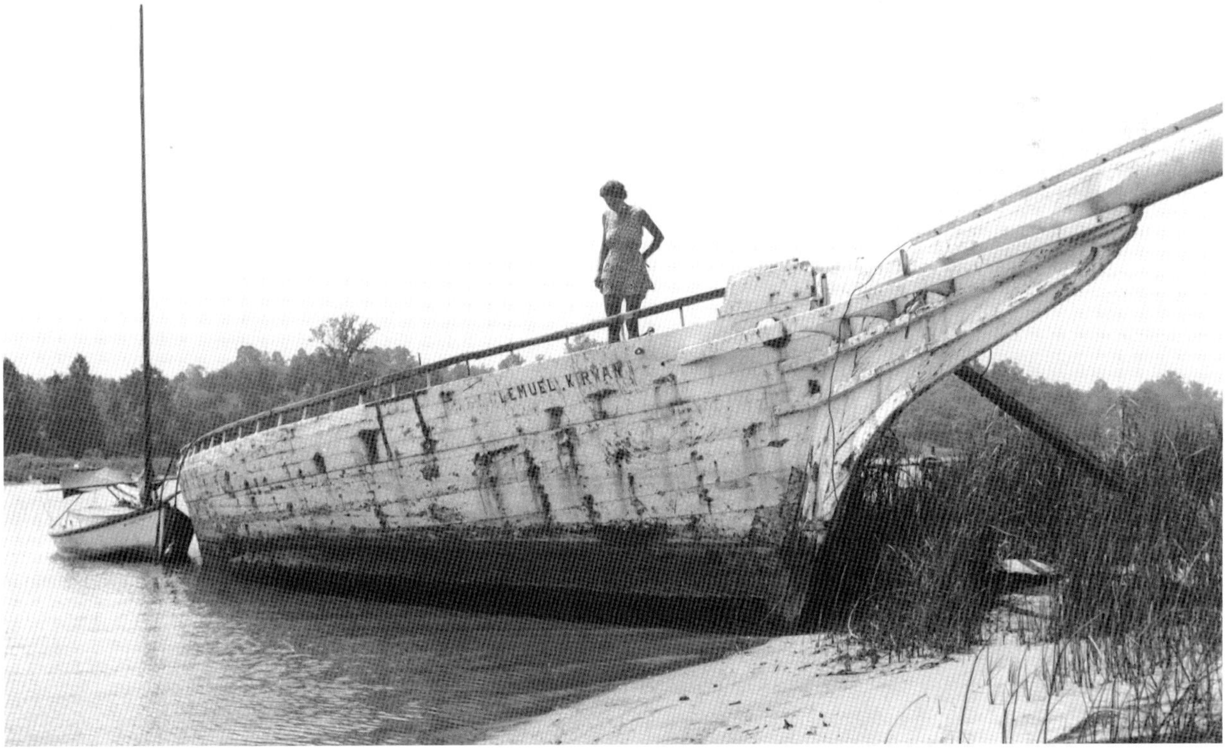

About 1937 the bugeye *Lemuel Kirwan* was blown ashore on a sandspit of Deep Creek, making off the Magothy River, Maryland, and there she remained. The author visited the hulk on August 9, 1938, with his wife, Adele, shown on deck, in his sloop *Trade Wind* tied up alongside. The vessel had been stripped completely; even the hawsepipes had been removed. The *Kirwan* was built at Chester, Maryland, in 1901.

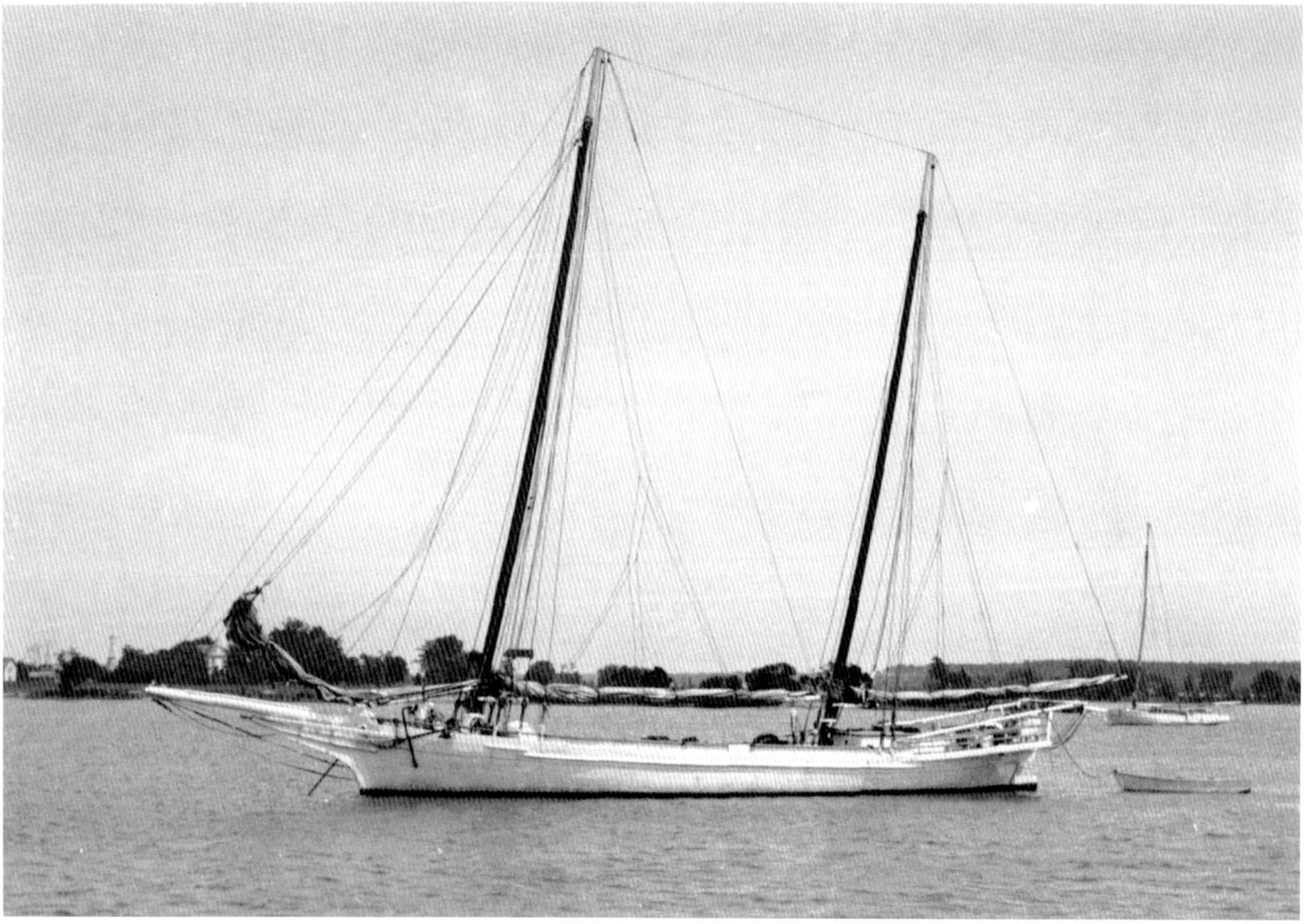

✪ The quaintly named *Little Jennie* at anchor with her yawl boat trailing astern, at Oxford on June 17, 1936. She was built in 1884 in Calvert County, Maryland, and was owned by James H. Mills. Of 22 gross tons, she measured 57.9 × 16.3 × 4.6 feet. Her handsome appearance belies her age in this photo. She survived until 2002, when she was broken up in Maine.

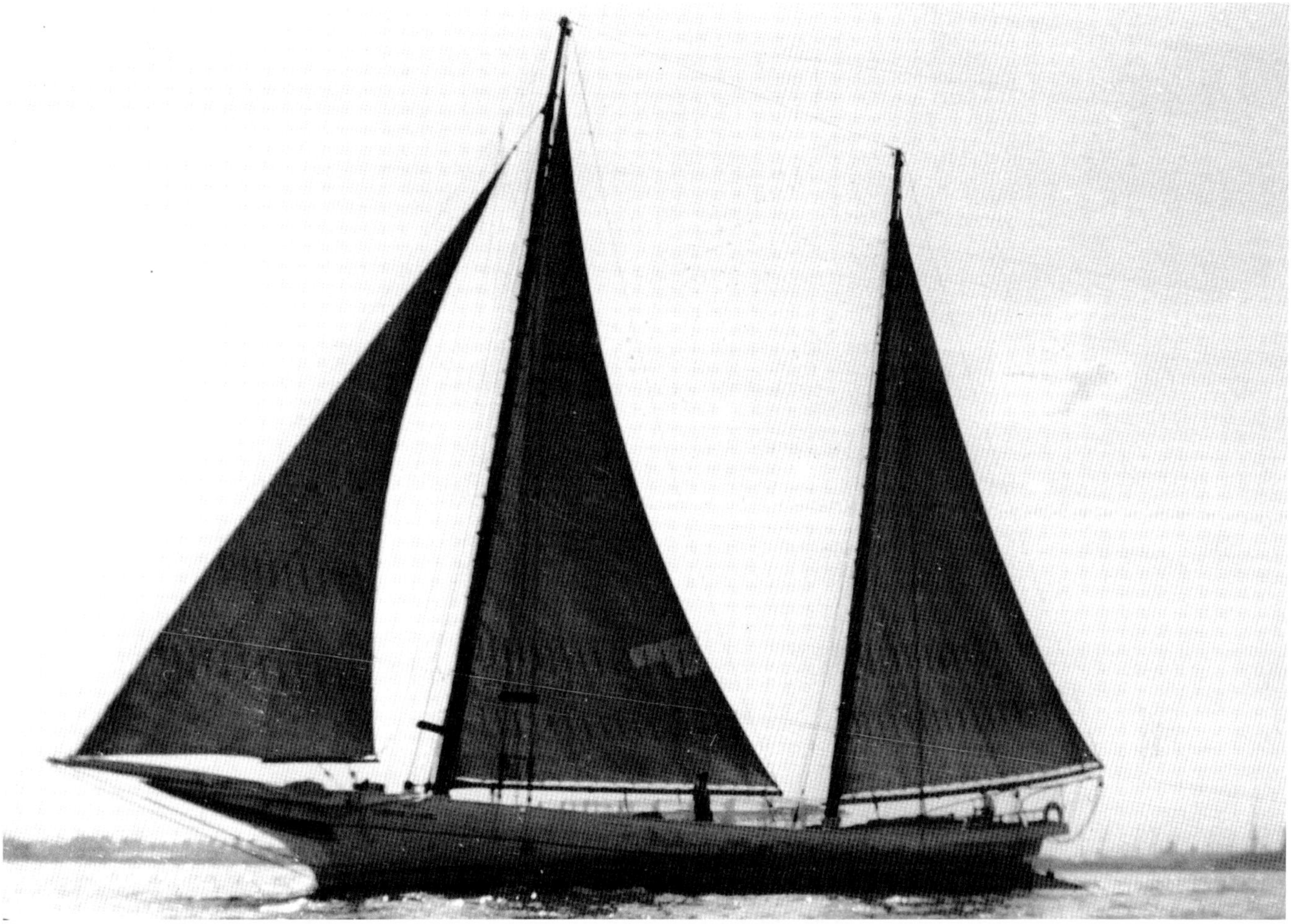

Bugeye *Lizzie J. Cox* stands out of Baltimore harbor in July 1933, after discharging a cargo of wheat. Her stay in that port is indicated by her grease-streaked side. The yawl boat is down at the stern to help push her to Chesapeake Bay where a better slant of wind will take her down to her home port of Cambridge. The *Cox* was built by John Branford at Fishing Island, Maryland, in 1905.

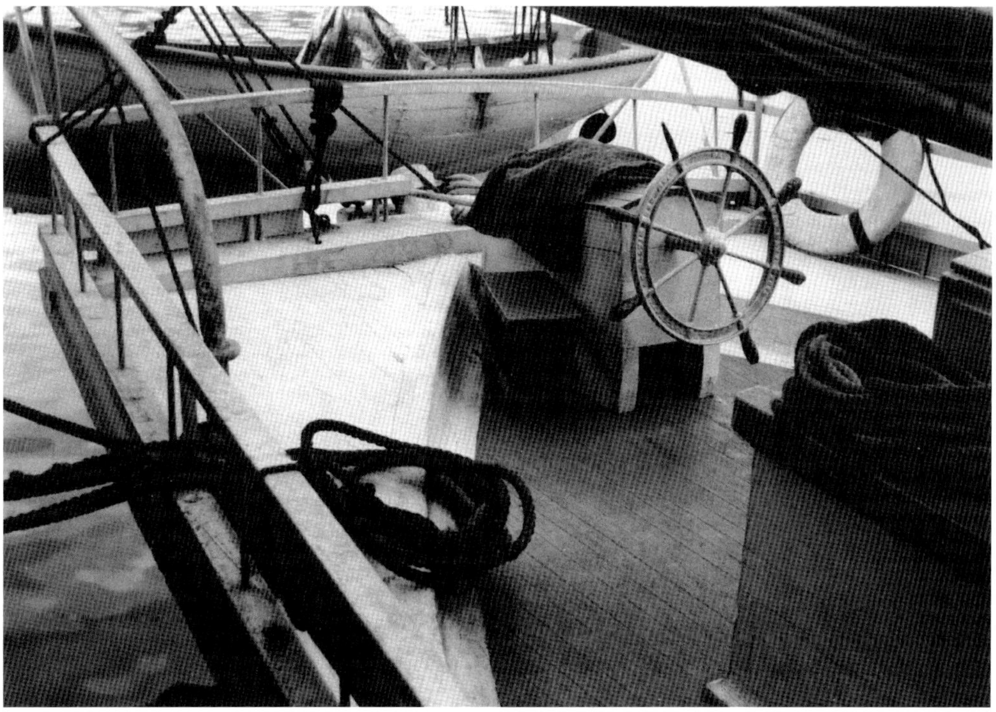

The patent stern and after deck of the *Lizzie J. Cox* are in shipshape order with lines coiled and everything in its place. After being sold to Connecticut, the *Cox* foundered in the harbor at Bristol, in that state, on August 21, 1938, three years after this photograph was taken.

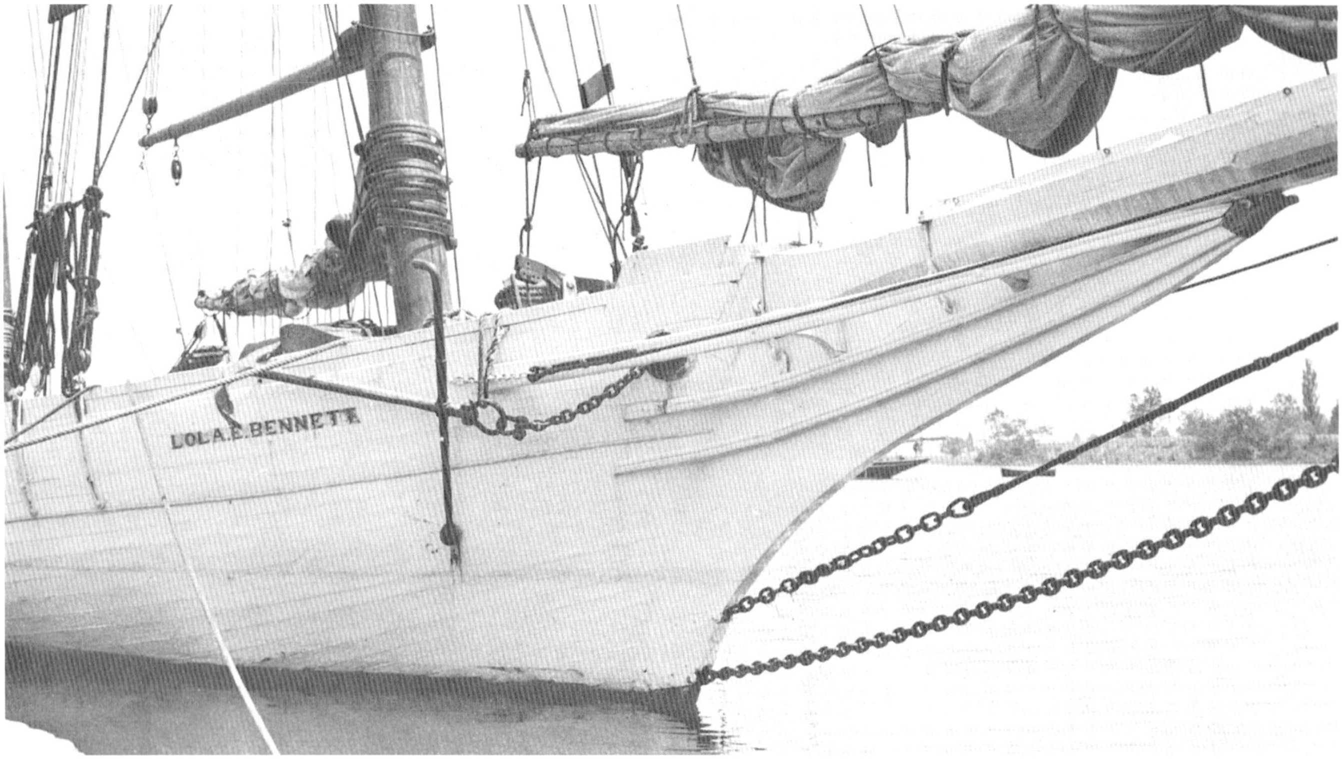

In this June 15, 1936, view the bugeye *Lola E. Bennett* is tied up at St. Michaels, Maryland, at the present site of the Chesapeake Bay Maritime Museum. The extra boom attached to her foremast supported a block and fall for unloading cargo in port. She had been built by John Branford at Inverness, Maryland, in 1902.

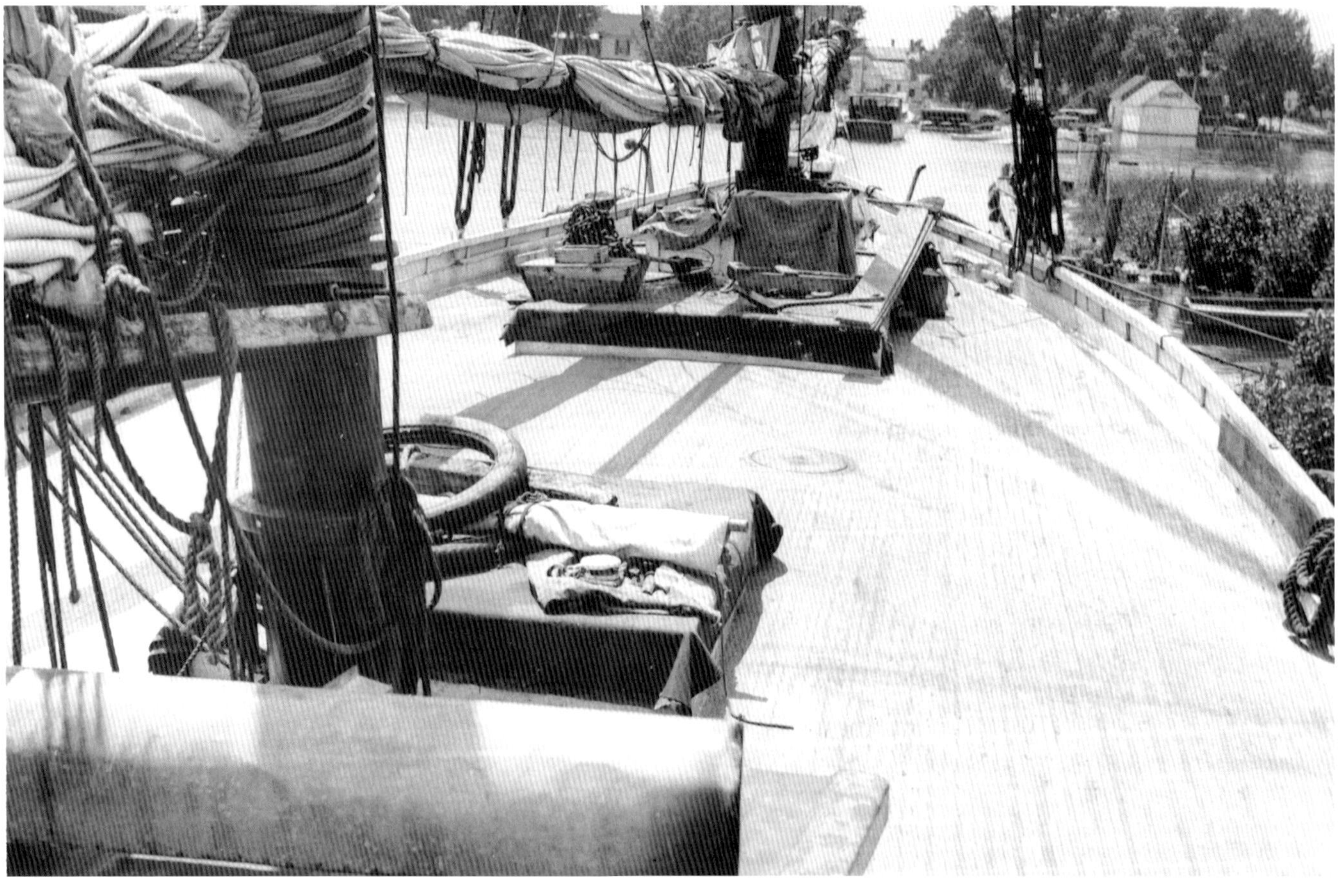

✪ Looking forward on the deck of the *Lola E. Bennett*.

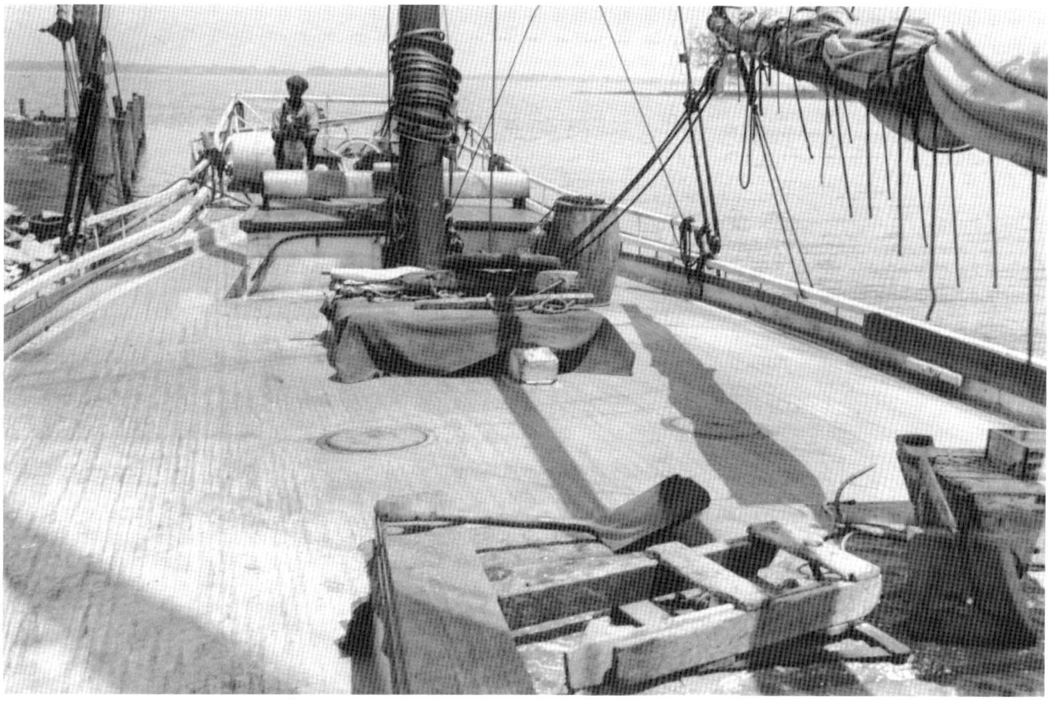

Looking aft on the deck of the *Lola E. Bennett* with the Miles River in the background. Her captain restrains his dog on the top of the after cabin as the author takes the photograph. Two manhole covers set in flush with the deck can be seen between the two hatches. Grain was poured through these openings into the hold. About 1940 the *Bennett* was sold to Florida and in 1948 was transferred to British owners in Nassau. She was still listed in the *Navy Mercantile List* of 1963 as a sailing craft.

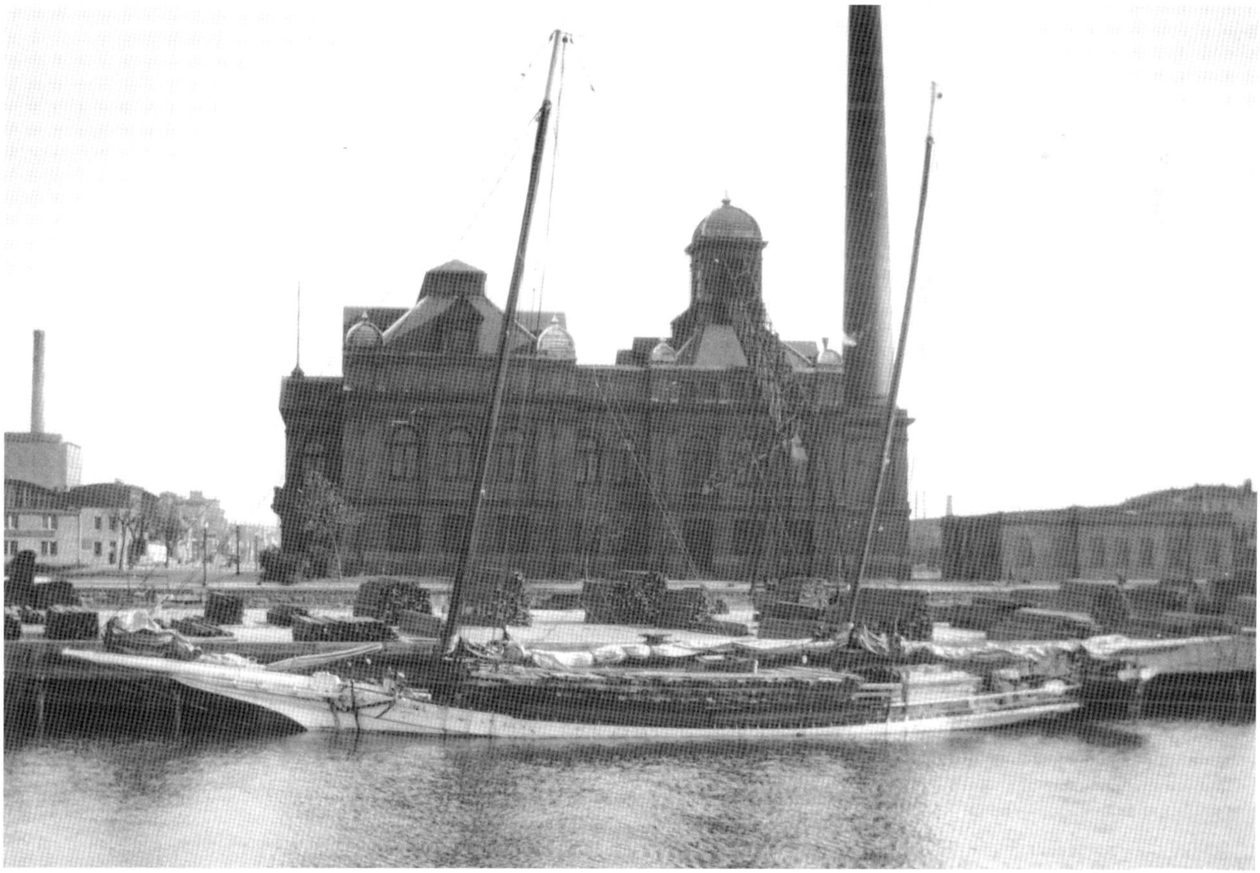

✪ The heavily laden bugeye *Maggie E. Maguire* unloads lumber at Pier 6, at Baltimore on November 12, 1933. Measuring 38 gross tons and 71.0 × 19.6 × 5.2 feet, she was built at Cambridge, Maryland, in 1905.

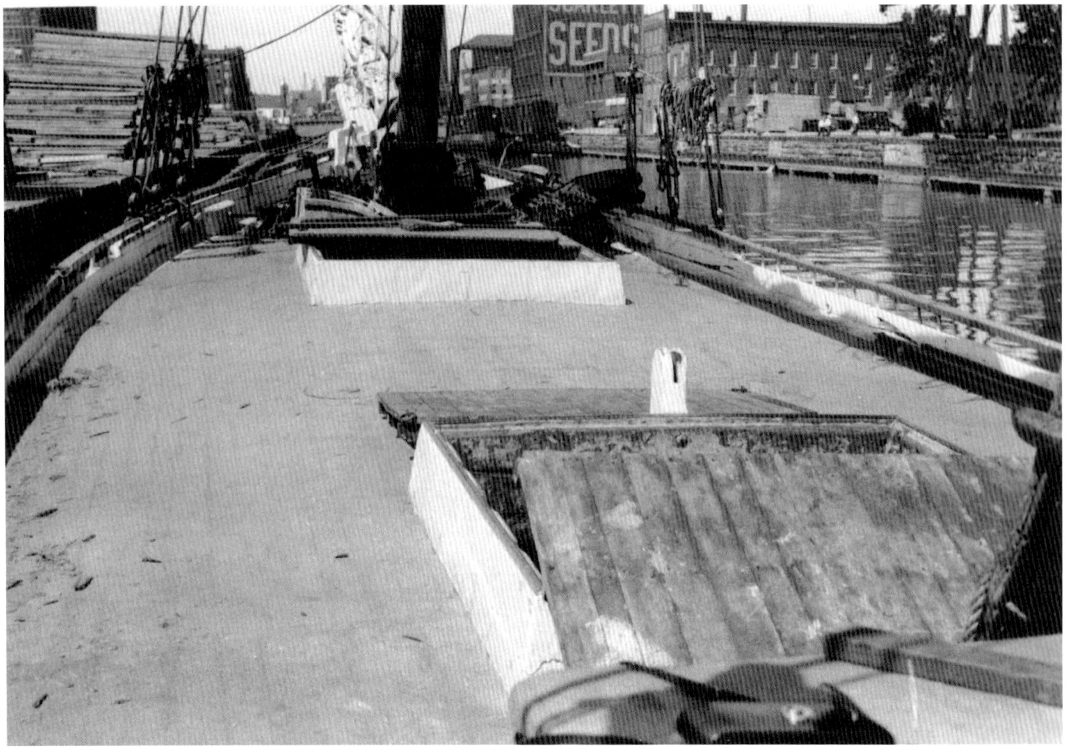

❂ A view on the deck of the *Maggie E. Maguire,* looking forward, at Baltimore in July 1936.

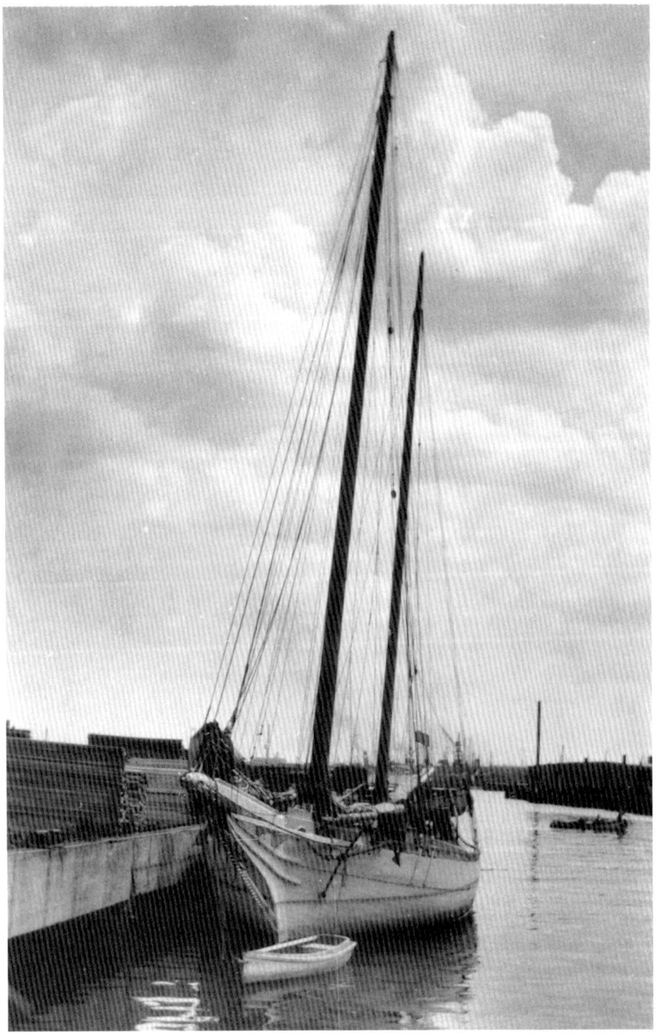

Majestic at Pier 6, Pratt Street, Baltimore, light and ready to return to her home port of Colonial Beach, Virginia, August 29, 1933. She had been built by John Branford at Fairmount, Maryland, in 1901.

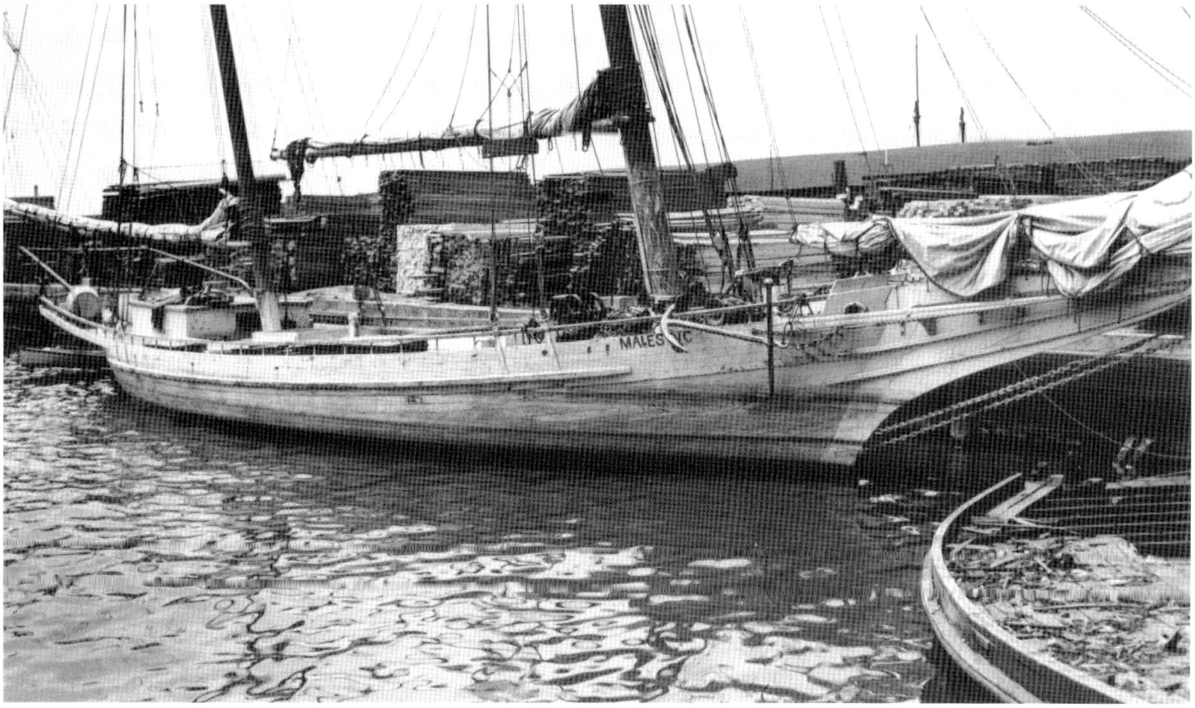

Majestic at Philpot Street lumber pier, Baltimore, after discharging her cargo on June 23, 1936.

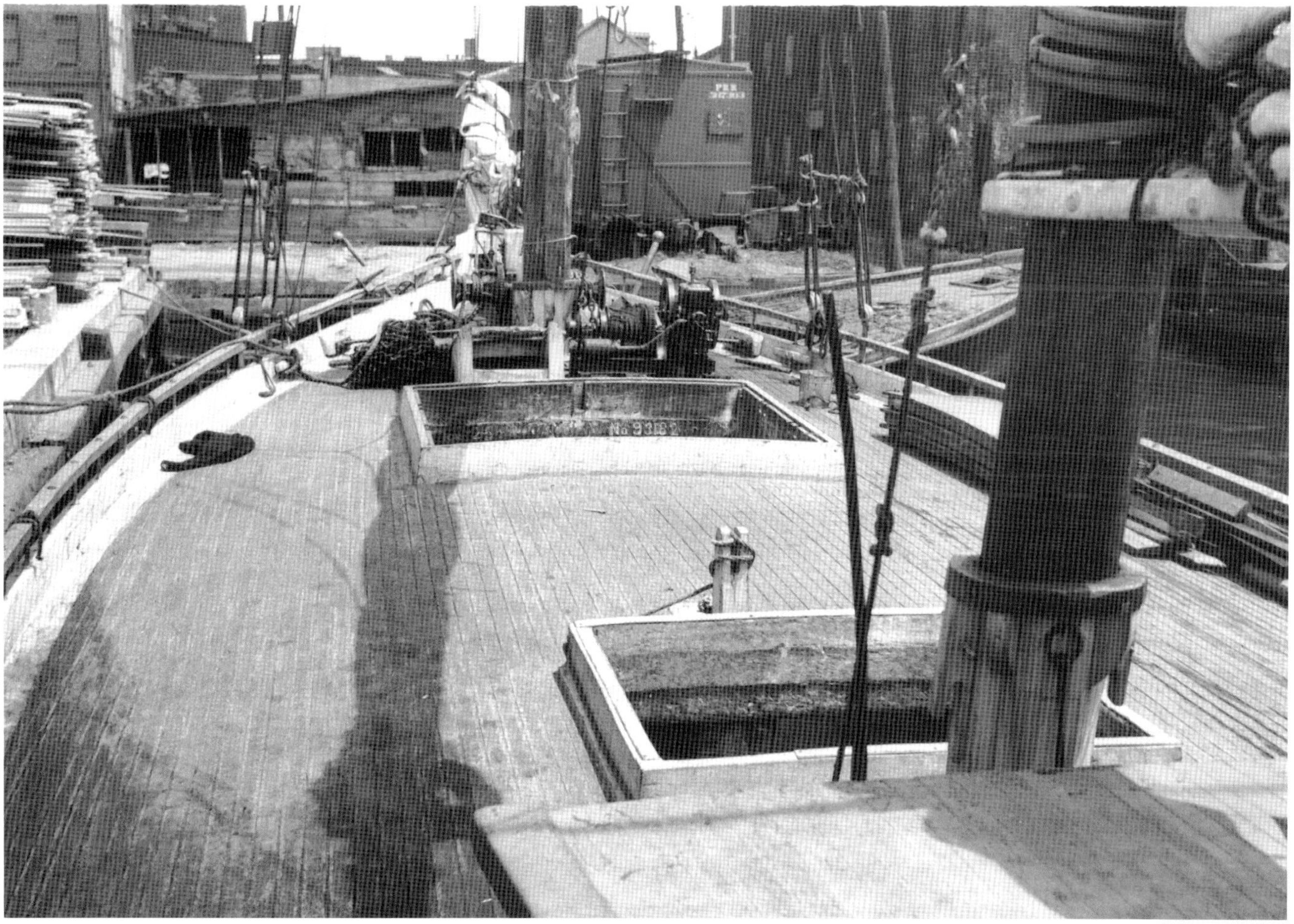

Looking forward on the deck of the *Majestic.* The small gasoline donkey engine on deck adjacent to her foremast was used to lift cargo out of the hold. Just forward of the main hatch can be seen the sheave for the centerboard pendant. About 1938 this vessel was converted to power, and five years later she stranded at Key West, Florida.

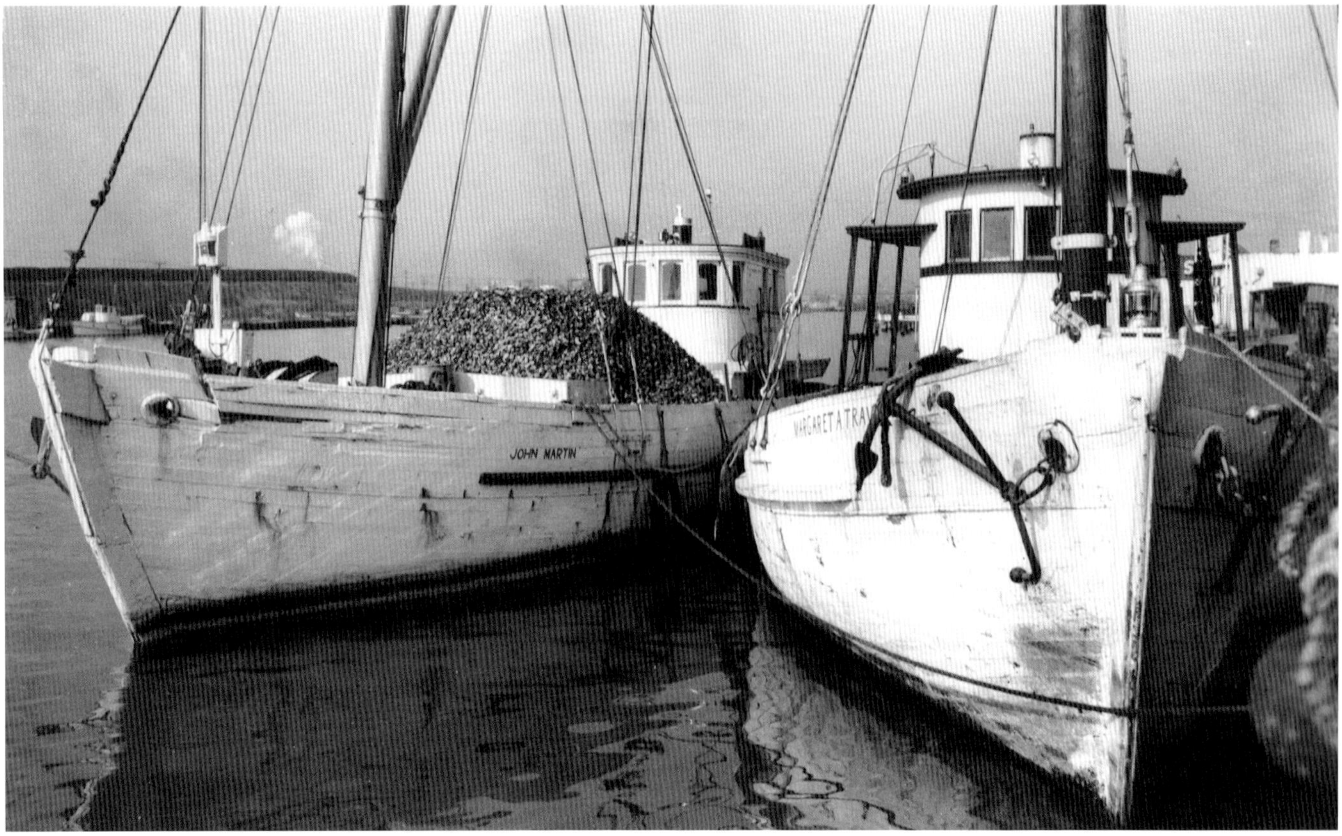

The round-sterned bugeye *Margaret A. Travers,* built by Otis S. Lloyd at Salisbury, Maryland, in 1901, was converted to power about 1935. Here she rests from her chores as an oyster buyboat at the boat harbor, Newport News, Virginia, December 15, 1947. At about 1:45 a.m. on December 16, 1952, while bound from Urbanna, Virginia, to Baltimore with 44,000 cases of canned beans, she hit a group of pilings near the mouth of the Patuxent River, Maryland, used as a target for testing planes from the nearby Naval Air Test Center. The impact knocked the pilothouse off and jammed the vessel's rudder. The craft circled the pilothouse, to which her master, Captain Jennings Burton, and a crewman clung, four times before she sank. The *John Martin,* at left, was a former schooner before being converted to power about 1933 and worked primarily as a buyboat thereafter. She was listed as "Abandoned" in the 1968 *MVUS.*

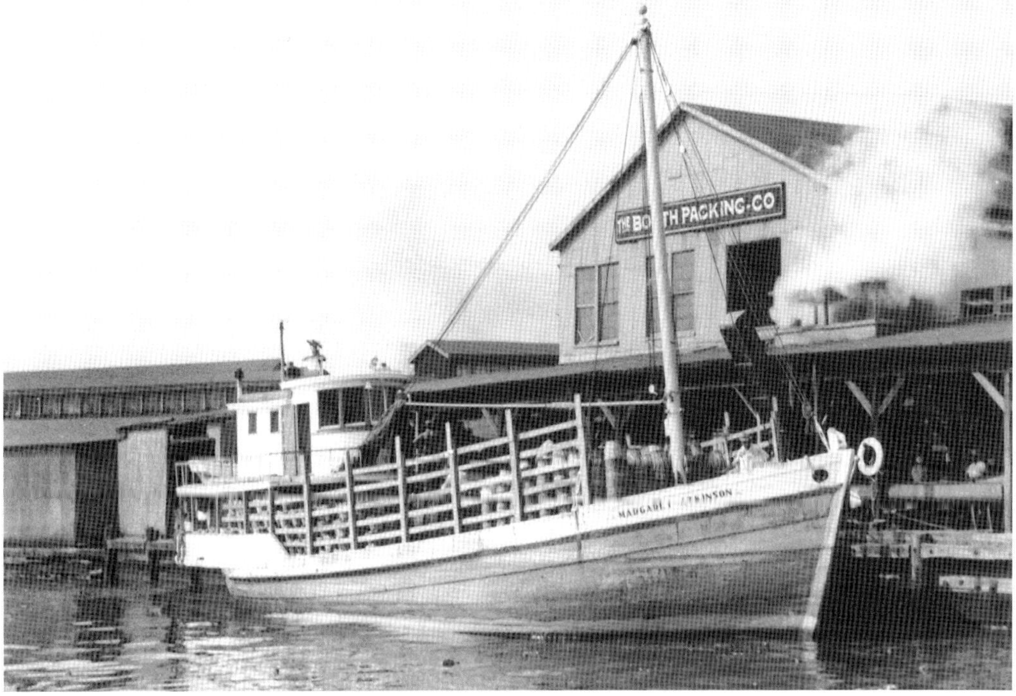

Margaret Atkinson, a bugeye built in 1904 at Lankford, Maryland, appears as a full-fledged powerboat with pilothouse atop her cabin. *MVUS* indicates she was powered from the outset of her career, although she probably retained her sailing rig. Here she discharges baskets of tomatoes at Baltimore, August 25, 1947. ✪ She was reported as "Abandoned" in the 1974 *MVUS.*

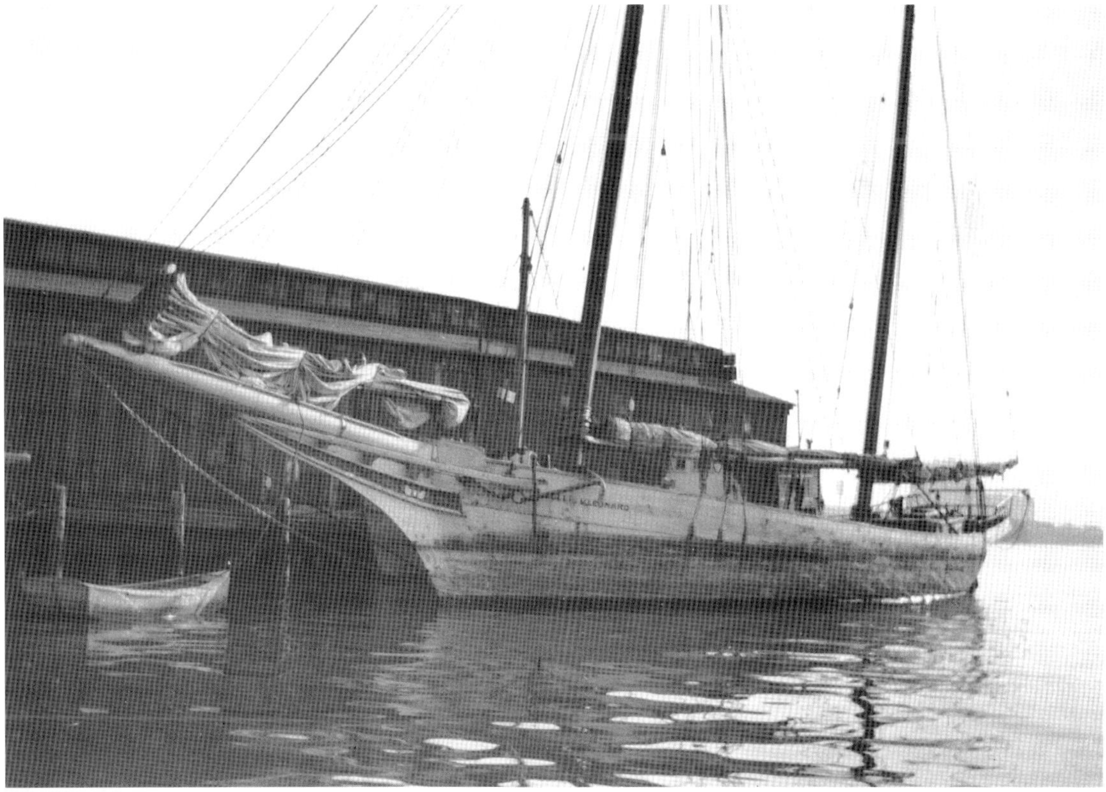

Mollie V. Leonard, second longest bugeye ever built and the largest in terms of tonnage, bears the appearance of a thoroughbred even though her grease-stained hull seems to belie the fact. Built by F. Cantwell at Cambridge, Maryland, in 1904, the 82.5-foot-long *Leonard* was the first bugeye to have a patent stern, which her owner invented. Although motorized at an early date (about 1917), she kept her sailing rig as in this view on January 20, 1934. She became an outright powerboat about 1940 and was sold to the Delaware Bay area in the early 1950s. In the 1963 *MVUS* she is listed as "Scrapped."

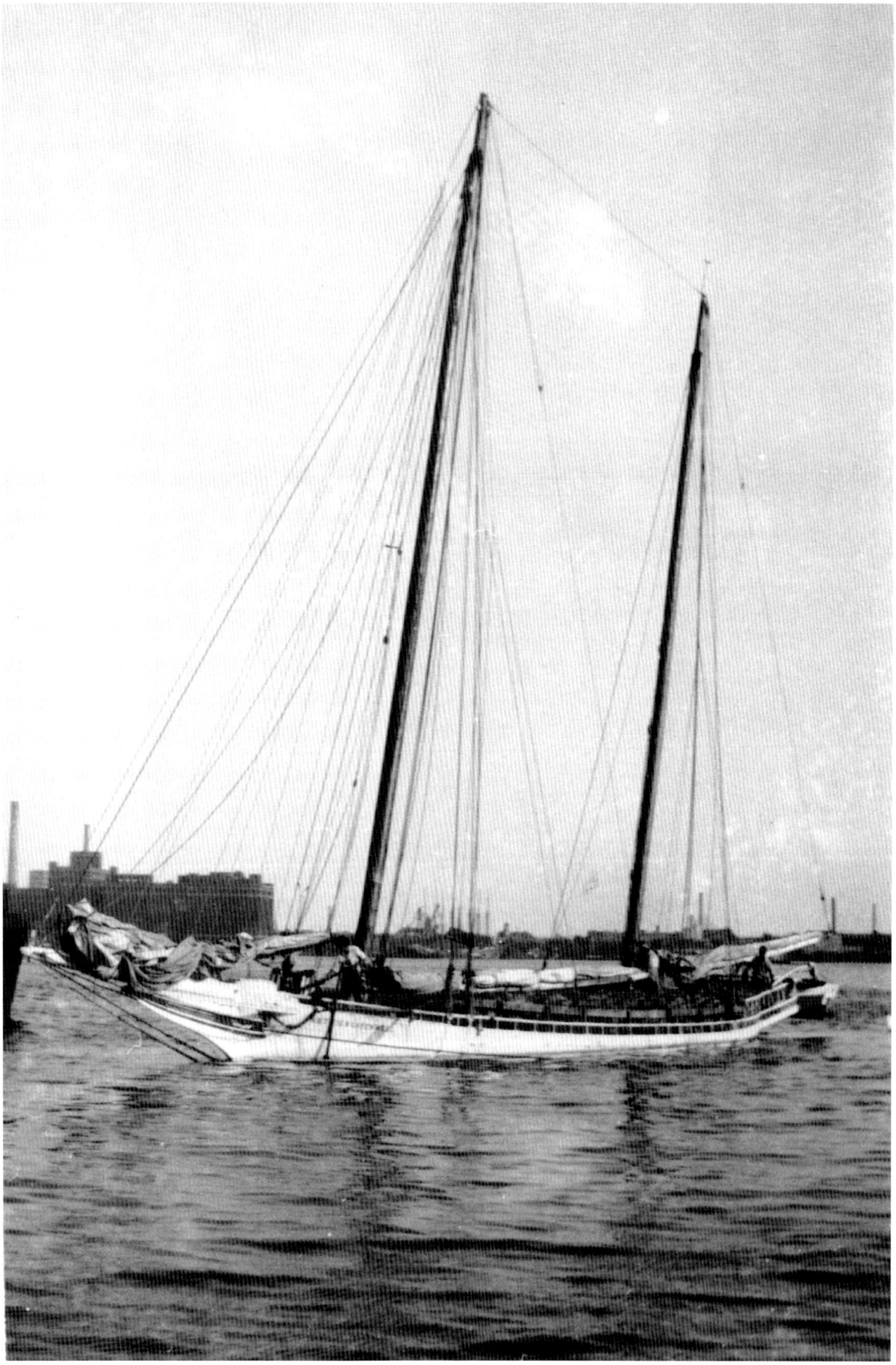

Nettie B. Greenwell has sailed up Baltimore harbor, dropped her jib and mainsail, and coasts into Long Dock with a load of watermelons on September 2, 1933. In the 1930 *Sun*-sponsored Chesapeake Bay Workboat Regatta, held off Piney Point, Maryland, the *Greenwell* won the bugeye event. Built by Joseph W. Brooks at Madison, Maryland, in 1897, this bugeye became a powerboat about 1936. On March 11, 1947, she was rammed by the Chesapeake and Ohio ferry steamer *Virginia* off Newport News and sank.

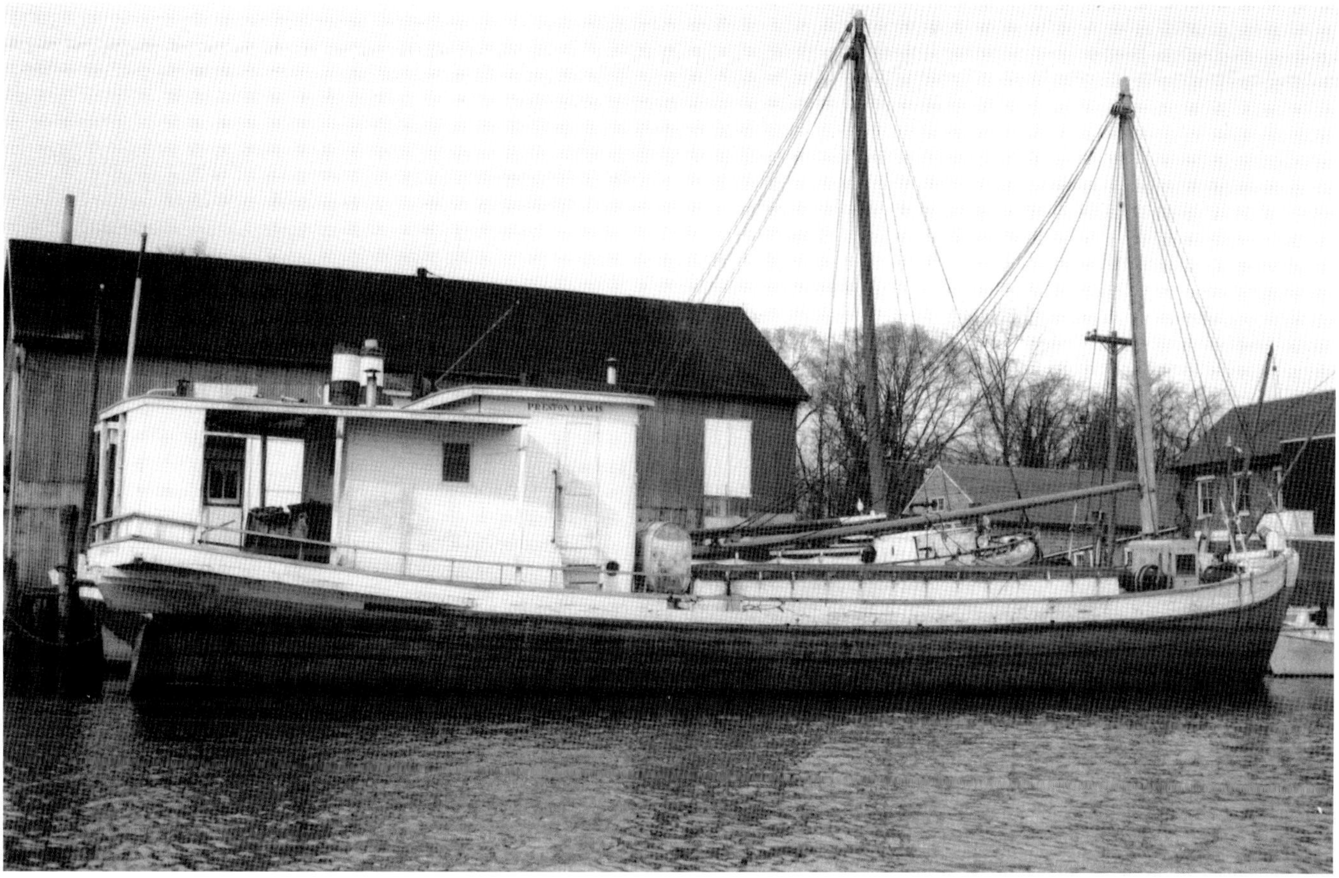

The *Preston Lewis* appears here as a powerboat, at Cambridge, Maryland, in 1947.

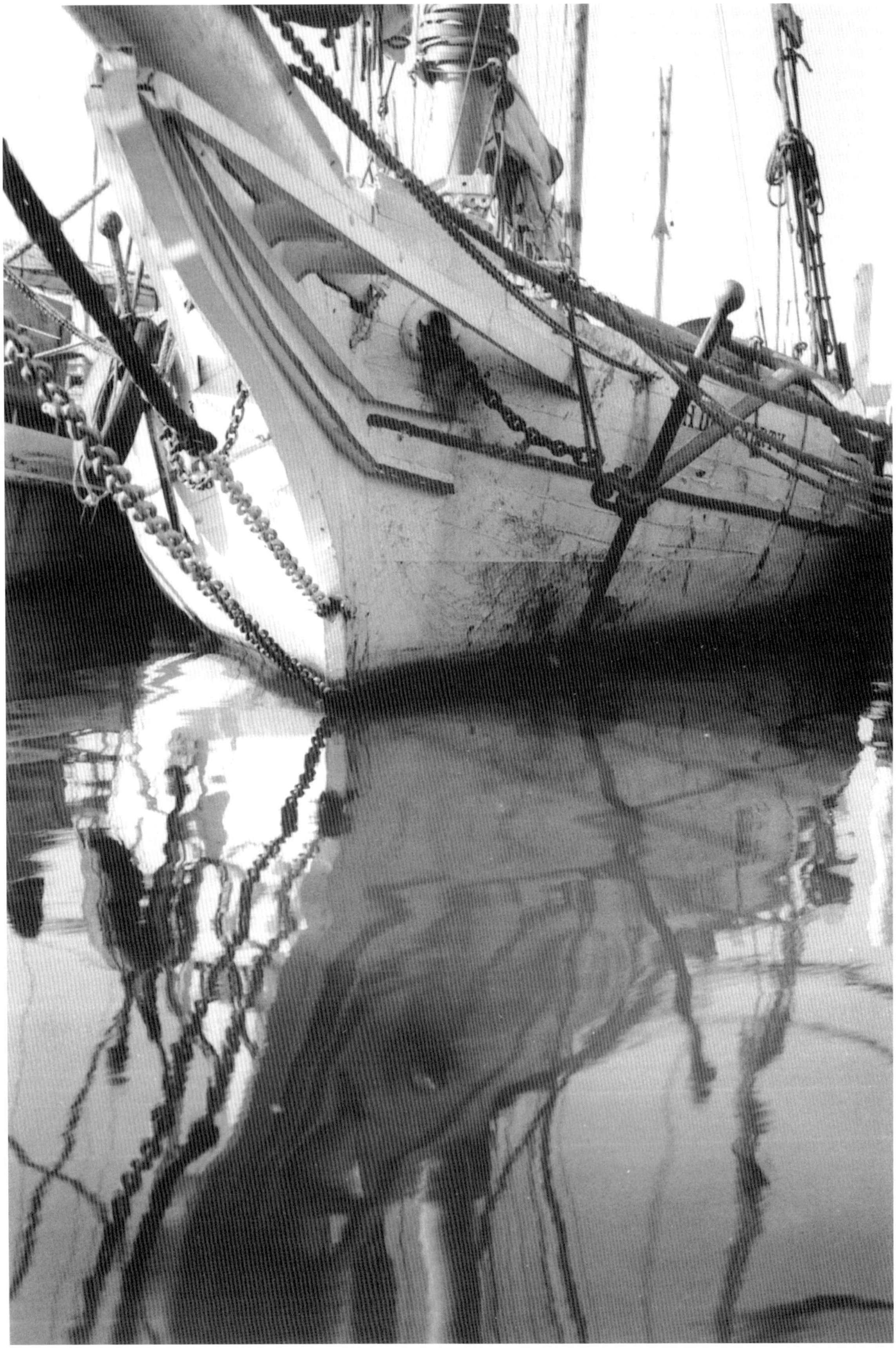

R.H. Daugherty has a day of rest in Annapolis, on February 10, 1952, after a rigorous week of oyster dredging on the Chesapeake. Annapolis was a favorite weekend mooring site for the dredgers working in that part of the Bay. The oystermen kept cars at that port, enabling them to return to their Eastern Shore homes for the weekend. Built at Madison, Maryland, in 1891, the *Daugherty* appears in the 1969 *MVUS* as "Abandoned."

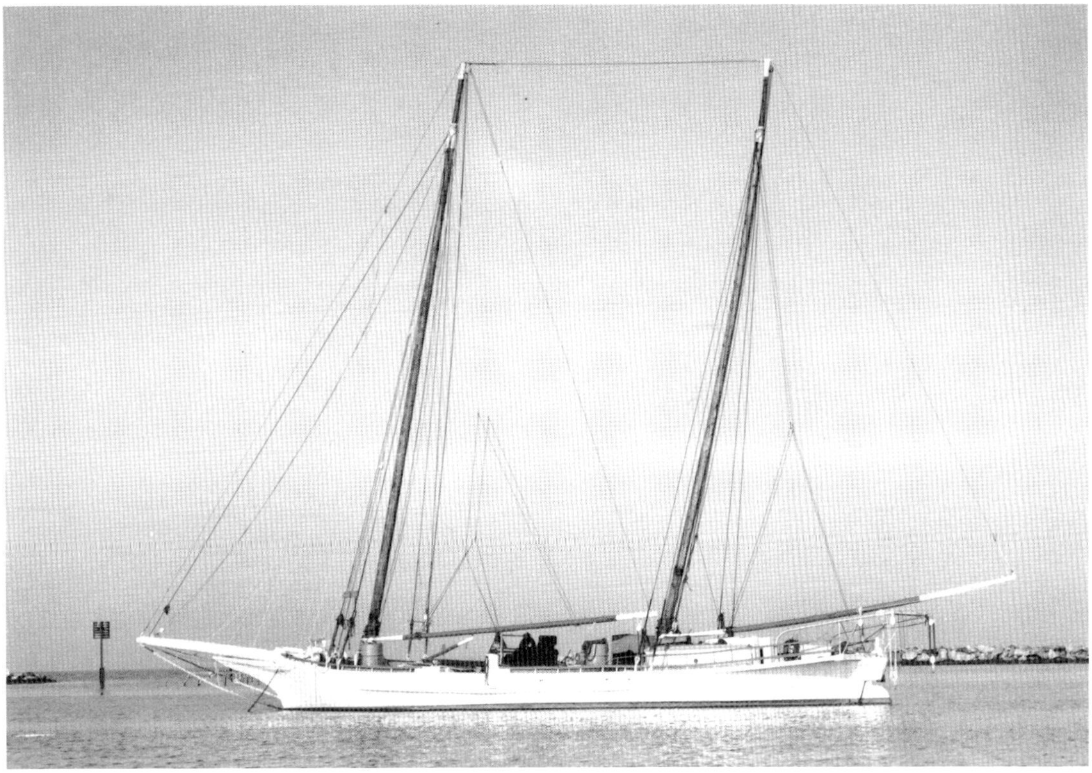

✪ A splendid profile of the *R.H. Daugherty,* taken at Deal Island on October 13, 1952. She certainly didn't show her 61 years in this view, and she sailed on the Bay for 17 more.

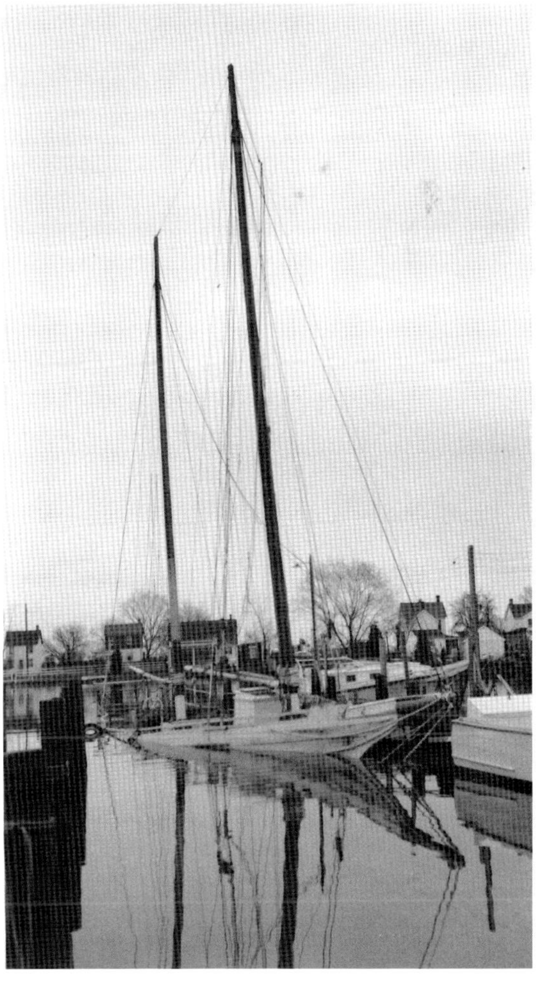

✪ The old bugeye *Richard Smith,* as she lay partially sunken at Oxford, Maryland, on March 9, 1962. She was built at Crisfield, Maryland, in 1881.

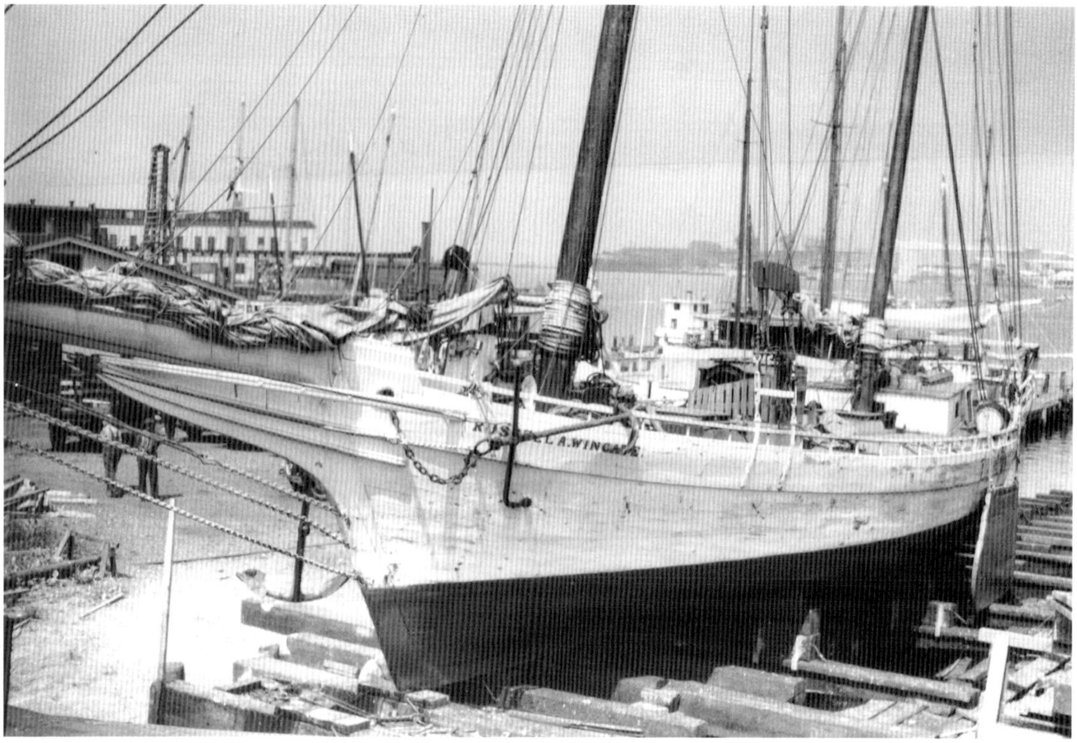

Russell A. Wingate on the marine railway of A. Smith & Sons, Curtis Bay, Baltimore, June 1939. Her centerboard has been removed and is suspended over the port side for painting. The white schooner anchored out in Curtis Creek is the *Ida Mae*. The *Wingate* was built by Joseph W. Brooks at Madison, Maryland, in 1901. About 1948 she was sold to Delaware Bay and motorized and was still in the 1972 *MVUS*. ✪ She was reportedly owned by a New Jersey historical society in the late 1970s, then abandoned in the 1980s.

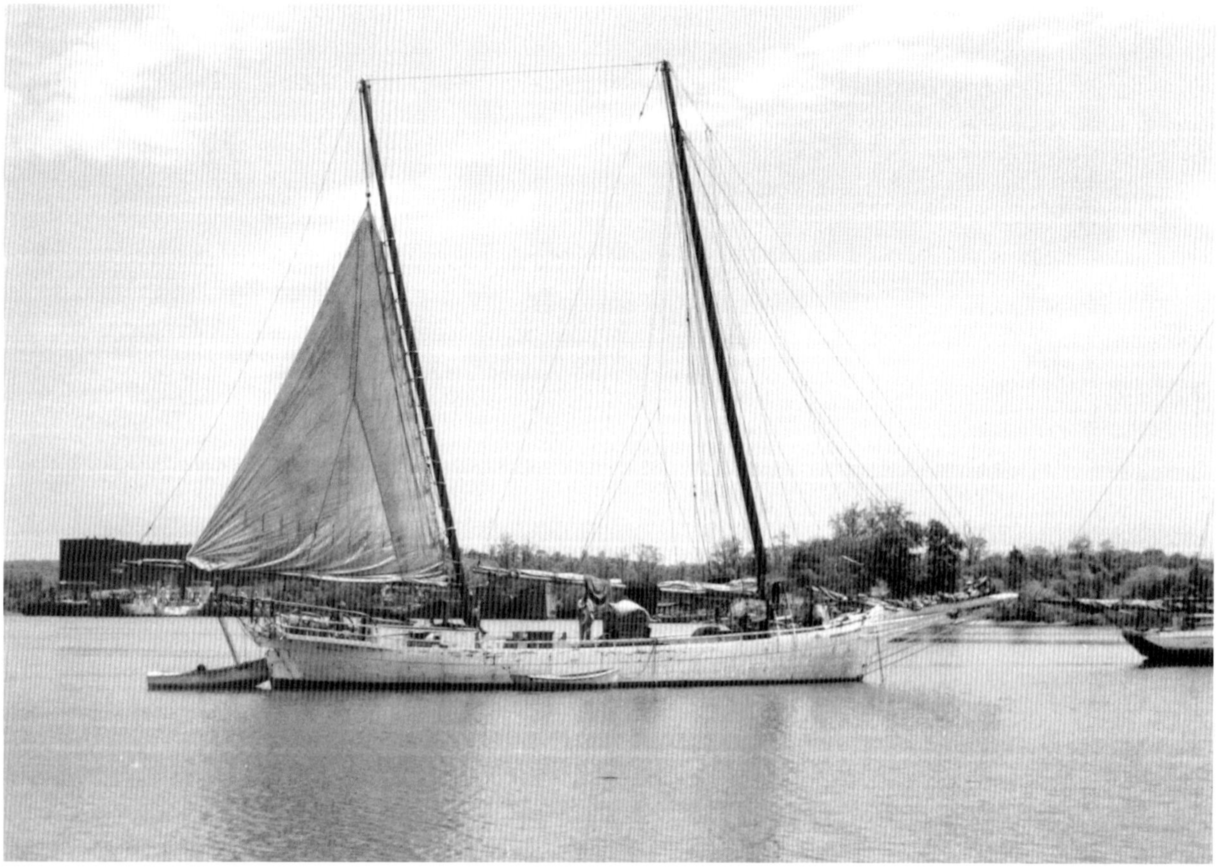

✪ A good profile view of the *Russell A. Wingate* at anchor, taken in June 1939.

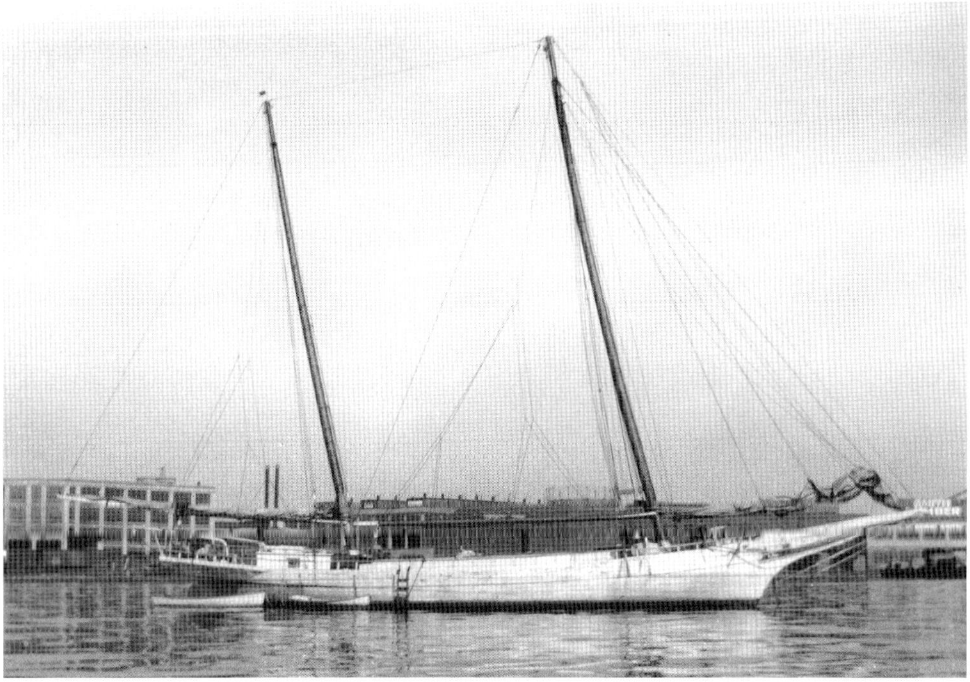

Russell De Hughlett, shown anchored in Canton Hollow, Baltimore, on October 29, 1933, had an auxiliary engine installed about 1926, but kept her sailing rig. Here her foresail is lacking, which might indicate that she is more dependent upon her engine. Or the sail could be at the sailmakers for repairs, or a new one is being made. She had been built in 1902 by Joseph W. Brooks at Madison, Maryland. Boston Street runs parallel with the harbor's edge, in the background. The building at the left was the American Can Company, from which many of the Bay sailing craft took cargoes of empty cans to canneries in the tidewater country. The *De Hughlett* was sold to the Delaware Bay area about 1934 but later returned to the Chesapeake before being recorded as "Abandoned" in the 1937 *MVUS.*

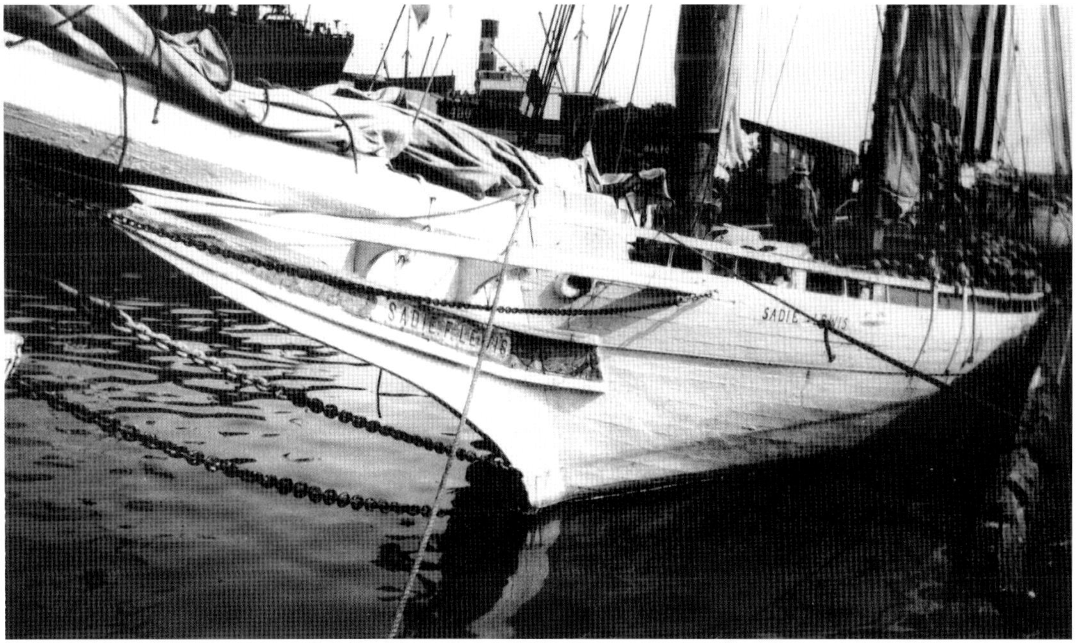

Sadie F. Lewis alongside Long Dock with a cargo of watermelons on September 13, 1935. Apparently, she was successful in disposing of them as indicated by the grease line above her waterline. The melons on deck were taken out of the hold for viewing by prospective purchasers. Proof that these smaller Chesapeake sailing craft went far afield is found in the 1905 report of the United States Life-Saving Service stating that the *Lewis* went aground near Ocracoke Inlet, North Carolina, while bound from Ocracoke to Portsmouth, North Carolina. Surfmen came to the bugeye's aid and released her. But she was to sail for many more years until listed in the 1939 *MVUS* as "Abandoned." The *Lewis* had been built by John Branford at Inverness, Maryland, in 1902.

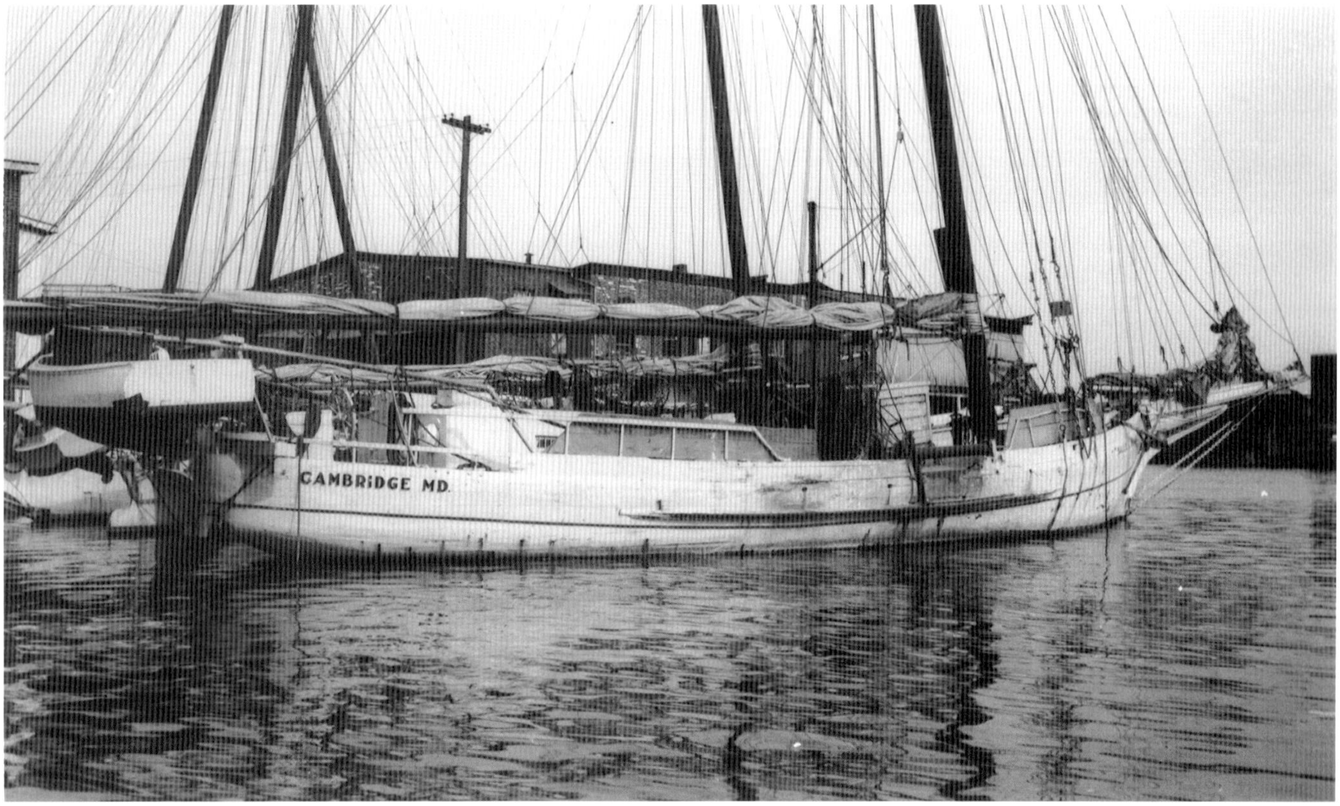

A rarity on the Bay was the round-sterned, one-masted bugeye *Sallie L. Bramble,* built by Otis S. Lloyd, Sr., at Whitehaven, Maryland, in 1890. This builder specialized in the construction of round-sterned bugeyes, usually with two masts, but he also built the round-sterned, one-master *Hattie Lloyd* in addition to the *Bramble.* J.B. Harrison of Tilghman, Maryland, also built a one-masted, round-sterned bugeye, the *Agnes,* in 1898. The *Bramble's* round stern is visible in this 1949 view in Cambridge where she was based as an oyster dredger. ✪ This vessel was broken up in 1991.

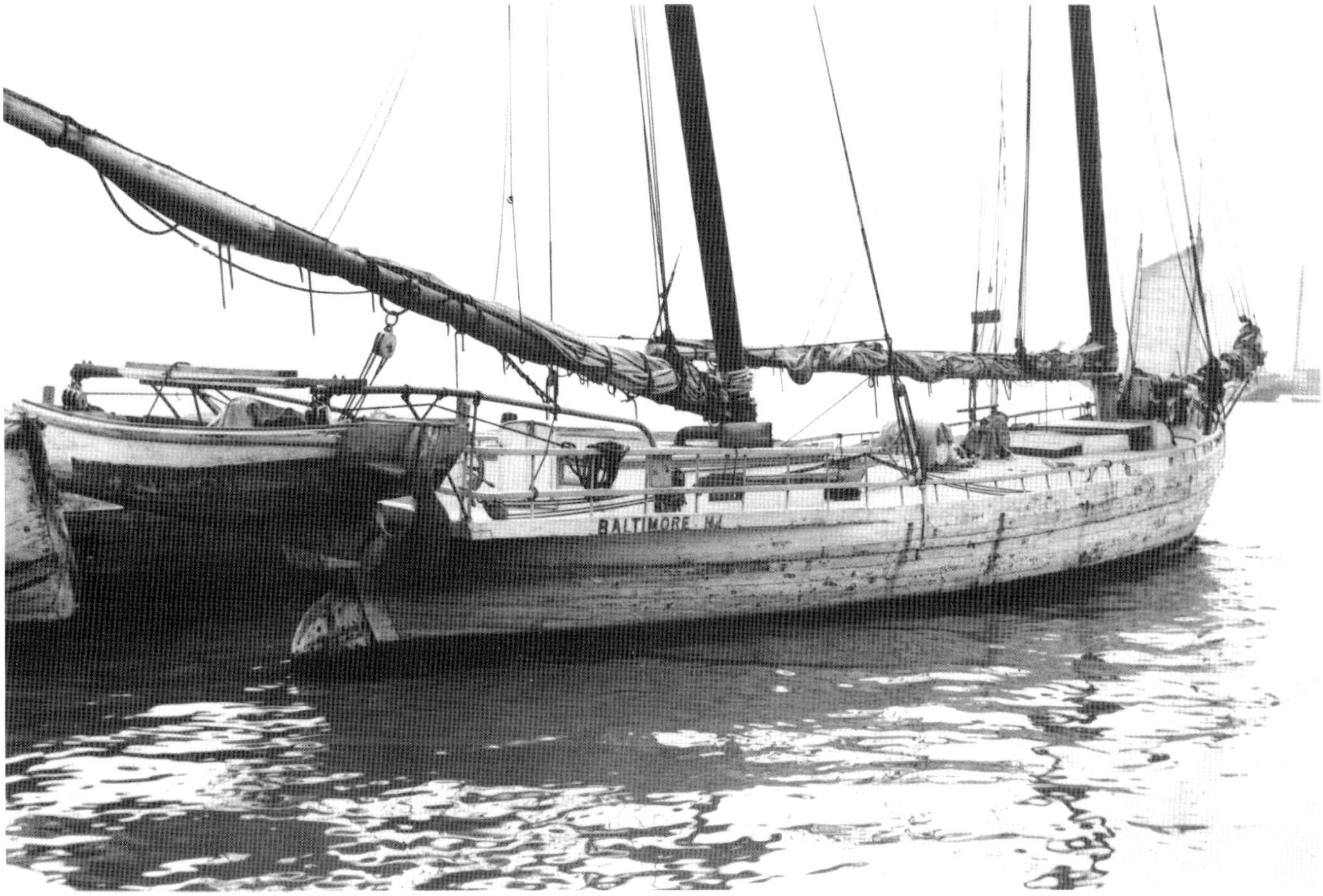

Sarah W. Wingate, built by Joseph W. Brooks at Madison, Maryland, in 1897, was to have a varied career in her later life. But here she is shown tied up at Montford Yard, Baltimore, on March 8, 1936. On September 9 of that year the *Wingate* was bound from Baltimore to Centreville, Maryland, with a cargo of fertilizer. Her owner, Captain Harry Dill of Centreville, was in command. As the bugeye approached Town Point, four miles below Centreville, Captain Dill was at the wheel and asked a crewman to go below and fetch him a glass of water. When the seaman returned he found Captain Dill slumped on deck, dead, and the *Wingate* hard aground, having strayed from the Corsica River channel. The crewman rowed ashore and walked seven miles to inform the captain's wife of the tragedy, attributed to heart failure. Shortly thereafter the bugeye was sold to Captain P. Jay Elbourn of Rock Hall, Maryland, and was motorized in 1938 and renamed for her new owner. On February 15, 1939, while bound from Oxford, Maryland, to Rock Hall with a cargo of oysters, the *Elbourn* encountered a fog and while feeling her way up the channel to her destination ran upon a bar, opened her seams, and sank. She was later raised and repaired and carried on as a powerboat until she was abandoned in Swan Creek, Rock Hall. She last appears in the 1952 *MVUS.*

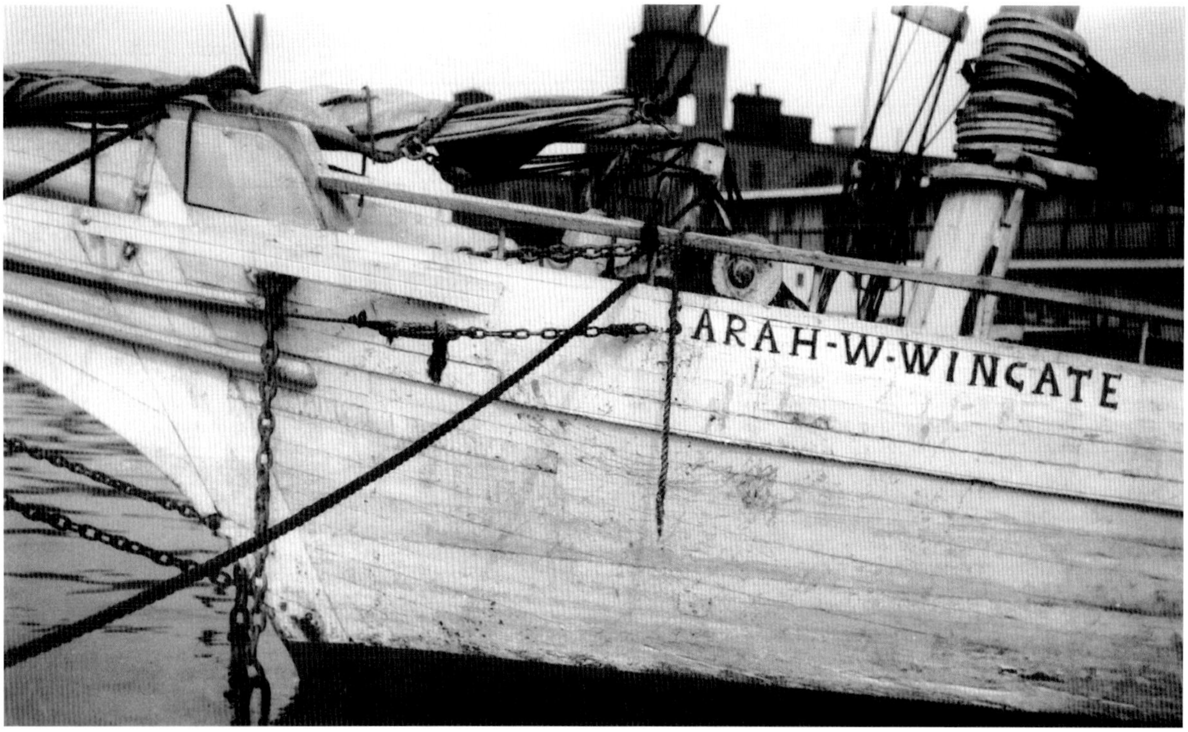

✪ Bow detail of the bugeye *Sarah W. Wingate,* also on March 8, 1936. She measured at 34 gross tons and 70.8 × 19.4 × 5.2 feet.

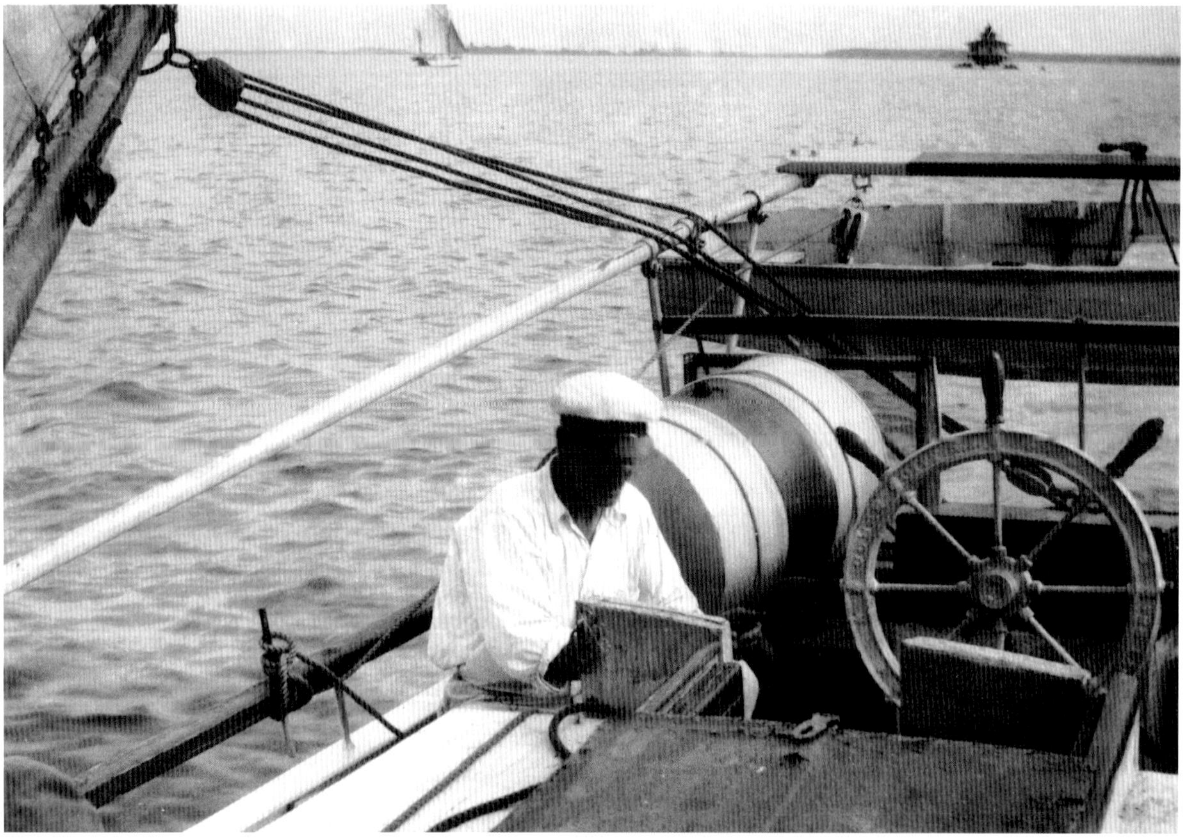

Onboard the bugeye *Susie E. Parriss,* Captain Charles W. Pope tends the wheel in light winds as he steers around the course in the Choptank River near Oxford during the 1933 Labor Day race with the bugeye *Hugh S. Orem,* cutter *Windrush,* and yawl *Spray.* At upper left is the *Spray* steering to round the Choptank River Light, upper right. The *Parriss* won the race as a result of the leading boat failing to properly round a marker and was awarded a turkey. Slung from the davits in this view is a flat-bottomed skiff in place of a yawl boat. The *Parriss* had an auxiliary engine at this time; gasoline was stored in the green and white metal drum lashed down behind Captain Pope on the patent stern. The *Parriss* was built in 1885 by W.P. Benson at Oxford, Maryland.

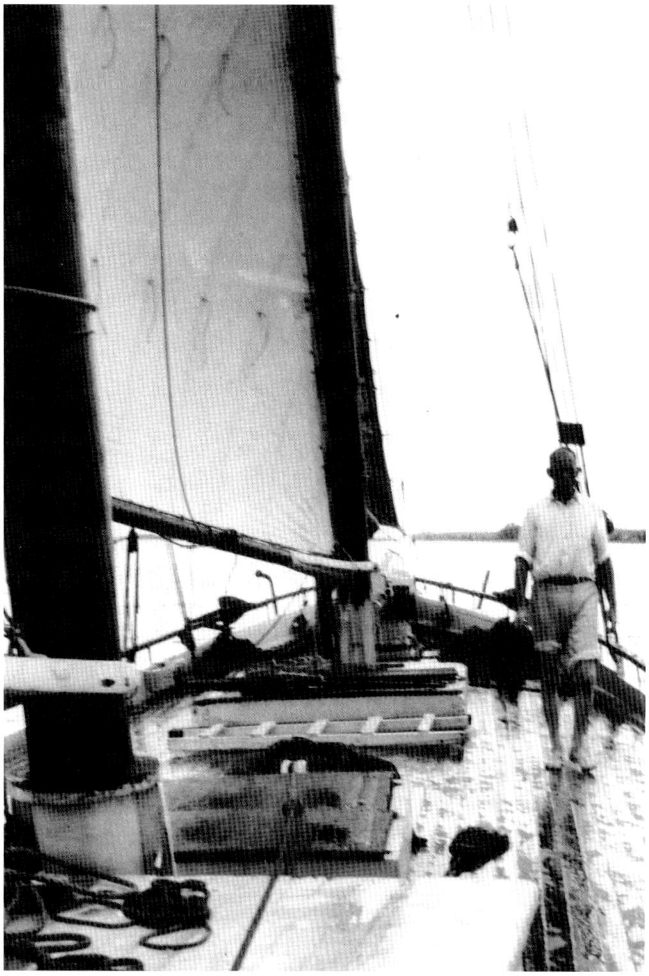

Looking forward on the *Susie E. Parriss* shortly after a rain squall passed during the 1933 Labor Day workboat race at Oxford. The line crossing the cabin top is the centerboard pendant. The *Parriss* encountered difficulties on April 16, 1935, while bound for Baltimore with a cargo of oysters. Off Greenbury Point Light she developed a leak and as her crew of five frantically bailed and pumped she flew distress signals, first sighted by the schooner *Kate H. Tilghman* that went to her aid. About the same time, the Annapolis-Matapeake ferry *Governor Albert C. Ritchie* came alongside and towed her to an Annapolis marine railway. This bugeye last appears in the 1947 *MVUS*.

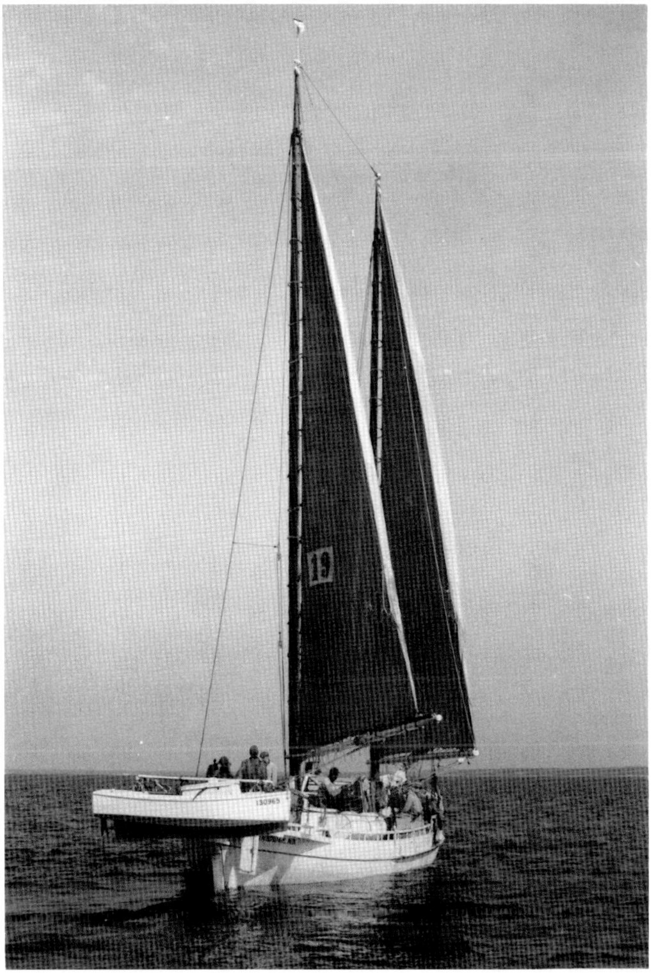

Thomas M. Freeman left the Chesapeake in 1965 to become a floating exhibit in New Jersey. That year she was acquired by the Historic Smithville Inn and taken up to Delaware Bay, out in the Atlantic to Mullica River to Port Republic Lake. She was transported by trailer to Lake Meone at Smithville and floated as part of that town's restoration. In this view, taken on the Choptank River on November 5, 1957, the *Freeman* was working for a living on the oyster beds. Her crew sits idly on deck as the port dredge makes a lick over the bottom.

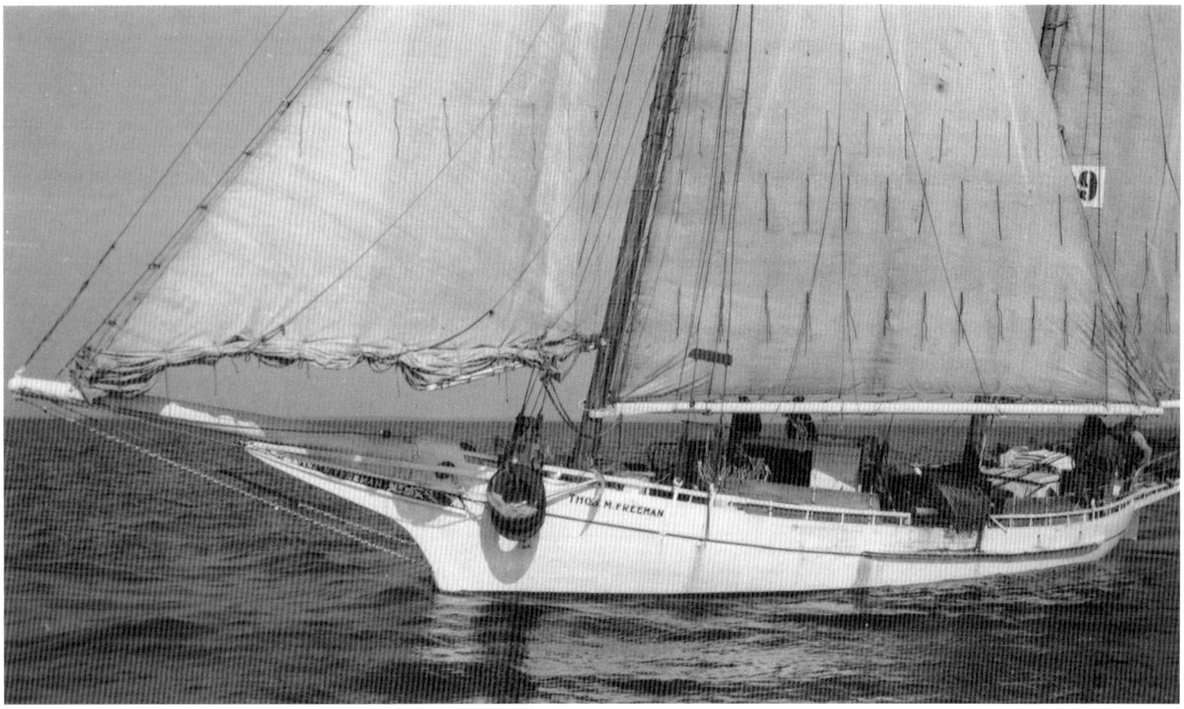

The rakish-looking *Freeman* ghosts along in light airs as she dredges in Choptank River. Reefs are in her jib and mainsail, not that she needed them at that time but perhaps because heavier winds were expected later in the day. The *Freeman* was built at Madison, Maryland, in 1889 by Joseph W. Brooks.

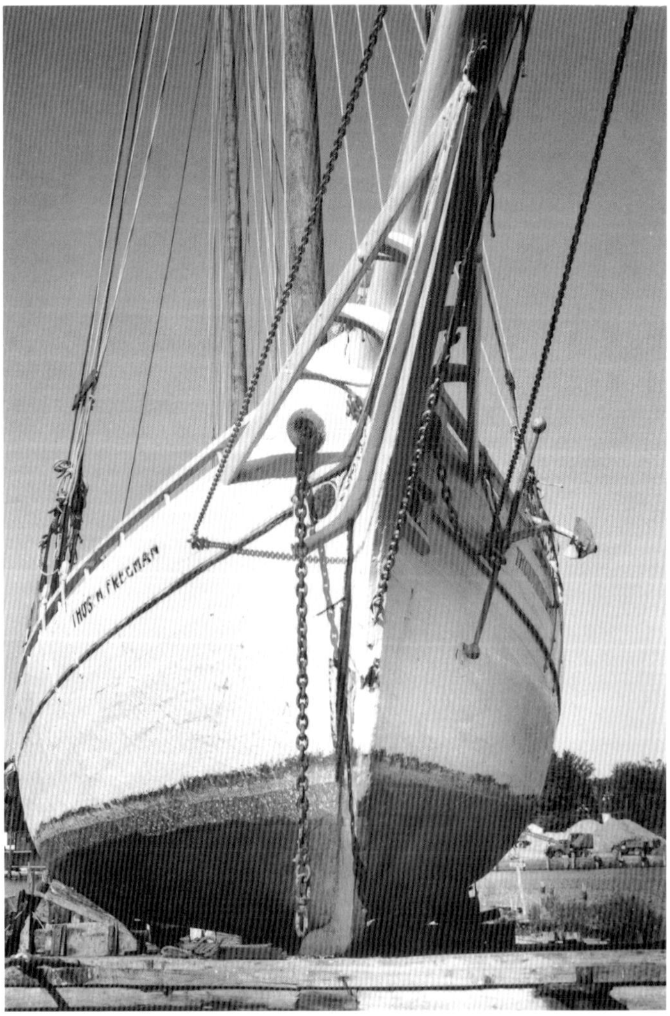

At a Cambridge shipyard the *Thomas M. Freeman* rests on a marine railway for repairs and overhaul on October 26, 1959, at the outset of another oyster season. One major repair was the replacement of her lower bobstay chainplate. The metal sheathing along her sides at the waterline, crinkled from wear, protected her planking from sharp ice, which invariably covers the Choptank in winter.

✪ A view across the bow of the *Freeman* in 1949. Her name was abbreviated on her hull, but it was spelled out in *MVUS*.

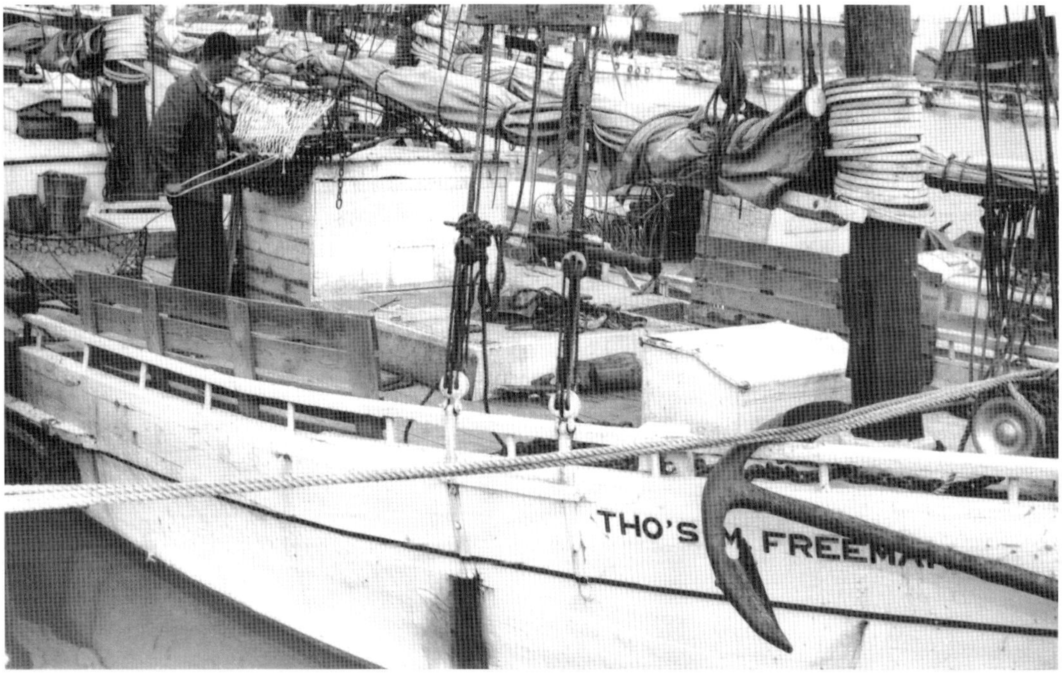

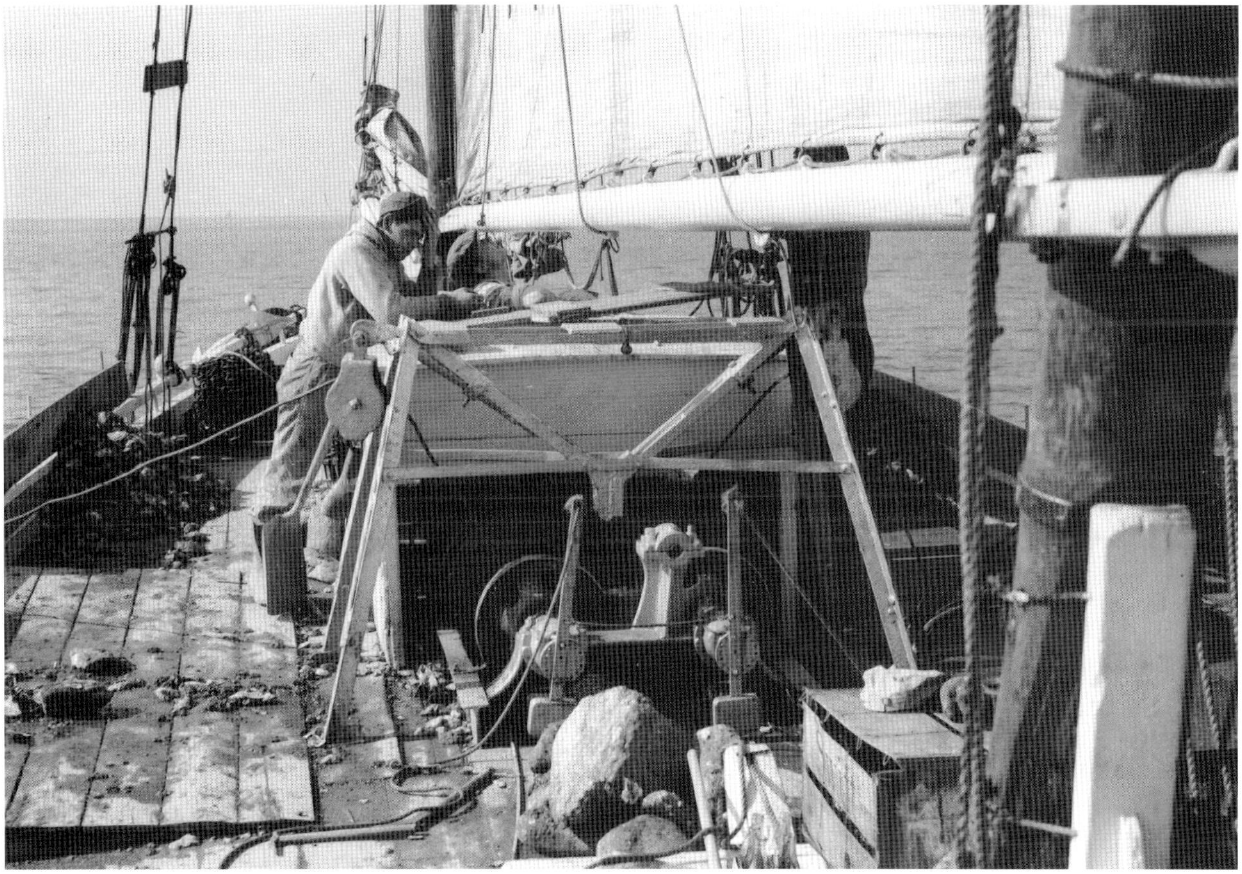

At noon on the Choptank a young deckhand on the *Freeman* rests against the winder engine before carrying on until sunset. Abaft the winder is the crane made up of a pair of A-shaped frames, one at each side of the winder and secured at its top by a strongback. Heavy iron swivel gin blocks, here shown, hung in each apex. The purpose of the crane was to lead the steel cable dredge lines in the proper direction to the head of the Hettinger winder. The vertical bars in lower center, attached to the winder, activated the clutches as the line at the tops of the bars was pulled by the man at the wheel when he wanted to bring the dredges onboard. The oysters are dumped out of the dredge onto the rough planks on deck to be culled. The legal-sized oysters, three inches and over in length, are placed in piles fore and aft on port and starboard sides. Boards have been placed above the log rail to keep the growing piles of oysters inboard as the day progresses.

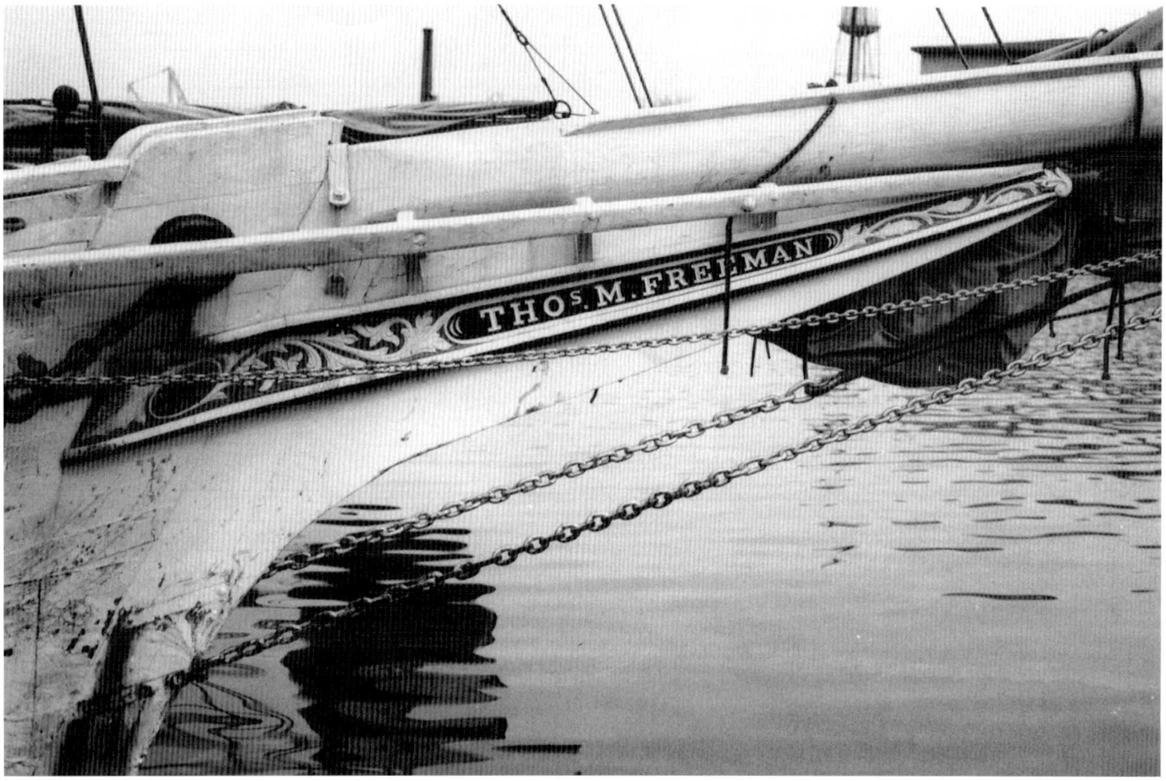

Delicately carved trailboard and sea monster figurehead of the bugeye *Thomas M. Freeman* at Cambridge on January 30, 1949.

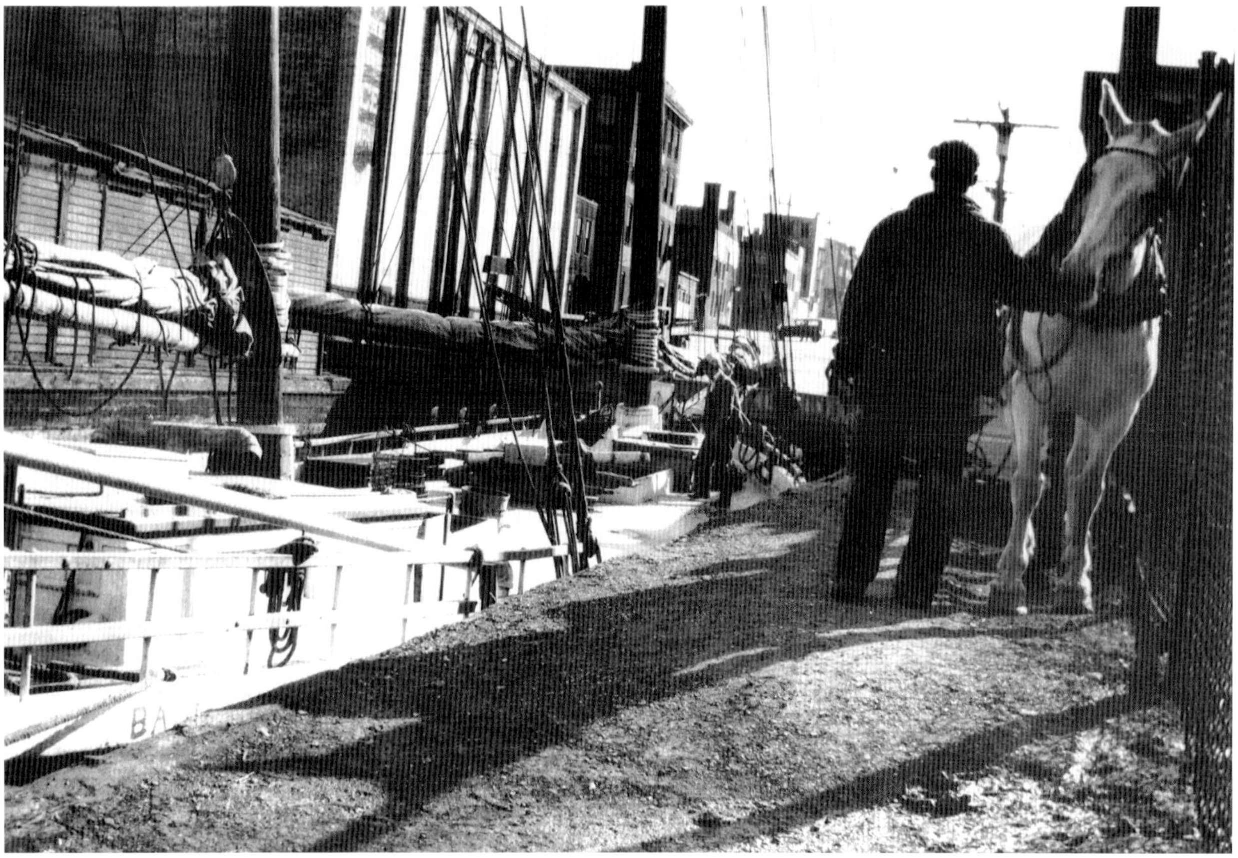

Oysters are lifted out of the hold of the bugeye *Vaughan* by a dependable one-horsepower "engine" at the foot of William Street, Baltimore, November 4, 1933. A block and tackle suspended from the head of the foremast is activated as the horse is led along dockside to lift the metal bushel containers of oysters into a wheelbarrow. *Vaughan* had been built by Joseph W. Brooks in 1899 at Madison, Maryland, and appears as "Abandoned" in the 1950 *MVUS.* She had been motorized earlier.

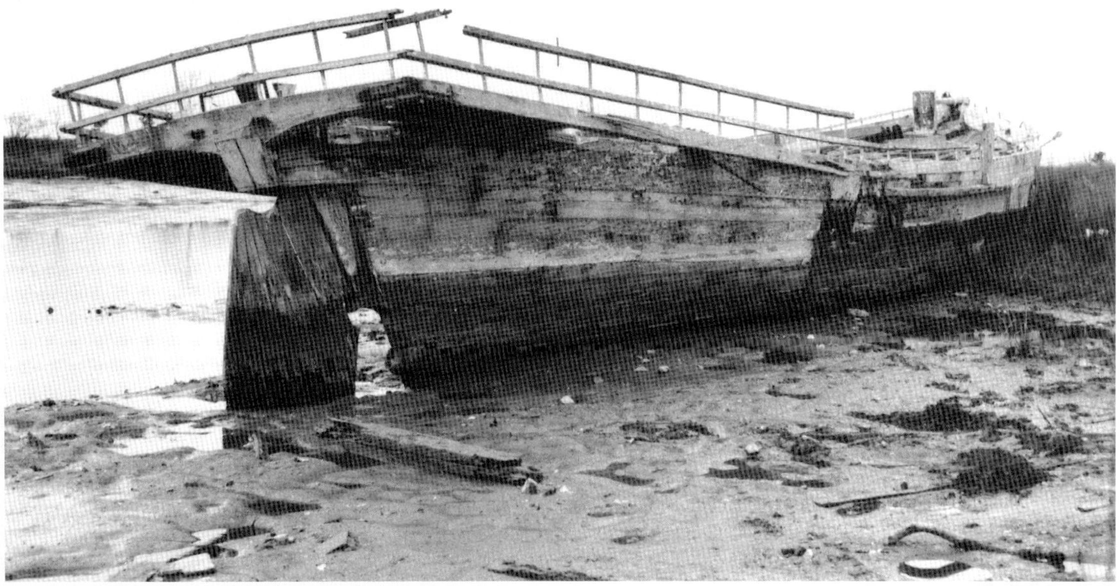

Wm. A. Egerton, built in 1881 at Manokin, Maryland, by W. Lawrence, came to her end not too far from her birthplace, at Chance, opposite Deal Island, and looked like this on February 15, 1950. She ended her days as a powerboat, having been motorized about 1931, as evidenced by the propeller aperture.

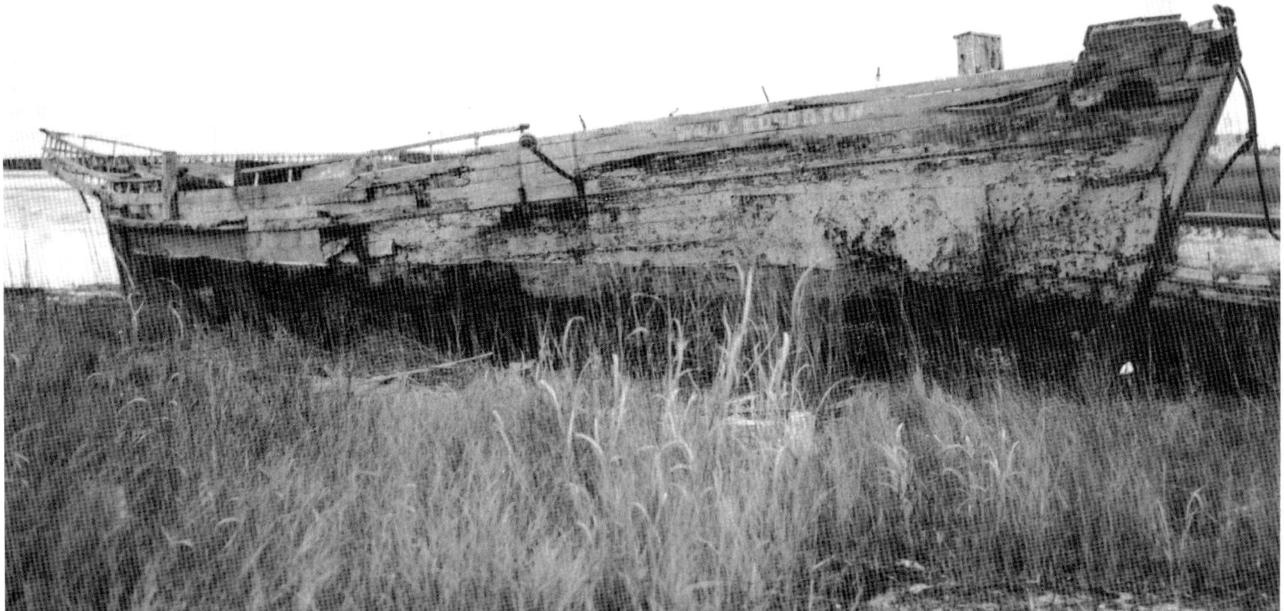

✪ Another view of the *Wm. A. Egerton* at Chance, Maryland, on February 15, 1950. Several of the Bay vessels abbreviated their first names in this way according to the official listings in *MVUS.*

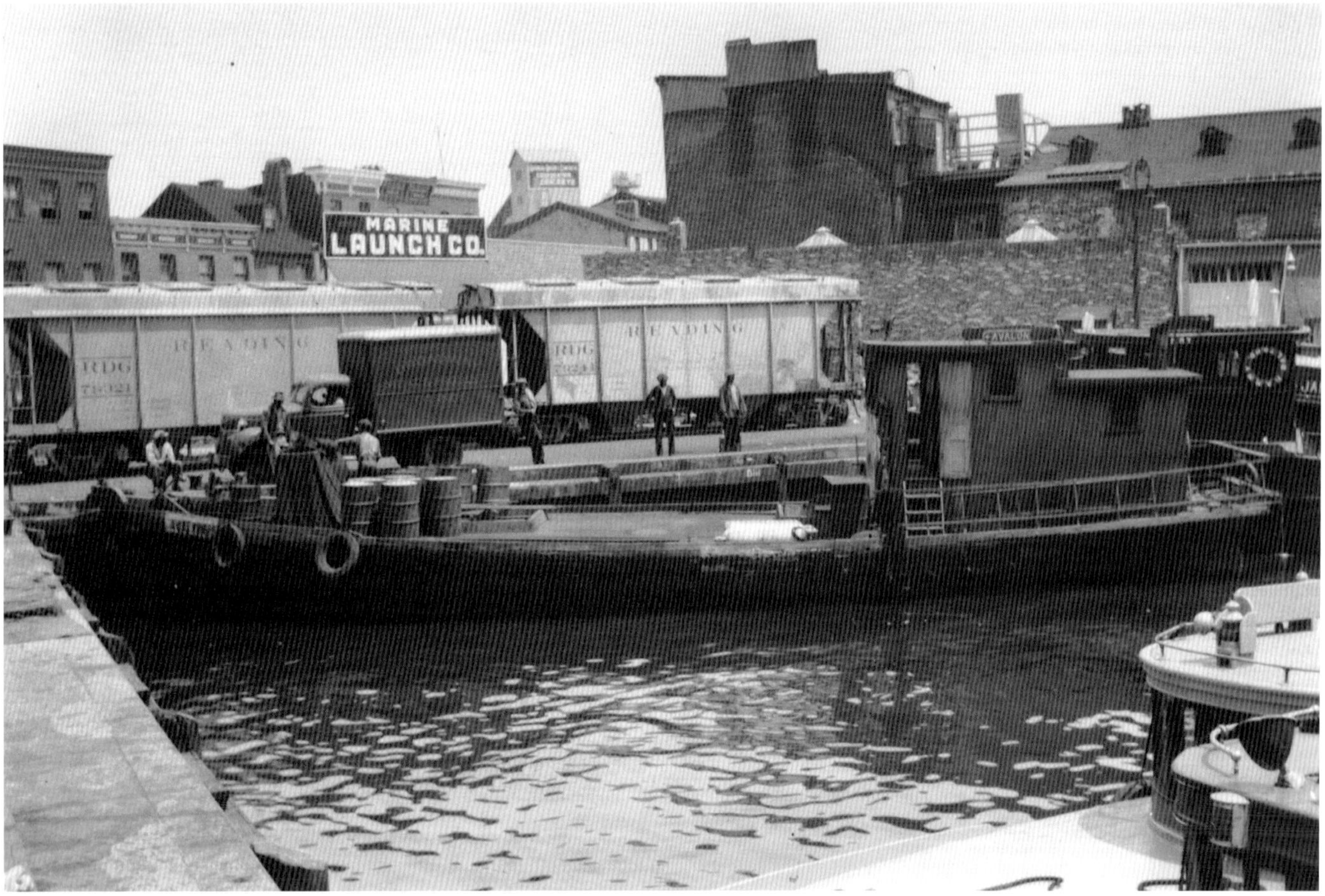

✪ *Avalon,* after her conversion to a powerboat, at Baltimore on May 31, 1947. One would never guess that she had once been a graceful bugeye.

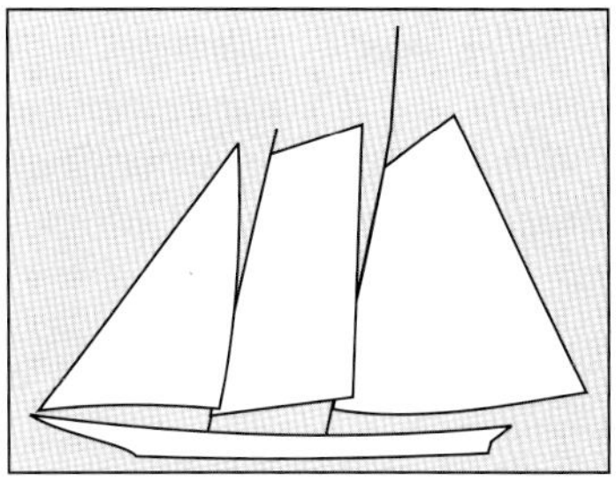

Pungies

There was always something exciting about the rakish sailing craft of the Chesapeake, particularly the pungies. By rakish I mean the lean hulls with pleasing sheer, fine lines, tall raking masts, with thin main topmast bowed slightly forward. The stout bowsprits, shaped by master shipwrights and unlike the pole-like spars found on most of the schooners and rams, had been fashioned so that their upper surfaces seemed to dip downward slightly as if being pulled by the bobstay chains, straining to hold the craft in leash. This combination of curves and daintiness imparted the feeling of swiftness even if a particular pungy was not noted for smartness.

Prior to World War I and up into the 1920s, any general view of Baltimore harbor would reveal the tall raking masts of the pungy tied up at a pier or anchored out in the stream. However, if the hull was not visible, perhaps obstructed by a building, that rig could be misleading, for some bugeyes, like the *Avalon* and *R.B. Haynie,* adopted the sail plan of the pungy. A definite identification of the pungy class was the hull: full flaring bows, long lean run, deep drag aft, sharp floor, flush deck, a log rail except abaft the main rigging where an open quarter rail would be found, and raking stempost and sternpost. She carried a simple schooner rig but who, with the slightest trained eye, could ever mistake a pungy for a schooner?

Even now I can see, in my mind's eye, a pungy drifting in to Baltimore on a misty early fall morning. The wind is ever so slight, just barely filling her sails, but her sleek lines and appearance suggest more motion than she is actually making. There is a curl of water around her stem and faint ripples follow in her barely noticeable wake.

This is toward the end of the reign of the pungy on the Chesapeake, in the early 1930s. There were five of these vessels still sailing the Bay—the *Amanda F. Lewis, James A. Whiting, Mildred Addison, G.A. Kirwan,* and *Wave,* ranging in length from 54 to 67 feet. Only two of these, the *Wave* and *Mildred Addison,* adhered to the traditional color scheme of pungy-pink wales and bronze-green bends.

As our pungy proceeds up Baltimore harbor she might anchor out in Canton Hollow, and when her anchor chain rattles through the hawsepipe, its echo resounds off the hulls of the other sailing craft anchored nearby. If she chooses to continue up the harbor to a dock, she creates a scene purely Chesapeake, harking back to the days when the even more rakish privateers and Baltimore clippers slipped up those very same waters more than a century before.

The pungy was a descendant of those privateers, a link in the chain of fast sailing craft that are believed to have originated in the West Indian sloops that later traded to Bermuda. From Bermuda the sloops came to the Chesapeake, where their hulls were admired and studied, improved upon, and emerged as schooners referred to as "Virginia-built" just prior to the mid-18th century. Their fame spread along the entire East Coast.

The Revolutionary War gave impetus to the construction of these fast, sharply built schooners for use as privateers and blockade runners. By the end of the war the "Virginia model" had been refined into an extreme clipper schooner. The demand for swift sailing vessels continued after the Revolution and the "Virginia model" was further developed. They became popular as pilot boats that required speed to be able to put their pilots first aboard a ship needing that service.

In the early 1800s the term "pilot boat" replaced "Virginia-built" in describing the fast schooners. Due to this speed the vessels readily found employment as privateers. They were owned privately and were manned and armed to capture enemy merchant ships in wartime. Many of these schooners had been built in the Baltimore area and were to be eventually referred to as Baltimore clippers. Whenever a swift vessel was needed—for instance, to out sail a pursuer in the slave trade—the Baltimore clipper could serve. Though fast, the clippers were not good cargo carriers due to sacrifices made in payload capacity to gain speed. This feature caused their decline in popularity; by the 1850s the type was passing out of favor. As clippers departed from the waterways a smaller version came into being—the Chesapeake Bay pungy that carried on into modern times.

The earliest mention of the term "pungy" this writer has found was made in a letter from a Baltimorean to the Eastern Shore of Virginia boatbuilder. Dated April 14, 1850, it reads:

> Mr. Boggs
>
> Dear Sir
>
> I want to know whether you will take a contract from me to build a pungy boat, say about 42 ft. keel, 7 ft. hole, (*sic*) 21 ft. wide and built of the very best materials, and finished in a thorough workmanlike manner. And if you can take up the contract let me know your lowest terms. I want her by the middle of September or October or sooner if possible. As soon as the vessel is finished you can have the money immediately. If you build her I will come and see you as soon as I hear from you. Direct your letter to me, care of Mr. I.B. Brinkley, Baltimore.
>
> Your respectfully
>
> Ben Somers

As with the name bugeye, the origin of the word pungy is obscure. In fact, I have never seen an explanation of the source. I have long felt that it came about through an association with the locality of Pungoteague on the lower Eastern Shore of Virginia. The aforementioned shipbuilder Boggs had a yard situated on a neck of land between Onancock and Pungoteague creeks. Perhaps the first of these extensions of the privateers and Baltimore clippers were built in that area. If a number of them were built there, "pungy" would have been a logical derivative of Pungoteague. It is believed the pungy type came into being in the 1840s or '50s and, judging from the Somers letter, the term is used in a familiar manner between persons at opposite ends of the Chesapeake, suggesting that it had been in use for a while.

There is no record that Mr. Boggs built the pungy, nor of its identity if it was built. It is interesting to note that the *Wave*, the last pungy to sail the Chesapeake—albeit as a yacht in those last days—was built in 1863 in Accomack County, which encompasses the Pungoteague section. In the mid-1920s the former owner of the *Wave*, John J. Crouch of Rock Hall, Maryland, also owned another pungy, the *Plan*, built in 1855 in Accomack County. Pungies were also built in other sections of the Chesapeake, in Maryland and Virginia, until the late 1880s. Gradually they were replaced by centerboard craft as the deep draft of the pungy restricted it from waterways that were silting up.

Some of the pungies migrated to other waters along the East Coast, principally the Delaware Bay area. In the 1908 *Transactions* of the Society of Naval Architects and Marine Engineers, Charles H. Cramp, vice president of that organization, stated, "At the oyster and fish markets in Philadelphia, more pungies than any other type were to be found, all commanded by Chesapeake men, all of whom used the name of 'pungies' as applied to them." These were probably transient craft with farm goods and watermelons from the Bay country via the Chesapeake and Delaware Canal. Pungies were also used as oyster dredgers on Delaware Bay, hailing out of Port Norris, New Jersey. The last in that area were the *Mary and Ellen*, built in Baltimore in 1881, and the *Ellen and Alice*, built in Gloucester County, Virginia, in 1877. Both were omitted from *Merchant Vessels of the United States (MVUS)* by the mid-1930s.

Pungies were seagoing and had some influence on the design of New England fishing craft. Sailing out of Baltimore, they hauled pineapples from the Bahamas. Their great asset in this trade was speed, a requisite in getting the fast-ripening fruit to market. Pineapples were picked green and allowed to ripen in the hold on the return passage. But pungies were "wet;" heavy seas swept their decks, protected only by a log rail. Their sharp-floored holds prevented proper stowage of fruit, in pens, so that only a few pungies participated in this trade. They included the *Francis J. Ruth*, the *Ruth A. Price*, and the *Amanda F. Lewis*. The bulk of the pineapple trade went to the faster two-masted schooners.

Pungies were engaged as oyster dredgers as late as the early 1920s. Later, they hauled oysters to the market, farm products, and other cargo. An excellent record of how the last commercial pungy on the Bay, the *Amanda F. Lewis*, kept busy resulted from a project the United States Army Engineers carried out in the early 1930s when the Bay sailing craft were required to fill out a form supplying such information as:

> 1. List all articles separately and state accurately the quantity and unit of measure of each commodity.
> 2. Fill in vessel data accurately, name, draft, registered tonnage, and trips being especially important.
> 3. Under point of origin and destination of shipment, please show, if possible, name of wharf or town, in addition to name of river.

Also on the printed form appeared spaces for the name of the shipper, name and kind of vessel, loaded draft, net registered tonnage, number of trips made, name of captain, managing owner, and address.

On January 31, 1934, Eugene Fallin, managing owner of the *Amanda F. Lewis*, of Coan, Virginia, Captain Gus Rice in command, submitted a form covering the activities of the *Lewis* for the calendar year January 1 to December 31, 1933, which shows the pungy made 23 trips during that period, broken down in this manner:

- 1 trip with 53 tons of wheat from E. Fallin and Brother from Coan, Virginia, to Baltimore
- 1 trip of 56 tons of coal from T.S. Southard Company, Norfolk to Coan
- 1 trip of 28 tons of cases of cans from the Continental Can Company, Baltimore to Bundicks, Virginia
- 1 trip with 80 tons of lumber shipped by T.C. Booker, Bundicks to Baltimore

- 6 trips totaling 440 tons of lumber from E. Fallin and Brother, Coan to Baltimore
- 3 trips with 95 tons of cases of empty cans from Metal Packaging Corporation, Baltimore to Coan
- 4 trips with 247 tons of fertilizer from Armour Fertilizer Works, Baltimore to Coan
- 2 trips with 165 tons of lumber from Colonna and Howard, Nomini, Virginia, to Baltimore
- 2 trips with 155 tons of cases of canned tomatoes from E. Fallin and Brother, Coan to Norfolk
- 2 trips with 251 tons of cases of canned tomatoes shipped by E. Fallin and Brother from Coan to Washington, D.C., all on a loaded draft of 9 feet

(On the form the *Lewis* is identified as a schooner.)

With such chores the *Amanda F. Lewis* kept busy until 1939. By then Captain Rice was feeling his years, and so was his pungy. So he sold her, closing out just about a century of commercial pungies on the Chesapeake.

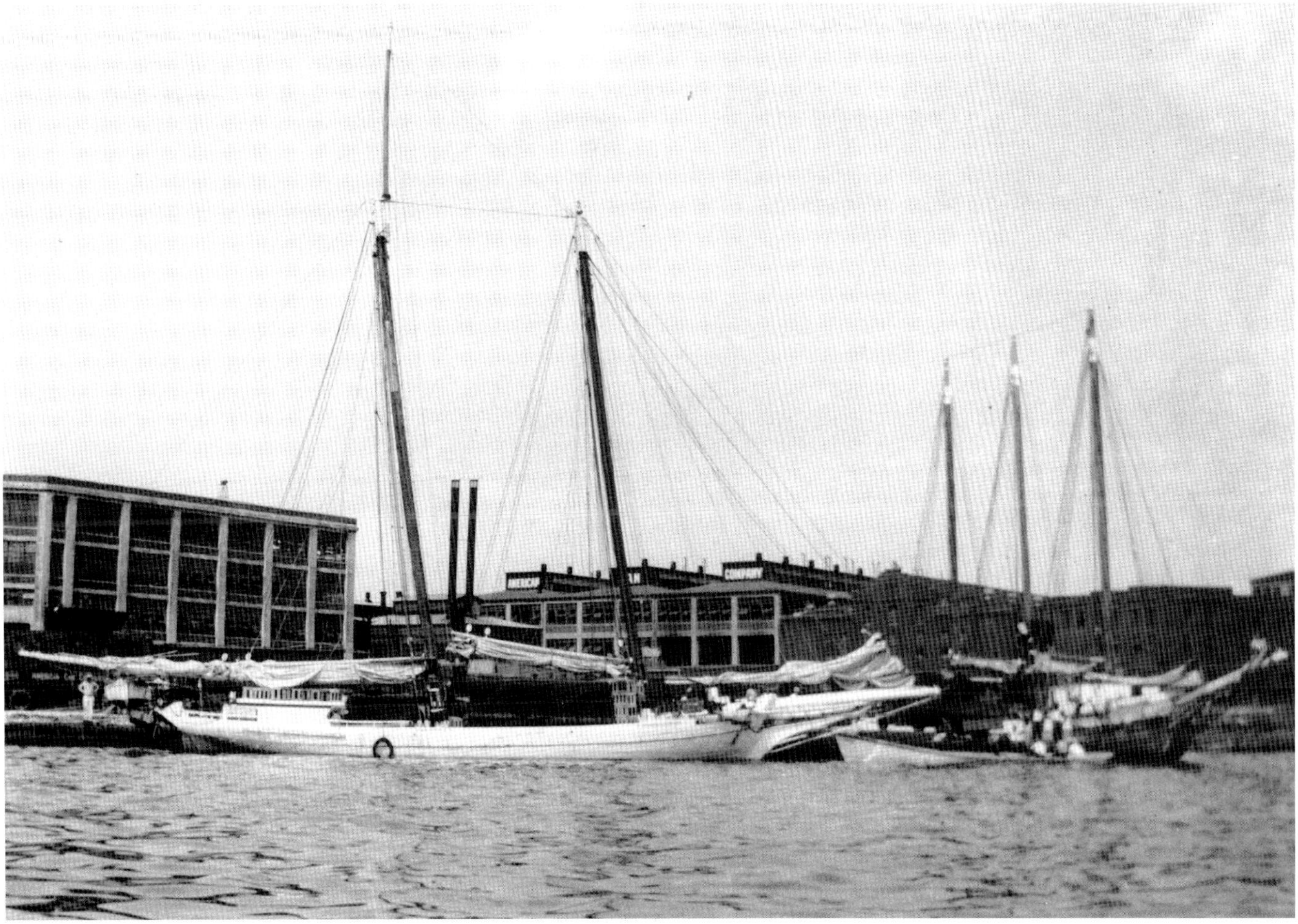

The pungy that stands out most prominently during the past 50 years of sail on Chesapeake Bay has to be the *Amanda F. Lewis.* Her lines have been taken off and plans prepared, models, paintings, and drawings made of her, many articles written about her, and her carved work preserved at the Smithsonian Institution, thus perpetuating her name for all time. During that period she sailed the length of the Bay and many of its tributaries in various roles—freighter, oyster dredger, menhaden seiner—and was maintained in first-class condition. On July 11, 1934, she is moored at the dock of Montford Yard, Baltimore, in front of the American Can Company on Boston Street, with tarpaulins covering cases of empty cans for delivery, no doubt, to the tomato cannery of Eugene Fallin, her managing owner, at Coan Wharf, Coan River, Virginia. This profile view shows off the classic lines of the pungy to advantage. Pleasant sheer, raking masts, and thin, tapering topmast were some of the characteristics of this class of Bay craft. Atop the gold ball on her foremast she carried a carved wooden anchor. The *Amanda F. Lewis* was built in 1884 at Madison, Maryland, by Joseph W. Brooks for Captain M.O. Lewis, who named the vessel after his wife and had her likeness carved in wood for a stern decoration. The bald-headed, three-masted schooner at right (not a ram in this instance) is the *William J. Stanford,* probably also taking on empty cans for a Potomac River landing. Standing on the edge of the dock in white duck pants, directly under the *Lewis'* main boom, is the marine artist Louis J. Feuchter, famous for his paintings of Chesapeake Bay sailing craft.

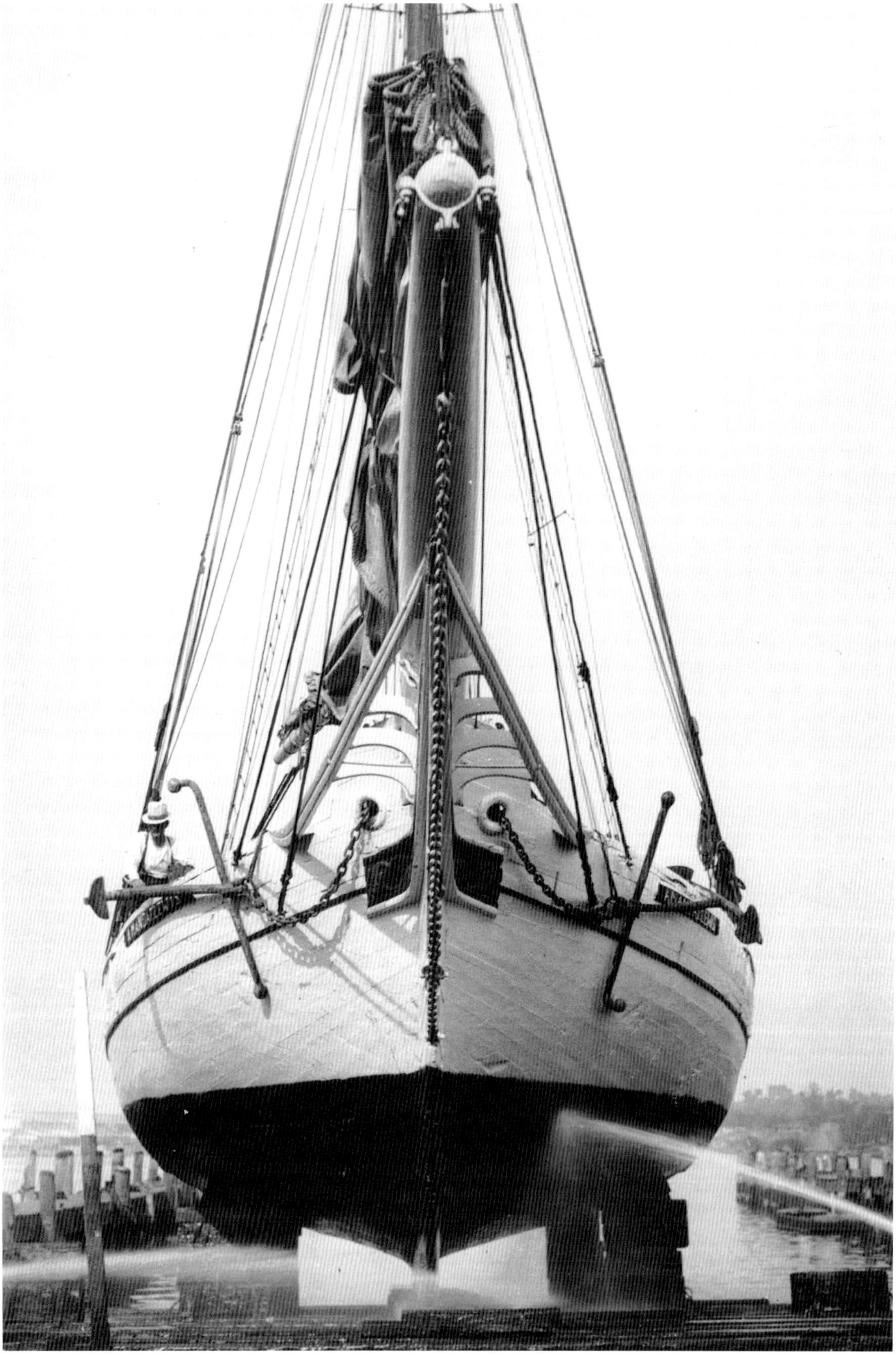

On a late afternoon in June 1939, the *Amanda F. Lewis* is hauled out on the marine railway of A. Smith & Sons, Curtis Bay, Baltimore, to have her bottom painted. As she clears the water, workmen hose off her bottom to remove any growth while it is still damp. Except for a few barnacles, the bottom is relatively clean, but regular paintings were necessary as a guard against the teredos, or marine worms.

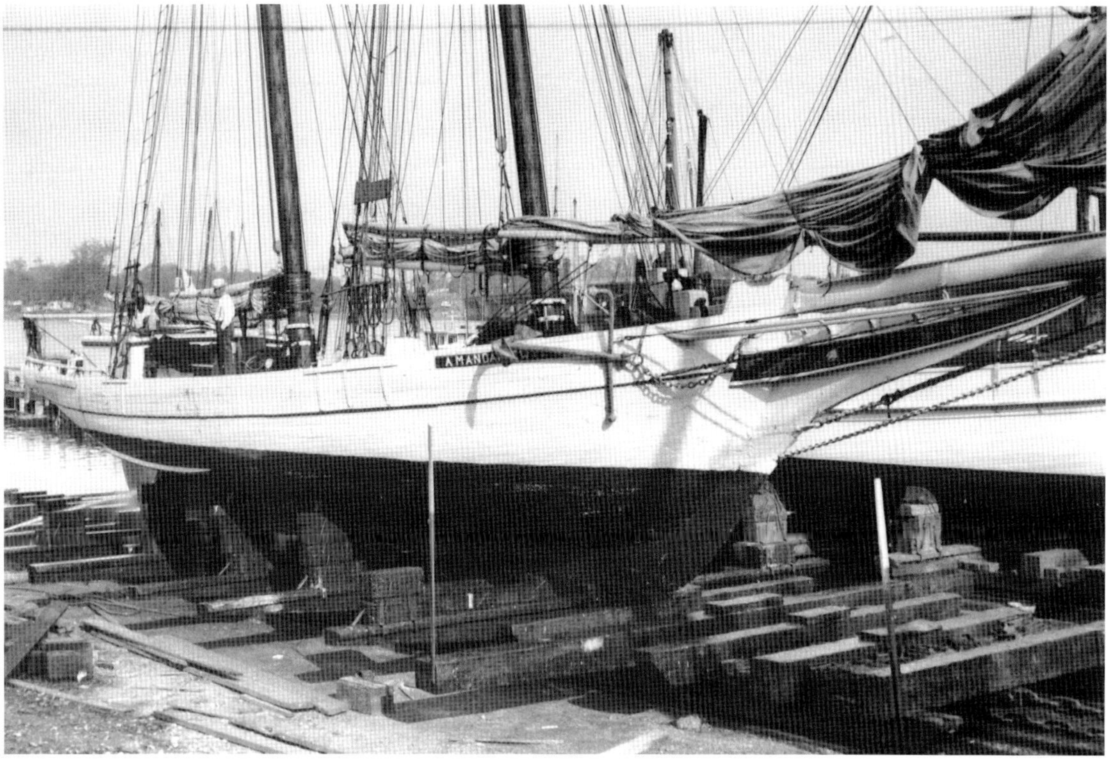

Free of the water, the pungy's racy appearance and lineage with the swift pilot boats and Baltimore clippers can be seen. Standing on deck near the main shrouds, with hands in pockets, is the *Lewis*'s master, Captain Gus Rice, a Chesapeake waterman of long standing who owned and skippered this pungy for many years.

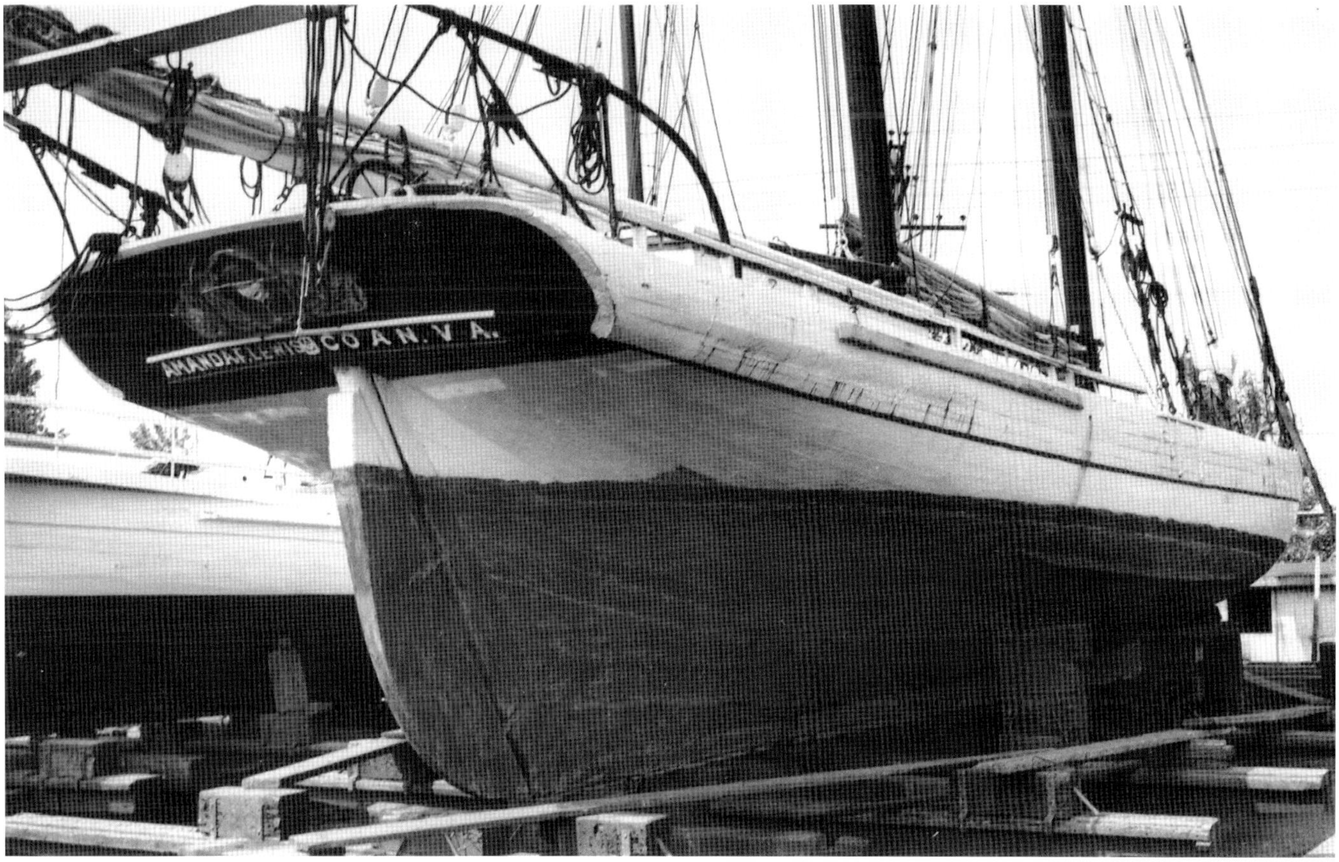

In this stern quarter view can be seen the deep drag aft, typical of the pungies, which eliminated the need of a centerboard required in the more shallow-draft Bay schooners.

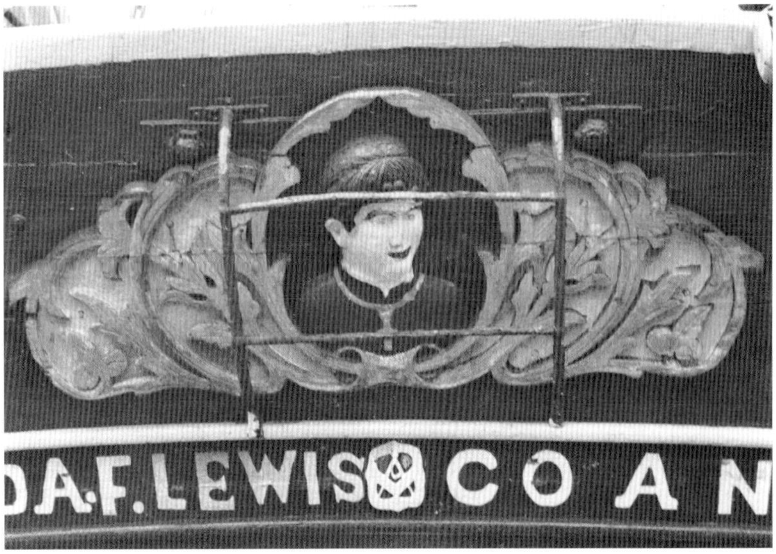

Stern carving of the pungy's namesake with metal rods to protect it against contact with yawl boats, piers, and other sources of damage. Beneath the carving, in primitive lettering, is the vessel's name and home port and the Masonic insignia that many of the Bay skippers proudly displayed on the sterns of their vessels. The *Lewis* was converted into a powerboat about 1939 at the Krentz Shipyard, Harryhogan, Virginia, after being sold to Captain William J. Stanford. She was last reported in 1949 as being owned in Haiti.

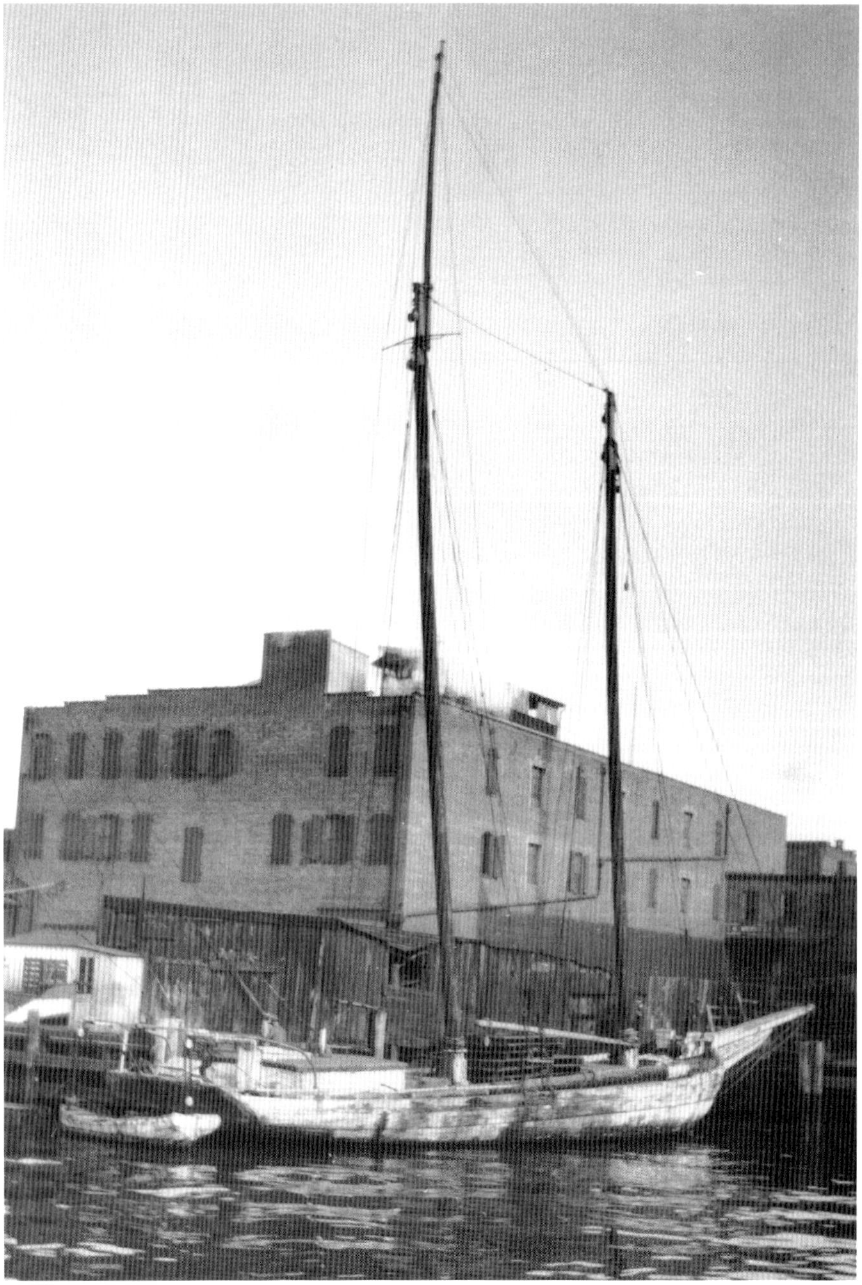

Pungy *G.A. Kirwan,* built in 1882 at Madison, Maryland, tied up at Webster's Wharf, Baltimore, November 29, 1933, after being out of service for a long period. Her main boom lies on the deck and her yawl boat is suspended just afloat so its bottom remains swollen and the weight is taken off the davits.

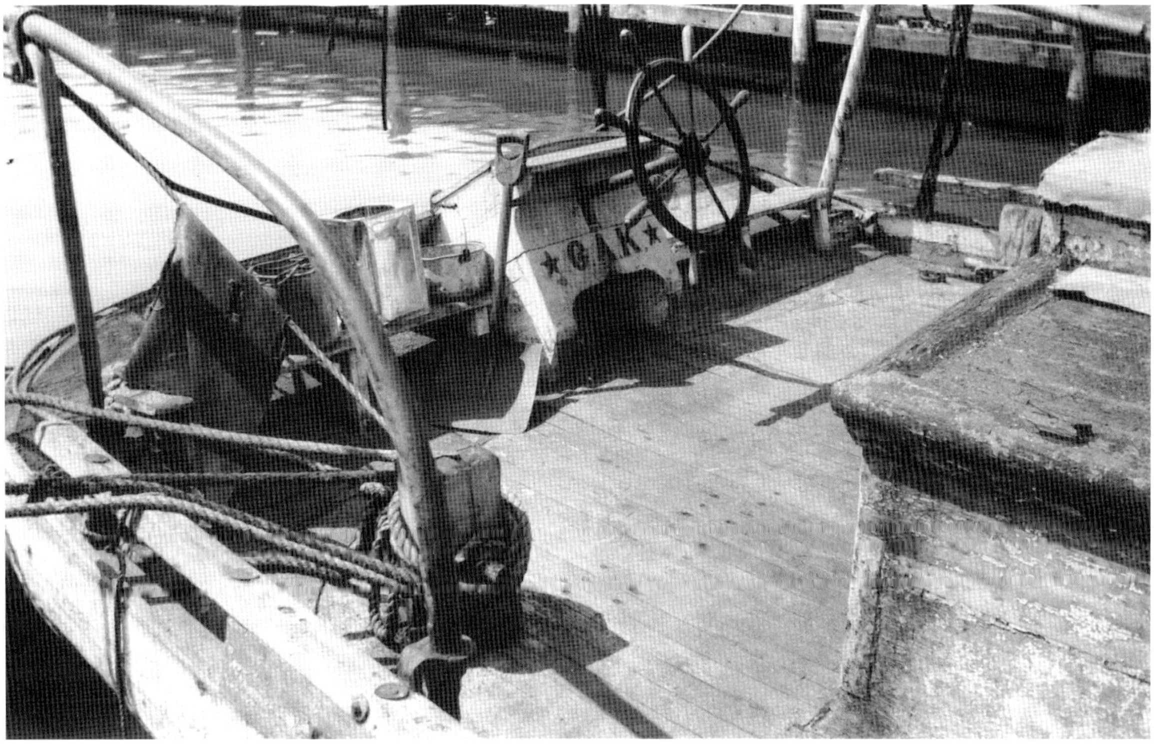

After deck and steering wheel of the *G.A. Kirwan,* November 11, 1934.

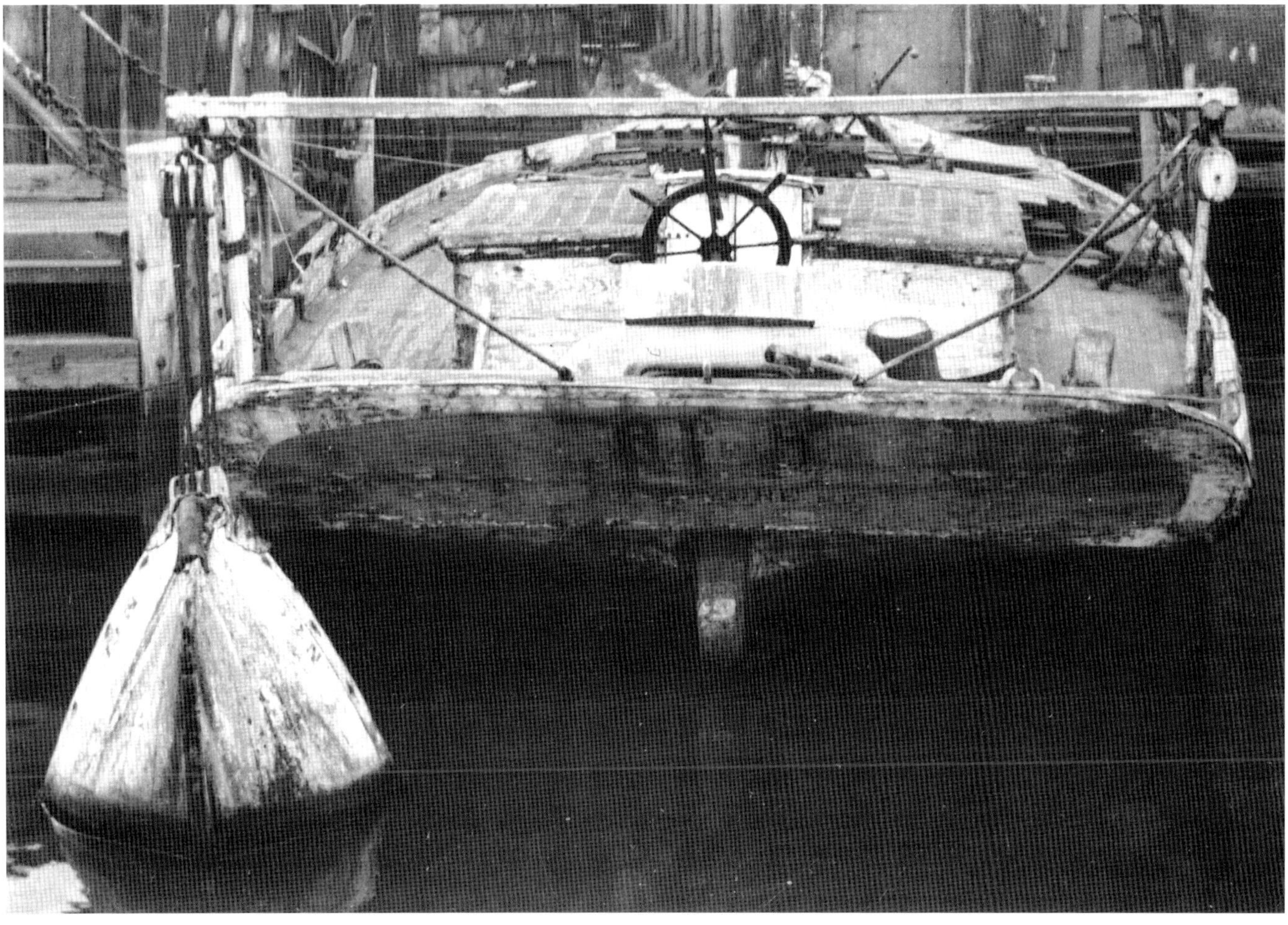

The starboard rope fall of the *Kirwan's* yawl boat parted, the craft filled, and was left suspended by the port block and fall, November 1934. This stern view reveals the pungy's deck outline.

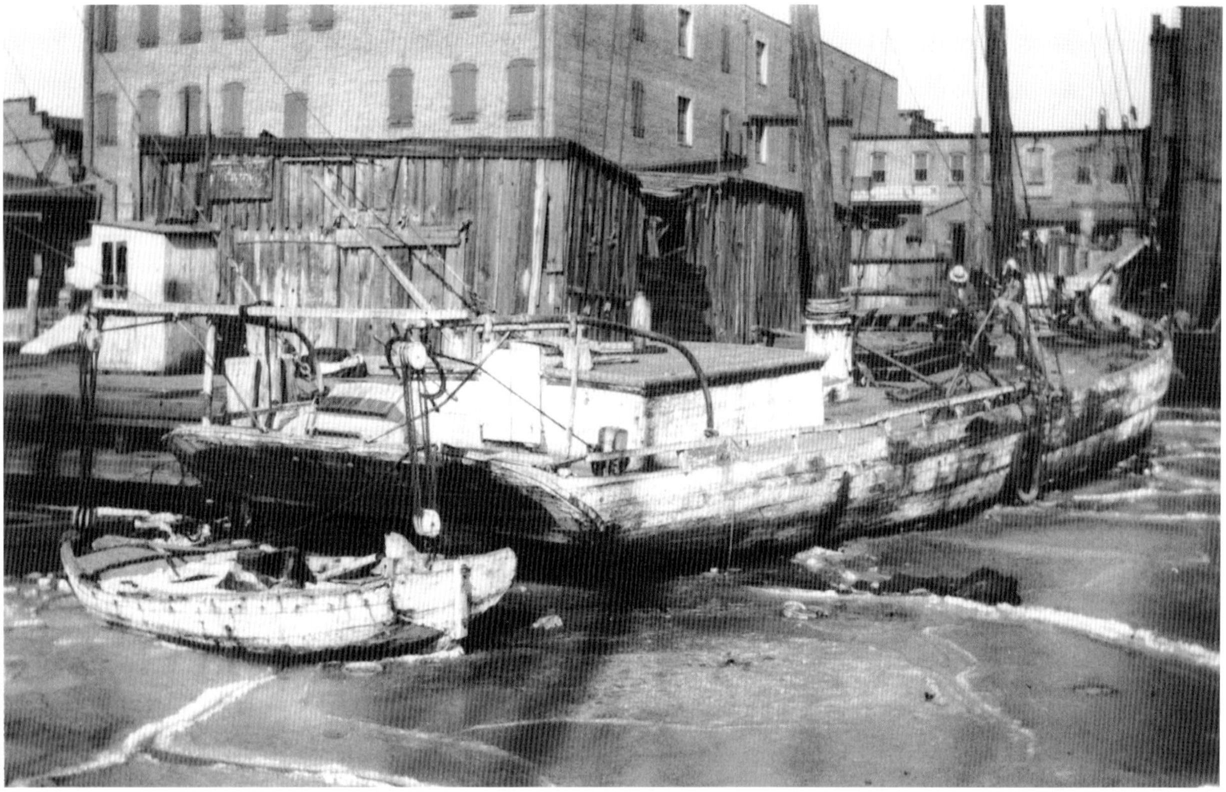

❂ The *G.A. Kirwan,* laid up in ice at Webster's Wharf during the winter of 1934–1935.

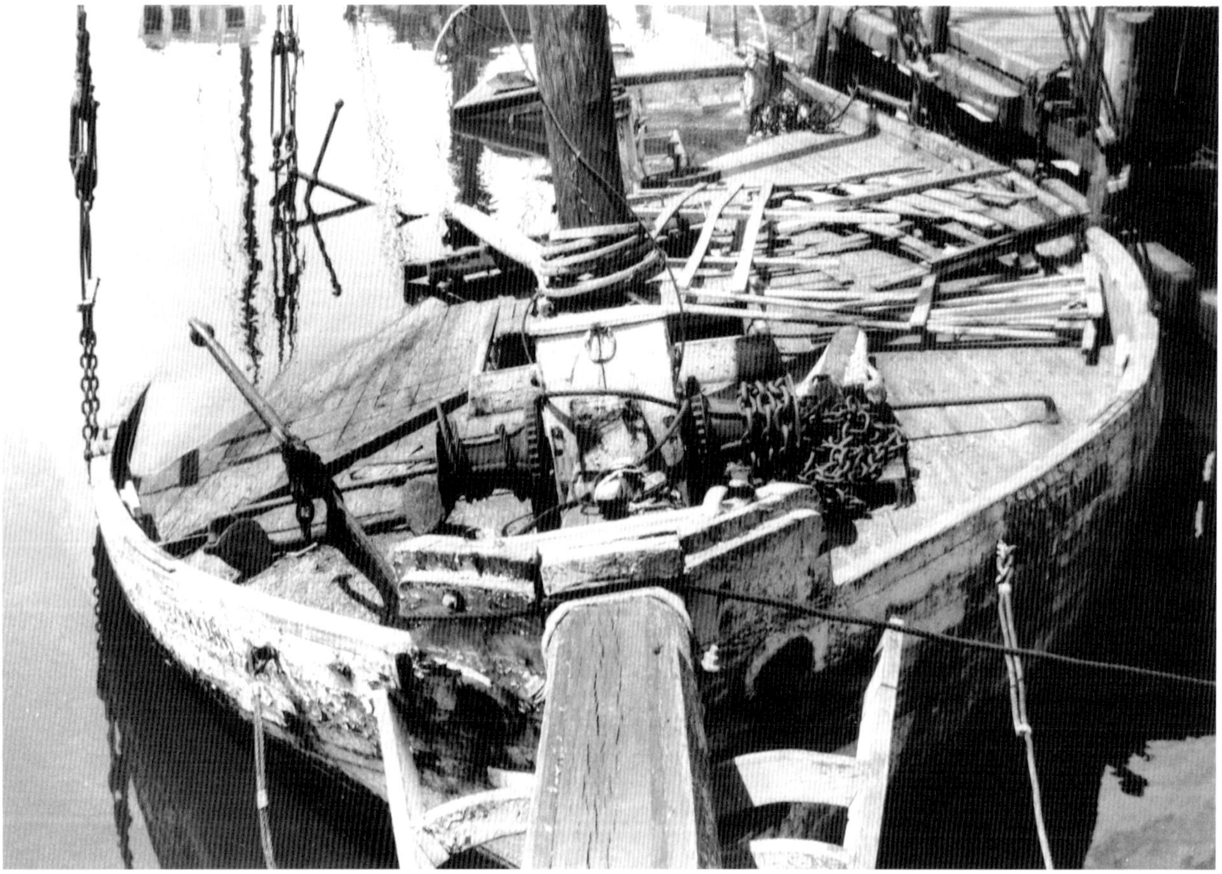

During high water in early 1935 the bowsprit of the *Kirwan* hung on the dock ahead and as the water receded her weakened stern settled and filled. Looking aft from the *Kirwan's* bowsprit at Webster's, May 26, 1935. She was raised the next year but only to be towed down the Patapsco River and abandoned on the flats off Point Breeze.

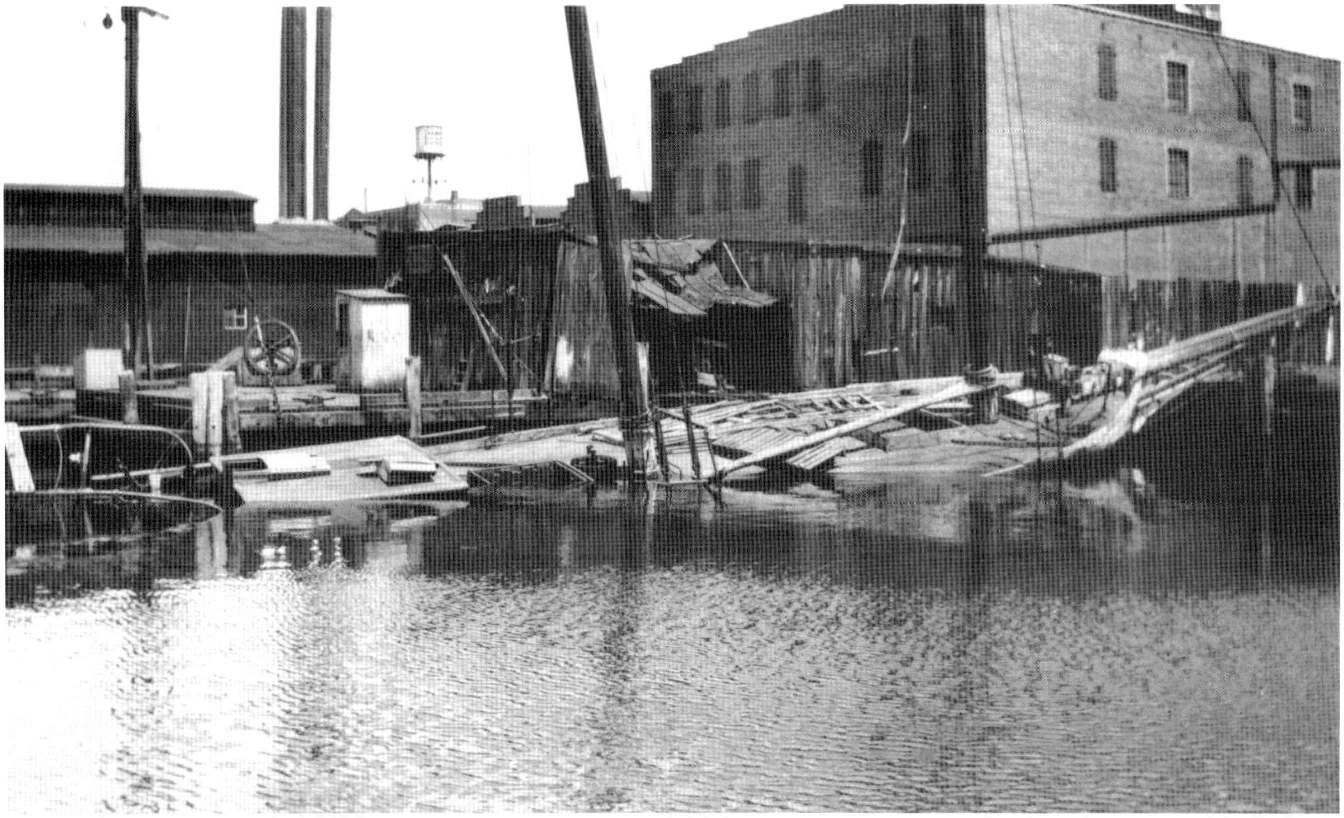

✪ An overall view of the poor *Kirwan,* at Webster's Wharf on May 26, 1935.

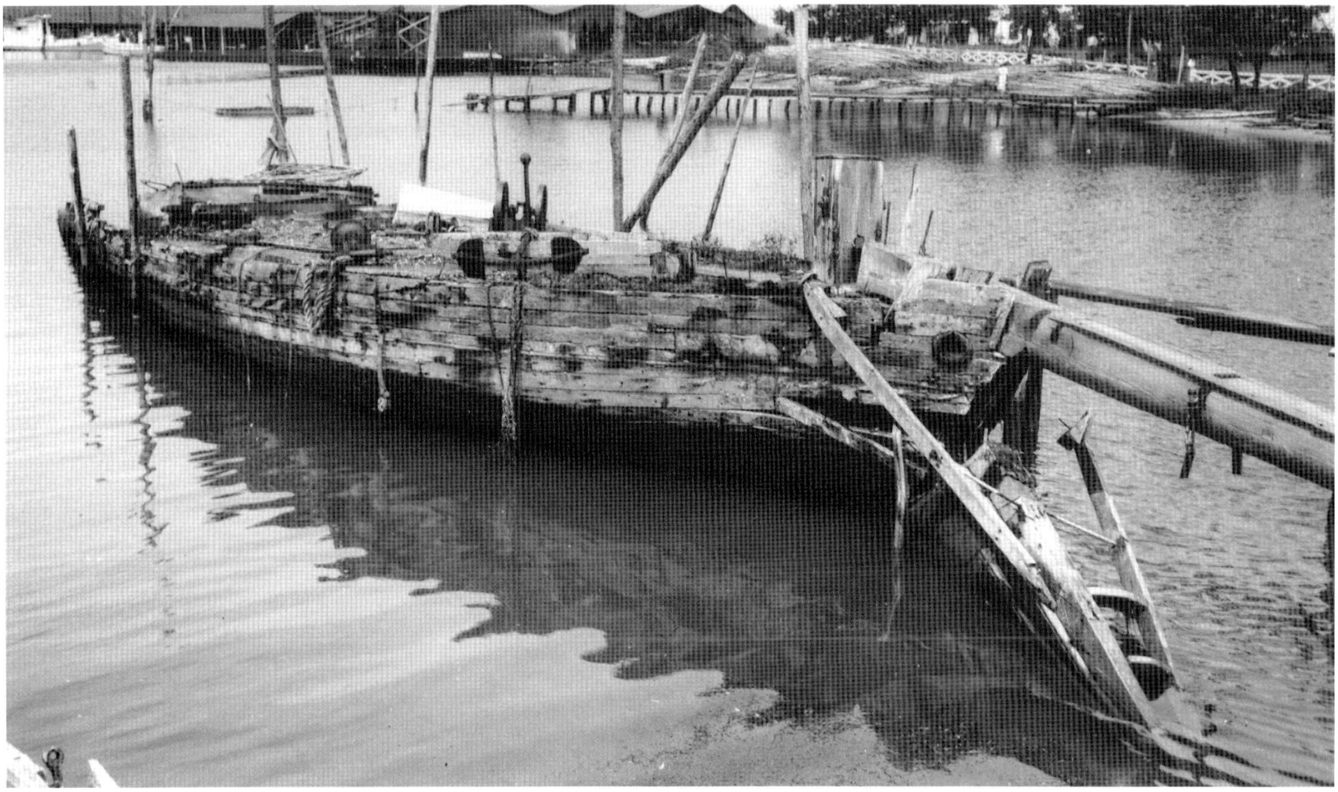

When the pungy *Jas. A. Whiting* was unfit for further use, she was pulled ashore on the property of her owner, A.J. Lewis, at Walnut Point, Coan River, Virginia, and left to disintegrate, as in this view of July 22, 1947. She had been built in 1871 in Somerset County, Maryland.

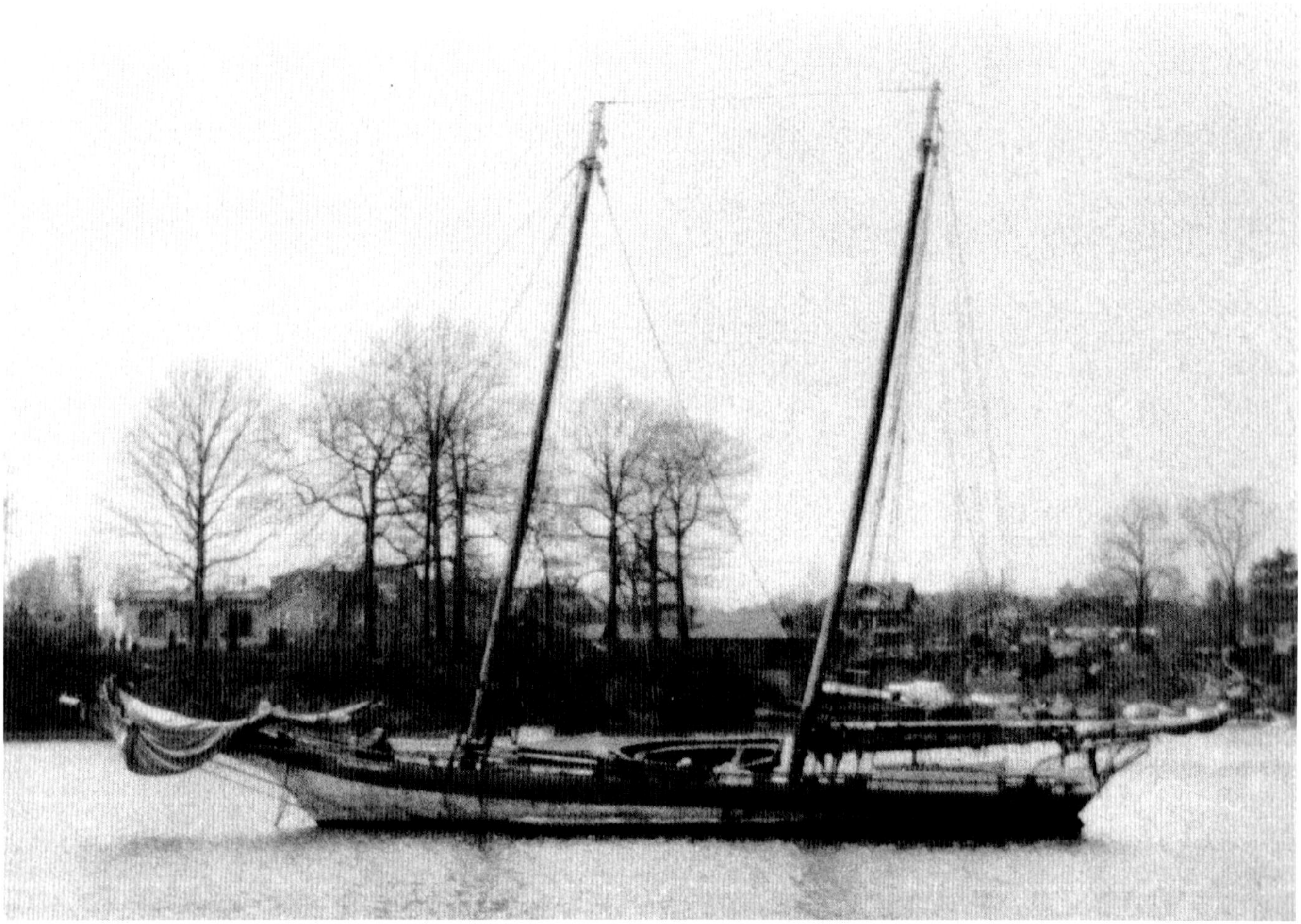

A pungy without a maintopmast just didn't look right. That slender spar, sprung forward slightly, added the touch that gave these sailing craft their racy look. Otherwise, they resembled many of the bald-headed two-masted schooners when, on their last stand, they removed their topmast and a general run-down condition followed. One of the last of the Bay pungies, the *Mildred Addison,* is anchored in Spa Creek, Annapolis, on April 22, 1934. Not long after this, about 1939, she was abandoned in those waters. This pungy started her career as the *Thos. H. Baily Browne,* having been built in 1870 in Somerset County, Maryland. The *Addison,* however, held true to the pungy style of painting—pungy pink and green—to her very end.

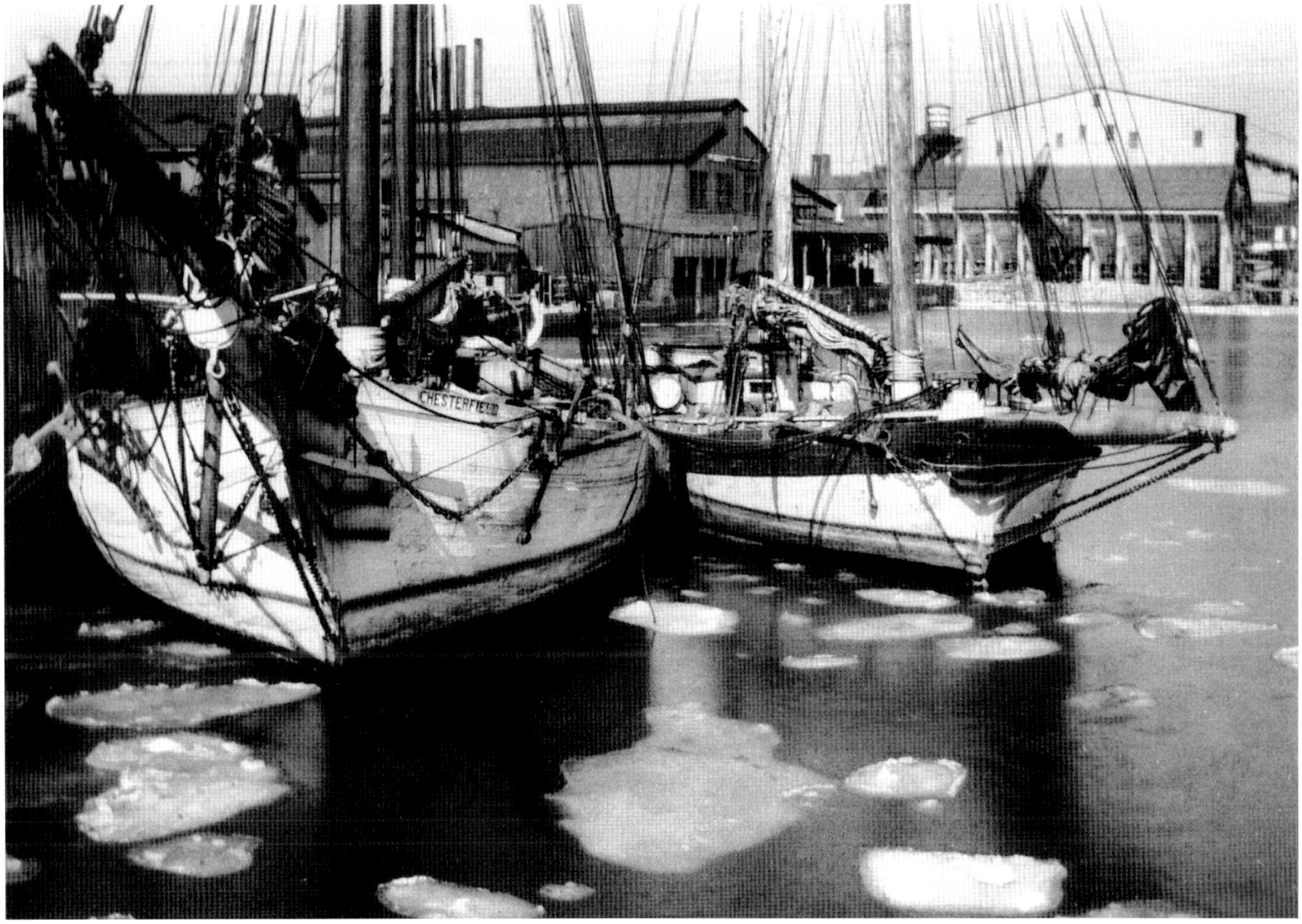

A sheet of ice over Baltimore harbor on February 18, 1934, imprisons the schooner *Chesterfield* and pungy *Wave,* tied up at the Lord-Mott Company, Fells Point packers. The yacht-like *Wave,* only 54.8 feet in length and built in 1862 in Accomack County, Virginia, and painted with dark green bends and "pungy pink" sides down to the waterline, was the jewel of the Chesapeake's sailing fleet. She was the pride of her owner, Captain John J. Crouch of Rock Hall, Maryland, who also owned the pungy *Plan,* built in 1855 and smaller than the *Wave.*

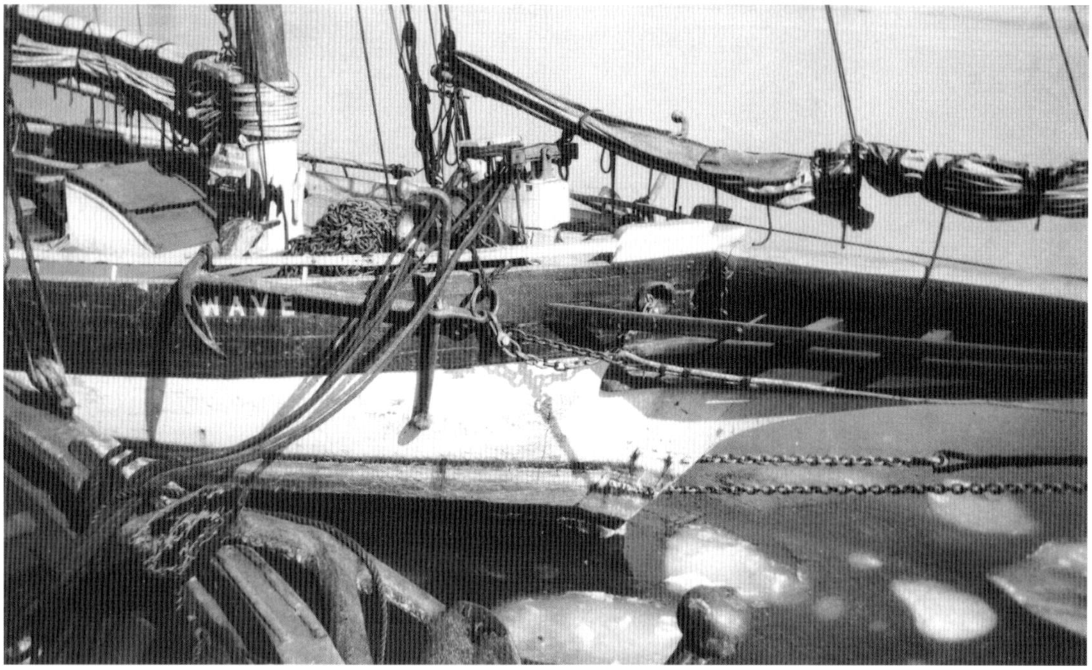

Starboard bow of the *Wave,* showing her ground tackle and deck arrangement.

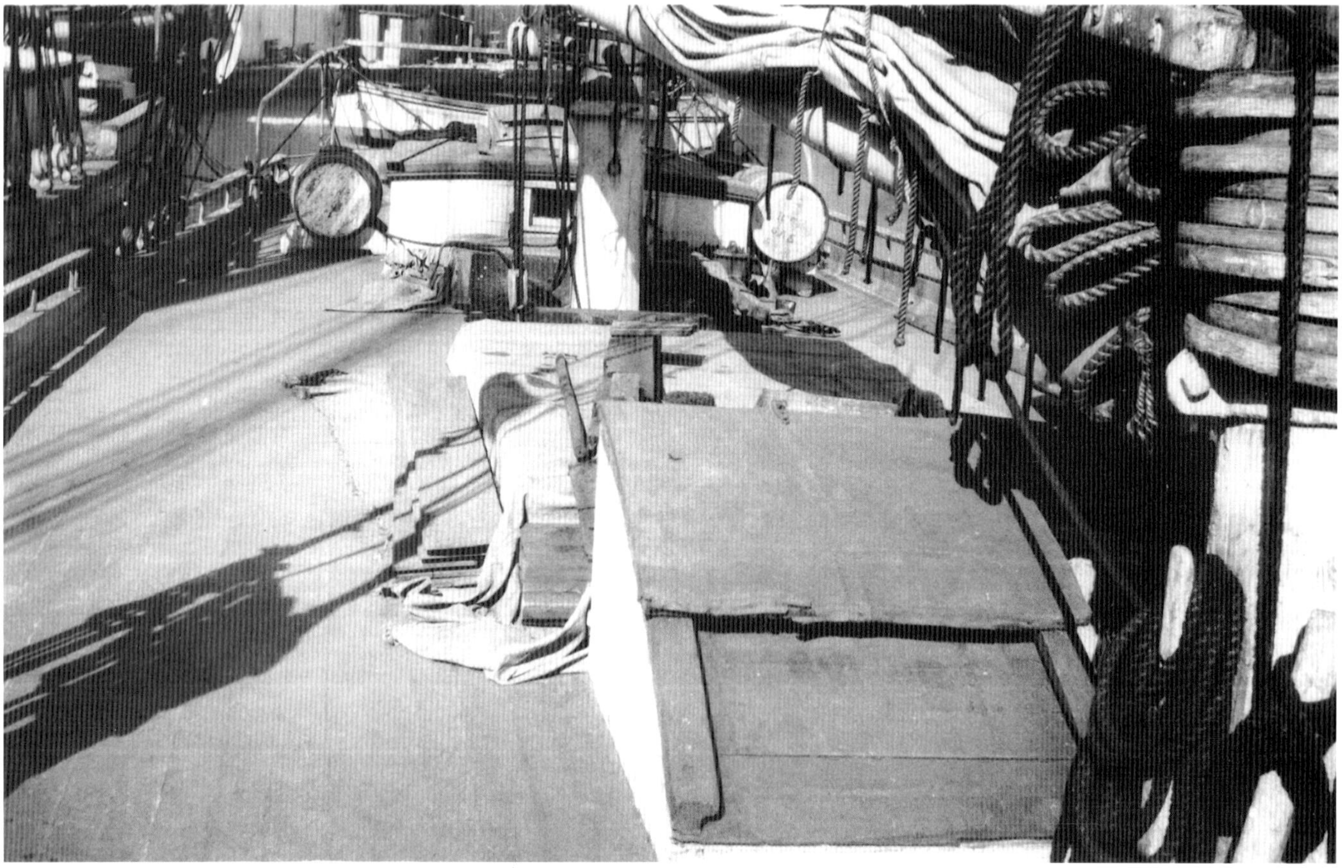

Looking aft on the *Wave's* deck, showing the log rail and pipe stanchions running along her sides. In the foreground is the forecastle hatch. Starboard of the after cabin is the water barrel or scuttlebutt, and on the port side a metal drum containing fuel for the motor yawl boat. With her sleek lines, it was inevitable that someone would recognize the possibilities of converting the *Wave* into a yacht. That is what happened in 1937. But it was the ruination of this fine boat after a high cabin, with seven big ports on each side, was erected running full length on the deck from the foremast to just abaft the main shrouds, with skylights atop this, companionways aft and amidships, and two huge wooden lockers on the afterdeck on either side of the wheel! She looked like an Ark or worse, since she had lost her sheer. In 1947 she was used as a floating residence at Washington, D.C. She found her way to the Great Lakes area in the 1950s and was reported abandoned there about 1959.

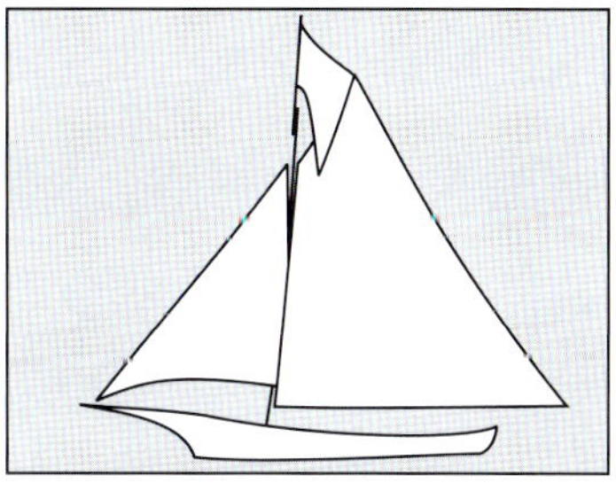

Sloops

Chesapeake sloops and schooners are the vessels about which it is most difficult to find historical information. This could be attributed to the fact that few maritime chroniclers were active before the turn of the century, or those types were so common that little attention was paid to them. We do know these craft were around in great numbers from the colonial period on because they are noted in newspapers of the period recording Chesapeake arrivals and departures.

Of all the craft discussed in this volume, the sloop and schooner are the only ones not indigenous to the Bay. But, except for the log canoe, they were here first and for a longer period than all the others.

Displaced in popularity by the bugeye and skipjack, the sloops dwindled in number and were not replaced by others of that type except in limited numbers. The last great surge of sloop building occurred in the 1880s at Taylors Island, Dorchester County, Maryland. Remnants of that fleet remained active over the oyster dredging grounds into the 1970s.

The term "sloop" refers to the round-bottomed, chunky hull with centerboard, with gaff mainsail and topsail, and one or more jibs. *Merchant Vessels of the United States* correctly classes skipjacks as sloops but they were never so designated on the Chesapeake.

All views of the dredging fleet in the early 1900s, at work or in harbor, particularly at Solomons Island, Maryland, reveal the single topmasts standing out and even outnumbering the bugeyes within camera range. Averaging between 30 and 60 feet in length, they were handy, handsome little craft, constructed by professional builders, more difficult and expensive to construct than bugeyes and skipjacks. The latter factor and the necessity of a larger crew to handle the large boom, mainsail, and topsail no doubt contributed to their decline.

As with schooners, sloops were brought in from other waters for use on the Bay. One of the largest of these was the 65-foot *Flora Elsie,* built at Islip, New York, in 1880, registered out of Washington, D.C., in her latter days, and listed as "Abandoned" in the 1932 *MVUS.*

One of the largest, if not the largest, built on the Bay in so-called modern times was the 74-foot *Charles M. Kelly,* constructed in 1892 on the Severn River, Gloucester County, Virginia, by George W. Marshall. She was abandoned in 1928. Those dimensions can be compared with the 45-foot *J. T. Leonard,* giving some idea as to the size of the boom that swung across the *Kelly's* deck. The *Leonard* was the last gaff-rigged commercial sloop on the Chesapeake and was active as an oyster dredger on the Choptank River up to the time she was acquired by the Chesapeake Bay Maritime Museum as an exhibit.

Around the head of the Bay, in the Havre de Grace area, the last of the scow sloops, the 56-foot motorized *Elsie,* was active until the mid-1940s. With her blunt bow, gaff rig with topsail, leeboards, and tiller steering, she was an unusual looking craft.

A group of the 1880-period sloops gave up the gaff rig to adopt the skipjack rig in the 1920s and could be seen in Cambridge harbor. One, the *Rebecca T. Ruark,* was still operating out of that port as this was written.

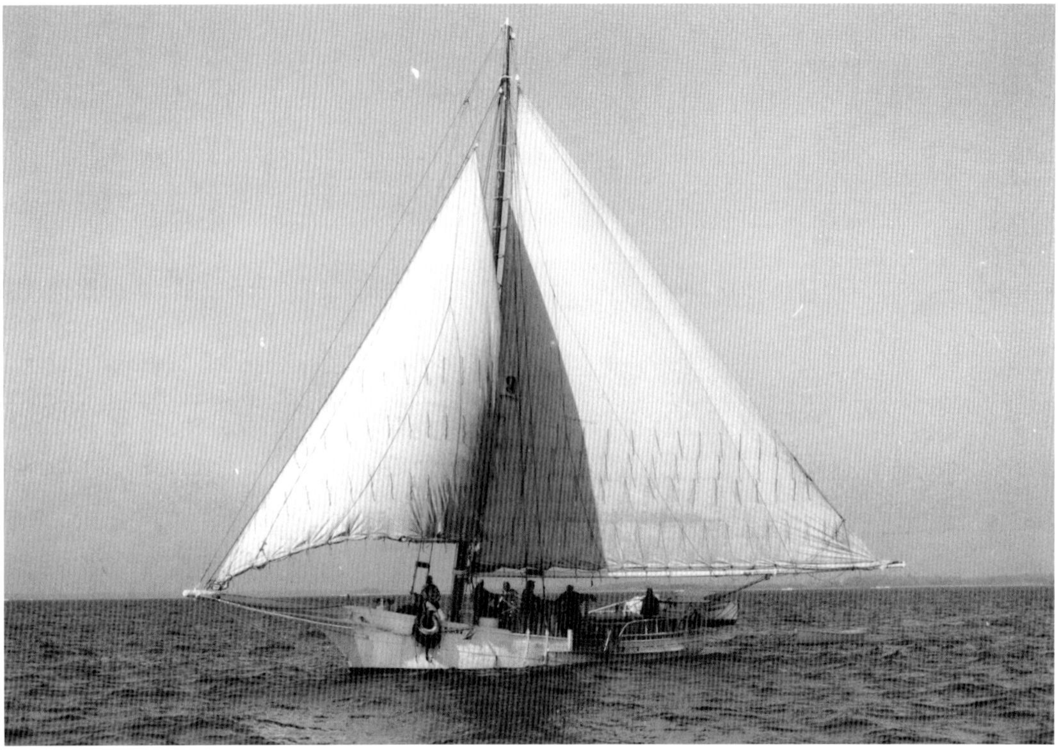

The skipjack rig on the hull of the former gaff-rigged sloop *Henry W. Ruark* would fool the average viewer into thinking that she was just another skipjack. But closer examination of that hull would reveal her true sloop characteristics. On this day of relatively light winds on the Choptank River oyster dredging grounds, November 15, 1962, her jib and mainsail each have a single reef. The other reef points on those sails indicate that there may be occasions to reef even more as the winter winds increase in strength and she has to reduce sail to carry on her work. The *Ruark* was built in 1884 at Taylors Island, Maryland, and measured 39 feet in length.

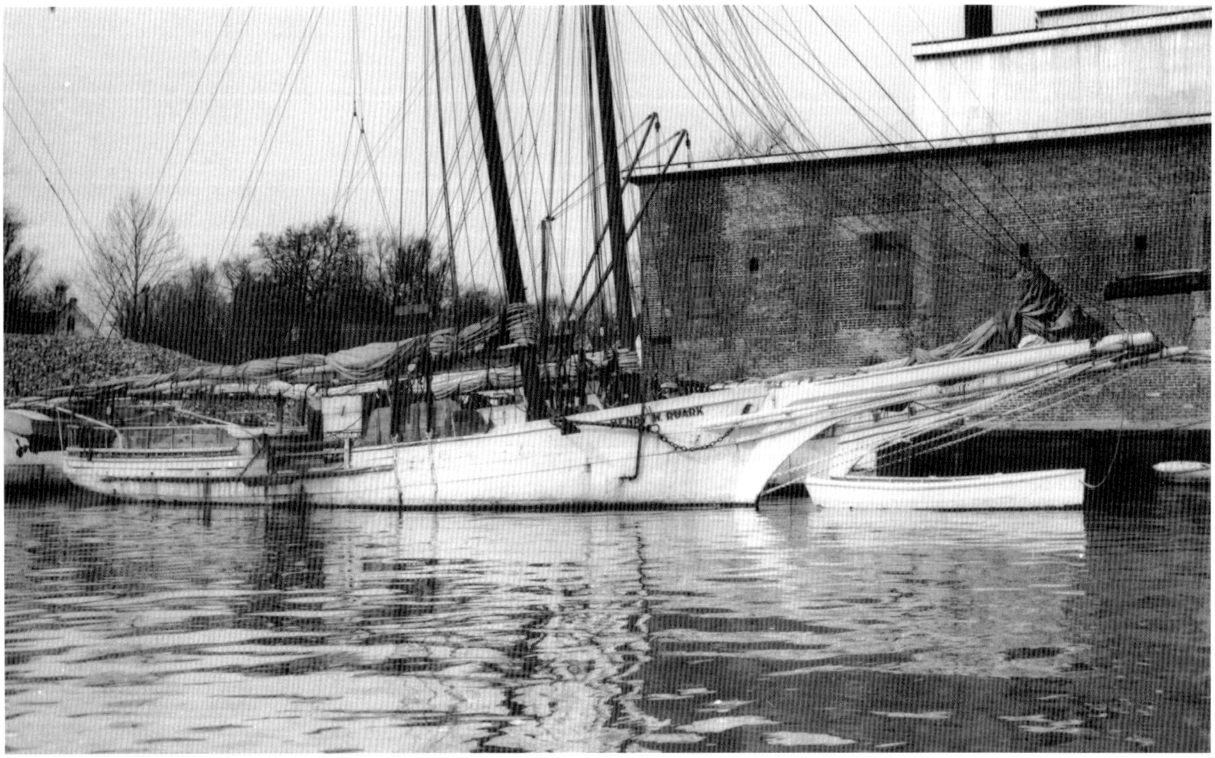

Even though the *Henry W. Ruark* may have lost some of her sheer in 65 years of oyster dredging, she still presents a pleasant appearance with her clipper bow and long bowsprit as she rests from her chores over a weekend in Cambridge in February 1949. ✪ The 1978 *MVUS* listed her as having been "Dismantled."

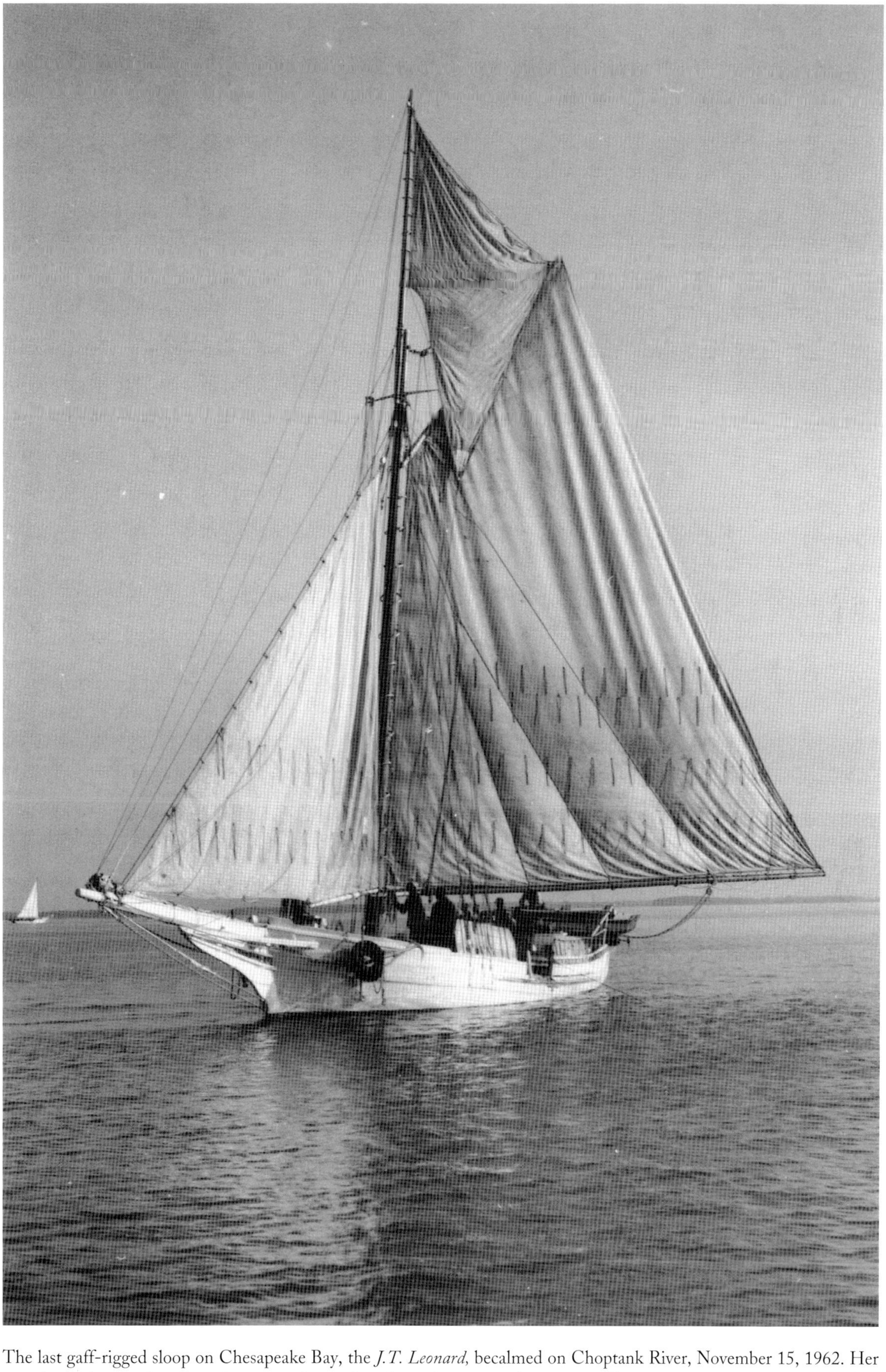

The last gaff-rigged sloop on Chesapeake Bay, the *J.T. Leonard,* becalmed on Choptank River, November 15, 1962. Her port dredge is over the side acting as a tether as the tide tends to sweep her upriver. The *Leonard* had been built in 1882 in Dorchester County, Maryland, and after a long life on the dredging grounds was purchased in 1966 by the Chesapeake Bay Maritime Museum, St. Michaels, for preservation.

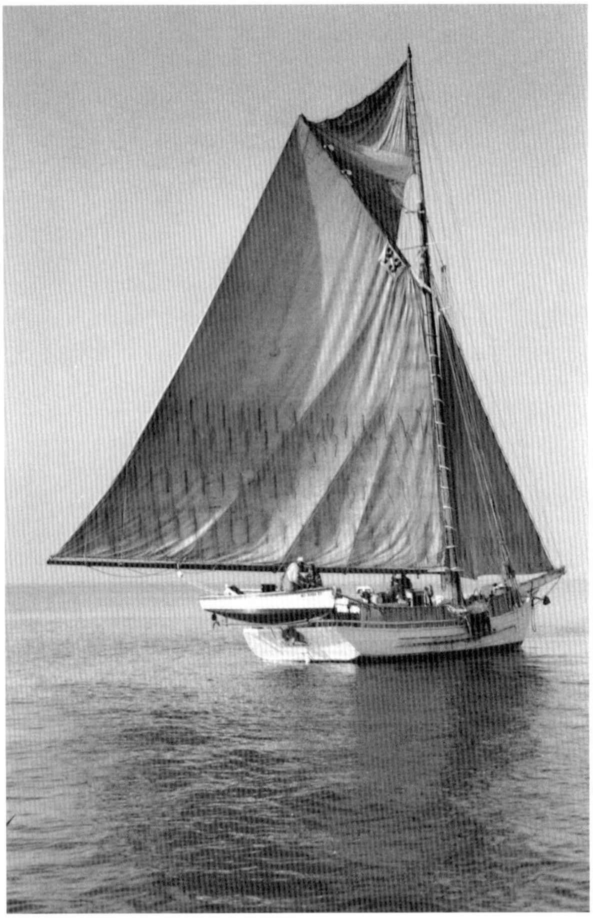

❂ Another view of the *J.T. Leonard,* becalmed on the Choptank River in November 1962.

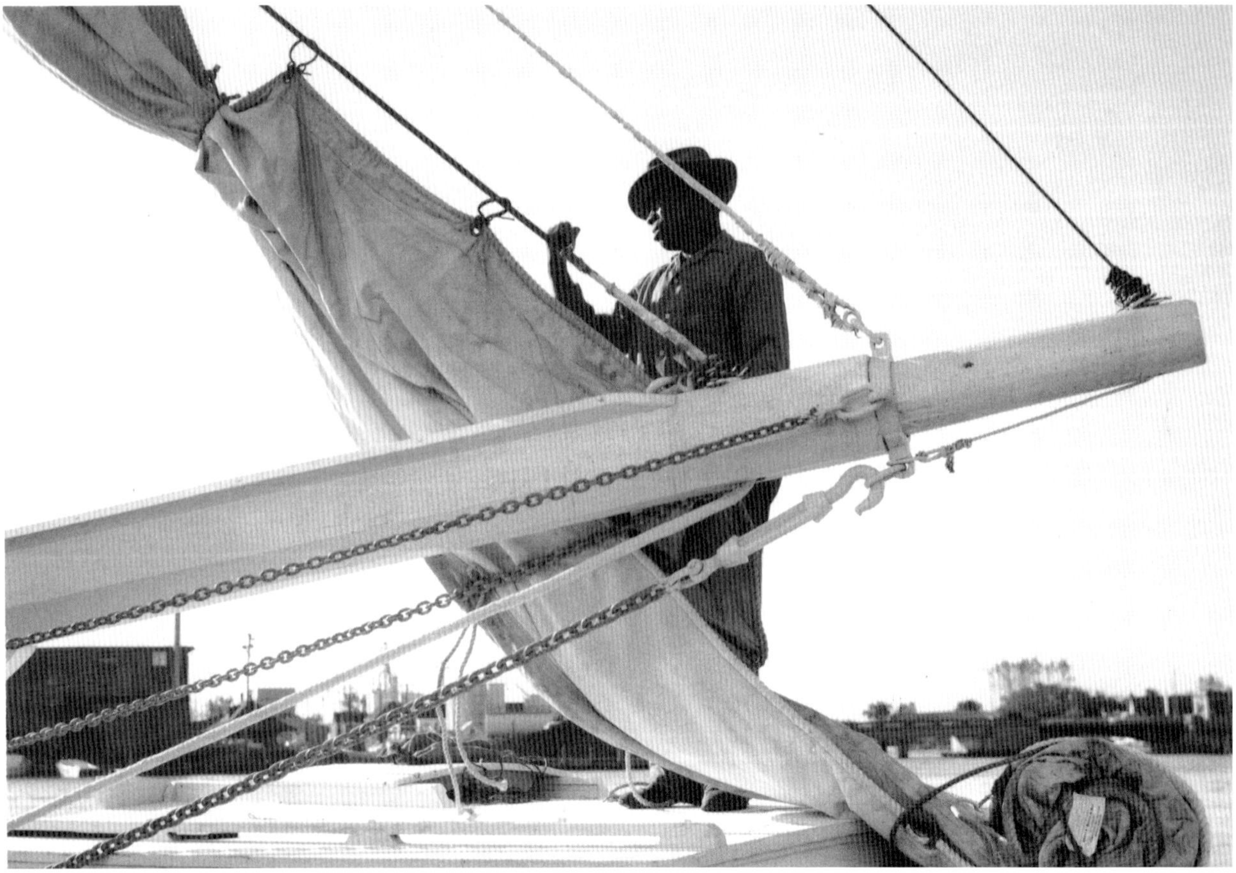

❂ Bending the jib onto the headstay of the *J.T. Leonard* on October 26, 1959.

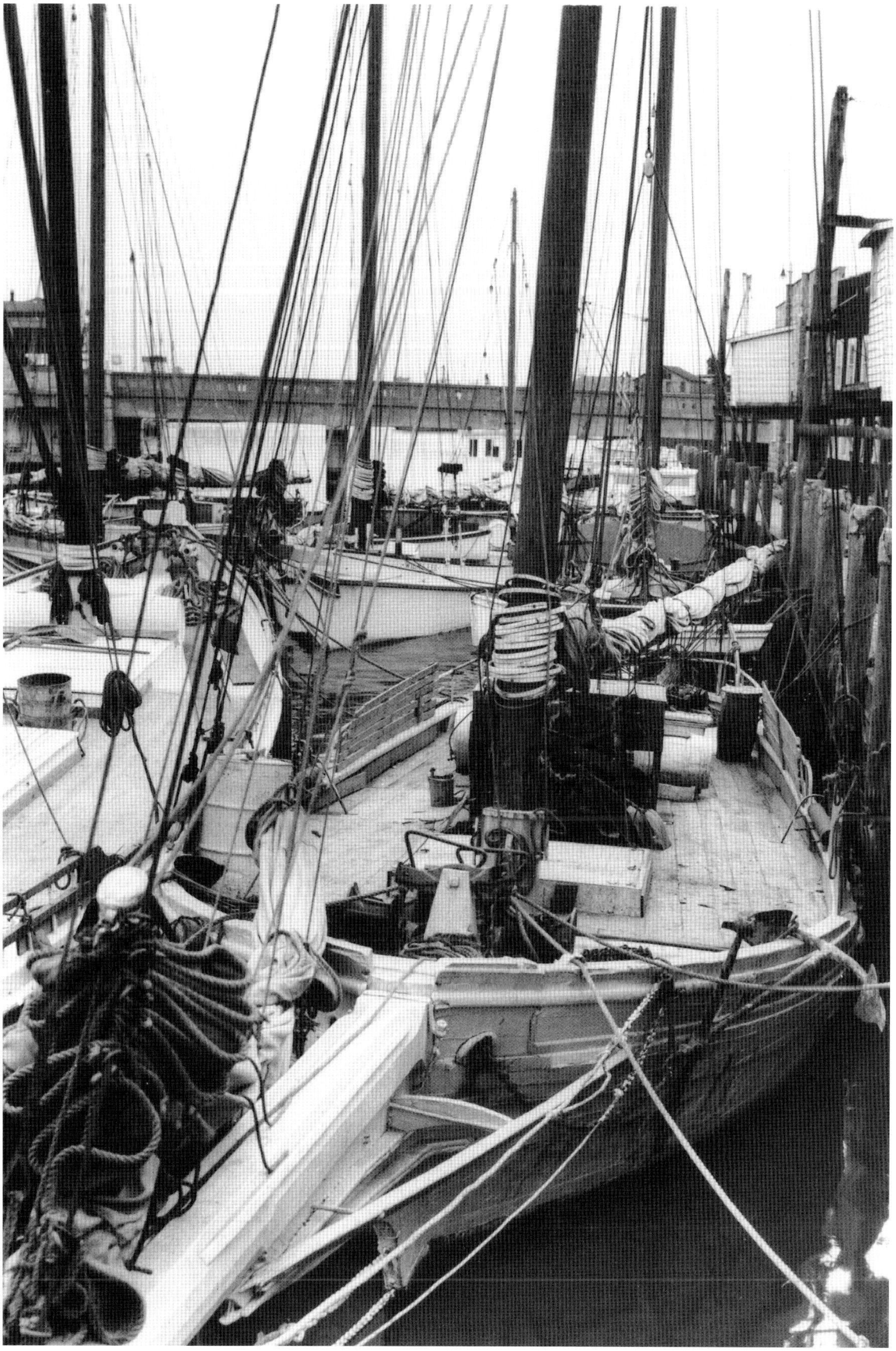

Former gaff-rigged sloop *Mamie E. Robinson,* shown here with the traditional skipjack rig, tied up with other dredgers and buyboats at Cambridge, February 1949. Looking aft along her deck, one sees a shipshape little craft that worked hard for a living from the time of her building at Taylors Island, Maryland, in 1884, until her abandonment at the headwaters of Cambridge harbor in 1953.

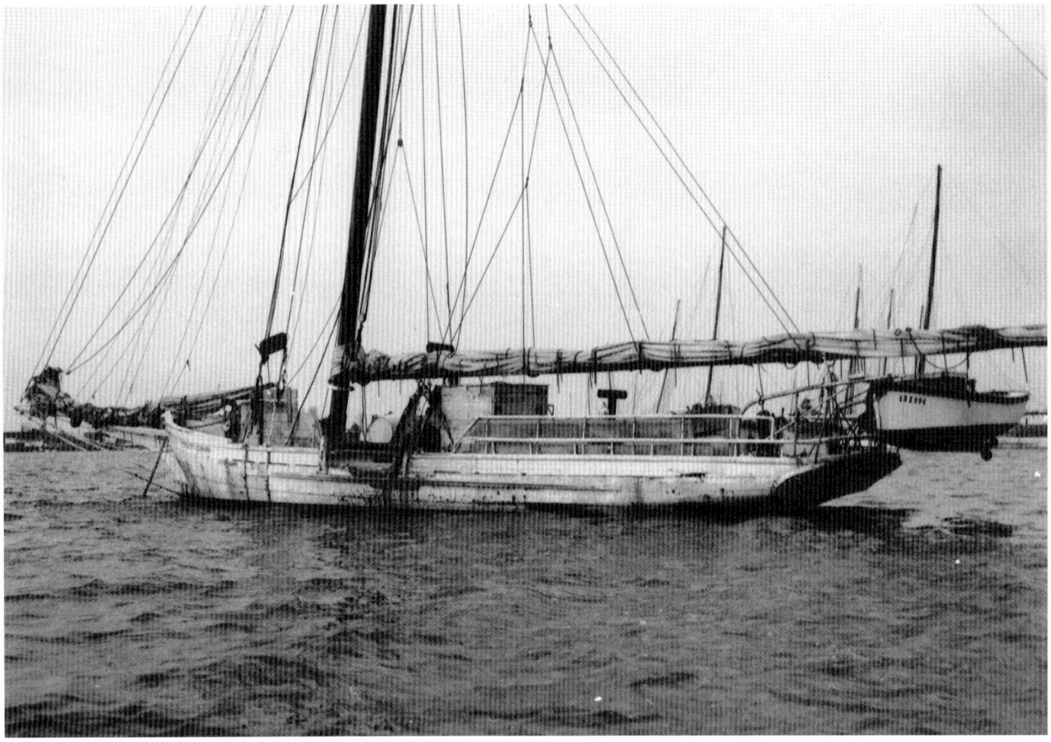

Mollie E. Leonard anchored in the upper harbor at Deal Island, Maryland, in February 1949. A former gaff-rigged sloop, she shed that more complex rig for the simpler skipjack rig as shown here. Built in Dorchester County, Maryland, in 1881, the *Leonard* left the oyster dredging fleet in the 1960s when she was sold with the thought of being converted into a yacht. The plan never materialized, and after being moved finally to a little shipyard on the upper James River, just below Richmond, she sank and was abandoned.

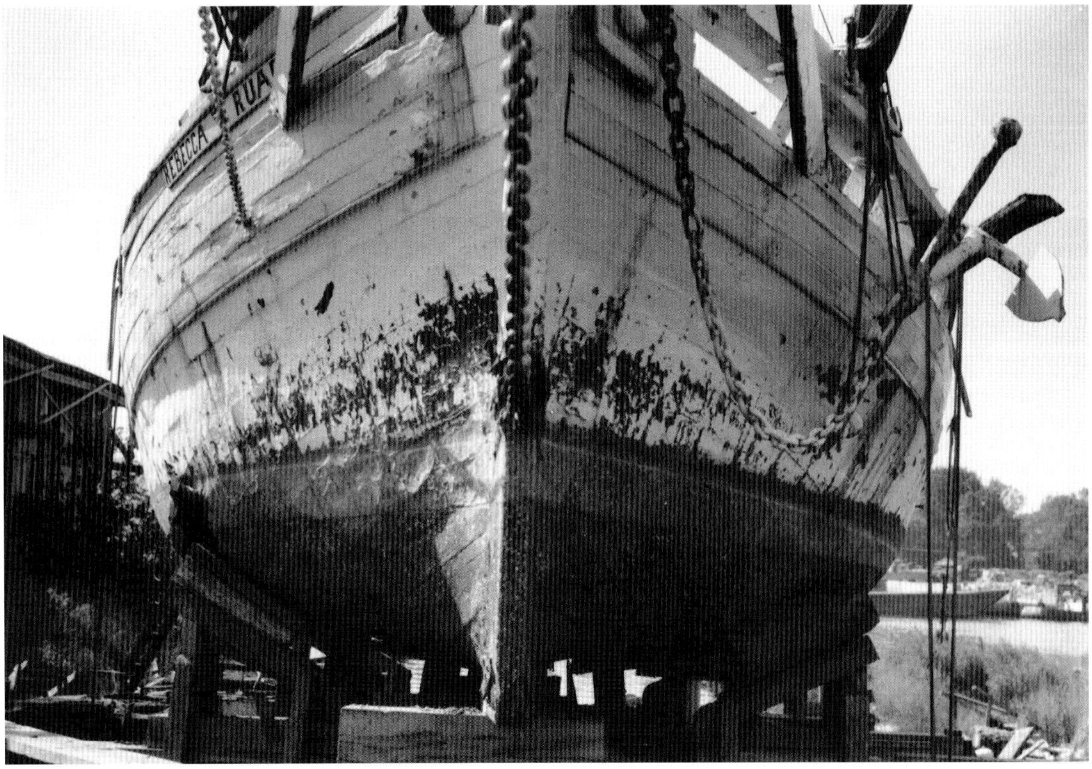

On a marine railway at Cambridge the *Rebecca T. Ruark* reveals her sloop hull of rounded bilges and full bow. At this time, October 1959, she carried the skipjack rig. The ice-scarred metal sheathing reveals her encounters with the floes in the Choptank River the previous season. But when she takes to the water again that and her barnacled bottom will be repainted. ✪ She had been built in 1886 at Taylors Island, Maryland, and was still active in 2005.

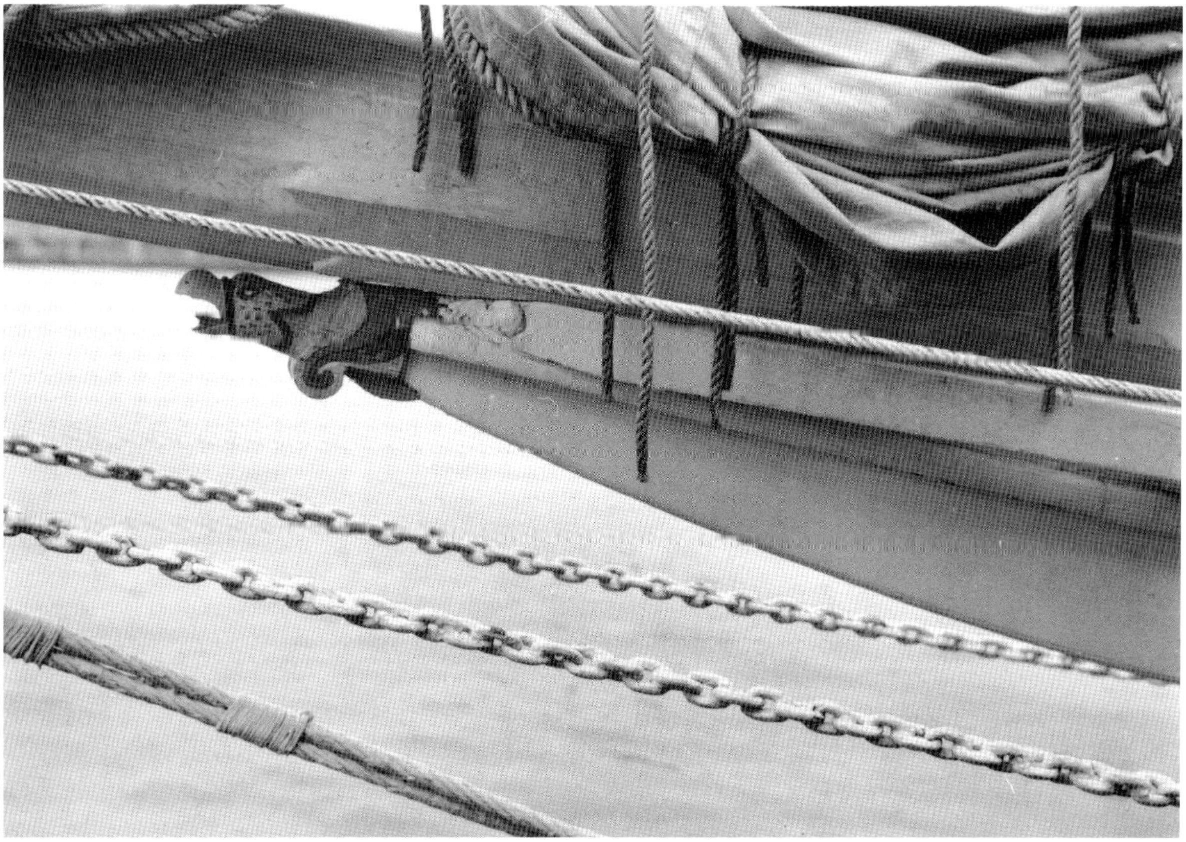

Garishly painted eagle figurehead, with broken beak, on the *Rebecca T. Ruark* at Cambridge, March 10, 1962.

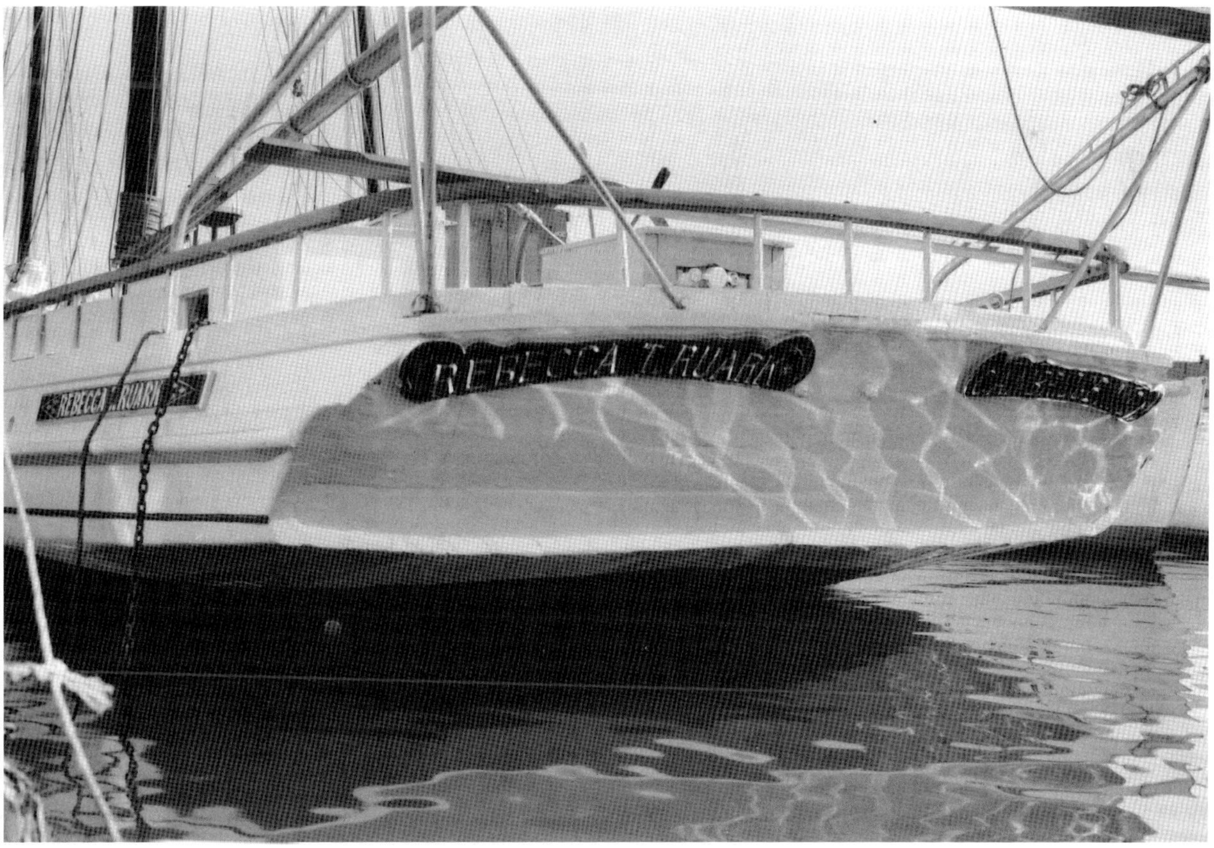

The last photograph taken by the author for this volume, at Cambridge, Maryland, shows the wide transom of the sloop *Rebecca T. Ruark* with its newly carved sternboards, and quarterboard on the port quarter, freshened up for the 1974 oyster dredging season.

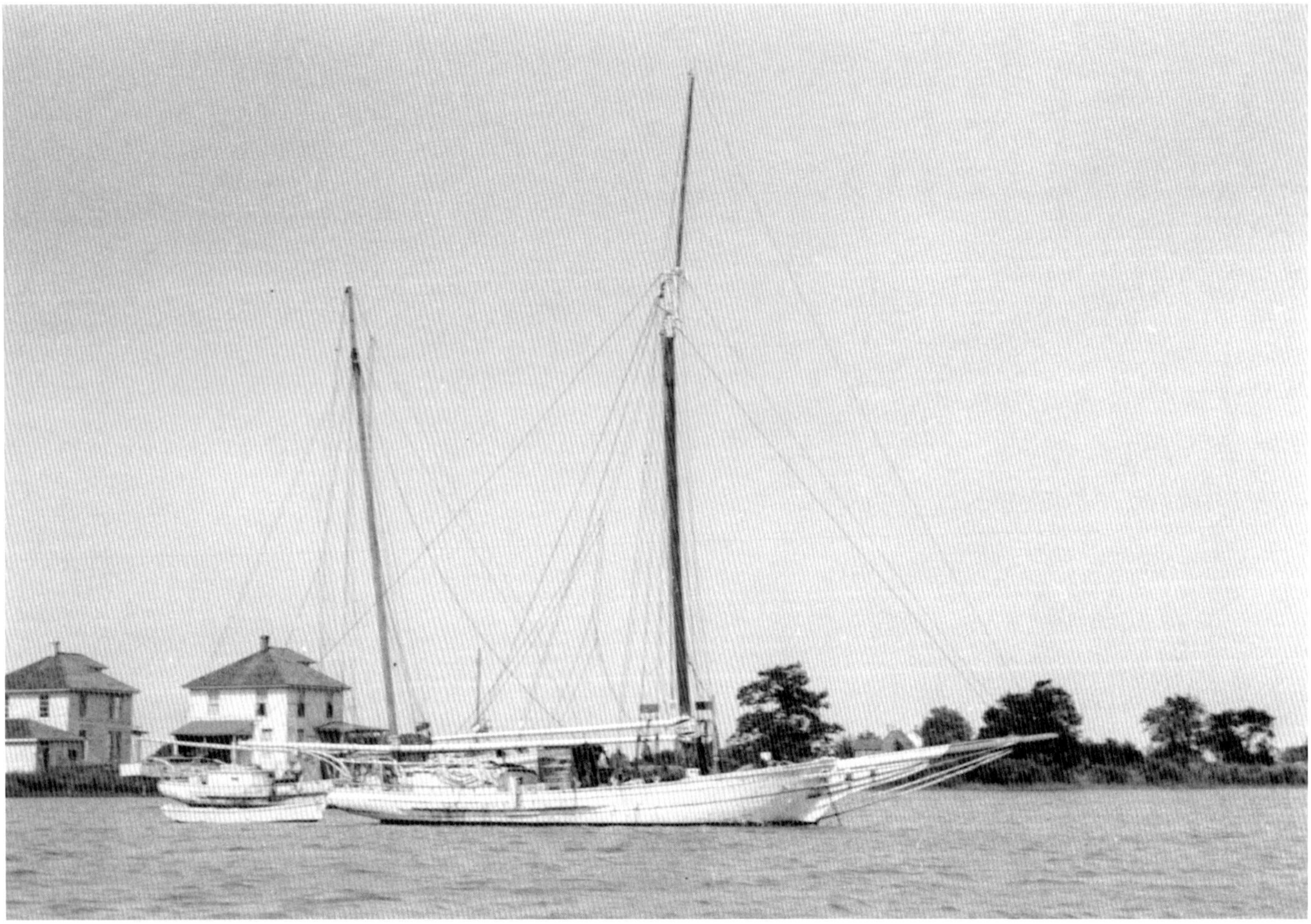

When this photograph of the sloop *William Wesley,* anchored in Town Creek, Oxford, on June 17, 1936, was made, the gaff-rigged sloop was virtually extinct in Chesapeake waters. In addition to the *Wesley,* only the *J.T. Leonard* and scow sloop *Elsie* carried that rig at that time. In the left background can be seen the former gaff-rigged sloop *Mamie E. Robinson.* The *William Wesley,* built at Crisfield, Maryland, in 1874, was abandoned in 1951.

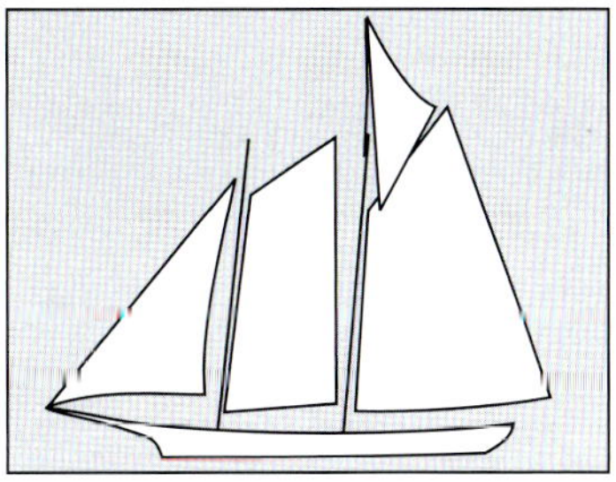

Two-masted Schooners

Two-masted schooners were the most numerous of the Bay sailing types for many years. Half a century ago they were found not only in the major ports like Baltimore, Norfolk, Washington, Annapolis, and Cambridge, but also back in the remote creeks while tied up at steamboat and cannery wharves, makeshift landings, or anchored in the stream. Schooners were the lifeblood of the Chesapeake country, but even then hard-surfaced roads, bridges, trucks, and powerboats were cutting into their trade.

It is surprising, though, how long the sailing craft were able to hold out on the Chesapeake as compared to waterways in other sections of the country. The topography of the Bay area was an important factor. Thick forests still covered much of the land in the lower Bay region, supplying what seemed like endless cargoes of lumber. The rivers penetrated into the forested sections and were navigable close to the cutting sites. Rails had not come to many of those sparsely settled places, and the roads were poor. The rivers were the outlets. The cheapest means of shipment was by water, under sail. A two-masted schooner with a crew of two, captain and deckhand, would have its hold filled, and then the boards stacked on deck, overhanging the sides and atop the after cabin, until water sloshed through the scuppers onto the deck.

Schooners were the workhorses of their day, much as the tractor-trailer rig is today. They were called upon to haul anything, anywhere, but lumber became their mainstay. The more remote areas for water transportation in the Chesapeake regions were in King William and King and Queen Counties, Virginia, at such locales as Rosespout and Aylett. To reach these places by water required a sail down the Chesapeake, then up to the head of the York River and continuing up the Mattaponi River. Rosespout, or Pointers Landing, was 29½ miles above West Point located at the juncture of the Mattaponi and York Rivers, and Aylett was 33 miles up the river. In 1925 the channel at the approaches to Rosespout had a depth of 6 feet and width of 40 feet, and at Aylett the Mattaponi River was 5½ feet deep. Freshets made it difficult to maintain those depths. In the late 1920s the 58-foot steam tug *James E. Lee,* of the York River Towing Company, would offer round-trip tows between West Point and Courthouse Landing, 13 miles above the mouth of the Mattaponi River, for $25. Most schooners would sail if the wind was directly fair and enabled the vessel to negotiate the corkscrew bends of the river, with the help of the yawl boat. Channels there were unmarked as they remain today, but the schooner captains had excellent knowledge of the out-of-the-way ports and waterways. The river continued 55 miles beyond Aylett but was shoal and crossed by fixed bridges. However, it was used for rafting and logging at high stages.

Several small drawbridges spanned the Mattaponi downstream from Rosespout, and the schooners passed through the openings with little room to spare. The last bridge was at West Point, and once in the York River the schooners had clear sailing to Baltimore—but not always fair. Head tides and winds opposed them, but the captain took advantage of every favorable slant to get his cargo to port as quickly as possible so that he could make a quick turnaround to return down the Bay for another payload.

The bulk of the lumber trade was from the western shore of Virginia and northeastern North Carolina. The sailing craft favored the western shore of the Bay from Hampton Roads to the Patuxent River for anchorages. From Cove Point to Annapolis the shore offered little shelter in the event of a storm, with the exception of an indentation at Herring Bay and the Severn River mouth at Annapolis. Favorite anchorages for vessels leaving Norfolk or the James River, before entering the Chesapeake, were Hampton Creek or Mother Hawkins Hole, the latter behind Old Point Comfort. Farther north, sheltered anchorages could be found behind New Point Comfort, and off Gloucester Point in the York River; inside the mouth of the Piankatank River; in Carters Creek, off the Rappahannock River; behind Fleeton in Cockrells Creek, off the Great Wicomico River; Cornfield Harbor behind Point Lookout, and farther up the Potomac at Piney Point; and Drum Point in the Patuxent River. On the Eastern Shore of Maryland, shelter could be found behind Black Walnut Point at the mouth of the Choptank River and behind Bloody Point at the entrance to Eastern Bay. All of these anchorages offered protection against northerlies. When the wind came fair for going up the Bay, the fleets would leave the anchorages and proceed on their passage. The shipping column of *The Baltimore Sun* for April 8, 1925, reported: "Delayed on the Bay and tributaries for 10 days by adverse winds, a fleet of 22 Chesapeake freighters came into port Monday night and yesterday, loaded with lumber." On November 6, 1935, *The Sun* said: "Captain Joe Pope, master of the powerboat *Lydia,* reported last Sunday that the schooners *Josephine Wimsatt, Thomas B. Schall, Minnie and Emma, Columbia F.C., Ida B. Conway,* and *Chesterfield* were windbound in Carters Creek, Virginia."

Sometimes ice, covering rivers, creeks, and even the upper Chesapeake, held up the sailing craft. *The Sun* on February 26, 1935, reported: "After being frozen up in the Wicomico River for four weeks the Schooner *Bohemia* arrived in Salisbury last Monday, according to word received yesterday from Captain E.B. Riggin, master of the vessel. Captain Riggin reported the upper part of the river clear but much ice was at its mouth." Sometimes whole fleets of schooners would wait weeks in icebound anchorages before a thaw permitted them to move on. At such a time there was always the risk of supplies and water running short and the men would have to borrow from one another. Or they could walk ashore and replenish their groceries at a nearby country store. To pass the time of day they visited from vessel to vessel and played cards. In the upper Bay, from Seven Foot Knoll toward Havre de Grace, the Coast Guard cutter *Apache* came to the aid of many sailing vessels, furnishing supplies and water, breaking ice around them, or taking them in tow.

When the schooners had reached Baltimore with their lumber, the focal point for discharging was Pier 6, at Pratt Street and Back Basin. Back Basin! How important that name was forty and more years ago! The main part of the harbor area, bounded by Locust Point and Canton and leading up to Light and Pratt Streets, was known as the Northwest Branch of the Patapsco River, or the Basin. A little indentation off that, just east of Jones Falls, was Back Basin. Here, in the heart of the lumber district, bugeyes, pungies, rams, and schooners rigged with as many as four masts tied up along the granite dock paralleling Lancaster Street and disgorged their cargoes of pine from Virginia, North Carolina, and Georgia.

Lancaster Street was the main street of Back Basin. Lumber mills were situated within that area bounded by Lancaster Street, East Falls Avenue, east on Fawn Street past the foot of Albemarle Street to Central Avenue, and south to the Waterfront again. The lumber that was not stored on the docks was loaded onto wagons and taken to the neighboring lumberyards. At the foot of nearby Philpot Street were more storage piers. Other mills could be found along Boston Street, out of the Back Basin district, and the sailing craft would unload at piers closer to those plants.

Within walking distance for the crews of the vessels was Pratt Street, and here were the ship chandlers, pawnshops, cheap restaurants and rooming houses, tobacco shops, bars, barber shops, shipping agents and brokers, sail lofts, and other establishments catering to the Back Basin fleet. In the mid-1920s there was a considerable drop in the lumber trade to Back Basin. The South was becoming better equipped with finishing mills, and more lumber from the southern states beyond the Chesapeake was being shipped by rail. Cardboard box factories were also making inroads into the wooden box trade, eliminating another outlet for lumber.

Of all the Pratt Street businesses patronized by Bay watermen, the most prominent was Vane Brothers, ship chandlers. Shelf-lined windows on either side of the entrance displayed smaller nautical items such as lanterns, foghorns, compasses, portlights, and various fittings. These conveyed the impression that the store dealt primarily in small boat paraphernalia. Inside, that misconception was quickly corrected. There one saw anchors, steering wheels, bells and other big-vessel gear. The feeling of the sea was transmitted by sight, sound, and aroma.

Coils of clean Manila line, neatly arranged on shelves according to size, waited to be cut into lengths for halliards, sheets, or mooring lines. Containers of nails and spikes of all sizes, carpenters' and caulkers' tools, caulking cotton and oakum, and marline were prominent. A grocery section had shelves stacked with canned and bottled goods, canned milk, fruit juices, and barrels of molasses and salt pork stood on the floor. Flour and baking powder, sugar, and other ingredients were all waiting for some ship's cook to convert them into energy-producing meals atop a wood-burning stove in some schooner's galley far down the Chesapeake. These stores, purchased by the captain or some other member of the crew, along with kerosene, gasoline, and oil, would be sent to the anchored schooner down the harbor by Vane's powerboat *Charles Wright.*

Obscured behind the name Vane Brothers were two other brothers, Claude and Charles Hughes. Claude bought half ownership in the chandlery in 1920 after he had sold his schooner *R.E. Powell,* of which he was captain, to marry and settle down ashore. Charles joined the firm later. They eventually acquired the business outright from W.B. Vane, but retained the original name. Claude retired in 1957, but Charles carries on with his son at a new location.

Vane Brothers was more than just a supplier to the sailing men. The Hughes brothers assisted the seamen in legal and financial matters, social security and income tax problems, and port charges. Vane's was also a sort of clearinghouse for those watermen who had something to sell or trade. Theirs was a mailing address for many seamen; letters were kept in a receptacle to await the return of the sailing vessels from down the Bay.

As sail declined in the 1930s and then practically disappeared in the '40s, except for the oyster dredging craft, so did a way of life disappear for Vane Brothers. From its site overlooking the harbor on Pratt Street the firm moved to another waterfront location at the foot of Broadway. Neighbors of Vane's who either moved or went out of business included Myer Saperstein, who, among other services, supplied crews for sailing vessels; Captain's Ship Chandlery Company, and sailmakers William L. Godfrey and Stevenson-McGee. All these depended on the Back Basin fleet. Other Pratt Street nautical businesses in that era were Charles F. Eareckson and Company, sailmakers; W.B. Tull, marine engines; and William H. Whiting and Company, marine hardware. In 1973 the colorful brick buildings themselves were leveled to make way for the Inner Harbor urban renewal development.

Other chandleries, sailmakers, and blacksmiths plied their trade along Boston Street, within rowing and walking distance of the anchorage in Canton Hollow. Included were R.H. Dougherty, and W.L. Muir and Son, blacksmiths;

Charles E. Osenburg, welder; and S.W. Webster and Brother, ship chandlers.

Many of the schooners, pungies, bugeyes, and smaller craft were owned by individuals who served as captains. Some vessels were owned by merchants in the tidewater country who used the craft in their seafood business or to freight their merchandise to and from the city market. Other craft were owned in shares and included schooners of two to four masts, and rams. One of the better known owners on the Bay was Captain Tom Webster, a native of Deal Island who, it has been reported, had between 50 and 100 schooners, pungies, bugeyes, and skipjacks. Many of these were used as oyster dredgers, others as buyboats and freighters. When Tom Webster died, his nephew, T. Preston Webster, took over the fleet and remnants of this once vast floating empire operated until the mid-1930s.

The lumber schooner passed out of the Chesapeake scene in 1948 when the *Bohemia* carried her last cargo. Schooners were part of the oyster dredging fleet until the late 1950s. Conversion to power, "dude cruisers," or abandonment had been the fate of the others.

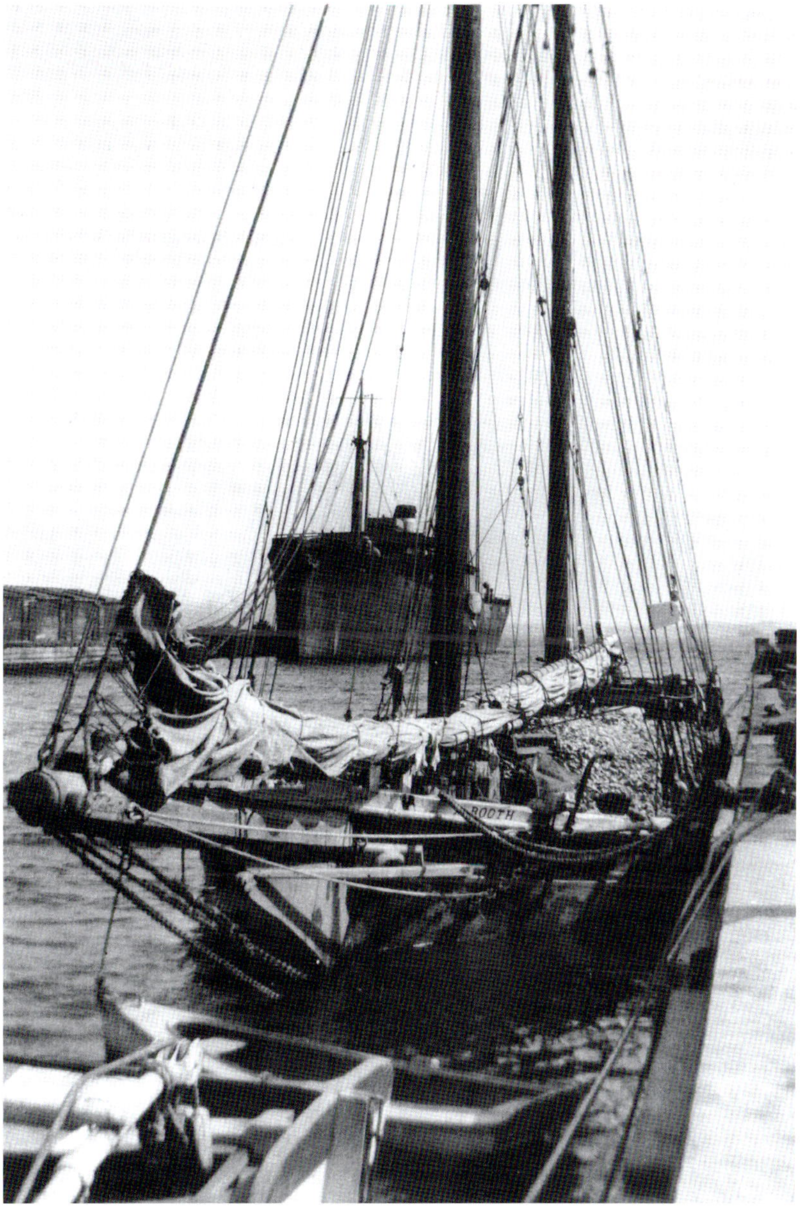

A. Booth was named for, and originally owned by, a Baltimore packer, having been built by T. Kirby in 1880 in Talbot County, Maryland. Here, on May 13, 1934, she waits to discharge a cargo of oyster shells at the Potomac Poultry Food Company in Canton, Baltimore, for grinding up as chicken feed. The vessel's only claim to fame is her placement as first in the alphabetical listing of sailing vessels in *MVUS* for 32 years. During World War II she was converted to power and in 1947 was operating out of Miami, Florida, finally being sold to Costa Rica. The steamer in center background is the *Munaires,* built at Newport News, Virginia, in 1917, and laid up with a tarpaulin over her stack.

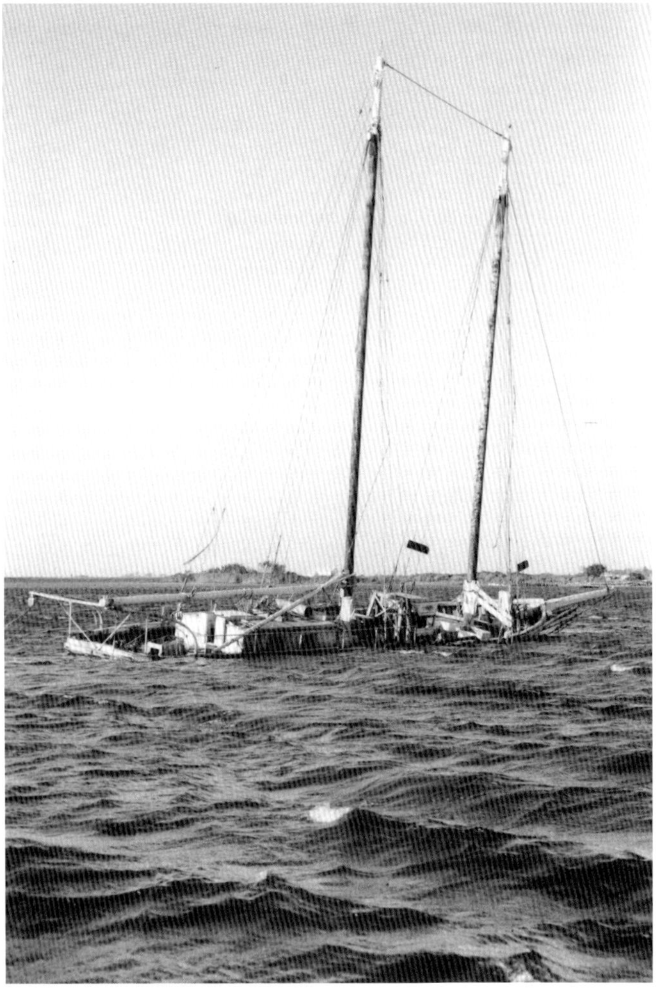

The last commercial schooner to ply the Chesapeake was the *Anna & Helen*, built at Dorchester, New Jersey, in 1911. She came to the Bay about 1941 and was employed as an oyster dredger. In 1957 she sank at Crisfield, Maryland, and there she stayed.

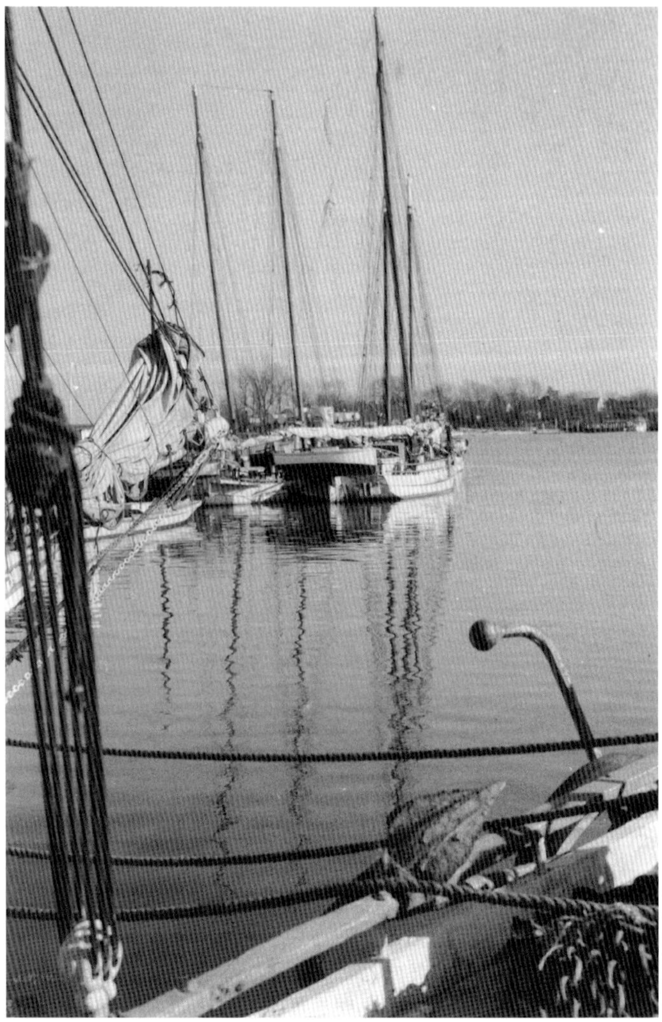

✪ The schooner *Anna & Helen* near the end of her working days, at Annapolis, on February 10, 1952. Skipjack *Mamie Mister* and bugeye *R.H. Daugherty* are tied up with her.

Recorded as "Abandoned" in the 1932 *MVUS*, the *Annie Belle*, built at Federalsburg, Maryland, in 1875, appeared like this on May 30, 1934, at Cambridge, Maryland. Having been rebuilt at West Point, Virginia, in 1918, she still retained some sheer and proves that there is beauty even in an old hulk. Those deadeyes still attached to the chainplates would soon be salvaged as relics if such a wreck existed today.

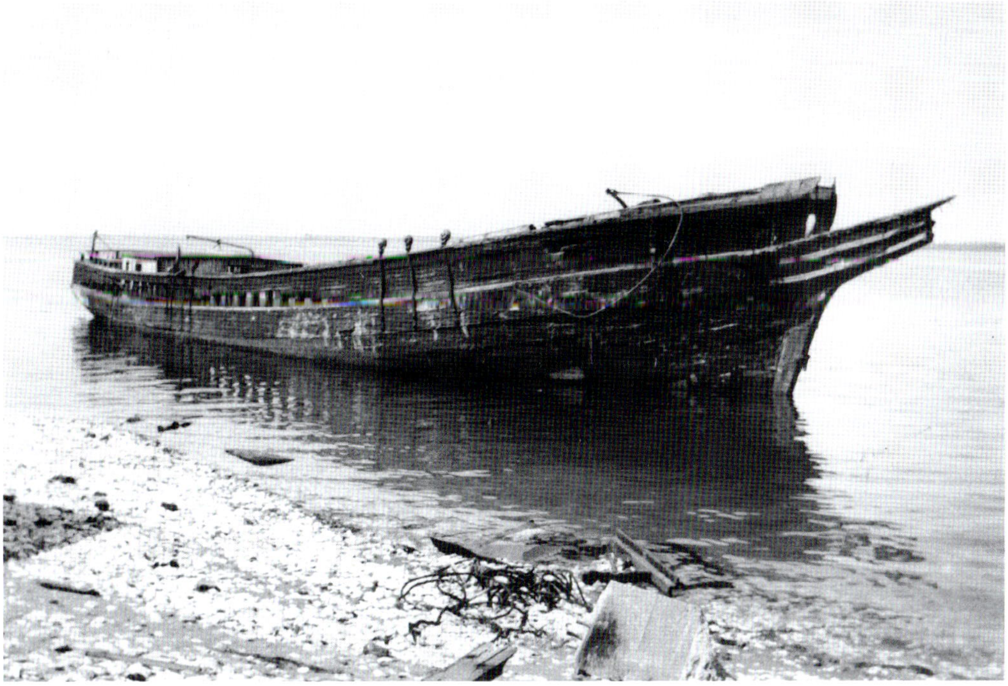

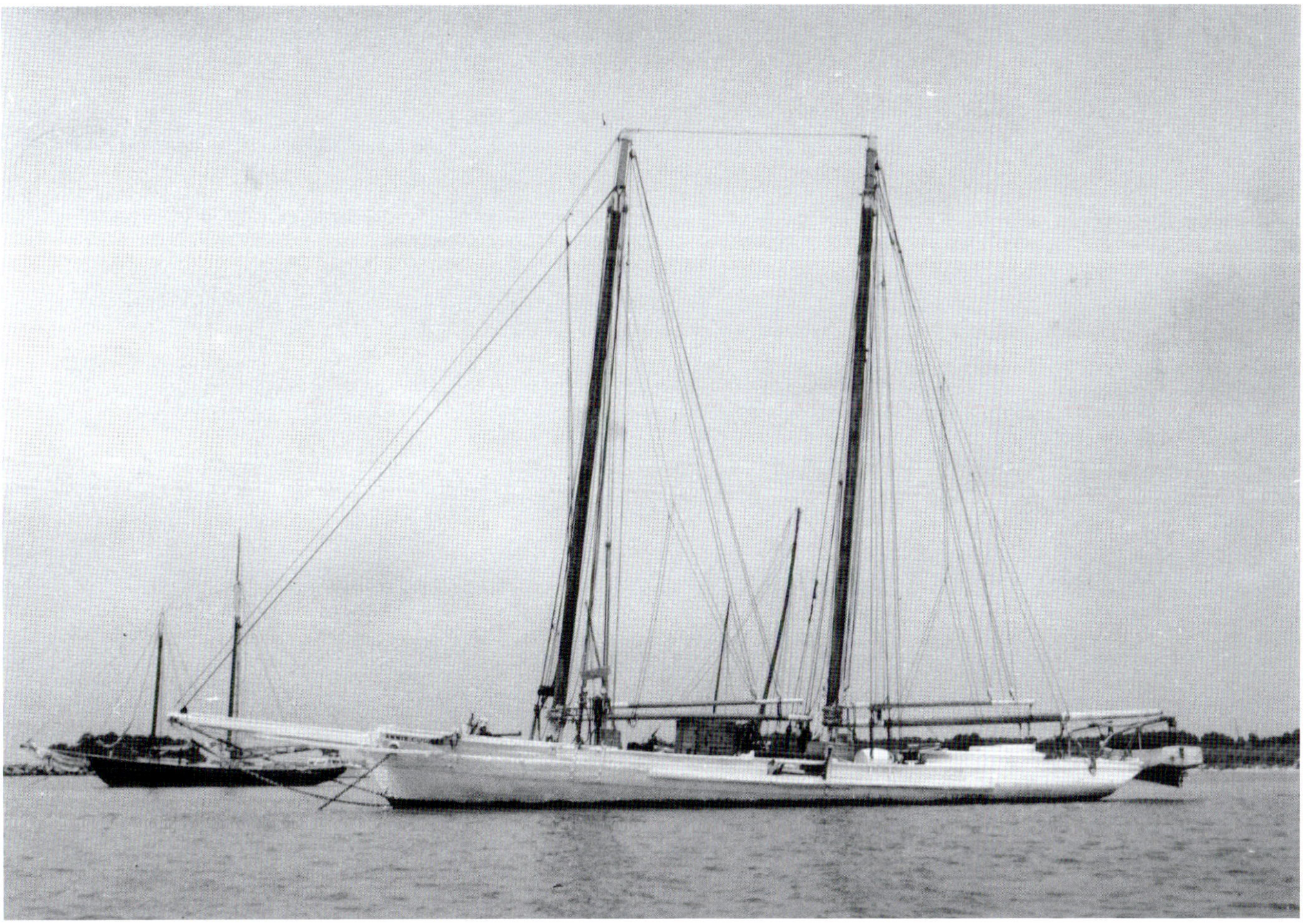

Sometime in the 1940s the *Annie C. Johnson* lost her longhead and it was never replaced, giving her the appearance of a spoon-bowed schooner, certainly not a characteristic of a Chesapeake Bay vessel. At anchor on July 12, 1950, at the upper end of Deal Island, the *Johnson* idles away the summer waiting for the oyster dredging season to start on November 1. The black vessel in left background is the former Nova Scotian fishing schooner *Lady Churchill*, converted to yacht use in 1949 by Hampton yachtsman Fred Kaiser and on a cruise, with the author, of lower Bay ports. The *Johnson* had been built in 1891 at Solomons, Maryland, sold to Delaware in 1953, and listed in the 1957 *MVUS* as "Abandoned."

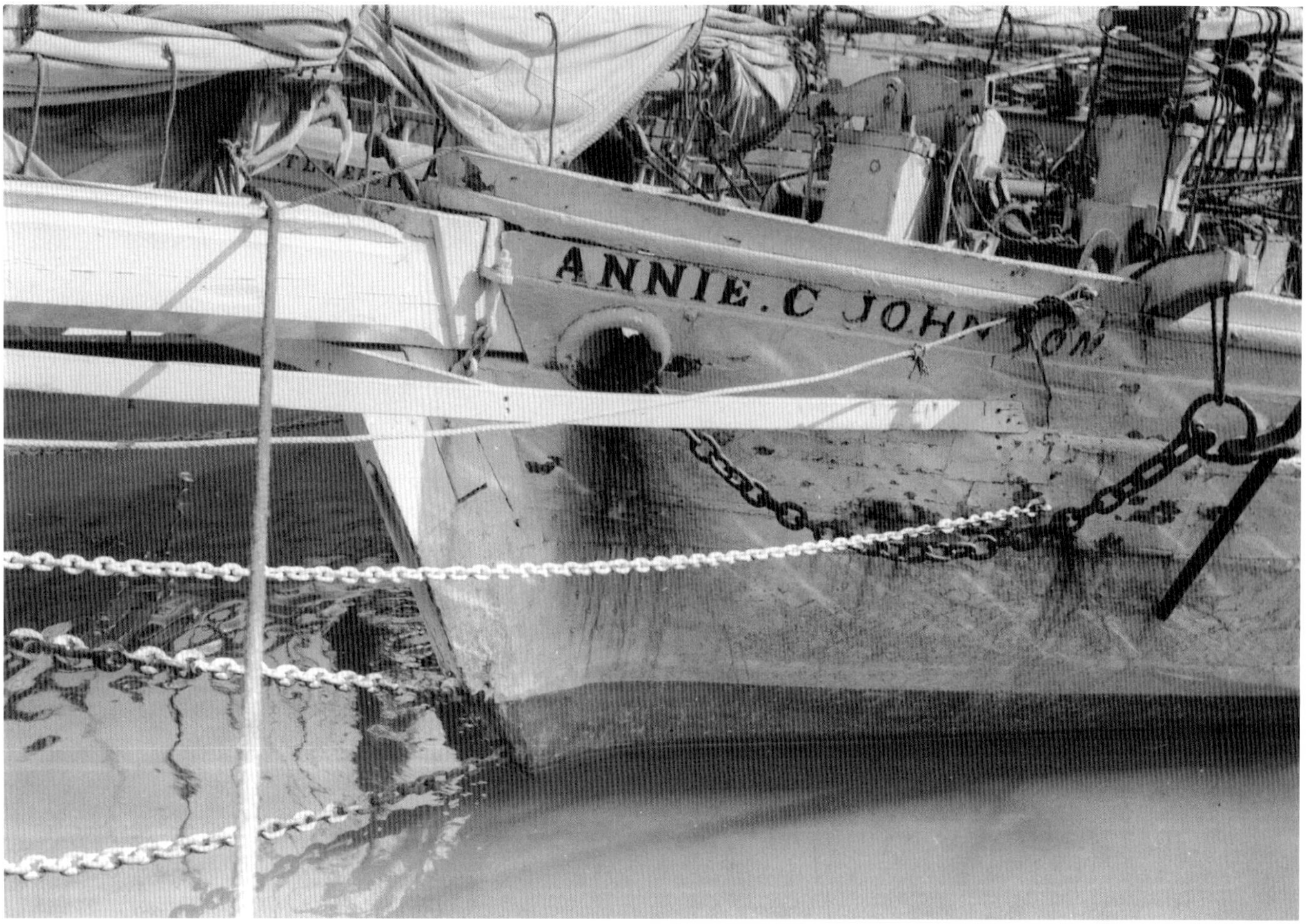

✪ Close-up of the bow of *Annie C. Johnson,* after removal of her longhead in the 1940s. Taken at Annapolis on February 10, 1952.

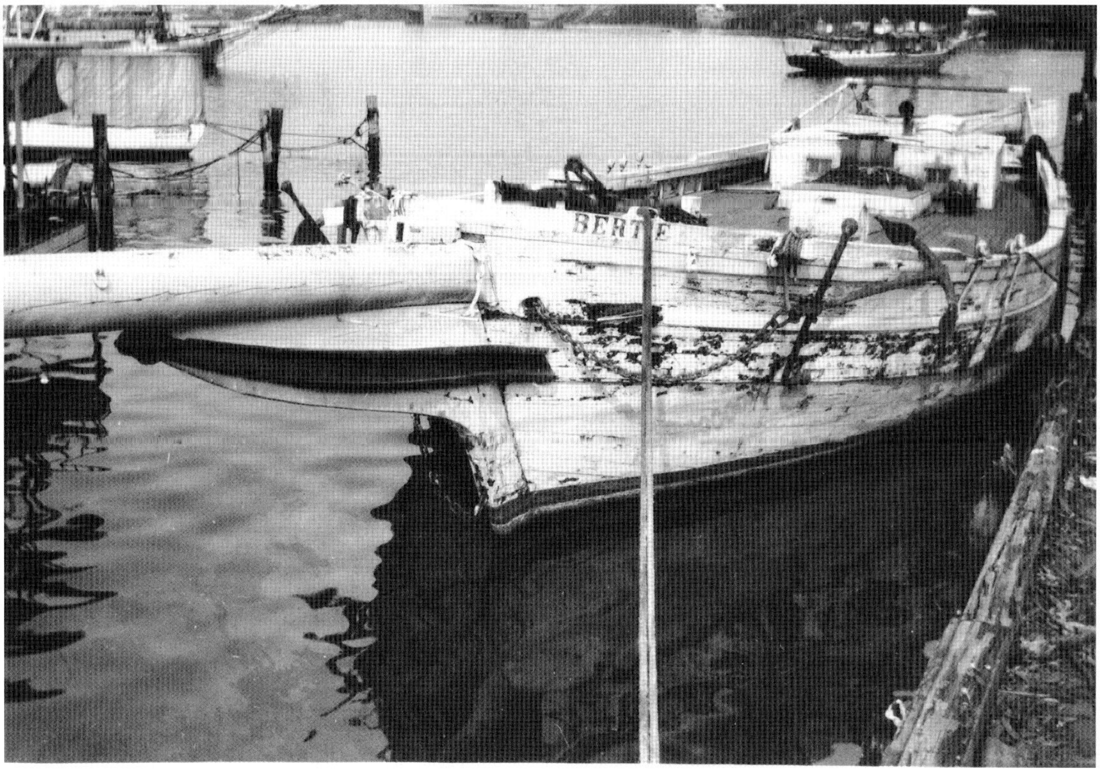

The schooner *Bertie* has been stripped in this view at A. Smith & Sons Shipyard on June 30, 1935, but already she was motorized. The engine is in her after cabin on the top of which is a ventilator. But she would eventually get rid of her longhead, bowsprit, deadeyes and chainplates, acquire a pilothouse atop her cabin, and lose all contact with sail. At upper left background is the bow of the schooner *Smith K. Martin* and at upper right the hogged hull of the schooner *Stephen Chase. Bertie* was built in 1875 in Dorchester County, Maryland, and after more than half a century under sail on the Bay and a brief period as a powerboat, was sold to New York before appearing in the 1956 *MVUS* as "Abandoned."

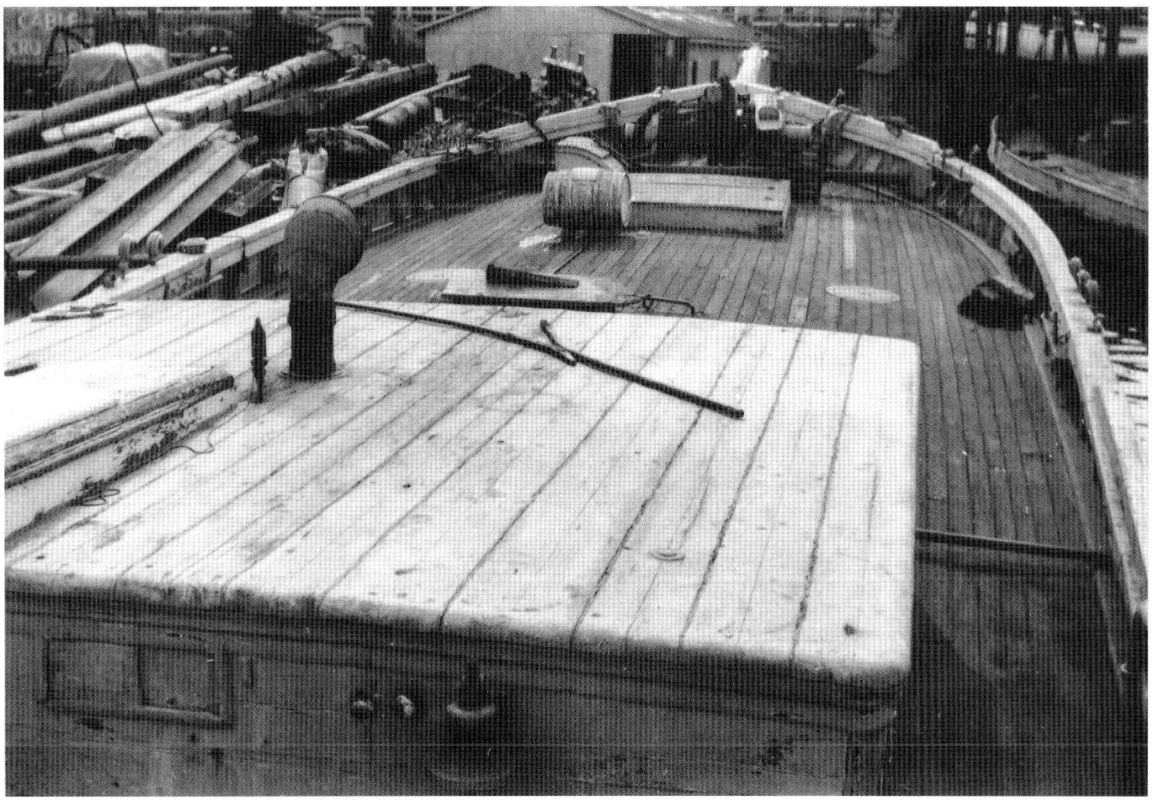

✪ Deck view of *Bertie* on the same day, looking forward. Despite the absence of masts, she still looks like a schooner here.

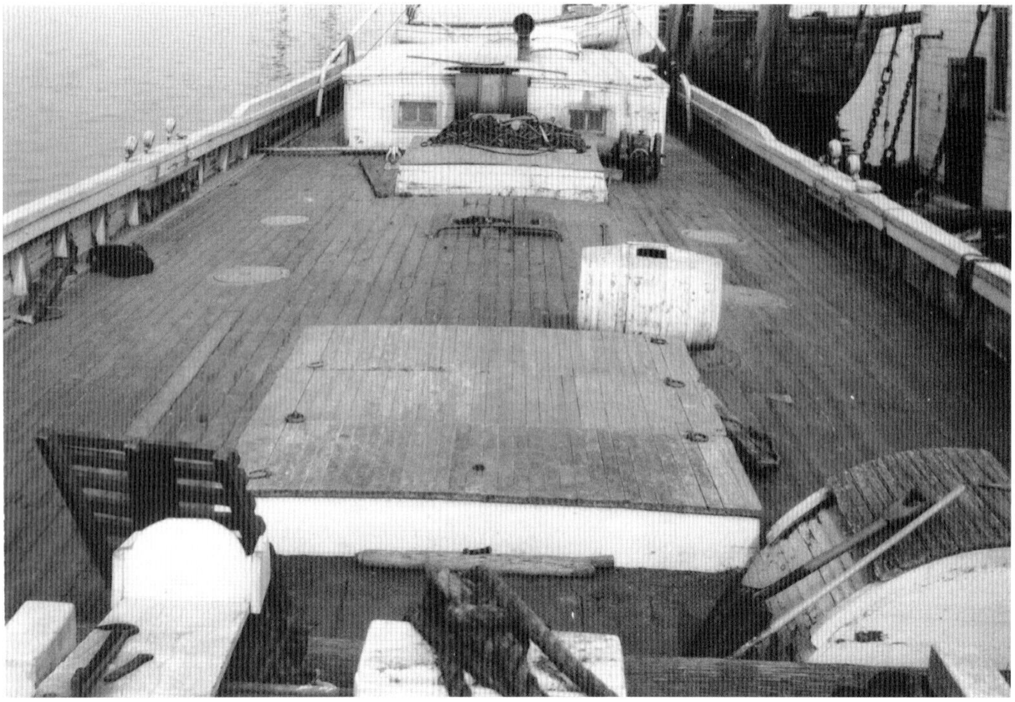

✪ Another view of *Bertie*'s deck, looking aft. Her yawl boat, wheel, and other equipment are still in place.

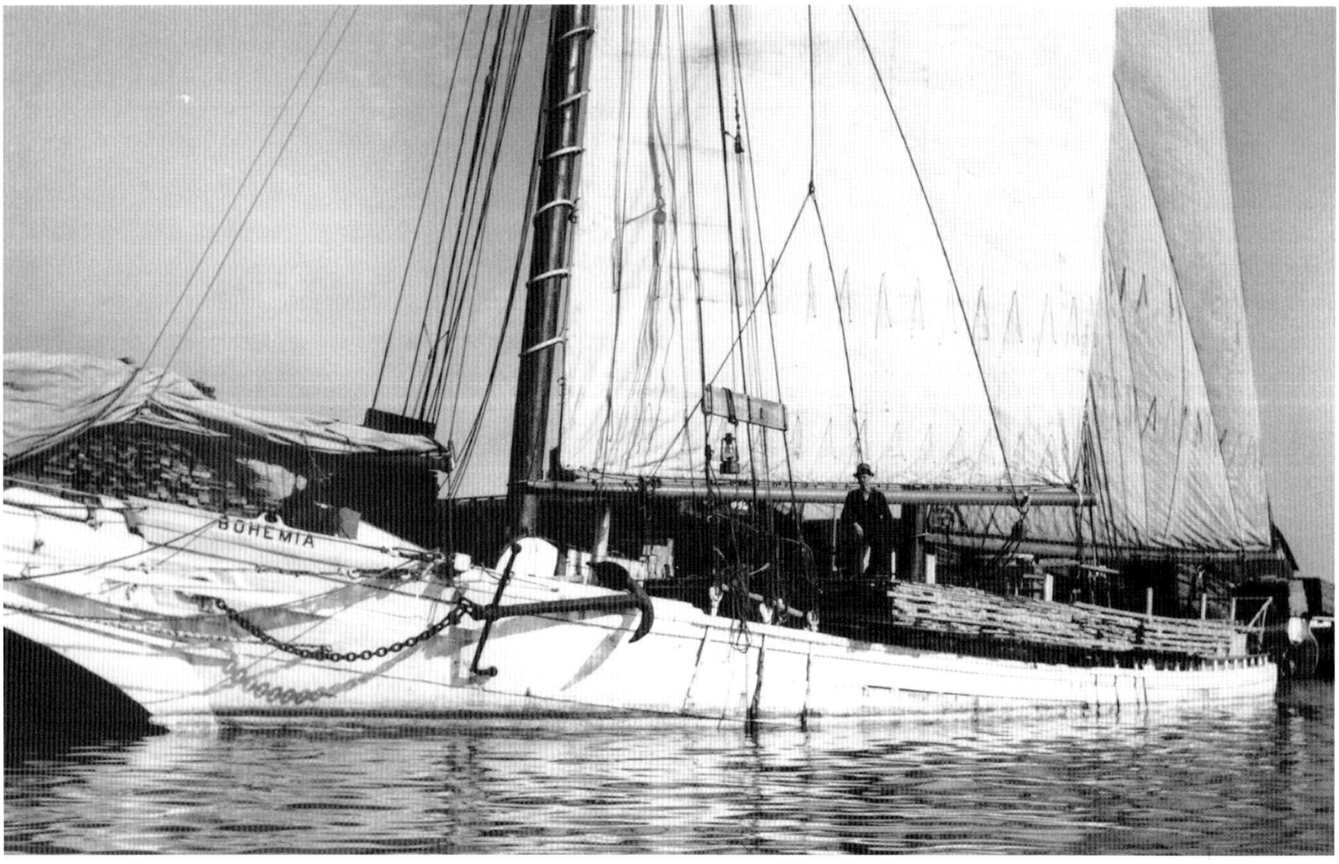

Bohemia was the last schooner on the Bay to carry lumber, a cargo that had kept her steadily employed in her latter days. Tied up at the Philpot Street Wharf of the F. Bowie Smith Lumber Company, Baltimore, she dries her sails. For 32 years she was owned and skippered by Captain Edgar B. Riggin, who poses for his portrait sitting on the deckload of lumber, September 1936. In 1947 Captain Riggin sold the *Bohemia,* and for a couple of more years she freighted on the Bay. She sank at her anchorage in the eastern branch of the Elizabeth River, Norfolk, on January 2, 1950. In March she was raised and taken to Sarah Creek, Virginia, for possible conversion into a powerboat, but this never materialized and she was abandoned there. *Bohemia* had been built in 1884 by T. Kirby at St. Michaels, Maryland.

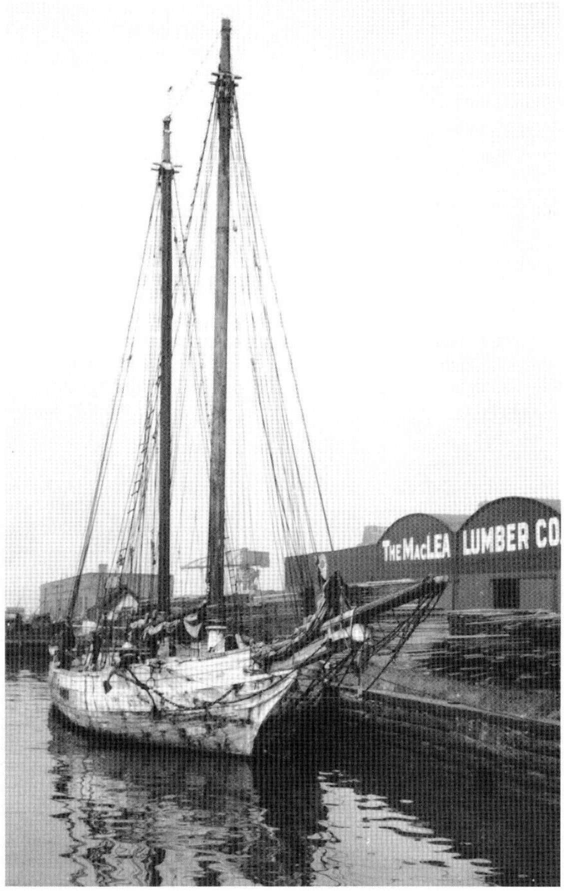

✪ The schooner *Bohemia,* after unloading her lumber cargo at Baltimore on September 8, 1948. Her laden waterline is revealed by the harbor dirt and oil staining her hull.

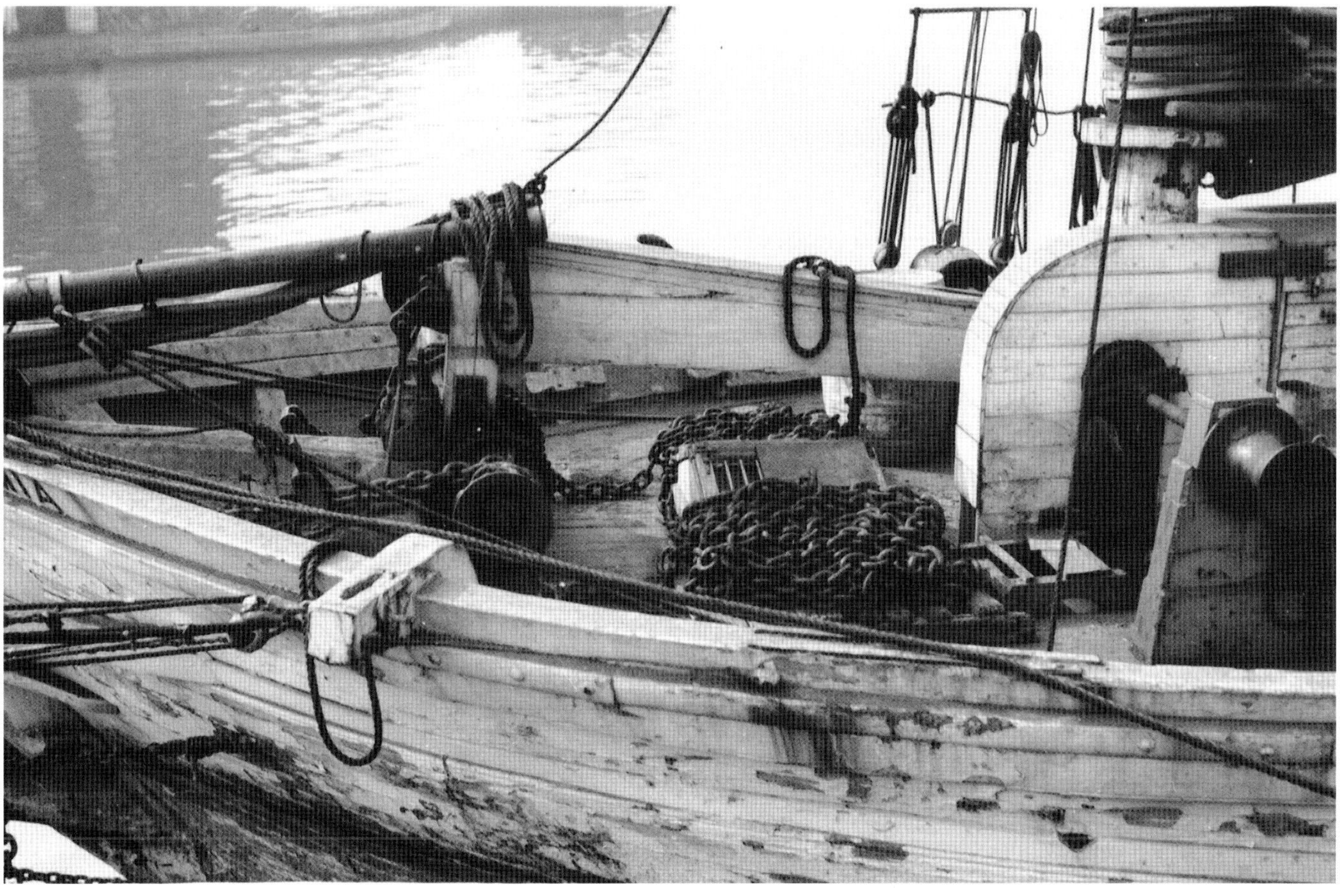

✪ Foredeck of the *Bohemia,* at Baltimore on September 8, 1948. Note her windlass, halyard winch, forecastle skylight, and port cathead.

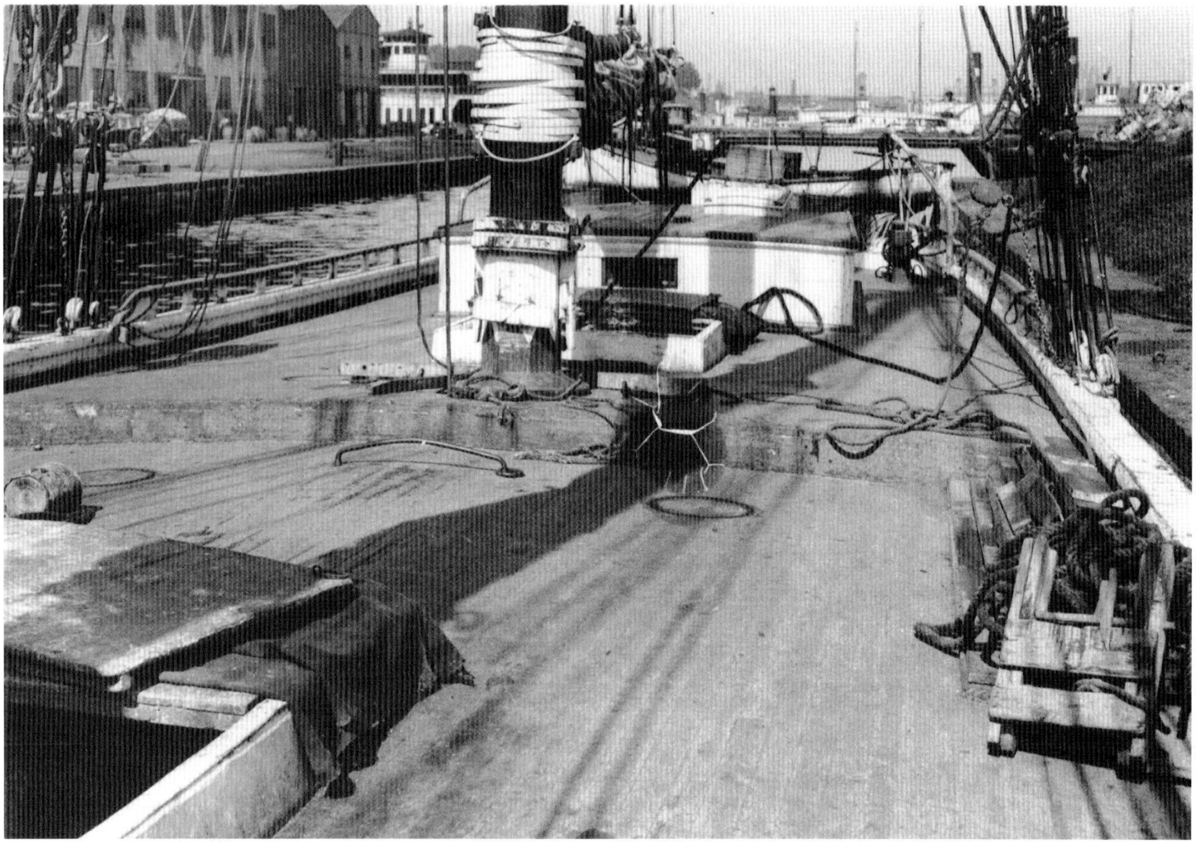

✪ Deck view on *Bohemia,* looking aft. It was unusual for such a small schooner built so late in the nineteenth century to have a break in the deck, as she had.

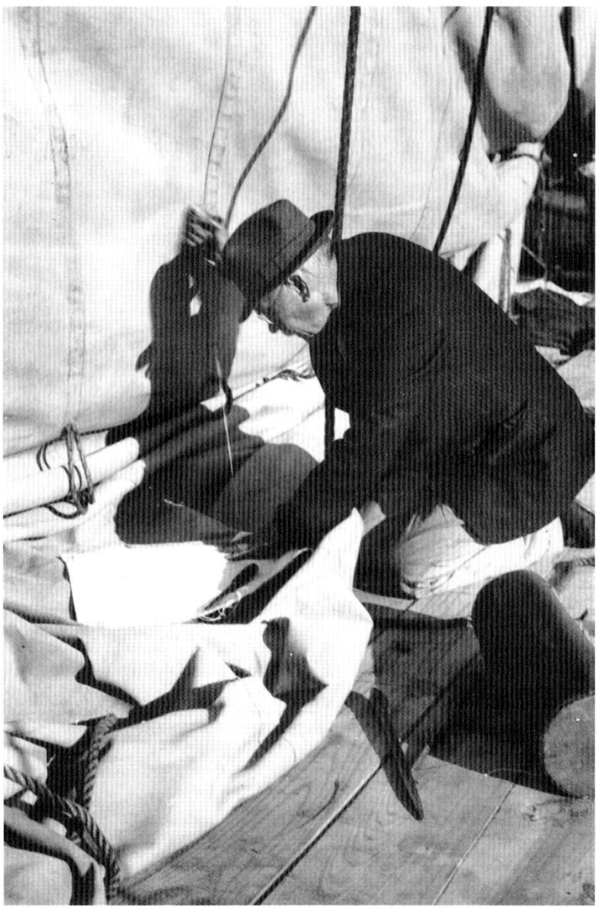

✪ George Gardner mending sails on *Bohemia,* 1936.

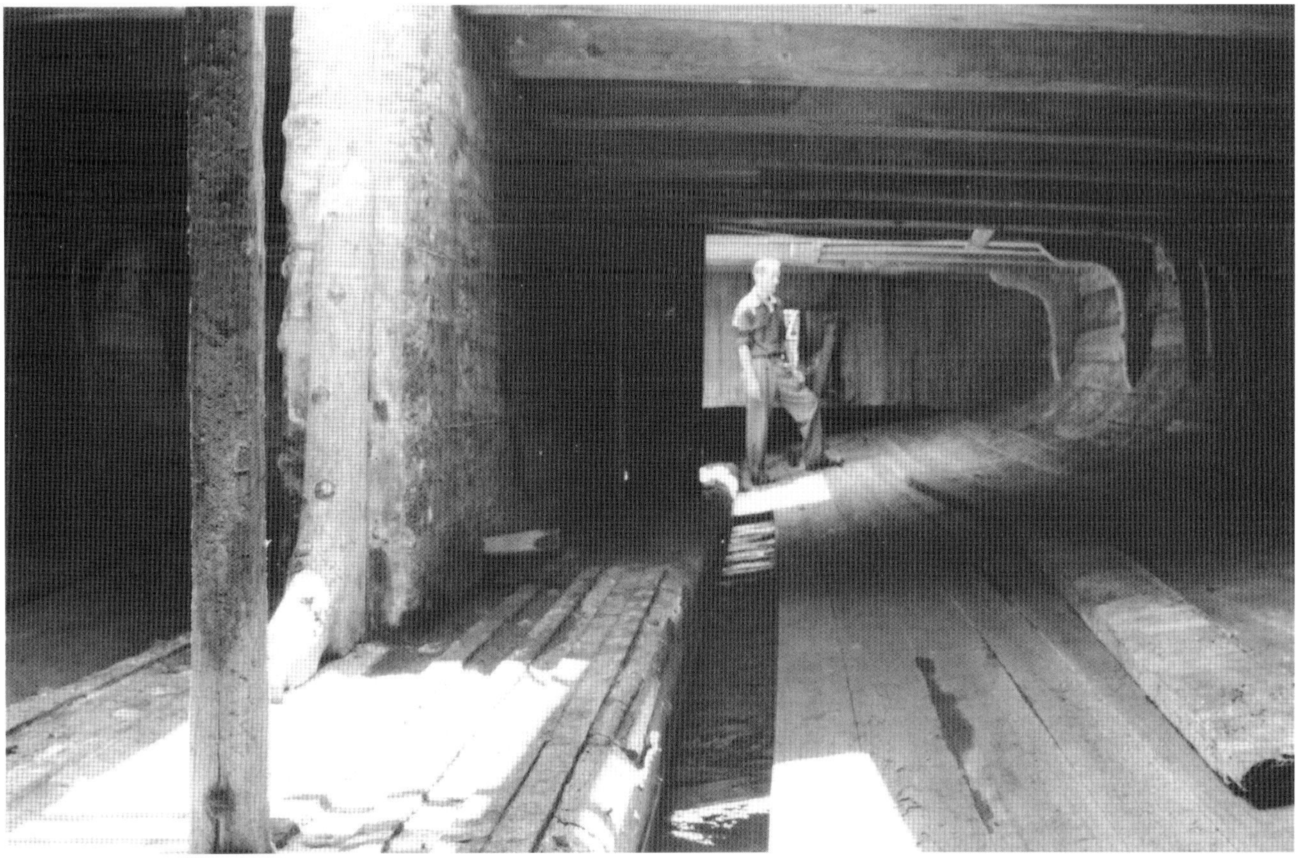

✪ *Bohemia*'s hold, looking forward. Her centerboard and mainmast were offset to either side of her centerline. John L. Lochhead, a former librarian of The Mariners' Museum, provides an idea of the scale of features in the hold.

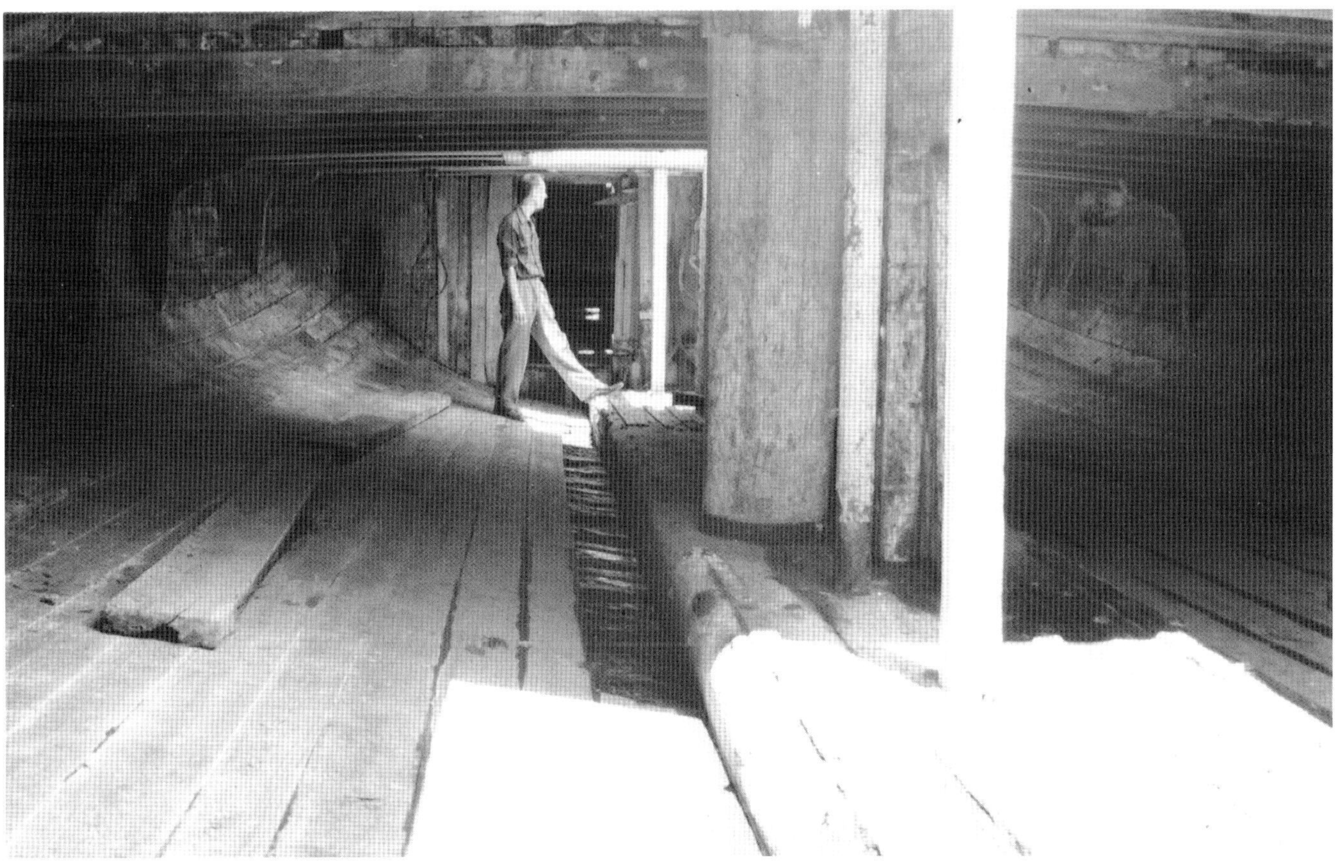

✪ *Bohemia*'s hold, looking aft.

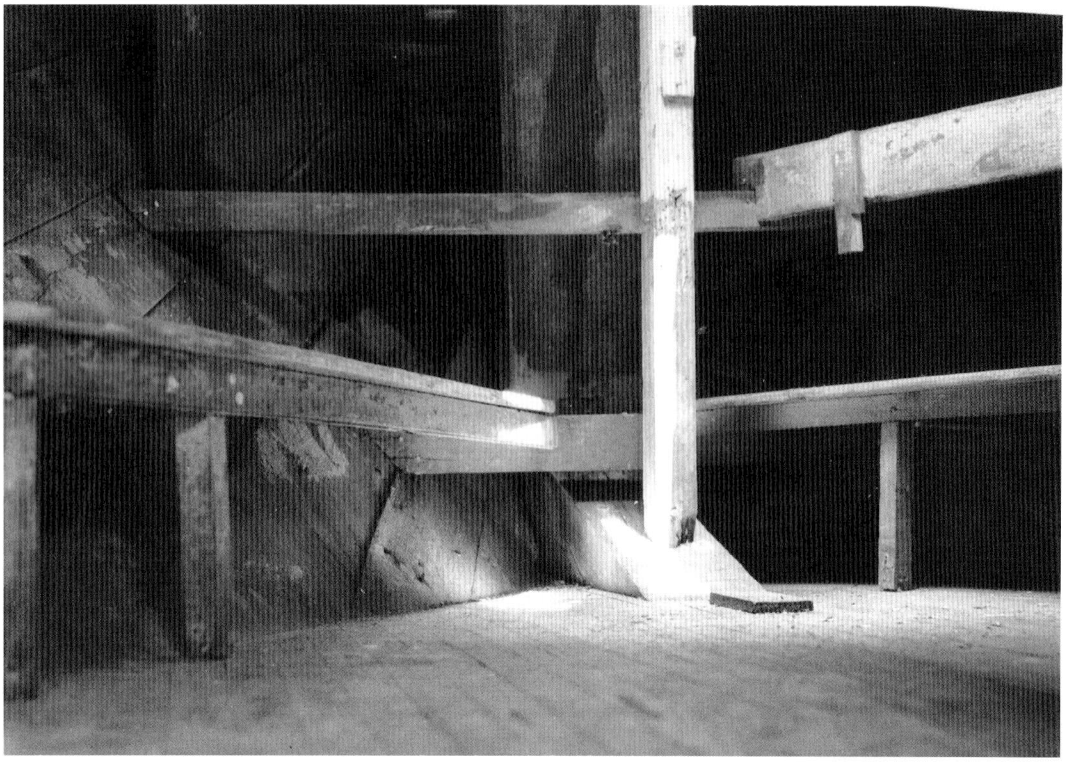

The fo'c'sle on the *Bohemia,* where her crew was housed, was dimly illuminated by daylight coming through a skylight on deck. Benches and hard board bunks offered few comforts, and in the early 1920s, when she was used to dredge oysters, as many as six men lived in this small area. The heavy timber in the background is the lower part of the samson post to which the windlass was mounted on deck. The interior planking is the ceiling. This was taken in 1953 after being raised following her sinking at Norfolk.

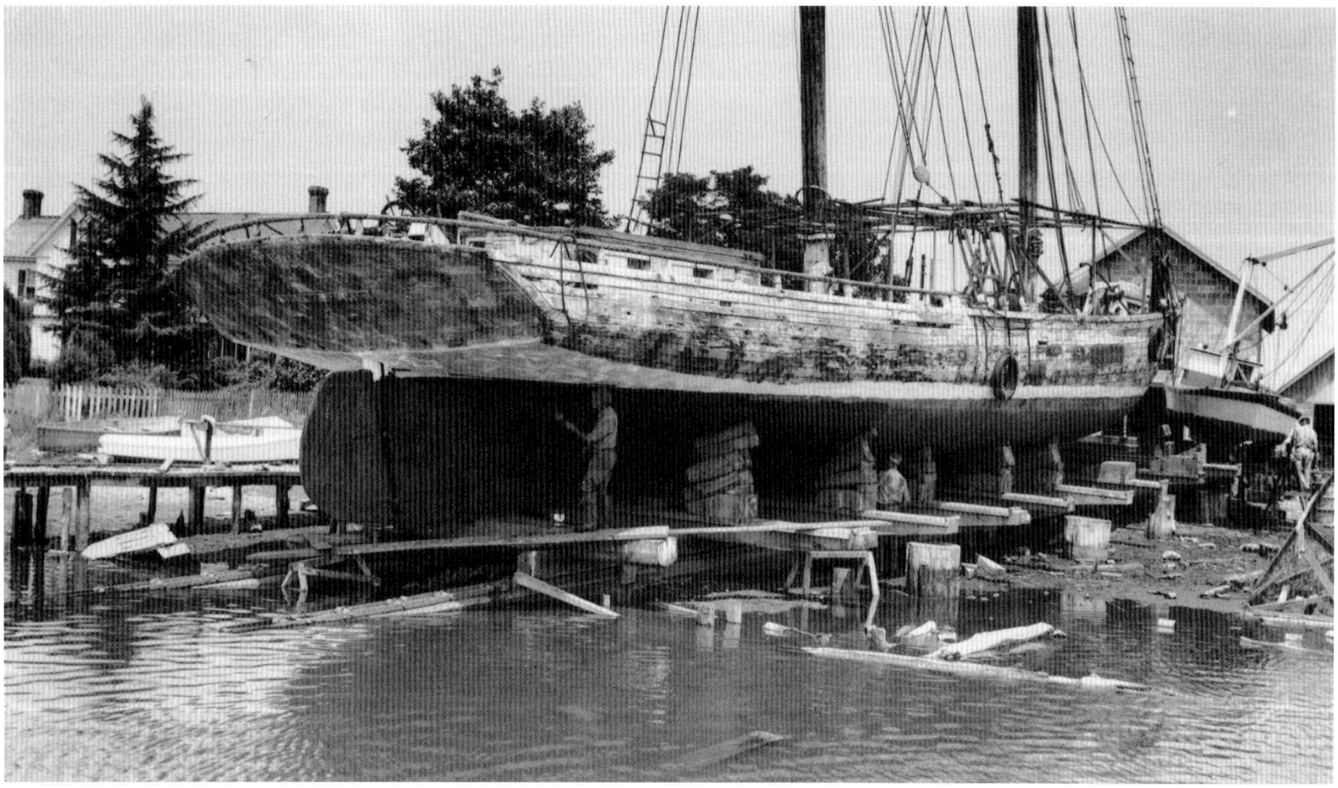

In June 1951, the *Bohemia* was hauled out at Brown's Wharf, Perrin, Virginia, to be examined for possible conversion to a powerboat and to have her bottom cleaned and painted. This is as high as her hull was painted, the plan being to paint the balance when she was waterborne. But she was never converted, so she was placed on a mud bank and neglected.

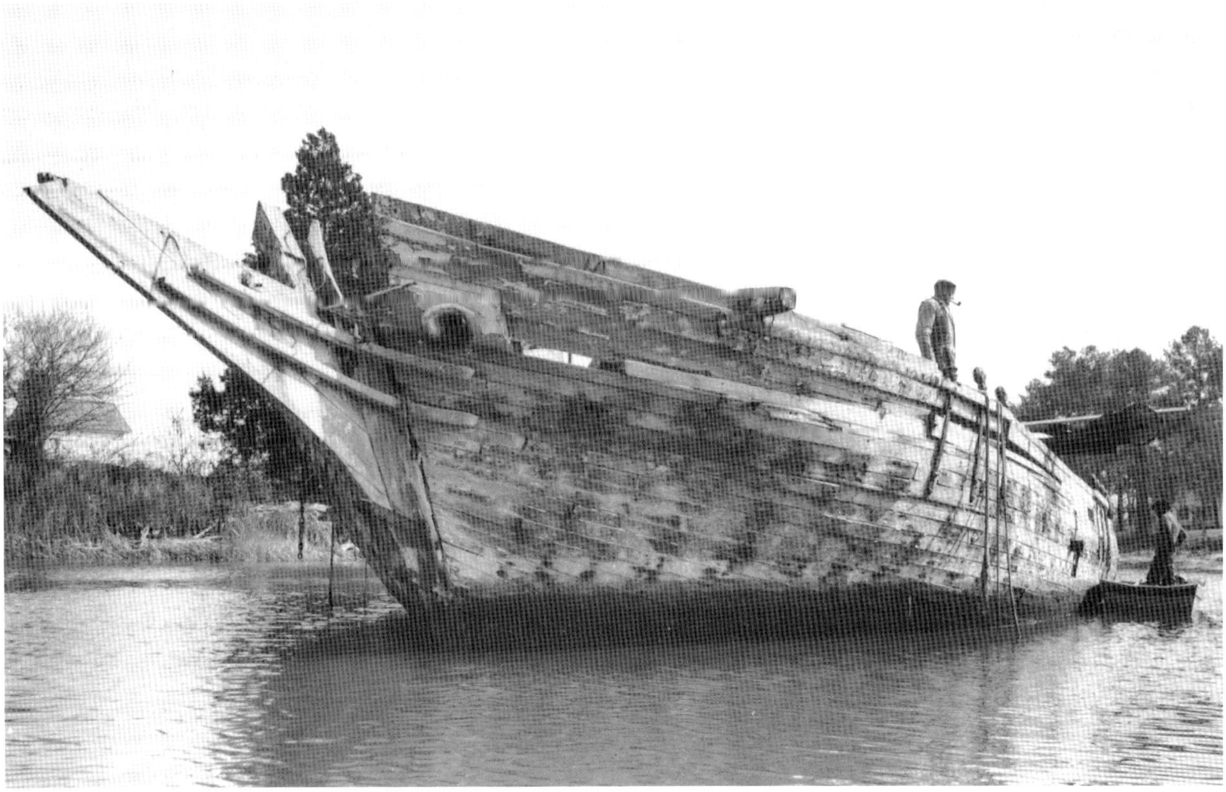

Bohemia as an abandoned hulk in an arm of Sarah Creek, Gloucester County, Virginia, February 1959. She has totally disintegrated now. ✪ At the time this photo was taken, a crew from The Mariners' Museum was using a chainsaw to cut through the deck, side planking and ceiling to expose some of the schooner's knees. They were able to get four of these in good condition, which were brought back to the museum. After being cleaned up, the knees were placed in arches for decorative and also educational purposes.

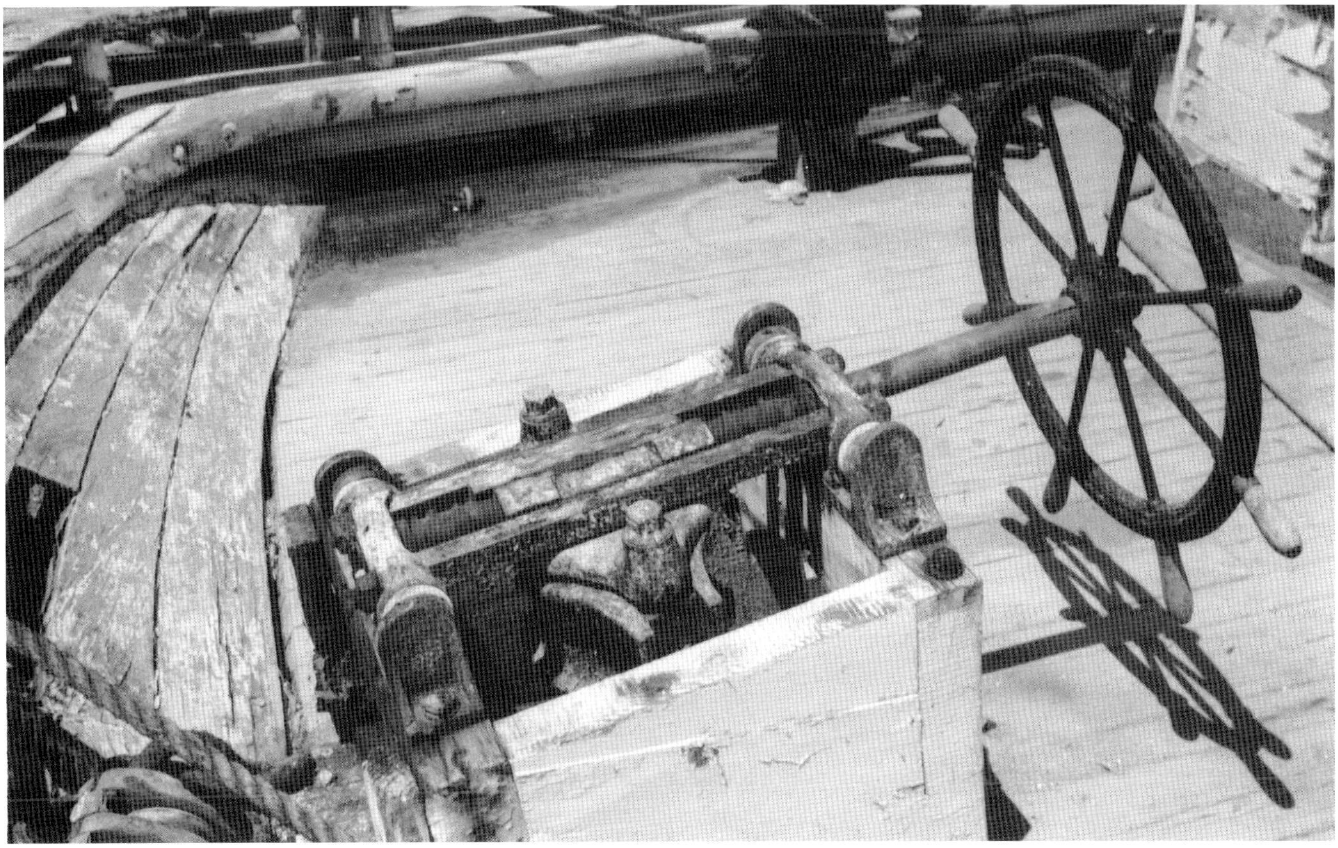

✪ *Bohemia*'s wheel and steering gear.

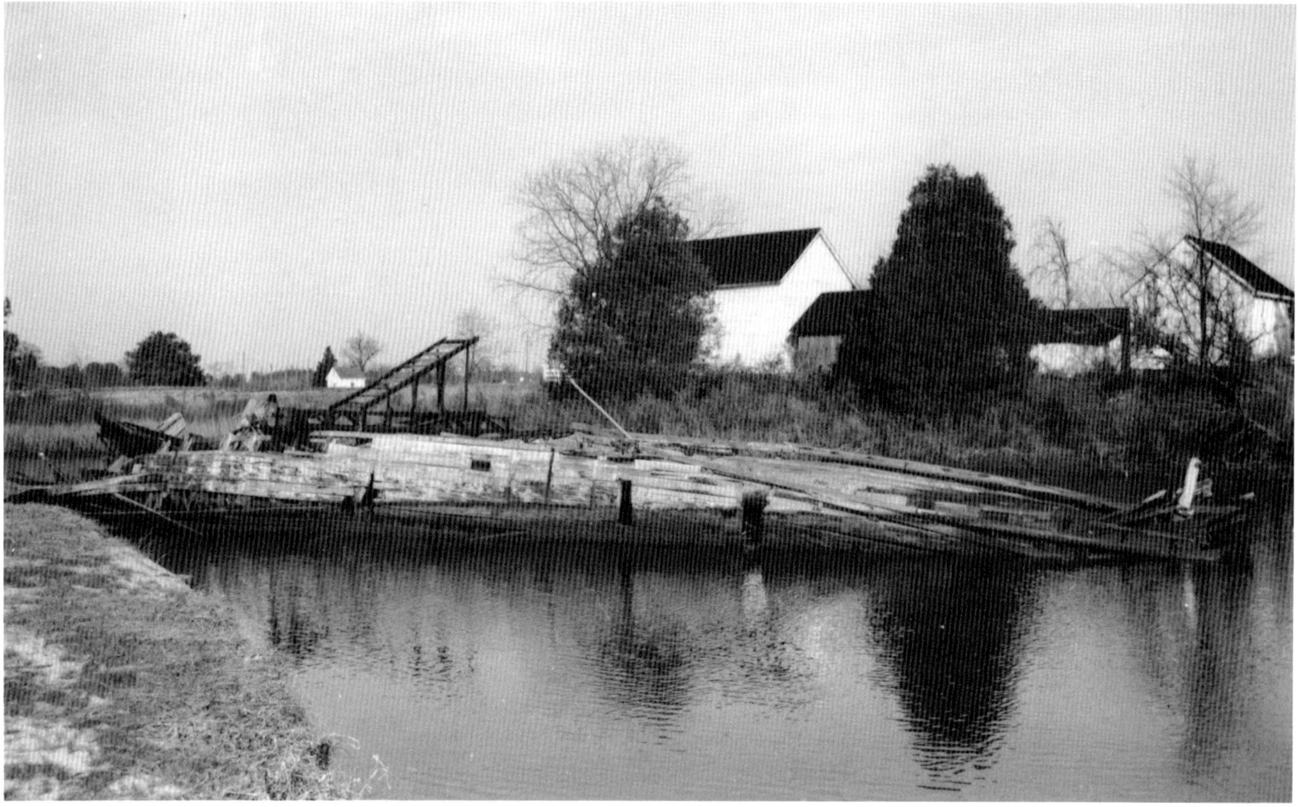

✪ The old schooner *Bohemia* settles in Sarah Creek. Taken on December 1, 1966.

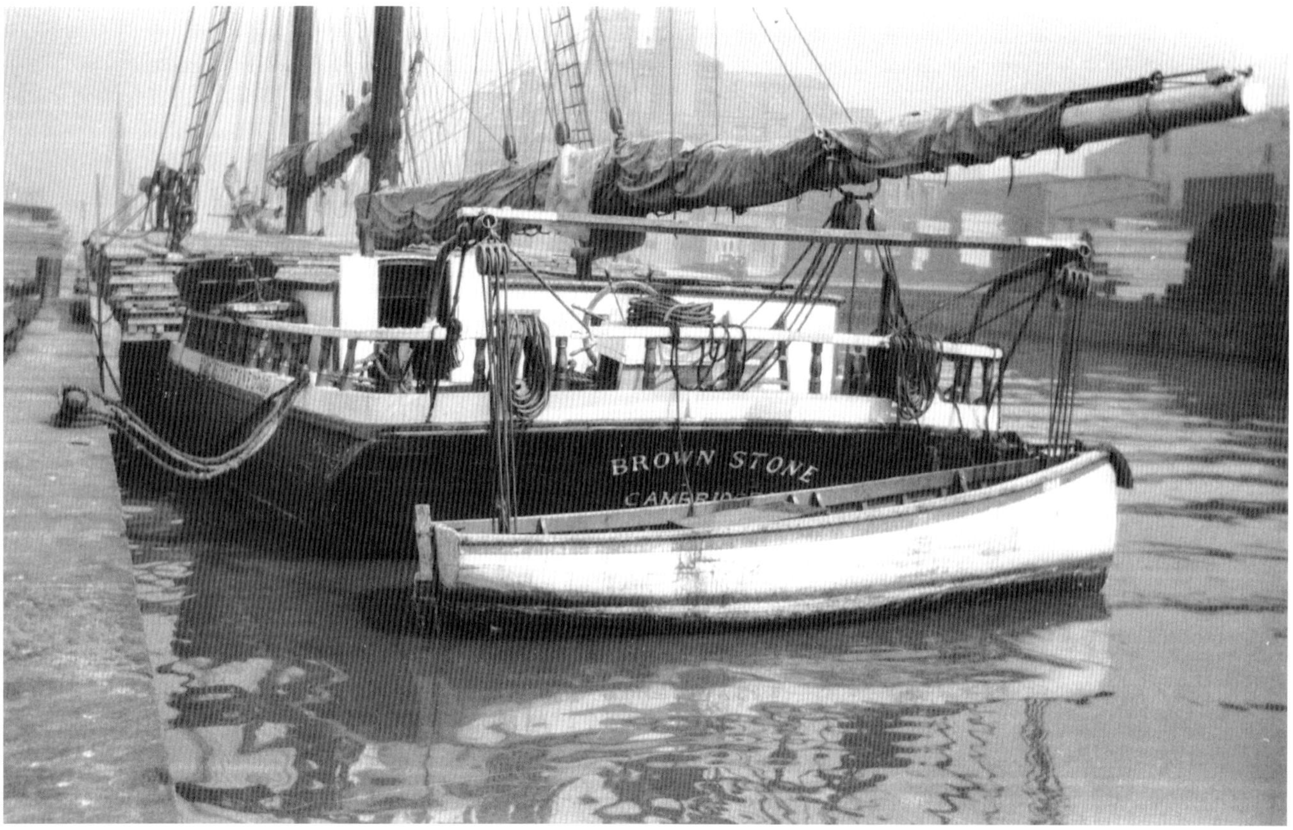

Burdened with a full hold and deck load of lumber, the *Brownstone* sits low in Jones Falls, alongside Pier 6, Pratt Street, Baltimore. The old schooner has covers on her sails, an unusual procedure on Bay vessels in this period, 1935. This schooner was built in 1869 at Hartford, Connecticut, as the *Fannie G. Warner*. Later names were *Gill and Baird, Manchester and Hudson,* and finally *Brownstone,* so named because of her cargoes of brownstone used to construct buildings in New York City. She came to the Bay about 1920.

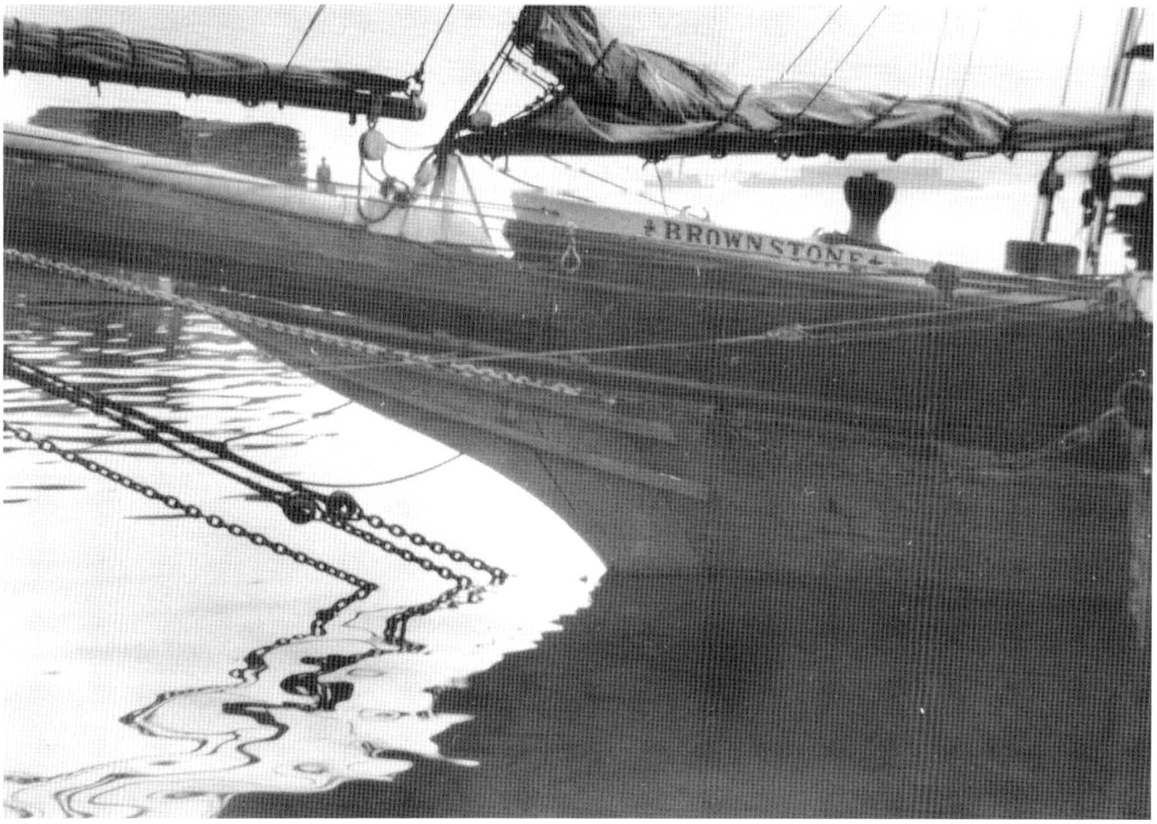

The bow of the *Brownstone,* at Baltimore on November 11, 1935, has a strong, swift appearance. Note the background vignettes: a lone figure on the dock and a passing steam tug and barge.

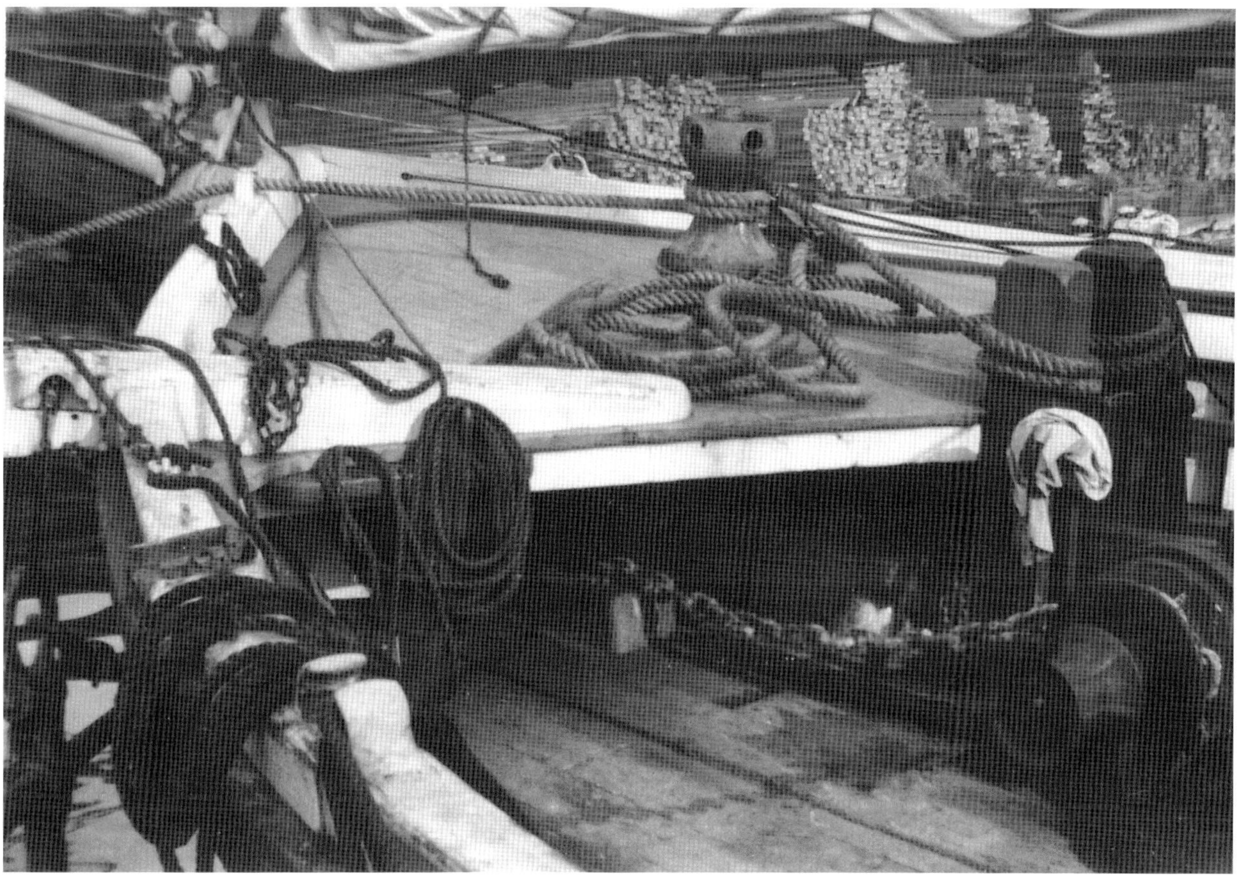

The raised foredeck of the *Brownstone* is fitted with a vertical capstan in addition to her horizontal windlass/winch.

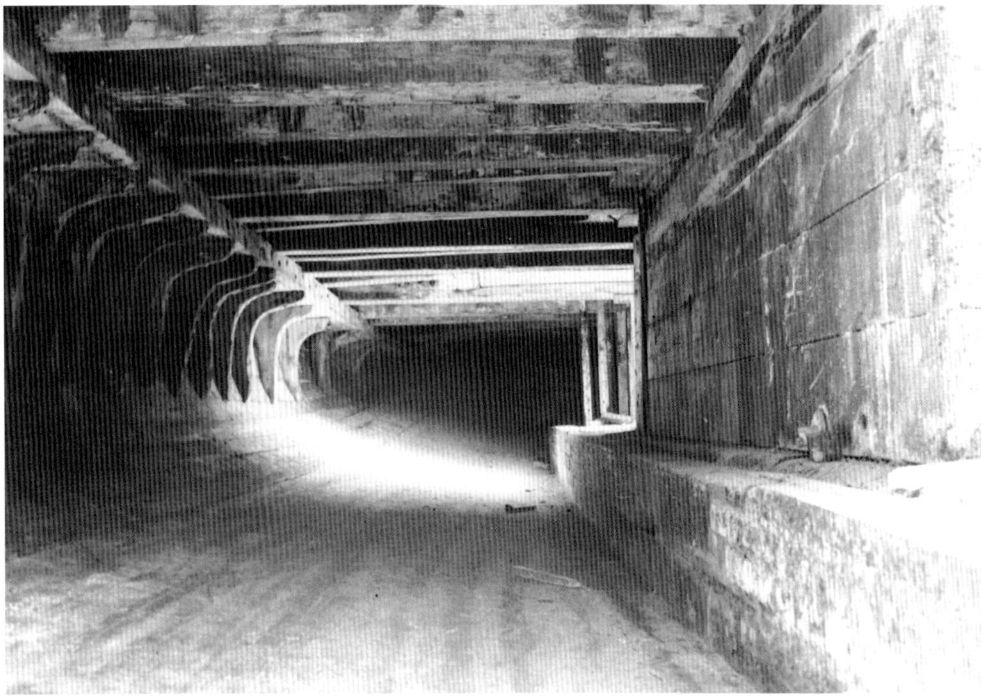

Looking aft in the hold of the *Brownstone,* May 15, 1936. The centerboard well is at right; natural-growth knees, supporting the deck beams, are at left. The hold has been swept clean after the lumber has been discharged, and she may carry fertilizer or cement, which must be kept dry, or coal, gravel, or slag as a down-the-Bay payload. Or she might have to go light to her destination to take on another load of lumber. *MVUS* calls for a crew of five for this schooner, but a captain and deckhand, who also served as cook, could easily handle her on the Chesapeake. A gasoline donkey engine did the heavy lifting.

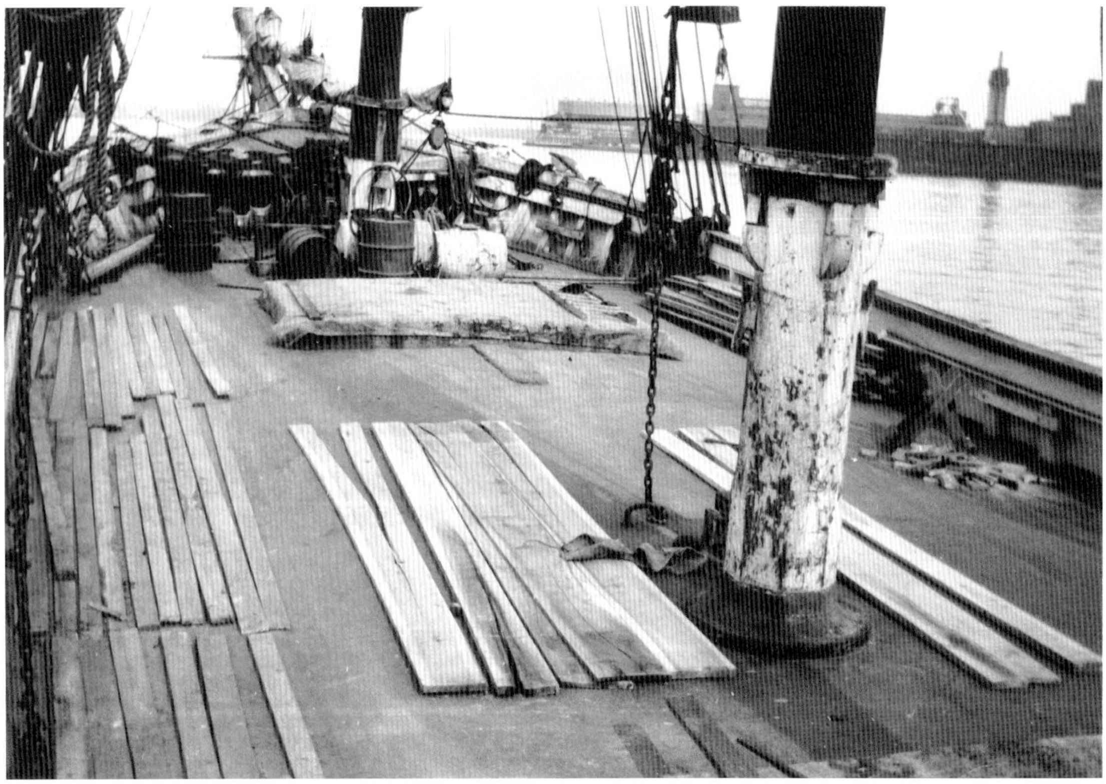

Looking forward on the wide deck of the *Brownstone* at Baltimore, May 12, 1934. Unlike most of the Bay-built schooners, this vessel had a raised foredeck and capstan. A large, seagoing schooner, she coasted in her early days before coming to the Bay. *Brownstone* foundered on Black Walnut Bar, at the southern end of Tilghman Island, Maryland, on February 15, 1943.

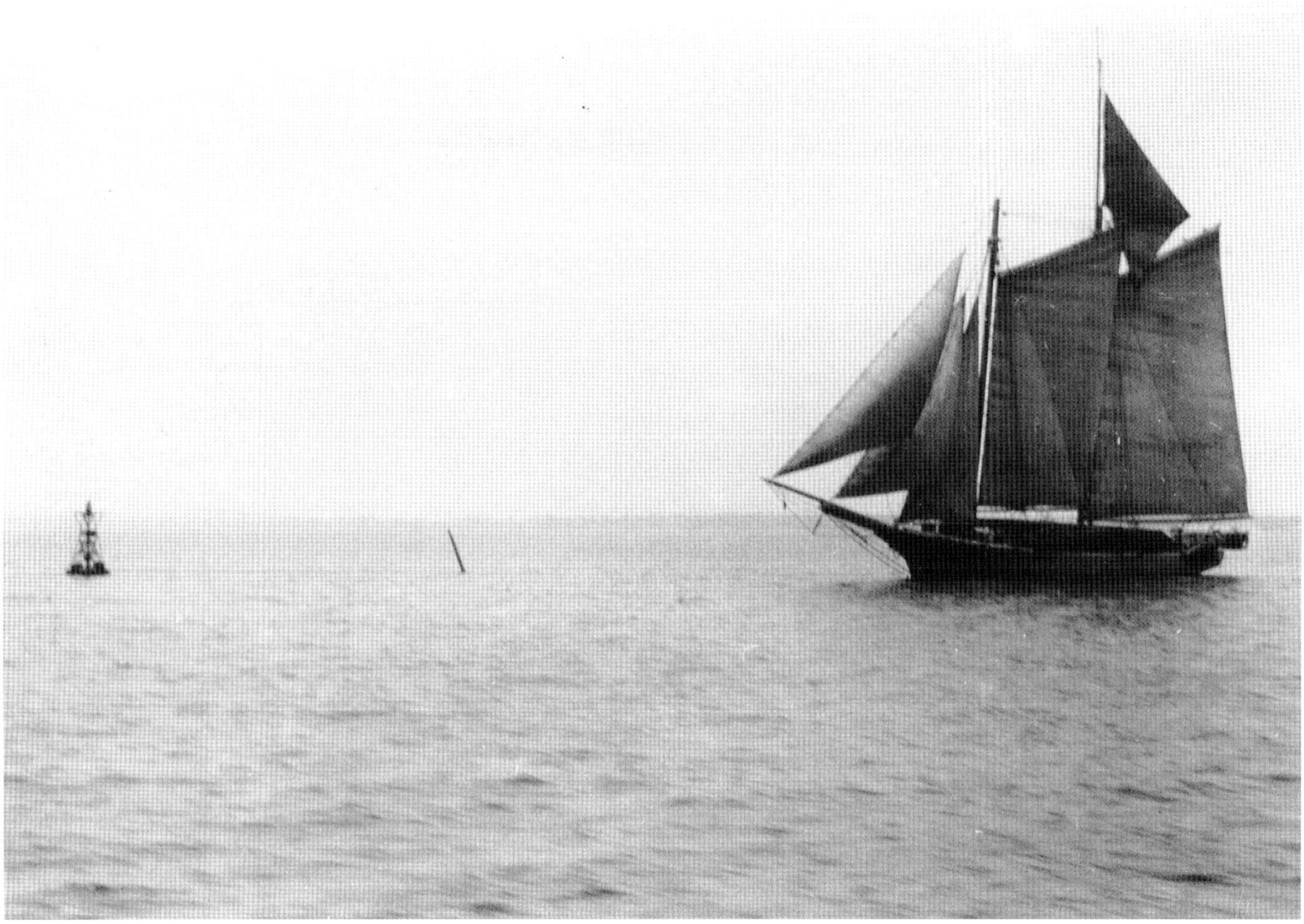

This is possibly the last time the schooner *Carrie L.* sailed down Chesapeake Bay, on July 22, 1933, and the last photograph taken of her as a seaworthy vessel. A month later she was cast ashore at Sharps Wharf, Rappahannock River, Virginia, during a severe hurricane and became a total loss. Heavy canvas tarpaulins cover her high deck load of what probably consists of cases of empty cans—a bulky cargo but not a heavy one. *Carrie L.* had been built in Church Creek, Maryland, in 1882.

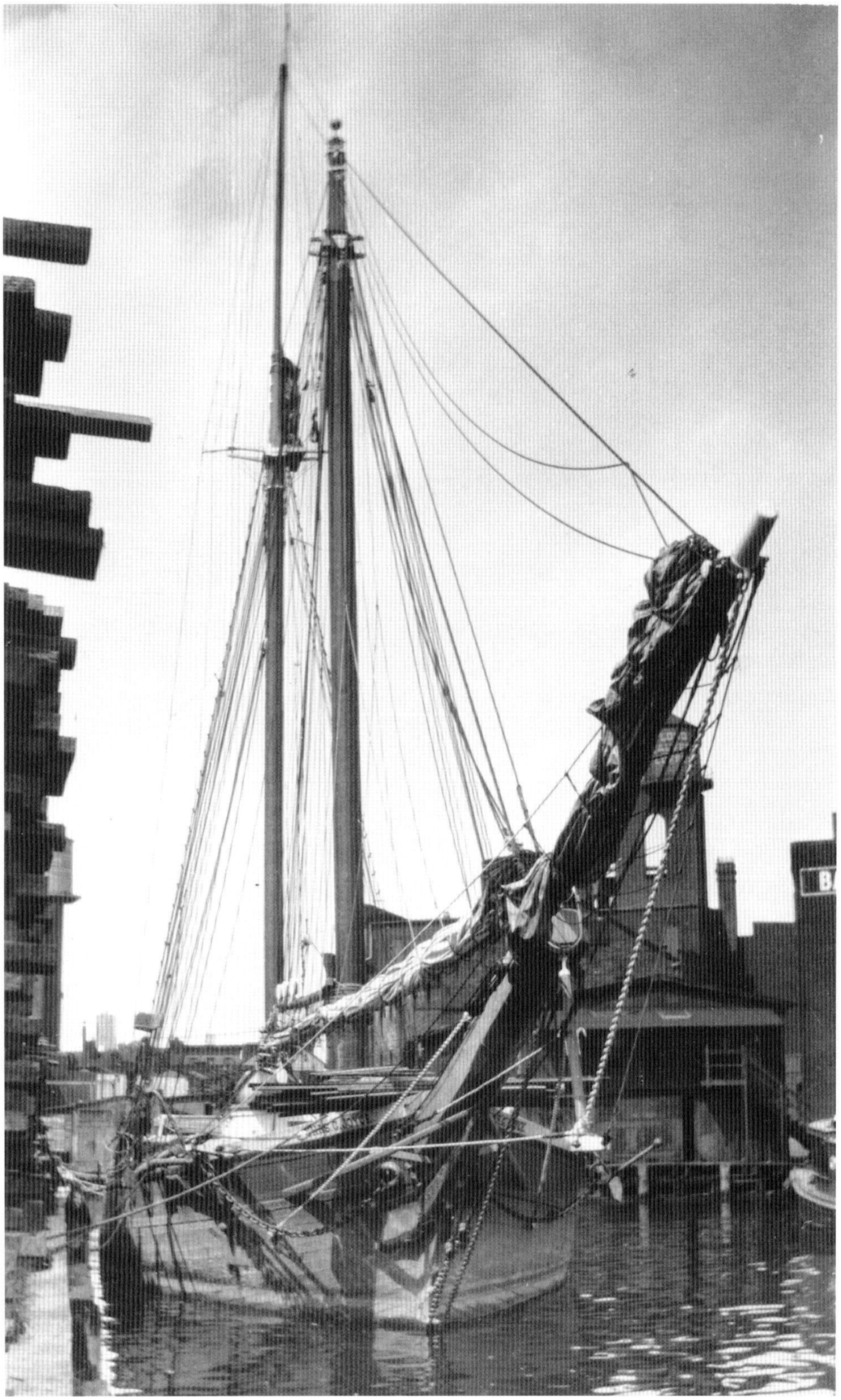

Charles G. Joyce at the Philpot Street lumber piers, Baltimore, August 12, 1933. Built in 1882 in Baltimore by William E. Woodall, this vessel had but two owners during her 61-year career on the Bay. She was jointly owned by Woodall and her master, Captain Thomas W. Wheatley, from the time of her building until 1913. During that time Captain Wheatley had been her only master except for two trips when the schooner went to the Bahamas for pineapples. In 1913 she came under the ownership of C.C. Paul and Company, which employed her as a Bay freighter, chiefly in the lumber trade. That firm sold her about 1943 and she was taken to sea by her new owner. On May 13, 1944, she foundered at the McArthur Causeway Docks, Miami, Florida.

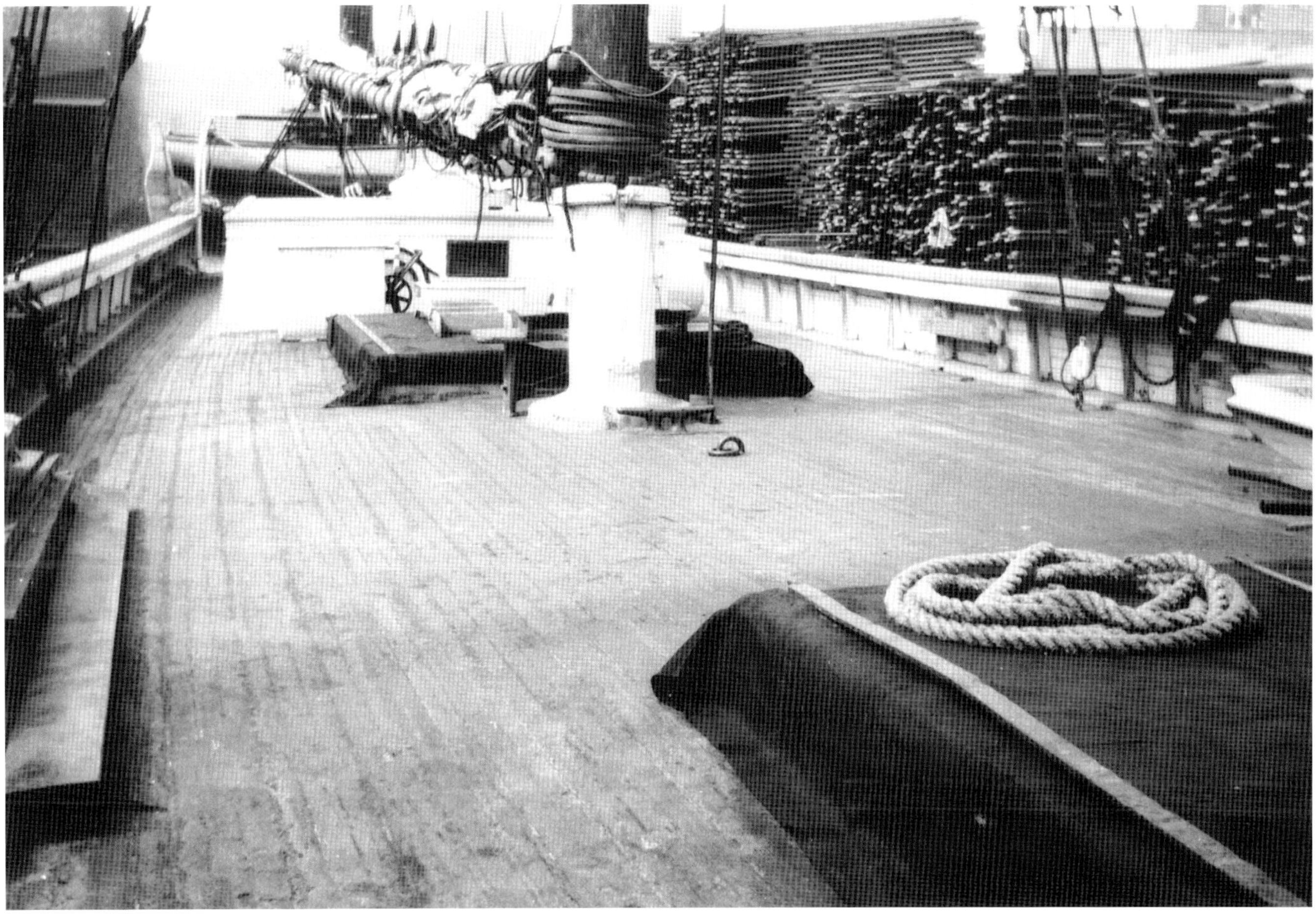

Looking aft on the deck of the *Joyce* in Baltimore on June 19, 1934. During her entire career on the Bay this schooner was properly maintained and was in excellent condition when she left the Chesapeake. The iron rod to port of the mainmast is connected to her centerboard and is controlled by a block and fall attached to the head of that mast. Forward of the after cabin is a gasoline engine and pump.

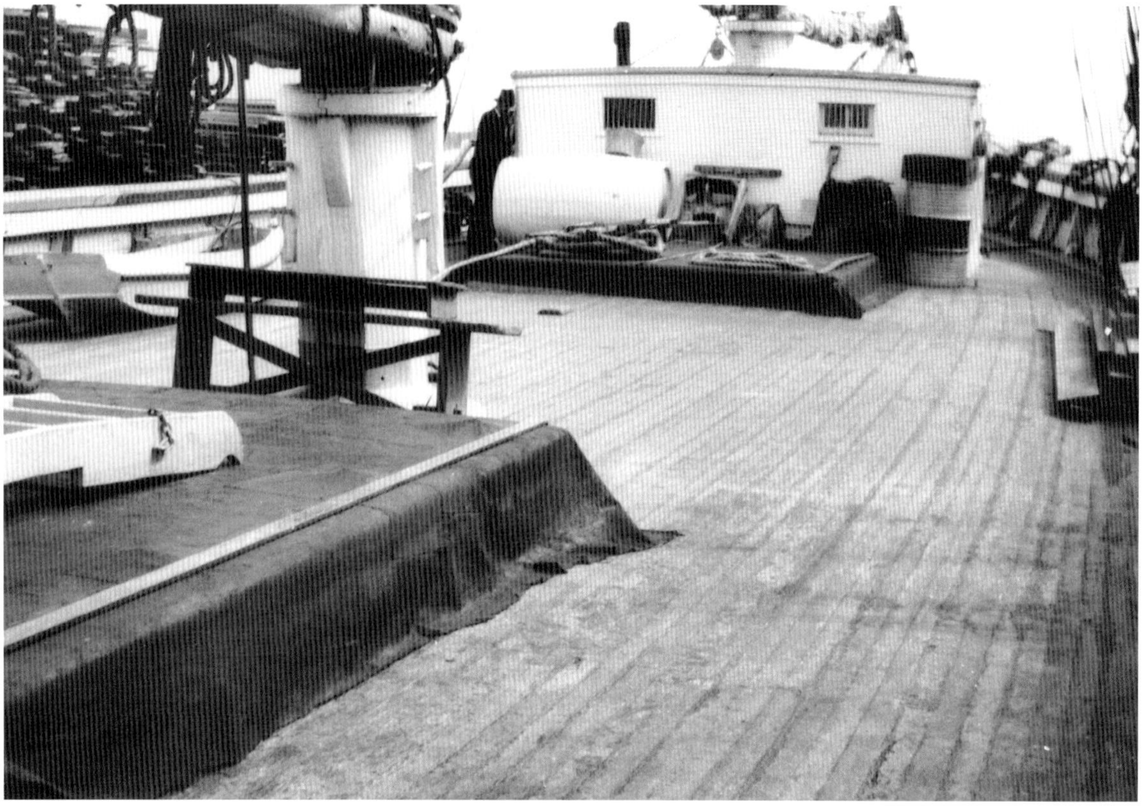

✪ Looking forward on the *Charles G. Joyce,* also on June 19, 1934. Her centerboard rod and mainmast are closer in this view.

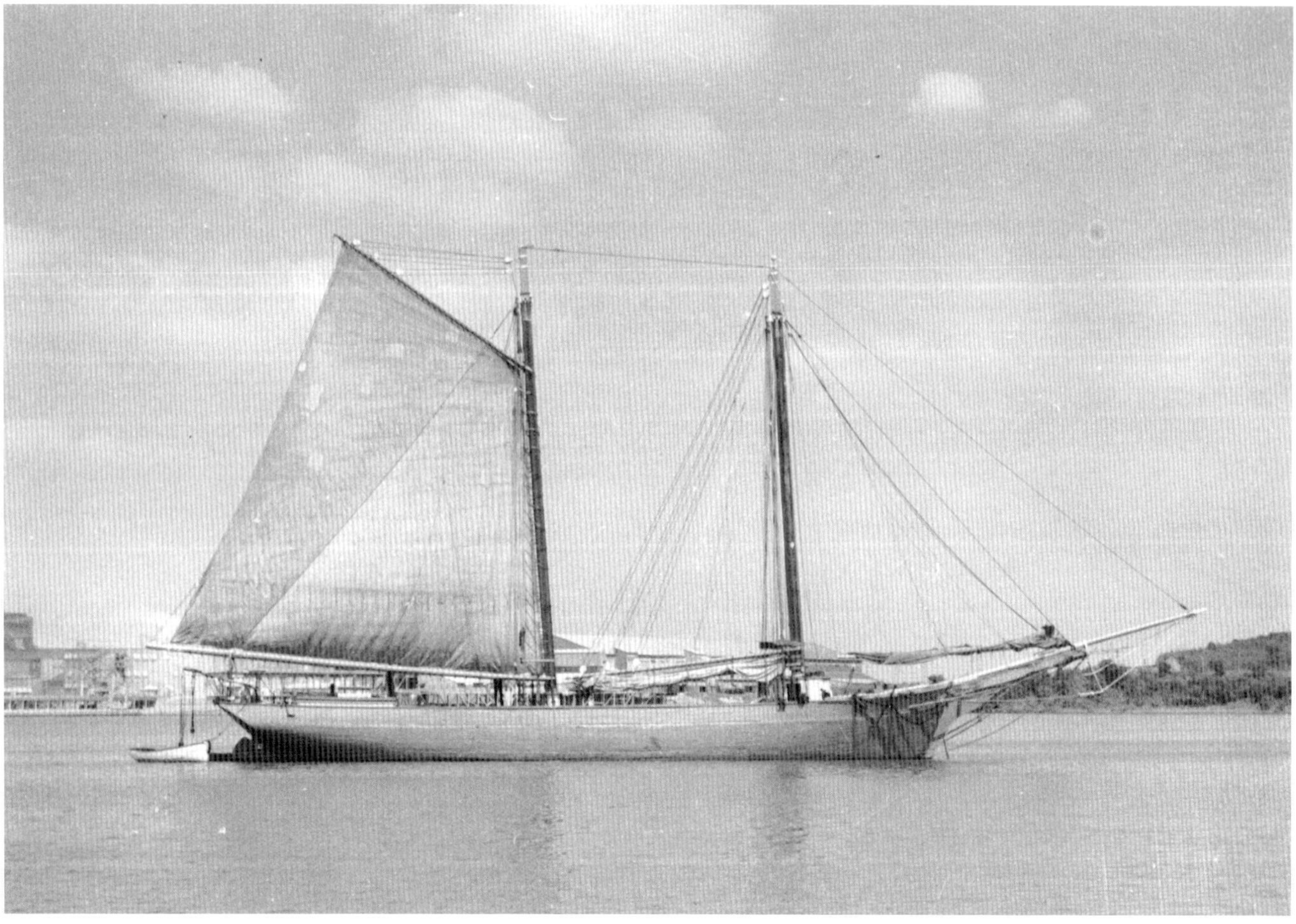

✪ The schooner *Charles G. Joyce* at anchor in June 1940. She presents a fine appearance.

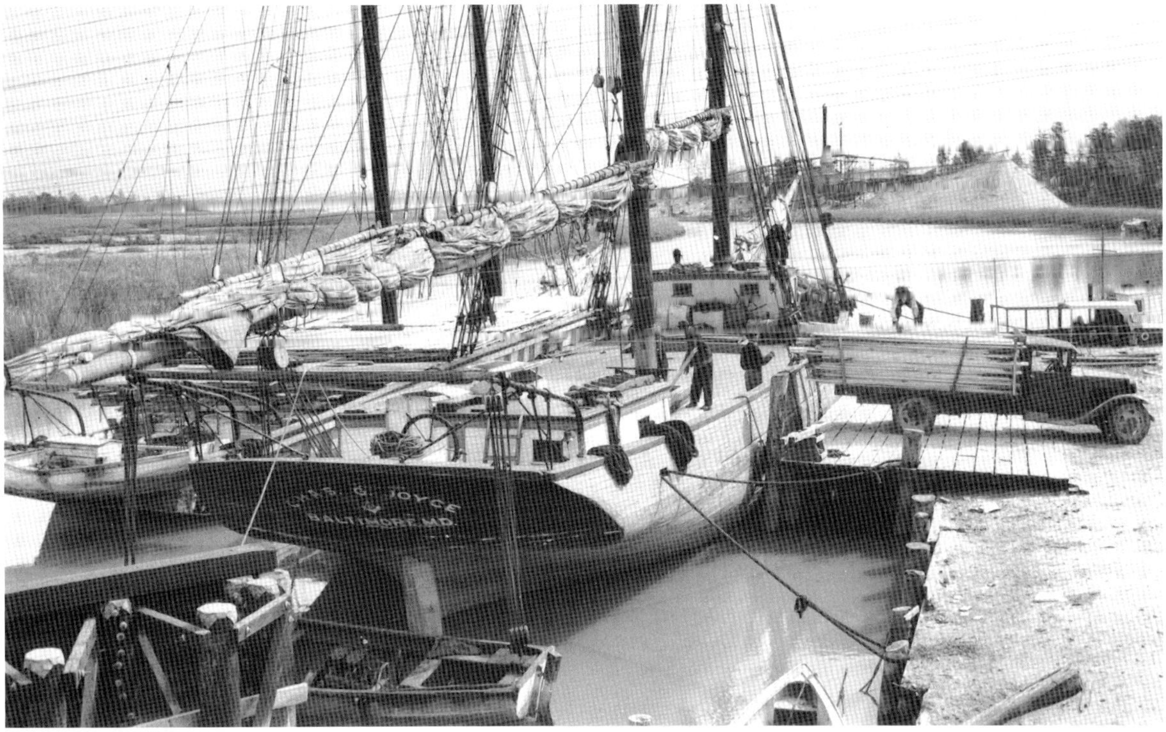

On April 23, 1941, the *Charles G. Joyce* loads lumber in Hoskins Creek, Tappahannock, Virginia. The lumber has come by truck from the mill in the background. Three men are passing the boards down to several men in the schooner's hold. When that is filled the hatches will be secured and the lumber loaded on deck. Note how the lumber has been piled high on the schooner to her port, the *Sarah C. Conway.*

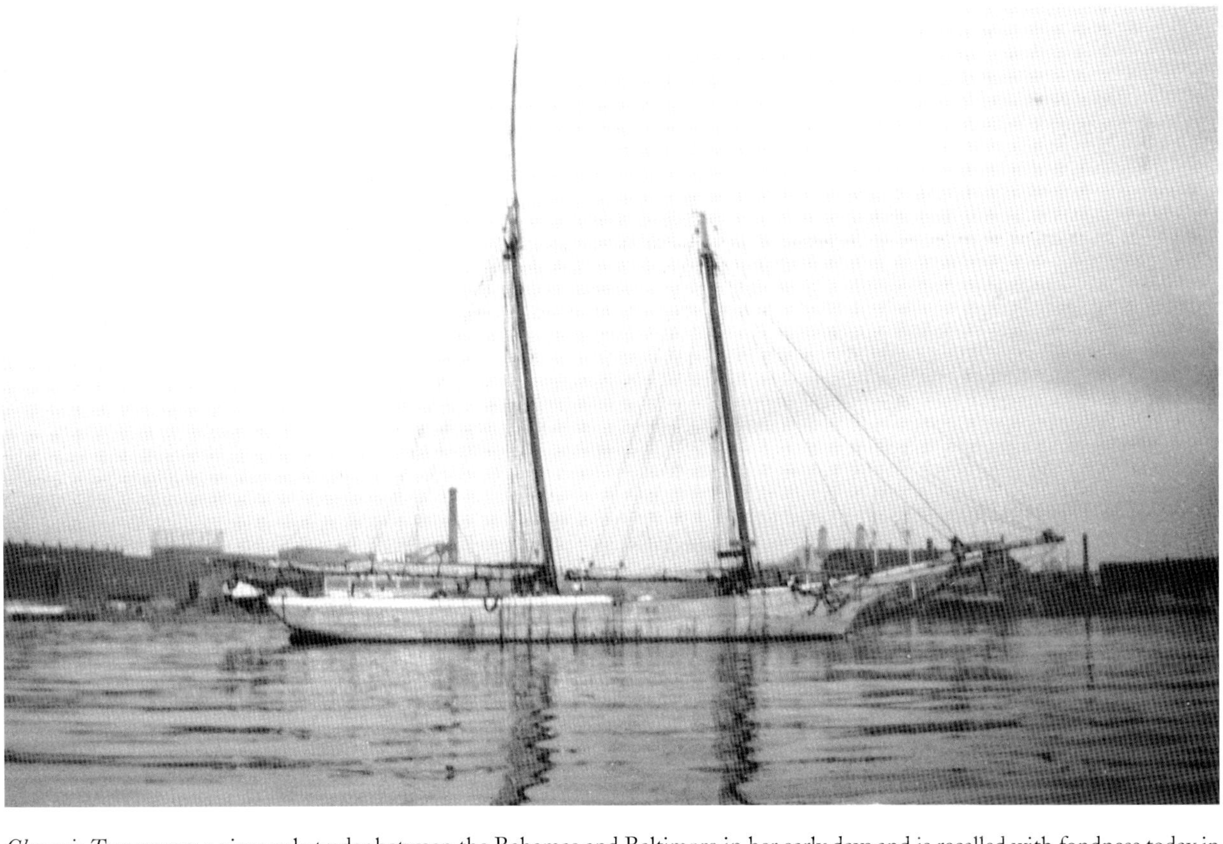

Clemmie Travers was a pineapple trader between the Bahamas and Baltimore in her early days and is recalled with fondness today in Mathews County, Virginia, where she was built in 1885. On Christmas Eve, 1933, she rides at anchor in Canton Hollow, Baltimore, her crew considering themselves fortunate to be in port at that time. The *Travers* stranded at Norfolk on September 18, 1936, ending a career of slightly more than half a century.

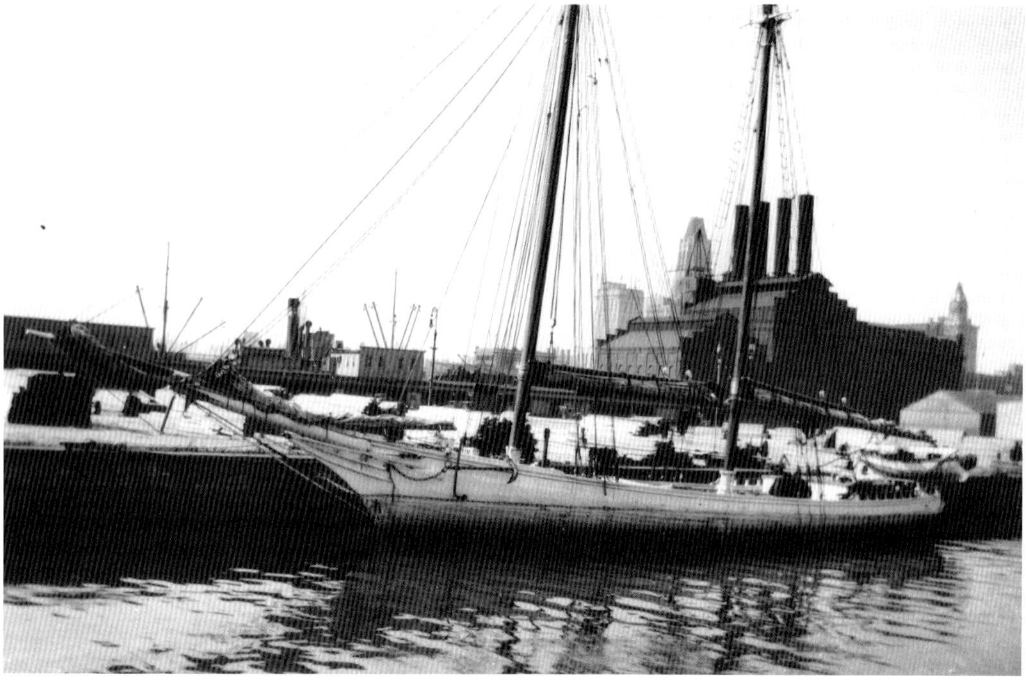

With acres of lumber piled on Pier 6, Baltimore, behind her, part of which she just discharged, the *Columbia F.C.* rides light on October 19, 1933. Baltimore's skyscrapers, the Baltimore Trust Building, First National Bank Building, and the Hearst Tower, loom behind the four-stacked Pratt Street power station. The steamer in left background is the Bull Line's *Frances. Columbia F.C.,* the F.C. meaning Fishing Club, was built at Mundy Point, Virginia, in 1874 and she was owned there by the Courtney family who sold her about 1946 to Captain Tom Henry Ruark of Deltaville. He continued freighting her on the Bay until selling her to Cuban interests in 1948.

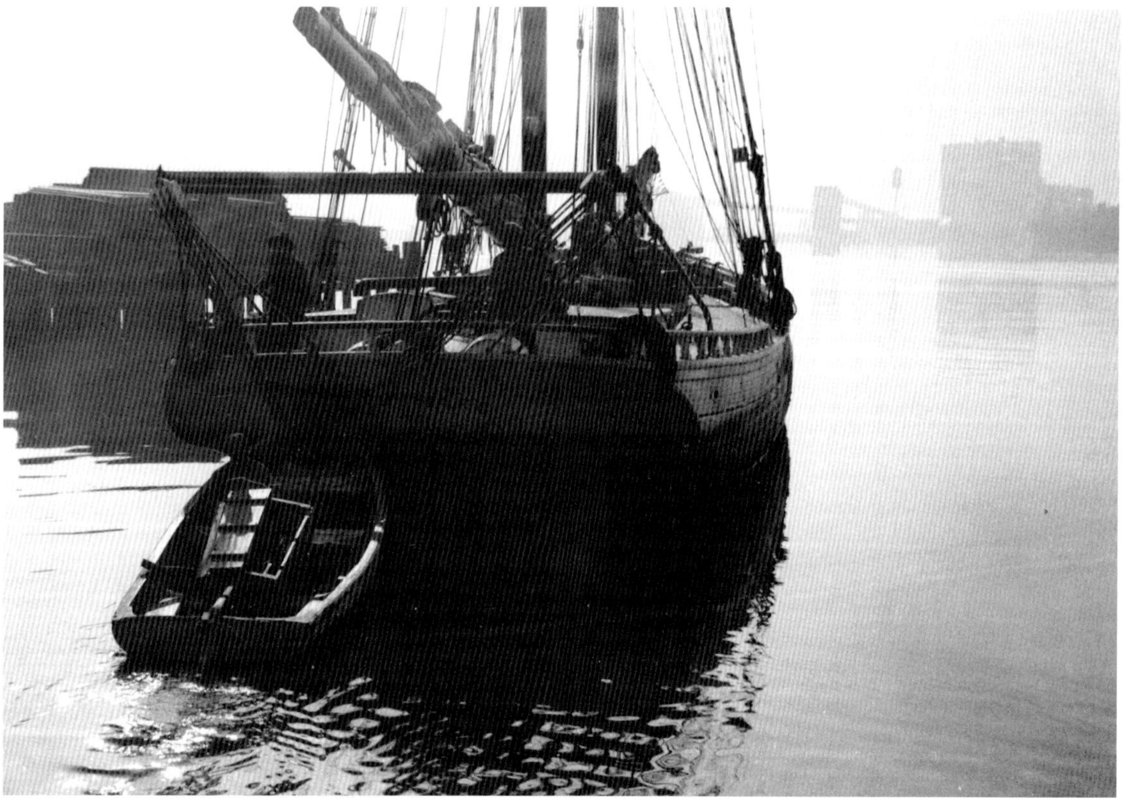

On the calm morning of January 21, 1934, the *Columbia F.C.* slips away from her Canton pier, Baltimore, pushed by her yawl boat suspended from the stern davits. She will head down the Chesapeake again to the Coan or Yeocomico Rivers, her home waters in Virginia, for another lumber cargo. The captain steers as a deckhand coils the mooring line on deck. Another crewman tends the engine in the yawl boat.

Sometimes the motor yawl boat was used to pull rather than push Bay sailing craft. A line would be secured to the martingale stays and then fastened to a ring bolt at the bow of the yawl. In this fashion the *Columbia F.C.* moves away from Coan Wharf, Virginia, in March 1939.

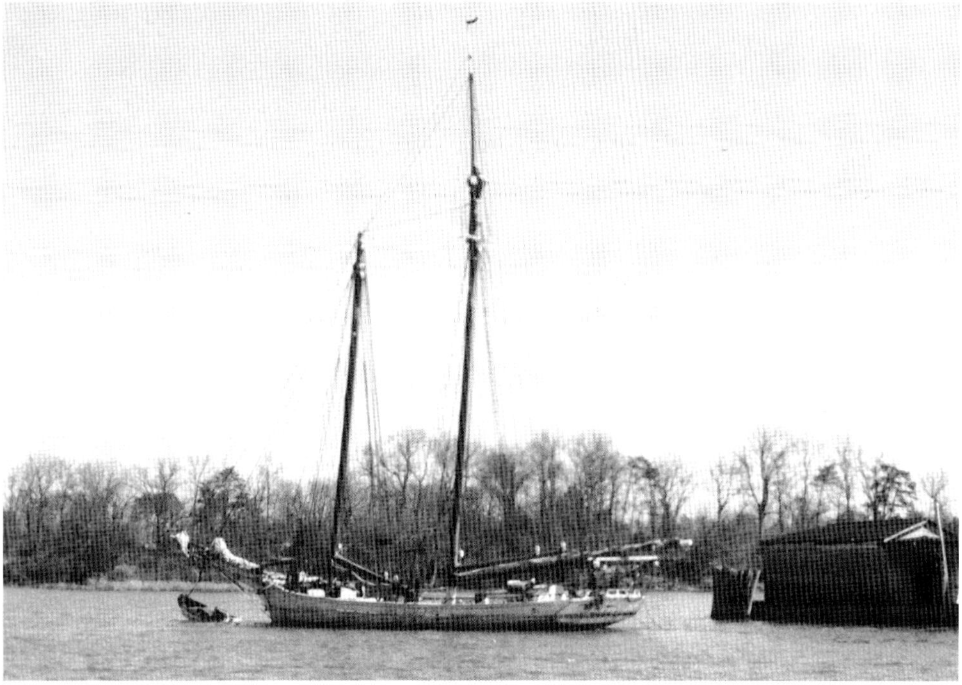

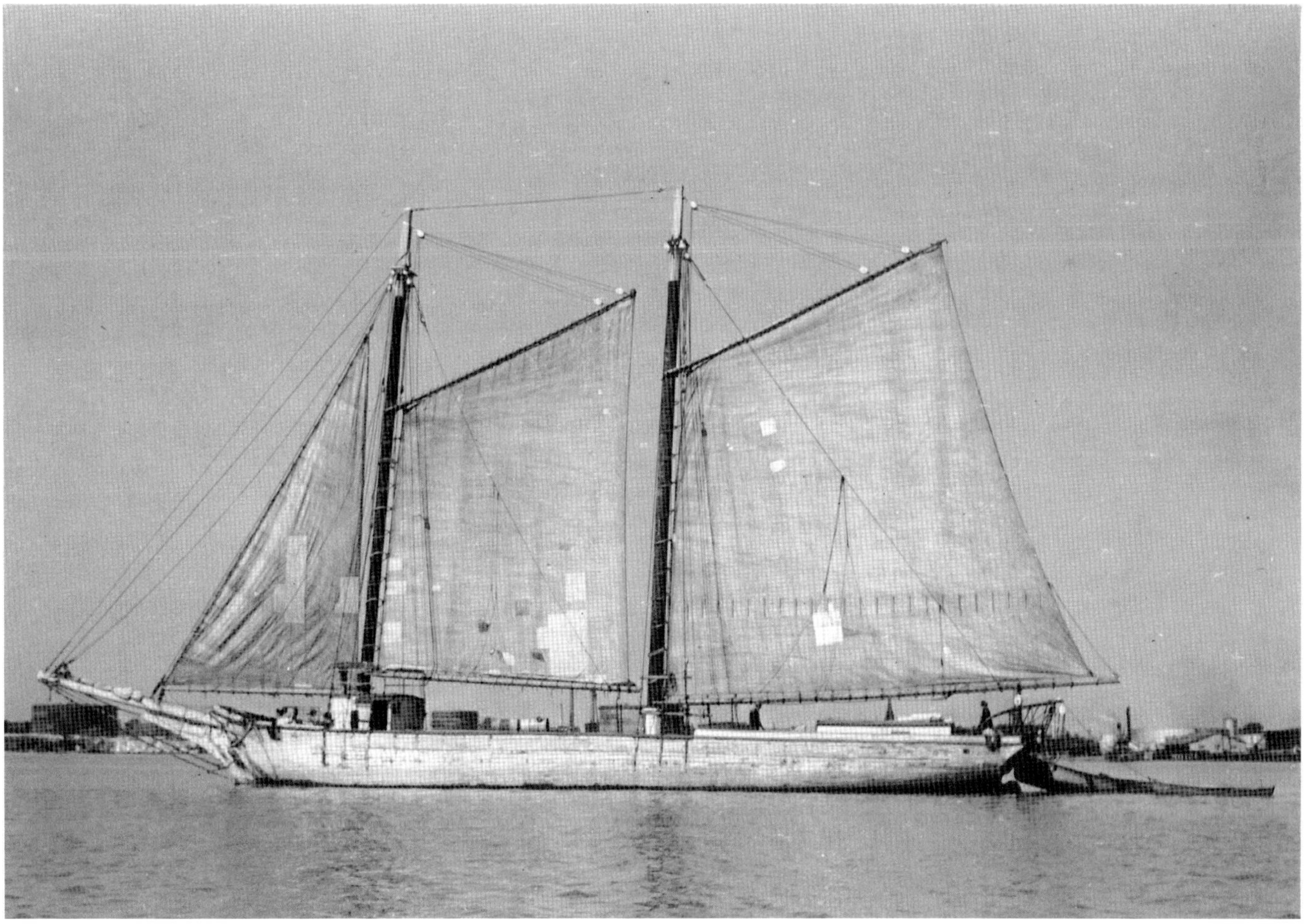

How the *D.J. Ward* managed to get around Chesapeake Bay as well as she did in her last days remains a mystery. Her hull was weary and hogged, her sails patched, and she was manned only by aged Captain Brinsfield and his wife, the latter at the wheel as the vessel moves up Curtis Bay, Baltimore, in October 1940. Her foresail is reefed even though there is scant wind to move her while the yawl boat keeps way on her. *D.J. Ward* was built as the *Elijah S. Adkins* at Salisbury, Maryland, in 1910. She was listed as "Abandoned" in the 1941 *MVUS,* coming to a final resting place on a mud bank off Mill Creek at Solomons, Maryland.

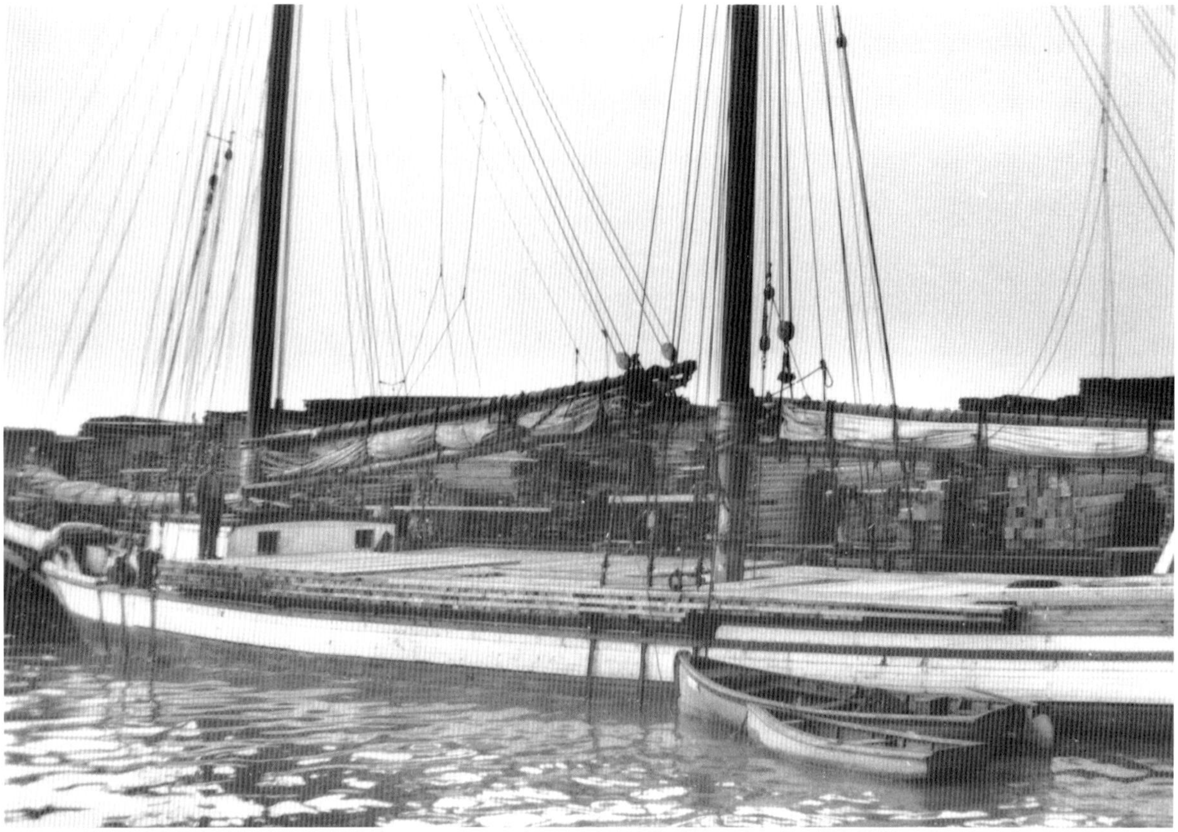

✪ Heavily laden, the *D.J. Ward* waits to unload lumber at Baltimore in the 1930s.

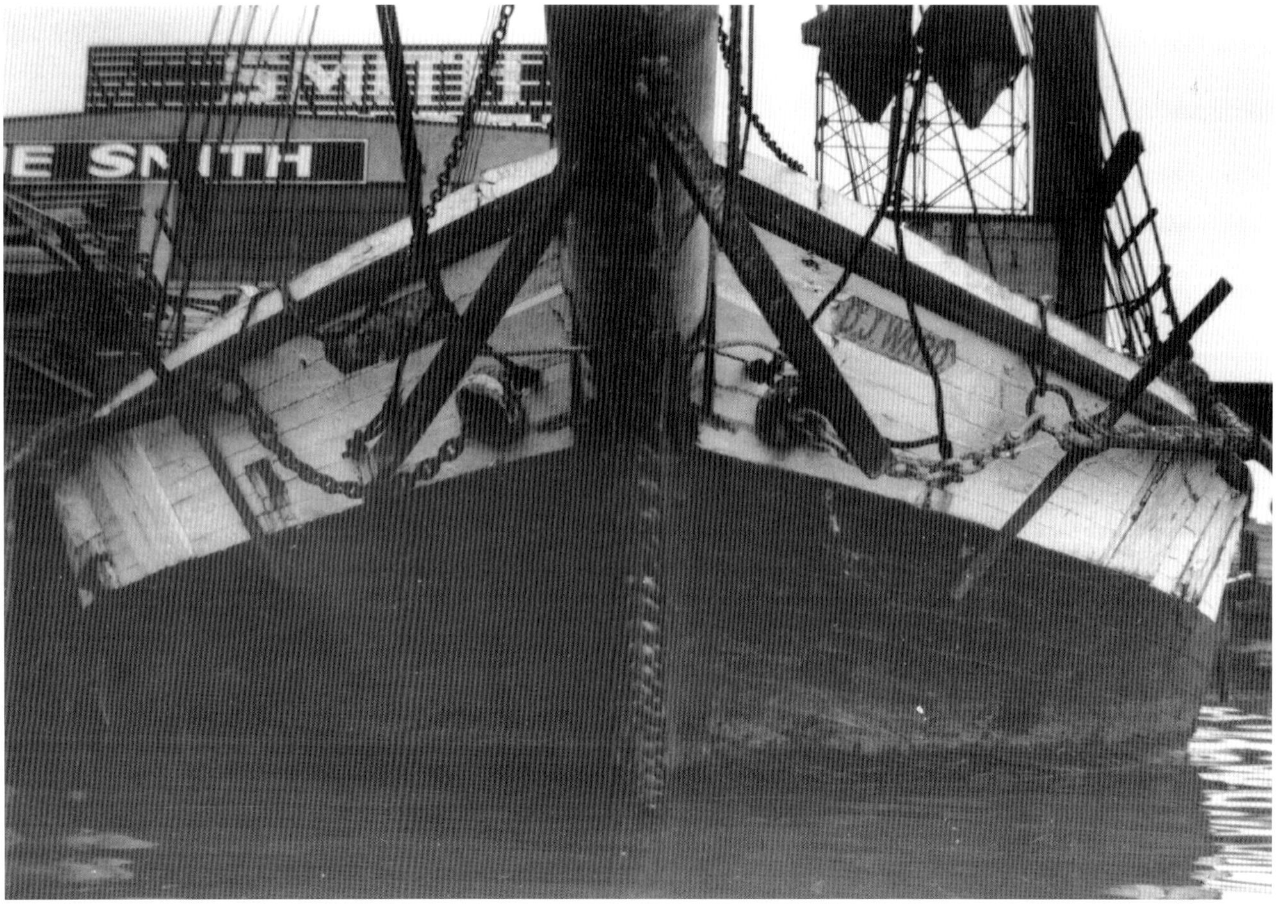

✪ The bow of the *D.J. Ward,* down to her marks with lumber.

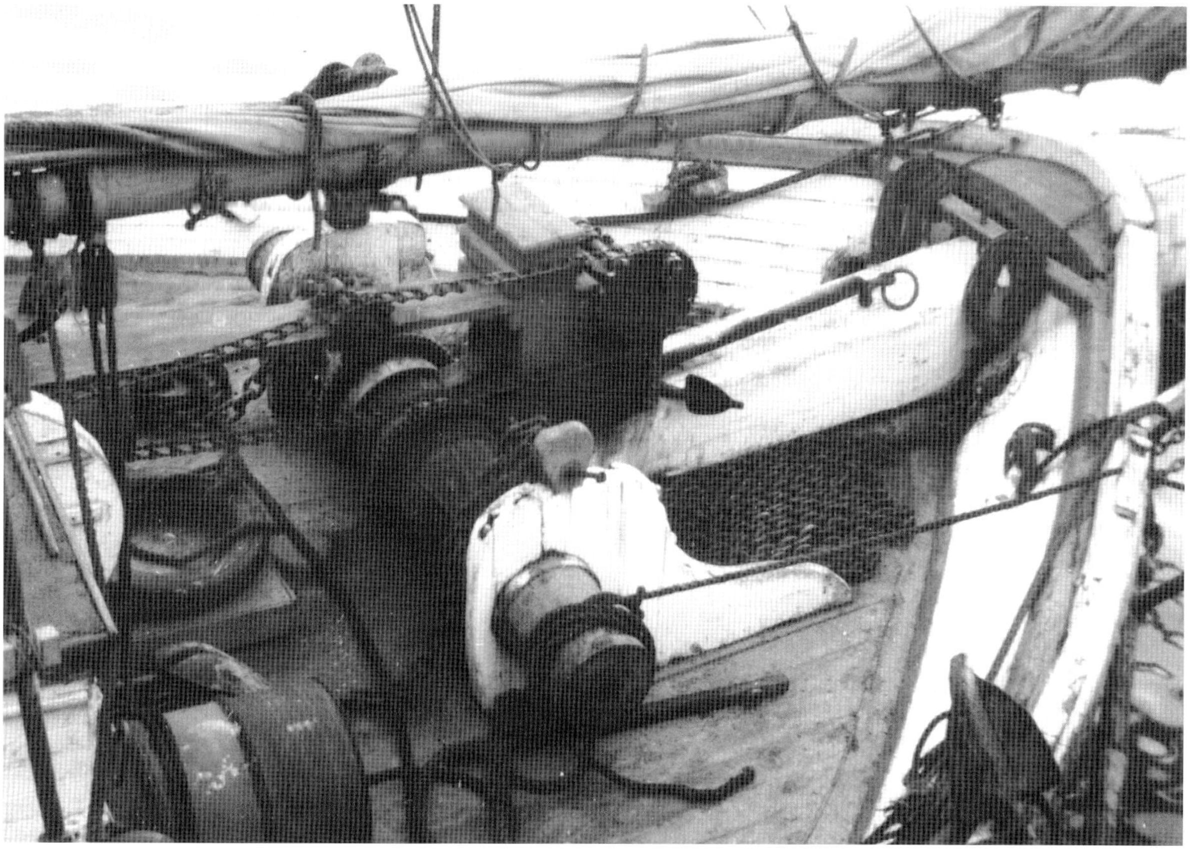

✪ The foredeck of the *Ward*.

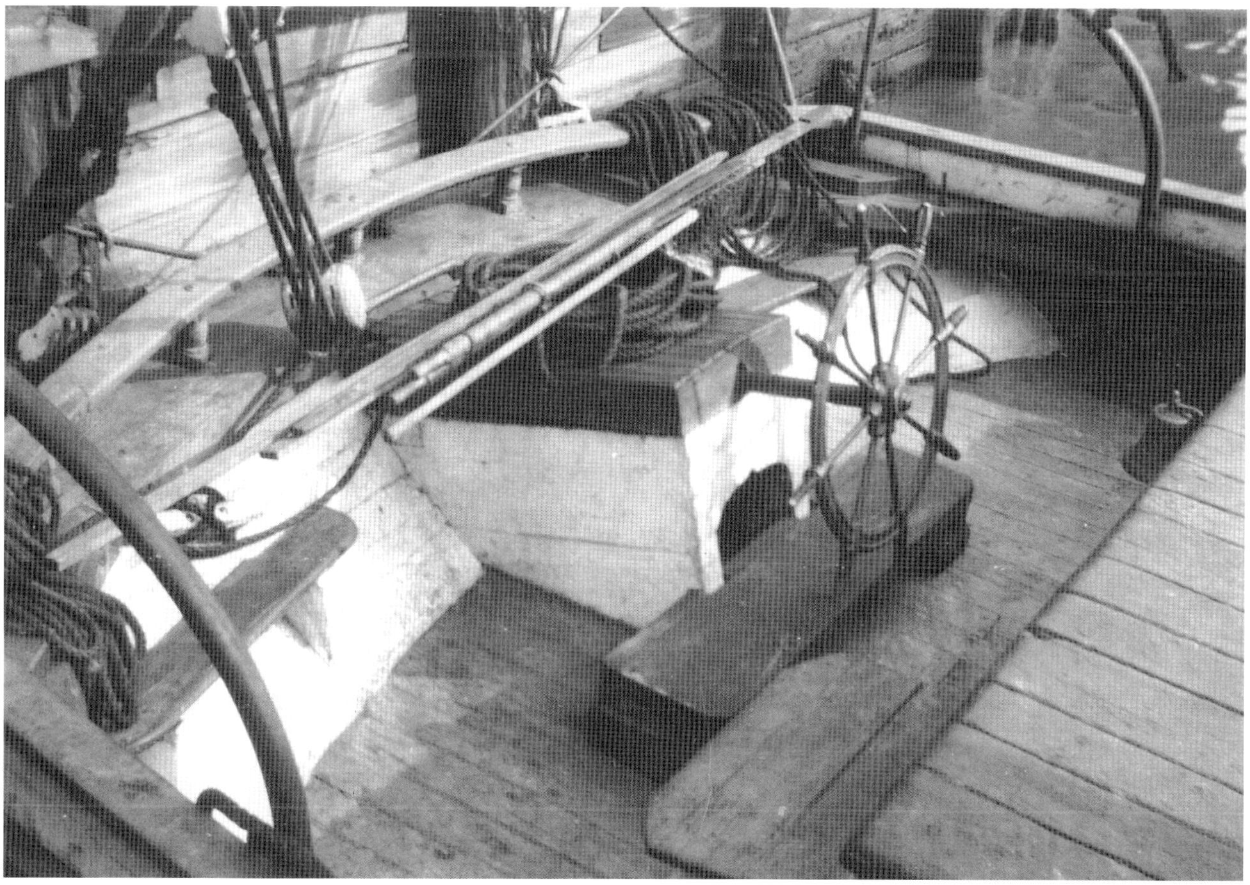

✪ The helm of the *D.J. Ward*, looking aft.

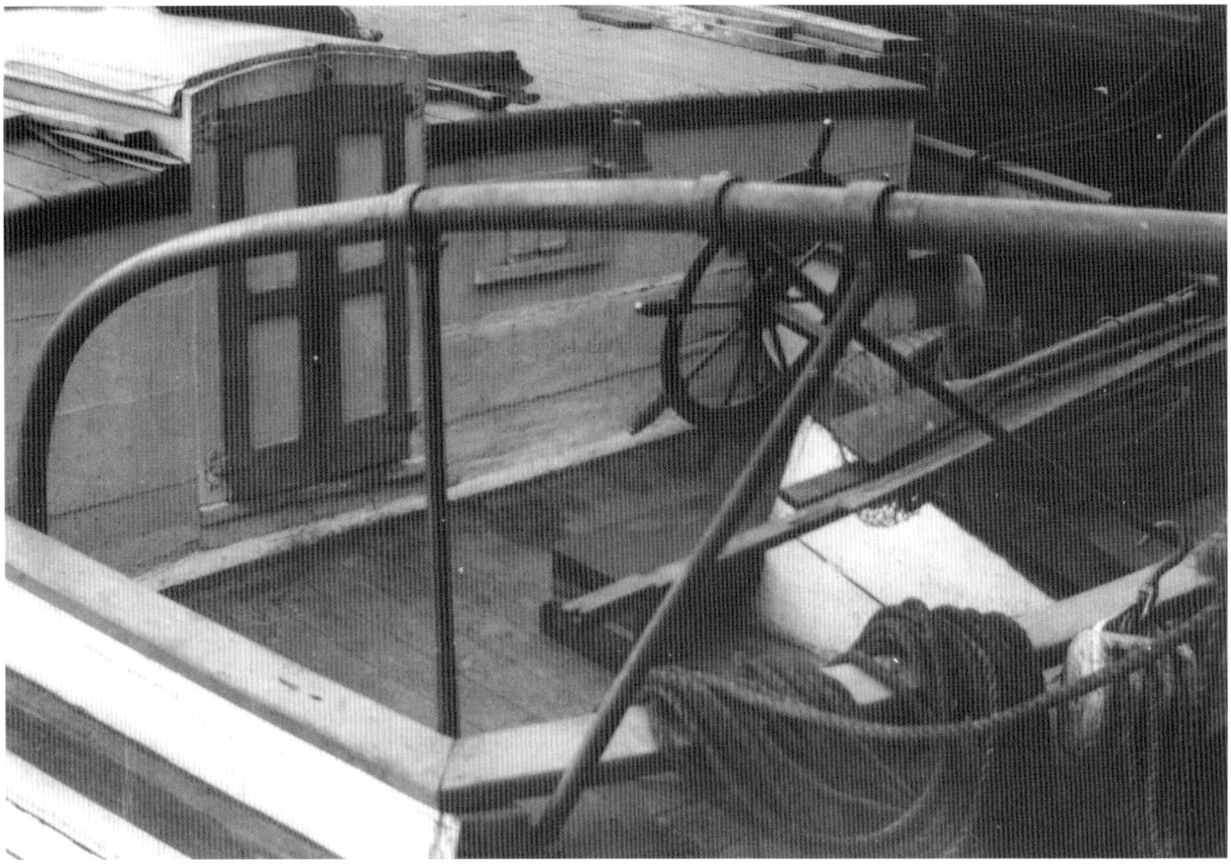

⚙ Looking forward on the *D.J. Ward,* showing her aft house.

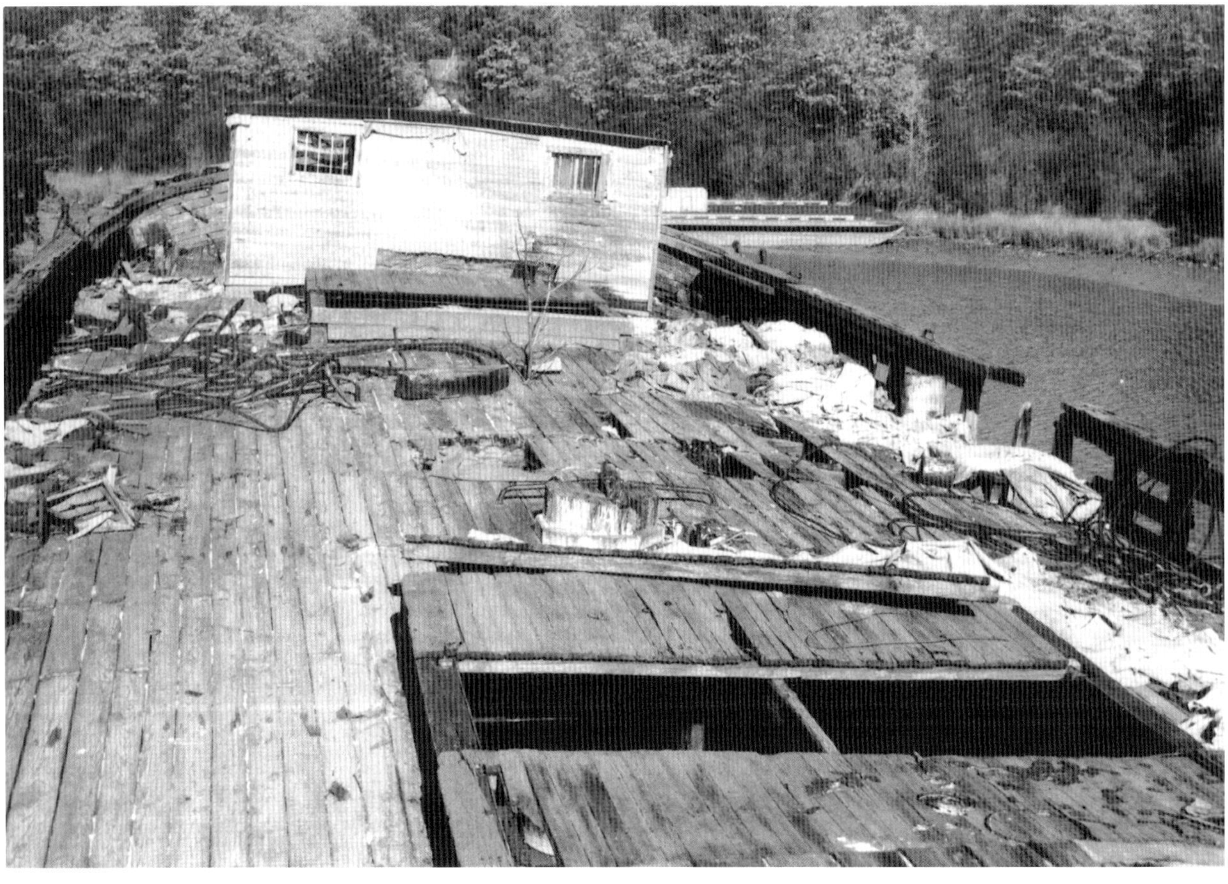

⚙ Hull of the old schooner *D.J. Ward* looking forward, at Solomons, Maryland, in December 1949.

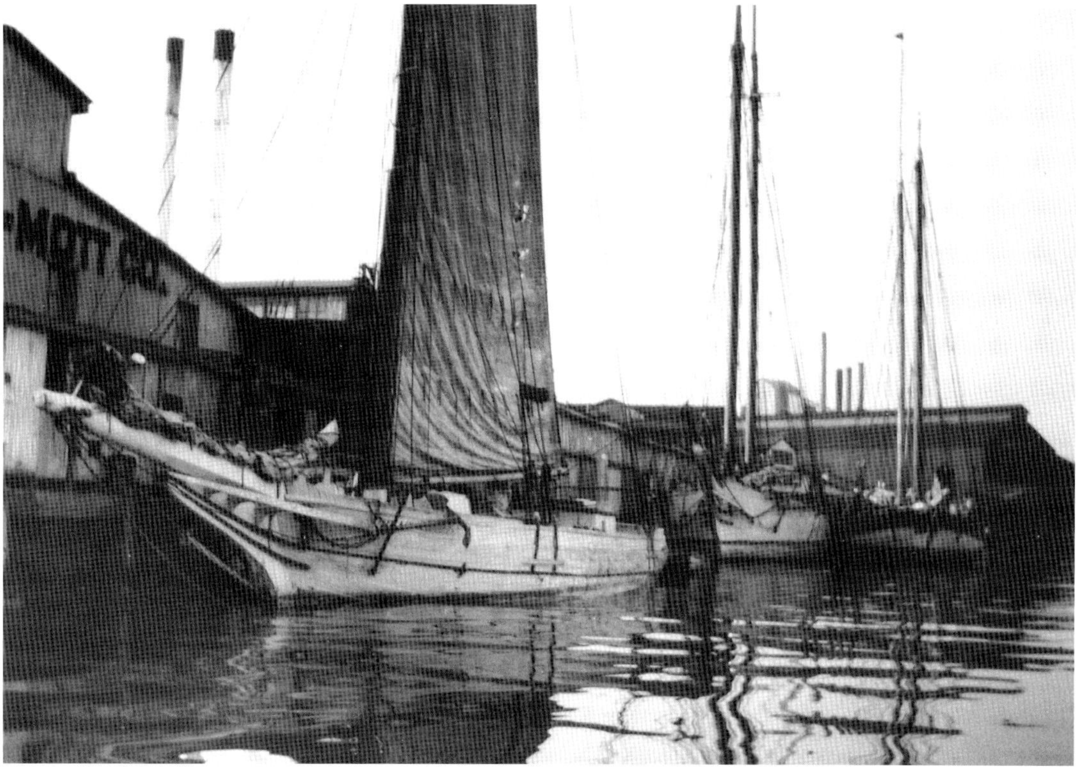

Three Chesapeake sailing veterans, their ages totalling 178 years, line up for their portrait on Christmas Eve, 1933, at the Lord-Mott packing plant in Baltimore. In the background are the schooner *Chesterfield* and pungy *Wave,* right. Left foreground is the *Effie A. Chase,* built in Pocomoke City, Maryland, in 1881. *Chesterfield* had been built in Talbot County, Maryland, in 1878.

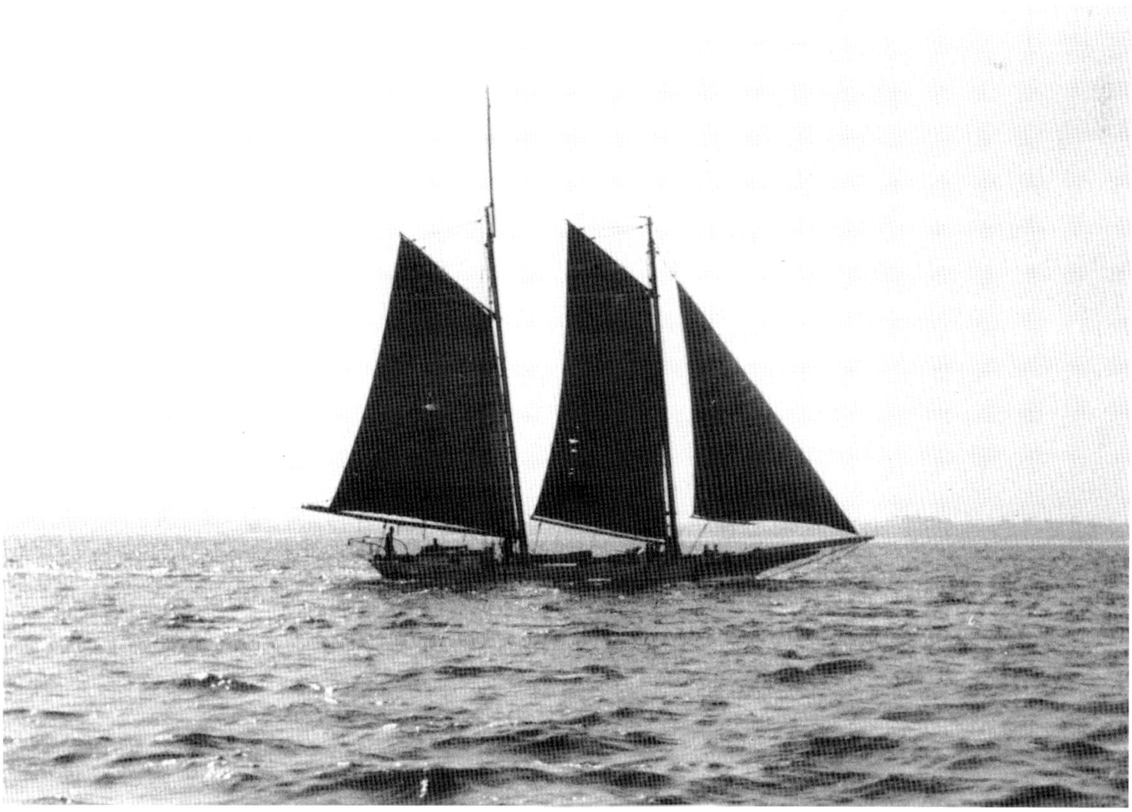

Effie A. Chase sails up Chesapeake Bay on November 17, 1933, with daylight showing through her sails as white spots. Only 53 feet in length, the *Chase* started her career as a sloop but was converted to schooner rig about 1893. She sank off the Arundel Boat Club near Hanover Street Bridge, Baltimore, in 1939 and was dismantled there.

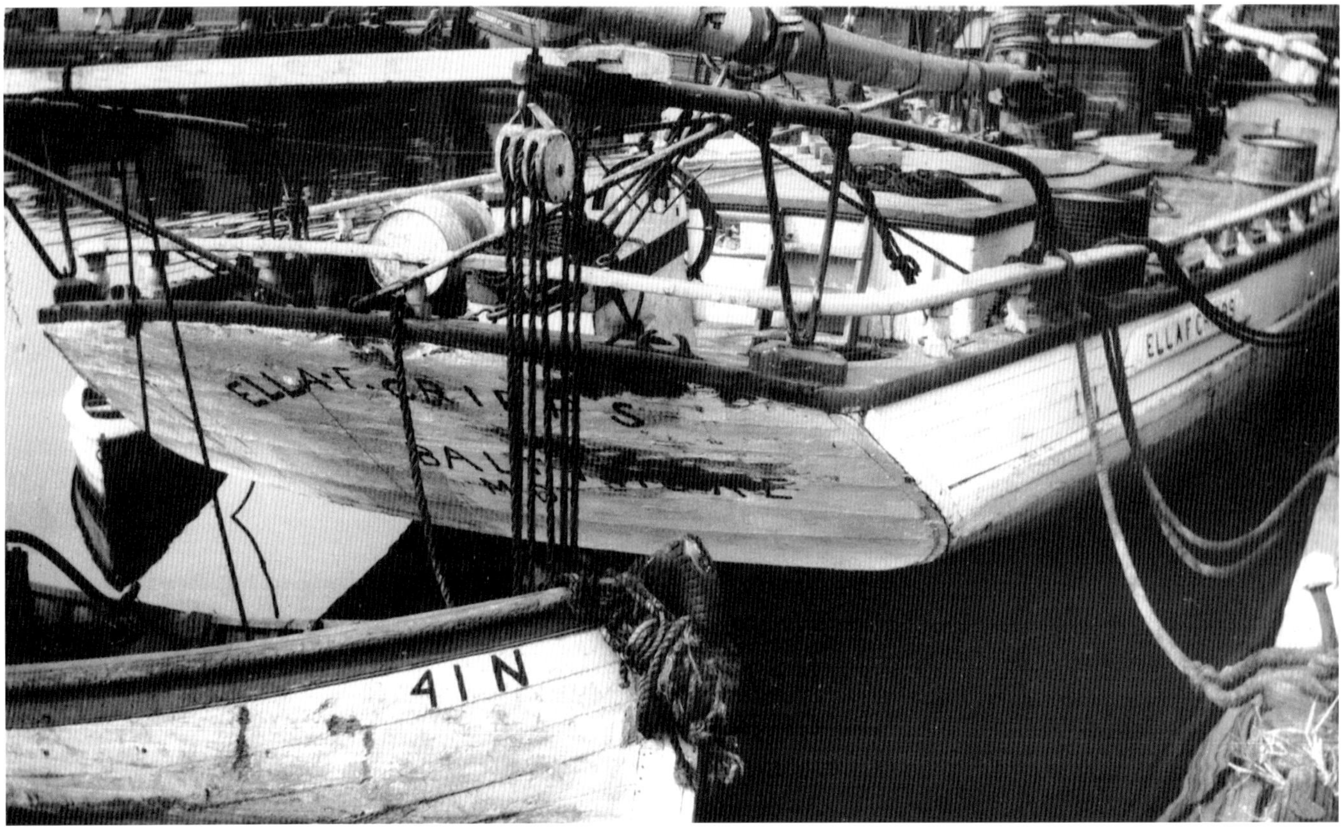

Preparing for the opening of the oyster season, the *Ella F. Cripps* ties up at Montford Yard, Baltimore, on October 27, 1935. She had been built in 1900 by Kirby & Sons at St. Michaels, Maryland, and after conversion to power in the 1940s and sent north, she returned to the Bay about ten years later as her purchaser planned to re-rig her as a schooner for use in the "dude cruise" trade. But after the tragic loss of the ram *Levin J. Marvel* in 1955 the idea was shelved. The *Cripps* was taken around to Back Creek, Eastport, Maryland, and abandoned. Several years later a marina was built at her grave site and she was broken up.

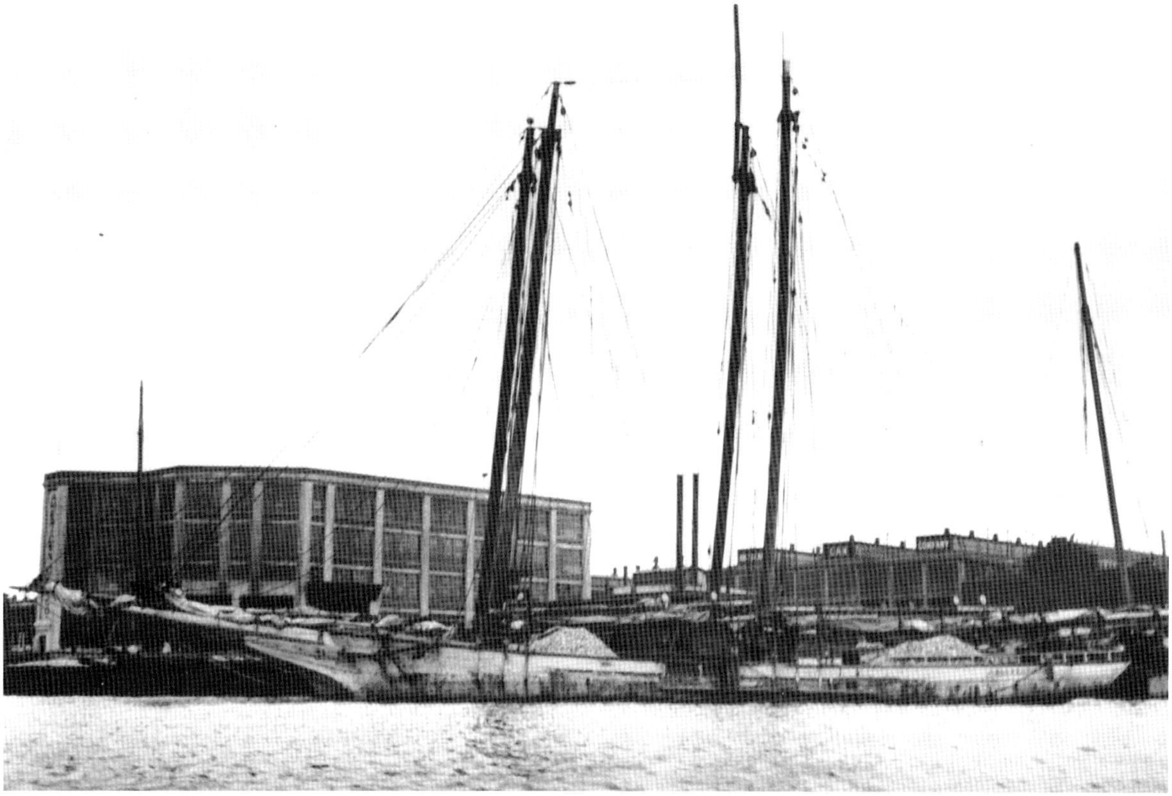

✪ The *Ella F. Cripps* at Baltimore on April 19, 1936, with a full load of oysters. *Minnie May Kirwan* is moored inboard of her.

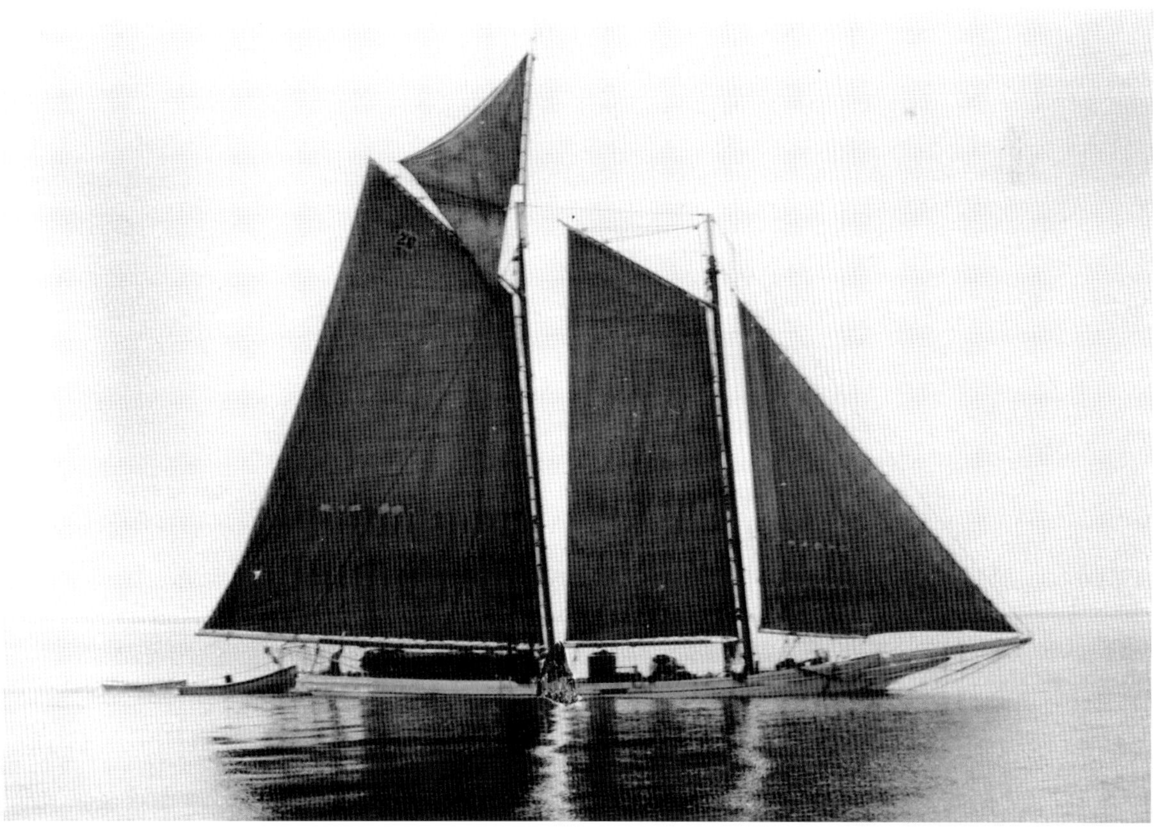

In the summer seasons, when she was not dredging oysters, the *Emma Temple* would carry watermelons, or wheat as in this view of July 22, 1933, to Baltimore. Deeply laden and in a flat calm on the Chesapeake, the schooner is nudged along slowly by her yawl boat. At the peak of her mainsail she carries her oyster dredging license number (26) of the previous season. The *Temple* was built in 1876 in Queen Anne's County, Maryland.

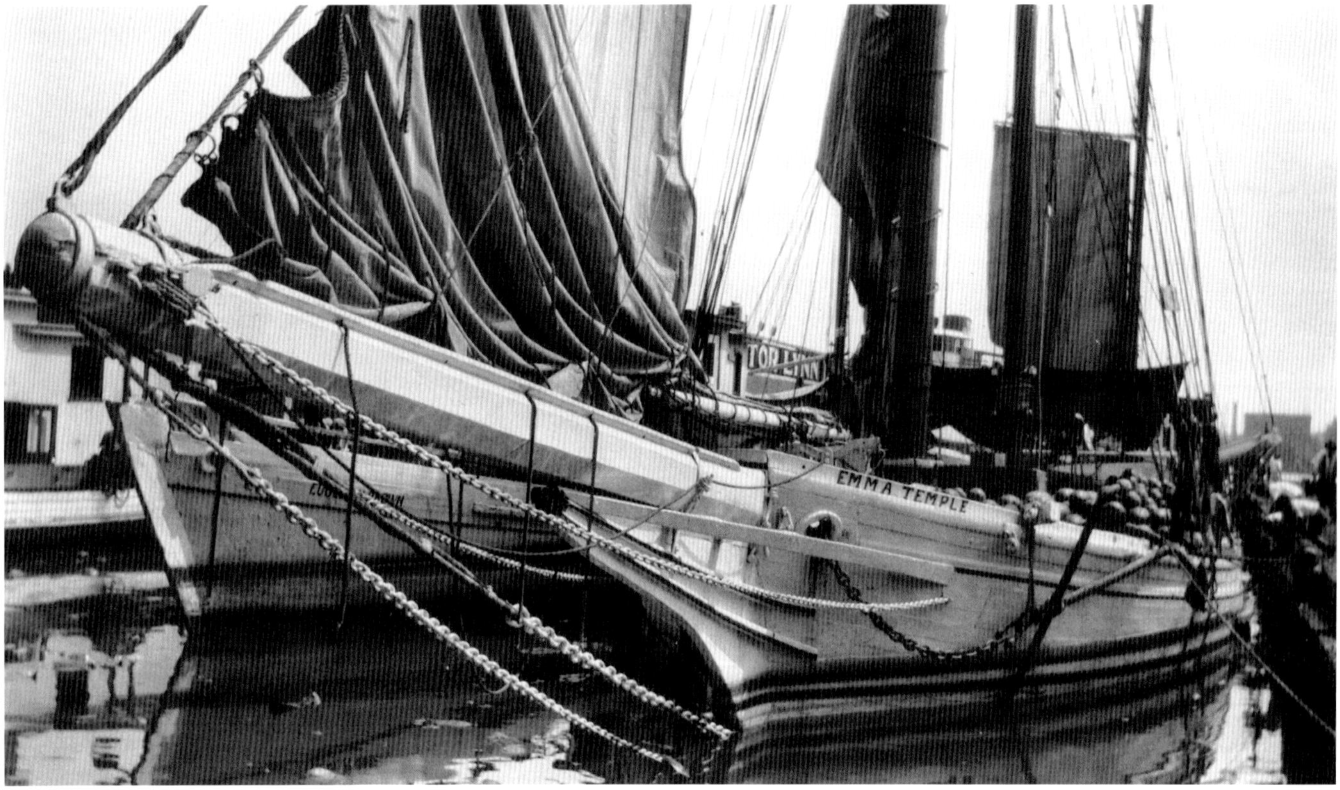

From the time the previous photograph was taken, the *Emma Temple* had discharged her wheat cargo, returned to Maryland's Eastern Shore, taken on a load of watermelons, and was back in Baltimore on August 12 peddling her produce at Long Dock along with the usual large fleet of sailing craft and powerboats on the same mission. Directly to her starboard is the powered former bugeye, *Eugene H. Brown,* built in 1892 at Oriole, Maryland, by B.P. Miles. It was on a passage to Baltimore with watermelons that the *Emma Temple* was to come to her end. About 2 a.m. on August 29, 1935, the schooner was coming up the Patapsco River with 7,000 melons from Salisbury. Onboard were her owner, Captain Samuel E. Shores, his son and mate Wilson Shores, cook Robert Price, and deckhand Earl Webster. As the vessel proceeded just outside the channel, those onboard observed a tug, which proved to be the *Columbia,* pushing a barge toward Sparrows Point. It was reported that the tug changed its course and the barge hit the starboard side of the schooner, causing her to sink. The *Temple's* crew was taken aboard the tug but the schooner was a total loss and remained an obstruction to navigation until lifted from the river's bottom later that year by a crane and removed to the Woodall Shipyard in Baltimore to be dismantled.

Fannie Insley tacks out of Baltimore harbor on May 6, 1934. This schooner was built in 1883 at Church Creek, Maryland, and was on the Chesapeake for more than half a century. On August 20, 1940, while bound to Crisfield with a load of oyster shells from the Rappahannock River, the *Insley* was dismasted, causing her seams to open, and she sank in 50 feet of water four miles northeast of Windmill Point Light. Captain Wilbur Willey and his crew of two were able to launch the yawl boat before the schooner went under, but were unable to start its engine. After drifting for two hours the men were picked up by the Baltimore Mail Line steamship *City of Baltimore* and taken to Cape Henry where they were sent ashore with a pilot. The wreck of the schooner was marked by a lighted bell buoy but she was total loss.

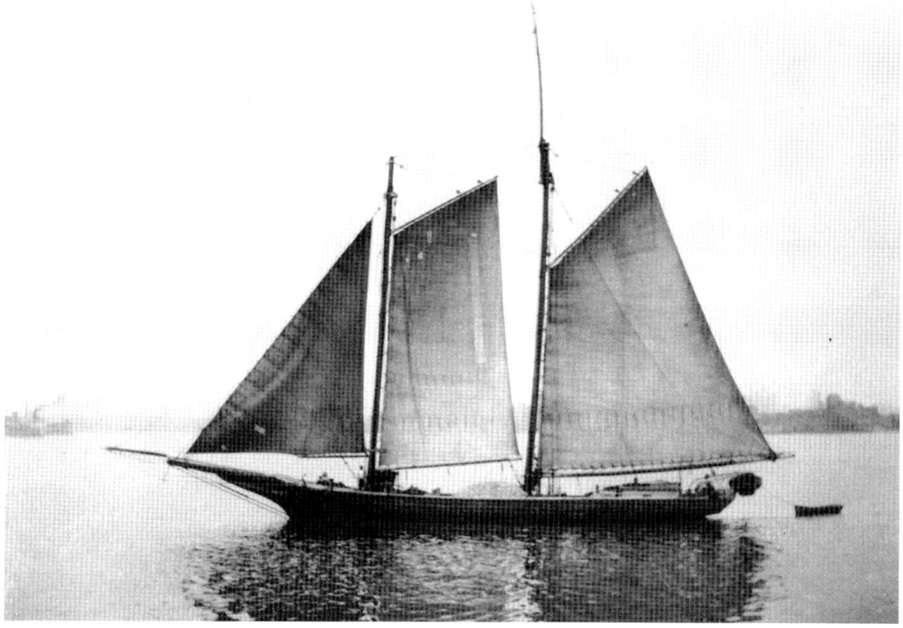

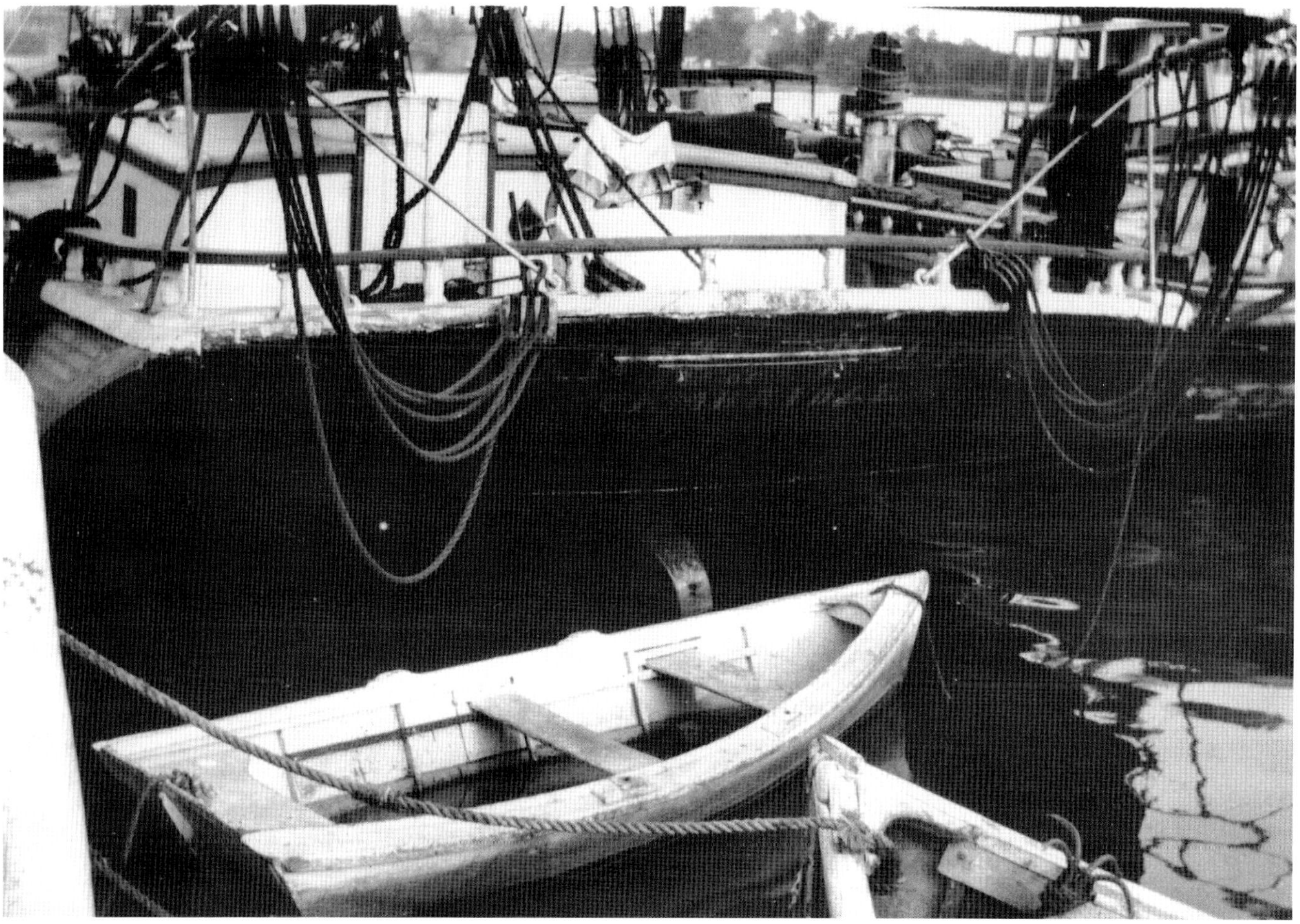

✪ The stern of the *Fannie Insley* at Curtis Creek, Baltimore, on October 12, 1935.

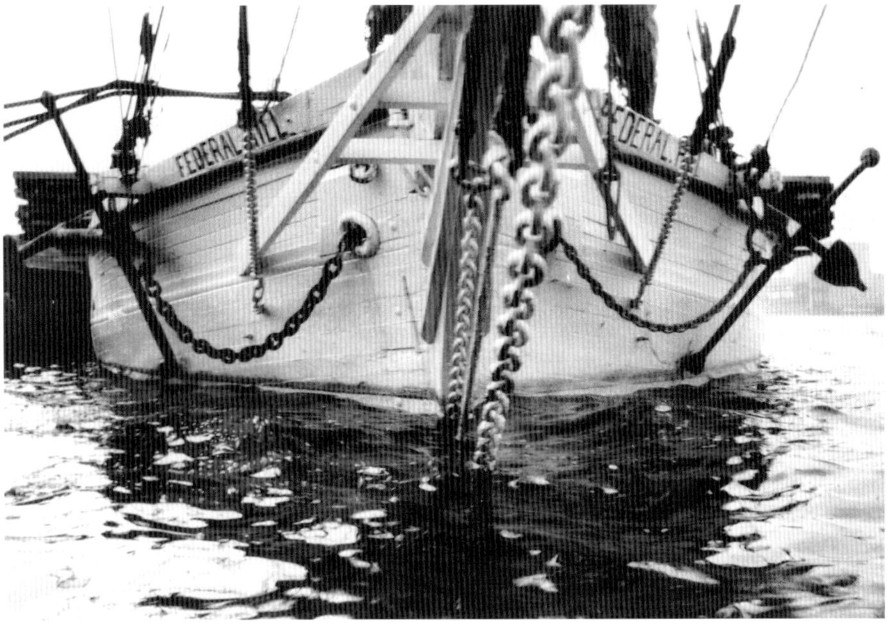

About 1933 the *Federal Hill* was rebuilt and emerged with her hull painted white and minus her jibboom and maintopmast. On January 14, 1934, she was at McGrath's pier in Baltimore with a lumber cargo jutting over either rail and her white paint getting stained by the oil-coated waters of the harbor. She enjoyed a few more years of activity but did not quite make the century mark. In the 1941 issue of *MVUS* she is listed as "Abandoned." She was actually cast aside in Monroe Bay, Colonial Beach, Virginia, the home of her owner, Captain William J. Stanford. Shortly afterward, Philadelphia lawyer Frank A. Moorshead recovered the small figurehead, a fine carving of an eagle's head, from the hulk of the *Federal Hill* and placed it on the longhead of his *Kessie C. Price,* a former Bay schooner converted to pleasure sailing.

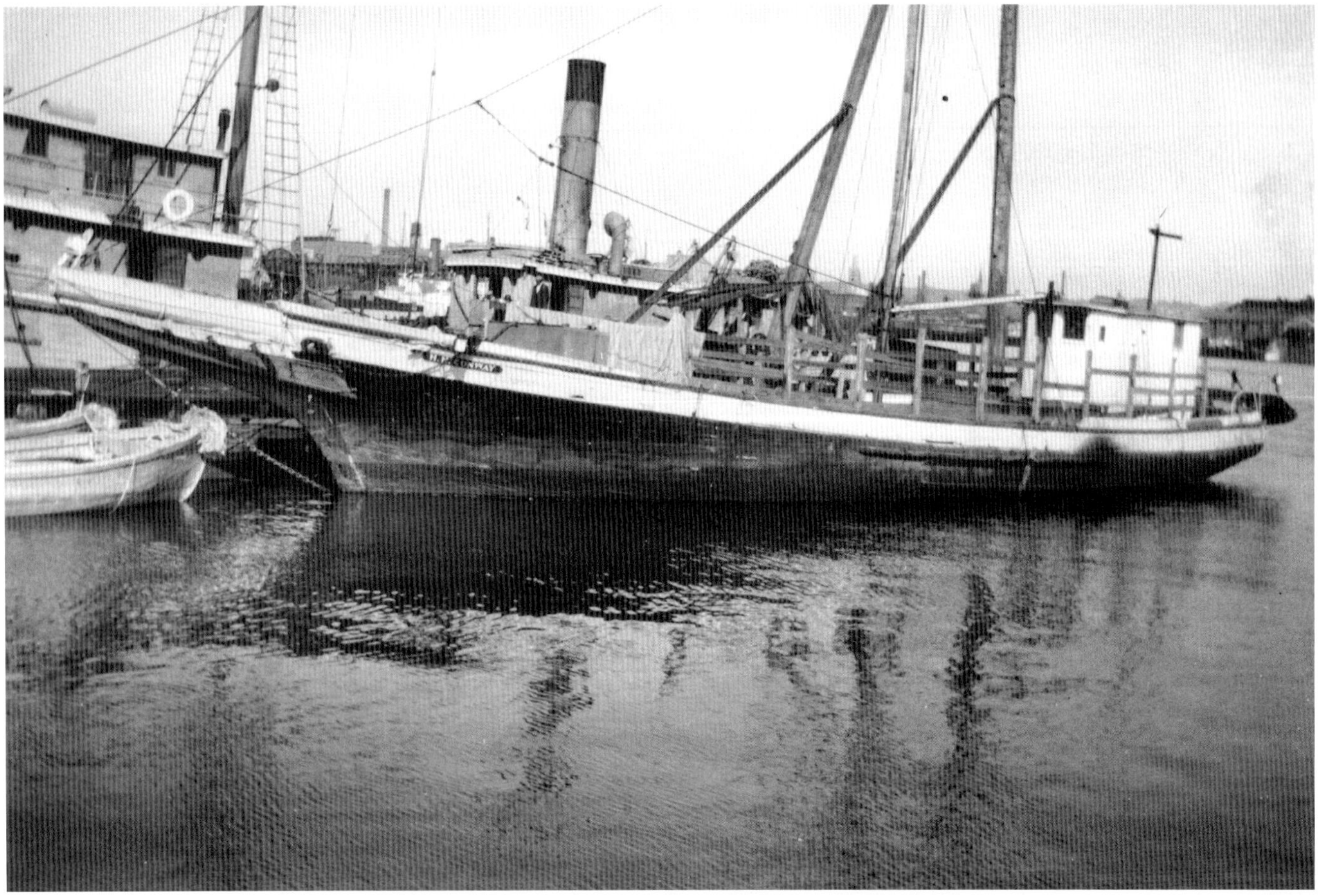

A good example of a hybrid vessel on the Chesapeake was the powerboat *H.H. Conway,* here at Redman and Vane's Shipyard, Baltimore, October 19, 1935. She is motorized but clings to her longhead, trailboards, and shortened bowsprit. Built as the *William Sommers* at Baltimore in 1883, she was acquired by Captain Harvey H. Conway of Cambridge and renamed after himself about 1921. She was sold about 1926 and taken to New Jersey as a schooner only to return to the Chesapeake as a powerboat when she was purchased by Captain Conway's son, Harvey, Jr., in November 1935, at a U.S. Marshal's sale. Her new career on the Bay was to be brief, for she was dismantled in October 1937. Menhaden steamer *Annie Dow* is in the background.

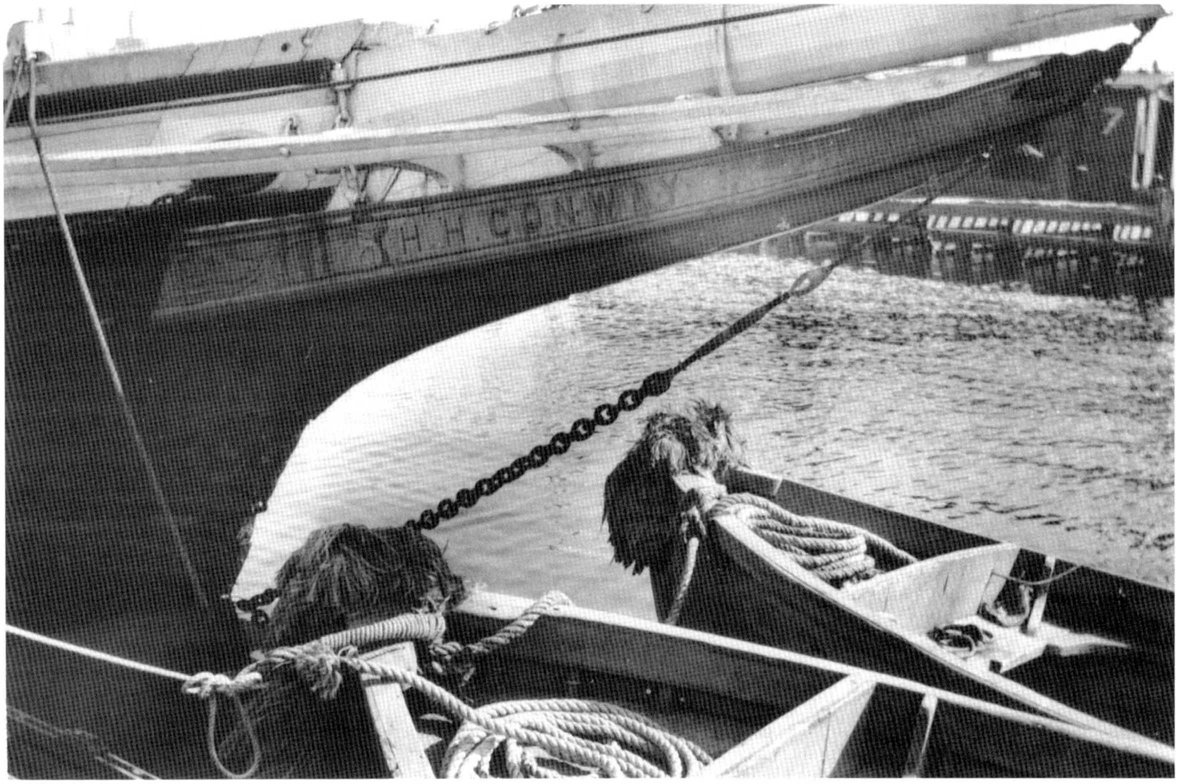

✪ A close-up of the starboard bow of the *H.H. Conway,* on October 19, 1935.

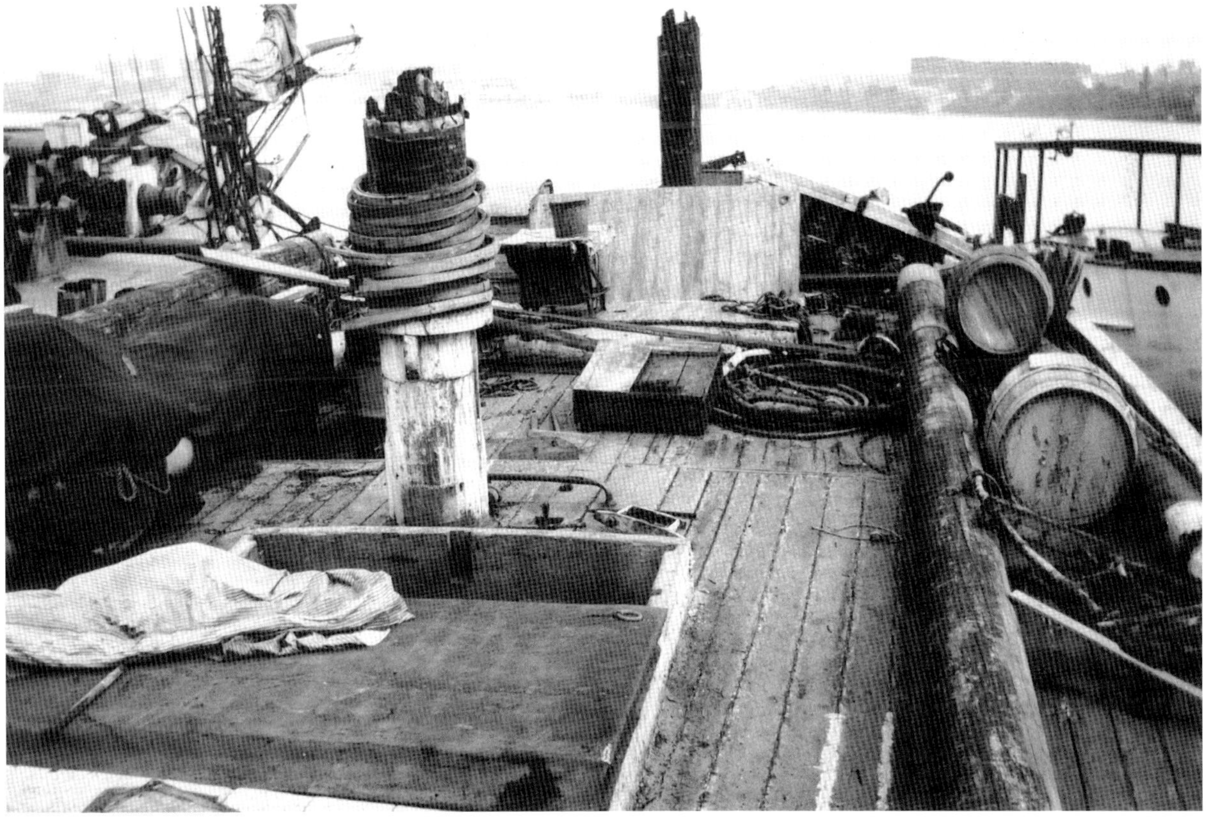

A bridge across the Chesapeake and Delaware Canal knocked the masts out of the schooner *Hannah and Ida* and on October 13, 1935, she lay at A. Smith & Sons Shipyard, Curtis Bay, as her owner, Captain H.A. von Pflugk transferred salvageable gear from the hulk to his newly purchased schooner *Fannie Insley,* at left. It appears that even the galley stove will be put to new use; it sits atop the fore hatch awaiting removal. The following month the *Hannah and Ida* was sold to new owners in Marley Park, Marley Creek, an arm of the waters in the background, for use as a breakwater. She had been built in 1872 at Mauricetown, New Jersey.

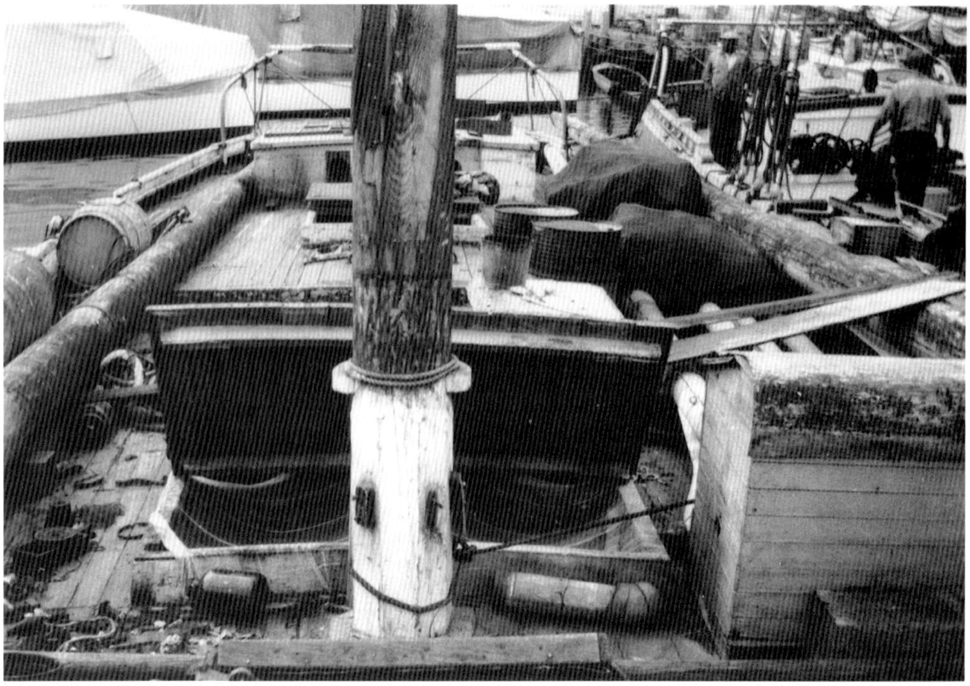

❂ A deck view of the ruined *Hannah and Ida,* looking aft on October 13, 1935.

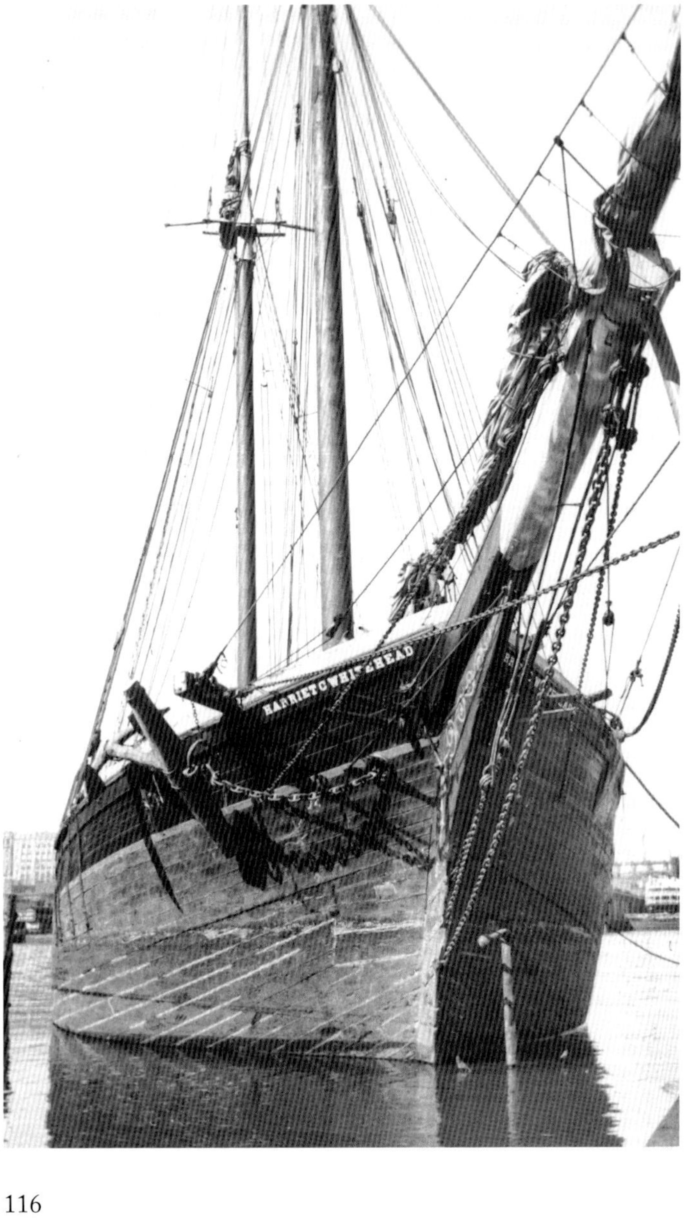

The largest schooner on the Bay, not registered there but a steady trader to Baltimore with lumber from North Carolina and lower Virginia, was the *Harriet C. Whitehead,* built in 1892 at Waterford, Connecticut, by Benjamin Stark, Jr. Measuring 119.6 feet in length and of 222 gross tons, she carried two topmasts and sailed coastwise before being sold to Elizabeth City, North Carolina, about 1924. Here she approached the ways of the Chesapeake Marine Railway on Key Highway, Baltimore, July 14, 1933, for an overhaul. Note the two-plank-wide bow ports for loading long timbers, the only set this author has ever seen on a two-masted schooner. Another distinction the *Whitehead* had among the Bay sailers was the huge anchor with a wooden stock on her starboard bow, certainly the last such anchor in use on a sailing vessel trading on the Chesapeake.

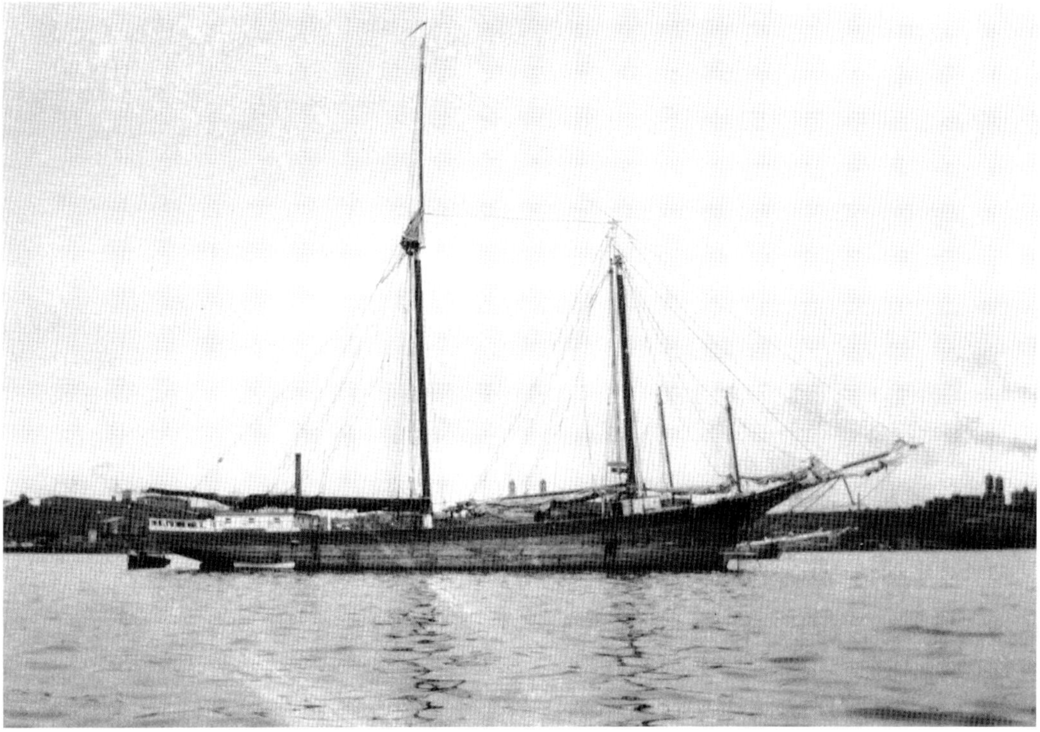

✪ The *Whitehead* at anchor before being hauled out in July 1933.

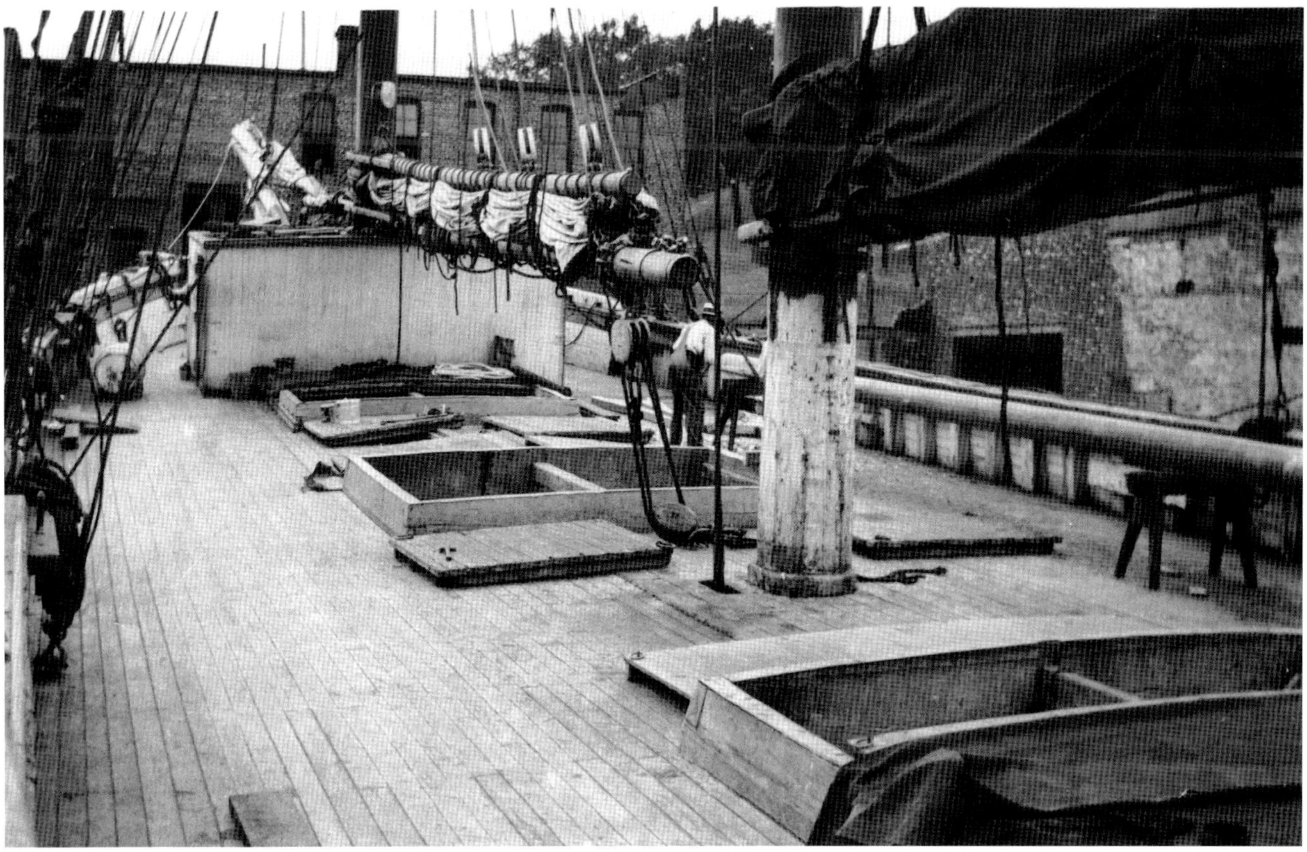

Looking forward on the wide deck of the *Harriet C. Whitehead* as she was hauled out on the marine railway. Her 31.3-foot beam is evident in this expansive deck. The carpenter is planning to fit jaws on the spar sitting on the sawhorses, probably to replace one of her booms, which seem to be in position. In 1936 this schooner was rebuilt at Deagle's Shipyard, Deltaville, Virginia, and resumed her lumber trade to the Chesapeake. When World War II placed a demand on vessels of all types, however, she was motorized and renamed *East Coast.* She operated out of Miami until she foundered on May 27, 1944.

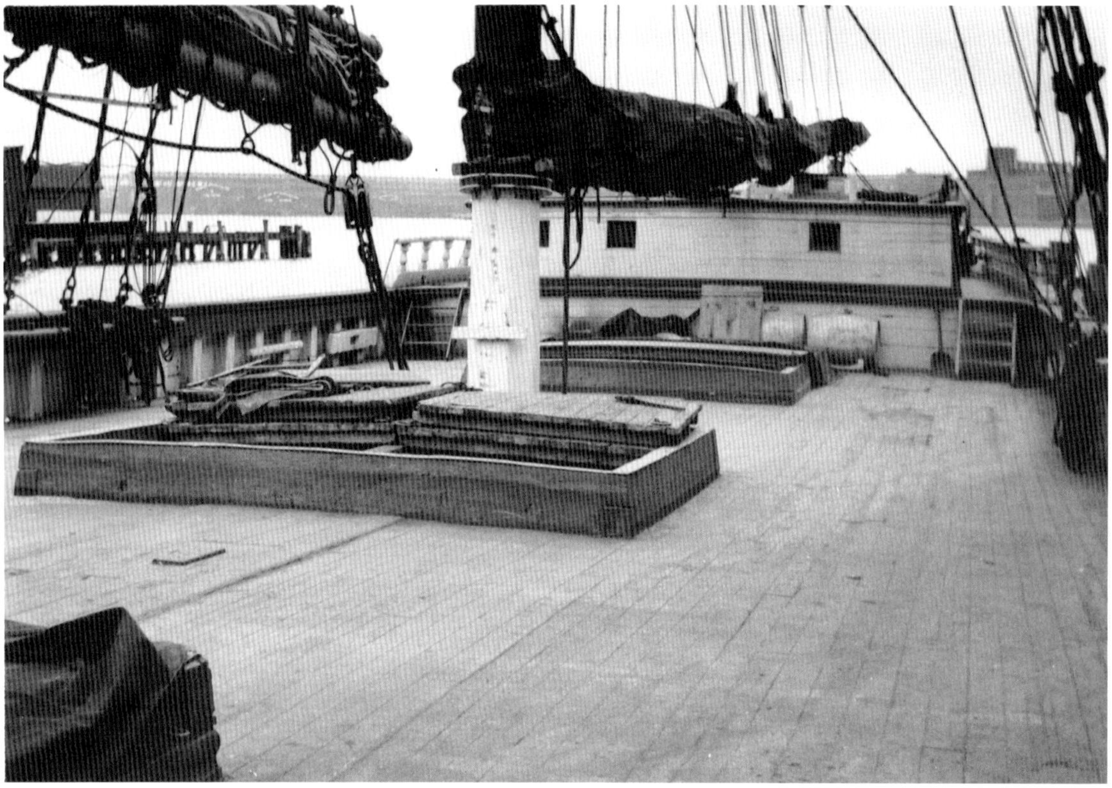

✪ Looking aft on the deck of the *Harriet C. Whitehead.* She presents a clean appearance as she enters the shipyard. Her hatch covers are off and stowed.

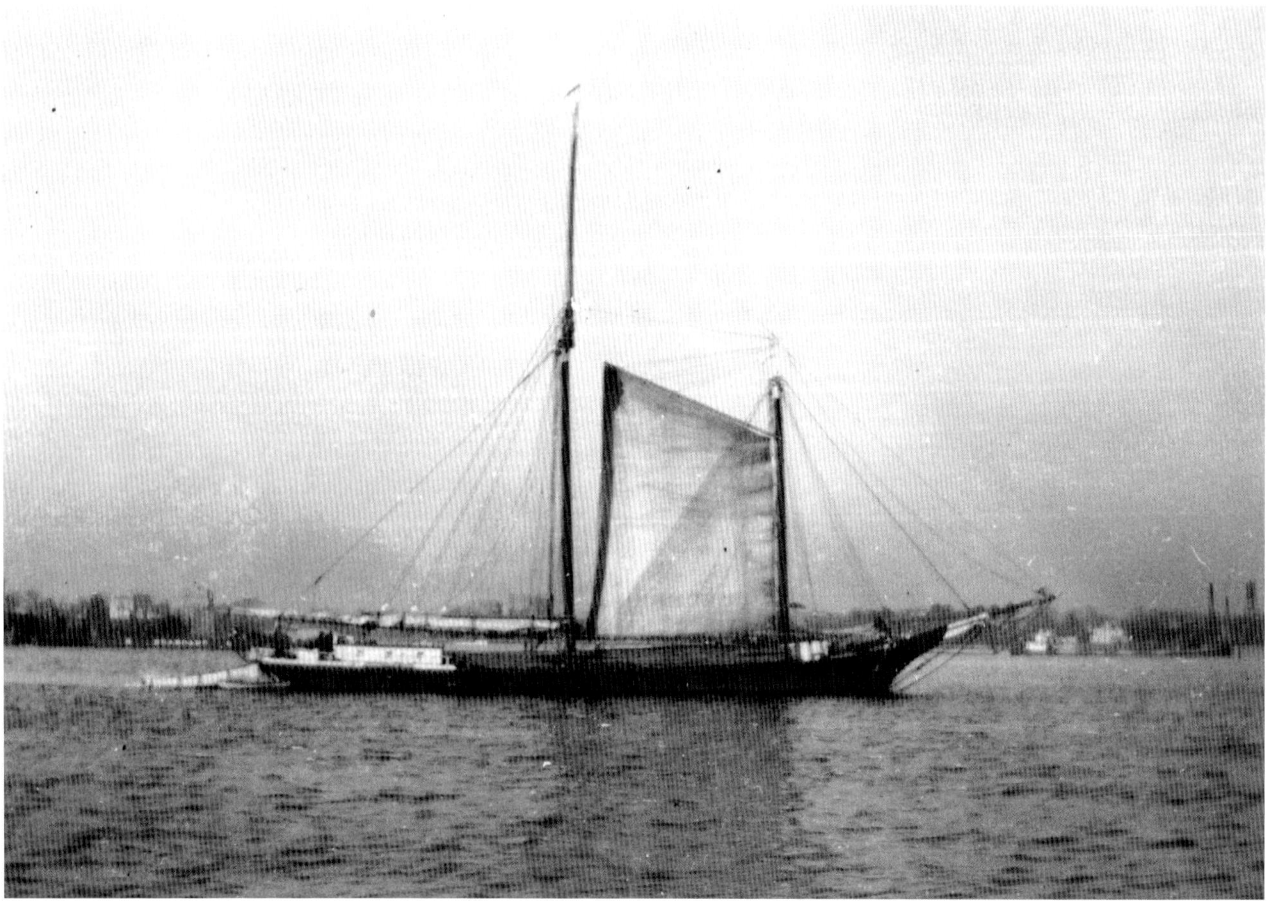

✪ At Norfolk, Virginia, the *Whitehead* is pushed by her yawl boat with her foresail still set.

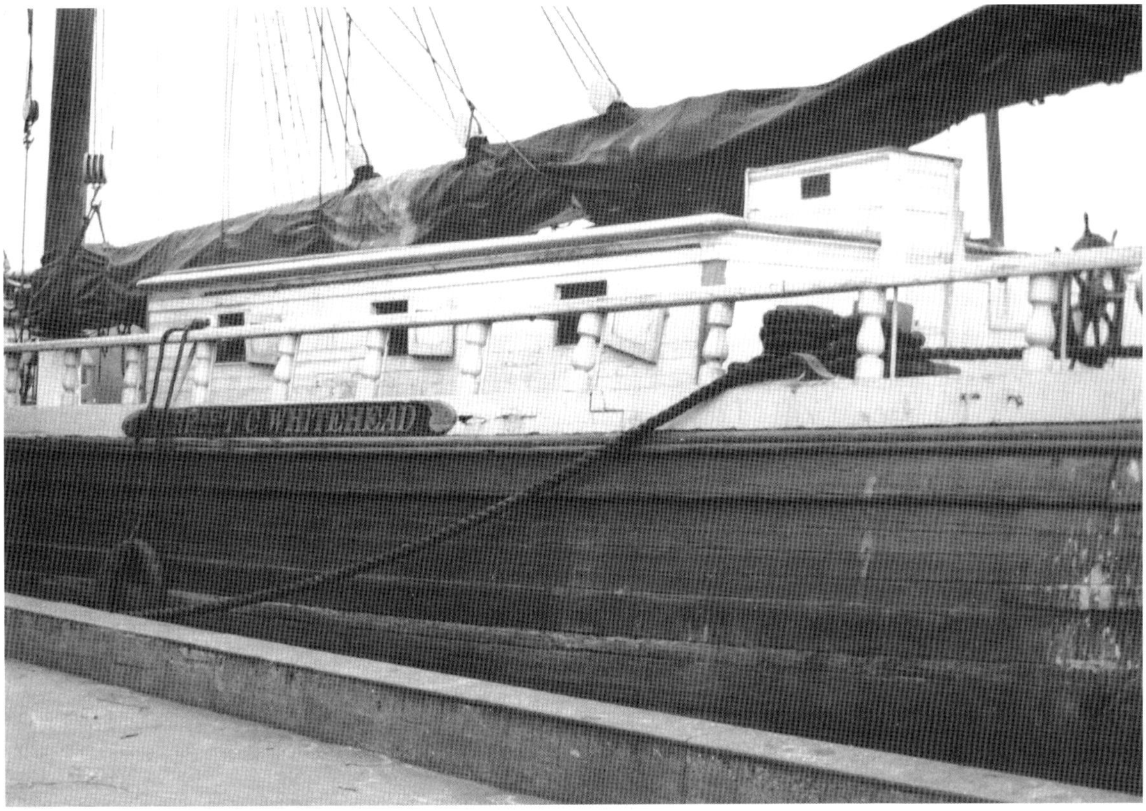

❂ Detail of the after house of the *Harriet C. Whitehead,* with her airports open. Her nameboards are mounted on her rails aft, not an unusual location for them on a schooner.

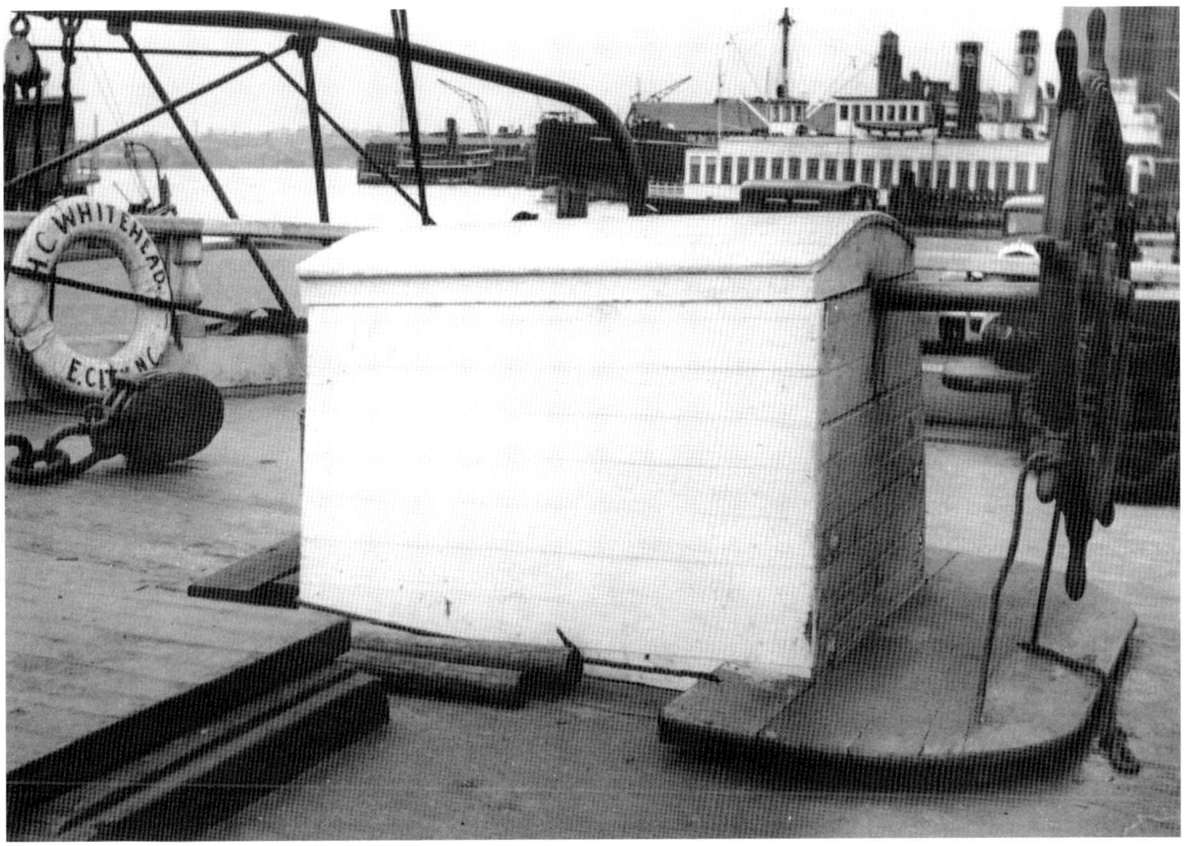

❂ A close-up of the steering wheel on the *Whitehead,* at Baltimore on July 14, 1933. Note the steam tug and Locust Point ferry in the background.

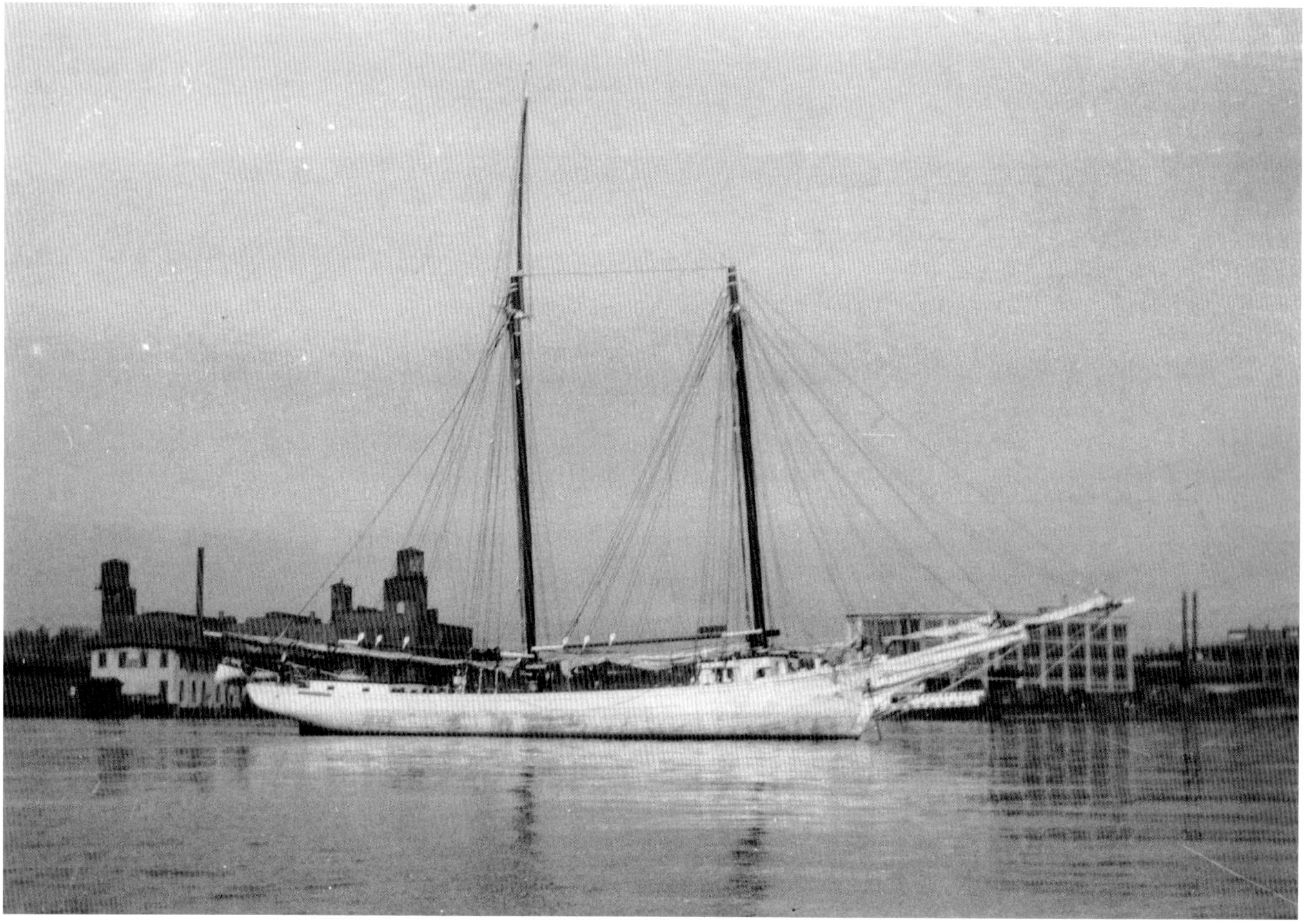

A fine example of a Chesapeake Bay schooner was the *Harvey H. Conway,* built as the *Richard Cromwell* by Joseph W. Brooks at Madison, Maryland, in 1888. In 1930 the *Cromwell* was purchased by the well-known Cambridge shipowner who renamed the schooner after himself. The *Conway* is mirrored in Canton Hollow on October 29, 1933, and presents a strong, seaworthy appearance. She was no stranger to offshore trips, for as the *Cromwell* she engaged in the pineapple trade from the Bahamas to Baltimore, as long as that trade lasted.

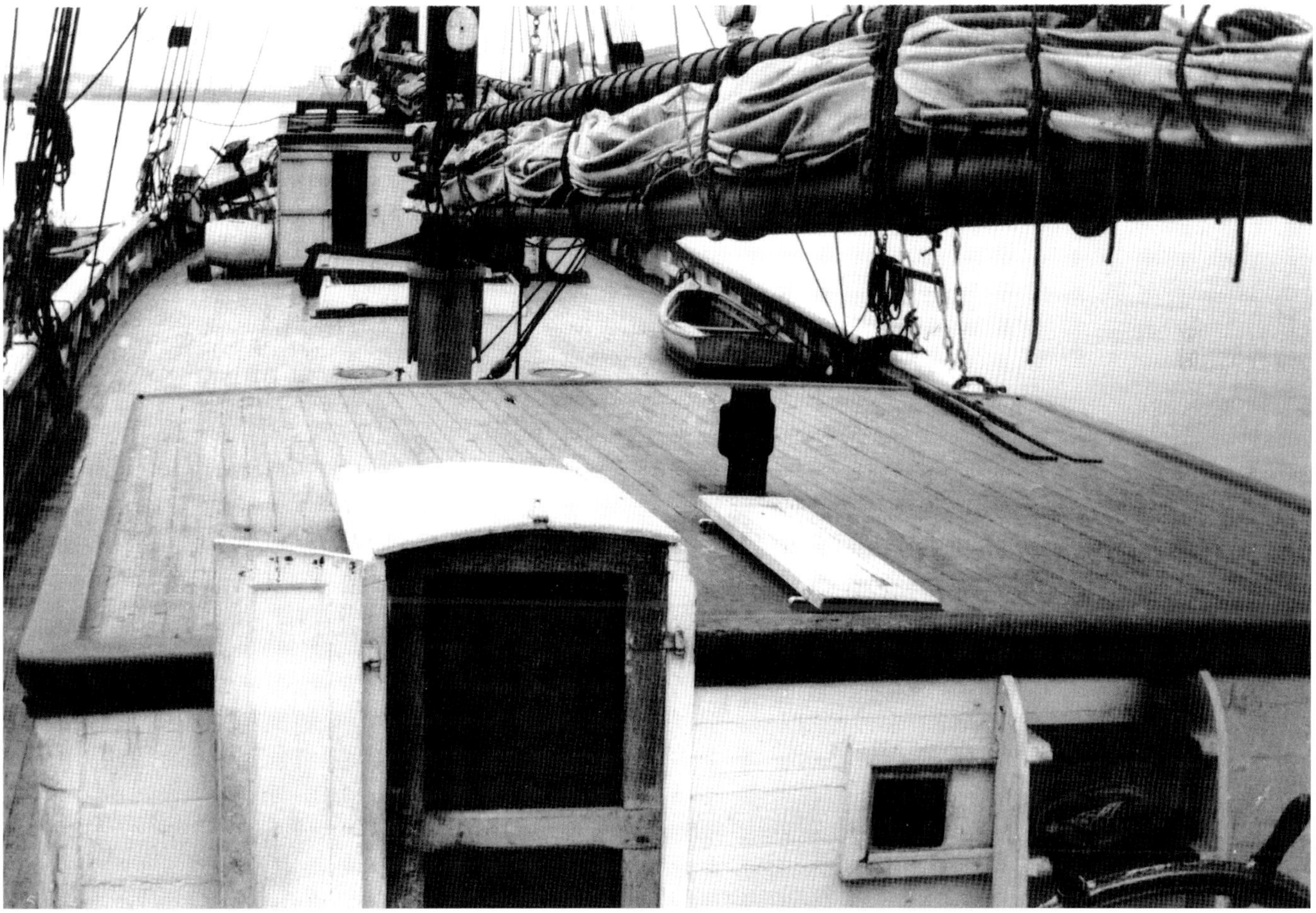

From the yawl boat in the stern davits one can look forward over the cabin top and along the deck of the *Conway* while tied up at McGrath's pier, Baltimore, November 3, 1935. The square opening in the after side of the cabin housed the compass, convenient to the steering wheel, lower right. On a tread of the ladder nearby was the lead line to take soundings. Entrance to the cabin was by opening the two-leaf door (one leaf off its hinges and resting on the cabin top), the screen, and then pushing the companion slide forward. The blackened stove vent protrudes from the top of the cabin. The *Conway* has given in to some modern fittings, such as the turnbuckles replacing the deadeyes on the shrouds. The position of the row-boat on deck was a natural one for this craft when not in use. Port of the forward house, close to the galley, is the scuttlebutt or water barrel; on top is a square hole covered with a canvas flap. A horseshoe, favorite good luck charm for Chesapeake watermen, has been mounted on the after side of the forward house.

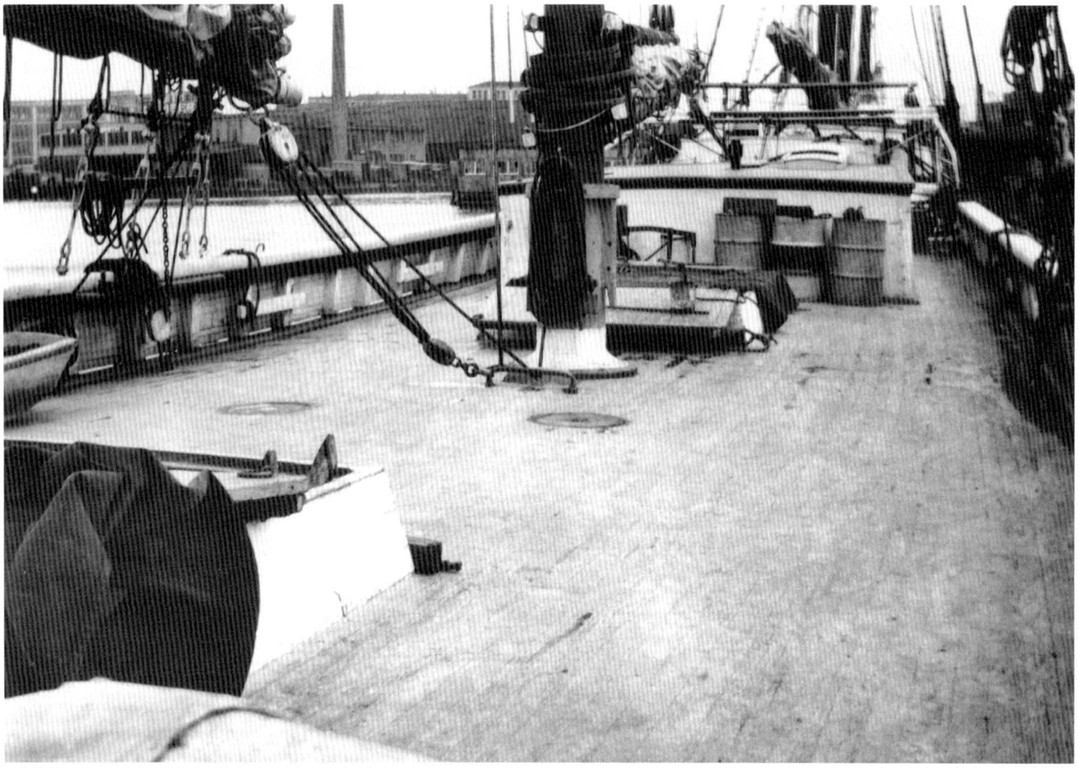

✪ The deck of the *Harvey H. Conway,* looking aft, has a shipshape appearance.

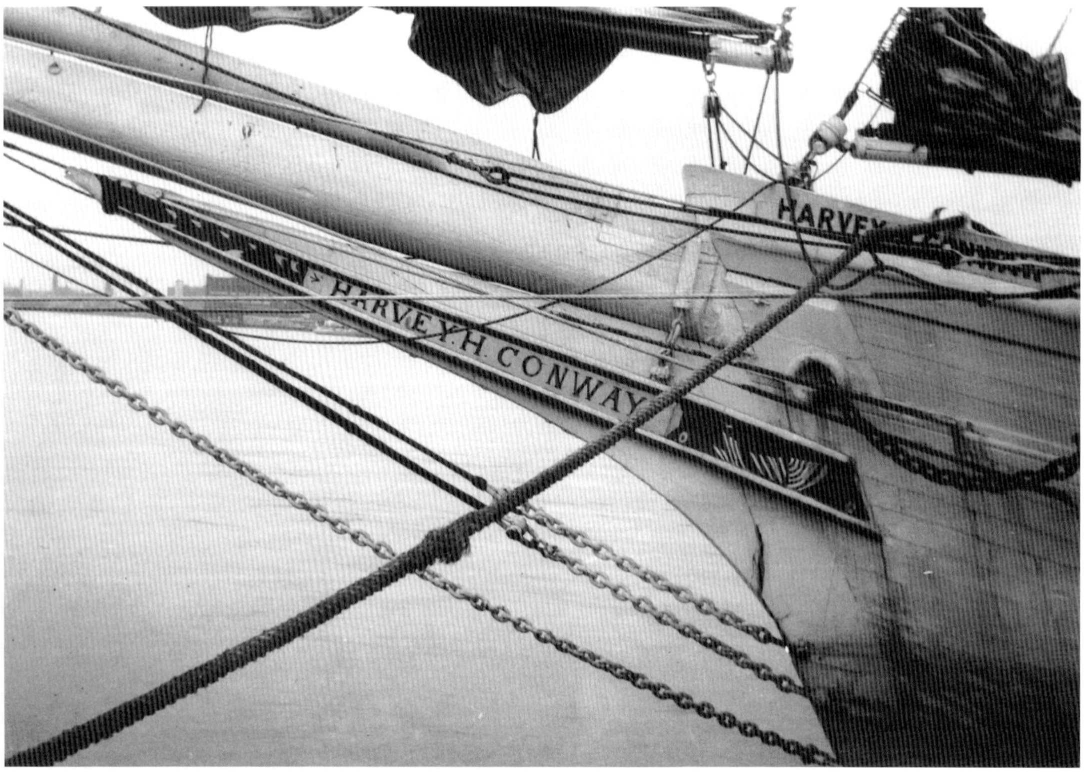

Carved trailboards and eagle figurehead on the *Harvey H. Conway* at Baltimore. When this vessel was renamed, that section of the trailboard bearing *Richard Cromwell* was cut out and replaced with the carved plaque *Harvey H. Conway.* The lower end of the board bears the traditional draped American ensigns, crossed staffs, and cannon. About 1938, after a brief period as a powerboat under the *Conway* name, she was sold to the Phillips Packing Company, Cambridge, Maryland, and renamed *Jeff!* That firm had the custom of naming its fleet of powerboats after comic-strip characters, others being *Mutt, Andy,* and *Popeye.* On October 15, 1954, after being sold from the Chesapeake, the *Jeff* foundered in the Atlantic 30 miles due east of Miami en route from the Dominican Republic.

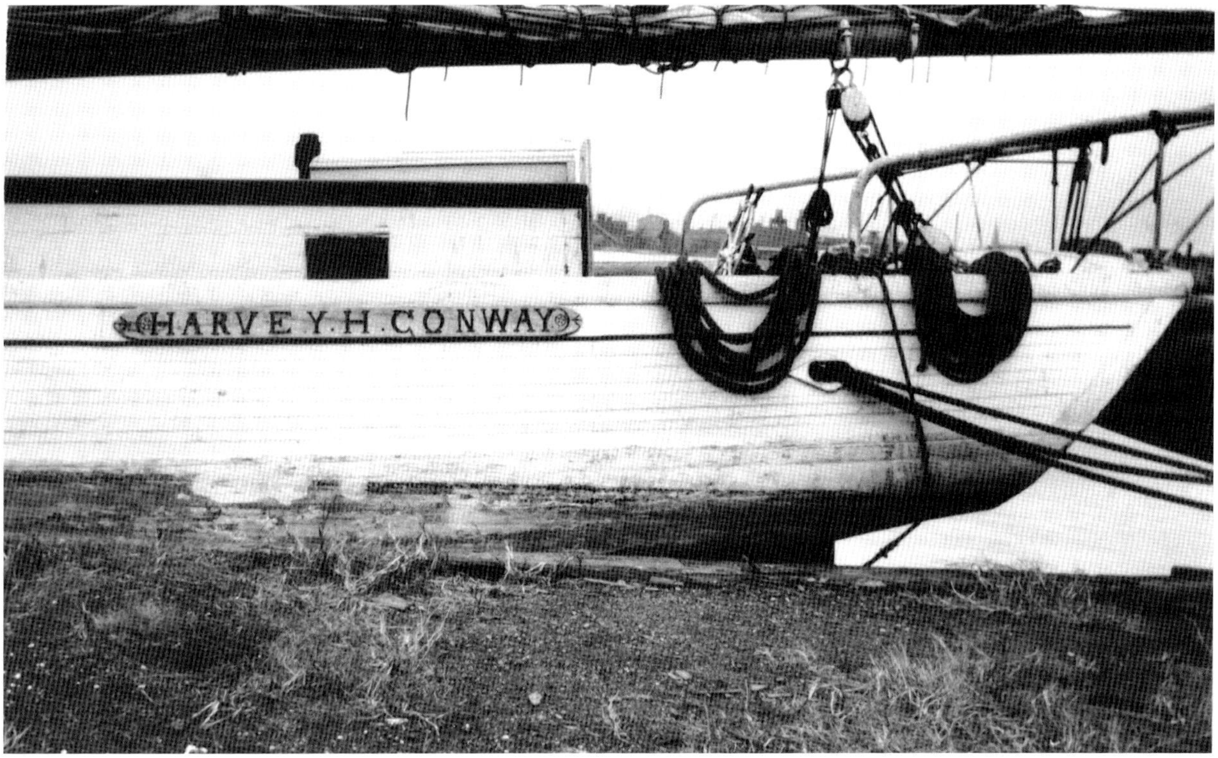

✪ A stern detail of the *Conway,* also with rail-mounted nameboards on her aft quarters.

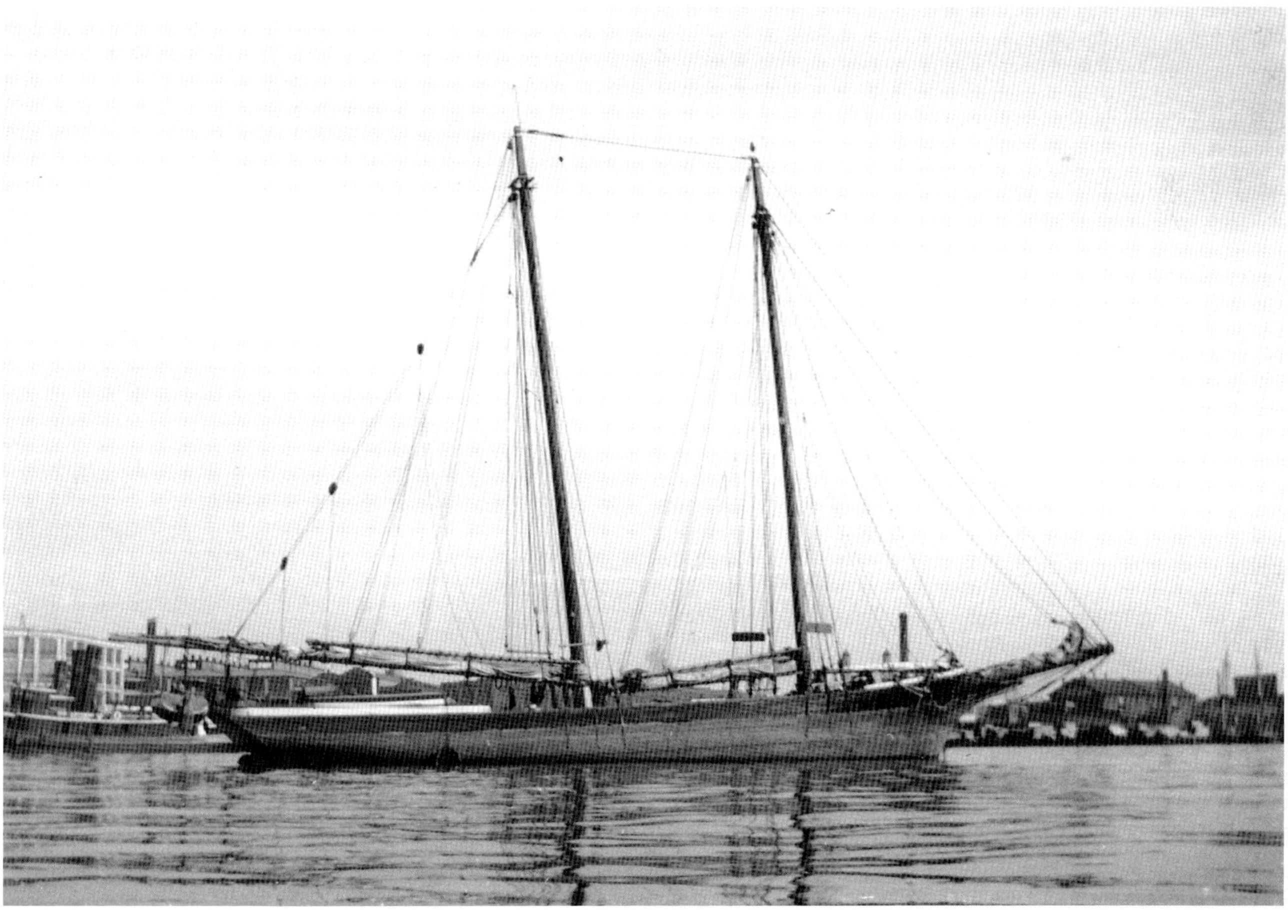

Hulda anchored in Canton Hollow, Baltimore, October 29, 1933. She was probably a nice looking schooner when she was built in Cambridge, Maryland, by J.H. Johnson in 1888 but after 45 years of trading on the Bay she has lost her sheer and has been stripped of her maintopmast and jibboom. *Hulda* carried the figure of a large acorn on the top of her foremast.

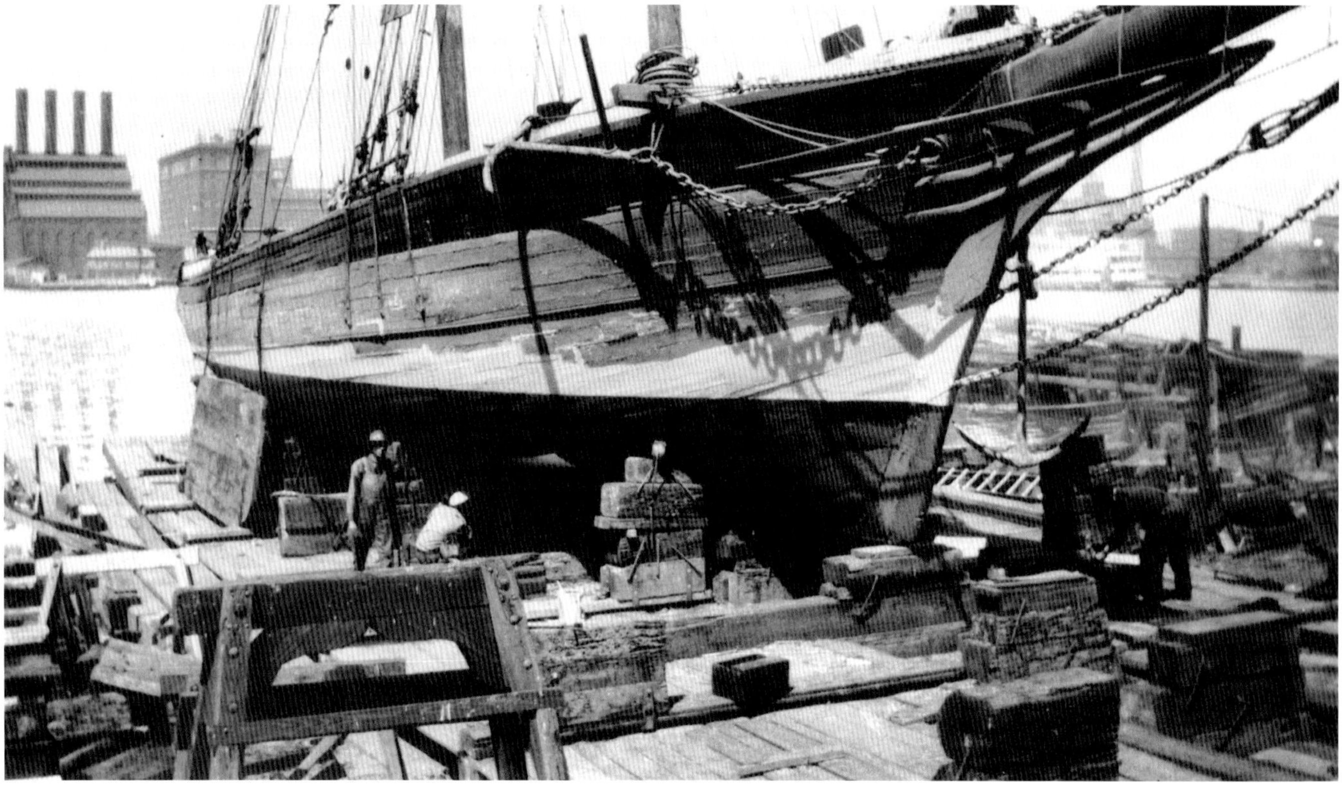

On May 1, 1934, the *Hulda* sits on the blocks of the Chesapeake Marine Railway, Baltimore, for a bottom painting. She shows her age in the uneven planking and metal patchwork in her hull, and this paint job was to do her little good. In 1937 she was abandoned.

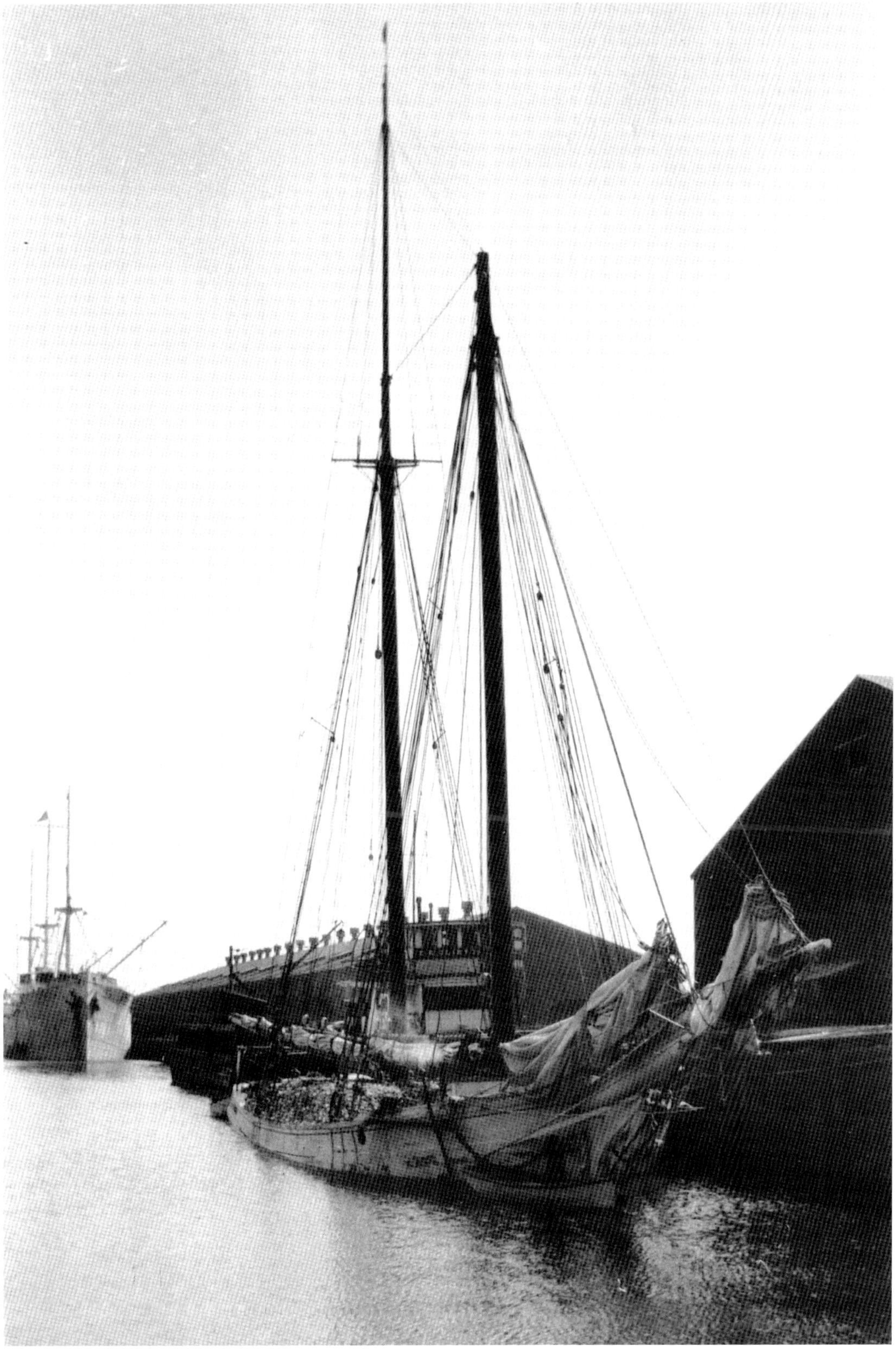

One of the last schooners serving as a freighter on the Chesapeake was the *Ida B. Conway,* which was converted into the powerboat *John J. Crouch* in 1947. Built as the *William Layton* at Nanticoke River, Maryland, in 1873, she was taken into the Conway fleet and renamed *Ida B. Conway* about 1924. On May 13, 1934, she is at the Potomac Poultry Food Company, Canton, Baltimore, with a cargo of oyster shells that has the waters of the harbor coming through the scuppers onto the deck. Following a brief period bearing the name *John J. Crouch* as a powerboat she was sold and renamed *Carol Ann.* In 1962 she was acquired by Captain Arthur Eubank who employed her as a freighter hauling, principally, soybeans. He sold her in September 1972, to owners in Anguila, British West Indies, for Caribbean trading, but her life there was brief as she sank off Puerto Rico in May 1973—exactly one century old.

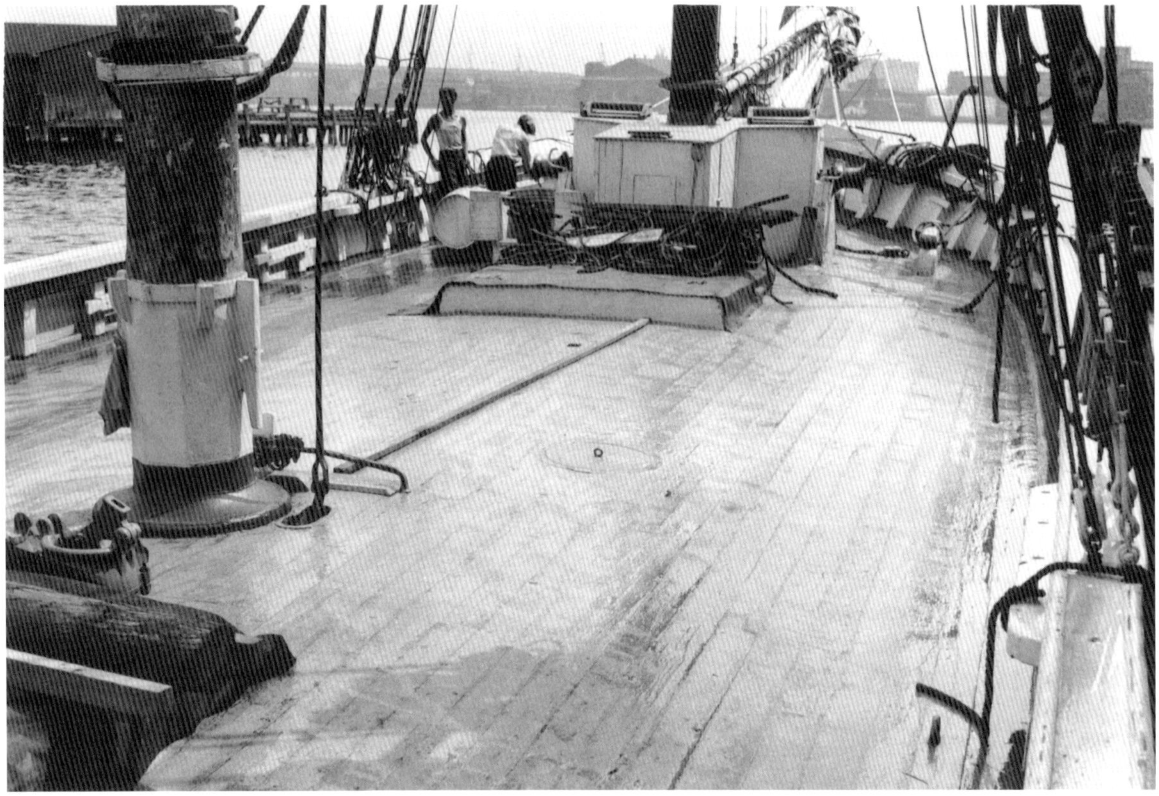

❂ Looking forward on the deck of the *Ida B. Conway,* moored at the foot of Broadway, in Baltimore, on August 22, 1936.

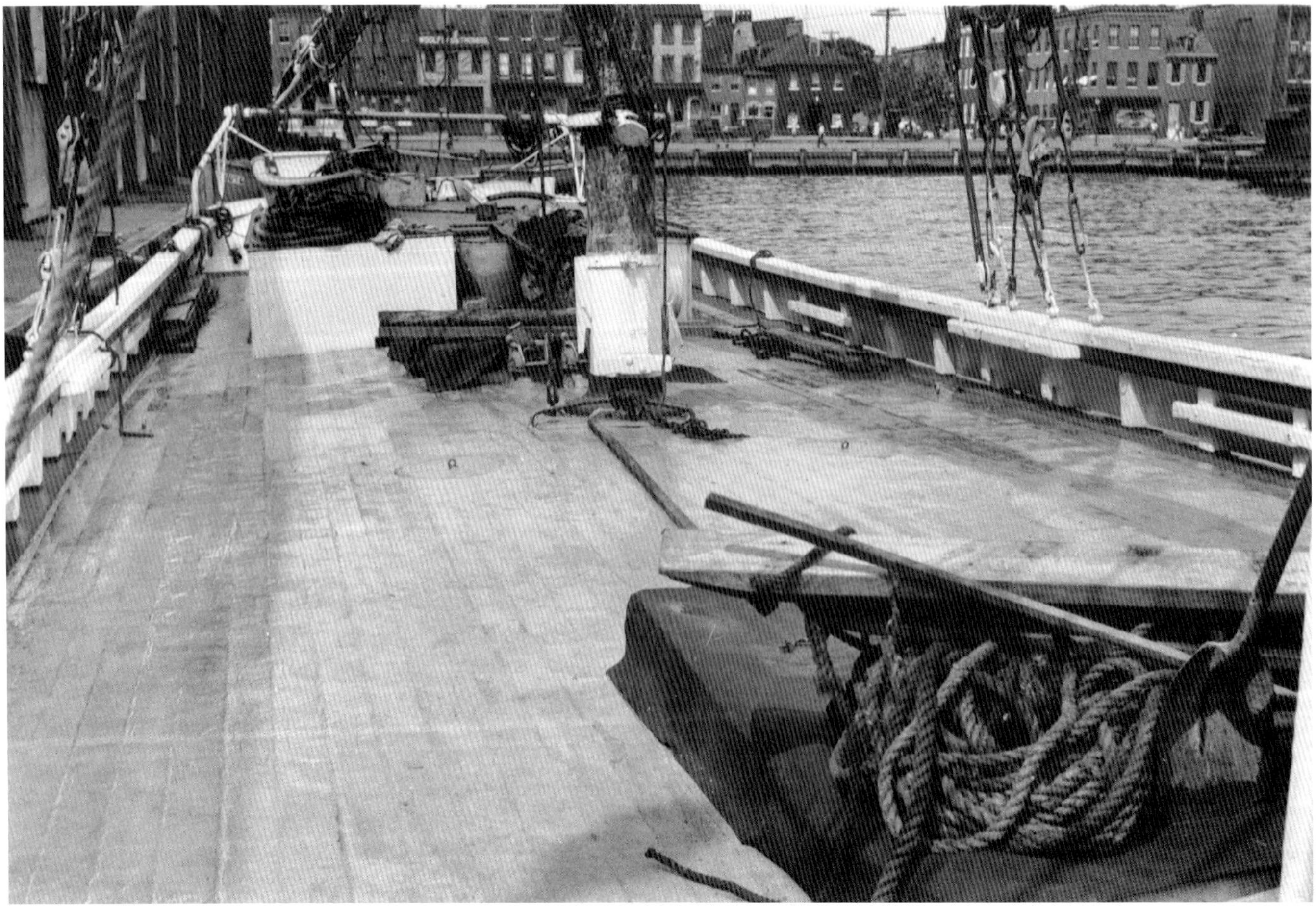

❂ Looking aft on the *Conway* on the same day.

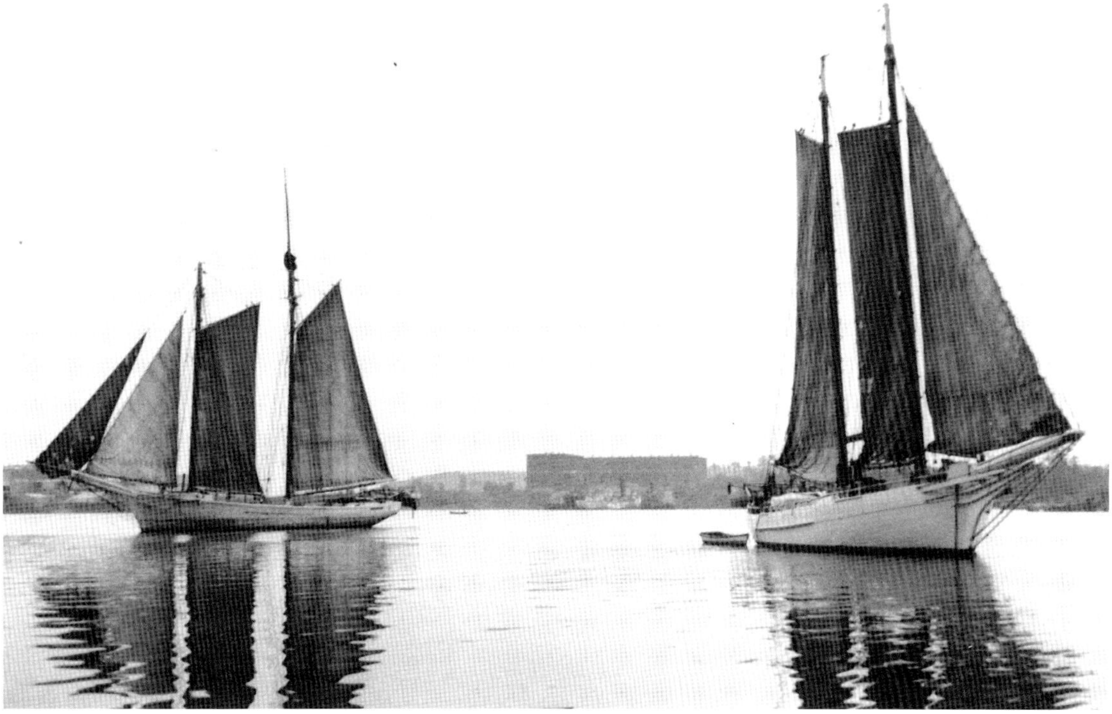

After a summer squall passed on August 31, 1938, the *Ida B. Conway*, left, dries out her sails while anchored in Curtis Creek off A. Smith & Sons Shipyard. The round-sterned bugeye *Catherine,* right, also takes advantage of the drying spell. Few of the Bay vessels bothered with sail covers at that late date.

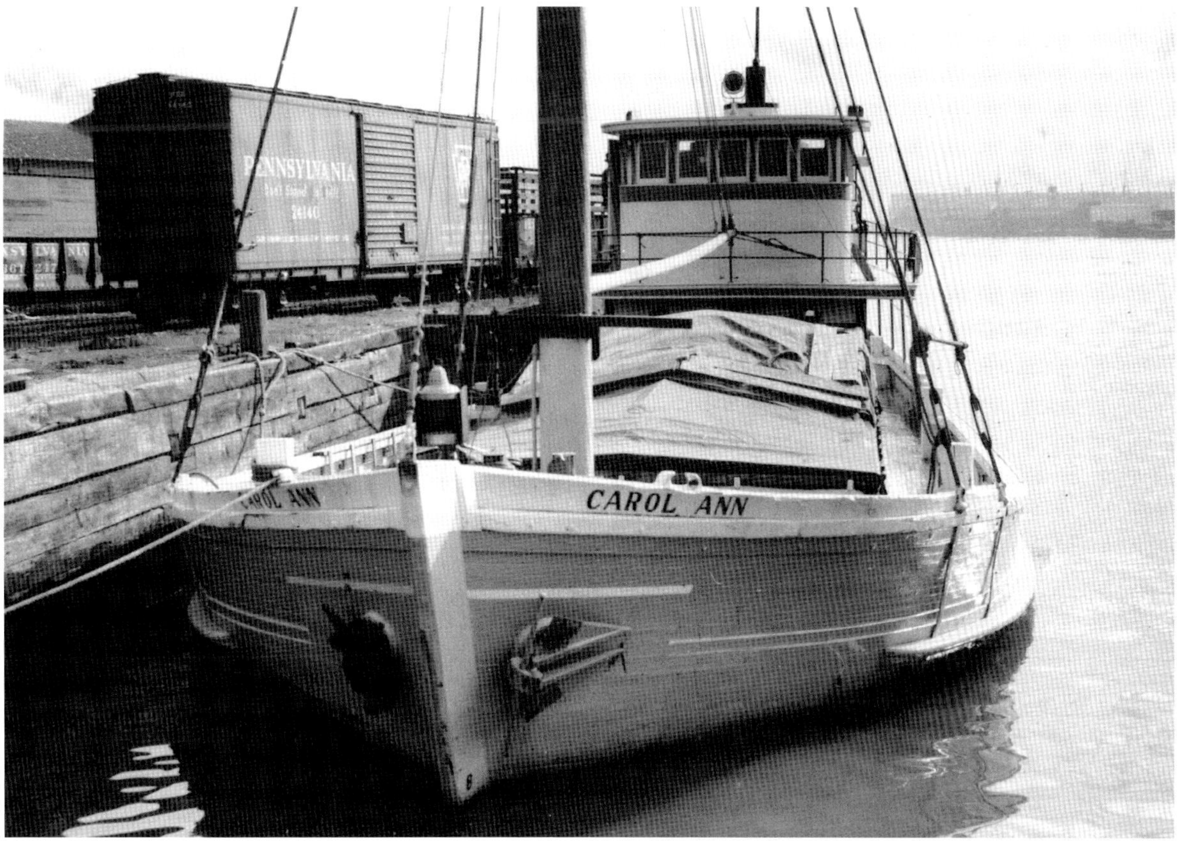

A vast change from the sailing *Ida B. Conway,* the same vessel makes a nice looking powerboat as she is loaded down with soybeans at a pier off Thames Street, Baltimore, on December 27, 1958. Her captain resented this picture being taken since the vessel is so deeply laden that water has come up on deck from the harbor. The insurance firm would not appreciate that method of loading.

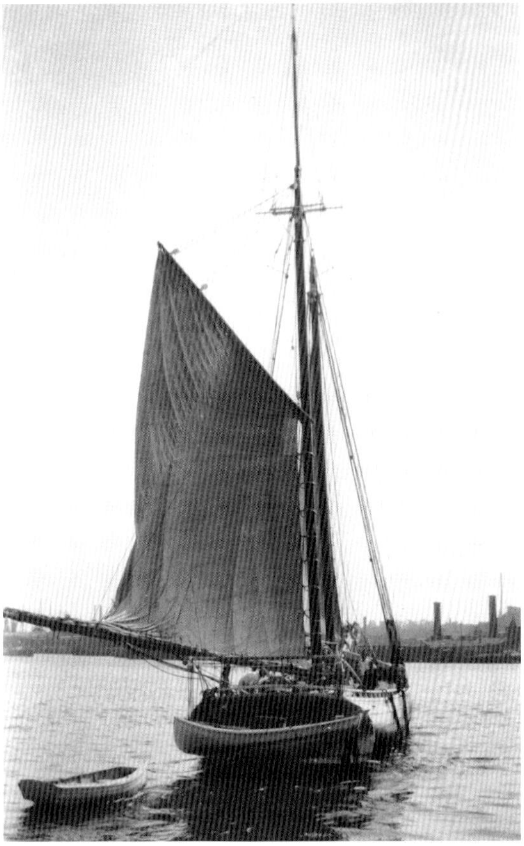

On September 22, 1933, the *Ida Mae* sails away from Long Dock, Baltimore, under forestaysail and reefed mainsail. Captain Barney Wessels, at the wheel and peering under the main boom to make sure the way is clear, will move out into the middle of the harbor and then sail down river with a fair wind, probably to anchor in Canton Hollow. The yawl boat hangs low from the stern davits and the skiff is towed astern, no position for these small boats if the schooner was planning to sail directly down the Bay.

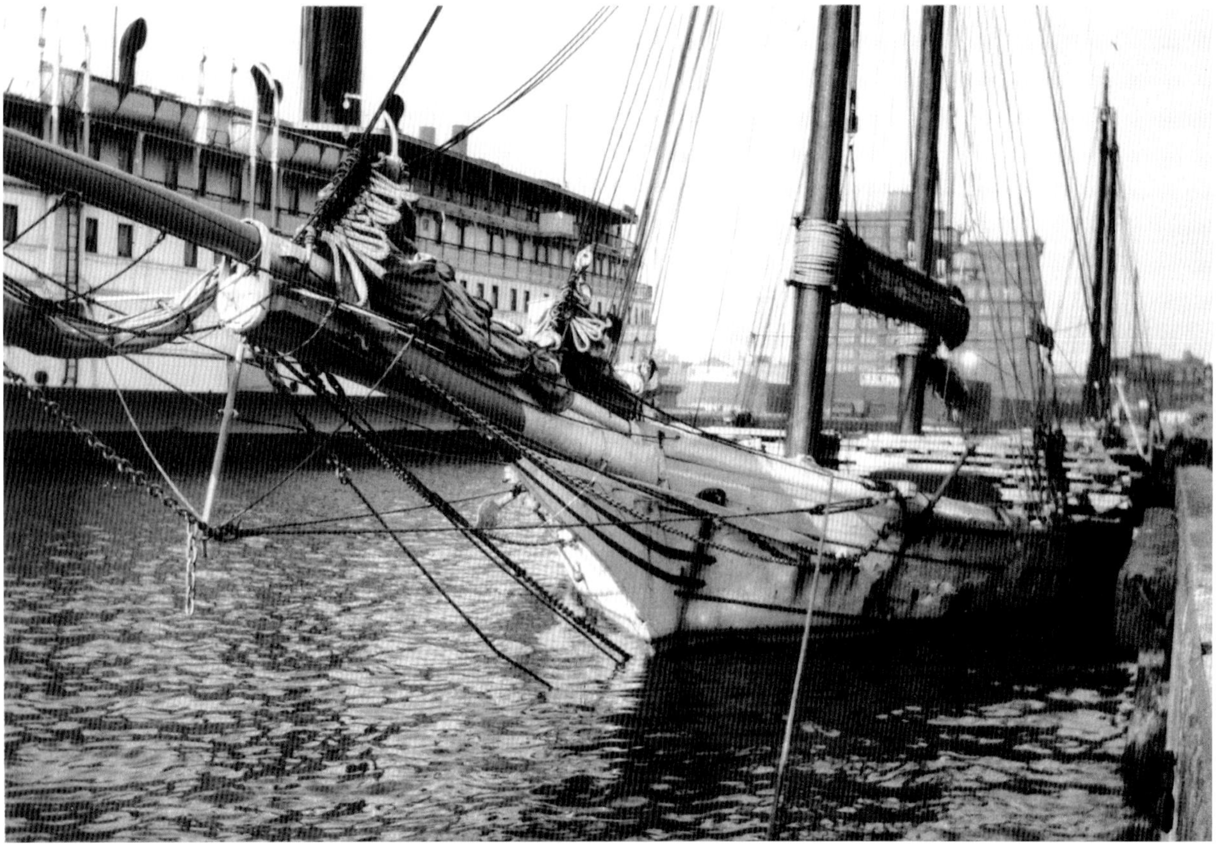

Christmas Day, 1933, finds the *Ida Mae* far from her home of Morattico, on the Rappahannock River, Virginia, as she rides alongside Pier 6, Pratt Street. But at least she is in port and not windbound in some down-the-Bay anchorage, and she has safely delivered a good-paying cargo of lumber. Astern of her is the schooner *J.A. Croswell.* The steamboat at left is the Old Bay Line's spare boat *State of Virginia.* In the background is the Candler Building on Pratt Street.

The clean lines of the *Ida Mae* are revealed at A. Smith & Sons Shipyard, Baltimore, May 1939. Her yawl boat lies on the railway at lower right to relieve her stern of that extra weight while out of the water. As her main gaff has been painted, her mainsail has been unbent. In 1946 this schooner was sold north for use in the "dude cruise" trade. The 1956 *MVUS* records her as "Out of Documentation." *Ida Mae* had been built in 1898 at Urbanna, Virginia, from the hull of the schooner *Jacob S. Barnes* which, in turn, had been built in Baltimore in 1870. Oldtimers said "the *Barnes* was one of the fastest schooners on the Bay in her time."

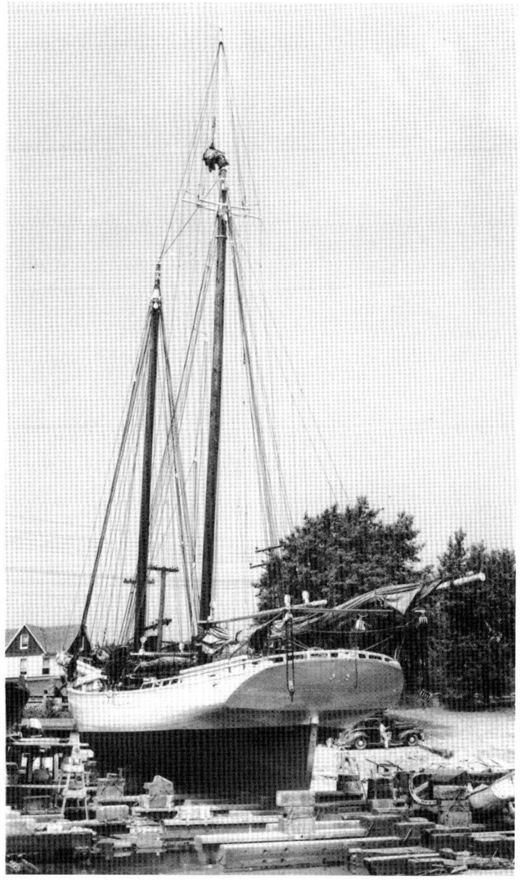

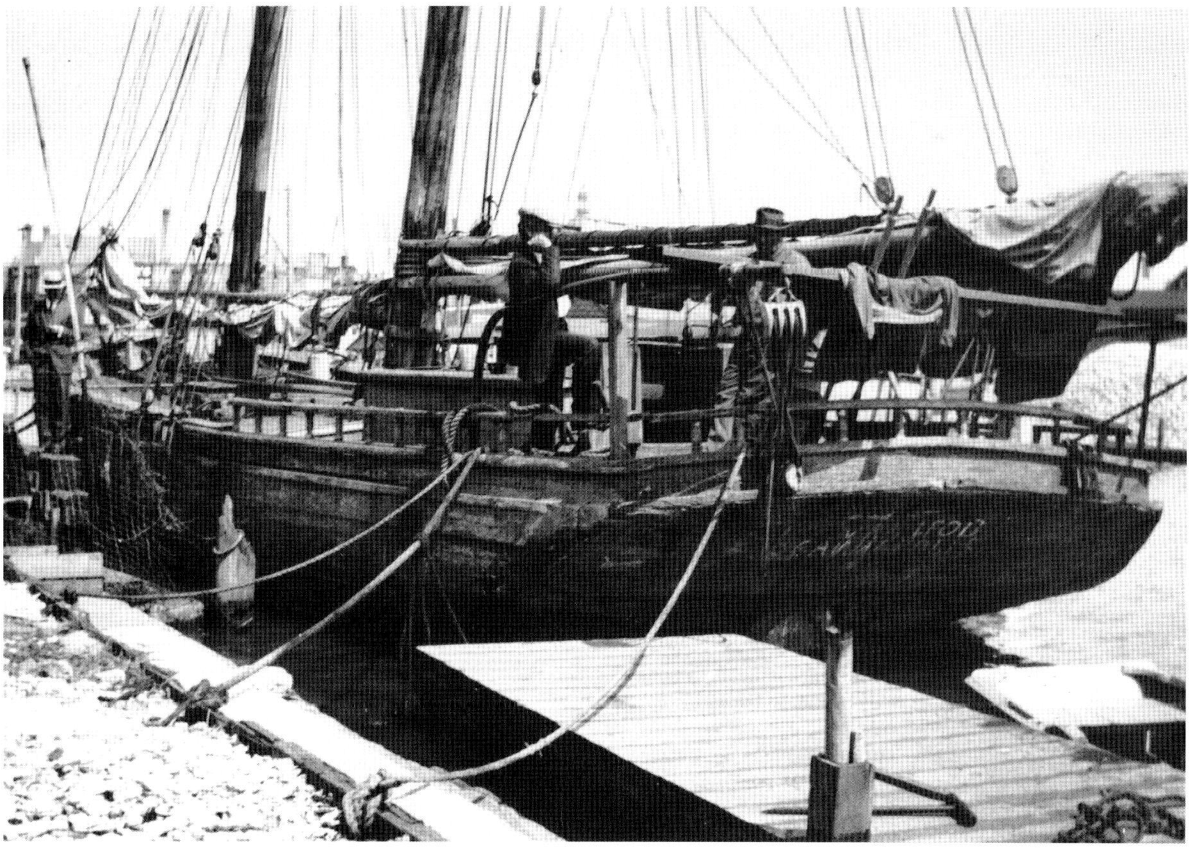

The *Iva Manifold,* built at Herring Bay, Maryland, in 1893, was long moored at the public dock in Annapolis, Maryland. On this Sunday, April 22, 1934, her captain chats with a friend on the quarterdeck, while at left center the marine artist Louis J. Feuchter studies some detail about the vessel's deck. The old schooner did not have much more time afloat; she was abandoned in 1936.

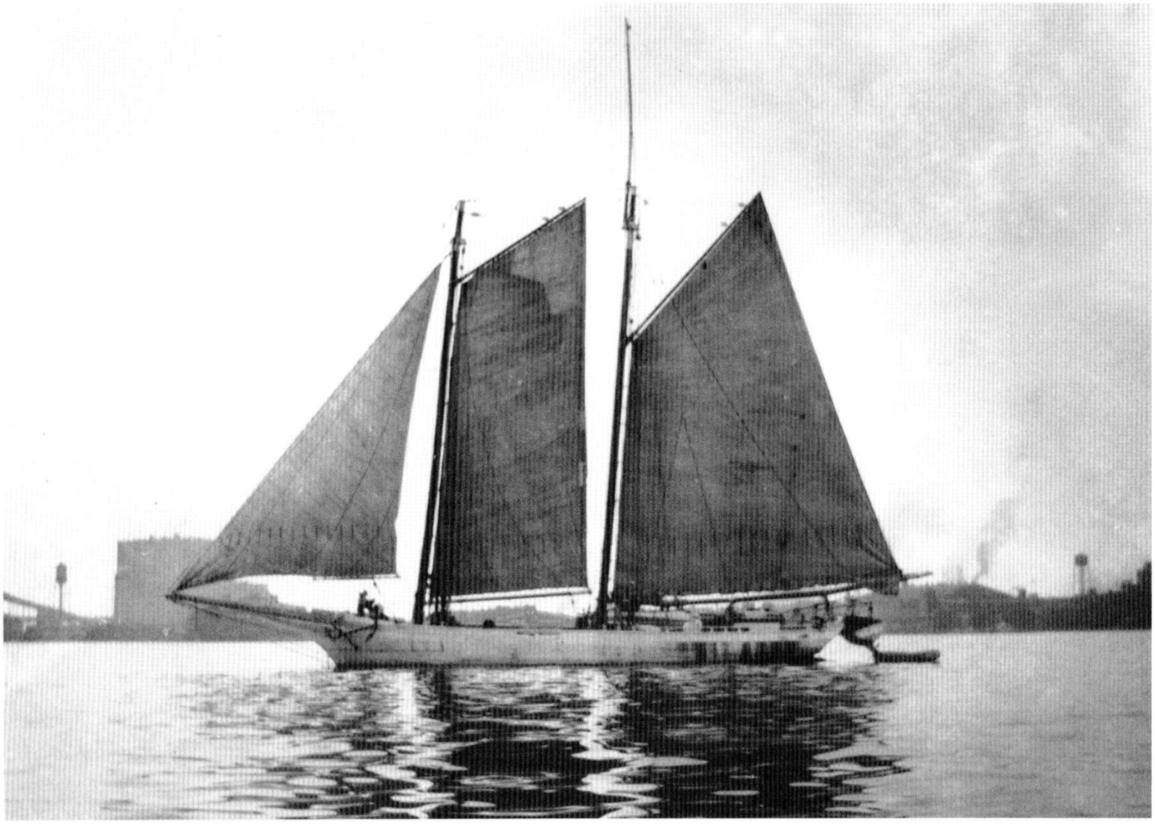

As her crew works on the foredeck to get the anchor aboard, the *J.A. Chelton* moves down Baltimore harbor with scarcely enough wind to fill her sails as she returns to Deal Island, Maryland, on July 19, 1934.

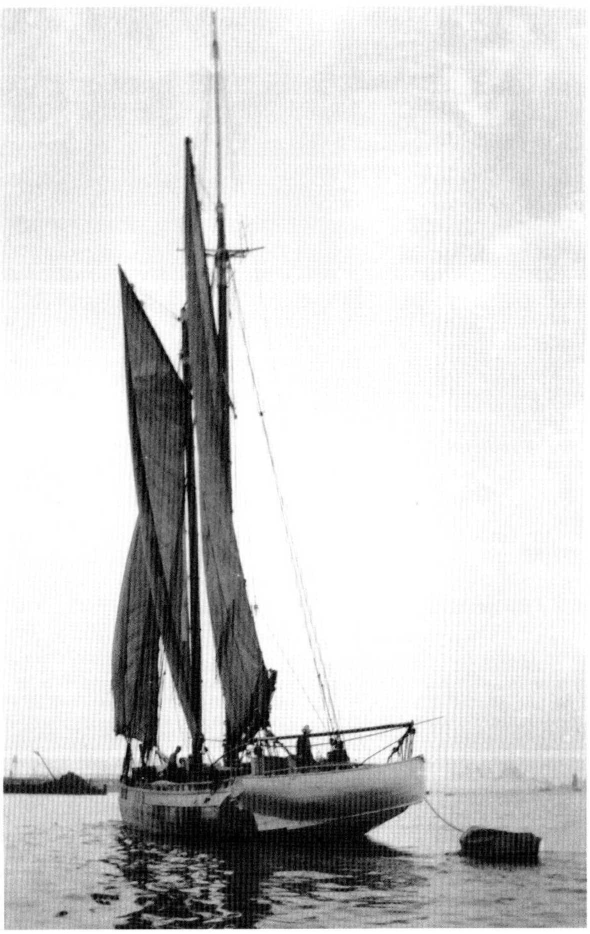

✪ The *J.A. Chelton* sails by on July 19, 1934.

❂ The schooner *J.A. Chelton* rests at Montford Yard, in Baltimore, on January 12, 1936.

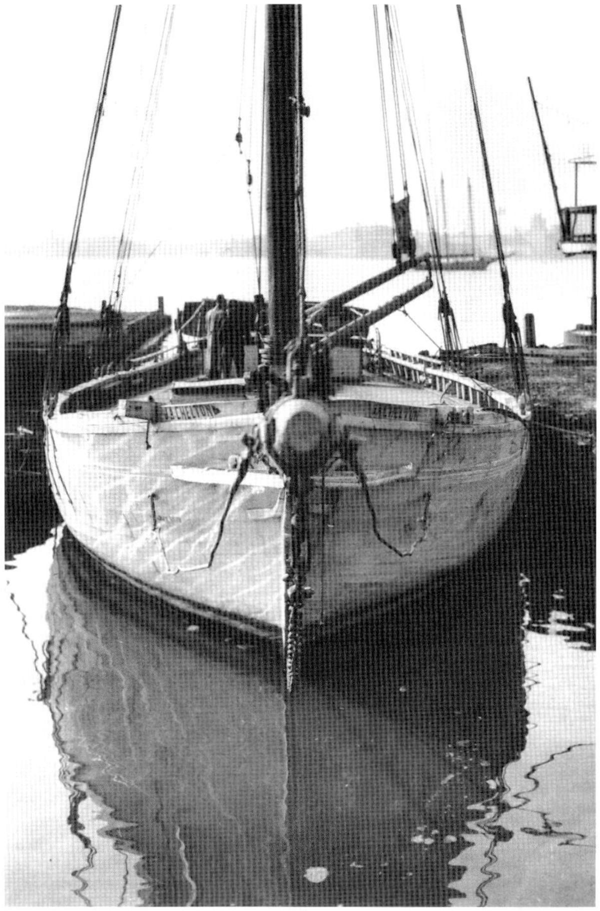

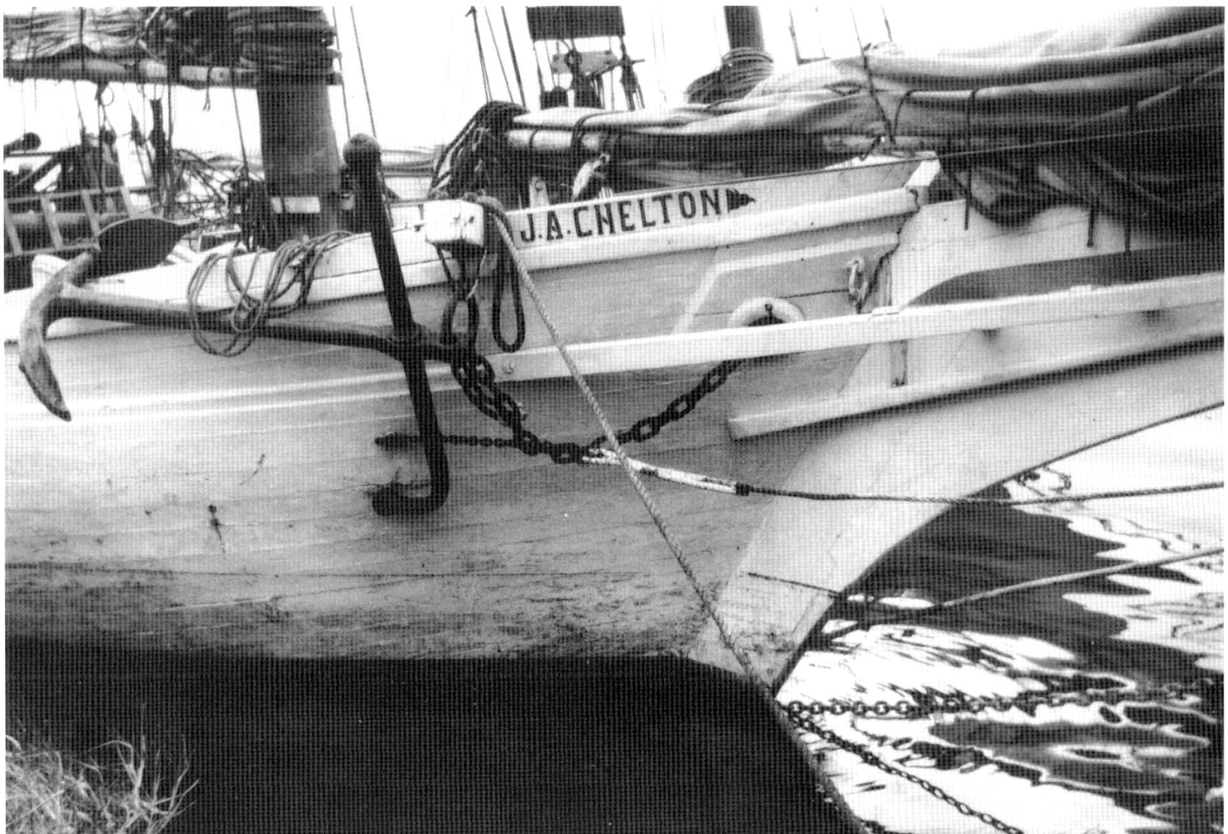

❂ Detail of the bow of the *Chelton* at Montford Yard on the same day.

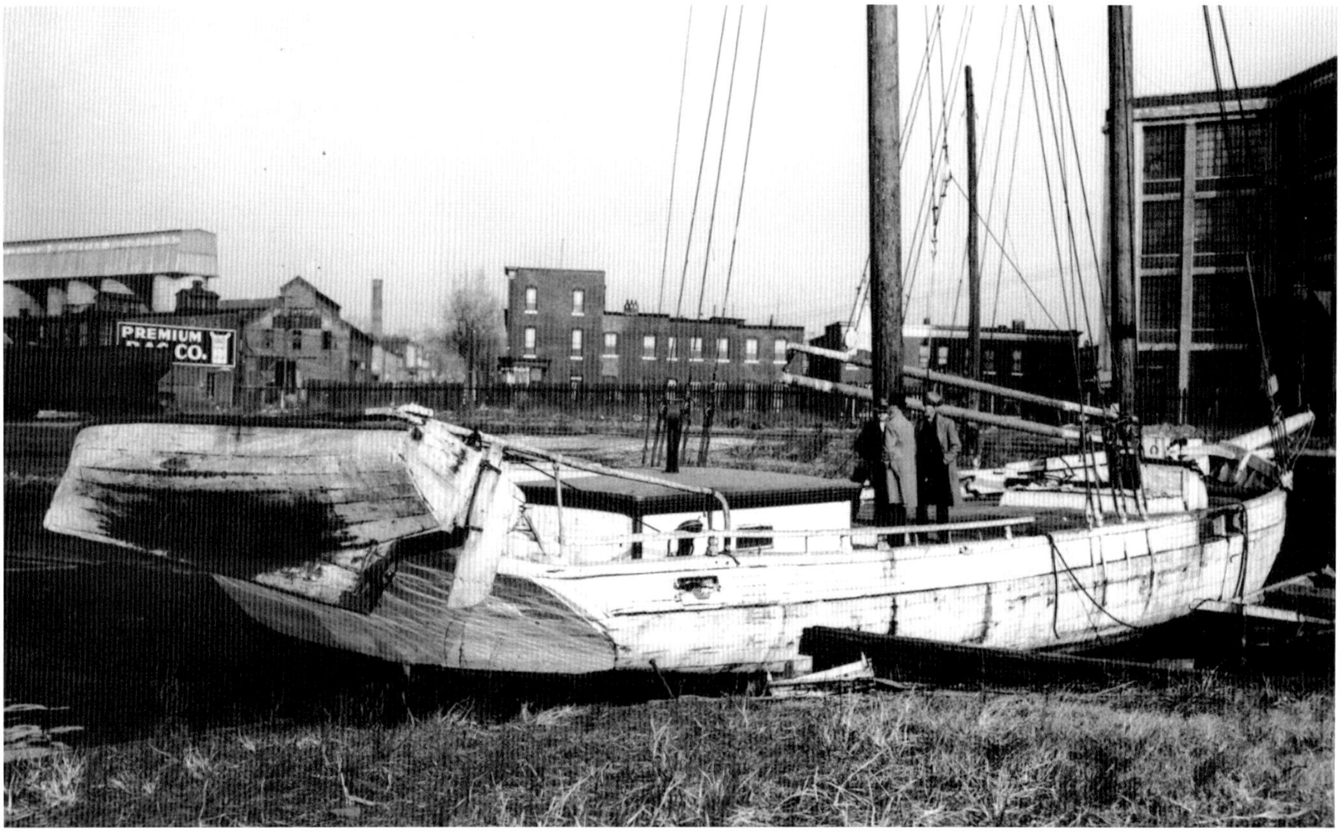

When this view of the *Chelton* was made on January 12, 1936, she had broken her "back" (her keel) and was tied up at Montford Yard, Baltimore, with a "For Sale" sign attached to her fore peak halyards. Perhaps her bad luck can be attributed to the upside-down horseshoe attached to the after side of her samson post, permitting her luck to run out. Many Bay craft carried horseshoes in that location for good luck, but the captains usually made sure that the opening of the shoe pointed upward. *J.A. Chelton* had been built in 1883 at Pocomoke City, Maryland, and served in the dredging fleet. But with no purchaser in view, the old schooner was taken to Jones Creek, near Sparrows Point, Maryland, and abandoned in 1937. Perhaps her yawl boat joined her, judging from the seepage of engine oil and grease through her open seams.

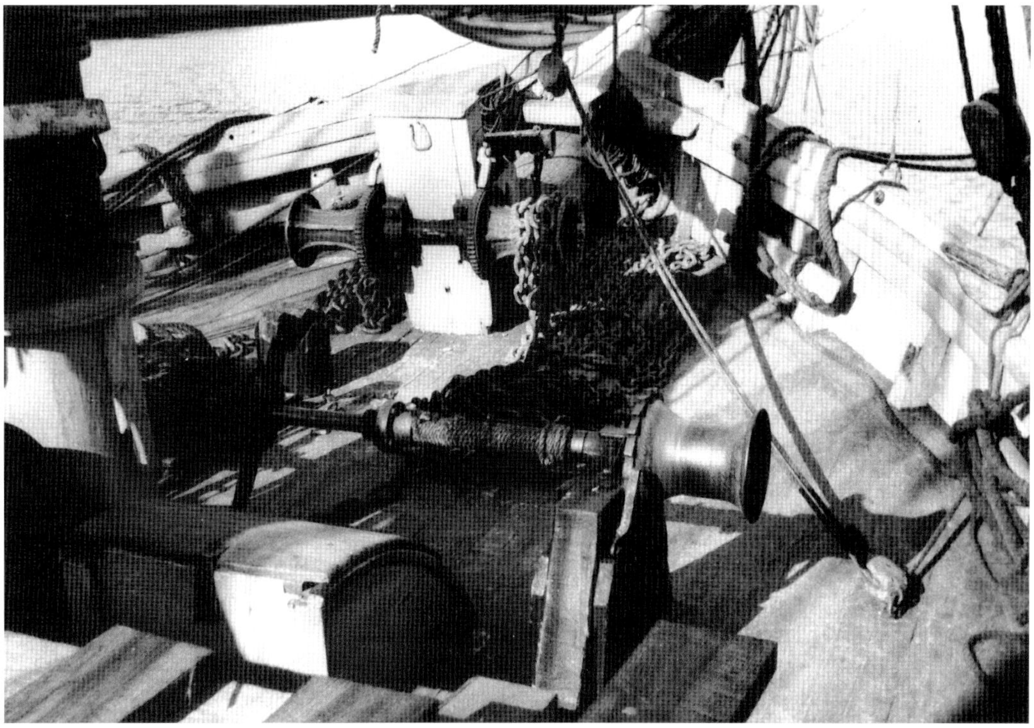

The foredeck on the schooner *J.A. Croswell* shows the arrangement of ground tackle and gear used in handling it and sails and cargo. Everything is snugged down on Christmas Day, 1933, in Baltimore, but would be easy to release if needed for fast anchoring down the Bay. The windlass, center, is fastened to the samson post, and is actuated by pump brake handles on the forward side of the post. In the foreground is the gypsy head turned by a chain drive from a gasoline engine abaft the foremast. In hoisting the sails the throat and peak halyards were led through blocks to the gypsy heads, one on either side of the foremast. The *Croswell's* horseshoe is secured properly to the samson post, but the vessel's luck ran out anyway when she foundered northeast of Poplar Island, in Eastern Bay, on October 6, 1936. She had been built in 1908 at Whitehaven, Maryland.

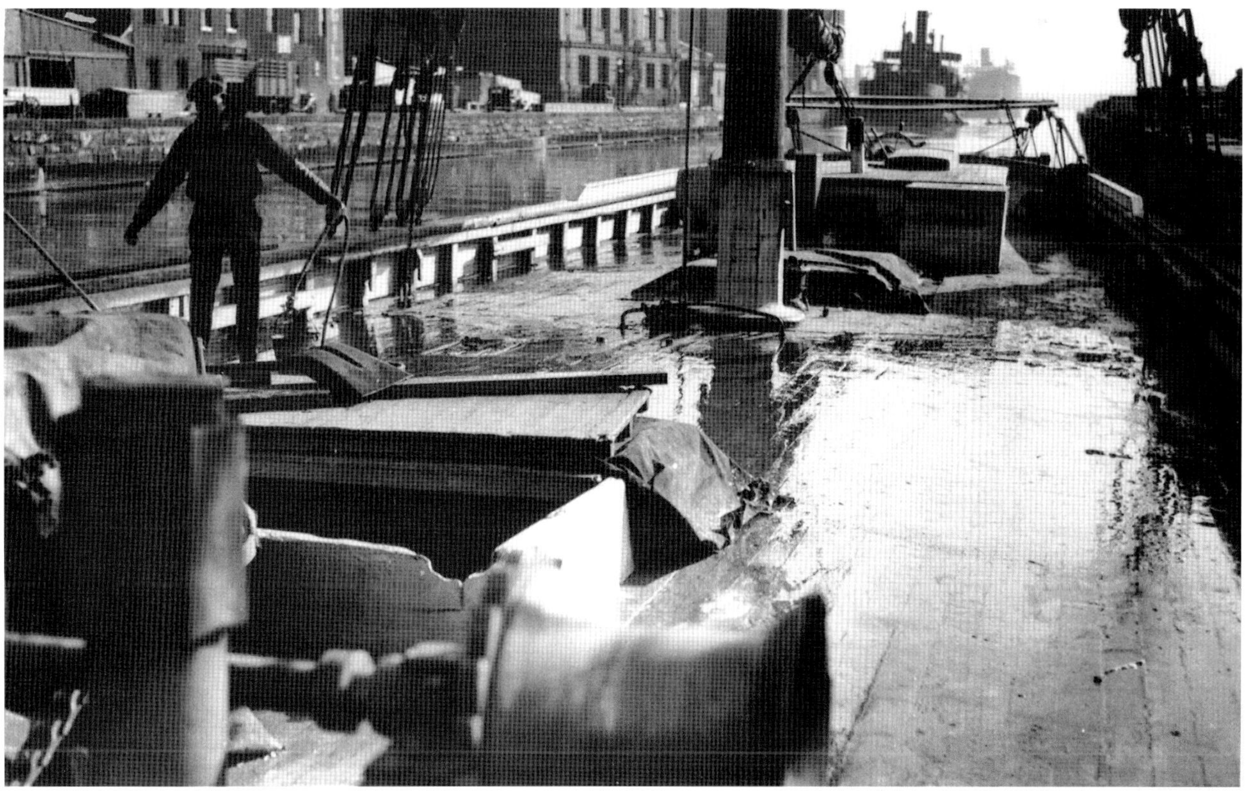

✪ The deck of the *J.A. Croswell* at Jones Falls, Baltimore, on January 14, 1936, looking aft.

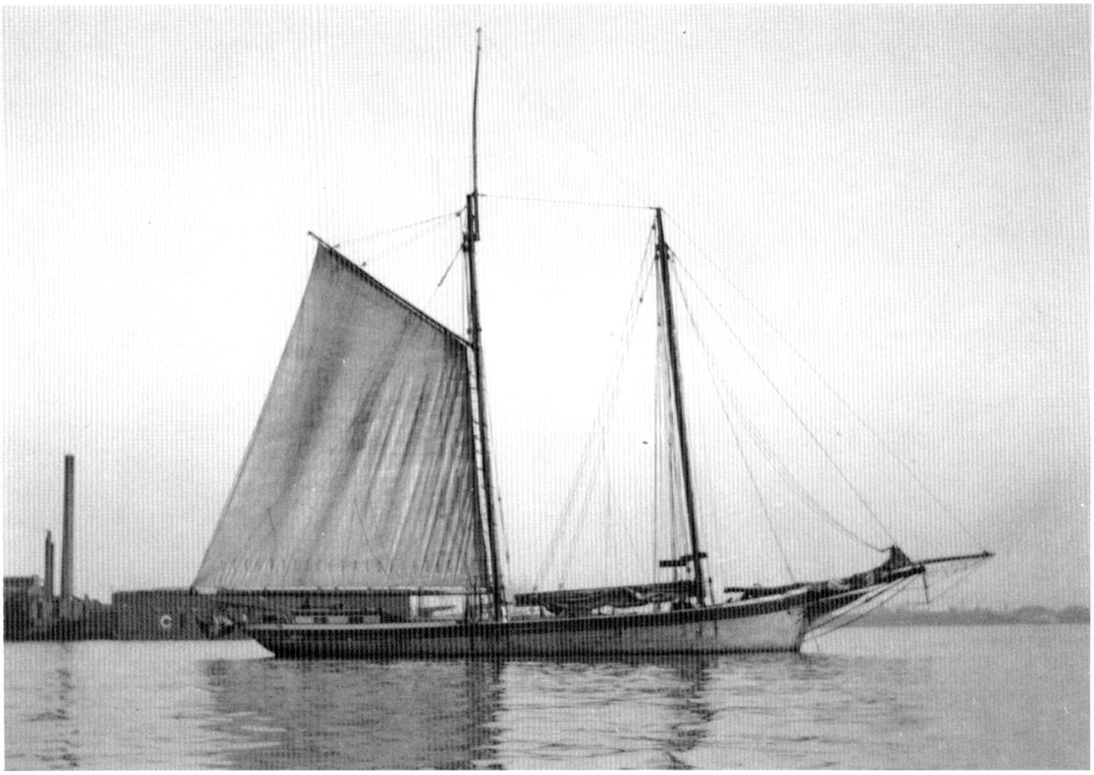

J. W. Knowles rides at anchor in Canton Hollow, Baltimore, January 20, 1934, light and ready to return to her home port of Kinsale, Virginia, on a branch of the Yeocomico River. Built in 1901 at Sharptown, Maryland, the *Knowles* is relatively young as Bay schooners go but her career is drawing to a close. Not long after this photograph was taken she developed a severe leak and was taken to Humphreys Shipyard, Weems, Virginia, for repairs. She was hauled out and caulked but when afloat again her hull continued to leak, so she was pulled onto the shallows at the shipyard and abandoned.

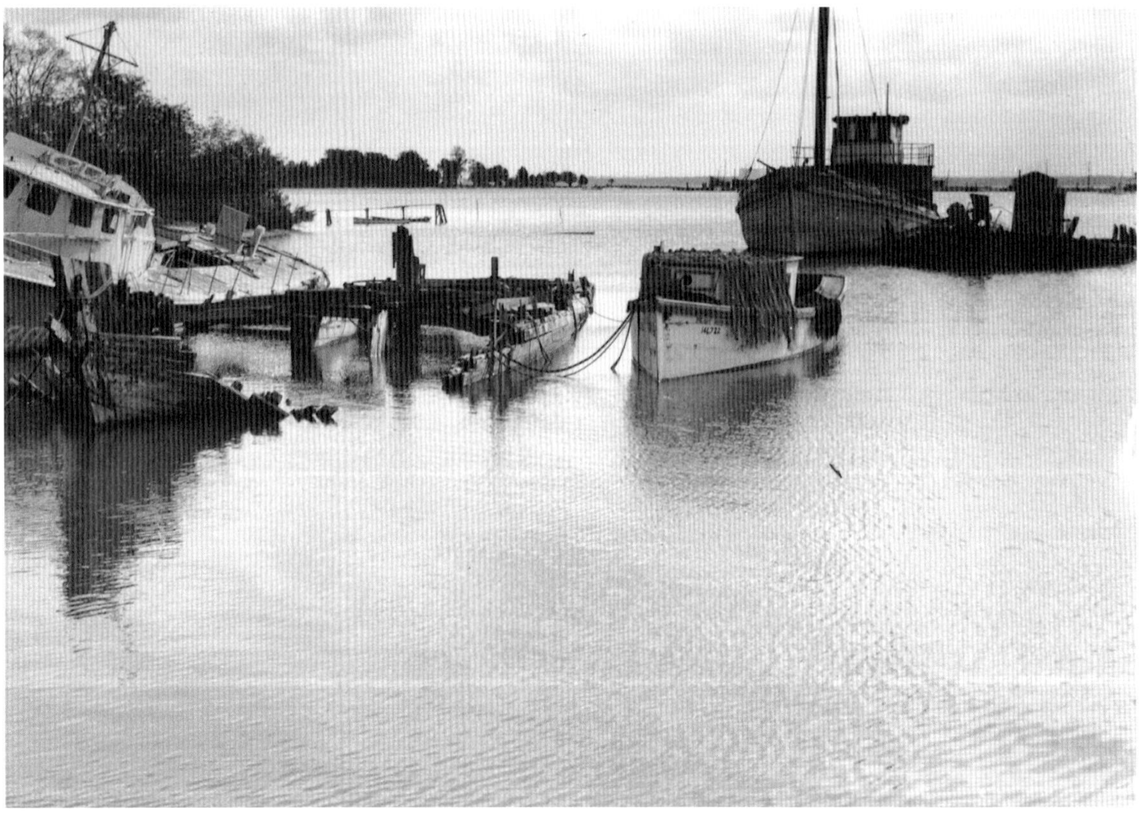

✪ Sunken hull of the *J. W. Knowles,* at Weems, Virginia, in September 1947.

The stately looking schooner *John R.P. Moore,* built in Snow Hill, Maryland, in 1877, rises slowly out of the water at Pier 6, Pratt Street, Baltimore, August 29, 1933, as her lumber cargo is piled on shore. Wide black bands above her light gray sides made her stand out in the Bay fleet, that is, until the bands were discolored by foul harbor waters when she was deeply laden. Wooden whiskers at the end of her bowsprit cap, used to extend the jibboom guys away from that spar to the end of the catheads on either bow, had been broken. Instead of being replaced, the starboard side of the whiskers was merely nailed together, making it about half its normal length. As a result, the *Moore's* head rigging looks awkward.

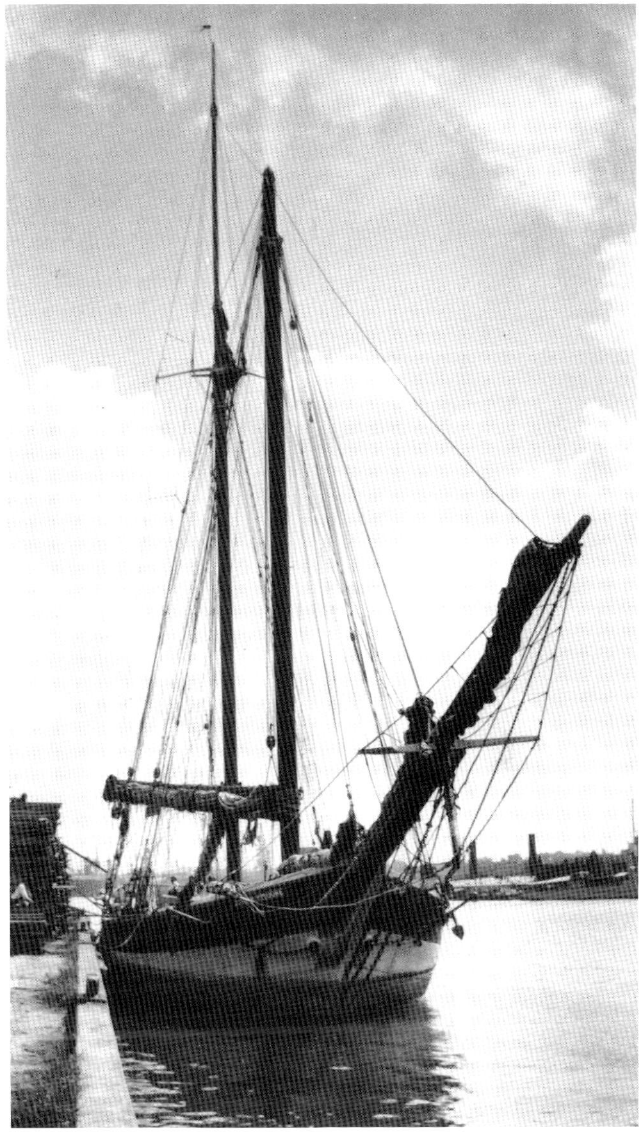

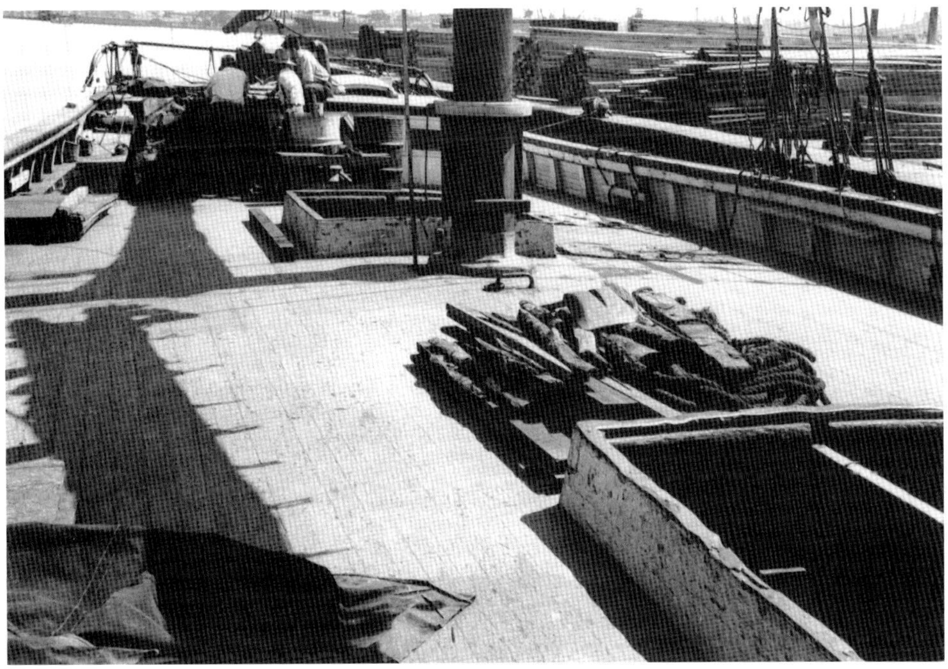

✪ On deck on the *John R.P. Moore,* looking aft, at Jones Falls, Baltimore, on May 20, 1936.

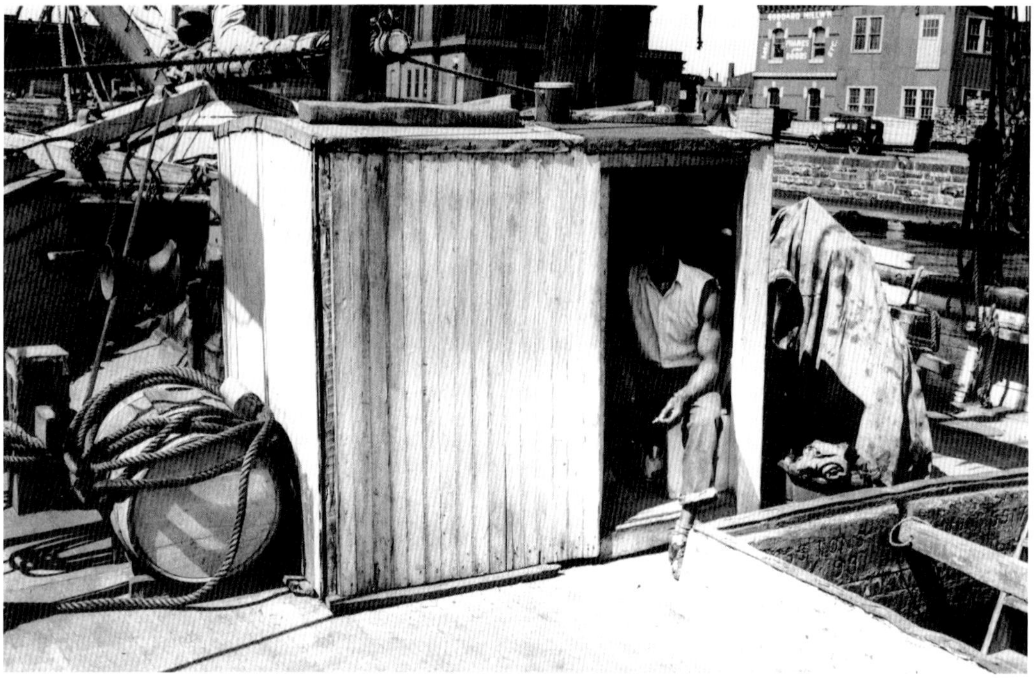

John R.P. Moore was one of the last, if not the last, Bay schooners to carry a caboose. This was a portable "summer kitchen" erected on deck when it got too hot below for cooking. Situated abaft, but just to the port of the foremast, it enabled this muscular cook to benefit from the downdraft from the foresail, making it a rather pleasant spot to be. To starboard of the caboose, covered with old canvas, is the gasoline donkey engine. To port, draped with the end of the mooring line, is the water barrel. As primitive as it may appear, this kitchen was capable of producing some mouth-watering aroma and tasty meals. In the oven of the iron cookstove, fired with wood, high, fluffy biscuits would be created. Country ham and eggs, the leftover grease placed in the plate and mixed with molasses, would be a breakfast to remember. A pot of bean soup would simmer on the back of the stove, while the coffee was ever hot. And when the man sitting in the doorway was not cooking, he and the captain would be sailing the schooner. Carved on the main beam of the fore hatch, lower right, appears GROSS TONS 99 No. 75917. That official number is retained through the life of the vessel and on an old abandoned hulk is often the only clue to the ship's identity. The *Moore* is tied up alongside Pier 6, May 1936, and is in Jones Falls, a stream that originates far north of Baltimore. In the city the stream is covered over with streets and emerges in the open again as it approaches the harbor. On October 31, 1938, the *John R.P. Moore* foundered in Norfolk harbor.

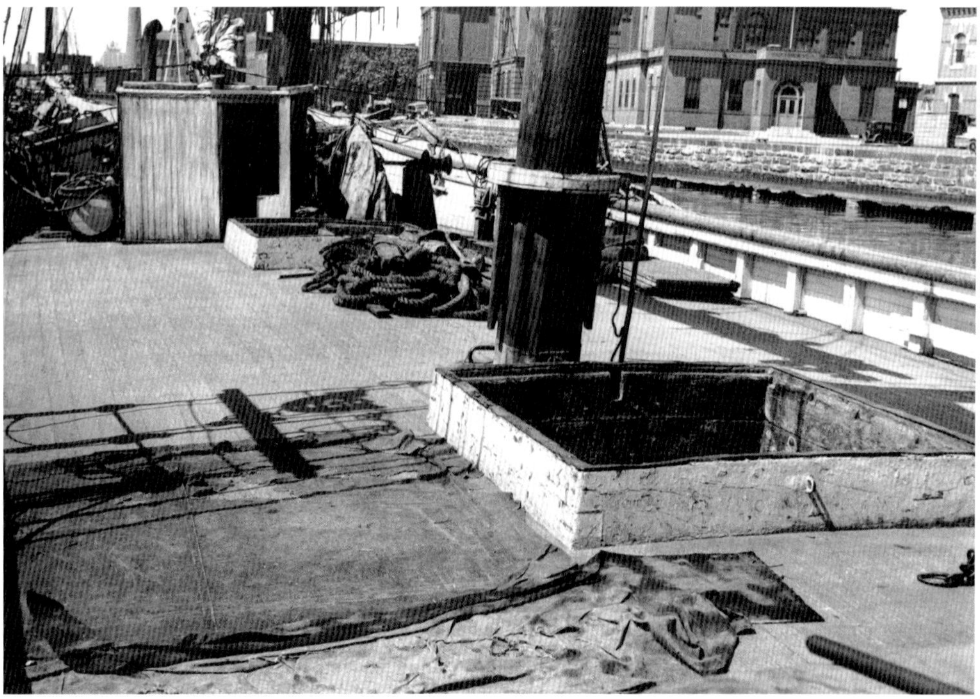

✪ The deck of the *Moore,* looking forward. Her hatches are open and her "caboose" is shown in this photo.

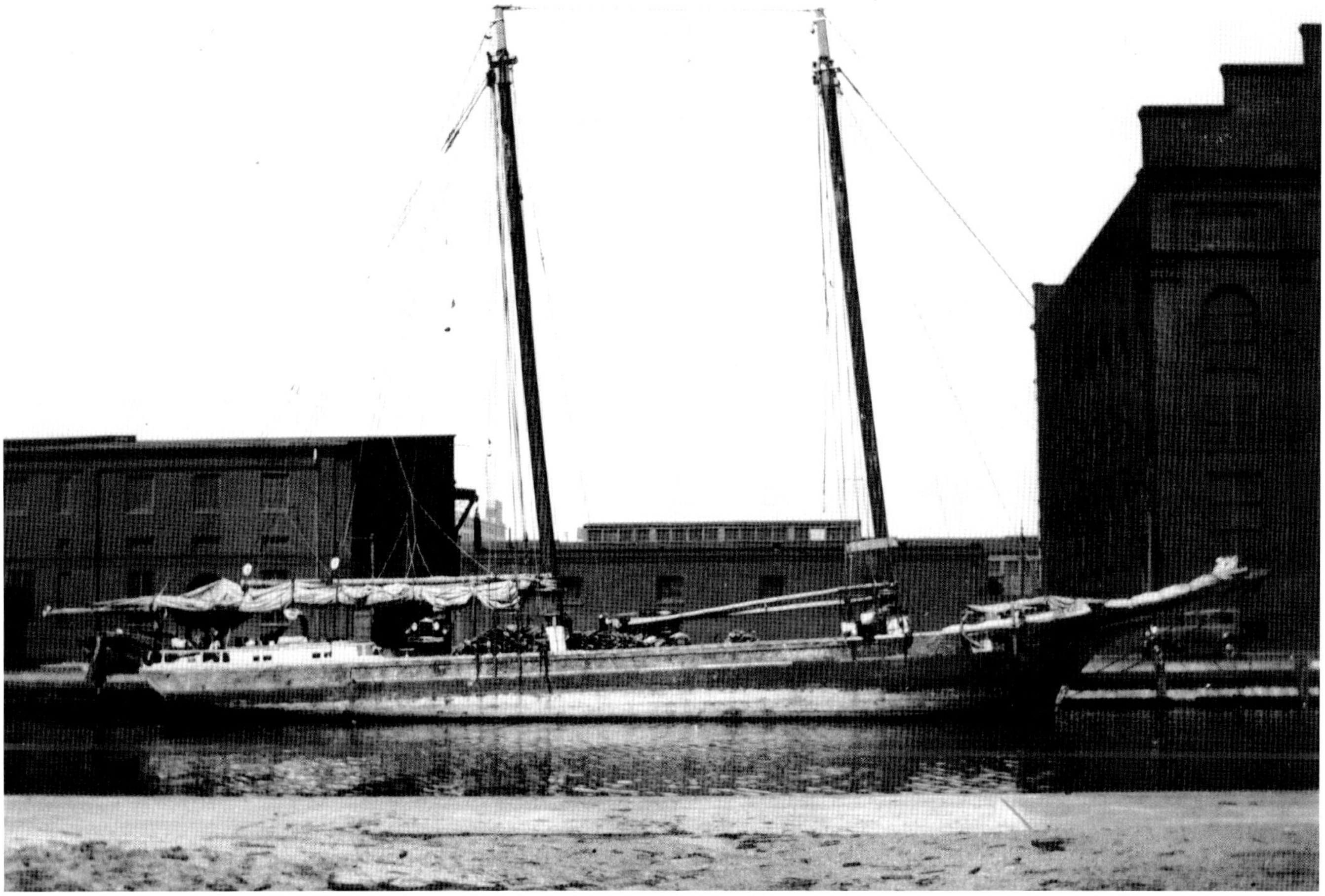

The oldest vessel sailing on Chesapeake Bay during the past 50 years, when the author was recording the history of the craft on film, was the two-masted schooner *John W. Bell,* built in New York, N.Y., in 1848 and rebuilt at Seaford, Delaware, in 1912. Often I looked at this venerable hull and my imagination would run rampant—here was a ship built only five years after the first extreme clipper ship, the *Rainbow* of 1843; the *Bell* was built the same year John Sutter discovered gold in California, an event that spurred on the construction of the great clippers; and the *Bell* outlived all those great examples of shipbuilding. In this illustration the *John W. Bell* is tied up at Long Dock, Baltimore. Cordwood is stacked on her deck. She has long passed that stage in the Bay schooners' careers when they had to resort to hauling cordwood. She is fortunate to even get a cargo such as this, for she was 86 years old when this photograph was taken on June 13, 1934. The schooner's foresail is not bent. It could be that a new one is being made or the old one repaired at the William L. Godfrey sail loft just around the corner on Pratt Street. About 1930 the *Bell,* then owned by C.C. Paul & Company, was tied up at a pier at the foot of Leakin Street, Baltimore, as it appeared her days of activity were over. Then, in 1932, a David MacMurchin of Houston, Texas, bought the schooner with the thought of taking her to St. Thomas, Virgin Islands, to trade. That plan never materialized, but the *Bell* did resume trade on the Chesapeake again, if only for a few more years. In 1935 she was anchored in Colgate Creek, making off the Patapsco River near the eastern limits of Baltimore, when she developed a leak and sank. Her stern, resting on a bar, sat high out of the water. Her bow was submerged. She disintegrated where she lay.

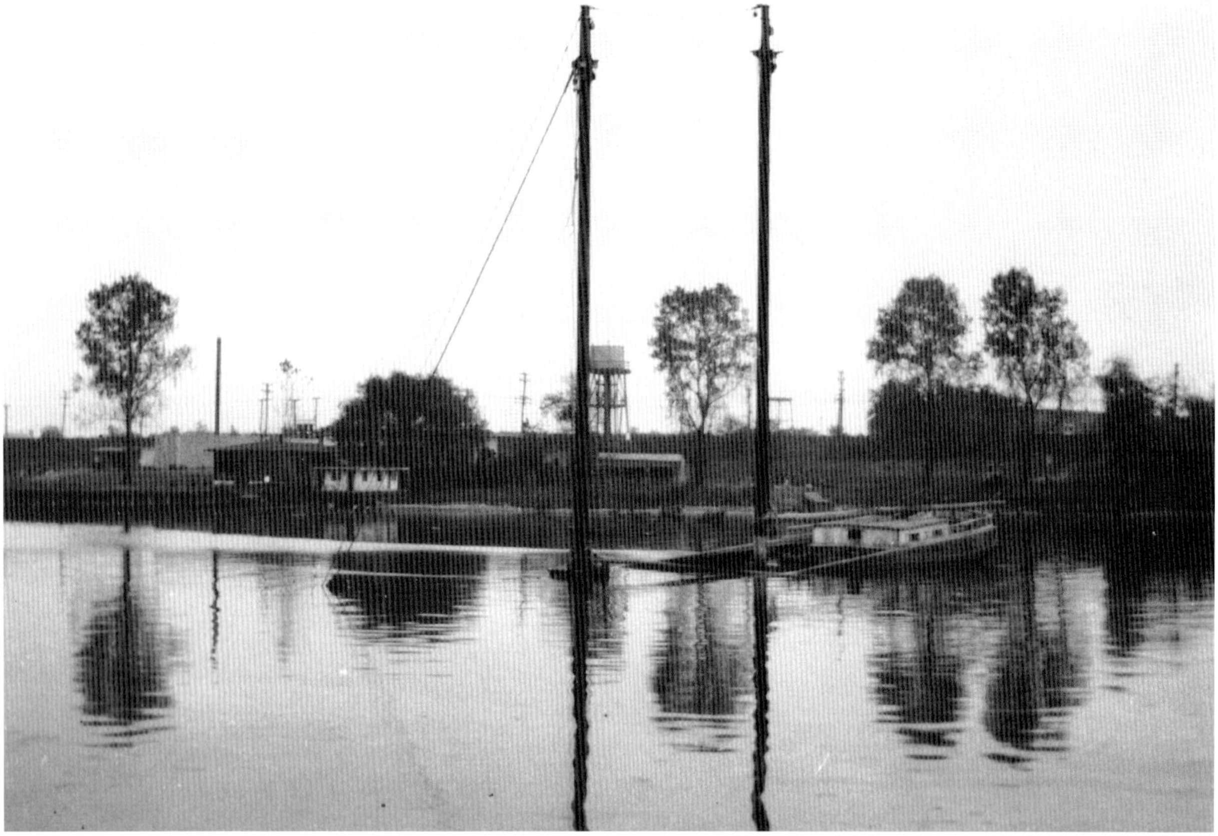

✪ The schooner *John W. Bell* lies submerged in Colgate Creek on October 13, 1935.

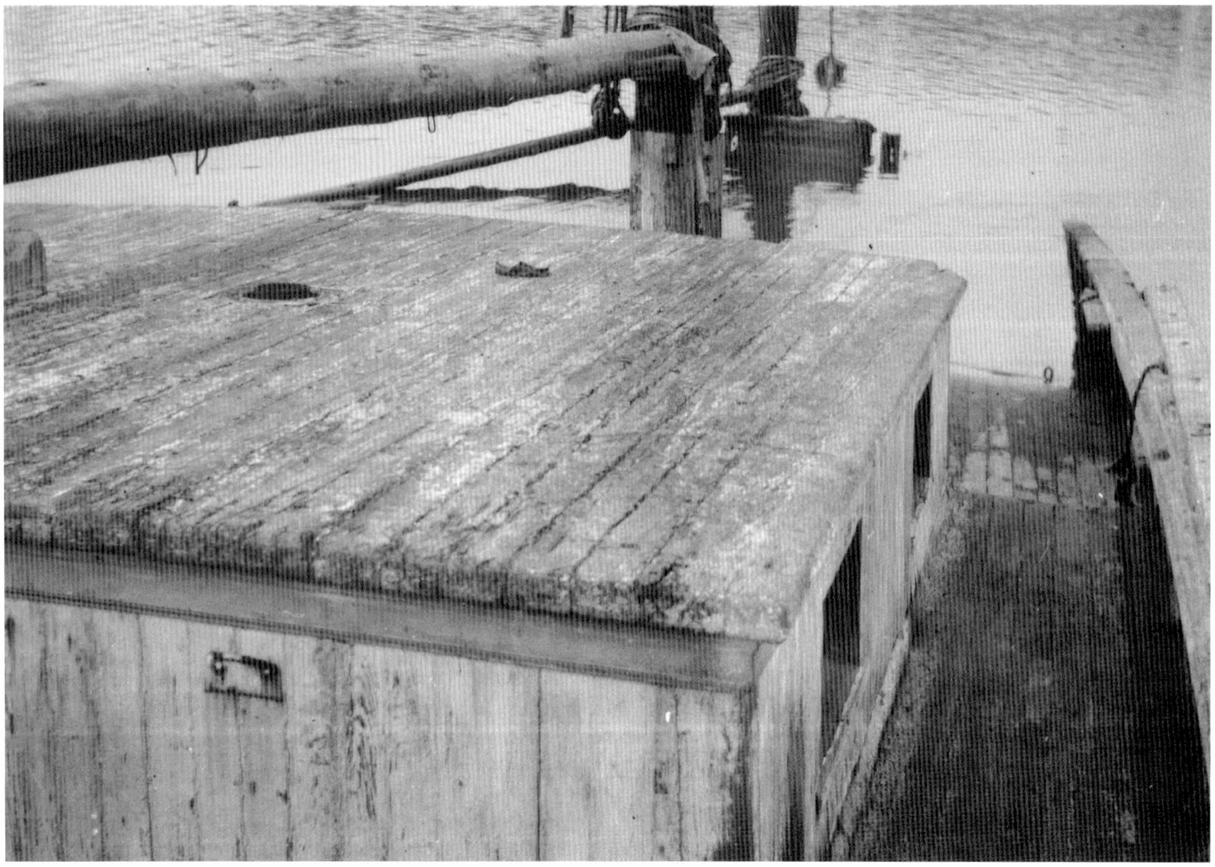

✪ On the *Bell*, looking forward at her sunken bow.

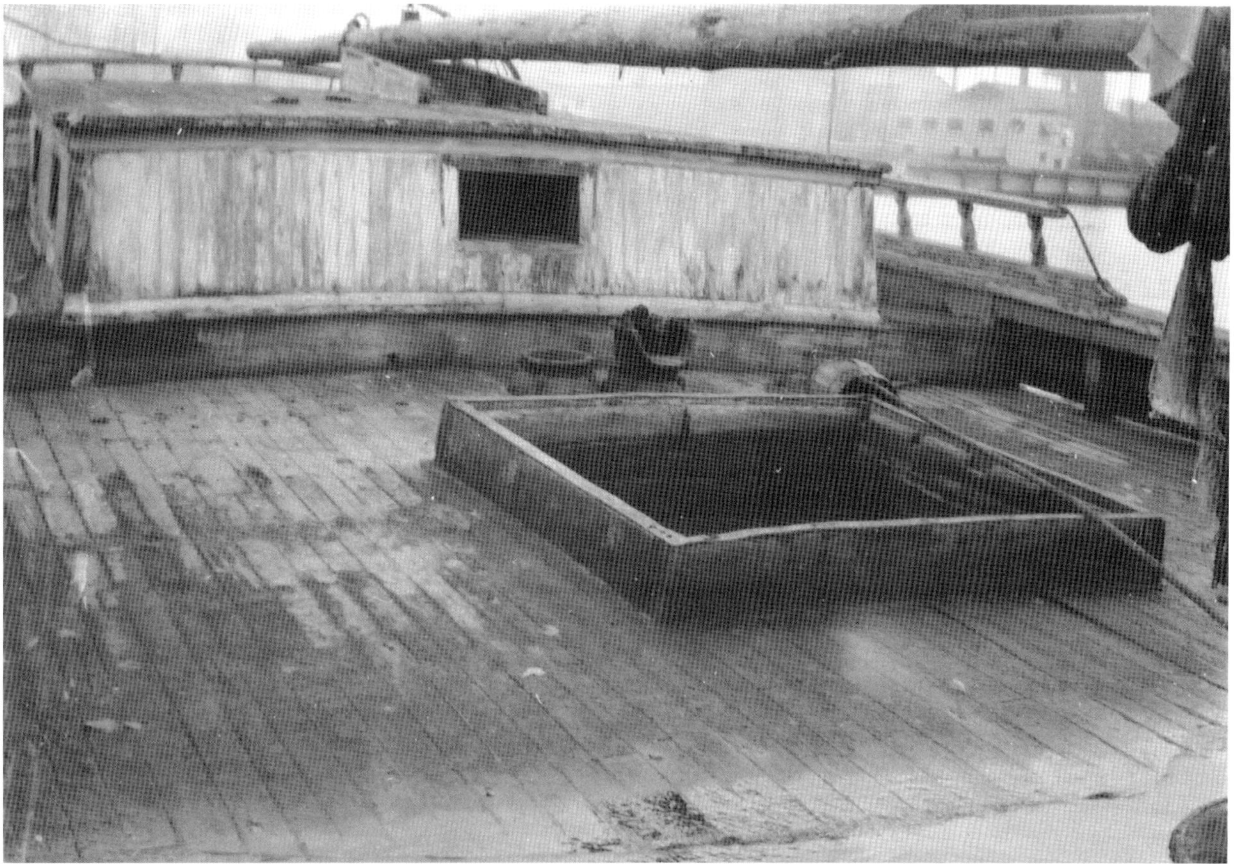

✪ Looking aft on the *John W. Bell,* on October 13, 1935.

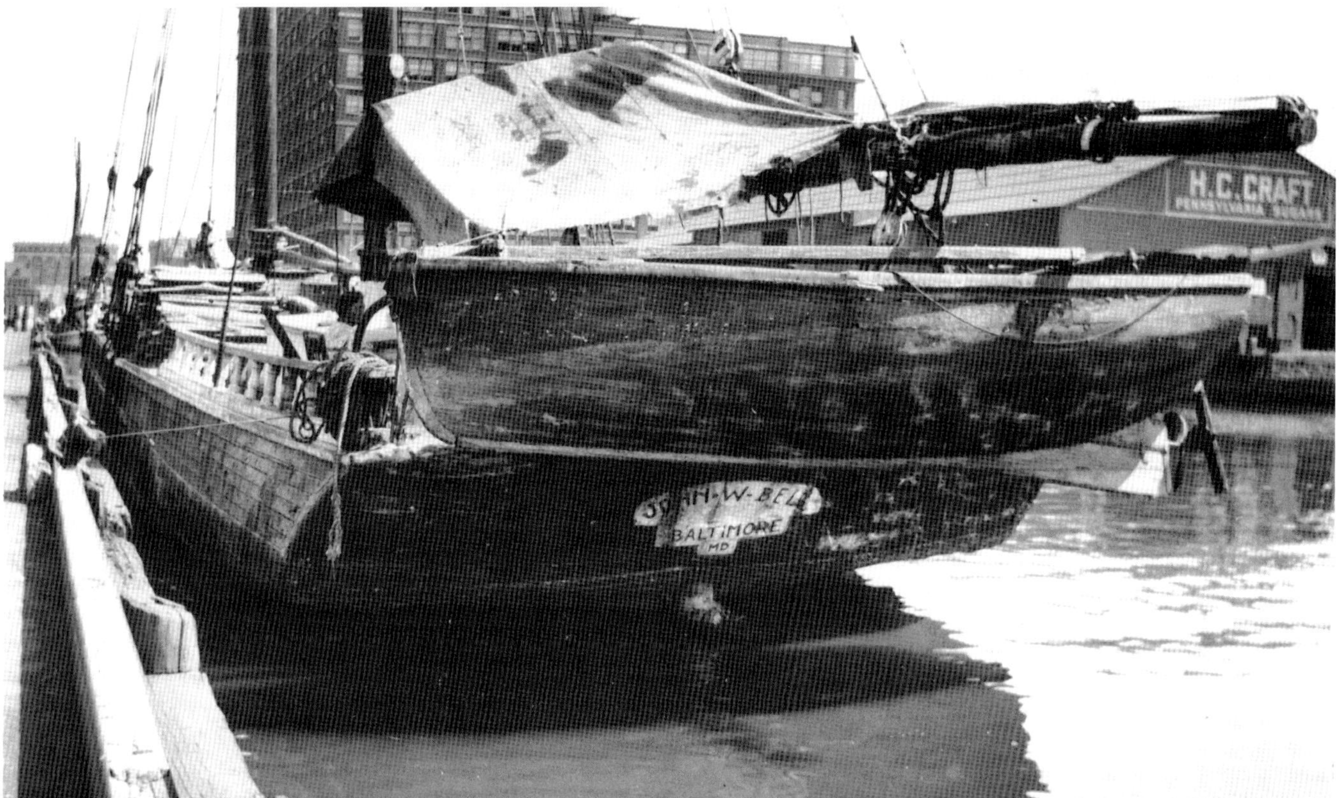

Stern view of the *John W. Bell* at Long Dock. A piece of old canvas is stretched over the after deck to keep the hot sun off the people sitting there. Note the light line holding her stern to the dock. The poor old schooner had surely come upon hard times.

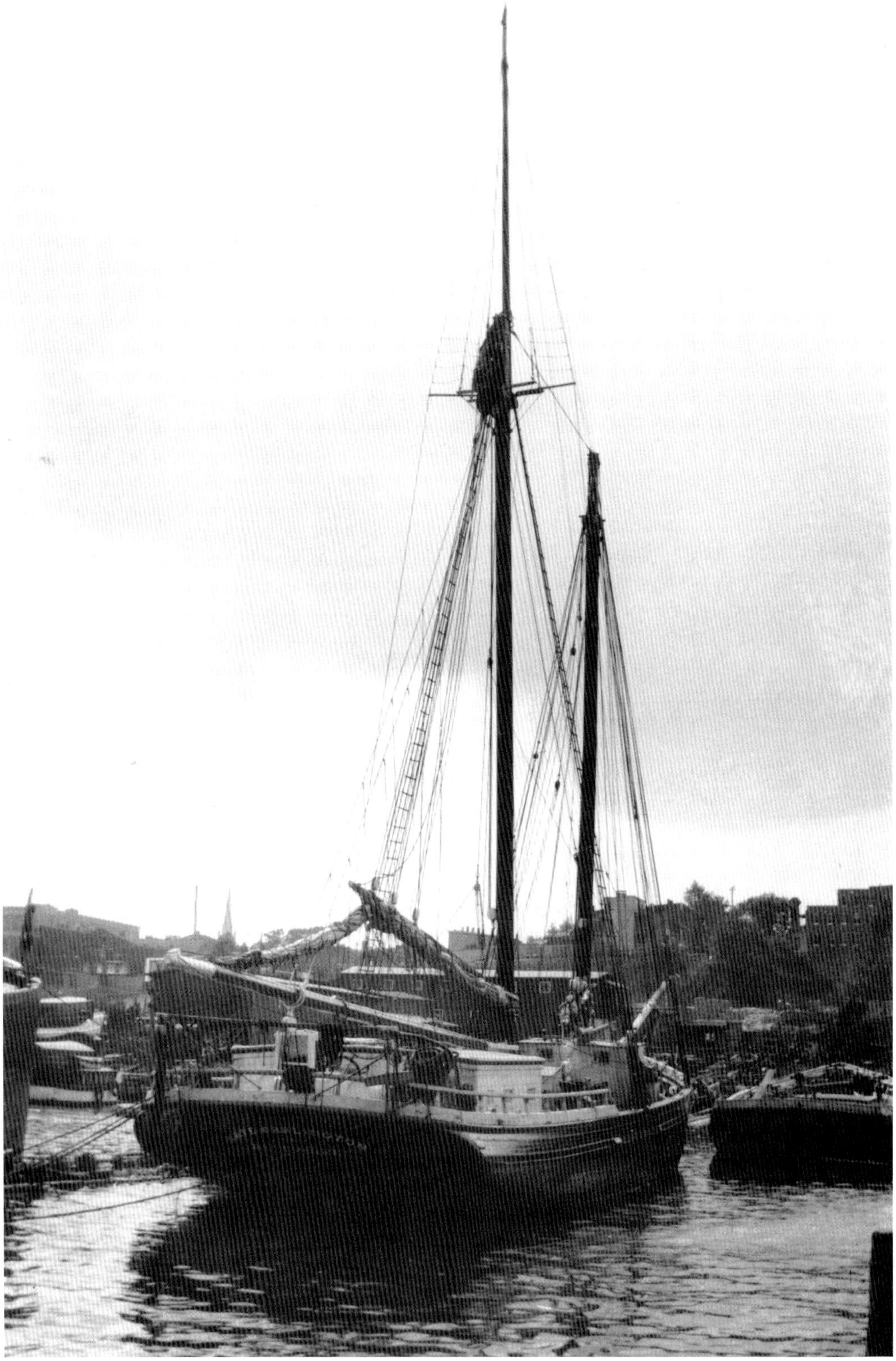

What might have been called the "grand old lady of the Chesapeake" in the 1920s and '30s was the schooner *Kate Darlington.* There was always a distinguished air about this vessel; it is evident in this portrait of her at Redman and Vane Shipyard, Key Highway, Baltimore, on September 2, 1933. The paint on her upper hull glistens even on that overcast day. Her mainsail has been lifted from the freshly painted main boom. Her future looks brighter than that of the recently dismantled schooner *Mary A. Fisher,* at right. To the very end the *Darlington* carried her maintopmast and jibboom, which added to her fine appearance.

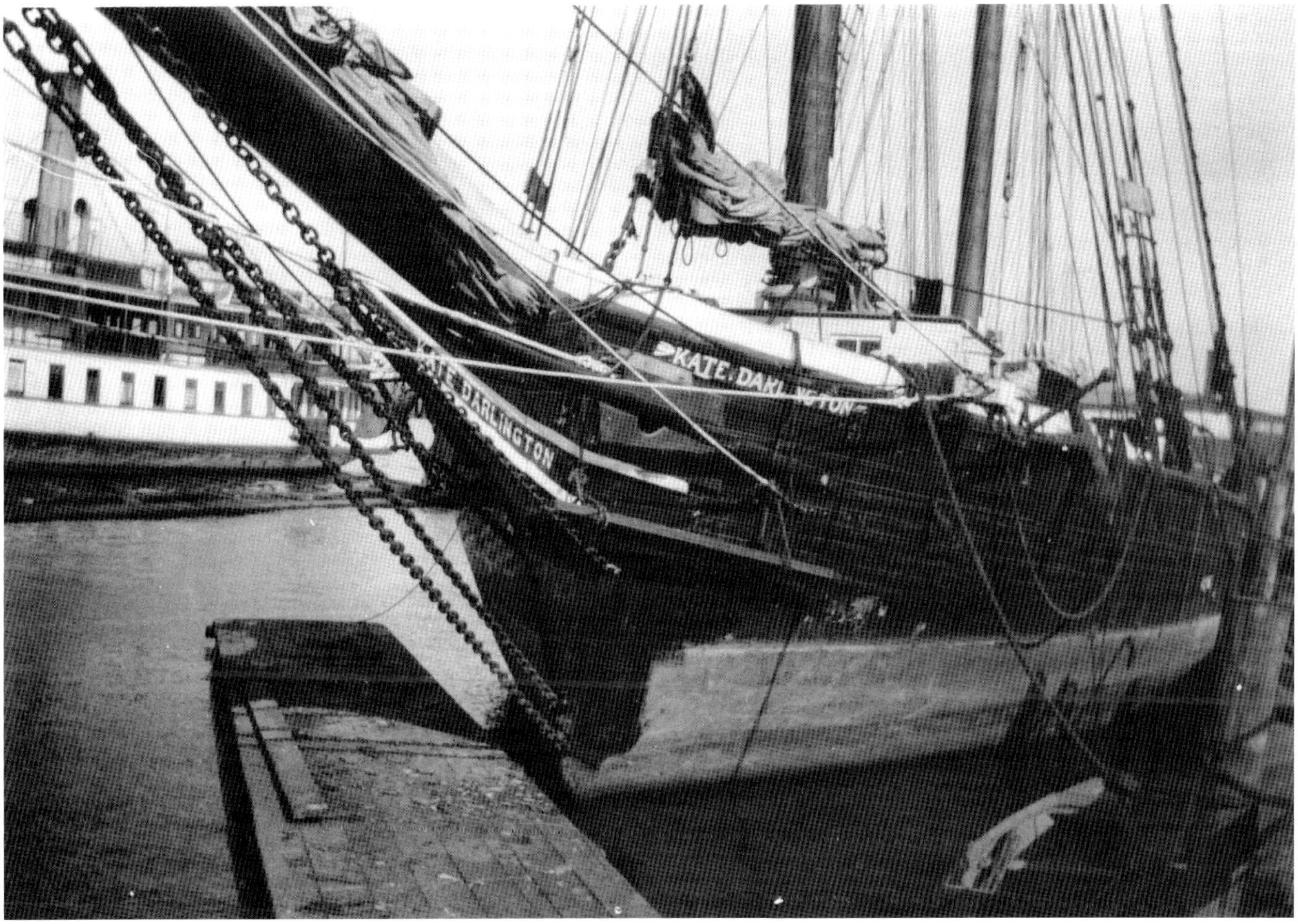

Still at the Redman and Vane Shipyard on October 28, 1933, the *Kate Darlington* has continuing work done on her hull from the raft in the foreground and the stage under her hawsepipe. This schooner had been built in Baltimore in 1889 by H. Brusstar and Brother and was fitted out with an eagle figurehead and carved trailboards, shown here. When the stripped hull of the *Darlington* was taken around to Port Covington on the Patapsco River's middle branch to be broken up for firewood by Mrs. Marie Mattson in 1939, the schooner still carried her trailboards. At left is the steamboat *Chippewa* that ran excursions to Betterton, Maryland, that summer.

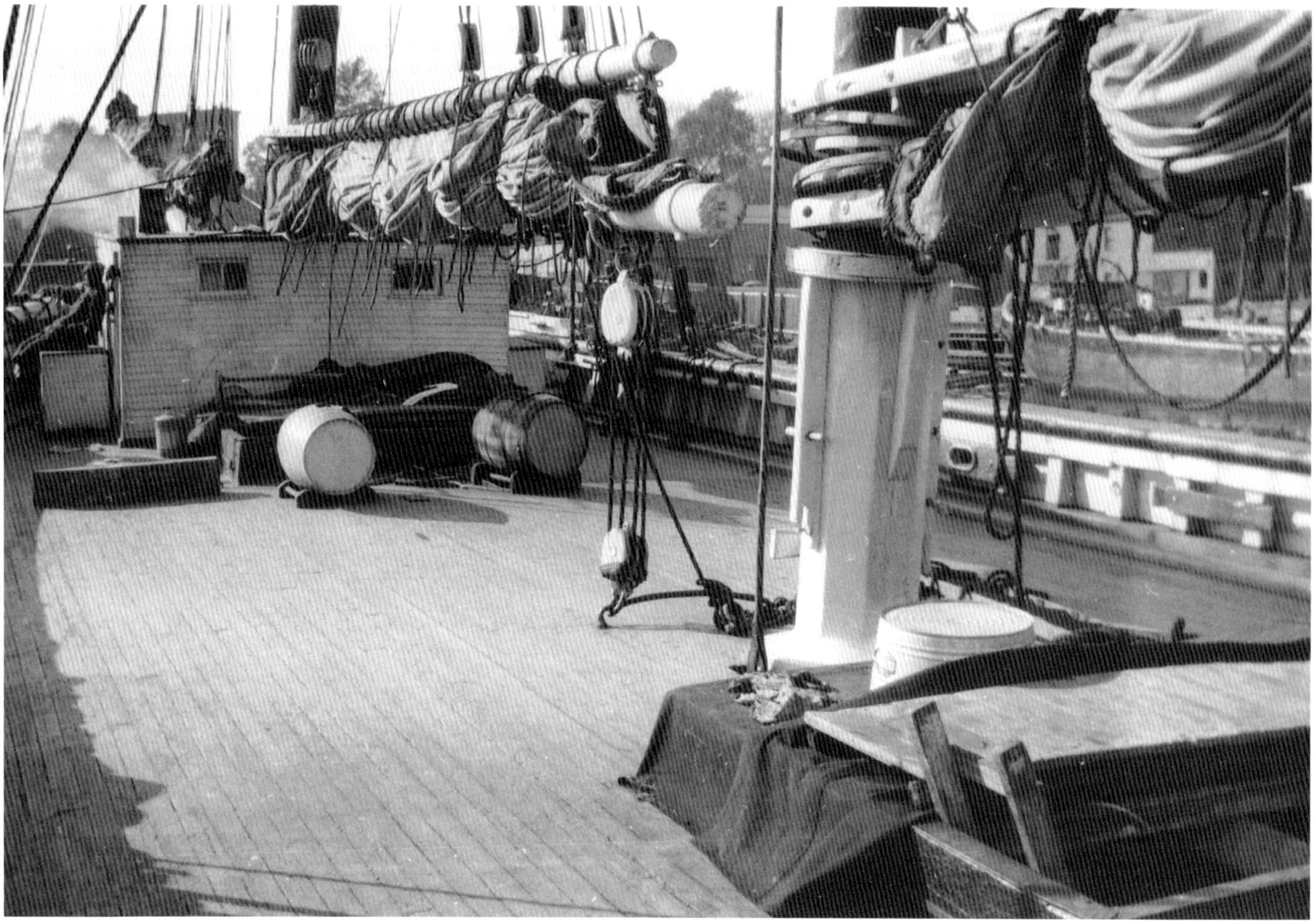

Looking forward on deck of the *Kate Darlington* at the Redman and Vane Shipyard, October 28, 1933, smoke emerges from the stack atop the galley in the forward house as the cook sits outside the door on a chest taking in the Indian summer sun. The donkey engine occupies the starboard half of the house. The traditional water barrels sit on deck—not their normal position, which is up against the house on either side of the fore hatch. The *Darlington* was a member of the pineapple fleet, so she knew the taste of the sea and looks built for it. But the Chesapeake was her usual trading route. In center background is wooded Federal Hill, overlooking Baltimore harbor. The *Mary A. Fisher* hull is still idled at right.

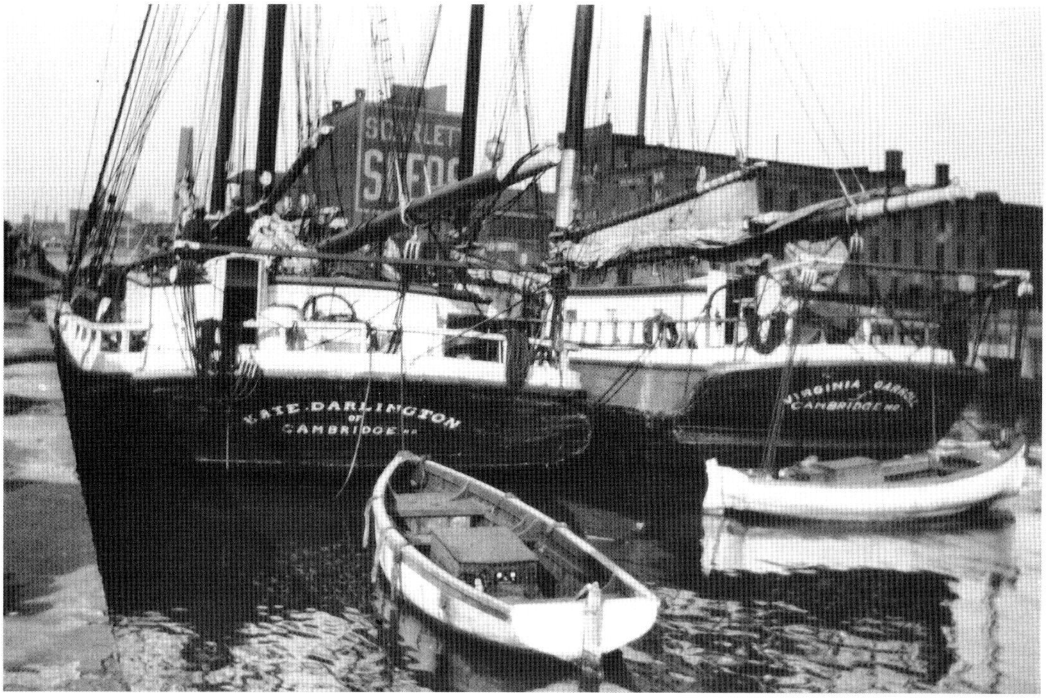

During the latter part of November 1933, while bound from North Carolina to Philadelphia with lumber, the *Kate Darlington* ran aground off Plum Point, on the western shore of the Bay above the Patuxent River. The schooner *Virginia Carroll* was sent down to take on the stranded schooner's cargo to lighten her for refloating. The powerboat *W.B. Blades* was dispatched to tow the *Darlington* off the bar. By December 5 both schooners were at Pier 6, Baltimore, the lumber cargoes being sent ashore. Baltimore's famed Shot Tower rises at upper left.

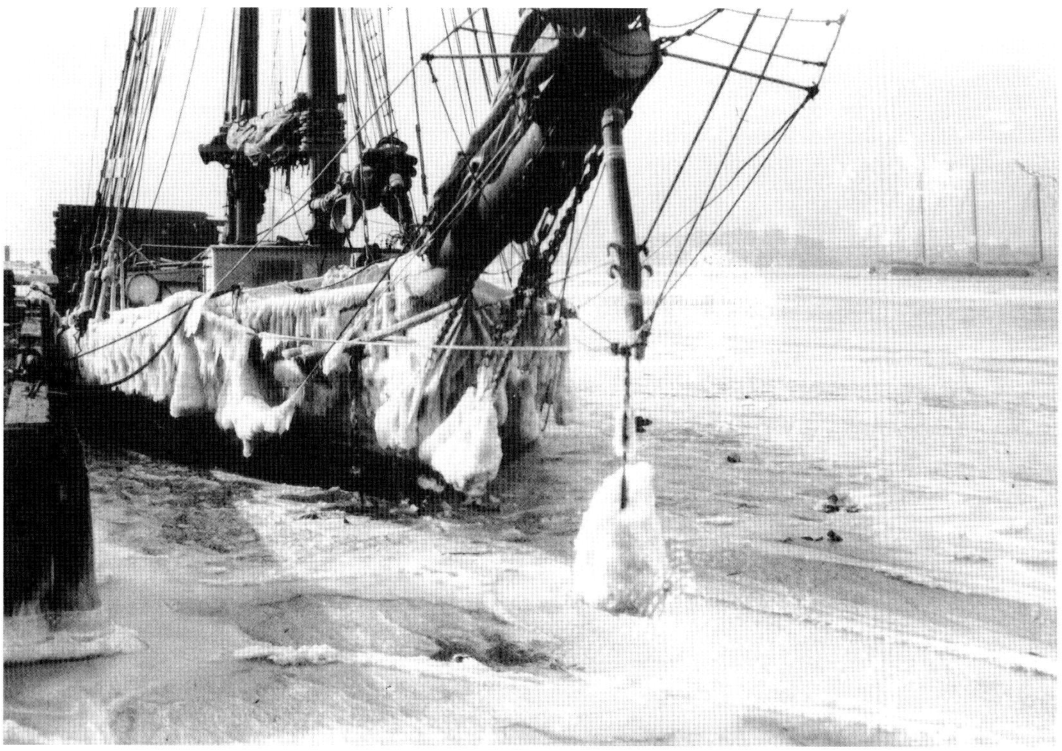

Looking like a fisherman in from the Grand Banks after a winter gale is the *Kate Darlington,* tied up at McGrath's, Baltimore, February 1936. She was at anchor in the Chesapeake when a storm bore down on her, and as she surged on her anchor chain the spray froze as it fell. The line at the bottom of the dolphin striker dipped in the Bay with each surge and froze as it was exposed to the air, building up the huge chunk of ice. Captain Marshall Pritchard was her master at the time. At right background is the ram *Kinkora,* frozen in Canton Hollow with her deck piled high with lumber. She was converted into a powerboat during World War II and sold to the Dominican Republic in 1943.

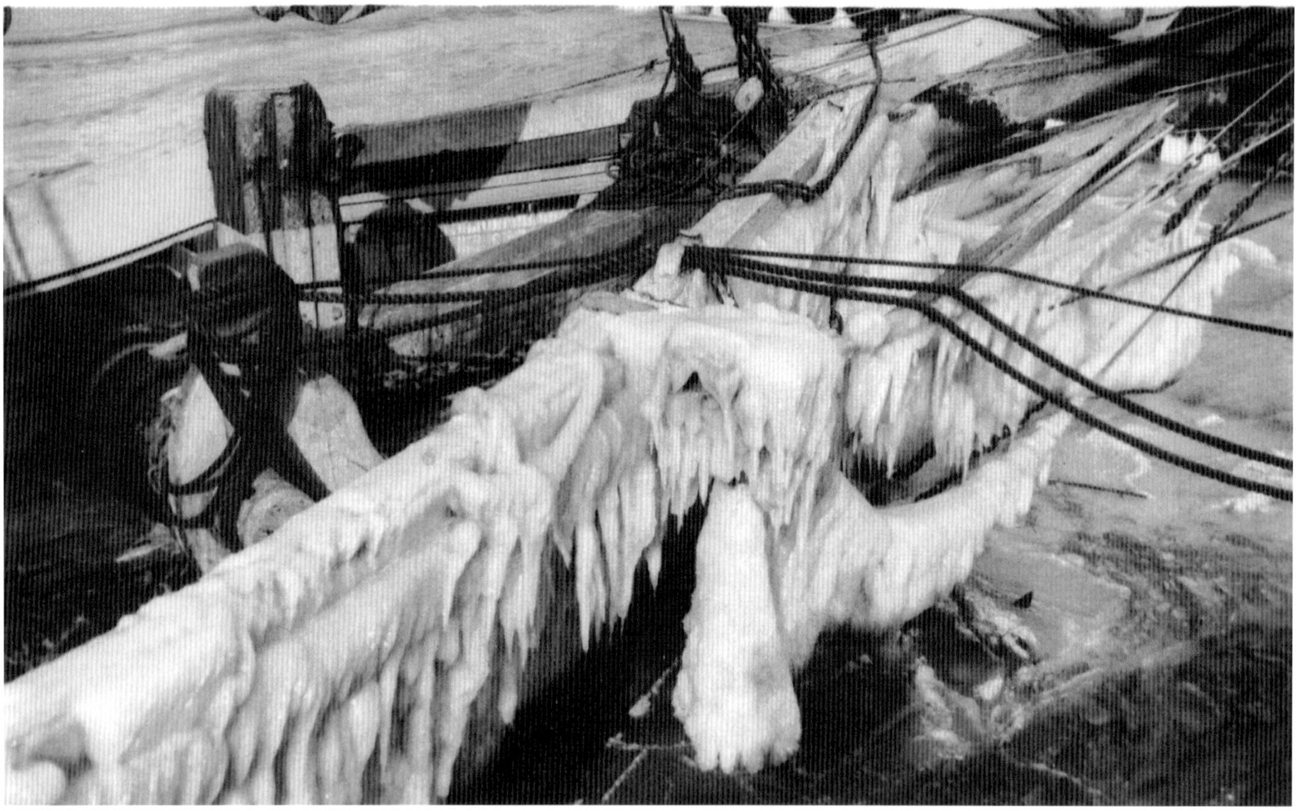

A close-up of the prow of the *Kate Darlington,* covered in ice in February 1936.

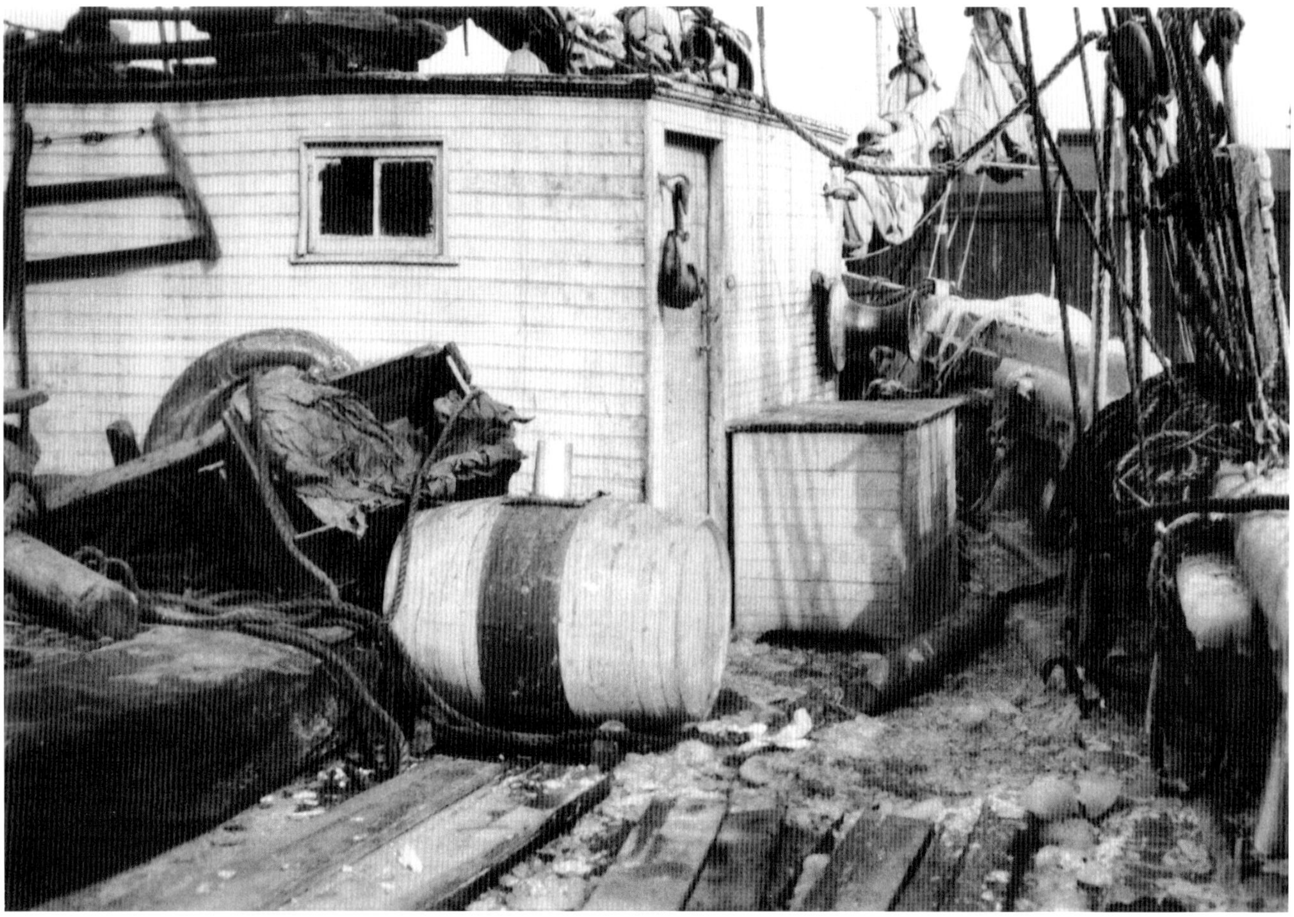

The forward deckhouse of the *Darlington,* looking forward, as it appeared in February 1936.

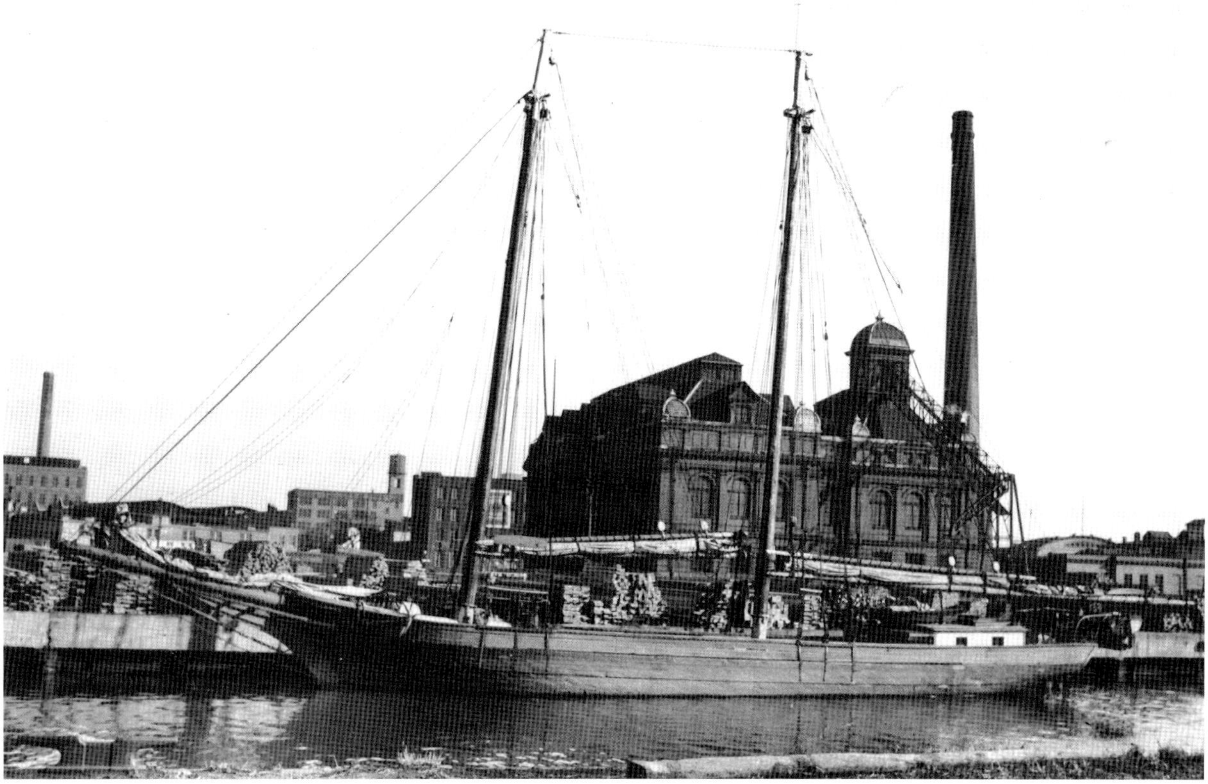

Masts and bowsprit of the *Kate H. Tilghman* look stumpy while tied up at Pier 6, Baltimore, on January 14, 1936. This "scalping" was usually preliminary to eventual abandonment. The huge brick building on Falls Avenue, background, was a pumping station. Adjacent to it in a low brick building was the city morgue.

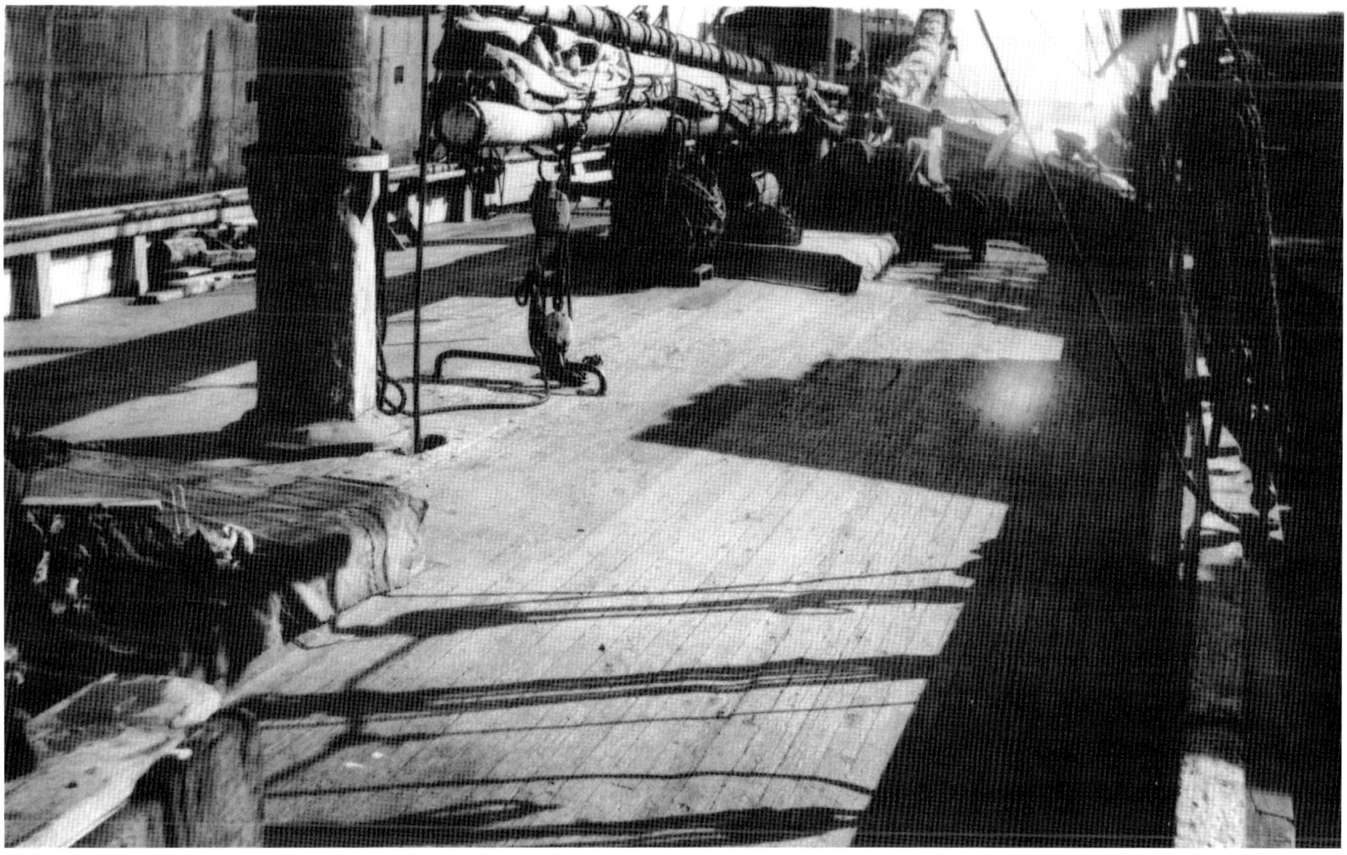

✪ On deck aboard the *Tilghman,* at Baltimore on October 25, 1935.

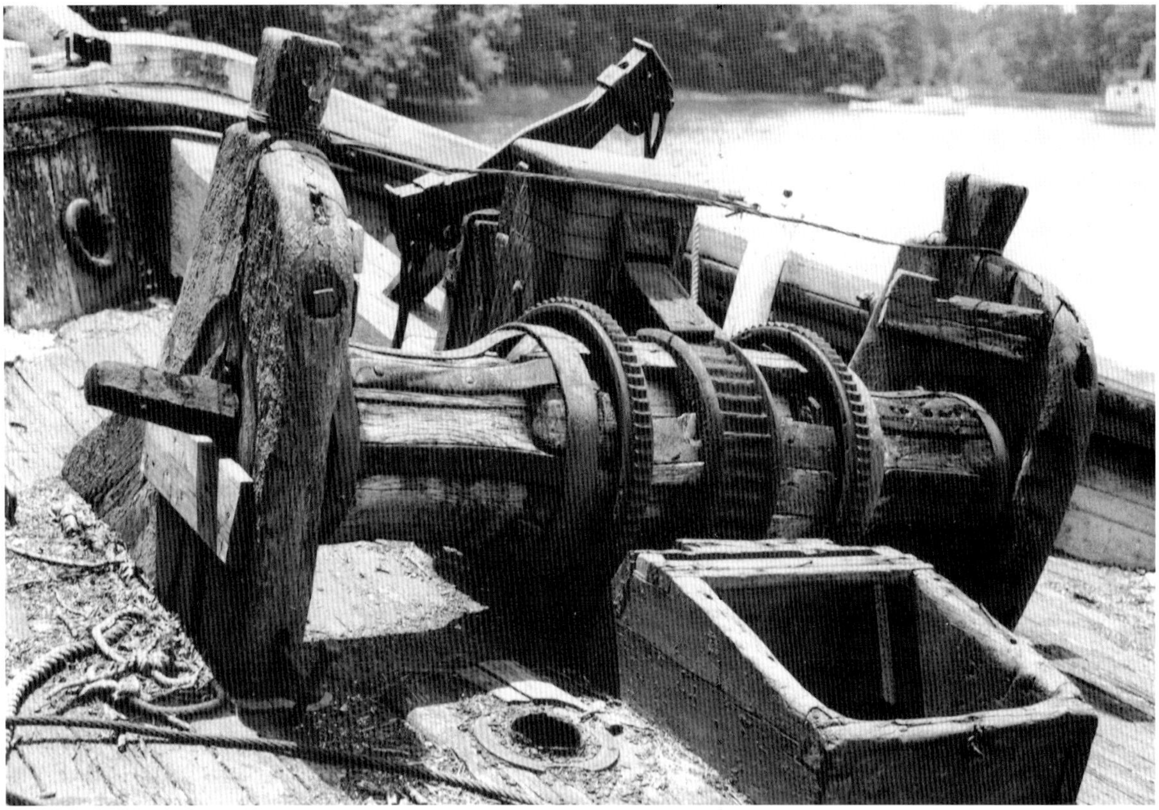

Abandonment was the fate of the *Tilghman,* in Urbanna Creek, Virginia, about 1945, and then total deterioration. By July 6, 1947, she had reached this state of dilapidation, sitting on the bottom, listing, and rot setting in throughout. Like some primitive monster, her log windlass waits for time to rust its metal and soften the wood, and then it will collapse. This was used to take in the anchor chain that would come in through the hawsepipe in upper left, or on the starboard side, and pass around the barrel of the windlass in three turns. Handles would be fitted into the rocker arm mounted on the fore part of the samson post, and the arms, in turn, connected to rods linked to ratchet ring gears on the barrel of the windlass. As the deck gave way with the years, the windlass eventually crashed to the bottom. The *Tilghman* was built in 1881 in Wicomico County, Maryland, and in her early career went beyond the bounds of the Chesapeake. On April 11, 1909, she stranded on Royal Shoal, seven miles west-northwest of the Ocracoke Life-Saving Station, North Carolina, while bound from the Rappahannock River to New Bern with a cargo of corn. With the assistance of the nearby lifesaving stations, she was floated the next day.

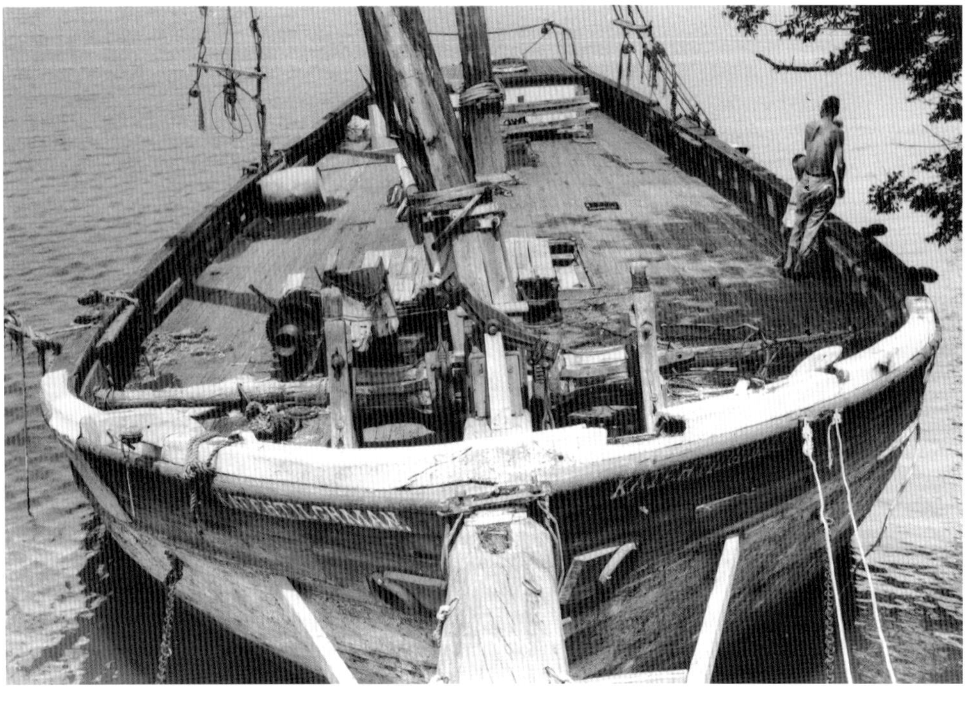

✪ The bow of the schooner *Kate H. Tilghman,* abandoned at Urbanna, on July 6, 1947, shows the wooden horse for her jib sheet traveler.

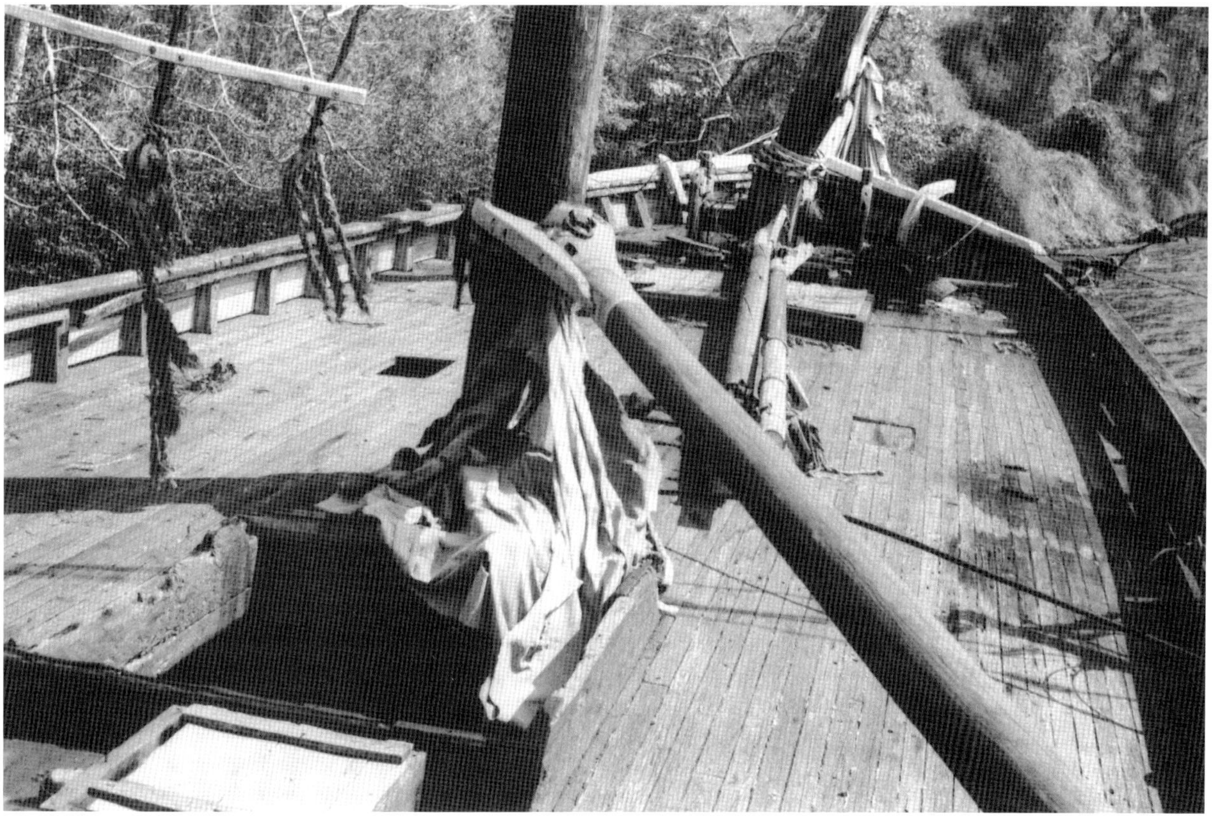

❂ Looking forward on the deck of the *Tilghman*, at Urbanna.

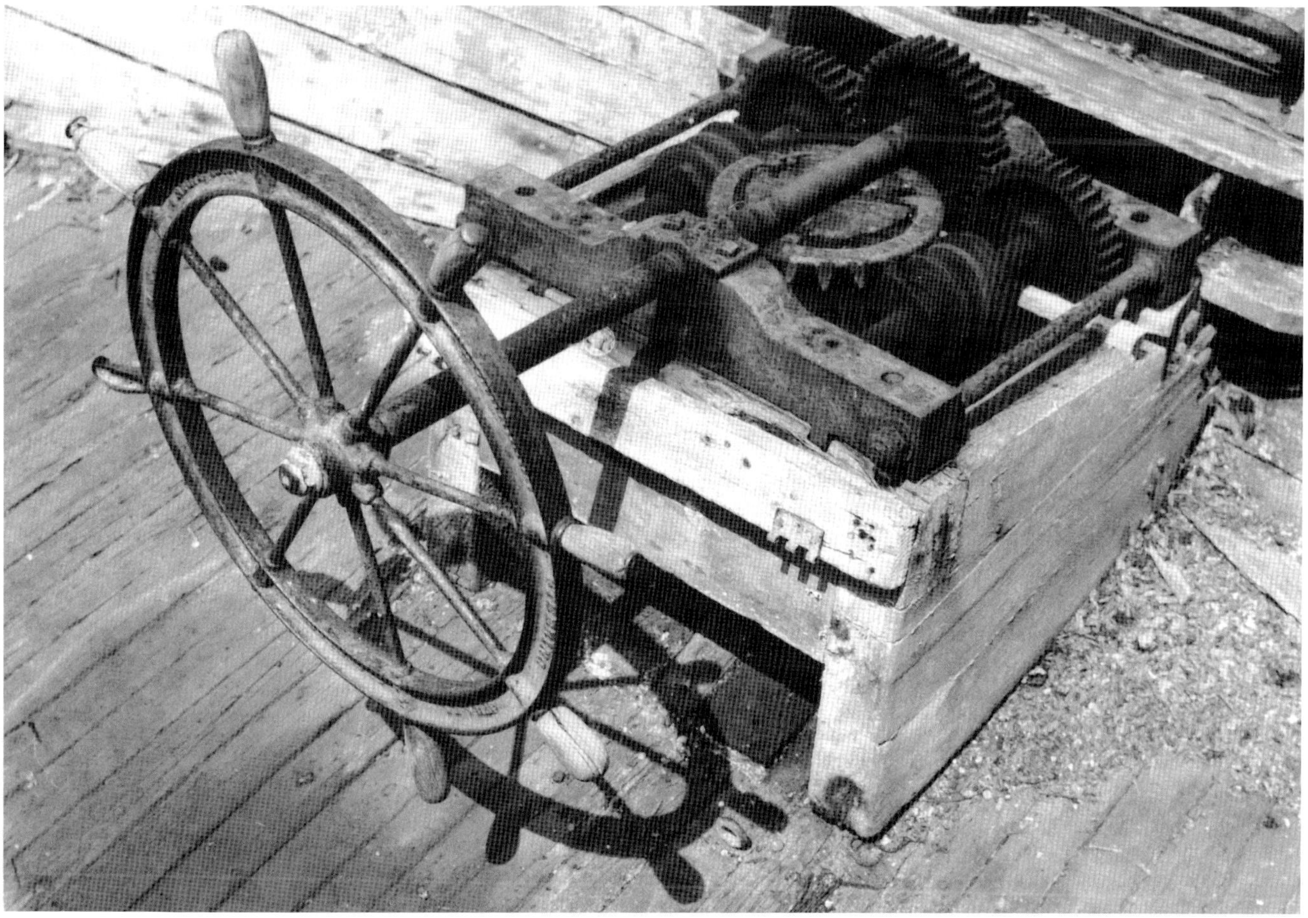

❂ The steering wheel of the old *Kate H. Tilghman* would be a great prize for a modern-day collector.

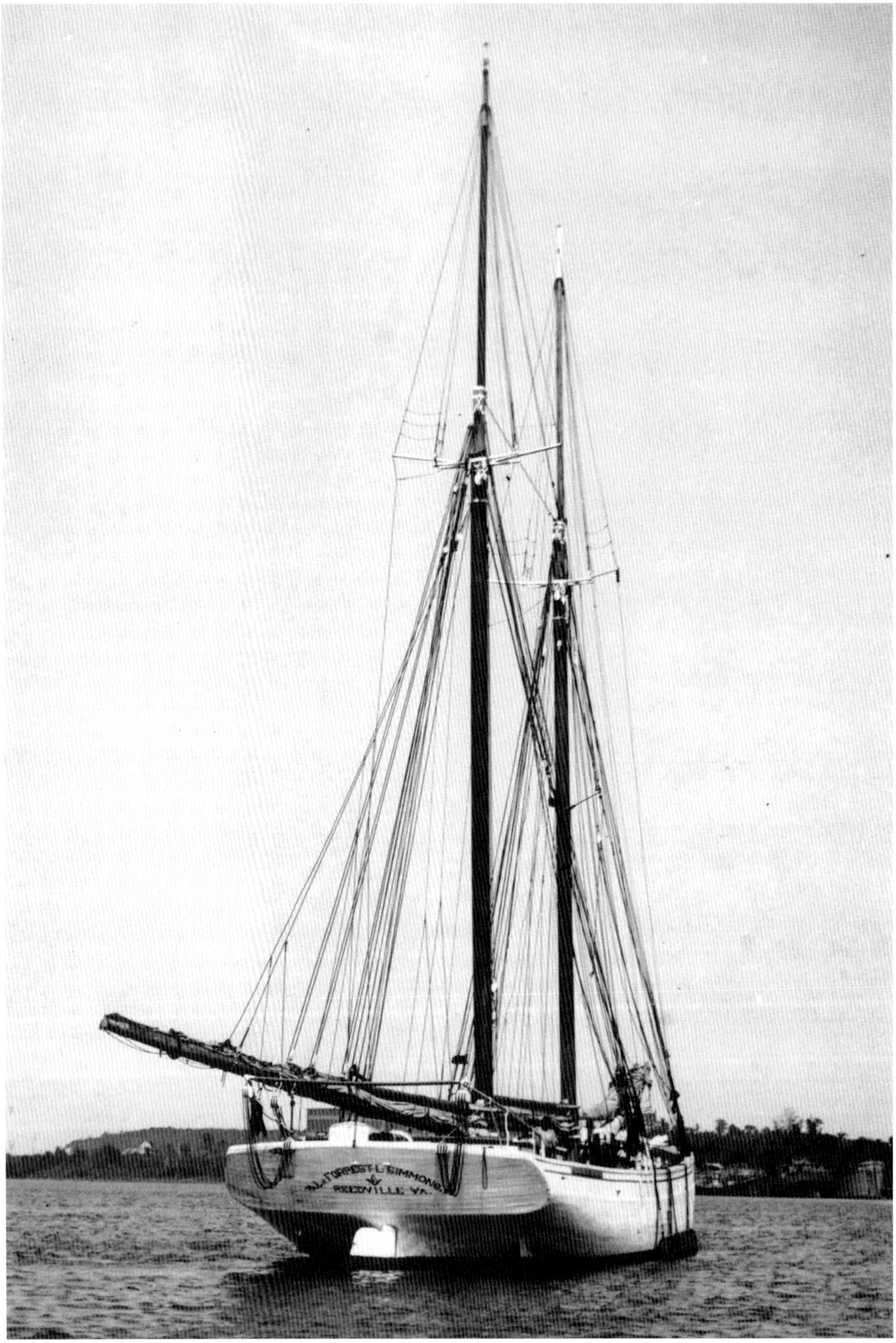

The last of her kind on the Chesapeake, a two-topmast schooner. The *LaForrest L. Simmons* rides at anchor in Curtis Creek, Baltimore, in August 1938, displaying one of the finest looking hulls on the Bay during her time. She had been purchased in Taunton, Massachusetts, in May 1936, by Captain Barney Wessels and sailed down to the Chesapeake the next month. Captain Wessels sold his schooner *Ida Mae* to Captain Raymond Evans. The *Simmons* was built in 1909 at Milford, Delaware, by W.G. Abbott. Here the *Simmons* has just come off the marine railway of A. Smith & Sons Shipyard and her centerboard is still suspended from her starboard side. Now comes the task of maneuvering that huge wooden piece back into its midships case with block and tackle.

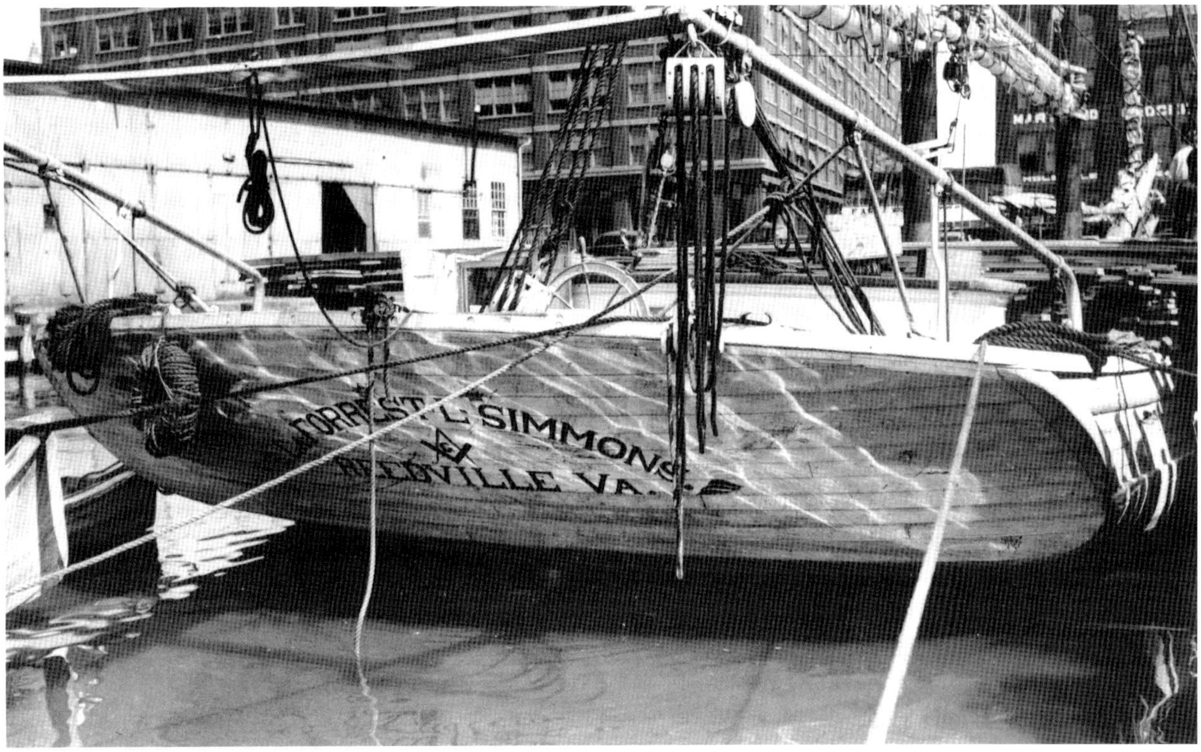

The wide transom of the *Simmons* displays her name, hail port of Reedville, Virginia, and the Masonic emblem. The rope-bound tire at left center takes the brunt of the pressure from the bow of the yawl boat when it is raised to that position to push the schooner. Her deck cargo of lumber is being transferred to the dock alongside as the schooner's jibboom almost overhangs the pavement along Pratt Street in July 1938. The large brick building in the background is the Candler Building.

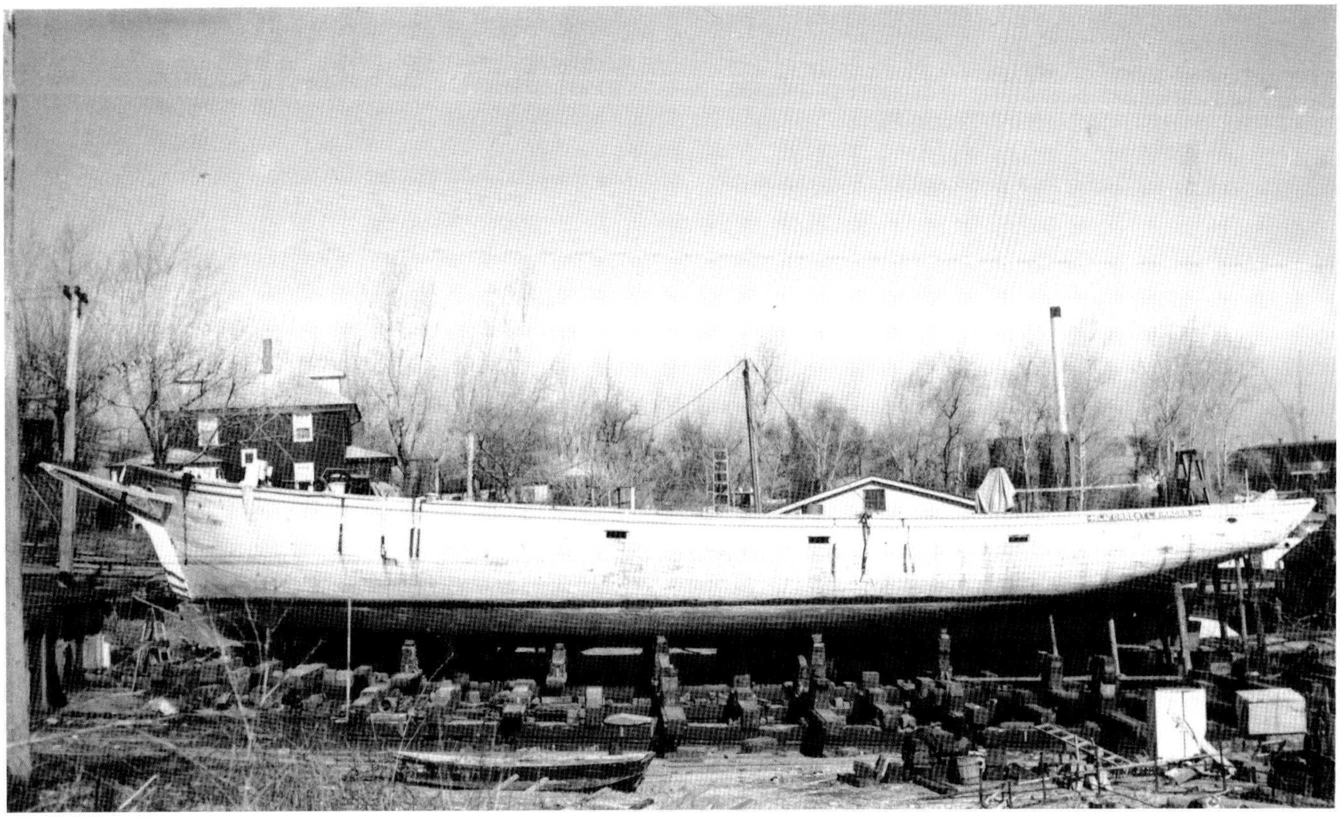

Captain Wessels was not to enjoy his new schooner very long. He died and the *Simmons* was acquired by the A. Smith & Sons Shipyard, which lost no time in converting the sturdy hull into a powerboat. Stripped of her masts and all rigging, the *Simmons* sits on the Smith railway in April 1939, as a diesel engine is installed where her after cabin once was. The graceful longhead with its trailboard and billethead was next to be removed.

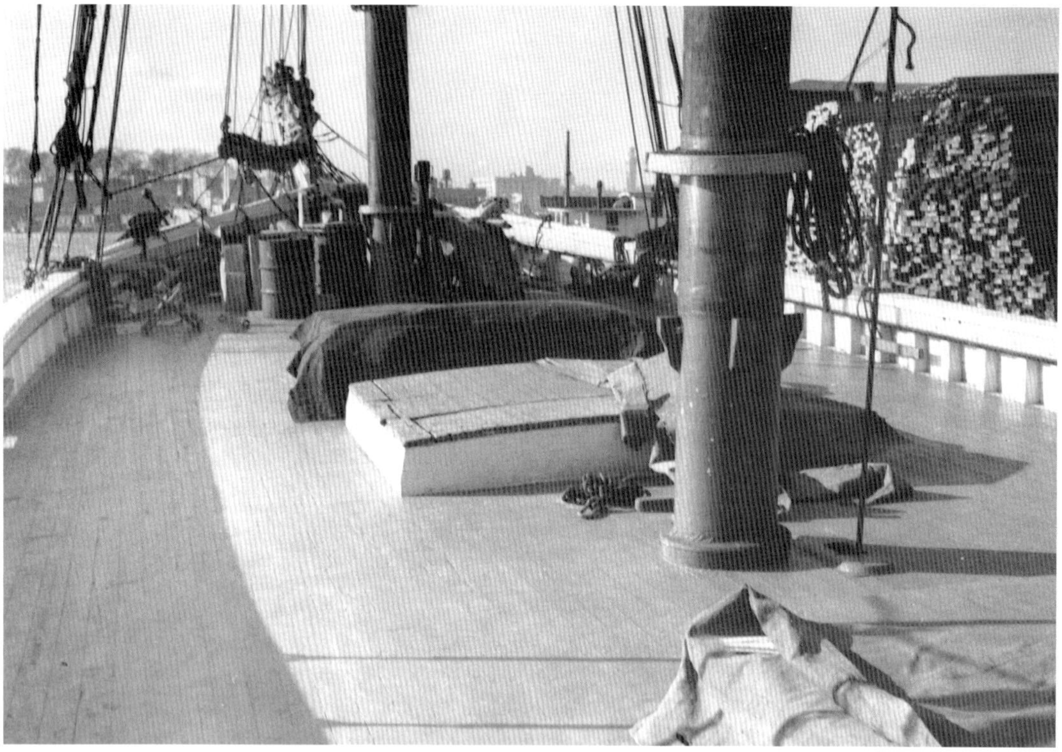

✪ The foredeck of the *Simmons* at Baltimore, probably in July 1938.

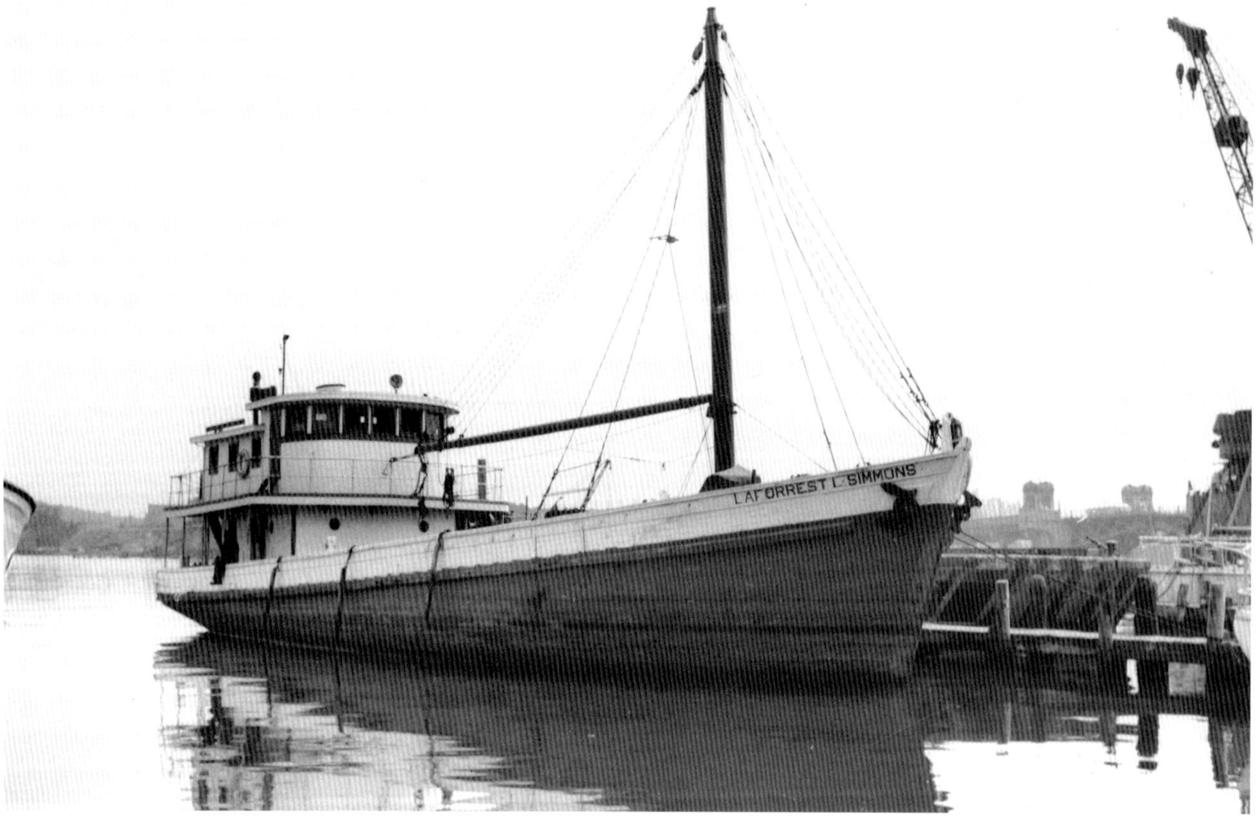

In September 1948, the powerboat *LaForrest L. Simmons* returns to the A. Smith & Sons Shipyard. By this time some of her graceful sheer has been lost due to her age and heavy engine astern, but she is a fine example of a powerboat conversion on a schooner hull. The excellent joinery work in her cabin shows why that shipyard was a favorite among Bay watermen. Retaining her original name, the vessel plied the Bay until she sank during hurricane Connie on August 12, 1955, one mile north of Sharps Island, Maryland.

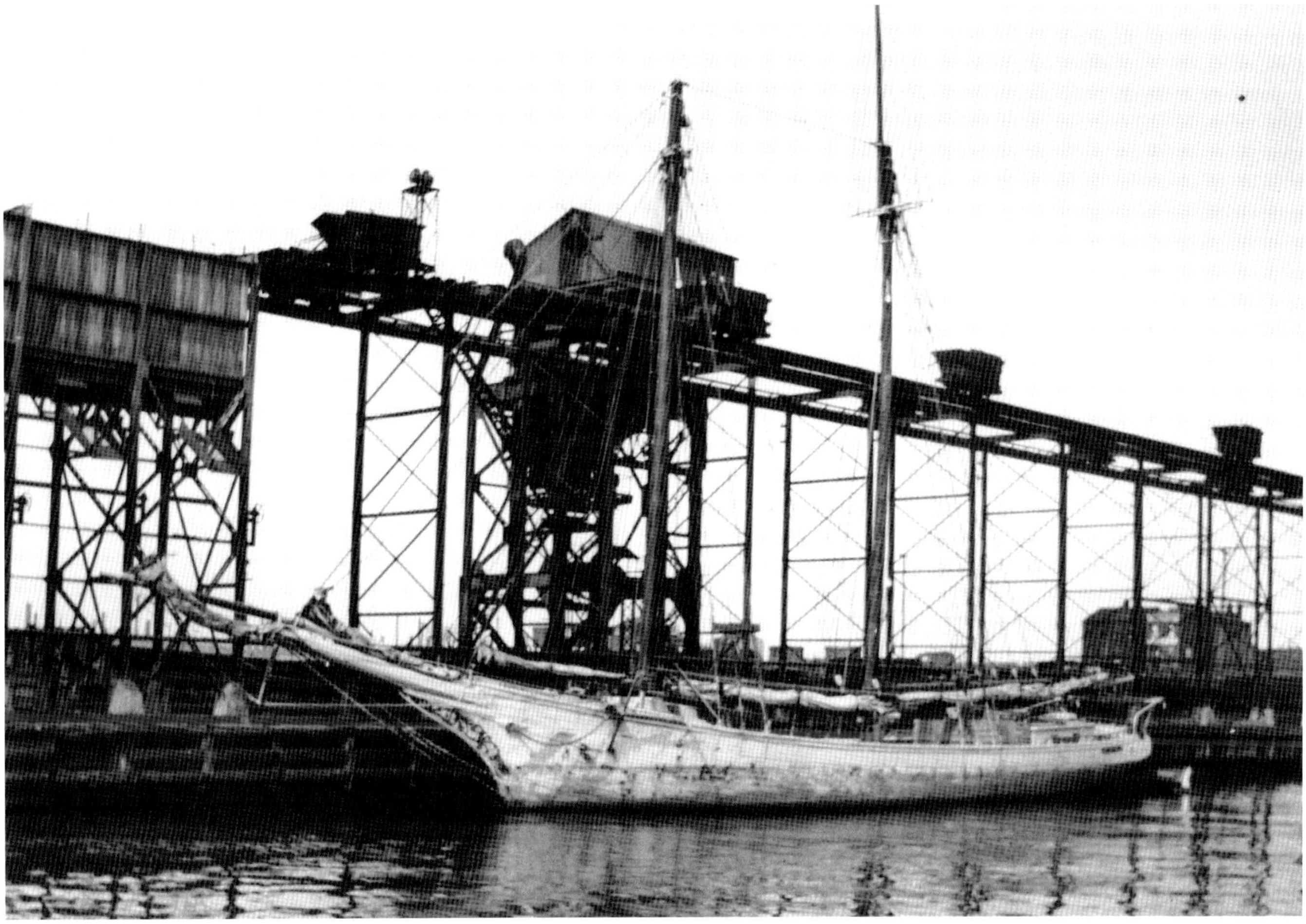

One of the Conway fleet of schooners not bearing his surname was the stately *Laurena Clayton,* here tied up at the coal piers on South Clinton Street, Baltimore, on April 29, 1934. What appears to be carved scrollwork along her longhead and stem is merely painted on, although she did carry a billethead. Perhaps, originally, she did carry carved trailboards and these may have been lost in a collision she experienced earlier. On March 25, 1925, the *Laurena Clayton*, in command of Captain John Conway, was bound up Tangier Sound under a press of sail. Captain Conway was at the wheel, and suddenly a powerboat attempted to cross the *Clayton's* bows. The powerful schooner ran down the powerboat, later identified as the *Southland,* which had left Crisfield about 3 a.m., and immediately Conway threw his helm up and his men lowered the yawl boat to go to the aid of two men on the sinking *Southland.* Heavy seas killed the small boat's engine and the rescue attempt failed; the two men were lost. The *Clayton* remained in the vicinity of the wreck until daylight and then proceeded up the Sound. It seems almost criminal to have converted a handsome vessel like this into a powerboat, but that was to be her fate. Renamed the *Effie M. Lewis,* she foundered in Delaware Bay on November 11, 1941. Thus ended the career of another schooner that once engaged in the colorful pineapple trade out of the Chesapeake. She had been built in 1892 at Cambridge, Maryland.

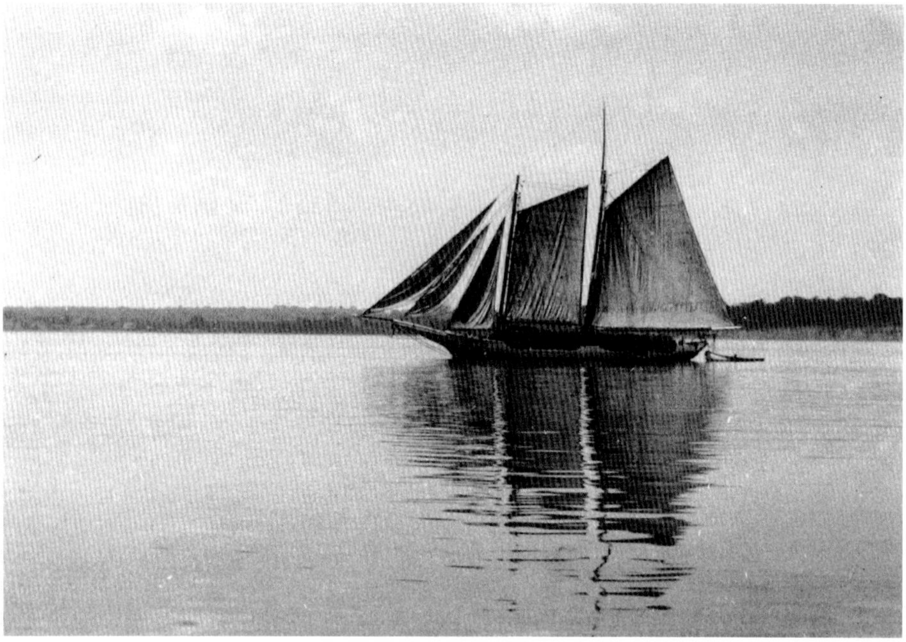

Another pineapple trader was the schooner *Levi B. Phillips,* here becalmed on Chesapeake Bay on July 18, 1934, but getting a slight assist southward from her yawl boat astern. Heavy canvas tarpaulins, held down by battens nailed to her waist, cover what are probably cardboard cases of empty cans destined for a tomato cannery in tidewater Virginia. During World War II she was among the Bay schooners to be sold south, she to the Peter Paul Candy Company to transport coconuts from Central America to Florida. In the 1946 *MVUS* she is listed as having been sold to Honduras.

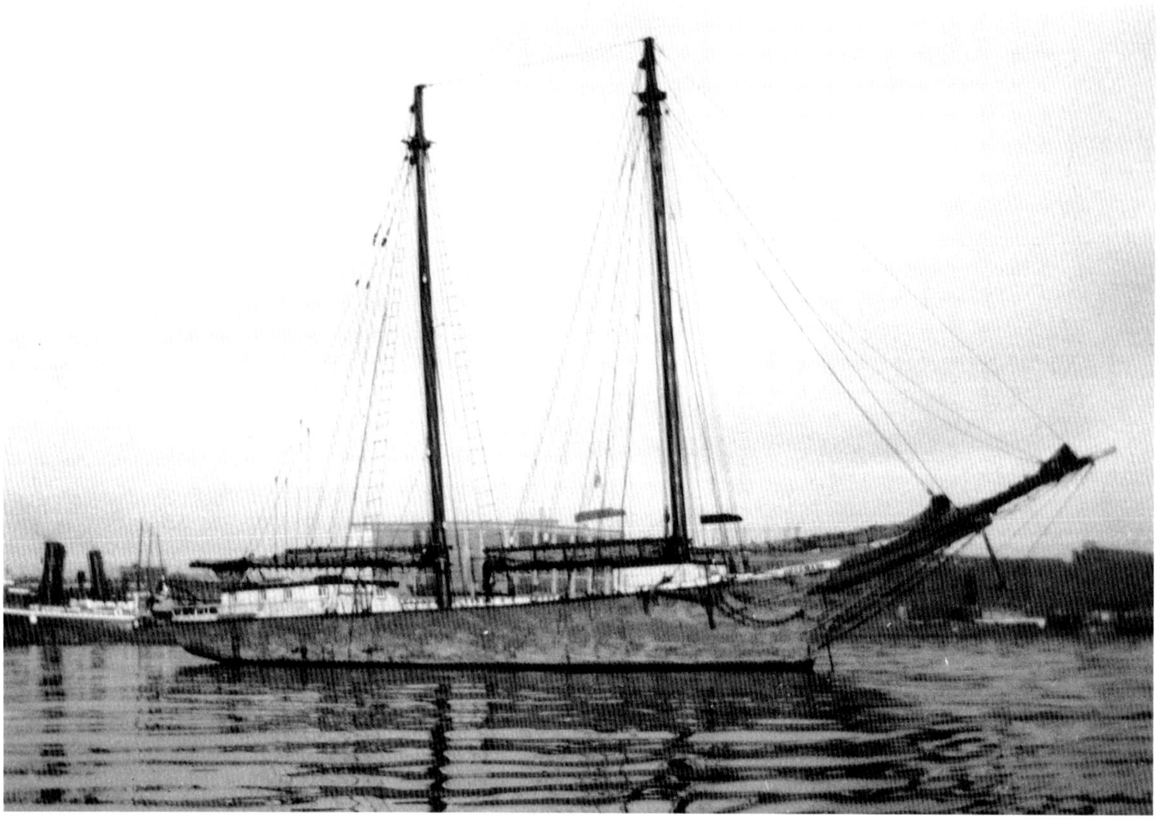

Lucy May was another of those schooners to come to the Chesapeake from "down-East" after a strenuous life along that coast. She had been built by George Emery in Cherryfield, Maine, in 1888. On November 18, 1910, while bound from Cherryfield with lumber for Boston, she ran into trouble along the Massachusetts coast and was assisted by the Pamet River Life-Saving Station. Then on May 11, 1921, she was towed into Southwest Harbor, Maine, in a waterlogged condition, by the schooner *Governor Grant.* About 1923 she came to the somewhat sheltered waters of the Chesapeake and became another "lumber lugger." Sheltered, yes, but still the unwary vessel could fall prey to its occasional tempests. On March 25, 1925, while bound from Laurel, Delaware, to the Rappahannock River, the *Lucy May* lost her masts in a heavy blow on the Bay. The Coast Guard cutter *Apache* was dispatched to her aid and towed her to an anchorage in the Rappahannock. The tug *James A. Clark* came up from Norfolk and towed the dismasted schooner to Baltimore on the 29th. At anchor in Canton Hollow, Baltimore, on October 24, 1933, those hard years are behind her. And only a few good years are ahead, for she was abandoned in the northwest branch of the Yeocomico River, Virginia, about 1937.

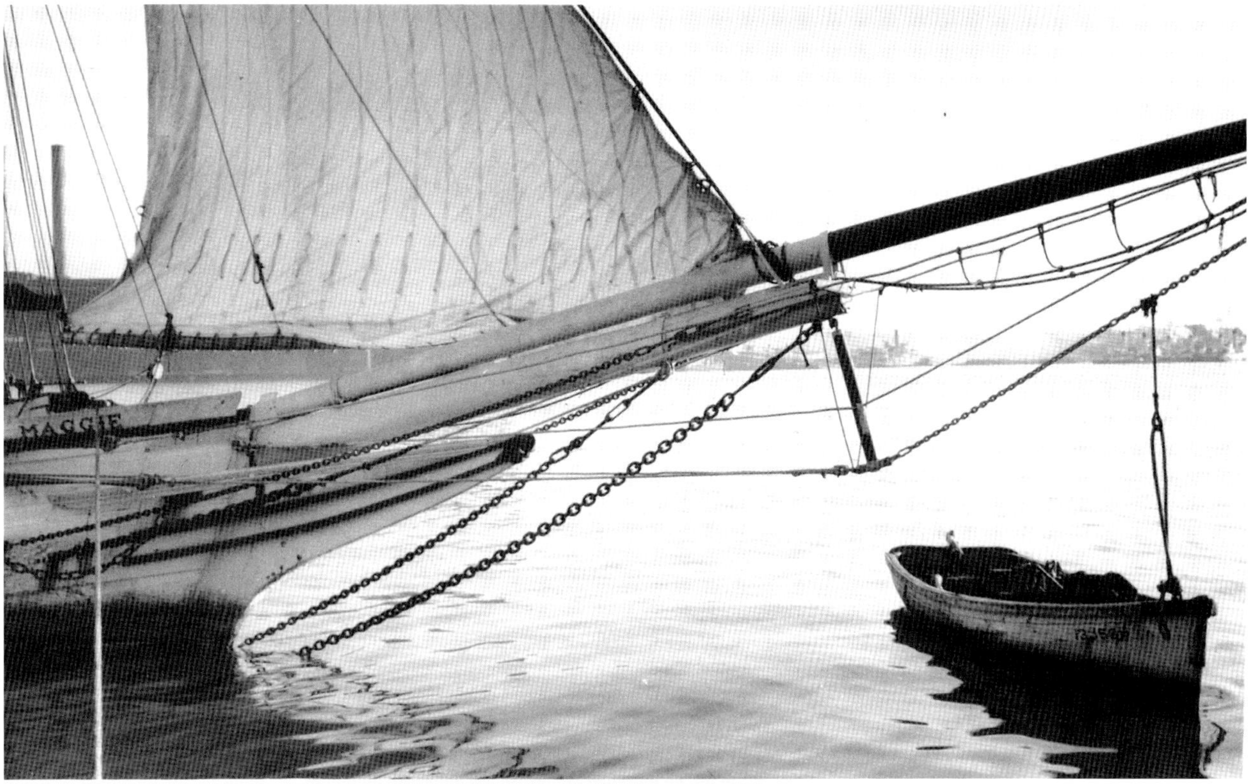

The schooner *Maggie* rests in Baltimore after discharging a cargo of lumber in 1937. She dries her new forestaysail as her yawl boat swings from the pendant to which that small craft is attached when it is employed to pull instead of push the schooner. In the background seen just in front of the dolphin striker, is the side-wheel steamer *Emma Giles*, recently laid up there at the site of the old Woodall Shipyard and eventually to be stripped to become a barge. *Maggie* was sold north in the 1940s to become a "dude cruiser" by her owner Captain Tom Henry Ruark, who later purchased the schooner *Columbia F.C.*

Maggie was seen abandoned at Rockland, Maine, in 1955. One reason some of these schooners lasted so long was that they were rebuilt at some point in their careers. *Maggie* had been built in 1871 in Dorchester County, Maryland. In the summer of 1914 she was at the Seaford Marine Railway, Seaford, Delaware, where she had been "retimbered, new centerboard well and keelsons placed, a great part of the planking renewed, new cabin and keel, caulked all over, refastened, and other repairs made." In August of that year she departed from the shipyard for Blakes, Virginia, to load lumber for Baltimore.

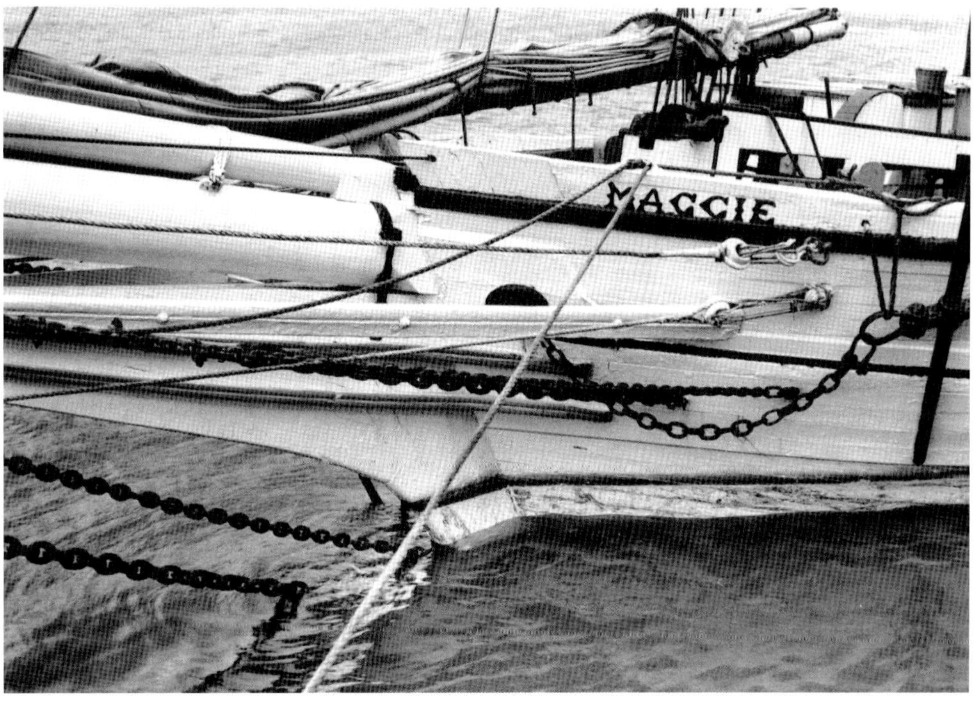

✪ Detail of the bow of the schooner *Maggie.*

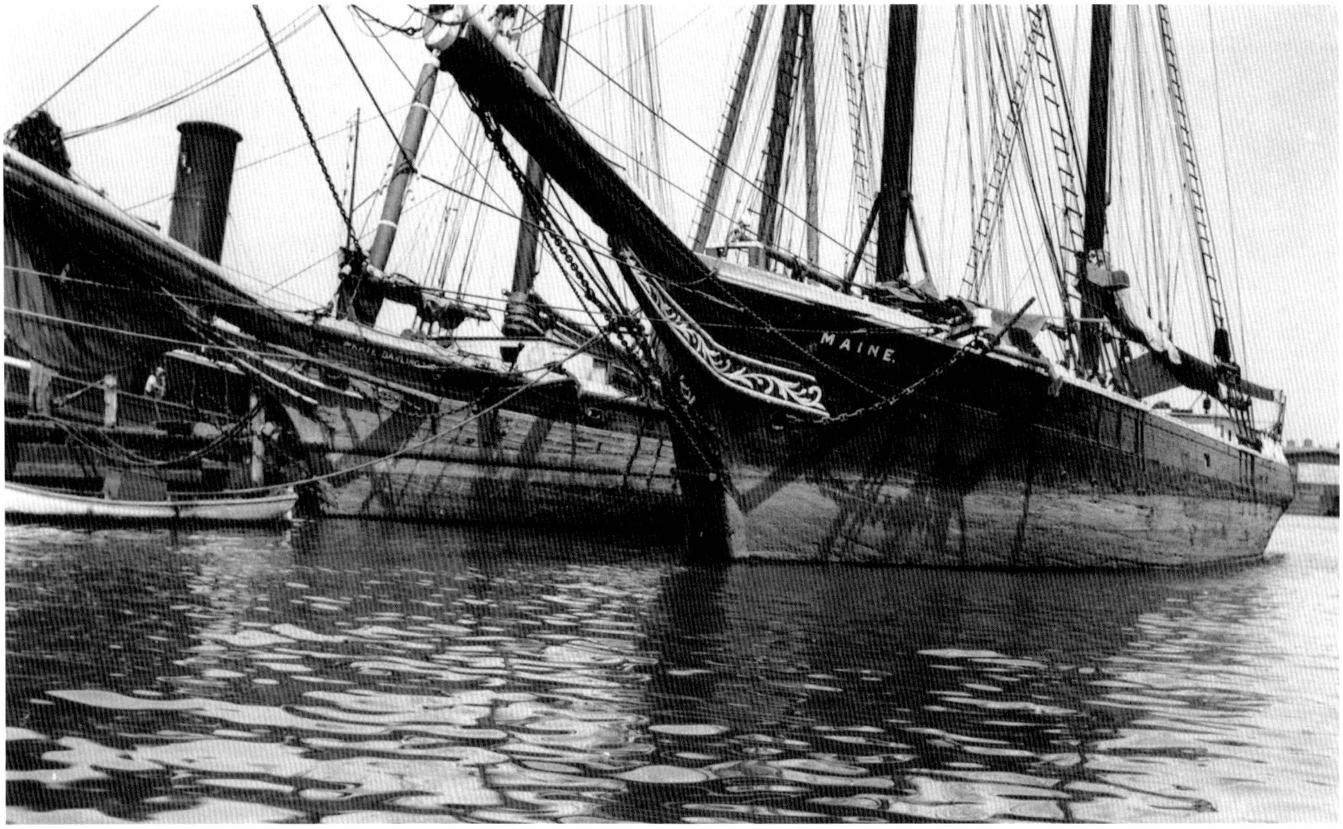

The huskiest two-masted schooner this author ever saw on Chesapeake Bay was the *Maine*, named for the state in which she was built, at Bath, in 1886 by the New England Shipbuilding Company. Her builders were used to constructing larger vessels, big schooners and square-riggers, and she was "sandwiched" in between the bark *William Baylies* and three-masted schooner *J.S. Hoskins*. She had been built for parties in Philadelphia and when she was delivered she sailed up the Delaware River and arrived off that city on September 19, 1886, under full sail. She engaged in the coasting trade out of Philadelphia until she was sold to Carter and Clark, Washington, D.C., by R.D. Wood and Company, in 1907. Thus started her career on the Chesapeake—with occasional runs outside, such as hauling canning machinery to the Isle of Pines, Cuba, in 1914. Importation of pineapples to Baltimore was coming to an end at that time, and canneries were being established in the islands. In May 1921, the *Maine* was bound out of Norfolk for Nassau, Bahamas, when she lost her mate overboard and returned to port. About 1933 she joined the sailing fleet of W.B. Vane, Baltimore, and was steadily employed freighting lumber from Virginia and North Carolina for Baltimore and Philadelphia. With another Vane-owned schooner to her starboard, the *Kate Darlington*, the *Maine* is tied up at the Redman and Vane Shipyard, Baltimore, August 1, 1936.

Looking aft on the deck of the schooner *Maine* in Baltimore, August 1, 1936. The "big ship" appearance of this two-masted schooner is revealed here—her high waist, poop deck, and high taffrail. And forward she had a forecastlehead and capstan. Another departure from the typical Chesapeake-built schooner style was the entrance into the after cabin directly from the main deck, as seen here. To starboard of the door is an icebox, its legs extended by boards nailed thereon so that a lard can may be placed beneath it to catch the constant drip of the melting ice. A canvas tarpaulin has been stretched over the main boom to keep the sun off the cabin roof.

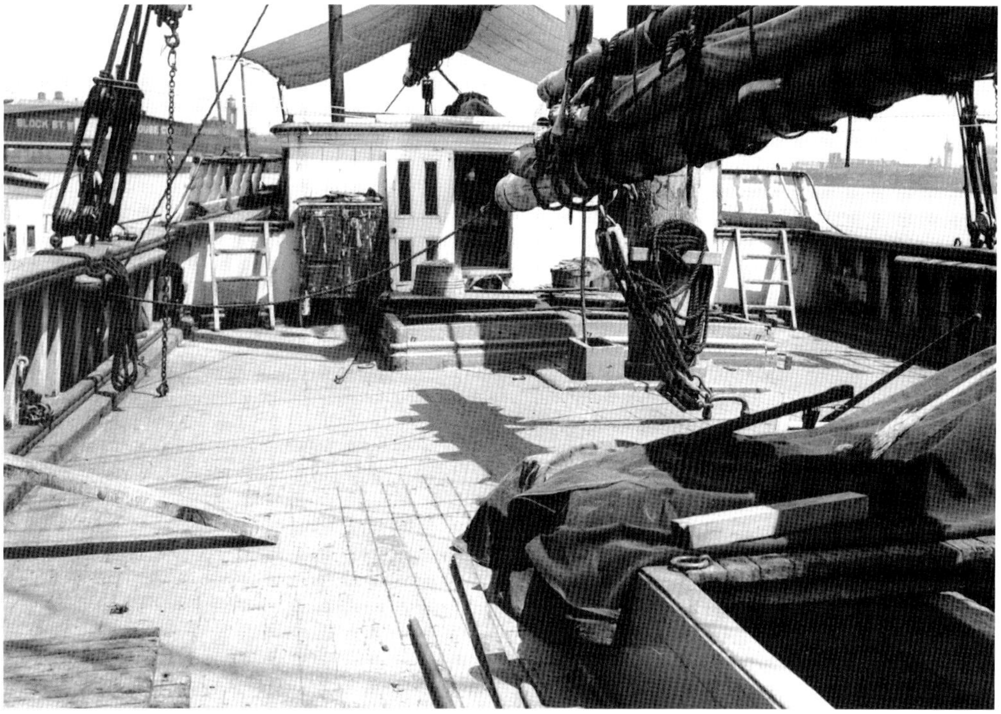

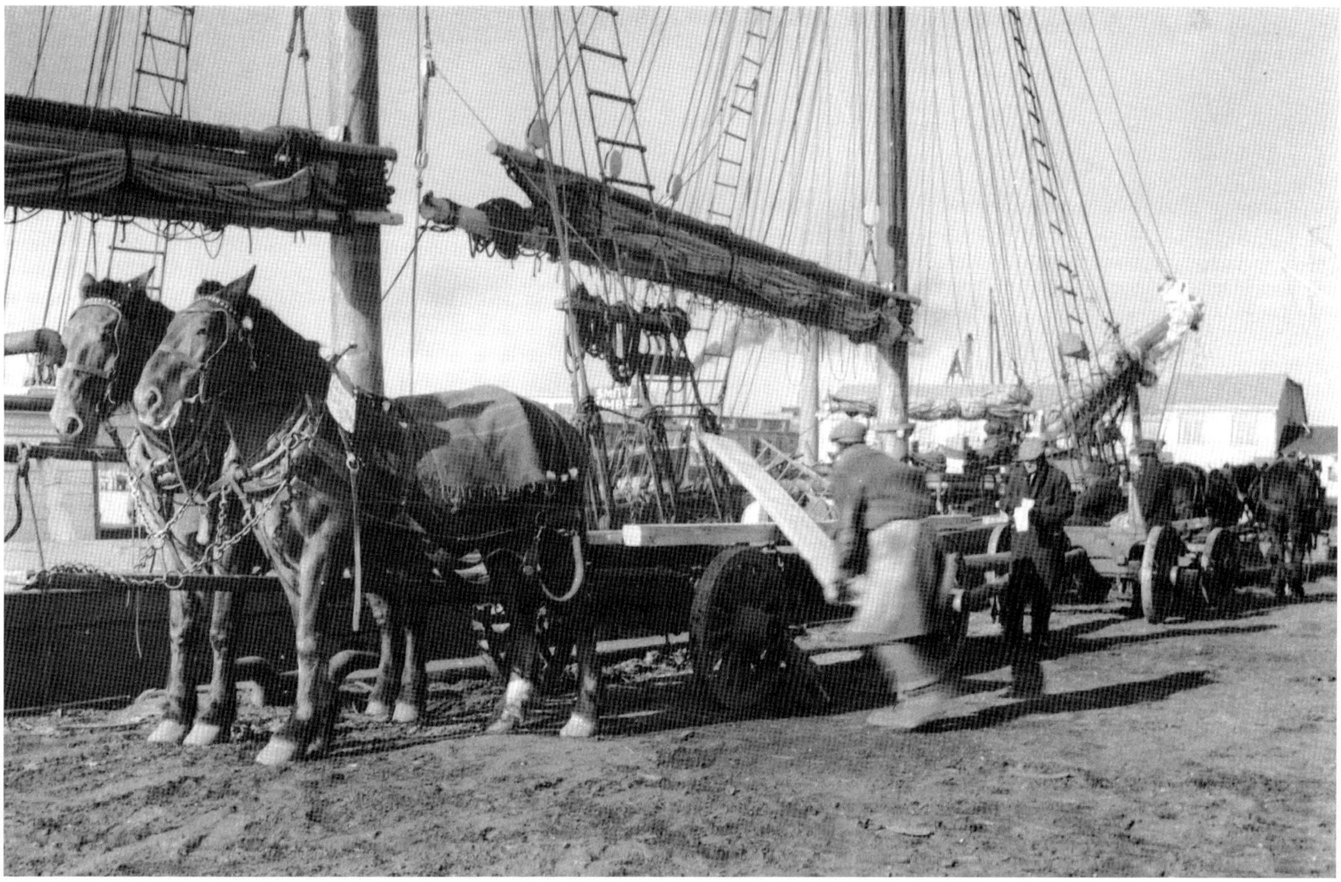

In December 1936, the *Maine* discharges lumber at McGrath's pier, Baltimore, a scene reminiscent of a century earlier with the horse-drawn wagons in the foreground. The pier-side lumber companies used these low wagons for deliveries to the mill and nearby customers. Big, well-kept horses, with highly polished brass harness fittings, patiently waited while the wagons were being loaded. A shiny brass plate attached to the collar is engraved "Canton Lumber Company," which was located nearby on Boston Street. A blanket has been thrown over the back of the closest horse, evidence of the good care given the animals. Sails and spars have been raised high above the deck to simplify the unloading of the schooner. The block and fall at upper left, directly in front of the mainmast, is used to raise and lower the centerboard.

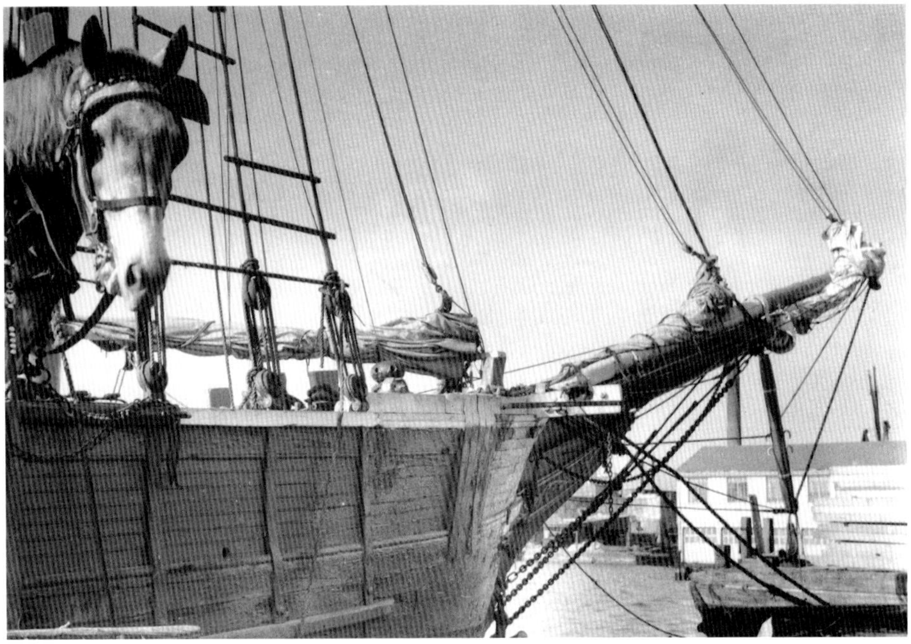

✪ One of the horses studies the photographer in this view of the bow of the *Maine.*

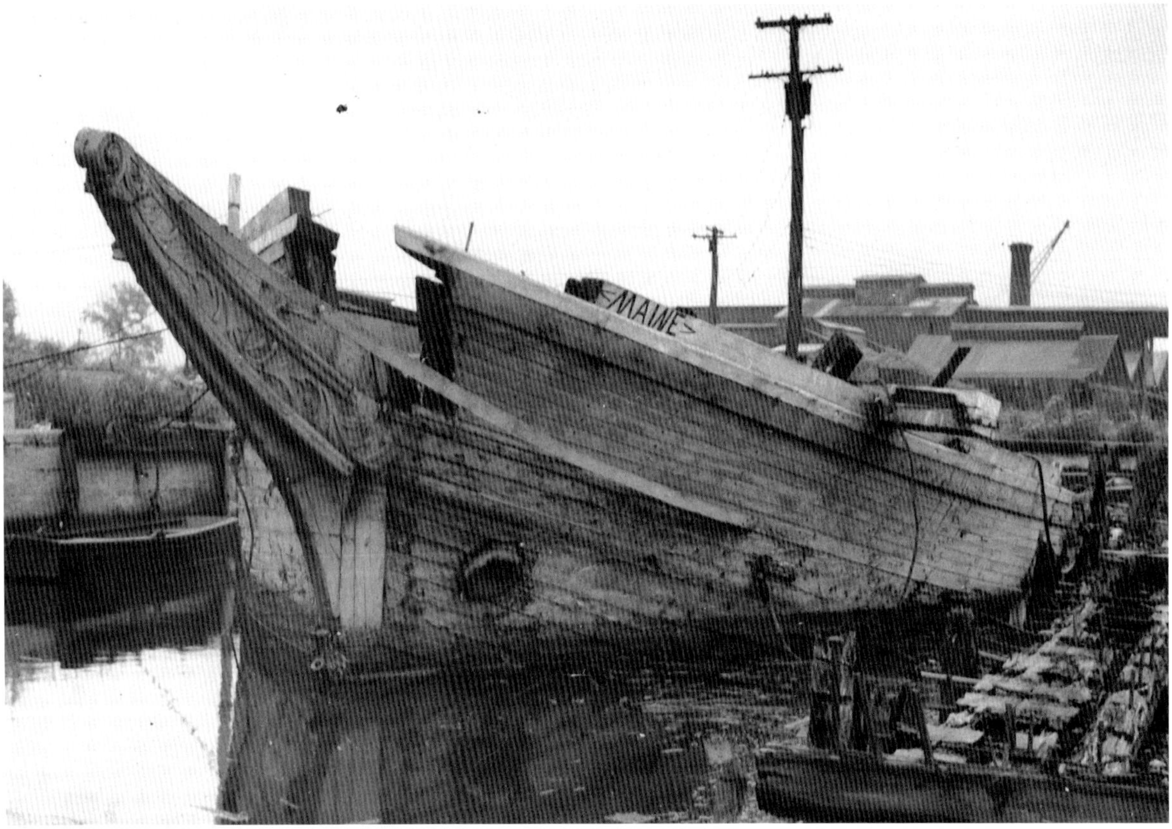

Ten years after the author visited the *Maine* and photographed her decks, the schooner was a waterlogged hulk sitting on the bottom of the Patapsco River at Port Covington, Baltimore. The *Maine* reached this stage not long after being sold to the Coulbourn Lumber Company in Elizabeth City, North Carolina, which also became her new home port. Sometime in 1945, in North Carolina but loaded with lumber for Baltimore, fire badly damaged the schooner's forward section. She delivered her cargo but after a survey it was felt the vessel was not worth the extensive repairs required, so she was dismantled and taken to the Port Covington graveyard and turned over to Mrs. Marie Mattson and helpers to be cut up for firewood. Mrs. Mattson lived in a shanty near the hulk and reduced the timbers of the old Bay vessels into small bundles of wood, held together by rubber bands made from old inner tubes, and sold them to neighborhood stores. On August 7, 1946, the bow of the *Maine*, with carved trailboards and billethead, reaches above the murky waters as if struggling for survival. But this was the end. The *Maine's* trailboards were massive carvings, their thickness equal to those found on the barks and larger schooners built by the New England Shipbuilding Company.

✪ Mrs. Marie Mattson took a rare break from her labors on September 7, 1948, to be photographed by Robert Burgess. She continued her colorful business from the 1930s into the 1950s.

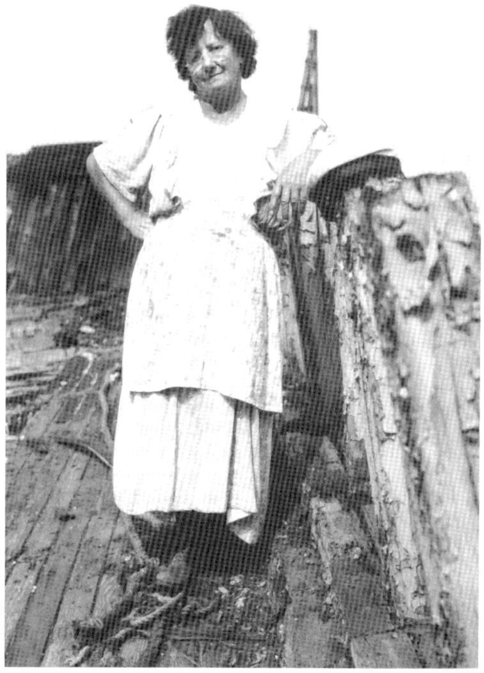

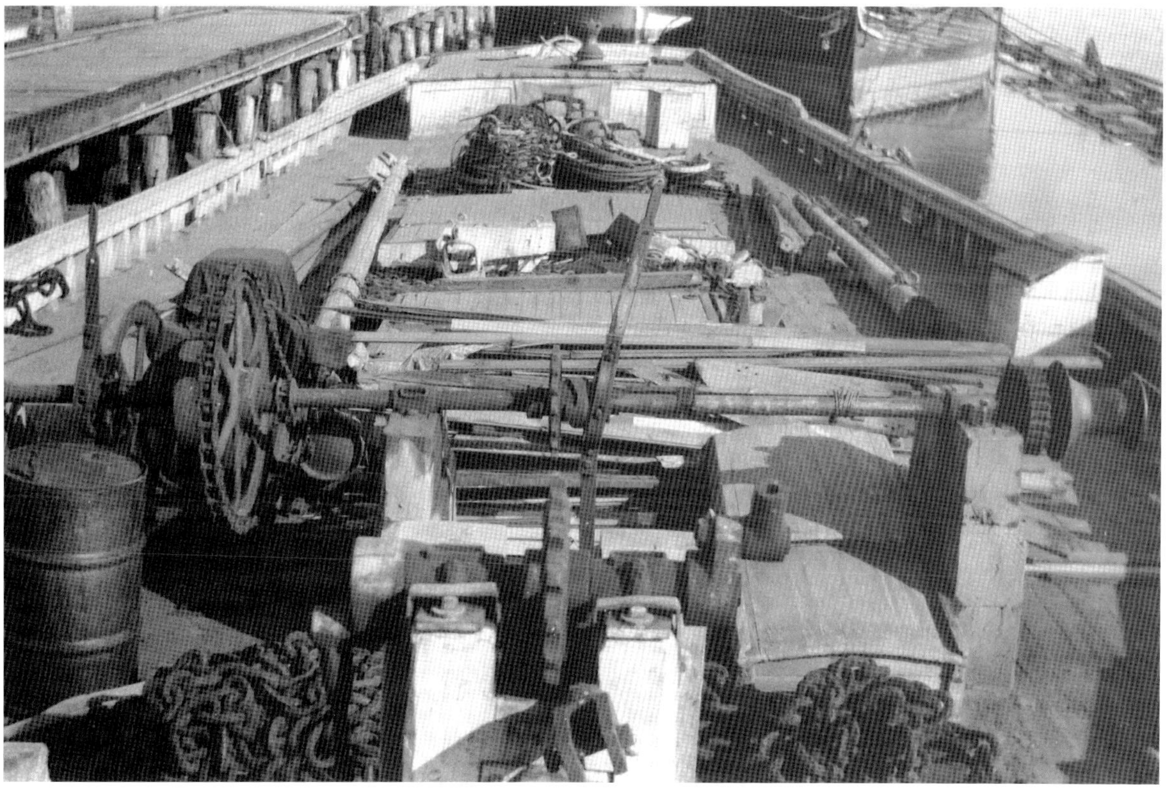

The *Mary A. Fisher*, built in 1882 at Sayreville, New Jersey, came to the Chesapeake about 1898. In those early years she went "outside." On August 21, 1903, she stranded on the New Jersey coast while bound from Norfolk to Little Egg Harbor. The nearby lifesaving station floated her after four unsuccessful attempts. Two years later, on August 5, she missed stays and ran ashore on Little Egg Harbor shoals, but was hauled off by the local lifesaving crews. That time she was bound from the James River, Virginia, with lumber for New York City. She was to last 30 years longer and on January 31, 1934, was tied up at the Booz Shipyard, Baltimore, stripped down to the bare hull, her spars, fittings, standing rigging, blocks, deadeyes removed, and the capstan placed on top of her after cabin. It was planned to convert her into a powerboat, but her hull was probably not worth the investment. In the 1935 *MVUS* she was listed as "Abandoned." At upper right can be seen the bows of the schooner *Kate Darlington* that was soon to follow the *Fisher* to those ranks.

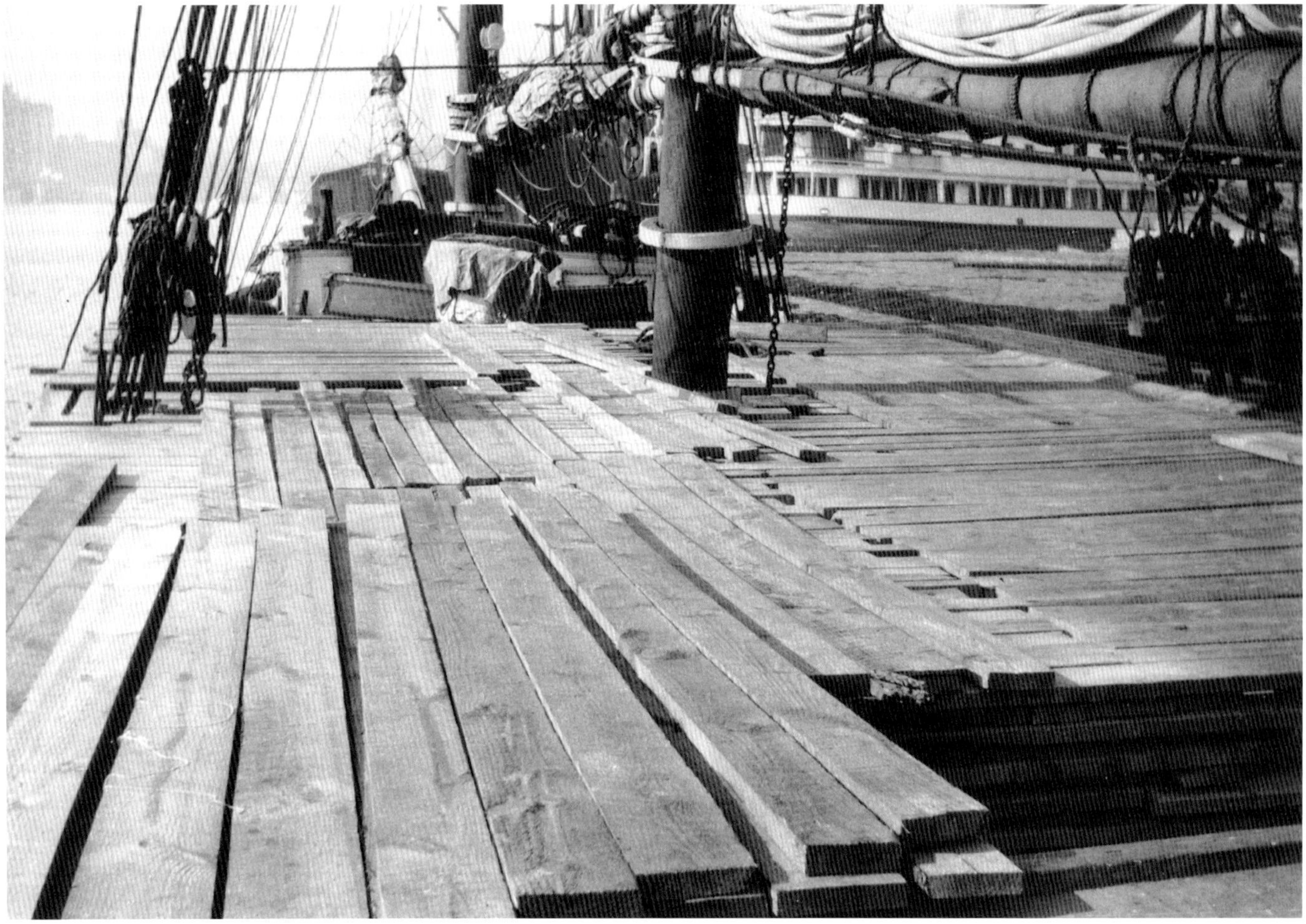

A schooner all but loses its identity when buried under a deck load of lumber such as this. But she is the *Mary Lee*, built in 1872 at Laurel, Delaware, as the *John McGinnis*. On this occasion the schooner had just been towed to Pier 6, Baltimore, by the Coast Guard cutter *Apache* on January 25, 1934. Twelve days out of Leedstown, Rappahannock River, Virginia, where she had loaded the lumber, she ran into heavy weather off Poplar Island, Maryland, but managed to anchor in Eastern Bay, off Claiborne. A leak developed and she filled but was kept afloat by her cargo. Safe in Baltimore, she was pumped out and her cargo unloaded. Then followed a long period of idleness, and about the fall of 1940 she was taken to the flats off Point Breeze, outside Baltimore harbor, and abandoned. Later she was refloated and taken to a ship graveyard at Spring Gardens, above Baltimore's Hanover Street Bridge, and cast aside once more. Illustrating the height to which deck cargoes were stowed, the top of the after cabin can be seen in the lower right corner. The boards are not lashed down for the inside run up the Bay as they would have been if the vessel had been running coastwise. Layers have been alternated, one fore and aft and the next athwartship. Beneath all this is the full hold. The lower foot blocks of the main rigging have been brought to the top of the deck load by means of a chain pendant, as at left center. The mast collar on the mainmast has also been raised and it was probably necessary to put a "lumber reef" in the mainsail, as that sail could not have been fully hoisted above the high deck load. The steamer at the upper right is the *State of Virginia*.

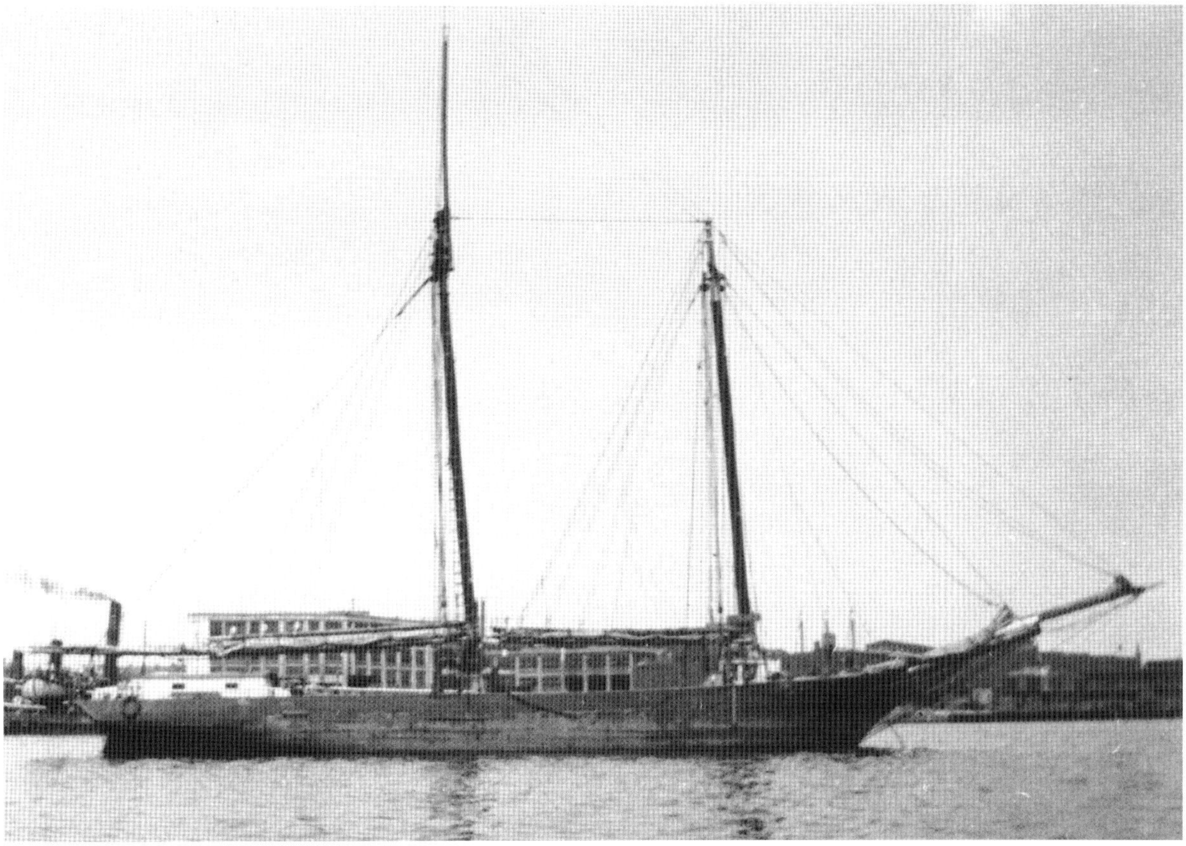

✪ The schooner *Mary Lee* at Baltimore, sometime in the 1930s. She was built as the *John McGinnis* at Laurel, Delaware, in 1872.

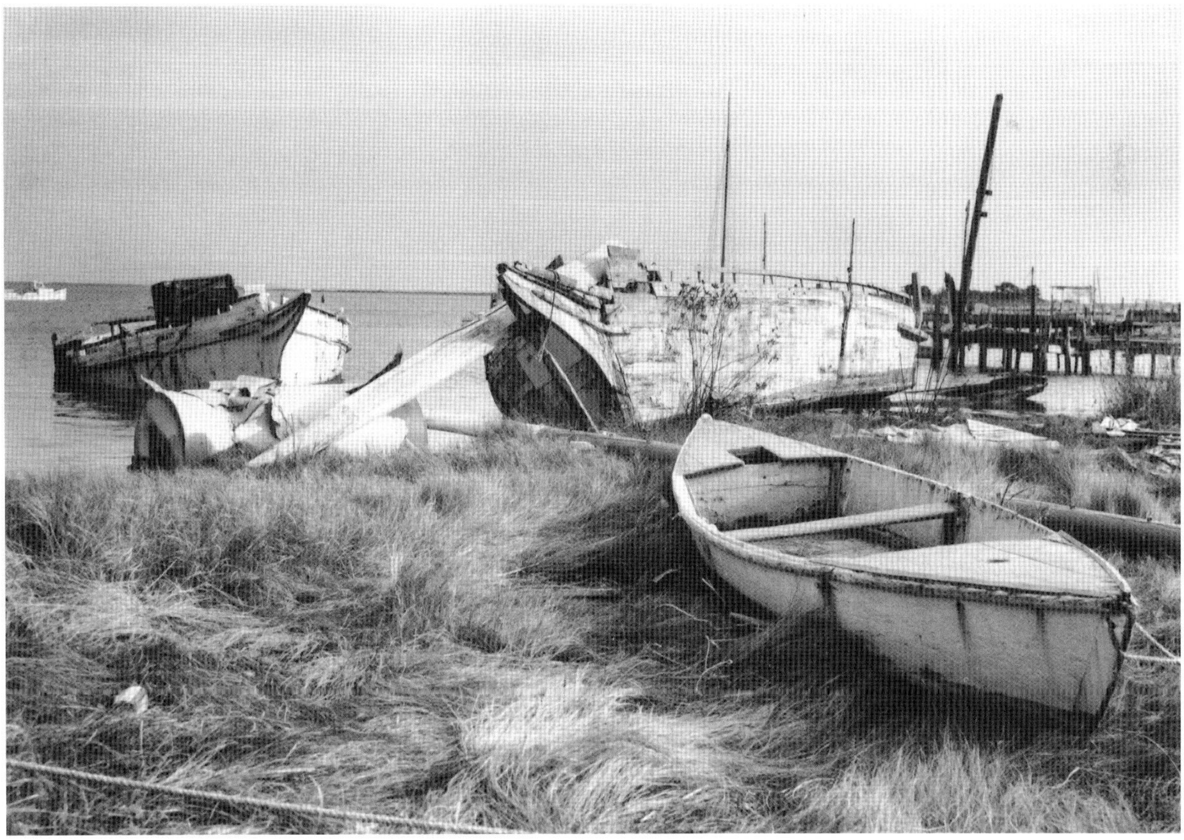

✪ The remains of the old schooner *Mary Lee* lie beached at the ship graveyard at Spring Gardens, Baltimore, in December 1958. The *Lee* is on the left in this photo, and the *J.L. Lawrence* is on the right.

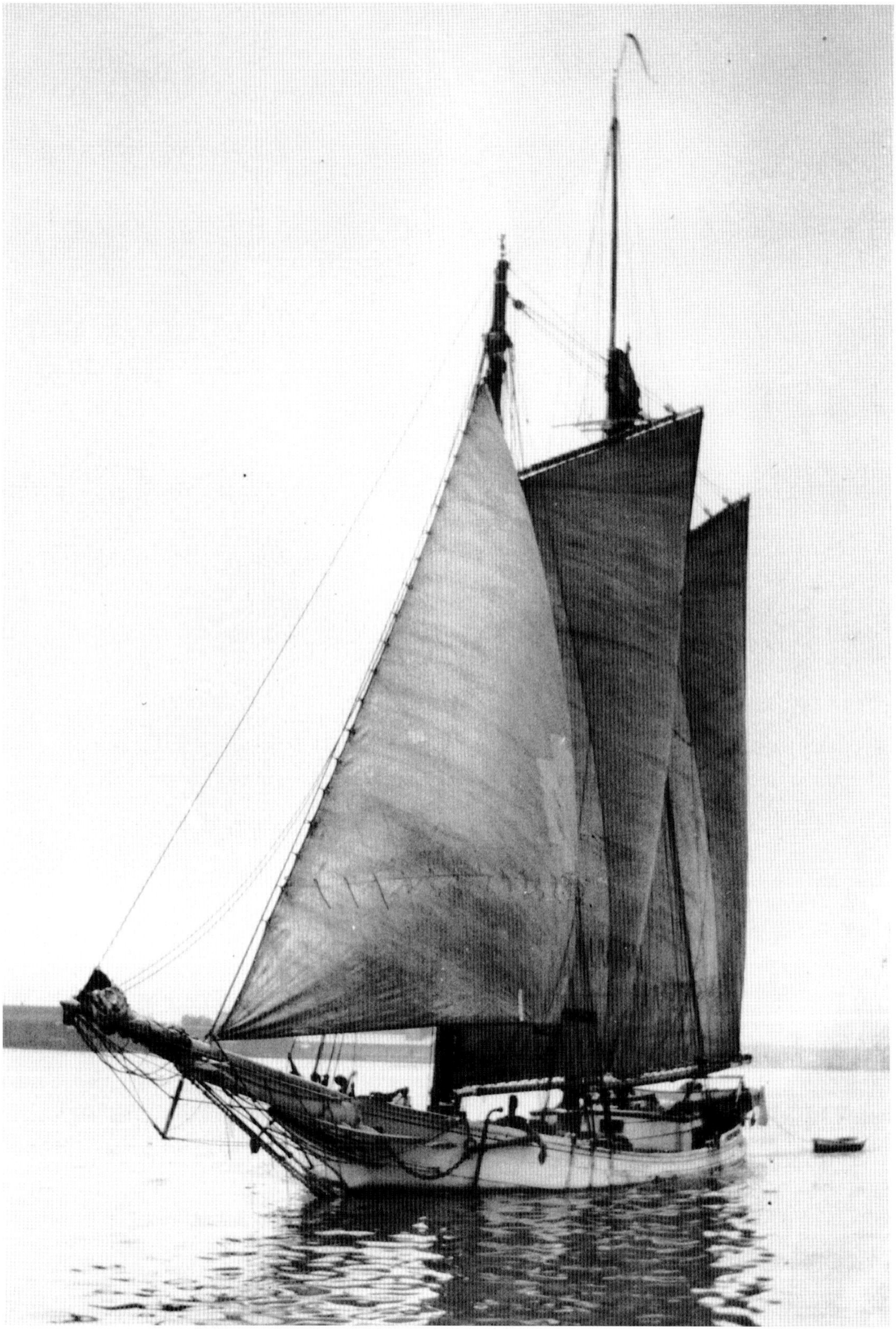

Mattie F. Dean was one of the more popular schooners on the Bay during the last days of sail there. Perhaps this was because she had been owned in several different areas of the Chesapeake and many people became attached to her. Built by Joseph W. Brooks in 1884 at Madison, Maryland, she was decorated as nicely as any schooner, with eagle figurehead, finely carved trailboards, quarterboards, and even a spread-eagle stern decoration. Atop her foremast was a carved wooden female figure with upraised arms. On May 6, 1934, she sails out of Baltimore harbor, her deck but scant inches above the water. In September 1928, the *Dean* brought a cargo of tobacco to Baltimore from St. Mary's County, Maryland, certainly one of the last such cargoes to move under sail on the Bay.

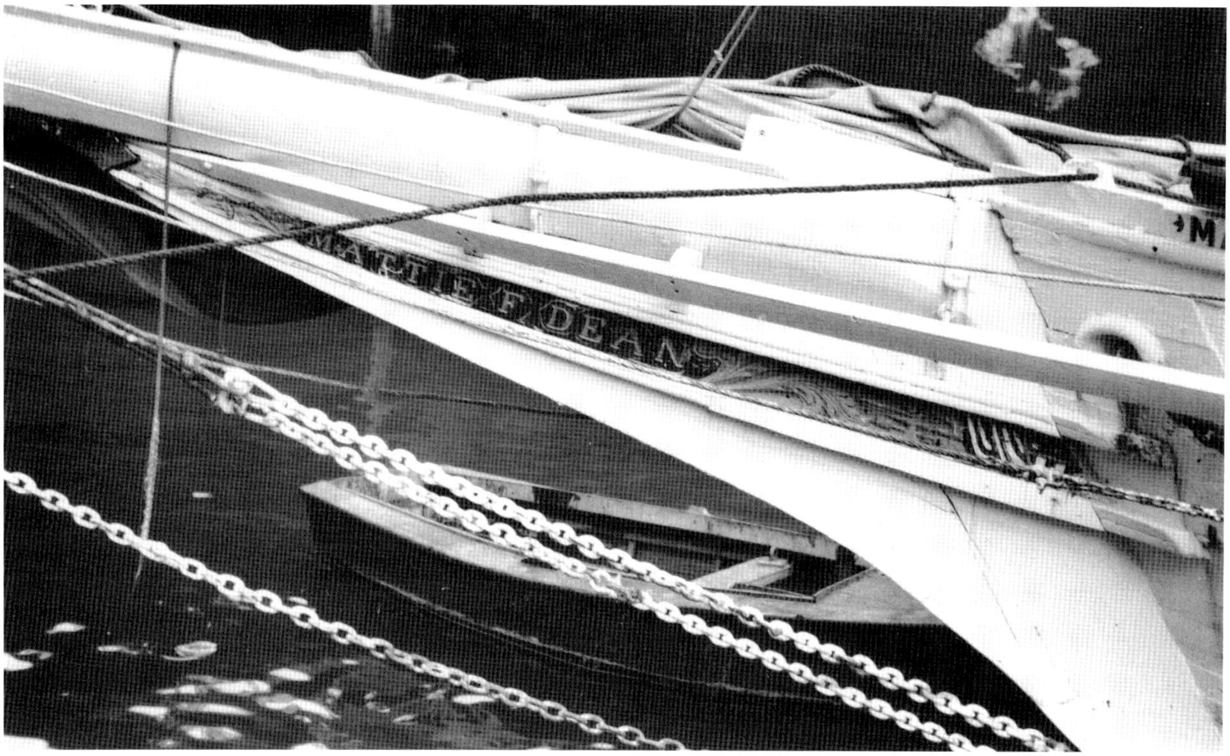

Eagle figurehead and port trailboard of the *Mattie F. Dean* tied up at Long Dock, Baltimore, with watermelons in September 1948. By this time she was owned at Chance, Maryland, and used as an oyster dredger.

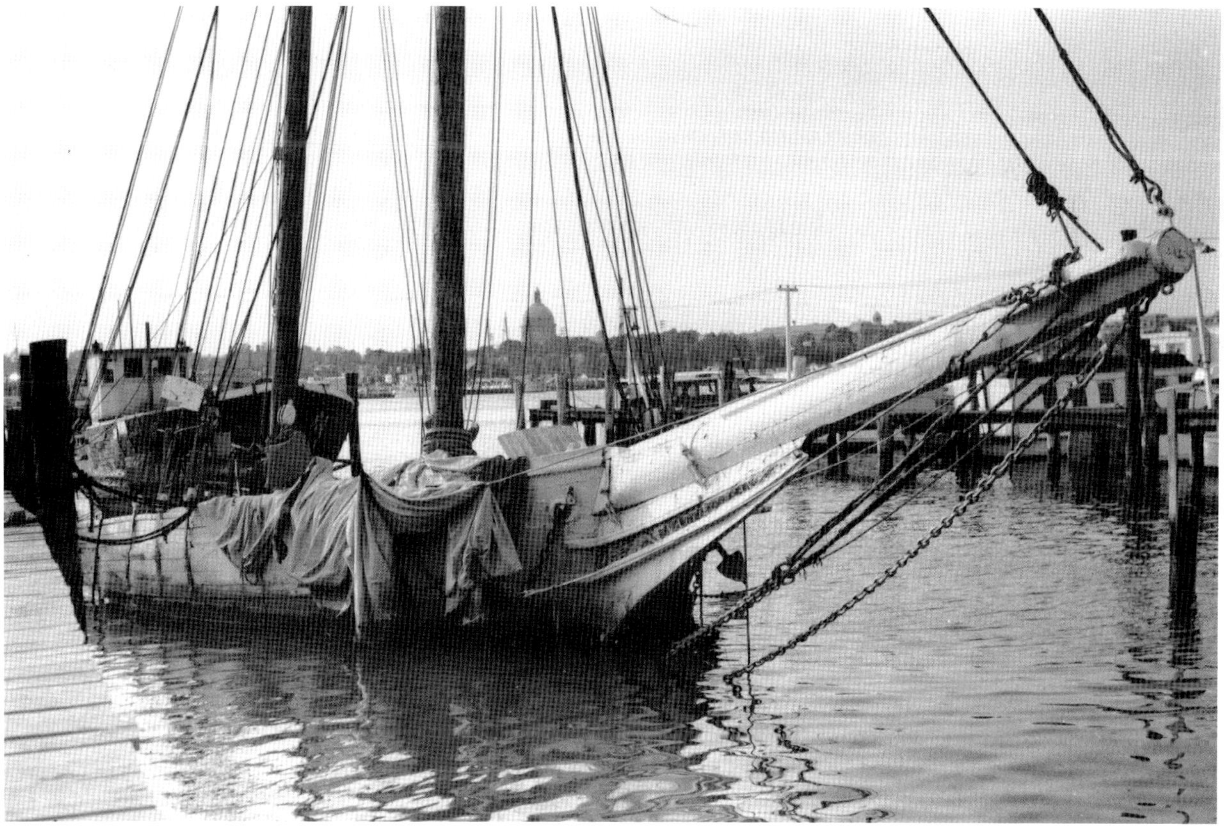

About 1953 the *Dean* was sold once again, to Harry Bureau of Annapolis, who continued to dredge oysters with her. The latter part of the next year she sank at his Spa Creek pier, as above. Astern of her is the powerboat *Ella F. Cripps,* a former Bay schooner. The *Dean* was raised about 1956 and she and the *Cripps* were abandoned in Back Creek, Eastport. Note that the *Dean's* eagle figurehead is missing. It was reported that it was placed on Burl Ives's yacht. In the background is the dome of the U.S. Naval Academy Chapel.

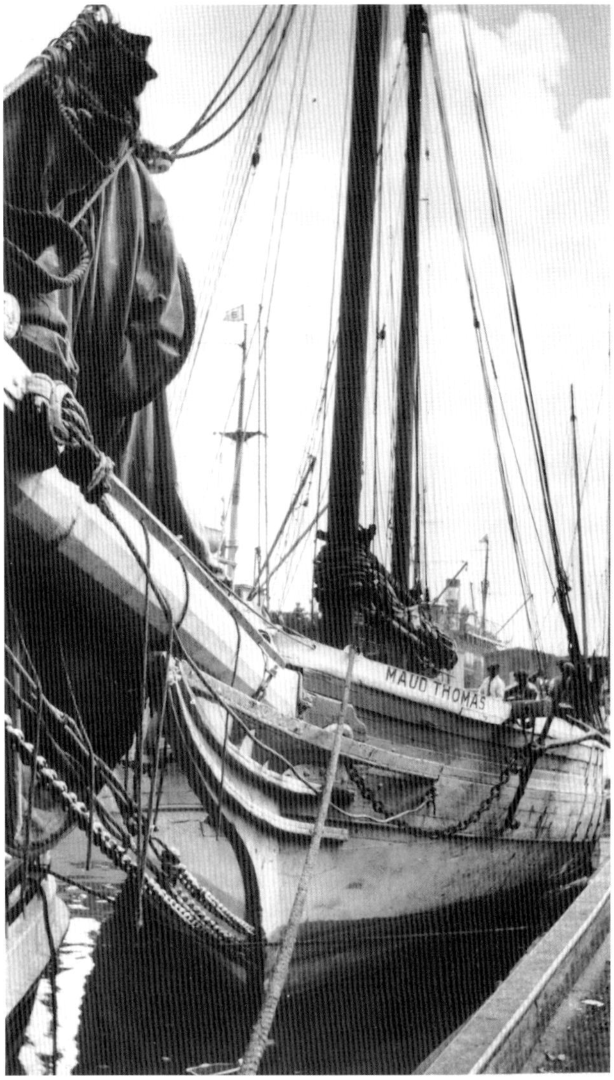

The name *Maud Thomas* was as synonymous with speed as that of any schooner on the Chesapeake in the 1920s. She won the schooner event in the Chesapeake Bay Championship Workboat Regattas for 1926, 1929, and again in 1931, the last year the races were held. Entered in the *MVUS* as having been built in 1908 at Whitehaven, Maryland, this vessel was actually rebuilt from the hull of the schooner *Judy*, built in 1881 at Baltimore. In the early 1900s the *Judy* was rammed and sunk by a steamer off Sandy Point, Maryland. She remained submerged for more than a year and was then raised and towed to Whitehaven where she became the new schooner *Maud Thomas*. The *Judy* had been noted for her speed, so perhaps some of that quality remained for the new vessel. On August 18, 1933, *Maud Thomas* was at Long Dock with watermelons. She remained in *MVUS* until 1945 and was then "Abandoned."

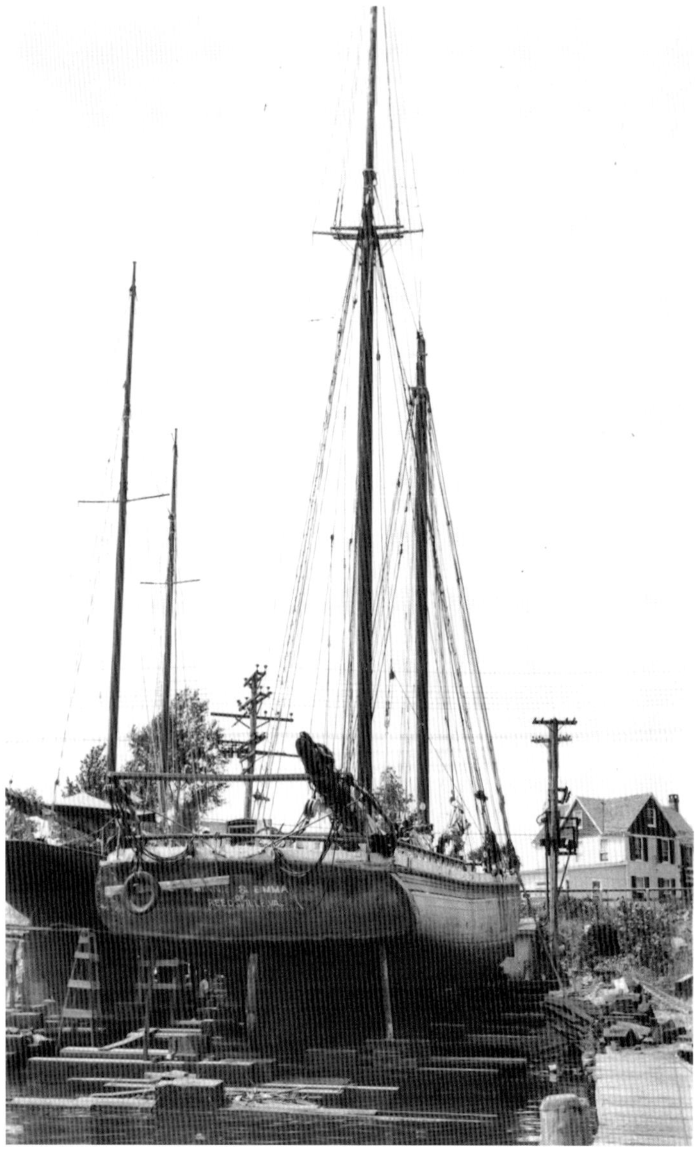

When the *Minnie and Emma* was photographed on the marine railway at A. Smith & Sons Shipyard, Curtis Bay, August 8, 1938, she was showing her age. Her sheer line was wavy and her stern had to be supported by timbers to prevent further sagging. She had been built in 1883 at Pocomoke City, Maryland, so there was good reason for the old schooner to lose her shape. Soon after this haulout *Minnie and Emma* met the fate of so many of her Chesapeake running mates. Her owner, Captain Hunter T. Belfield, ran her up into the head of a shallow slough off the Yeocomico River, Virginia, to die. The 1941 *MVUS* records her as "Abandoned."

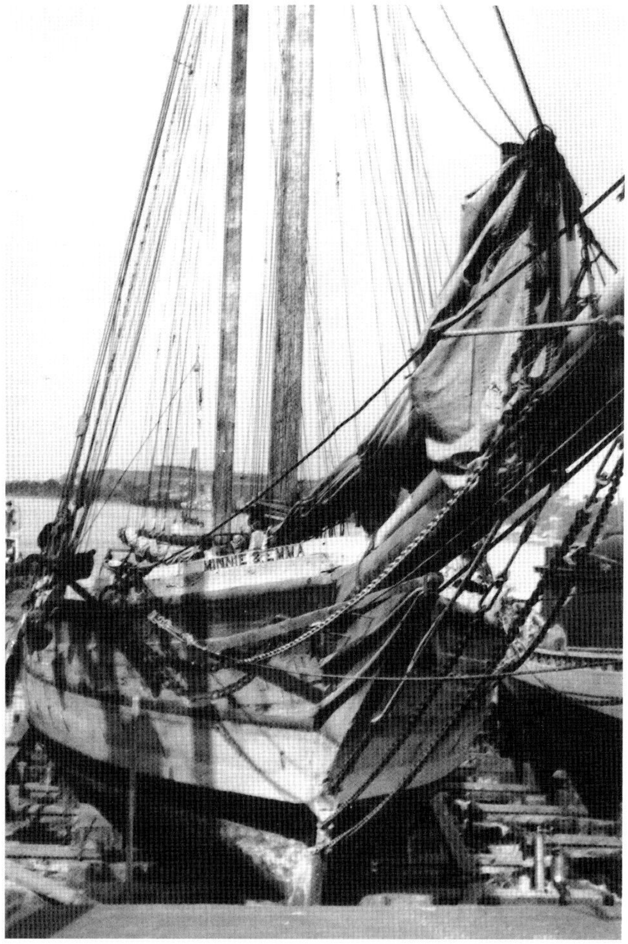

✪ A view of the bow of the schooner *Minnie and Emma,* at A. Smith & Sons Shipyard in June 1939.

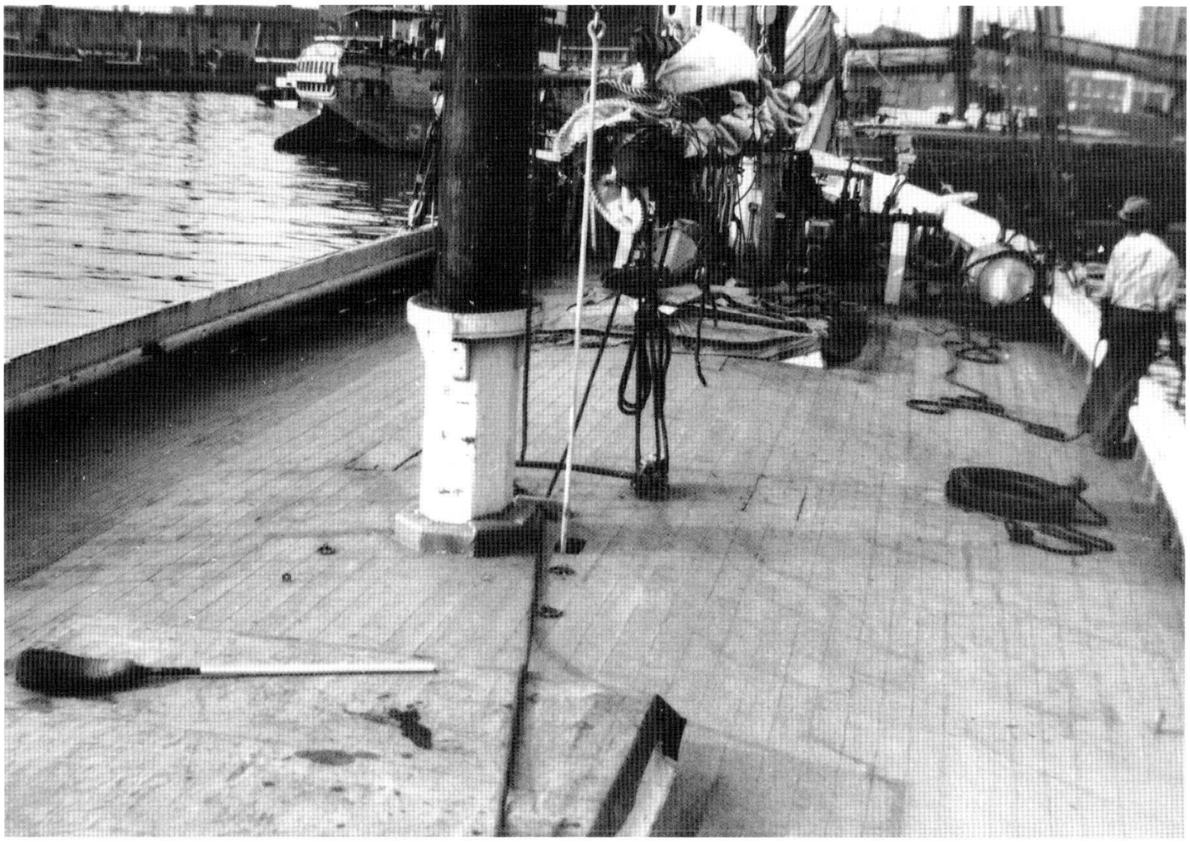

✪ Looking forward on the deck of the *Minnie and Emma,* on November 11, 1935.

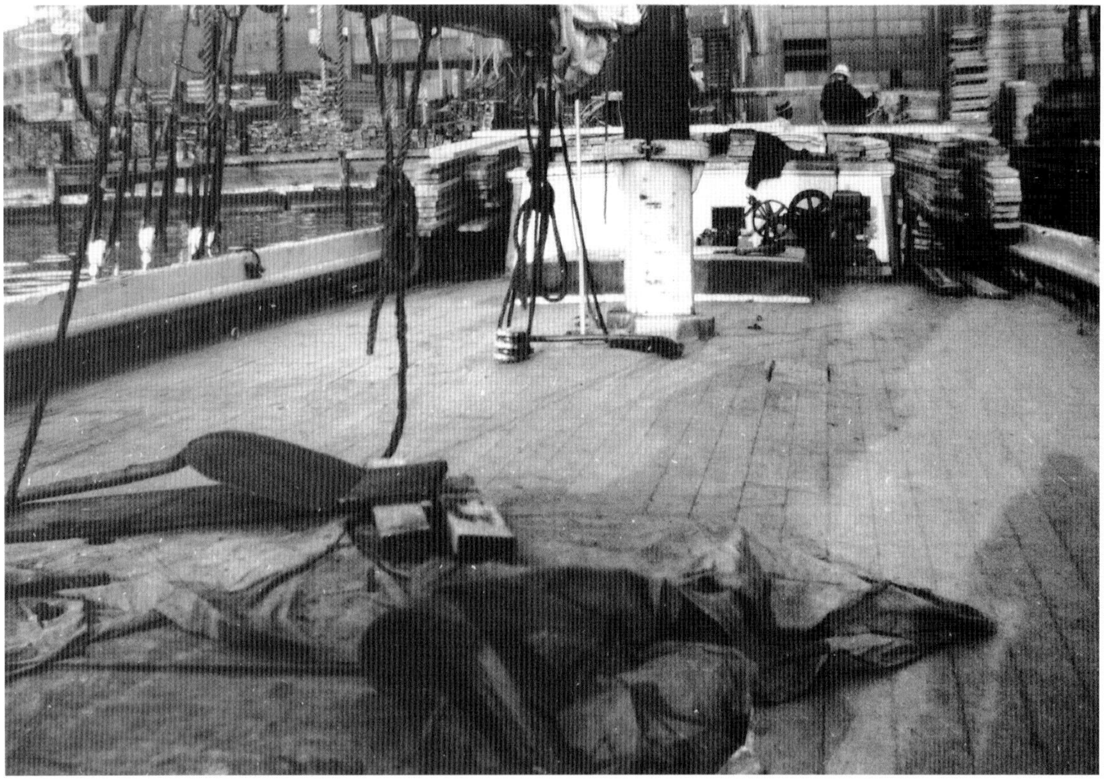

✪ Looking aft on the deck of the *Minnie and Emma.* Note the donkey engine next to her cabin.

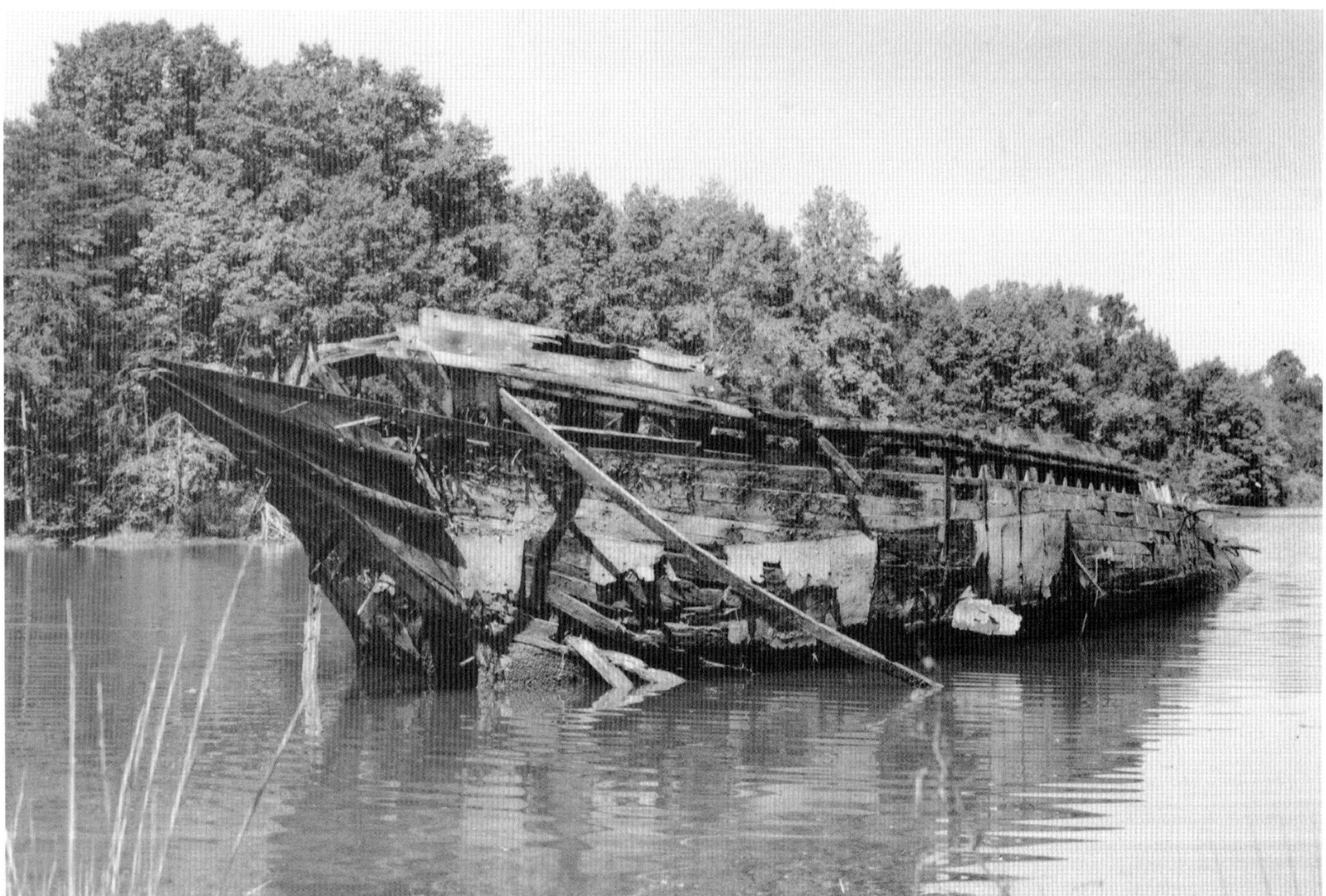

✪ The hulk of the *Minnie and Emma* rests in the shallows at Kinsale, Virginia, on October 13, 1957.

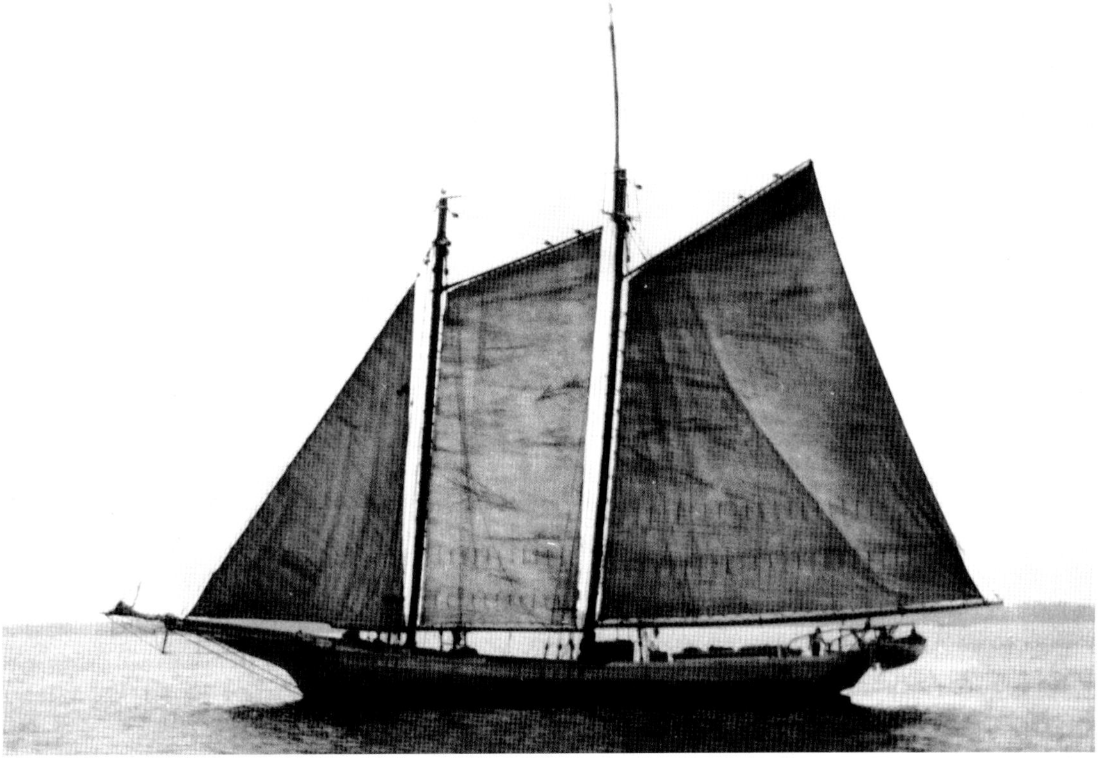

Minnie May Kirwan was to have a long and varied career. She was launched as the *G.W. Robertson* at Whitehaven, Maryland, in 1882. About 1907 she was renamed and was the *Kirwan* as long as she remained under sail. A handy little vessel of 75-foot length, about average for a Chesapeake schooner, she could be employed as an oyster dredger in season and haul watermelons or lumber. Here on June 22, 1934, she is bound down the Bay light, possibly to pick up a lumber cargo. Long owned by T. Preston Webster, who controlled a large fleet of schooners, bugeyes, skipjacks, and at least one pungy up into the early 1930s, the *Kirwan* was one he sold as the Depression was taking its toll. On November 29, 1935, this schooner was sold at a U.S. Marshal's sale for the Depression price of $550 to Captain Robert Shores of Chance. He had put his *J.A. Chelton* up for sale and found a good replacement in the *Kirwan* as a dredger. In April 1941, the *Kirwan* was at A. Smith & Sons Shipyard, Curtis Bay, where she was stripped for conversion into a powerboat and emerged under the name *W.C. Witt*. About 1956 she was renamed *L.R. Parker* and continued on the Bay until sold to Florida owners in 1968. She was engaged in diving operations off Florida until time caught up with her. In February 1970, she was observed abandoned on a coral shoal at Marathon, Florida.

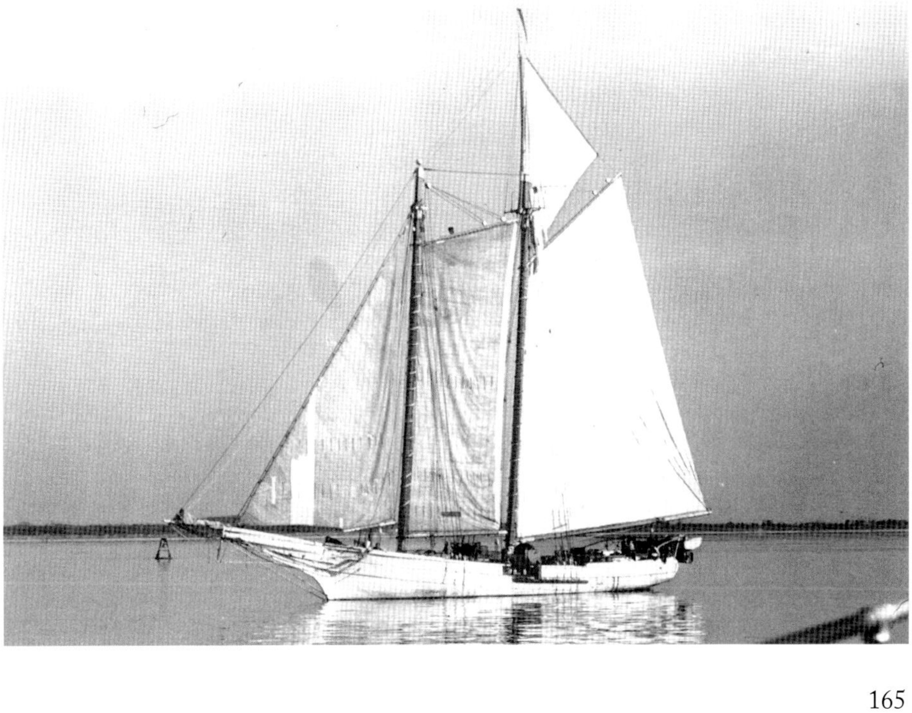

✪ Another view of the *Minnie May Kirwan*, with her main topsail set.

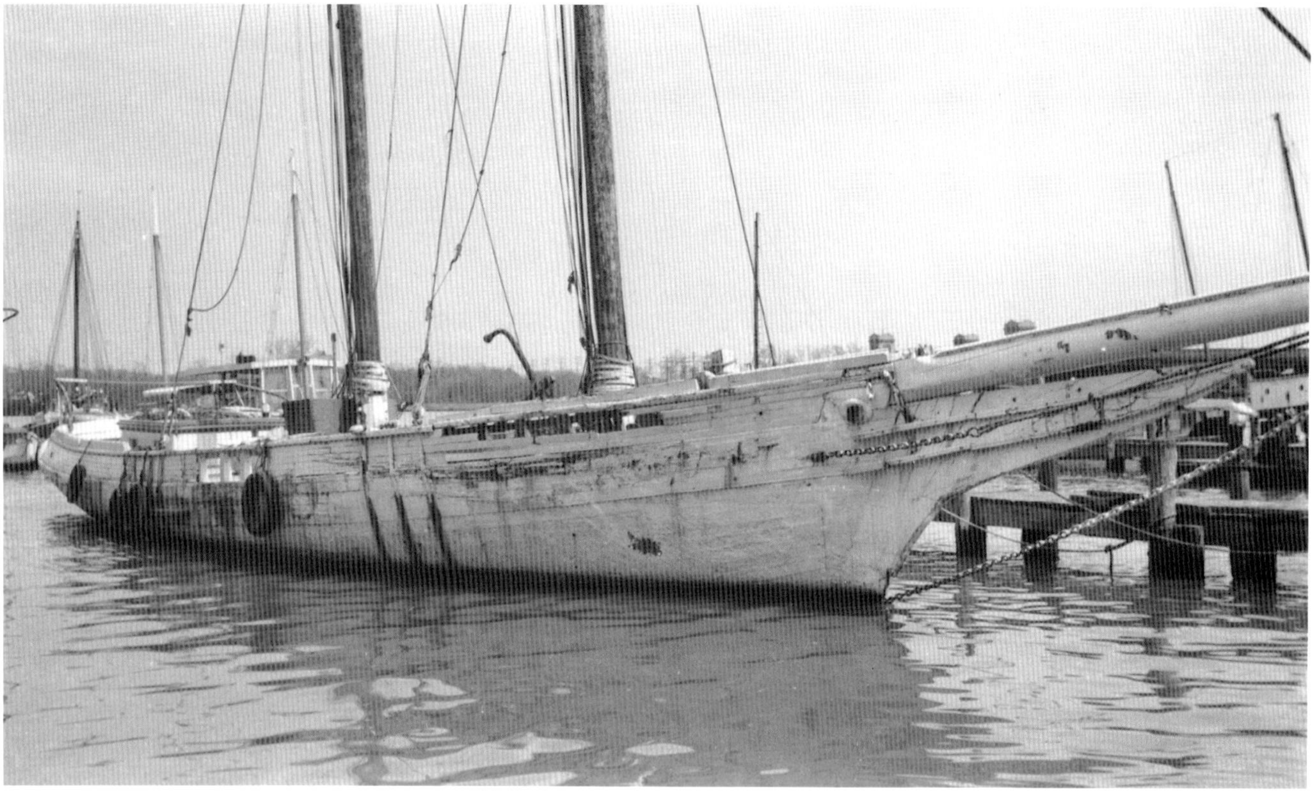

✪ Bow detail, *Minnie May Kirwan* at A. Smith & Sons Shipyard in April 1941. She measured 56 gross tons and 75.7 × 23.7 × 6.7 ft.

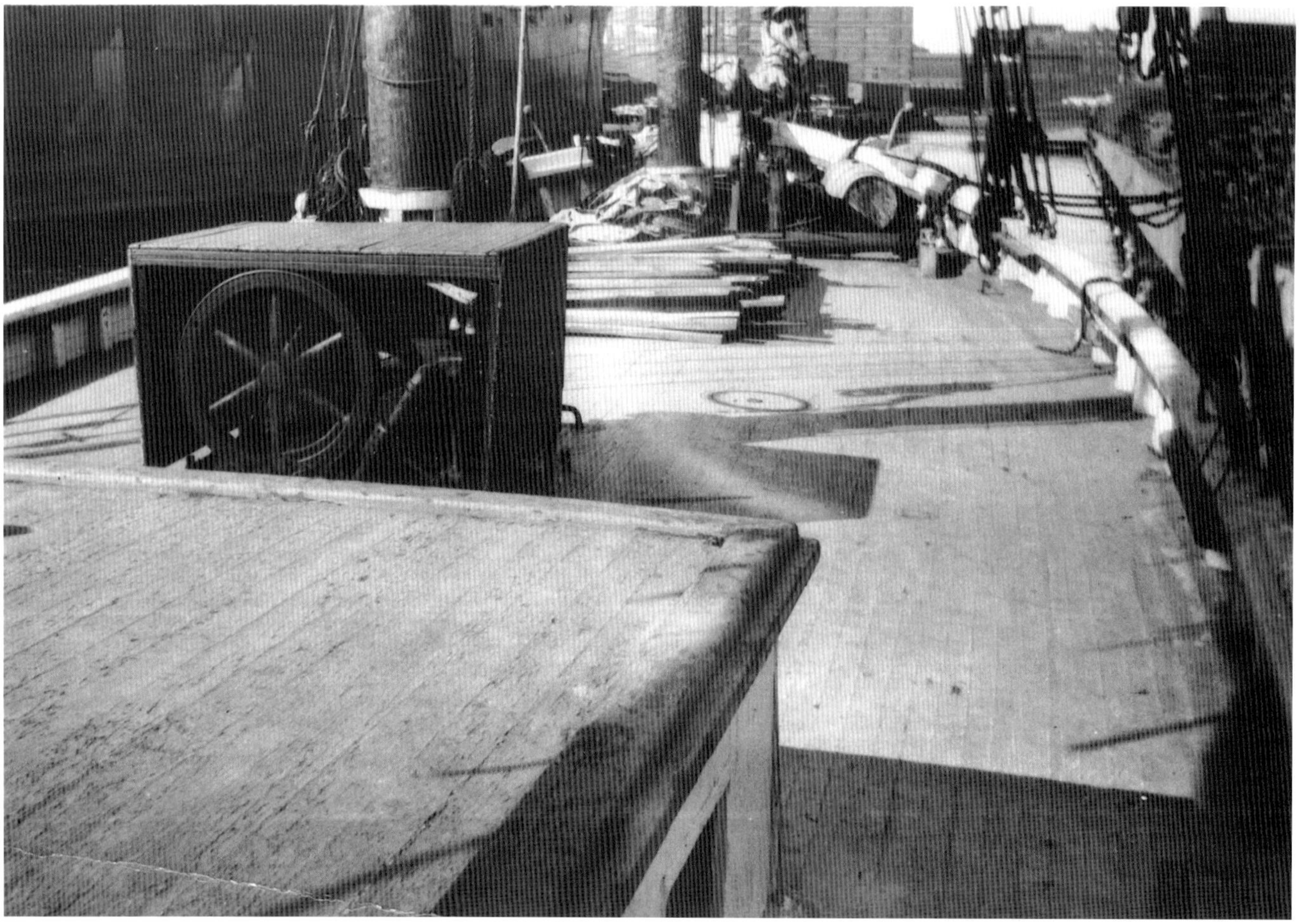

✪ On the deck of the *Kirwan,* looking forward, at Baltimore.

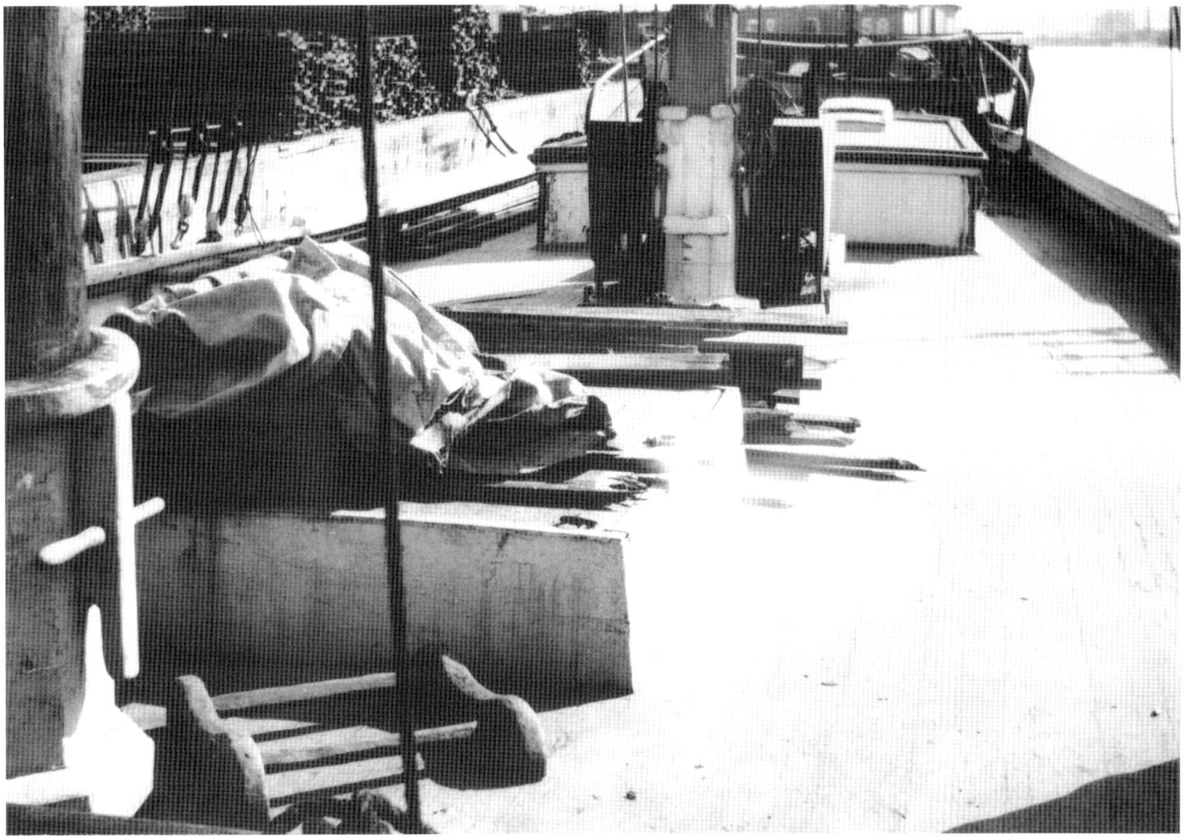

❂ On the deck of the *Kirwan,* looking aft. It appears that she has just unloaded a cargo of lumber.

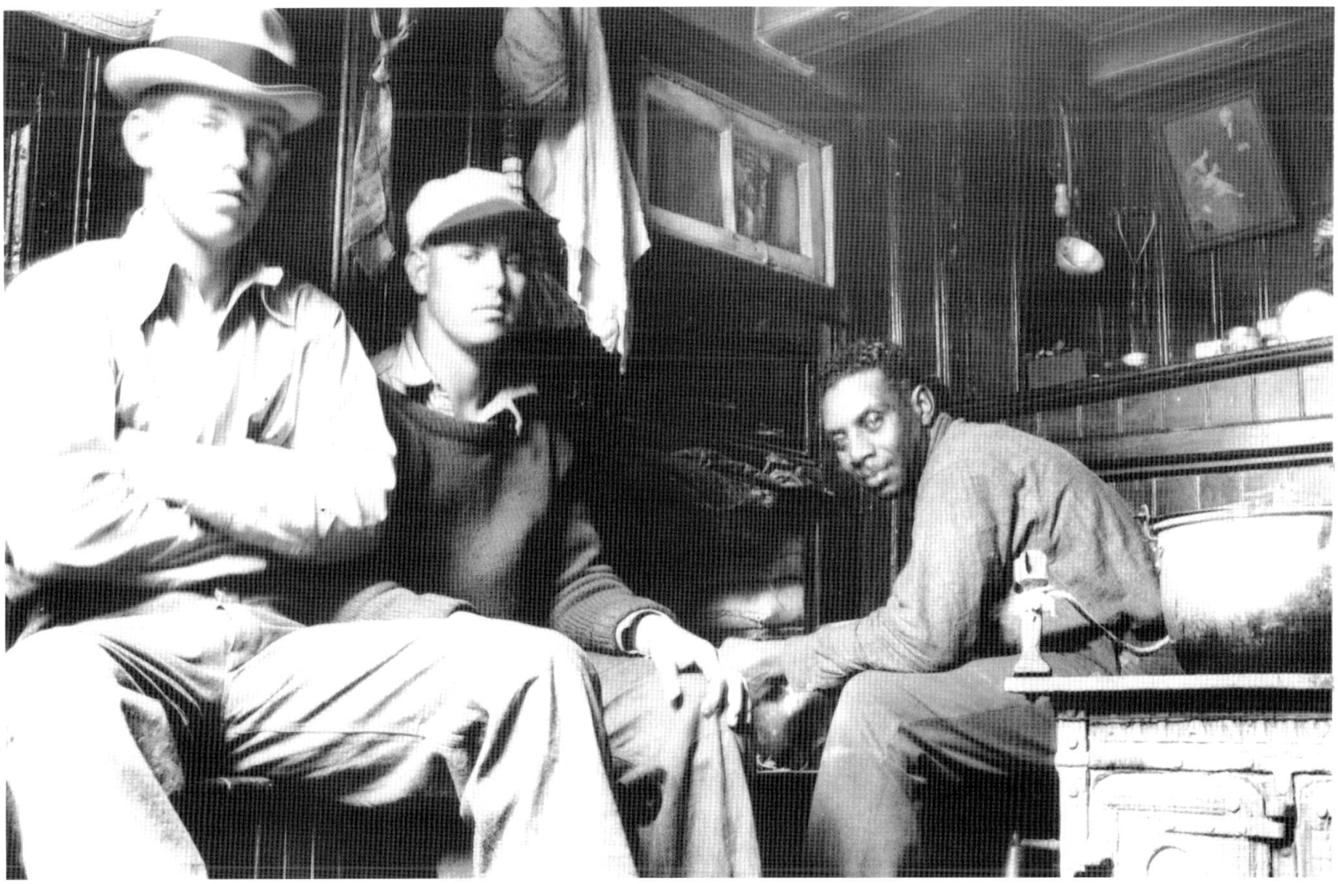

❂ The crew of the *Minnie May Kirwan* relaxes below deck on April 19, 1936. Pictured (l-r) are Lloyd Shores, Jerome Shores, and Percy Satchell.

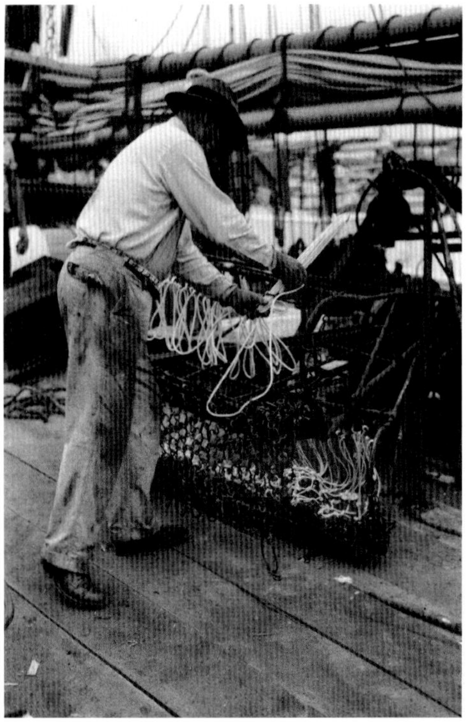

✪ Captain Bob Shores mends the dredge net on October 27, 1935.

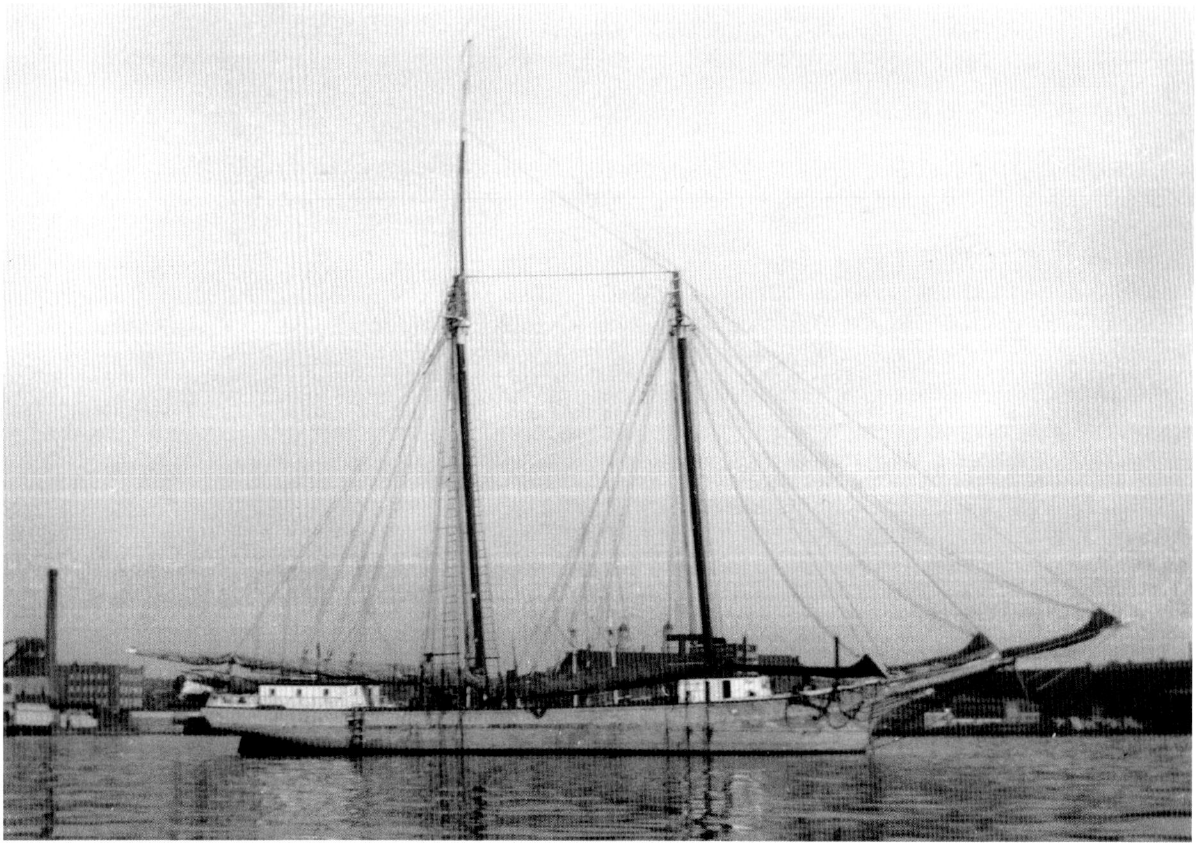

Minnie T. Phillips was one of the larger Bay-built two-masted schooners, measuring 100 feet in length, constructed at Baltimore in 1873 as the *George C.A. Travers*. On a voyage from Baltimore to Norfolk and return in 1910 she must have had everything in her favor. At 2 p.m. on May 28 of that year the *Travers* left Baltimore with 200 tons of guano and arrived in Norfolk at 3 p.m. the next day. The guano was discharged and 150,000 feet of lumber loaded in time for the vessel to sail for Baltimore on June 4. The round trip took less than eight days. About 1912 she was renamed *F.P. Murphy*, after Baltimore's oldest sailmaker, and in 1926 *Minnie T. Phillips*. She kept busy freighting lumber from Virginia to Baltimore into the early 1940s and was reported as "Abandoned" in the 1944 *MVUS*. With a little more than a decade of service still ahead of her, the *Phillips* swings at anchor in Canton Hollow, Baltimore, on October 29, 1933. Unusual for a Bay schooner, she had a complete set of sail covers at that time.

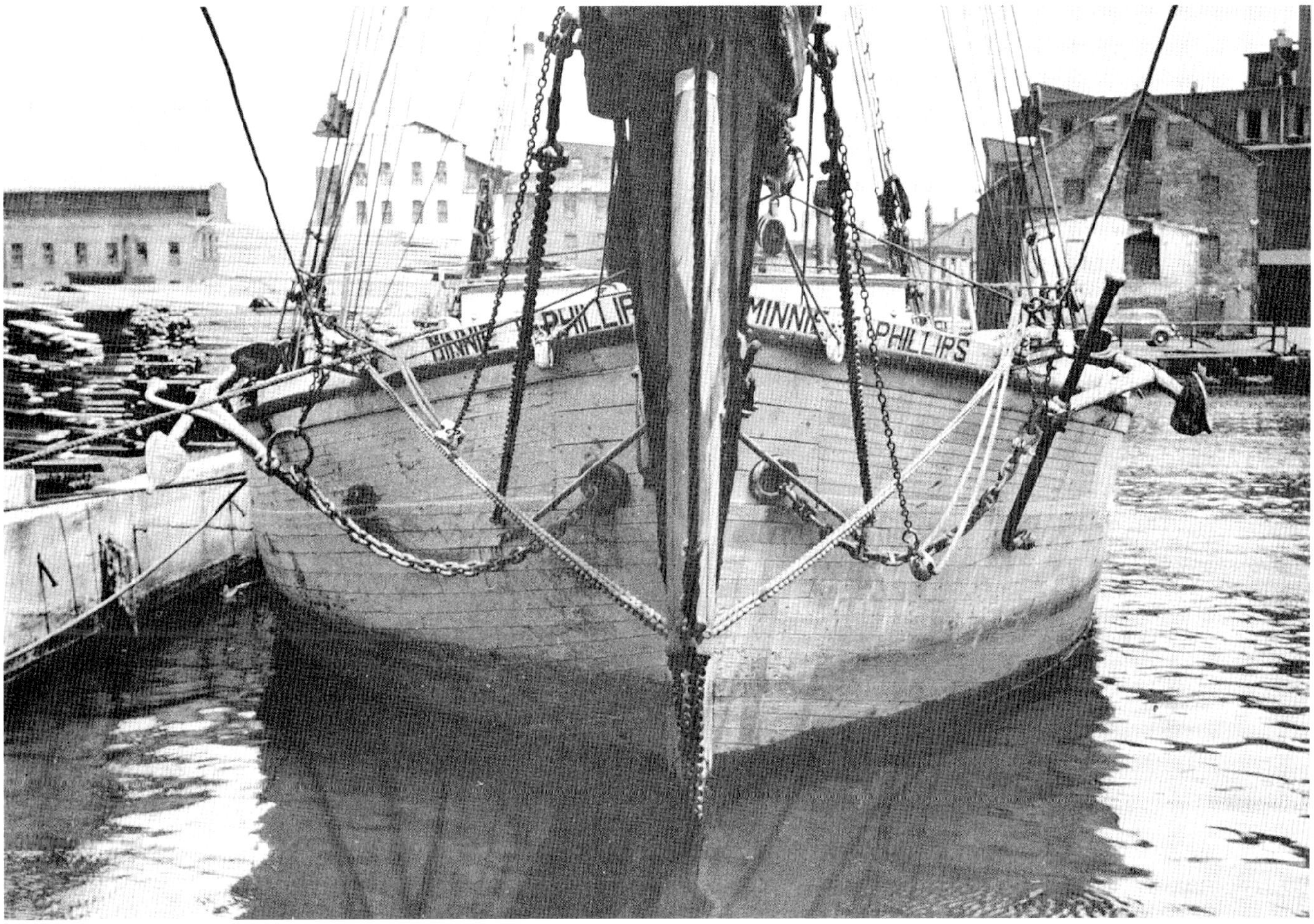

At the Philpot Street lumber pier, Baltimore, on June 26, 1936, the *Minnie T. Phillips* is viewed from under her jibboom. Her headgear and anchors are shown to advantage. Some of her planking seems to be new, probably as a result of replacement after she was rammed by the barge *J. W. Tillett* in tow of the tug *Margaret* in the latter part of February 1935, at which time she lost her bowsprit. Less than five years after this scene, she had been abandoned.

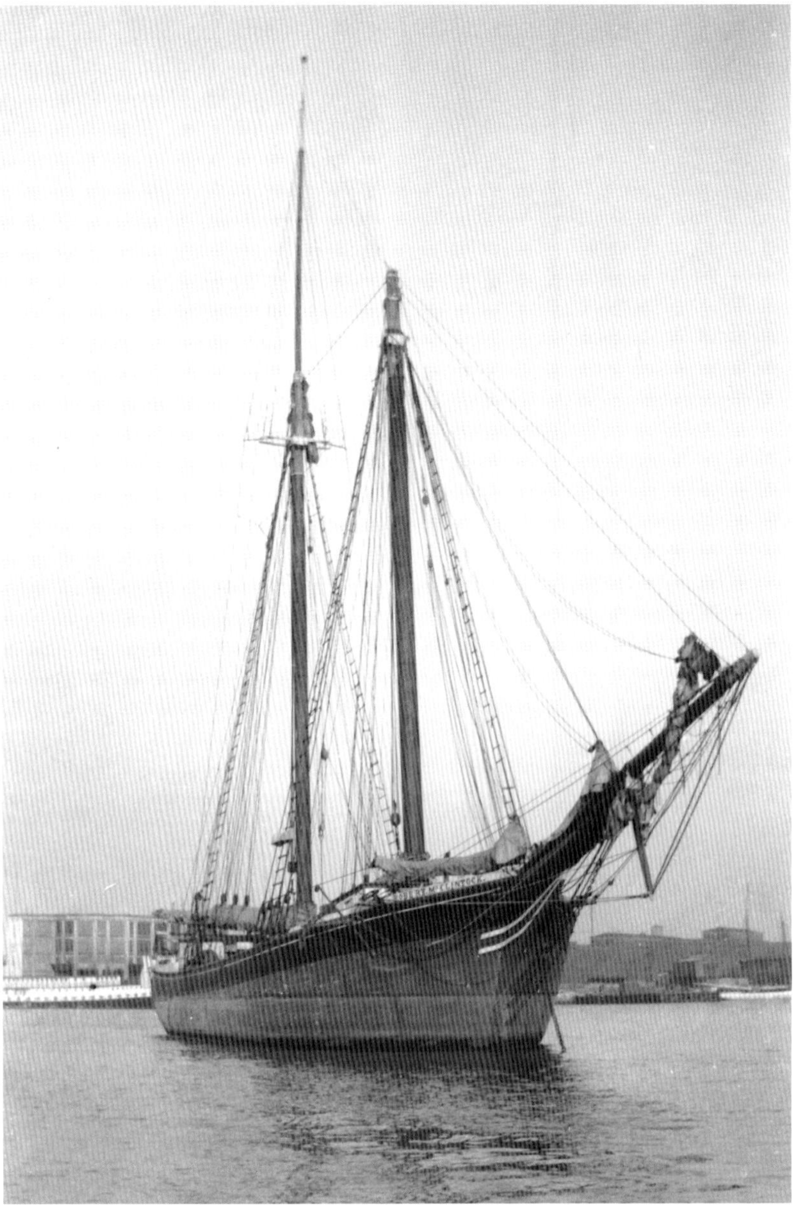

If this photograph were in color one would see a handsome example of a small coasting schooner painted a leaf green from the waterline up to deck line and black bulwarks with a white bead running the full length of the hull. Though small in comparison with the three- and four-masted schooners of the coasting trade, she was large—90 feet in length—for a Chesapeake schooner. This is the *Robert McClintock* at anchor in Canton Hollow, Baltimore, on October 29, 1933. Built in 1858 at Norwalk, Connecticut, by Richard and Bell as the *William F. Burden*, the 75-year-old schooner still retained a pleasing sheer and a stout appearance even though she had some ill experiences around the turn of the century. She had been hailing from various New England ports until 1891, when she was sold to Baltimore. In 1897 she was rebuilt at Whitehaven, Maryland, and renamed *Robert McClintock* after one of her owners. On November 28, 1901, she stranded on the east side of Fire Island Bar, New York, while beating out of the inlet in a strong northwest breeze. Surfmen from nearby lifesaving stations boarded her, ran an anchor out, and got her afloat. On August 19, 1903, she stranded on Fire Island Bar again but was refloated with the assistance of lifesaving crews. The first day of the next year she was frozen in off Patchogue, New York. Her master had discharged the crew and, fearing that when the ice began to move the schooner would overrun her anchor and stove a hole in her bottom, requested assistance from the Blue Point Life-Saving Station. Surfmen cut a hole in the ice, heaved up the anchor, and moved her to a safer anchorage. She had onboard a cargo of lumber from Georgetown, South Carolina, for Patchogue. Her most trying experience occurred on January 9, 1906, when she was disabled in a gale five miles south of the Popes Island Life-Saving Station, Virginia. With sails blown away, she displayed distress signals. Surfmen found her leaking badly and her mate and cook each with a broken leg. The injured seamen were transferred to the schooner *Maggie M. Keough*, which carried them to the marine hospital in Norfolk. The *McClintock* was maneuvered to a safe anchorage from which point a tug towed her into Norfolk. The schooner had a cargo of rails from Baltimore for Georgetown, South Carolina. During these trials a Captain Joseph Lewis had been her master but in early 1906 she was acquired by new owners. One of them, Captain Joseph M. White, became her master and part owner during the next 26 years and her outright owner from 1932 to 1942. The latter year she was sold and converted into a powerboat, trading out of Florida ports. In 1943 she was transferred to the Honduran flag. Her career away from the Chesapeake was brief, for she was reported aground on the east end of Yucatan Peninsula on January 14, 1944, and breaking up.

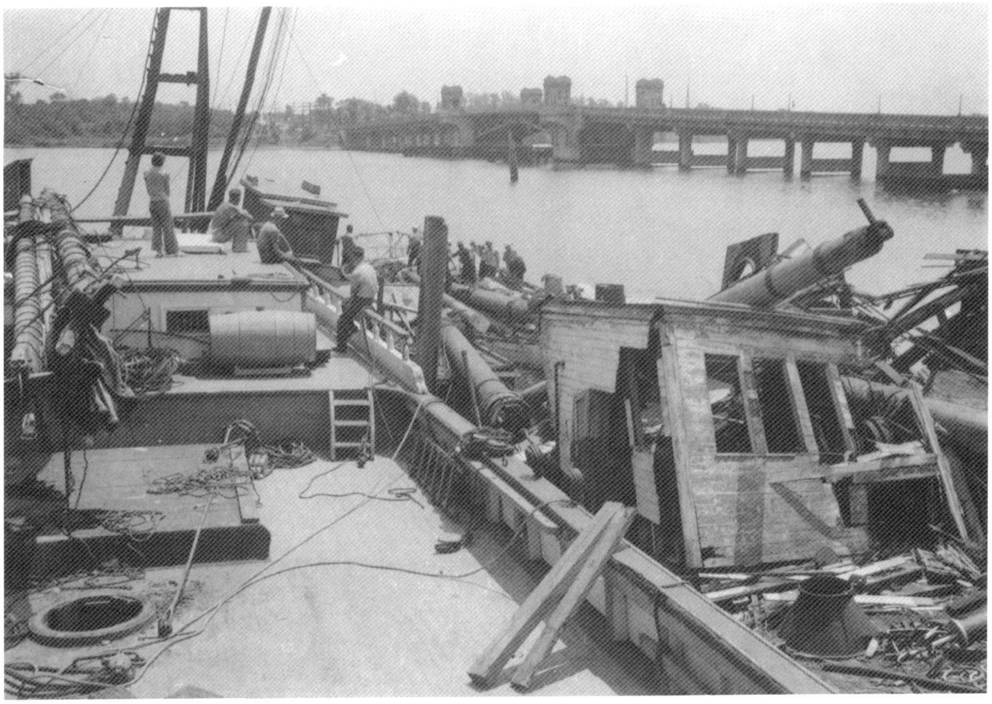

❂ The deck of the *Robert McClintock* at A. Smith & Sons Shipyard around 1937. The crane lifting out her mast has just collapsed on the adjacent bugeye, *Gorman C.*

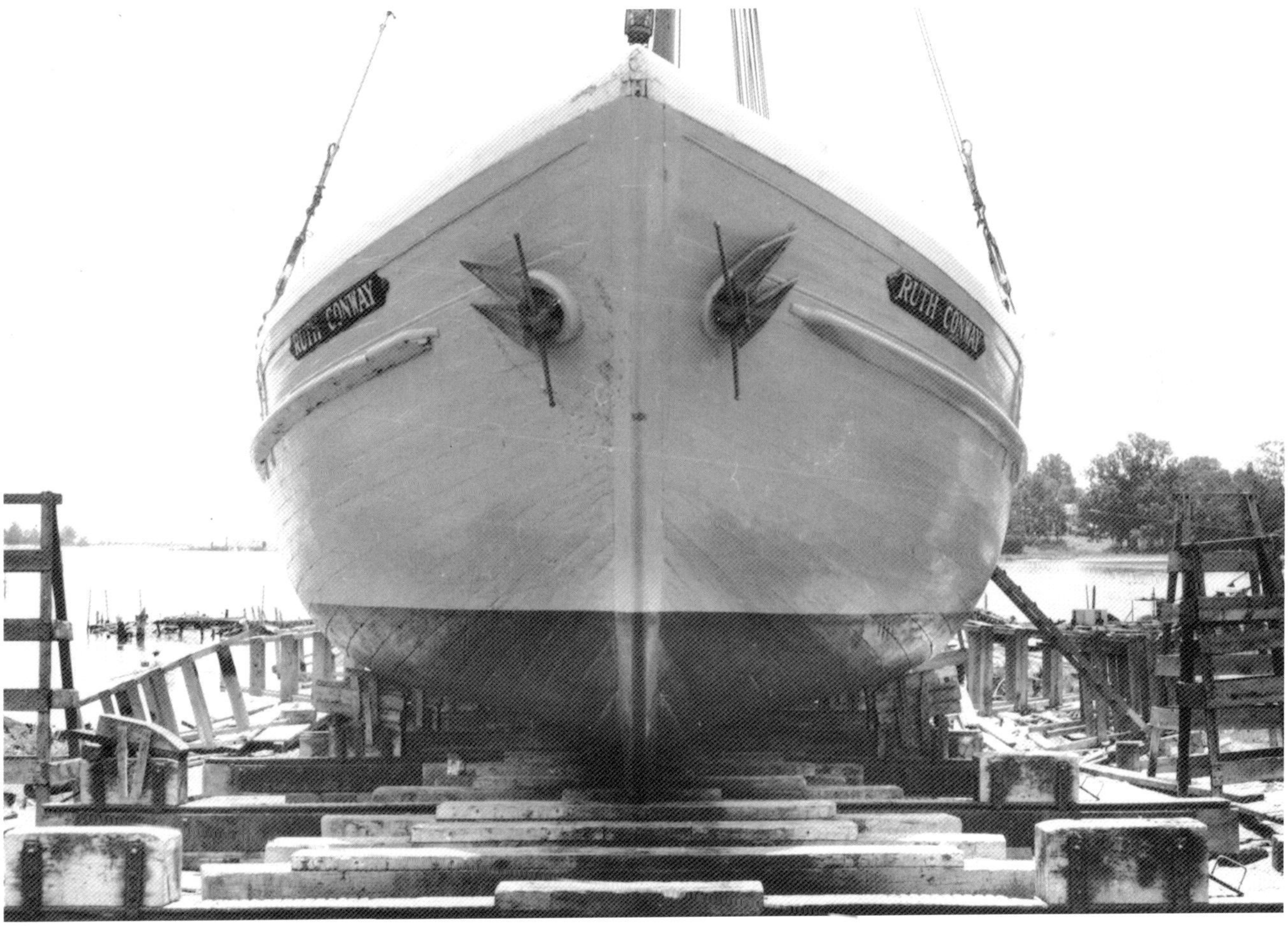

What is supposed to have been the last two-masted schooner built for commercial use in the Chesapeake area was the *Gracie May*, constructed in 1909 at Seaford, Delaware. This may be disputed since there was a schooner named *William F. Dunn* built in 1924 at West Norfolk, Virginia. In any event, the *Gracie May* only served 20 years under sail (the *Dunn* less than that, since she foundered at Tangier Island, Virginia, February 12, 1932) before being converted into the powerboat *Ruth Conway* in 1929 as part of Captain Harvey H. Conway's fleet. Freshly painted and glistening, the *Conway's* hull sits poised on the ways at Humphreys Shipyard, Weems, Virginia, in June 1958, almost half a century old but still retaining her lines.

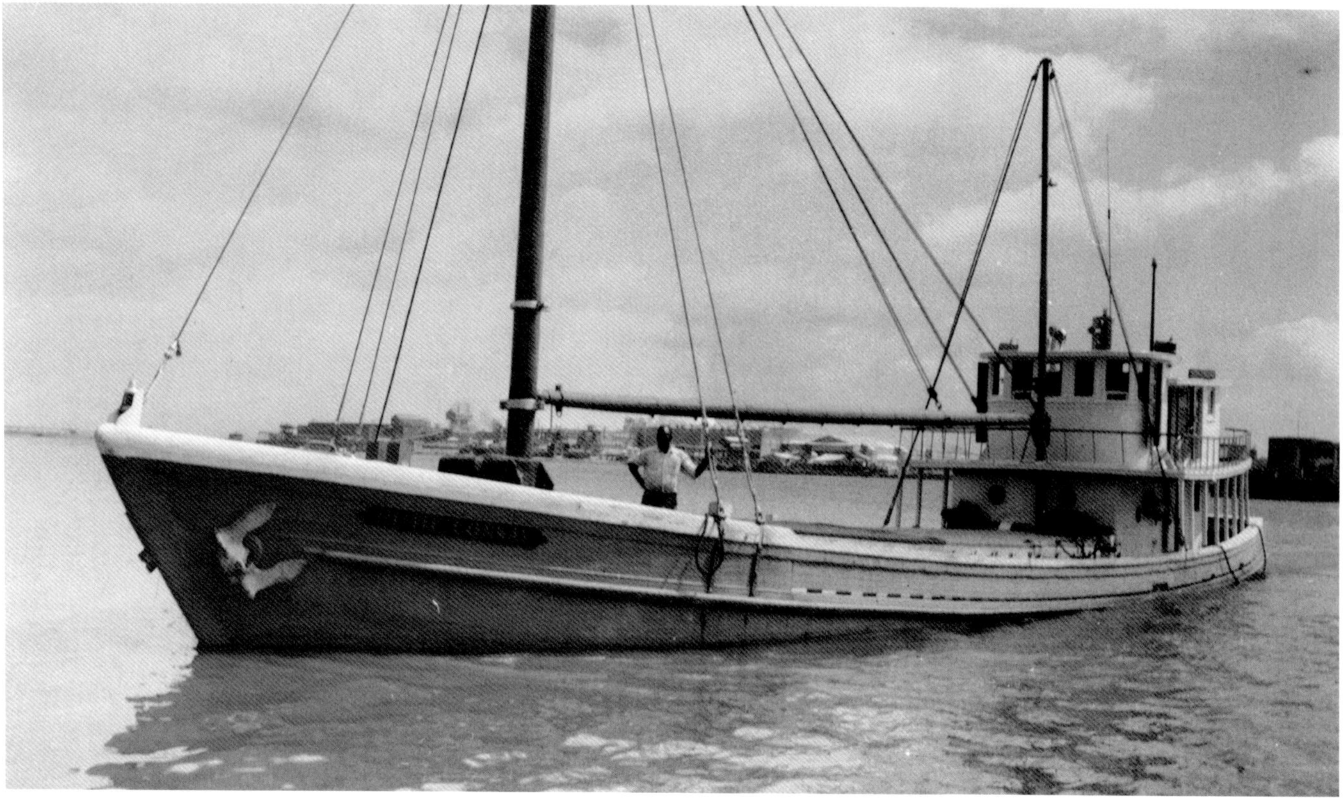

The old schooner hull is seen approaching the dock of A. Smith & Sons Shipyard, Curtis Bay, in the early 1950s and shows how well the Bay schooners could be adapted to power. She is low in the water, almost up to her deck, probably with a cargo of fertilizer, a product represented by the factories in the background. *Ruth Conway* had sunk in the Chesapeake and Delaware Canal in 1947 but was raised, overhauled, and put back into service again. She was not to be quite as fortunate the next time. On October 8, 1965, while bound from West Point, Virginia, to Norfolk with 7,000 bushels of corn, she developed a leak and sank about one mile northwest of Thimble Shoal Light and became a total loss.

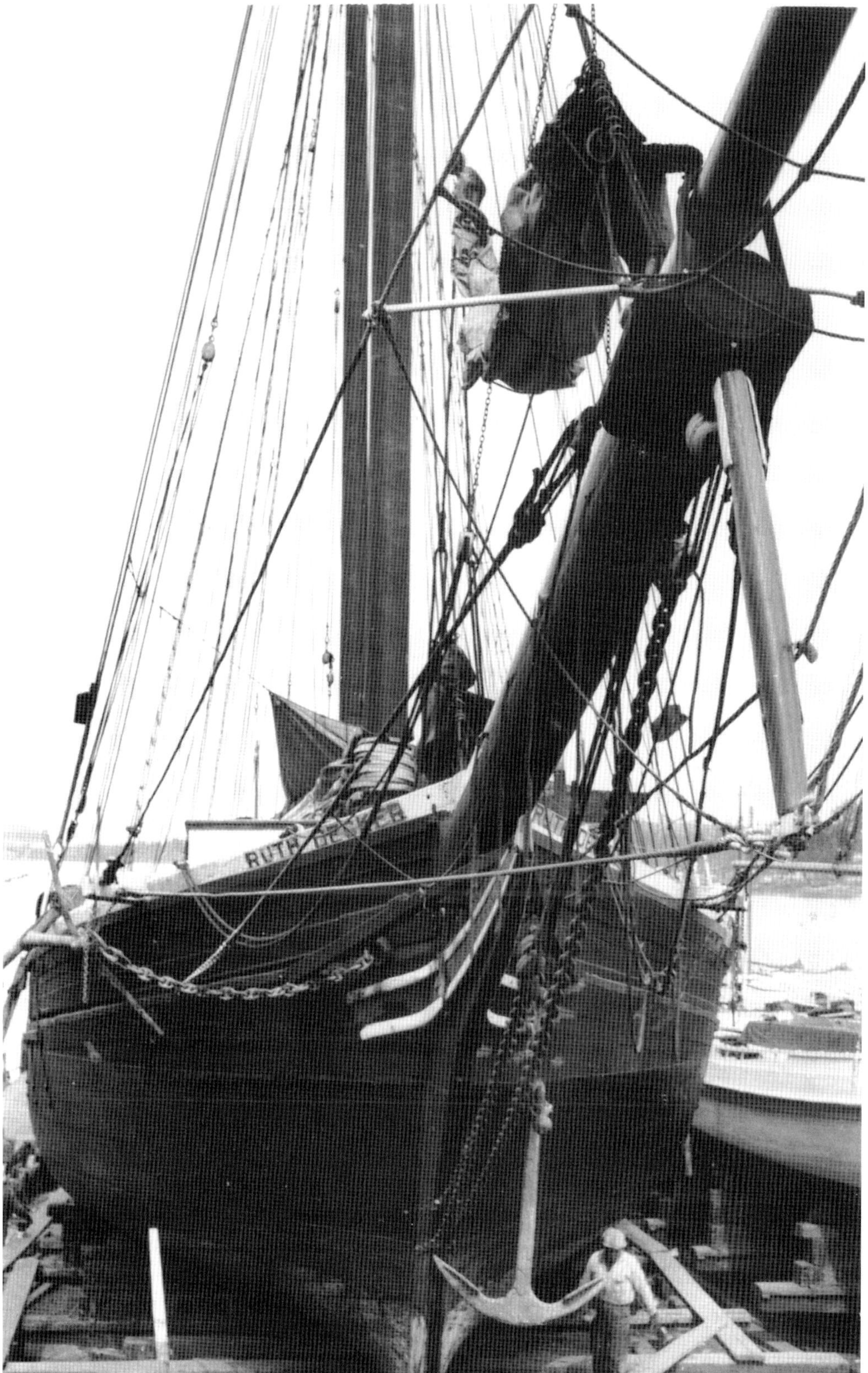

In July 1938, the *Ruth Decker* was getting a periodic repainting at the A. Smith & Sons Shipyard, Curtis Bay. Running the full length of her hull was a band of galvanized sheet metal about six feet in height. It eliminated further caulking of her seams, guaranteed her being more watertight, and protected her along the waterline from sharp ice. On October 14, 1938, while bound from Baltimore to Greensboro, Maryland, the *Decker* and the powerboat *E.G. Pope* collided off Sandy Point, Maryland, in a dense fog. The powerboat fared the worse, since she had a hole stove in her bow and had to be beached near Gibson Island. The *Decker* proceeded to Annapolis, where it was found that the schooner was damaged above the waterline but could continue to her destination. Perhaps the metal sheathing helped her. Built as the *Josie Crowley* at Port Jefferson, New York, in 1876, and renamed *Ruth Decker* in 1900, she was to be sold to Elizabeth City, North Carolina, about 1912, before arriving in the Chesapeake as a point of registry in 1923.

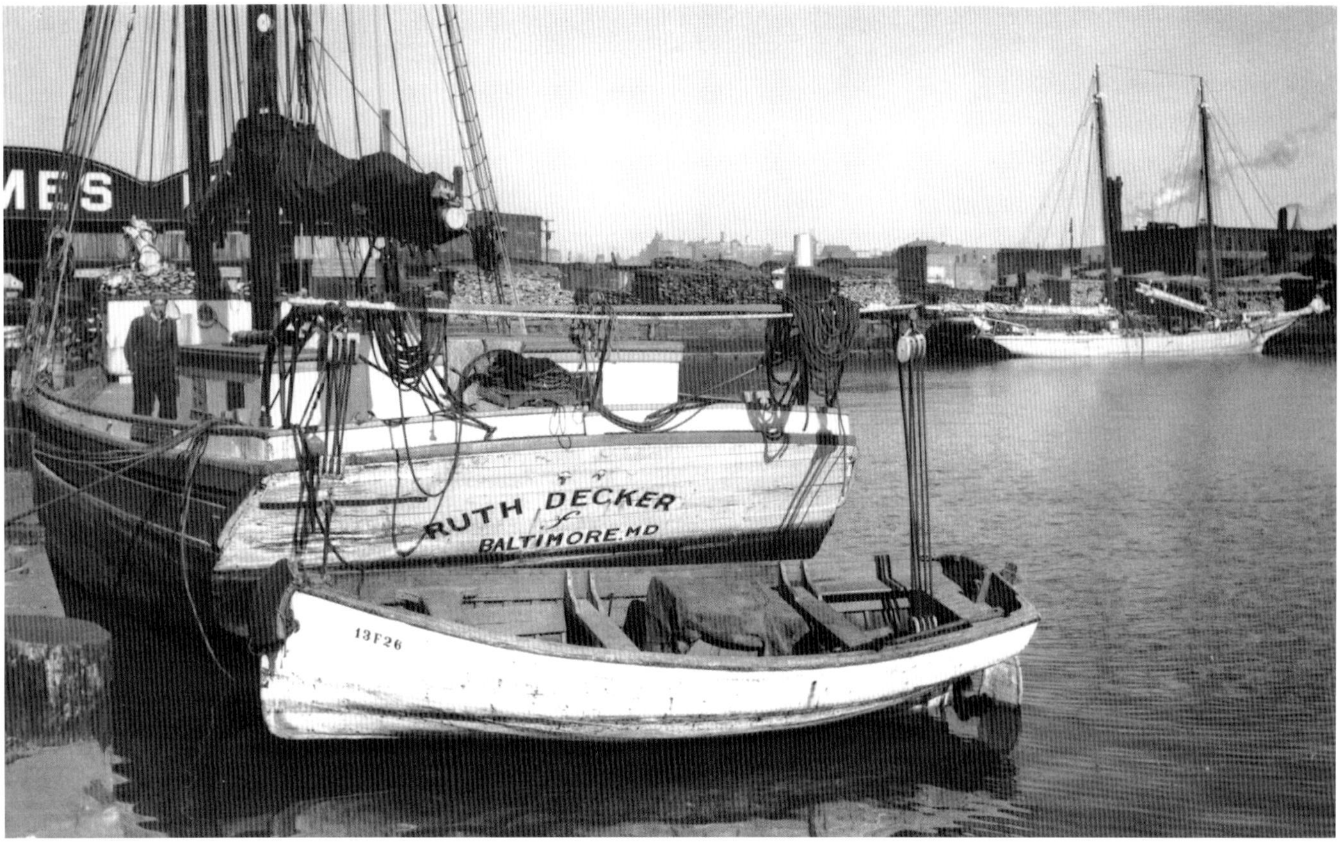

In February 1939, the *Ruth Decker* is moored at the city dock, Pier 7, in front of James Lumber Company, Lancaster Street. The yawl boat is prominent; these sturdy, powerful little auxiliary craft were of good design and measured about 26 feet long, with a beam of about six feet. How the sailing craft depended upon them is evident by the wear on the port side of the *Decker's* transom. At upper right is the schooner *Australia*. The *Decker* sailed on the Bay until 1945. In the early part of that year the head of her foremast carried away just below the crosstrees. This was never repaired. The next year she was stripped, with the plan of converting her into a powerboat, but this never materialized. In March 1948, trying to place her hull in position for use as a breakwater near Fort Howard, Maryland, she grounded offshore and eventually went to pieces there.

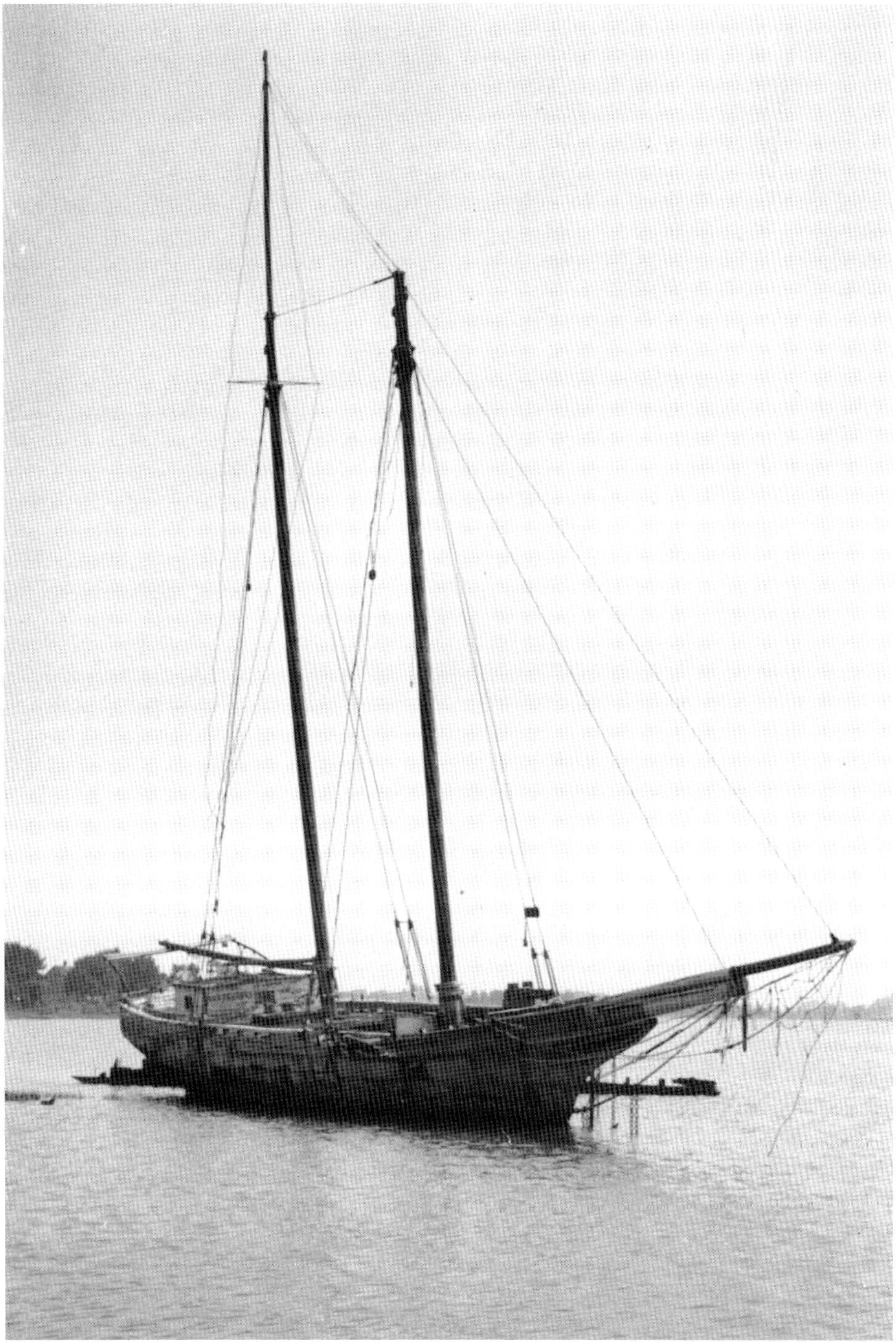

The schooner *S.J. Delan* was lucky in coming to an end where she did—in the haven of the Tred Avon River off Bellevue, Maryland. On June 17, 1936, she appeared like this after having been driven aground during the hurricane of August 23, 1933. But when one considers the number of times she touched bottom off the Atlantic coast in her career, she had to be fortunate. On February 2, 1877, while owned in Onancock, Virginia, she ran aground at Tucker's Beach, New Jersey, while bound to New York from Hog Island, Virginia, with a cargo of corn. She was refloated with no loss. On March 26, 1899, while bound from Hampton, Virginia, to Atlantic City, New Jersey, with a cargo of oysters, she struck her centerboard on the bottom while beating into Absecon Inlet, New Jersey. The crew from the Atlantic City Life-Saving Station refloated her. On May 31, 1902, the *Delan* stranded a half mile south of Atlantic City Life-Saving Station. Surfmen boarded her and found her leaking, her pump disabled and her steering gear carried away, but they refloated her. After repairs, she completed her passage to Atlantic City from Haverstraw, New York, with a cargo of bricks. On January 3, 1904, she dragged anchor in the vicinity of the Assateague Beach Life-Saving Station while bound from Chincoteague, Virginia, to Norfolk with oysters. She was hove clear of danger. On April 2, 1910, bound from Chincoteague to Abescon with oysters, she ran into difficulty north of Atlantic City but suffered no loss. While bound from Norfolk to Chincoteague with a cargo of coal on August 6, 1913, the schooner was disabled south of the Hog Island Life-Saving Station but was fortunate again in suffering no loss. About 1925 she was purchased by W.H. Valliant and confined her activities to the limits of the Chesapeake until her loss in 1933. Having been built at Islip, New York, in 1869, the 59-foot schooner enjoyed remarkable longevity considering the number of times she was in trouble.

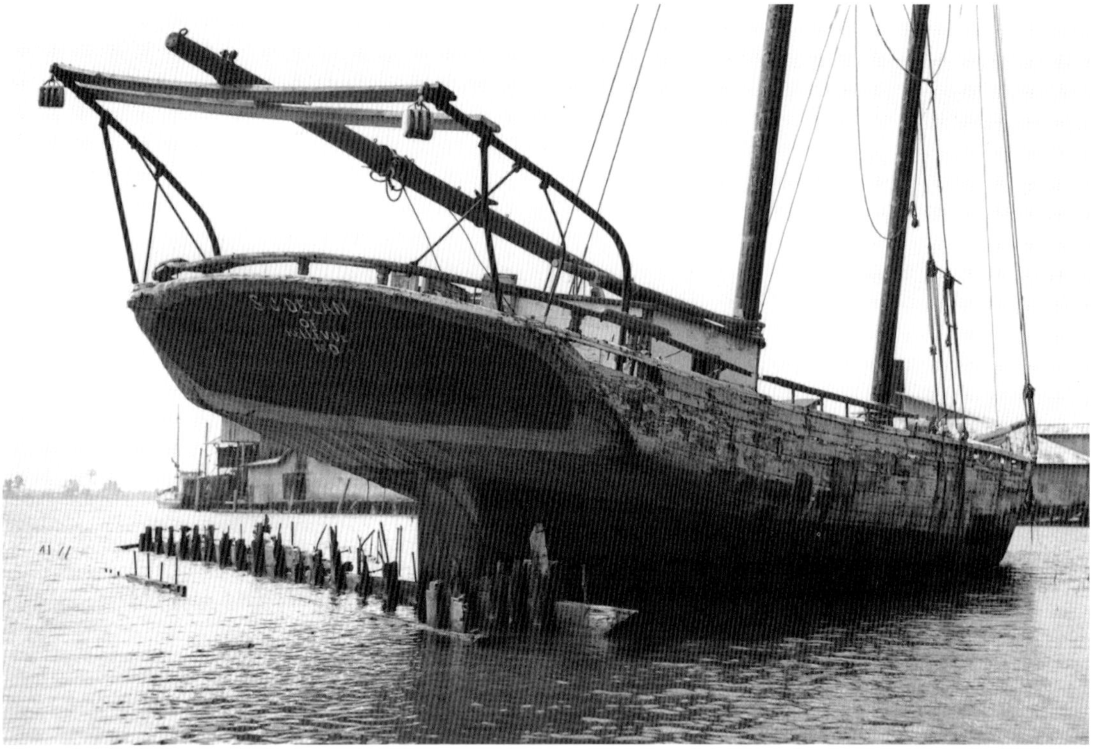

The wracked hull of the *S.J. Delan* after almost three years aground. Sixty-four years old when cast on the shallows, she was probably too far gone for an attempt at refloating even if that had been possible. In the background is the plant of her owner at Bellevue.

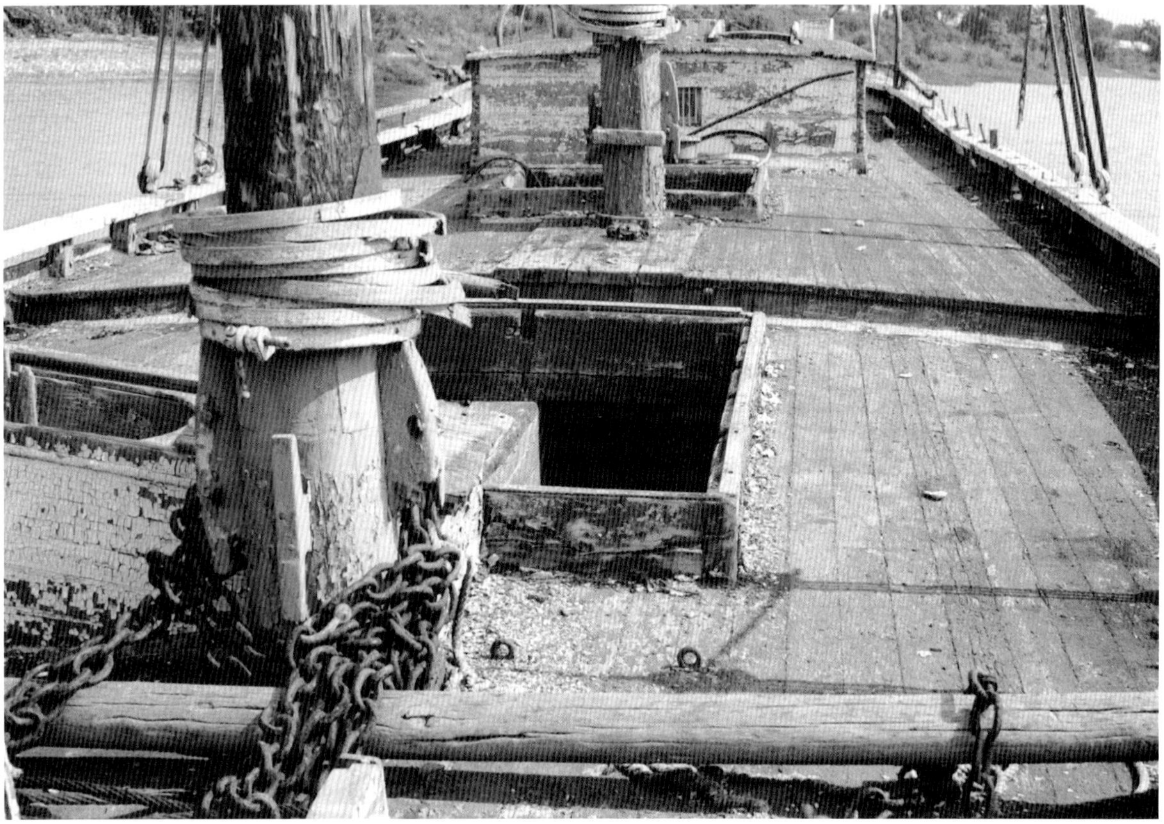

Come aboard an 1869-built schooner, dilapidated and abandoned as she is. This was the view looking aft on the deck of the *S.J. Delan*, littered with flaking paint. The loglike timber in the lower foreground was the "horse" for the jib sheet. Her raised quarterdeck is extended forward of the mainmast. This was a feature not usually found on Chesapeake-built schooners, although the *Bohemia* was fitted out this way.

Anchored in Curtis Creek, Baltimore, in April 1941, the *Sarah C. Conway* is reflected in the calm waters. She had been built in 1872 as the *A.H. Schultz* by William E. Woodall and Company at Baltimore and was one of the first vessels owned by Captain Harvey H. Conway of Cambridge, who was to build a large fleet by the time he died in 1937. He was given to naming his vessels after members of his family. The first was the renaming of the *William Sommers* (about 1921) to *H.H. Conway*, after himself. About 1925 he renamed the *Schultz* after his wife Sarah.

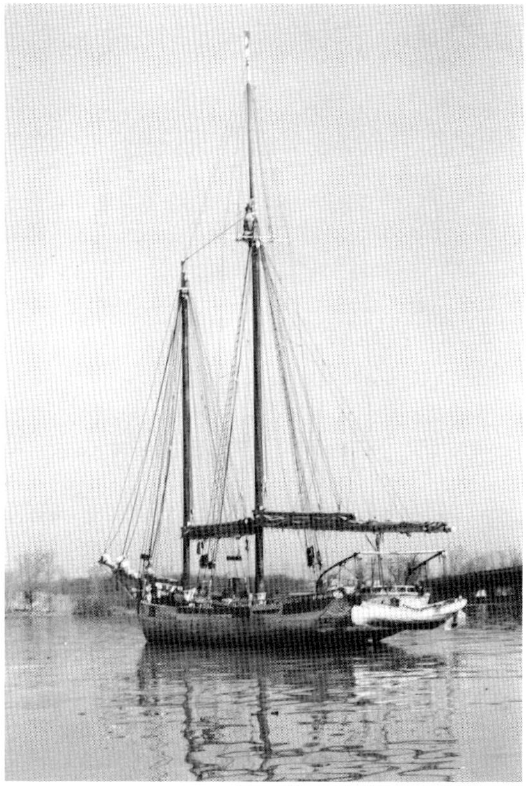

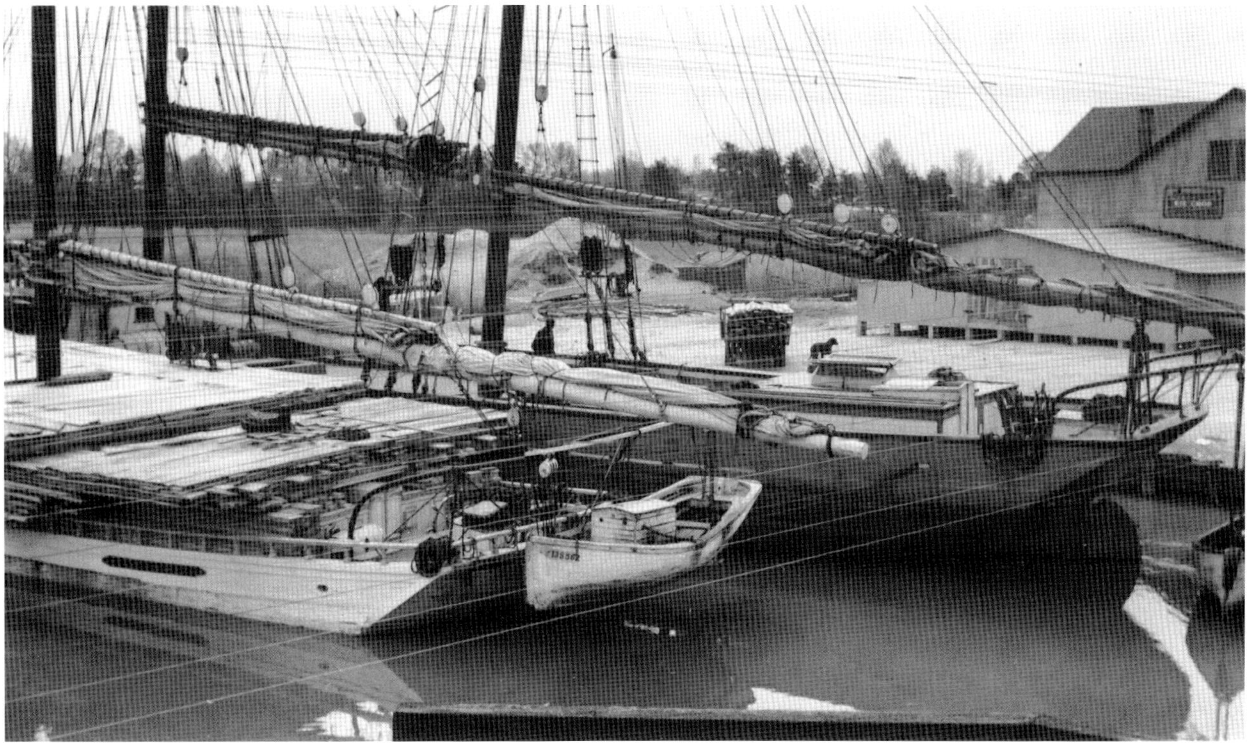

Sarah C. Conway, at left, at Tappahannock, Virginia, on April 23, 1941, loaded with lumber for Baltimore. To her starboard is the *Charles G. Joyce*. This view shows how high lumber was piled on the schooner's decks and cabin tops. It is even crammed along the quarterdeck on either side of the cabin. The crisscross layers of boards interlocked them and prevented them from shifting, an important factor when these cargoes were not lashed down. *Sarah C. Conway* became a powerboat about 1946. In May 1957, while loaded with grain and docked in her home port of Vienna, Maryland, she was rammed and sunk by a tug. For a year she remained completely submerged in the murky waters of the Nanticoke River. In May 1958, she was raised by cable and crane by A. Smith & Sons Shipyard and refurbished to carry on her career as a Bay freighter under Captain Calvert Evans. In 1971 the *Conway* was purchased by Captain Fred Savage of Ocean City, Maryland, and placed in the deep-sea clamming fishery off the Maryland coast. Her companion in her rigorous trade is the 1892-built former Bay schooner *Flora Kirwan*, skippered by Captain Ricky Savage.

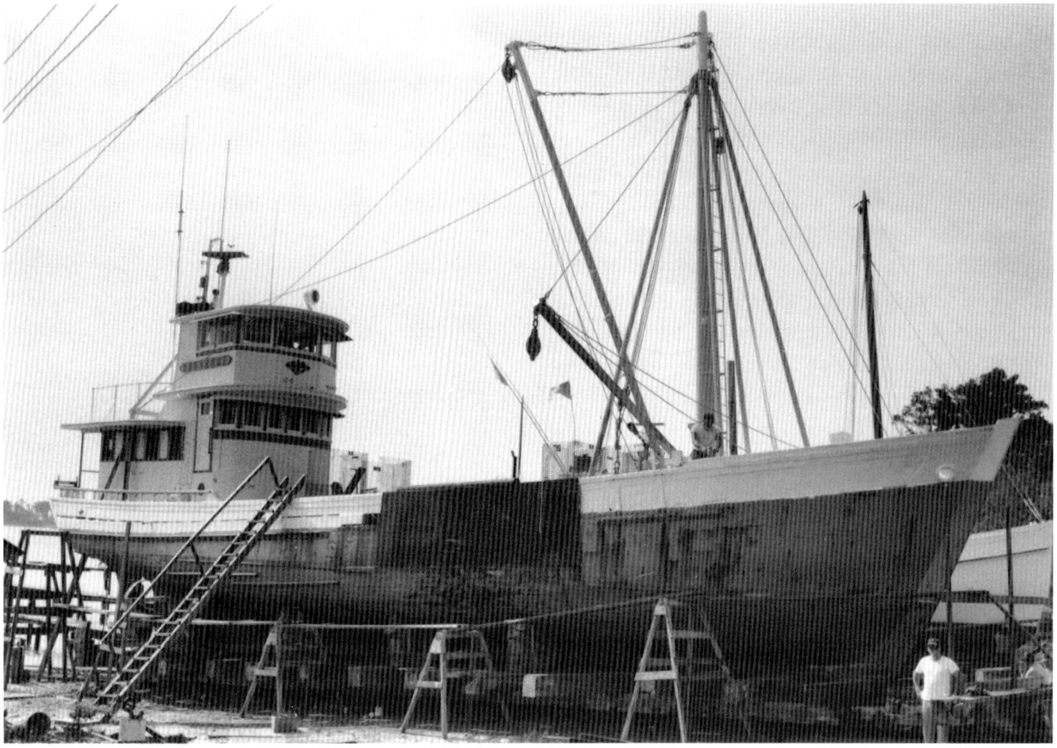

✪ The *Sarah C. Conway* on the railway, probably during her clamming days after 1971. Note the gear on her deck.

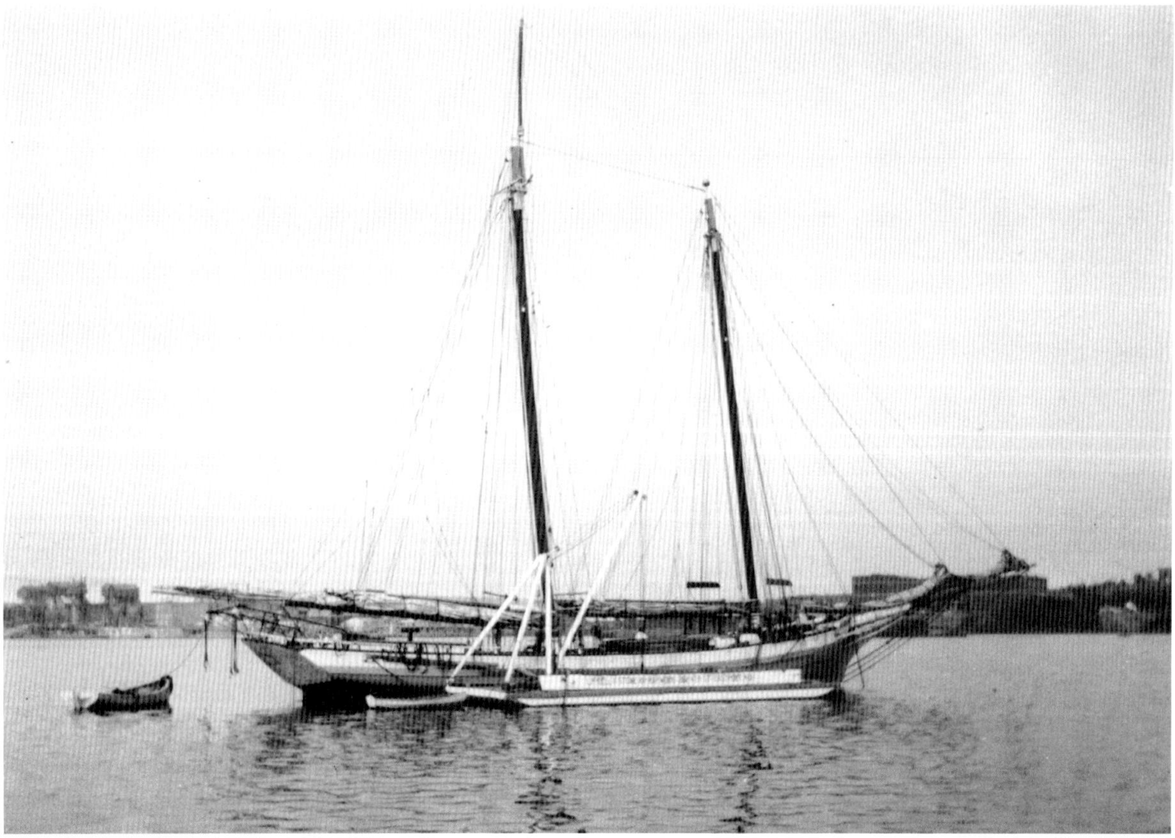

Smith K. Martin was a handsome schooner with a nice sheer and the largest transom this author recalls having seen on any Bay schooner of her size. Unfortunately, in this view of the vessel and anchored in Curtis Creek, Baltimore, on June 30, 1935, a scow moored to her starboard side detracts from her lines. The scow, however, was part of her activity around the Bay. On its side, in large letters, appears "L.P. Kelly Stone Rip Rap Work Eastport, Md." Apparently the *Martin*, owned by Kelly, transported stone to the work site and the derrick on the scow set the stone in place. About 1940 the *Martin* was reported to have been purchased by a former naval officer who took the schooner to sea and was never heard from again. Built in 1899 at Pocomoke City, Maryland, *Smith K. Martin* appears in the 1941 *MVUS* as "Forfeited."

There is no evidence that this schooner, the *Speed*, ever lived up to her name since she does not appear among the winners of any of the Chesapeake Bay Championship Workboat Regatta races held from 1921 through 1931. She even lost in the race of survival when she sank at her pier at the Charles L. Rohde Shipyard on South Clinton Street, Baltimore, and was covered by harbor ice on January 23, 1936. That was the final act in an episode that started on May 13, 1931. On that date the *Speed* was bound up the Bay for Baltimore with a cargo of oyster shells from Sharps, Virginia. Below Seven Foot Knoll Light, during a blow, the schooner developed a leak and rapidly sank, leaving three-quarters of her masts above the water and her jib still set. Before the schooner sank, Captain Albert Justus and his son, the only men aboard, manned the pump but could not keep ahead of the rising water. They launched the yawl boat, grabbed the ship's cat, and pushed off. The cat, having become confused, jumped back aboard the schooner and was lost. The schooner *Brownstone*, bound in from Salisbury, picked up the men and brought them to Baltimore where the captain said that he "believed a rat had bored a hole through the bottom to start the leak that sank the *Speed*." The schooner was raised and brought to the pier where she eventually sank again. Her remains were broken up during pier improvements in later years. *Speed* was built in 1882 at Russum, Maryland, and listed as "Abandoned" in the 1936 issue of *MVUS.*

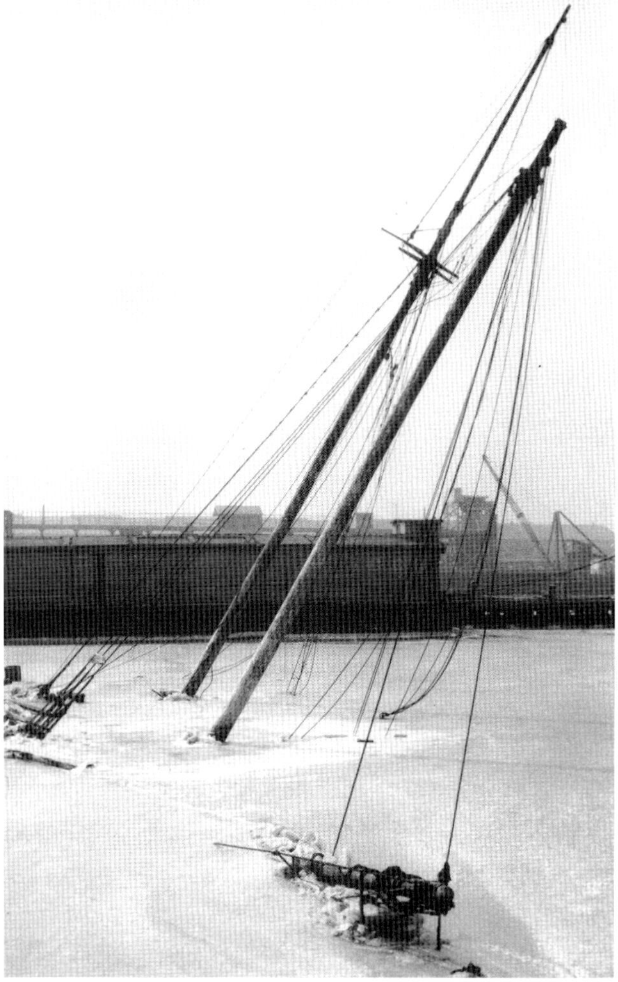

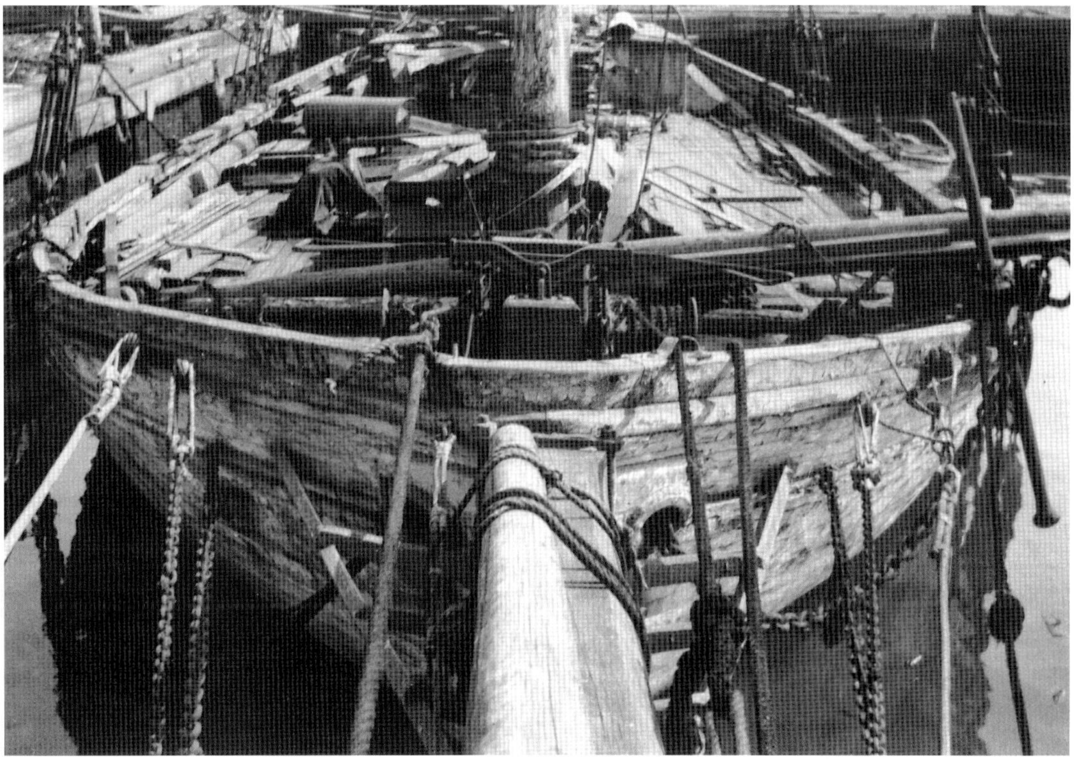

✪ The bow of the 67-foot *Speed,* at Baltimore. From the debris on deck, it appears that the photo may have been taken after her January 1936 sinking.

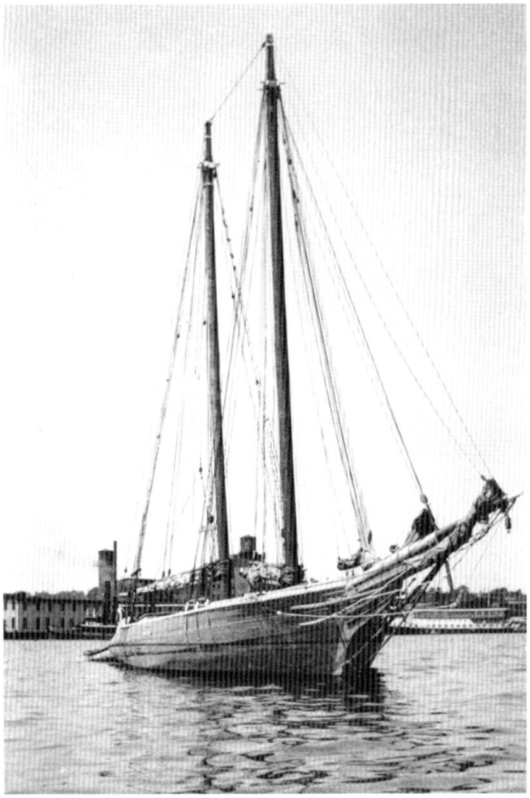

Stephen Chase has discharged her cargo of lumber at the Baltimore docks and dropped down to Canton Hollow to anchor, July 11, 1934. A little more than a week later she was heading down the Chesapeake with a cargo of cases of empty cans. She took with her a mark of her visit to the city, a black greasy ring around her hull, as shown, to indicate how deep in the water she was when she arrived with her lumber. In left background are tugs of the Cottman Company.

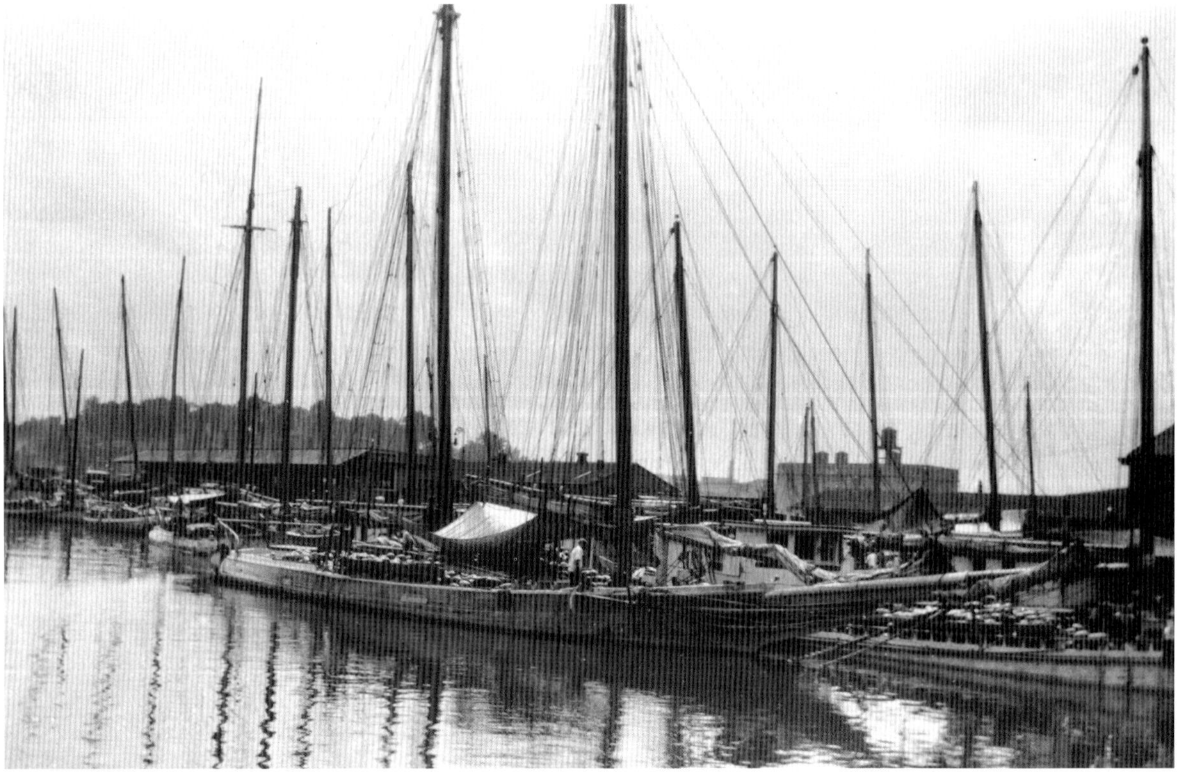

It's watermelon time around the Chesapeake and the sailing fleet is in Baltimore with its luscious cargoes on September 1, 1933. In the foreground is the *Stephen Chase*, as deeply laden as she had ever been with lumber. Schooners, bugeyes, skipjacks, and powerboats, three deep in some places, are tied up on both sides of the dock. As each boat offloads its melons onto huckster wagons or to individuals, it is maneuvered through the floating maze and the next boat takes its place alongside the pier. As the sailing craft disappeared, the trade fell to the powerboats. They, in turn, were displaced by trailer trucks, which delivered directly to the market. The pierside building in the middle background was the base of the Victor Lynn Lines. The trees surmount Federal Hill. *Stephen Chase*, long owned and skippered by Captain Major Todd of Chance, was built in 1876 in Dorchester County, Maryland. About 1940 she was abandoned in the Curtis Creek, Baltimore, ship graveyard.

✪ A nice view of the schooner *Stephen Chase,* under sail on the Bay. Her full sails and the "bone in her teeth" indicate that she was making good speed in the smooth water.

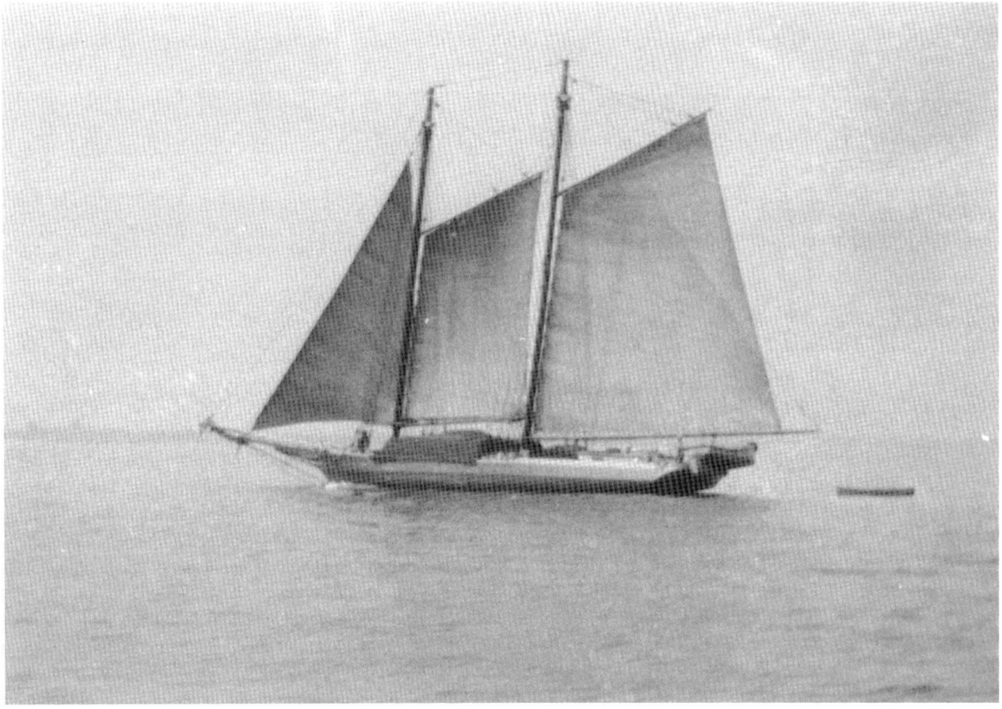

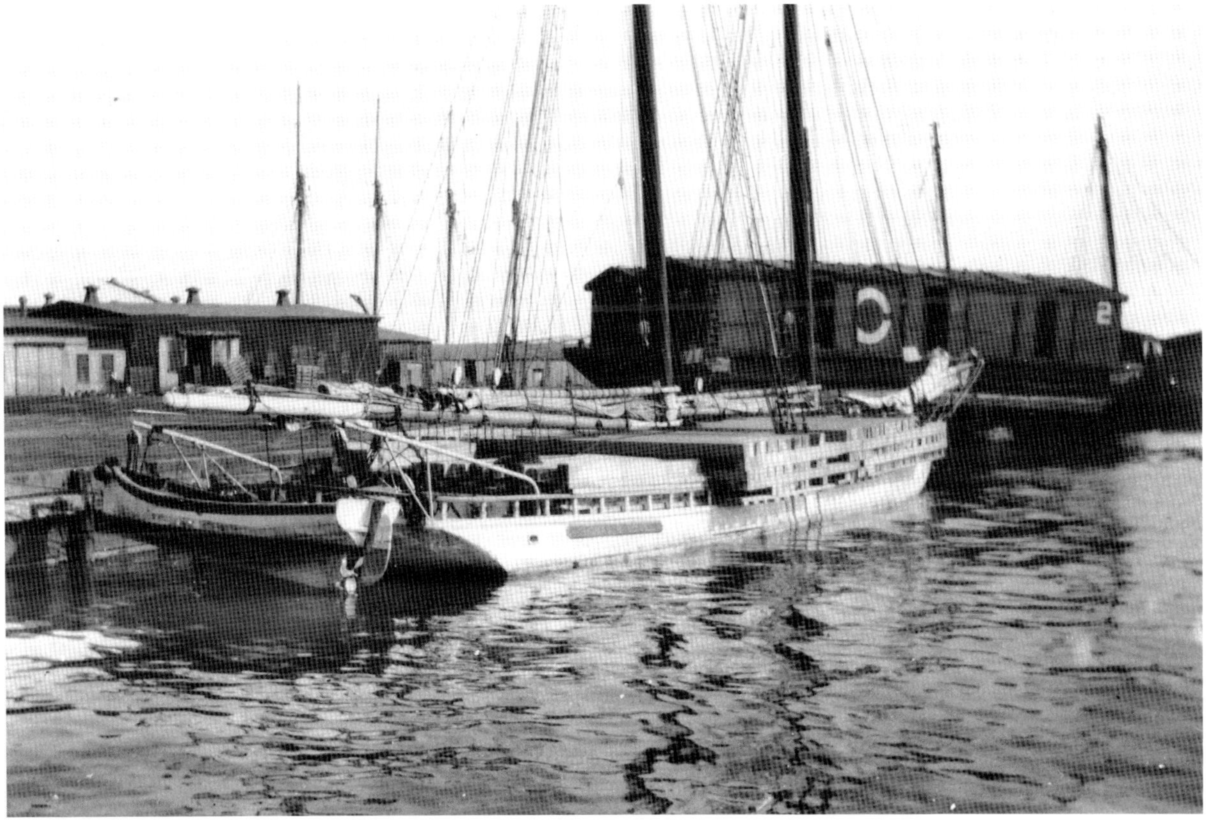

Big timbers make up the deck cargo of the schooner *Thomas B. Schall* in this colorful Baltimore waterfront scene at McGrath's pier on May 19, 1934. Since she was named for the owner of one of the largest pineapple packing plants in Baltimore, it is most likely that she took part in that trade. But about 1915 she came under the ownership of Warren J. Courtney, Mundy Point, Virginia, and with his *Columbia F.C.,* the *Schall* was kept busy freighting between Baltimore and the Northern Neck of Virginia. In the left background are the masts of the four-masted schooner *Doris Hamlin* tied up at the logwood pier of J.S. Young and Company. She had come in from Cap Haitien about a week earlier and the first week in June departed for Newport News to load coal for Fort de France, Martinique. To the right are the masts of the ram *B.P. Gravenor.* Of particular interest are the two covered harbor scows above the bowsprit of the *Schall.* They are sitting high and dry on the pier, one partly overhanging, having been deposited there during the August 23, 1933, hurricane. They could not be refloated and were dismantled where they lay.

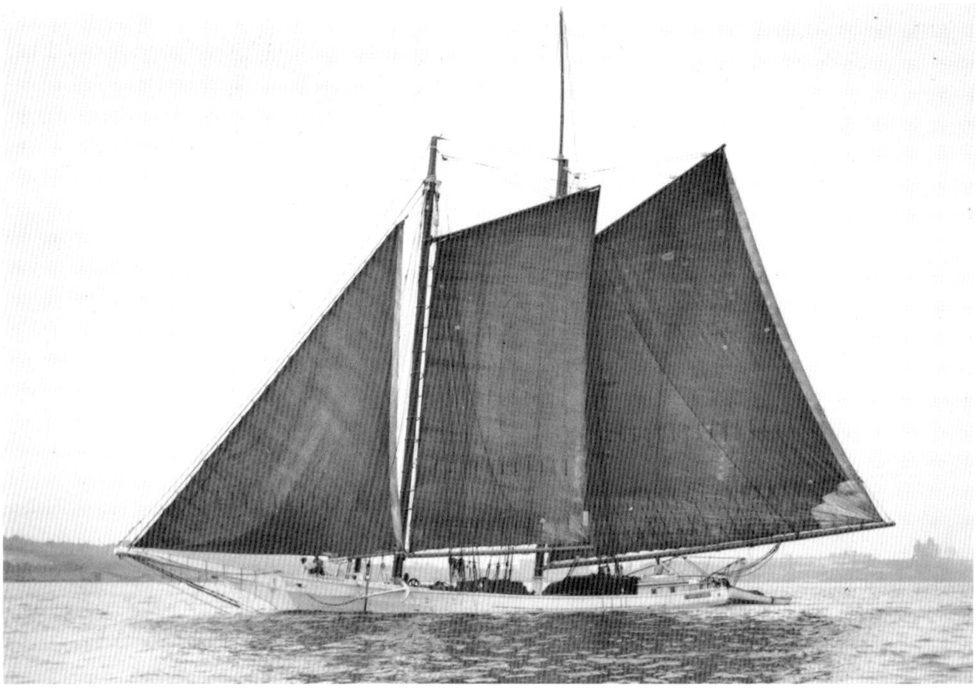

On August 31, 1938, the *Thomas B. Schall* sails and is pushed by her yawl boat out of Curtis Bay, Baltimore. She is leaving astern an area of fertilizer factories and, undoubtedly, it is a fertilizer cargo that has her down to her deck in the water. Some changes have come over the *Schall* in the previous couple of years. Her jibboom has been removed and her trailboards are missing. The longhead looks heavier and bulkier, not as shapely and fine as earlier, and is probably a replacement. As occurred so often when work was done around the longhead, if the trailboards were removed they were not necessarily put back even though they were in good condition. They could have been just cast aside. Built in Baltimore in 1882, by Thomas McCosker, the *Schall* plied the Chesapeake for 60 years. In 1942 she was sold and went to the Caribbean area, where she later capsized and was lost.

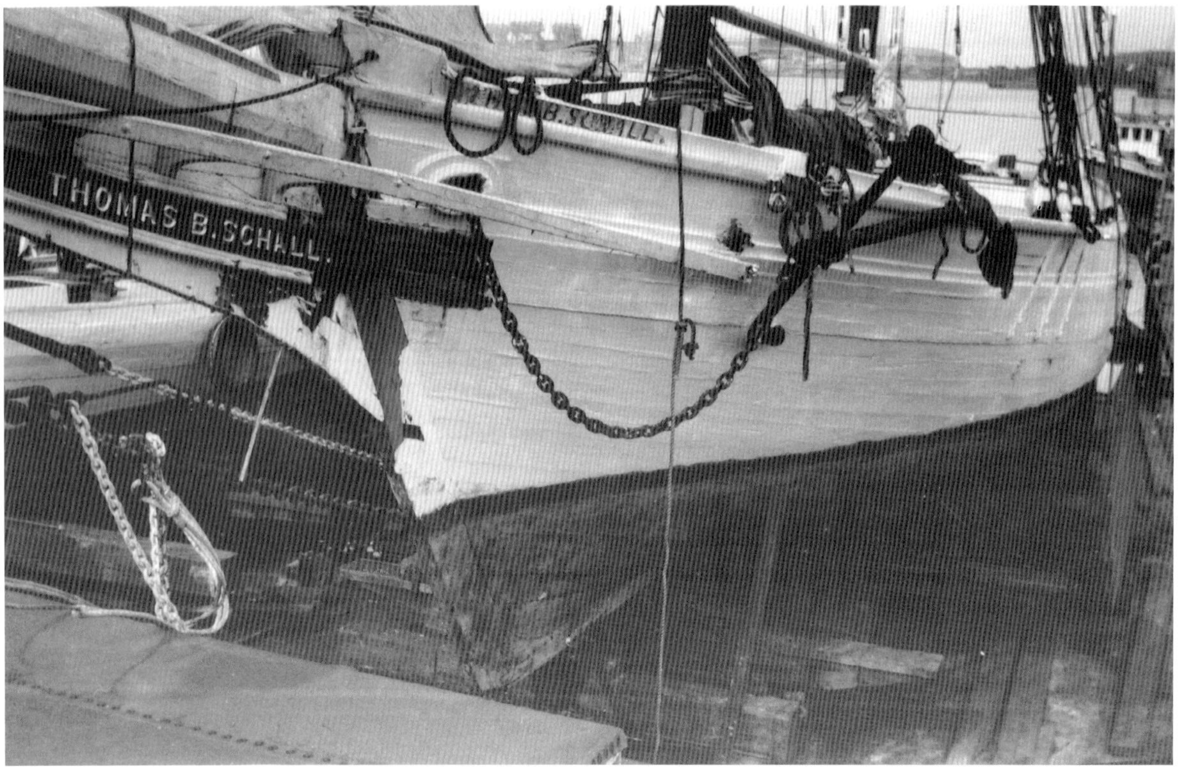

❂ A close-up of the bow of the *Thomas B. Schall,* at A. Smith & Sons Shipyard, Curtis Creek, Baltimore, before her trailboards were removed.

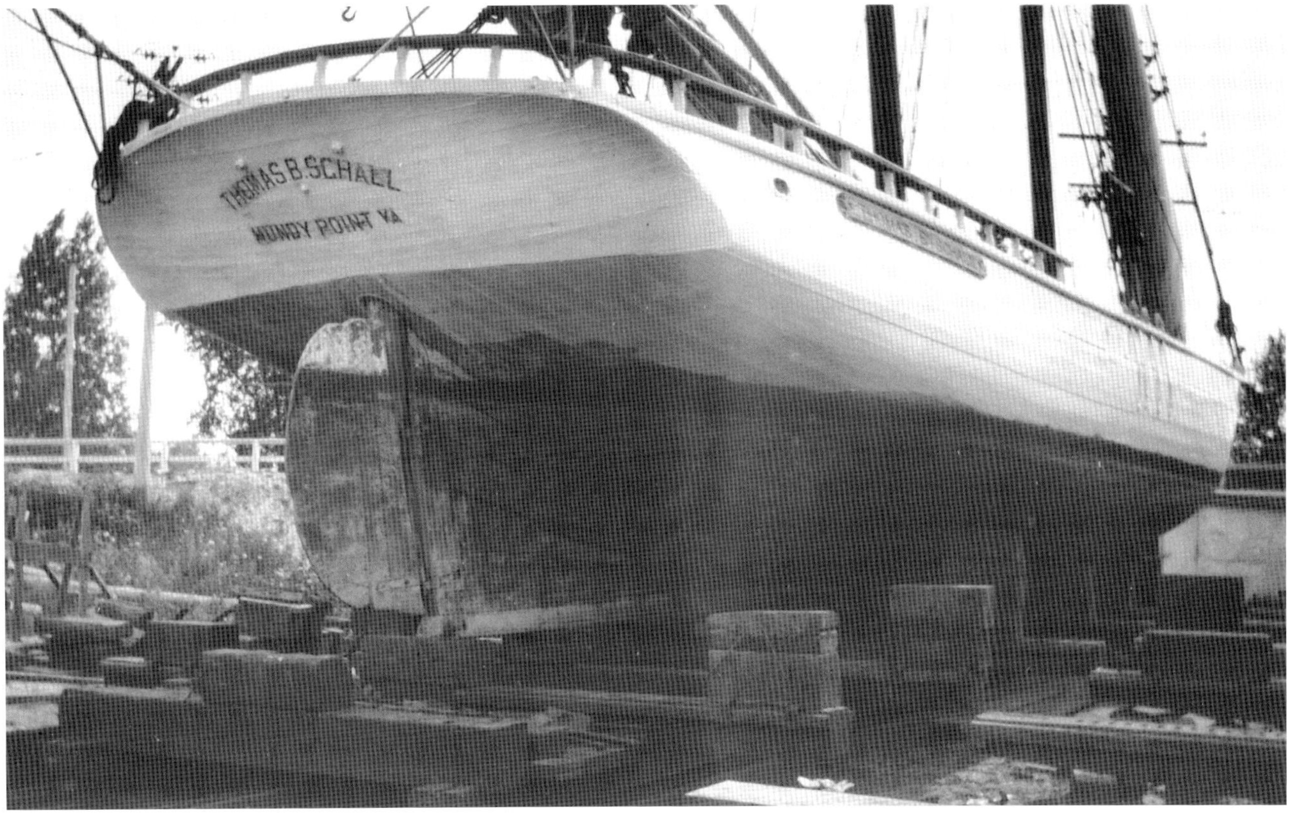

✪ The stern of the *Schall* on the railway at A. Smith & Sons Shipyard. Note the nameboard attached to her bulwarks, and her graceful lines.

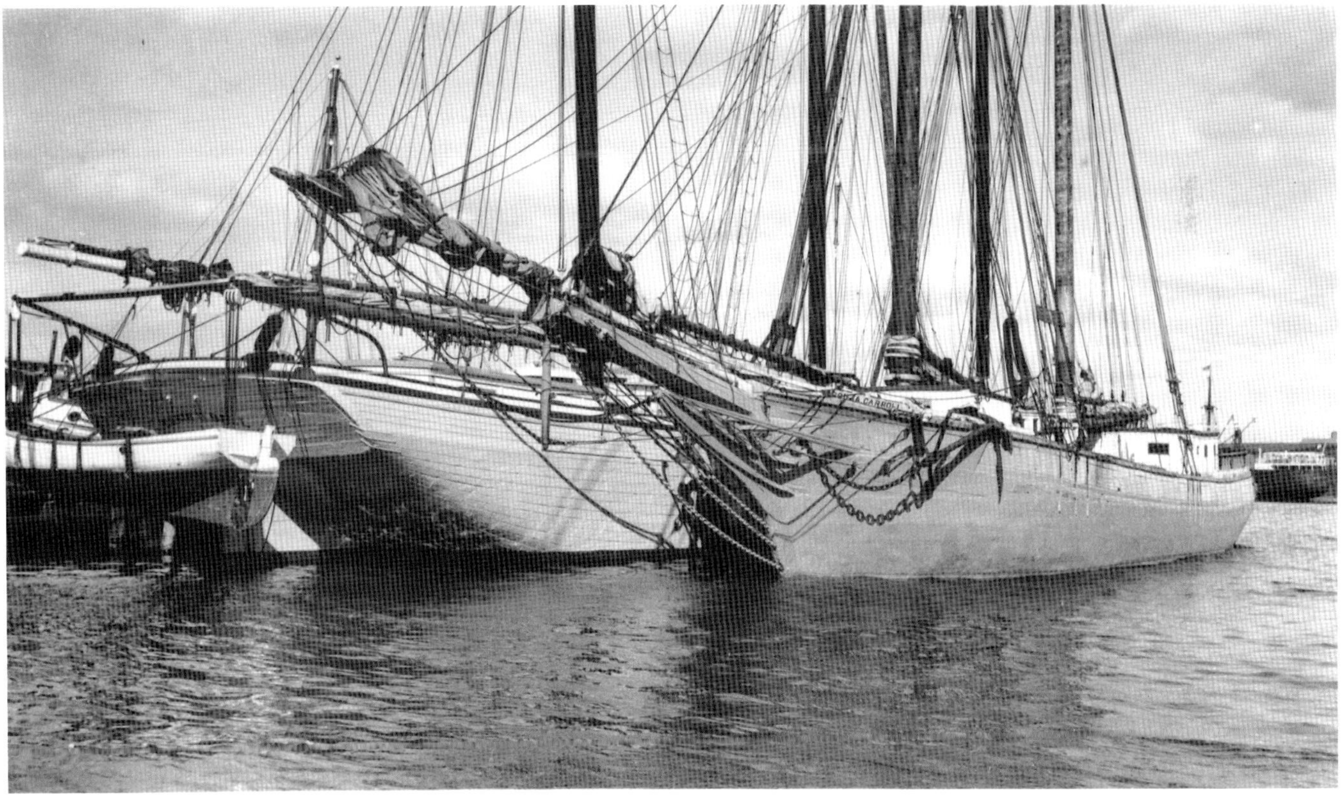

On a sparkling day in April 1936, the schooner *Virginia Carroll*, right, and the ram *Jennie D. Bell* take a brief respite from their lumberlugging chores at the Redman and Vane Shipyard. Judging from their clean waterlines and fresh-looking hulls, they have recently come off the marine railway. Like the schooner *Ruth Decker*, the hull of the *Virginia Carroll*, has been sheathed in galvanized sheet metal. Judging from the condition of some of the planking exposed, the sheathing contributes considerably toward keeping her afloat. She had been built as the *Hesper A. Waters* at Pocomoke City, Maryland, in 1884. She left the Chesapeake during World War II and was listed "Out of Documentation" in the 1946 *MVUS*.

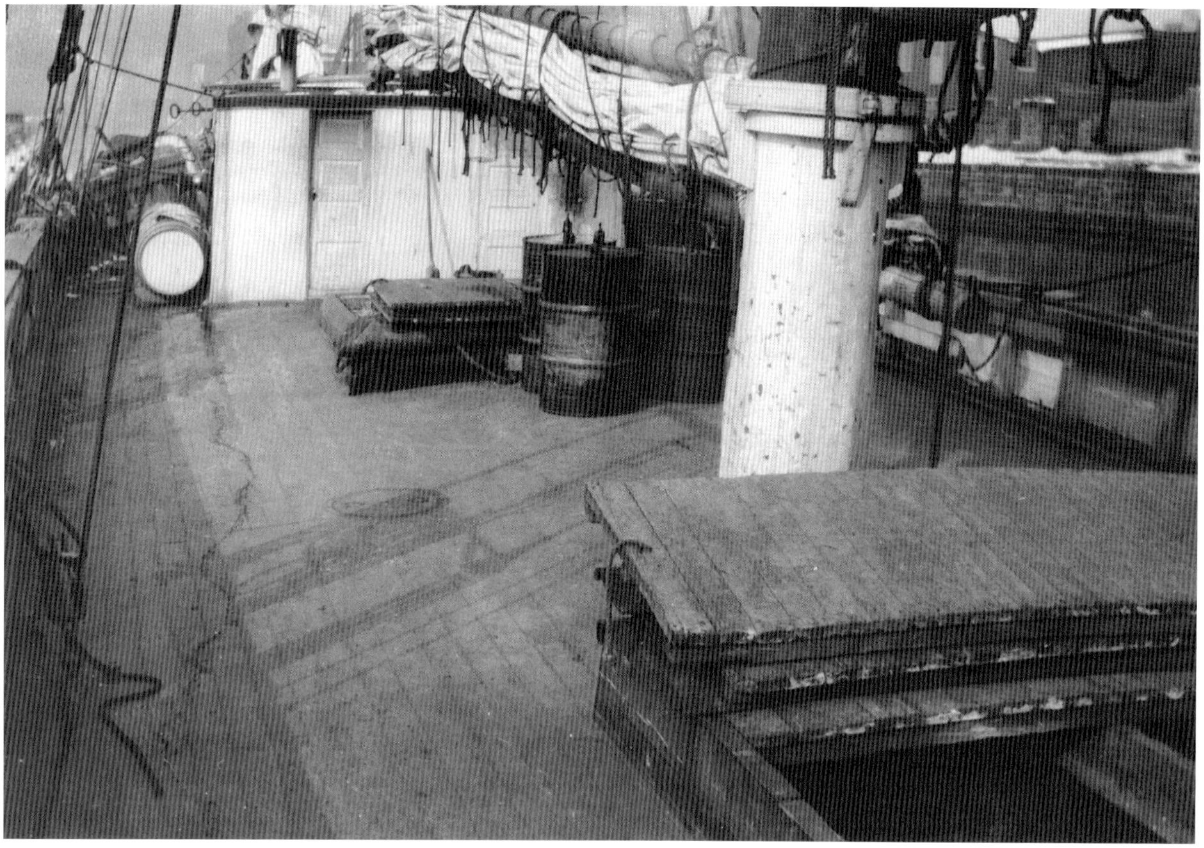

✪ On the deck of the *Virginia Carroll,* at Jones Falls, Baltimore, on February 1, 1935, looking forward.

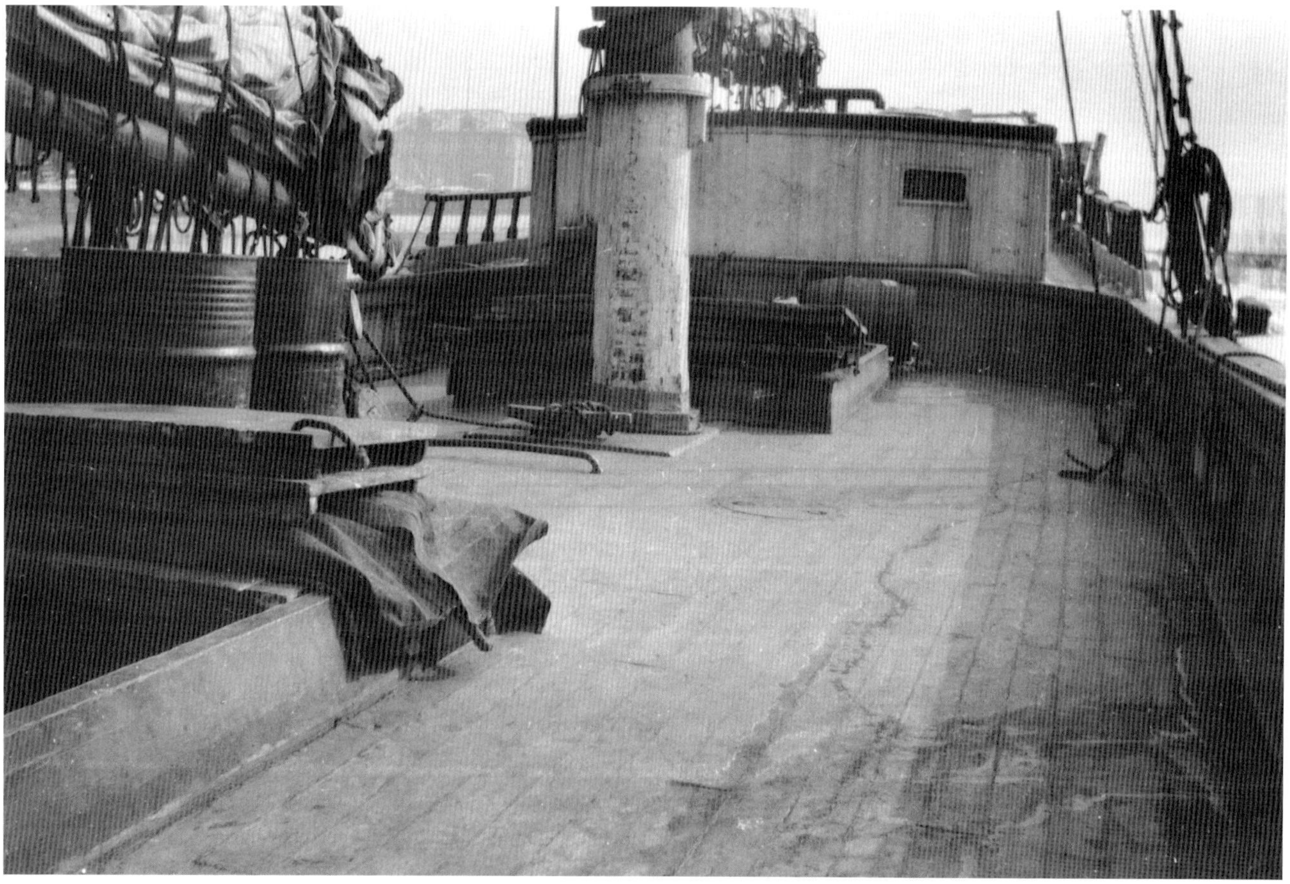

✪ On the deck of the *Virginia Carroll,* looking aft.

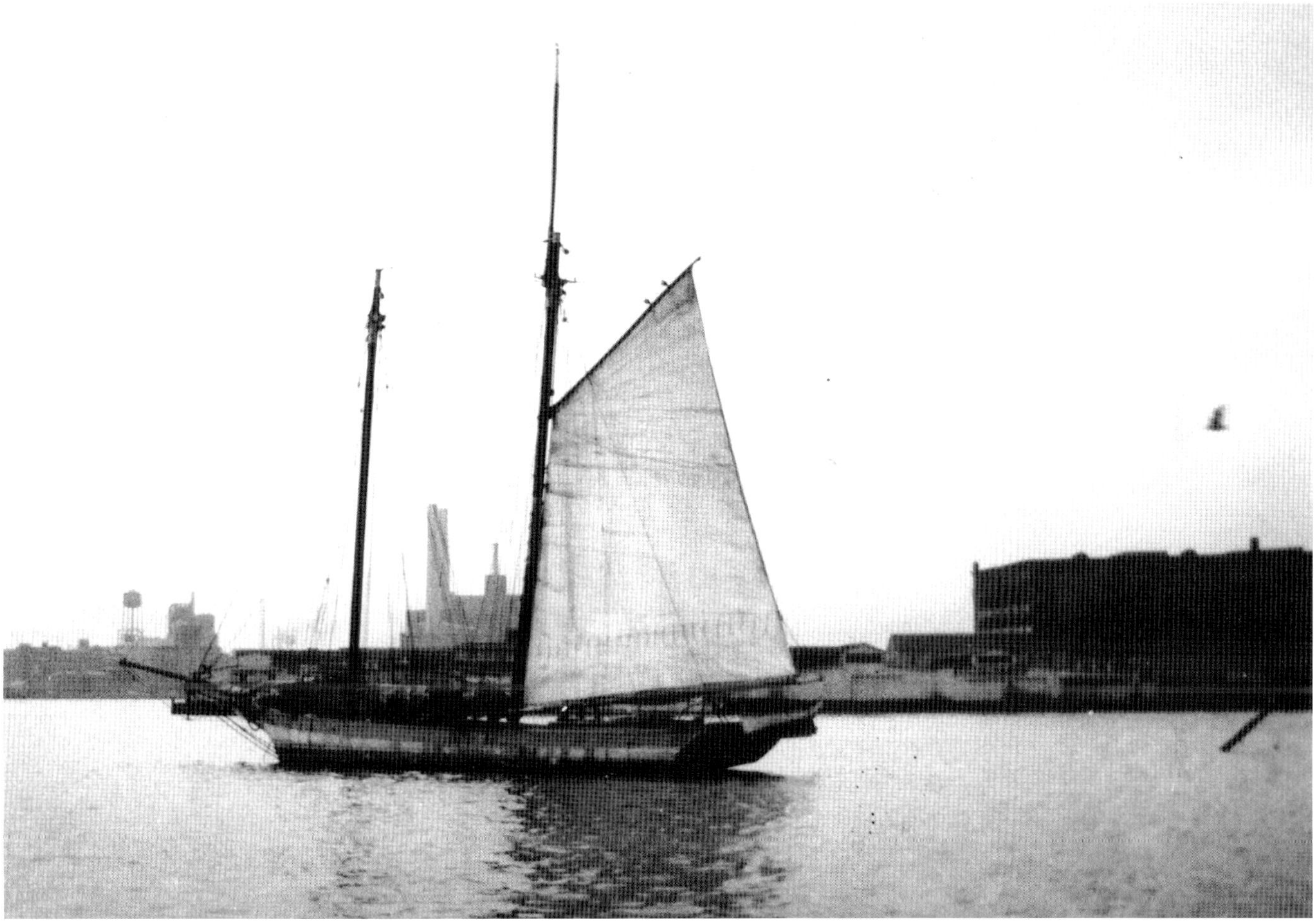

A close look at the mainmast of the schooner *W.H. French* will disclose a definite "break" about halfway up its length. For years it had been bound with galvanized sheet metal, as it was at this time while the schooner was anchored in Canton Hollow on November 11, 1933. The low buildings in center background are of the Lord-Mott Company, packers of oysters and tomatoes brought up from the country by the Bay craft. At the right is the Baltimore and Ohio Railroad's Tobacco Warehouse. Between the *French's* masts appear the stacks of the American Sugar Company refinery and at left the Procter and Gamble Company, Locust Point. The *French* had a black, one-armed captain, Alex Butler, who kept pace with the rest of the Bay skippers. *W.H. French's* life was short compared with some of the other Bay schooners. She was built at Berkley, Virginia, in 1895 and appears in the 1936 *MVUS* as "Abandoned."

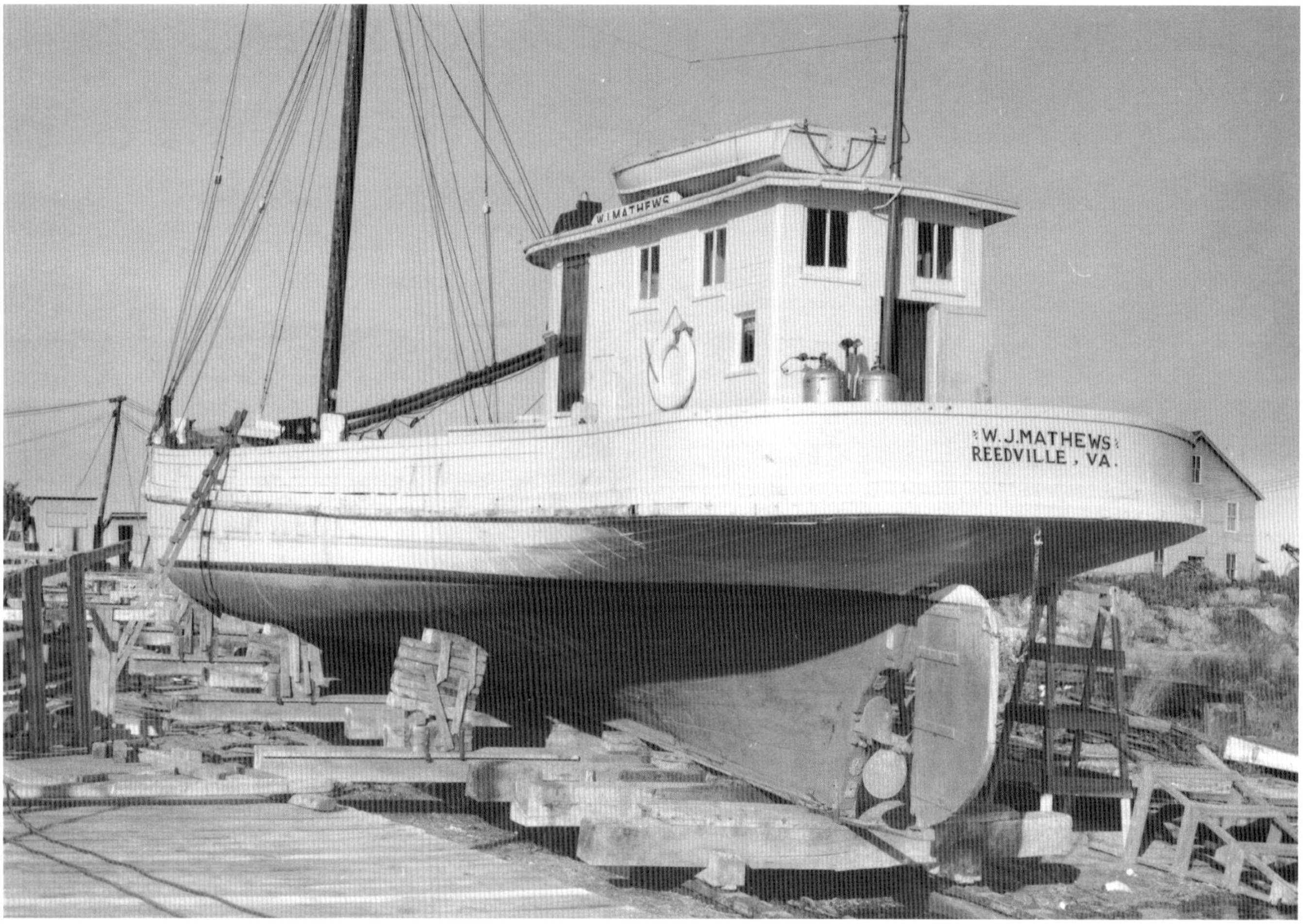

The schooner *Dorchester*, built at Madison, Maryland, in 1882, was regarded as a fine sailer. In 1914 the vessel was so completely rebuilt that she was given a new official number and was renamed *W.J. Mathews*. Judging from conversations with Bay skippers, she lived up to her reputation as a fast schooner. In 1934 she was converted to power and looked like the typical ex-schooner with pilothouse aft. She kept her foresail as auxiliary wind power. Some time in the 1940s she underwent another rebuilding and the hideous stern (see photograph) was added. She is shown here at Humphreys Shipyard, Weems, Virginia. The 1969 *MVUS* lists her as "Abandoned."

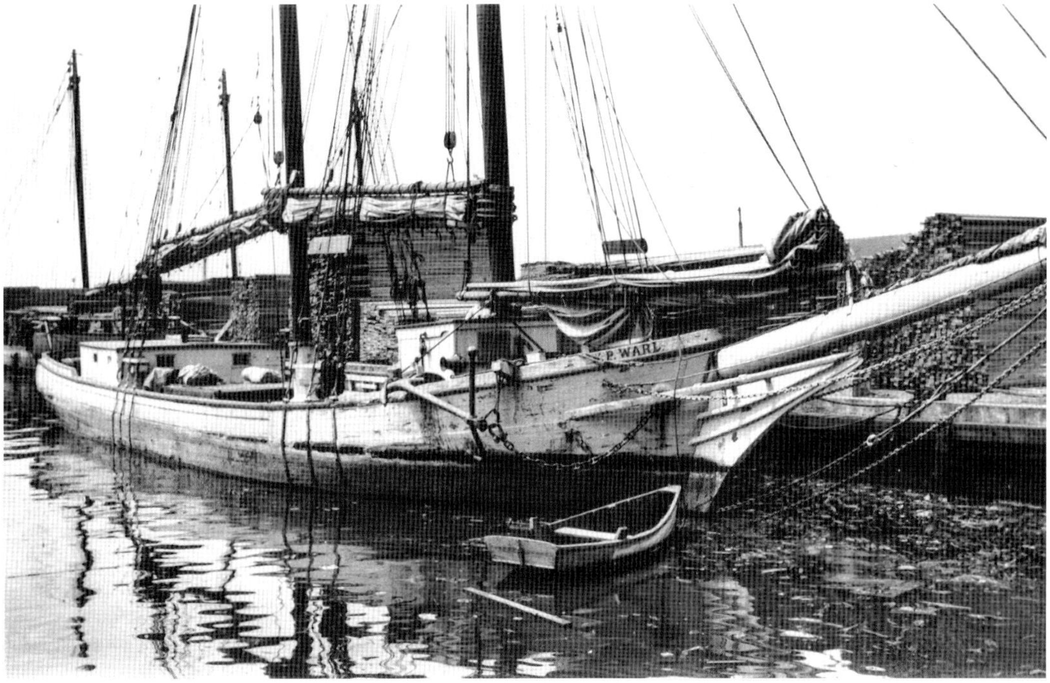

The pollution of Baltimore harbor has probably been a problem ever since the city became a port of any prominence. On May 3, 1936, this area at the Philpot Street lumber pier was a pocket of thick oil holding every manner of refuse to be found afloat in a harbor. The schooner *W.P. Ward* has become saturated during her stay there while unloading her lumber. Her skiff will never be the same. This schooner was built in 1882 at Cambridge, Maryland, as the *Thomas Richard*. About 1922 she was sold and given the name of her new owner. In 1945 she was converted to power and was operating out of Miami before appearing in the 1947 *MVUS* as having been sold to Honduras. At the end of the lumber pier are the masts of the ram *Edward R. Baird, Jr.*

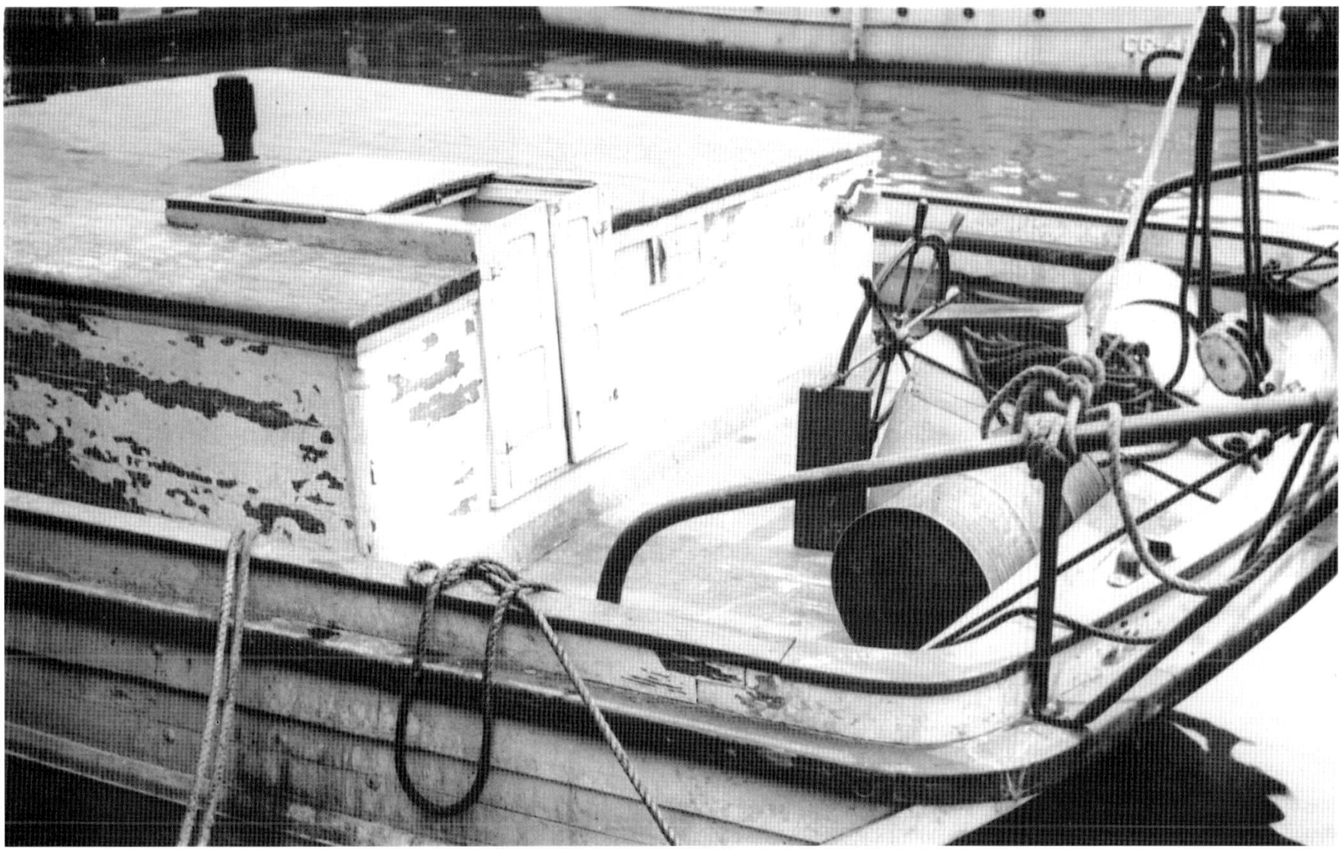

✪ Detail of the weather-beaten stern of the 51-year-old *Ward,* at Baltimore on May 3, 1936.

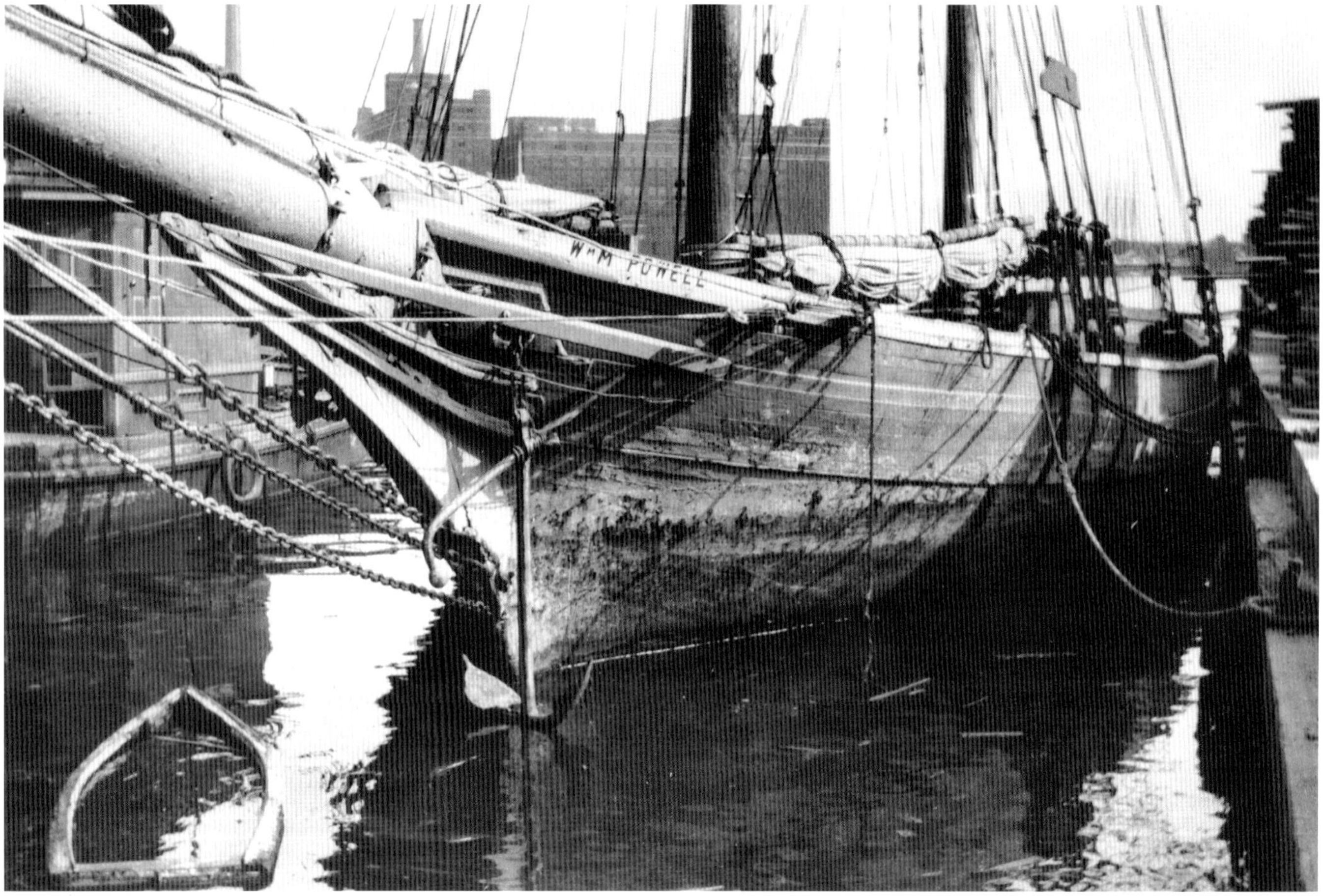

William M. Powell, her dinghy awash under her bowsprit and her hull scarred and stained, still looks jaunty as her fine example of a small eagle figurehead peers forward. She had been built at Finney, Virginia, in 1883, and here in July 1934, at Baltimore's Philpot Street lumber pier, she nears the end of her career. In the far right background can be seen the masts of the four-masted schooner *Katherine May*, which arrived in Baltimore in 1931 but never left that port again. She was abandoned in nearby Curtis Creek in 1936. The *Powell* met a similar fate in an arm of Indian Creek, Virginia, and appears in the 1937 *MVUS* as "Abandoned."

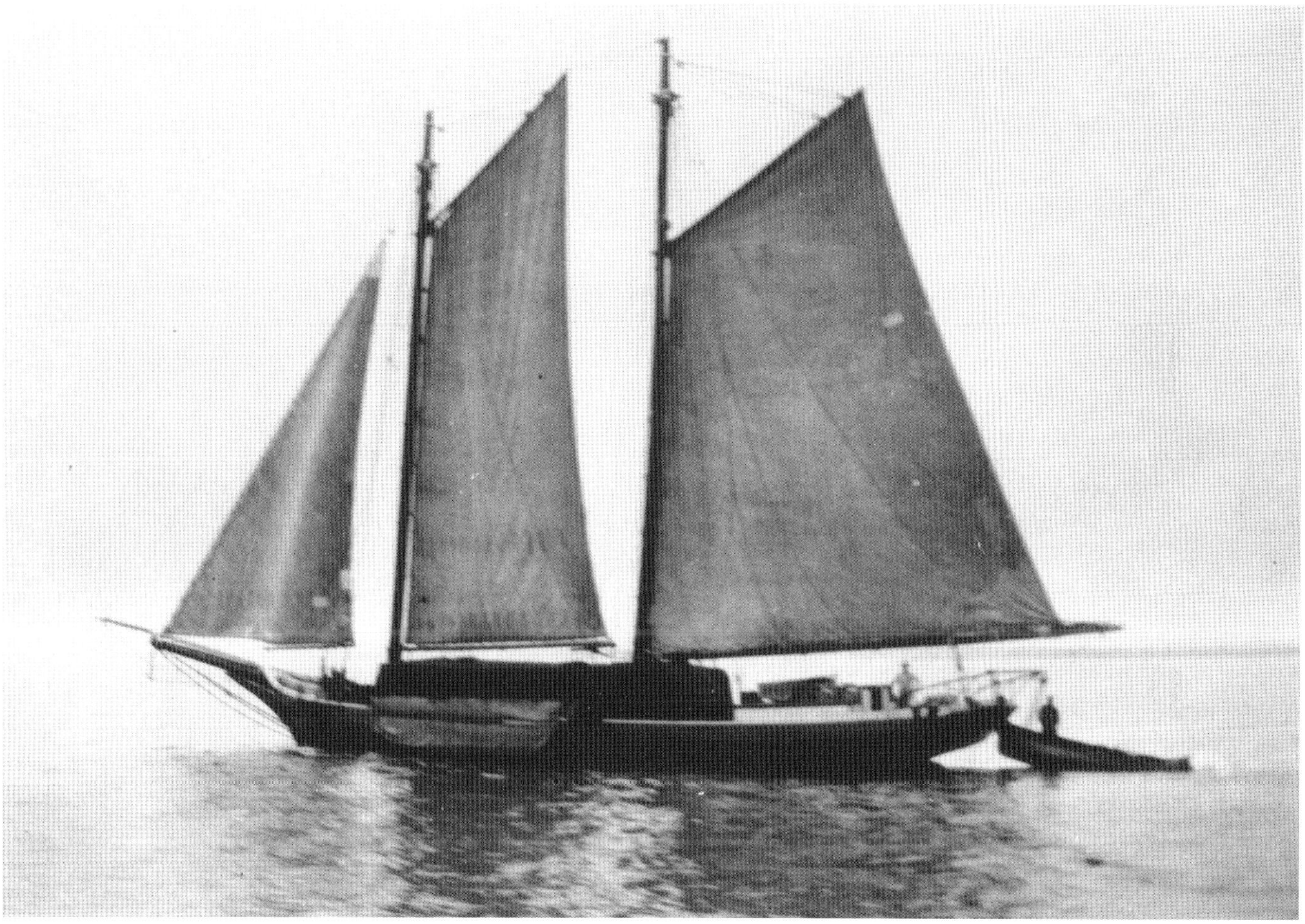

✪ The schooner *William H. Powell* under sail on the Bay, with her yawl boat pushing.

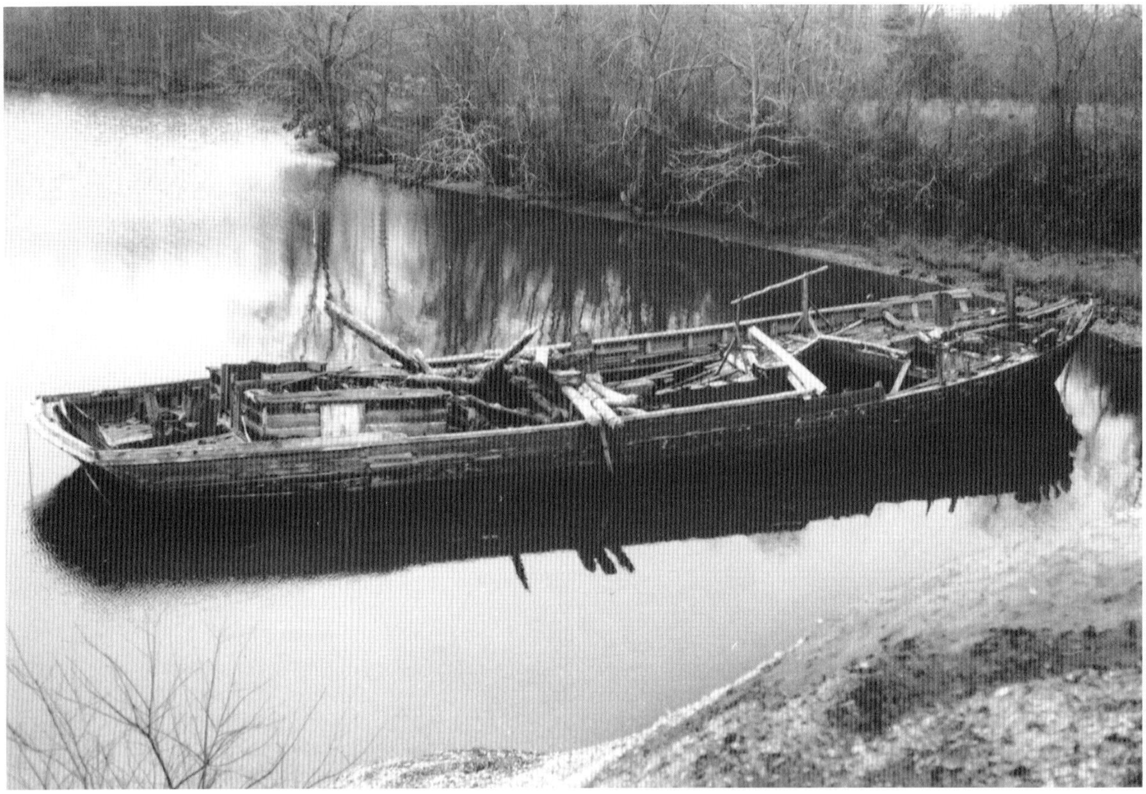

✪ The hulk of the *Powell* lies beached in Indian Creek, Virginia, in 1949.

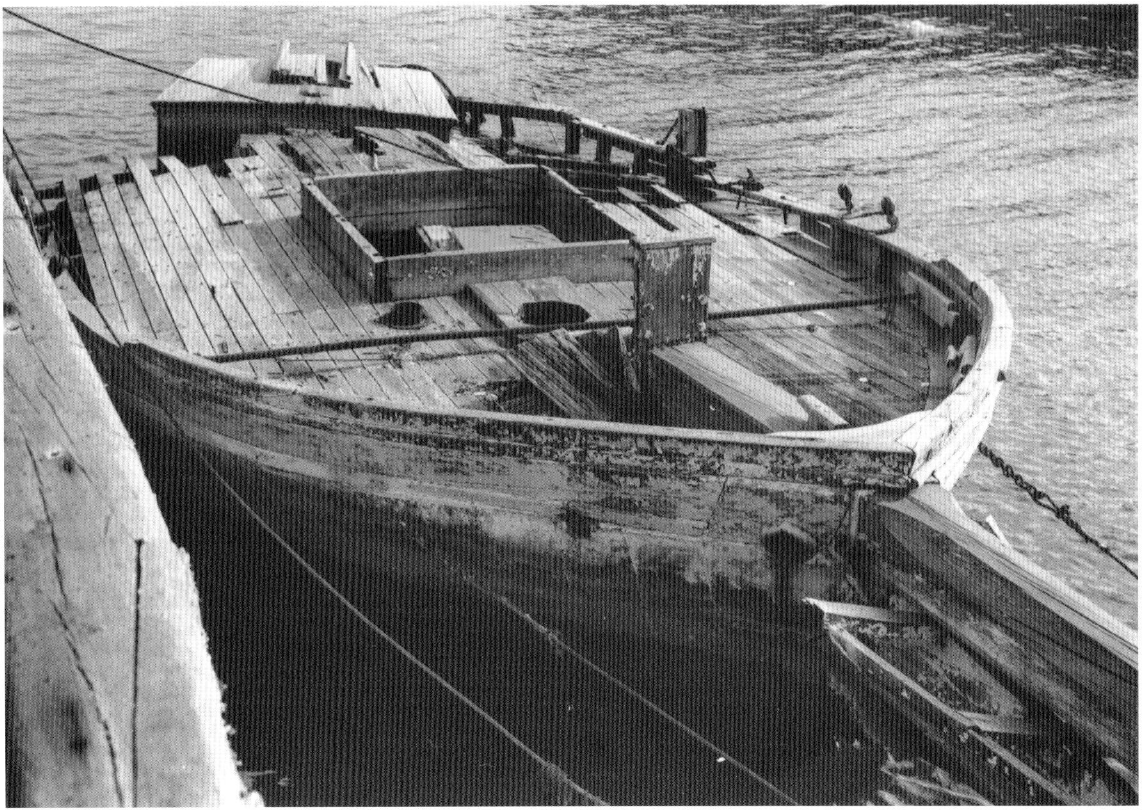

When the schooner *Zingara* was built in Anne Arundel County, Maryland, in 1852, she no doubt shared the Chesapeake channels with many famous Bay craft. She survived the Civil War, crossed the wakes of Baltimore's well-known coffee barks and barkentines, and served well up into the 20th century before being abandoned at her home port of Bellevue, Maryland, alongside the hulk of the schooner *Thomas A. Jones*, whose railing can be seen at lower left, in 1941. When this view was taken on February 19, 1948, the *Zingara* had broken in two and her after section sank lower in the water.

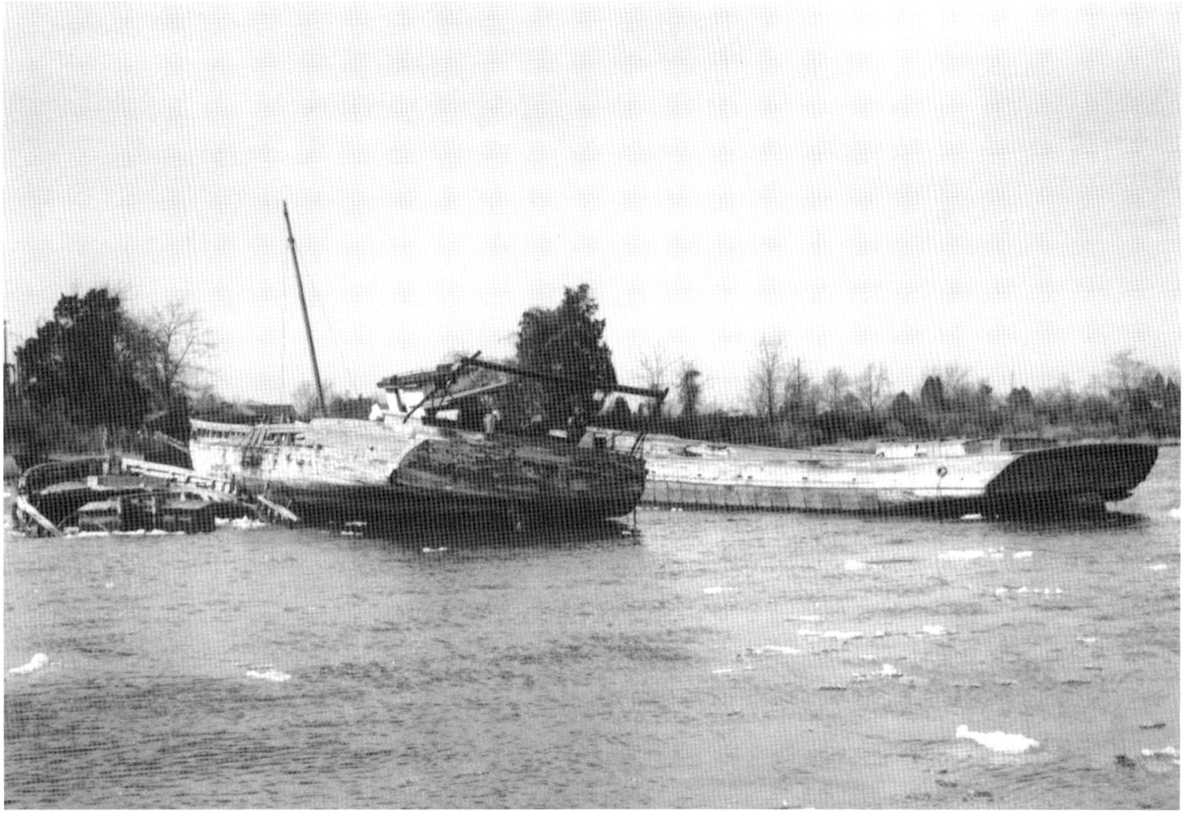

✪ An overall view of the *Zingara, Thomas A. Jones,* and *Florence* at Bellevue, Maryland, in February 1948. The *Jones* had been abandoned in the boneyard there since 1939.

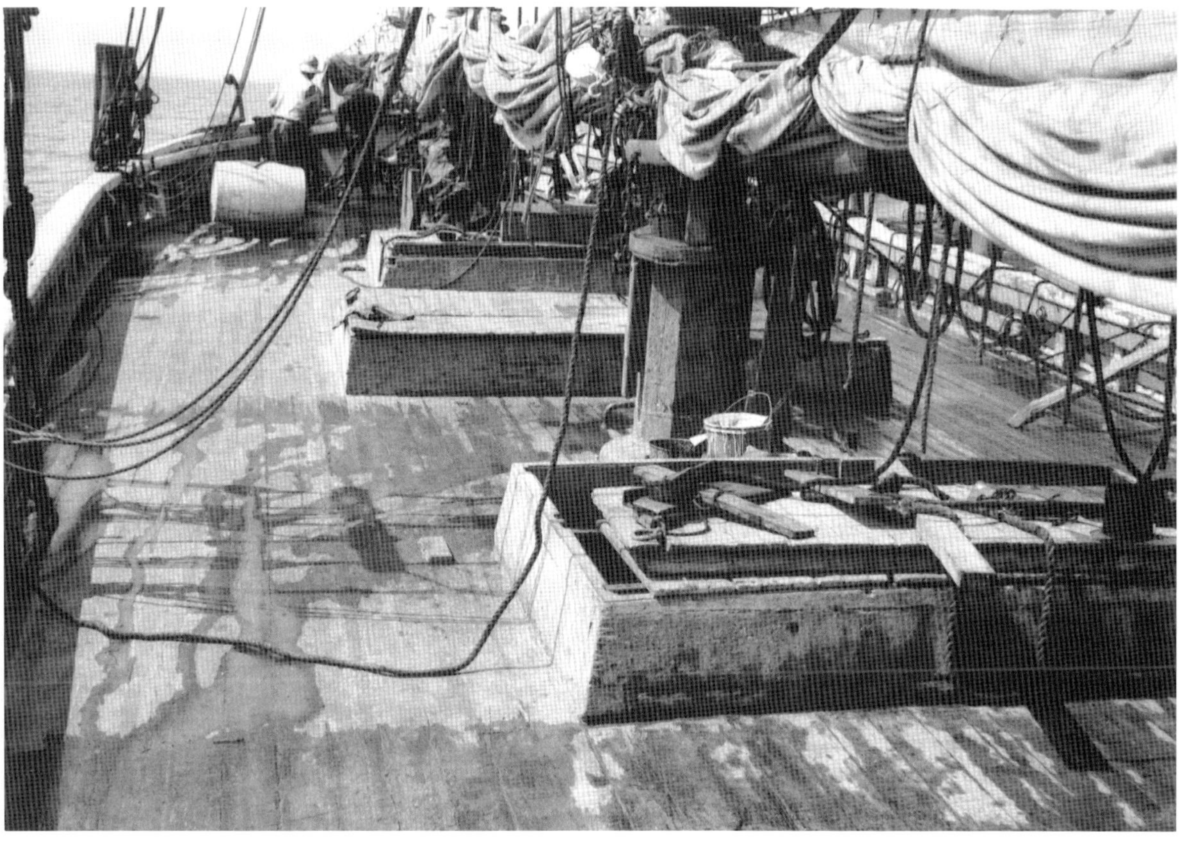

✪ Aboard the schooner *Thomas A. Jones,* looking forward, at Bellevue, Maryland, on June 17, 1936. She was built at Carters Creek, Virginia, in 1911. In 1936 she was owned by William H. Valliant of Bellevue, and she was home-ported in Cambridge, Maryland.

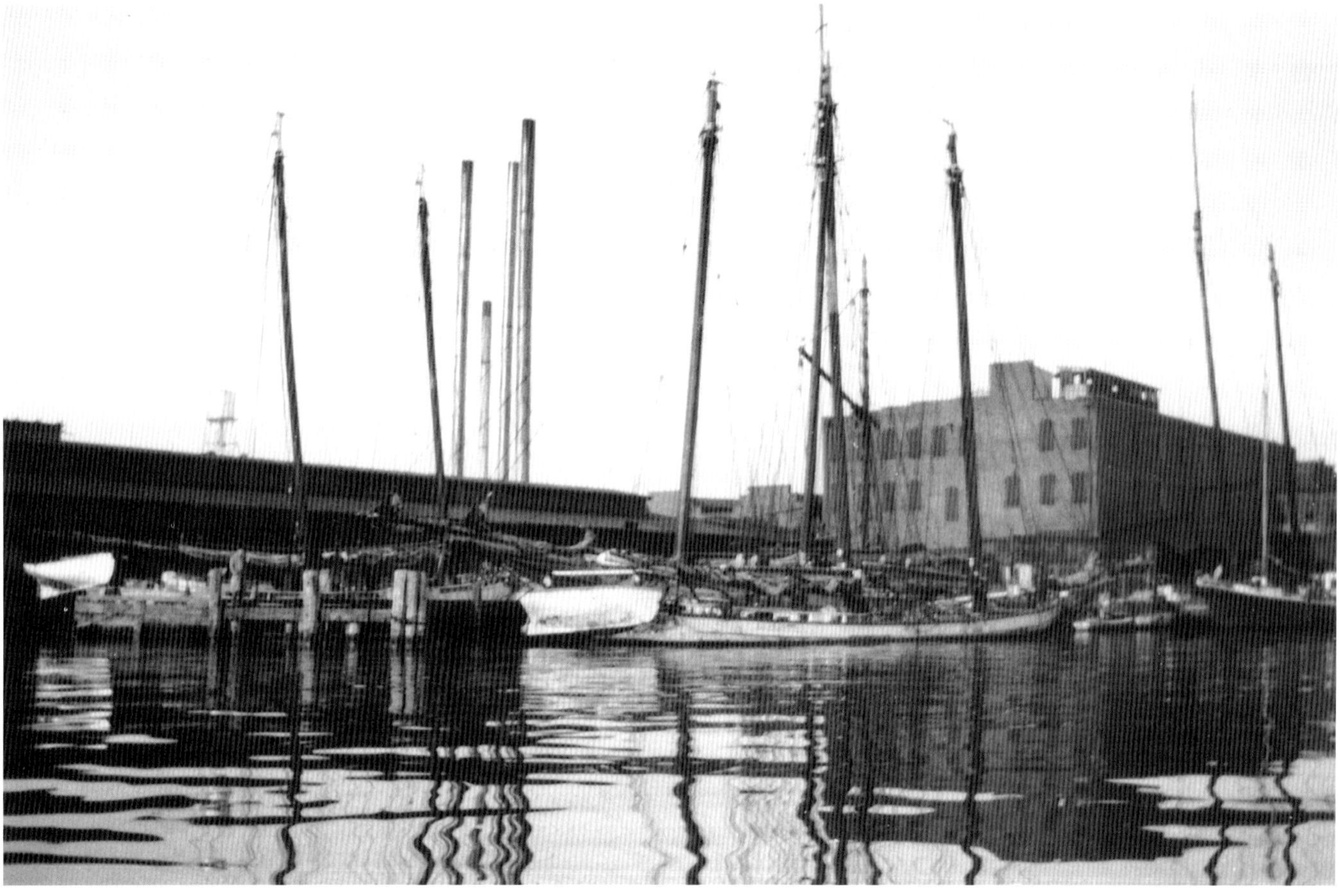

On Christmas Eve, 1933, some of the oyster dredging fleet ties up at a favorite gathering place in Baltimore, Webster's Wharf, on Boston Street near the foot of Wagner Street. At far left is the round-sterned bugeye *Catherine*. Across the pier from her is the schooner *Minnie May Kirwan* and to her port is the schooner *Ella F. Cripps*. At far right is the powerboat *Emmons* lying alongside the pungy *G.A. Kirwan*. All of the vessels but the *Emmons* are owned by a member of the Webster family. Tom Webster probably owned more Bay sailing craft at one time than any other Chesapeake waterman. This fleet eventually came into the possession of his nephew T. Preston Webster, and in the 1925 *MVUS* 14 schooners, four bugeyes, and one pungy appear under his name. Webster's Wharf was a most colorful site, even in the 1920s and early 30s. Minor repairs would be carried out on the craft, their crews would gather and discuss other vessels and conditions of the Chesapeake freight and oyster trades, and some of the older men would reminisce about their younger days.

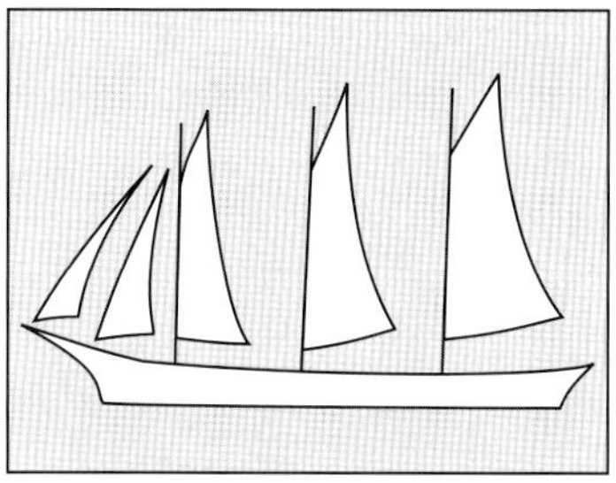

Rams

One may wonder what the feelings were within J.M.C. Moore when he first saw his new ship afloat on Broad Creek, Bethel, Delaware, in 1889. The shipbuilder, with his yard foreman Jacob Smith, had designed this new type of sailing vessel to negotiate the Chesapeake and Delaware Canal, connecting Chesapeake Bay with Delaware River. Limited in beam by the 24-foot width of the canal, but not in length, a strange looking craft had emerged from the builder's ways. Moore must have been influenced by a barge in the creation of his new craft, but he gave it the rig of a three-masted centerboard schooner, less topmasts. His vessel could transport large cargoes of lumber on a shallow draft from North Carolina to Philadelphia without going to sea.

Perhaps Moore wasn't too affected by the appearance of the new craft. Possibly he had seen something similar in his home waters and carried on from there. John Lyman, in his dissertation on "The Chesapeake Ram" in his informative *Log Chips*, a maritime history periodical he published from 1948 to 1959, suggests an interesting possible link to the origin of Mr. Moore's vessel. Lyman states that canal-type craft with schooner rig were used on the canal system of the Schuylkill River through which passed cargoes of coal to Philadelphia by way of the Delaware and Raritan Canal. Some of these craft came to the Chesapeake by way of the Chesapeake and Delaware Canal. One of these was the barge *Reading Railroad No. 34*, built at Hamburg, Pennsylvania, in 1863. She measured 106 feet in length, 19.1 feet in beam, and 8.6 feet in depth. That narrow beam must have been regulated by the width of the canal locks. About 1883 she was refitted as the three-masted schooner *Lulu M. Quillin* at Seaford, Delaware, a short distance from Bethel by land or water. Another Hamburg-built vessel of weird dimensions was the two-masted schooner *Robin Hood*, built in 1868 and measuring 100' × 17.4' × 8.6'. She was referred to as a "Long John." Both of these vessels were registered out of Eastern Shore ports, Seaford and Crisfield respectively. Perhaps such vessels influenced Moore in his design.

No view of the first of the type, the *J. Dallas Marvil*, has been found, but a similar vessel built two years later, the *Levin J. Marvel*, sailed the Chesapeake until 1955 and she was as homely a vessel as ever cleaved the waters of the Bay. Her cumbersome hull resembled a canal barge. The characteristics of these craft were wall sides, bluff bows, flat bottoms, little sheer, and no topmasts. There were some departures from these distinctive features in the later rams, however. A few carried foretopmasts and had more sheer and high bows.

The name assigned to the type was "ram." As with several other Chesapeake sailing types, pungy, bugeye, and skipjack, the exact origin of the term is unknown. Some have attributed it to a bystander's remark when he saw the first ram moving through the C & D Canal, "butting her way through the other schooners, acting just like a ram."

The centers of ram building were at Bethel, a small Delaware town on Broad Creek, which empties into the Nanticoke River, and at Sharptown, Maryland. Between 1889 and 1911, 26 rams were built. They ranged in length from 112.8 to 163 feet. Two were four-masters—the *Judge Pennewill*, built in 1906, and the *Albert W. Robinson*, built a year later. Two were also built in Baltimore and one at Madison, Maryland. So the design must have been successful if not eye-pleasing. They apparently were kept busy not only in the shallow-water trade but also coastwise, down to the Gulf of Mexico and West Indies, and even across to Europe, a voyage made by the *Joseph P. Cooper* during World War I. In April 1917, the rams *Edward R. Baird, Jr.*, and *Granville R. Bacon* had charters to carry coal from Norfolk to Bermuda. As late as 1933, 13 rams were still sailing the Chesapeake. Mostly they hauled lumber from North Carolina and lower Virginia to Baltimore and Philadelphia. Other cargoes were slag, road-building materials, and fertilizer.

Three men could easily sail a ram, with a gasoline donkey engine for handling the anchor and sails, and a powered yawl boat to assist in calms. A lot of freight could be hauled long distances at little labor expense. Records of one ram, the *Corapeake*, in the author's possession, give an insight into the maintenance costs and freight returns, which are average for the type and period. Owners of the Bay sailing craft were happy if they could obtain cargoes for a round trip but were satisfied to sail south light to load lumber for Baltimore.

Showing how the *Corapeake* fared on a round trip with freight, her statement for June 3, 1926, reveals that she transported fertilizer from Baltimore to Suffolk, Virginia, for \$356.73. After discharging, she sailed across the James River and loaded lumber on the Chickahominy River for Baltimore at $623.50. Expenses on the trip: towing, $29; loading, $85.10; discharging, $93.52; gasoline and oil, $28.15; commission, $39.00; totaling $274.77 and leaving net freight revenue of $705.46. The ram's share was $352.73; the same amount went to the owners in dividends. Out of the vessel's portion in bills paid by the captain were towing to shipyard in Baltimore, $5; new bearing, etc., $22.60; carpenter labor and material to make new jaws, $21; engine parts, batteries, etc., $19.15; canvas for staysail, $1.60; 14 days at Redman and Vane Shipyard, $74.90; mate's wages for 10 days, $20; cook, $18.66; grub, $28; totaling $210.91 and leaving the ram due $141.82, which added to the balance due on the freight from the previous trip, $11.20, amounts to $153.02. One hundred dollars was paid on a bill from sailmakers Marshall and Godfrey, and $53.02 on Redman and Vane's bill, leaving nothing in the coffers for the ram. So she gets another cargo and starts all over again, her owners hoping there are no mishaps to take away the slight profits she might garner. At that time the cost of a new mizzen sail from Marshall and Godfrey was $305.

In November 1928, the *Corapeake* was chartered to bring 61,293 feet of 4/4 lumber from Bridgeton, North Carolina, to the Canton Lumber Company, Baltimore, at $2.75 per thousand. Rather than negotiate the long haul from New Bern to Norfolk under sail and by yawl boat, by way of the Sounds and the Albemarle and Chesapeake Canal, the *Corapeake* took the Norfolk Lighterage Company tug *Dorothy* between the two ports at a charge of $100. Then the same tug towed her from Norfolk to Baltimore for $75. For a shift at Baltimore by Captain J.H. Mullen's tug *Easby*, from Curtis Bay to the Redman and Vane Shipyard, the *Corapeake* was charged $14.

In July 1929, the *Corapeake* was hauled on the marine railway at Booz Brothers Shipyard, with these charges: docking at 234 tons, $30; blocking, $10; 8½ days cleaning and painting, at $5 per day, $42.50; freshwater, $1; half a day to shift blocks, $2.50; half gallon red lead paint, $1.75; 3¾ days caulking at $7 per day, and 10 pounds of oakum at 20 cents per pound, totaling $28.25; 1 seam brush, 25 cents; 2 pounds of sheet iron, 16 cents; 1 pound of nails, 15 cents—totaling $116.56. More than a week's stay in a shipyard for a 134-foot vessel for $116.56!

The Depression was setting in about this time, but most of the rams managed to keep busy, perhaps spending a little more time at anchor in Baltimore's Canton Hollow than normal. During the 1930s, fire, strandings, and abandonment had taken their toll among the fleet. Then World War II came along and stirred some activity in their aging timbers as lumber was once more in heavy demand.

Early in 1941 the rams, schooners, and bugeyes had found a new discharge port near Baltimore, one which had not seen much use in previous years. This was an unloading dock on Marley Creek, in upper Anne Arundel County, the closest deepwater contact with Fort George G. Meade where the U.S. Army required some 40,000,000 feet of lumber for cantonment construction. Anchored in the creek or moored at the dock within sight of Old Annapolis Boulevard, just below Glen Burnie, the sailing vessels gave new life to the area. Marley Creek had once seen great activity when the production of charcoal iron at tidewater furnaces was a big industry in Anne Arundel County. Virgin pine forests afforded an abundance of cheap fuel. A short haul by cart to the creek for loading aboard the sailing craft that delivered the product direct to furnaces created a long-lived boom. Iron furnaces sprang up on the section's creeks; one stream with such a plant came to be called Furnace Creek. When the charcoal industry lagged, Baltimore afforded a ready market for Anne Arundel farm produce. Good roads were lacking and heavy tolls were charged at bridges leading to Baltimore. Other sailing craft than the rams found a livelihood moving fruits and vegetables to market. When good roads and faster vehicular traffic took most of the business, the sailing fleet moved out of Marley Creek not to return until it was called upon to carry materials for national defense.

After that last spurt of lumber hauling, the sailing fleet dispersed. Several of the rams left the Chesapeake, under full power or still as sailers. By 1945 only two remained in commercial use on the Bay, the *Jennie D. Bell* and the *Edward R. Baird, Jr.* Three others, *Levin J. Marvel, Edwin and Maud,* and *Grace G. Bennett,* remained on these waters as "dude cruisers." Within a decade only one ram was still active, the *Edwin and Maud*, renamed *Victory Chimes,* sailing out of Maine even today.

Facing page: Looking down at the butt of the bowsprit and along the length of the ram *Agnes S. Quillin* as she was tied up at the yard of her owner, Wathen and Company, at the south end of Andre Street, Locust Point, Baltimore, December 1, 1933. This ram was built by George K. Phillips and Company at Bethel, Delaware, in 1894. Few were the rams that did not have more than their share of mishaps during their careers. On December 14, 1907, while the *Quillin* and barge *Wicomico* were in tow of the tug *Defiance* on Chesapeake Bay, the four-masted schooner *William B. Palmer* ran into the ram's stern and deflected into the barge. But after repairs at Woodall's Shipyard, Baltimore, the *Quillin* was back in service again in early January. On January 6, 1915, the *Quillin* was aground, waterlogged, and leaking in Croatan Sound, North Carolina. The cutter *Pamlico* and a tug were sent to her aid. As the tug went alongside to pump her out, the cutter took both in tow and brought the ram in to safety. Another time, in April 1929, the *Agnes S. Quillin* was waterlogged in Tangier Sound for two days before the passing steamboat *Eastern Shore* came to her aid. Earlier the ram had displayed the reversed American ensign as a signal of distress, but the steamboat *Piankatank* had failed to respond. When the *Eastern Shore* approached the *Quillin* the steamer's quartermaster, Nevitt Muir, and two men were lowered in a boat to help Captain W.W. Jarrett and his crew. Back alongside the steamboat, a large wave upset the lifeboat as it was being raised and threw the men overboard. Rescuer Muir drowned. The ram was later towed to port and repaired to resume her lumber trade. She was to struggle on into the late 1930s. On November 8, 1938, bound up the Bay with a cargo of lumber for Philadelphia, the ram encountered a severe storm off Smith Point, Virginia, and became waterlogged. The powerboat *Popeye* went to her aid and took off the captain and two of his crew and put them ashore at Reedville. The Coast Guard patrol boat *Dione* took the *Quillin* in tow for Norfolk. That must have been her last passage on the Chesapeake, for the 1939 *MVUS* reported that she stranded at Smith Point on November 16, 1938, and she was removed from the register.

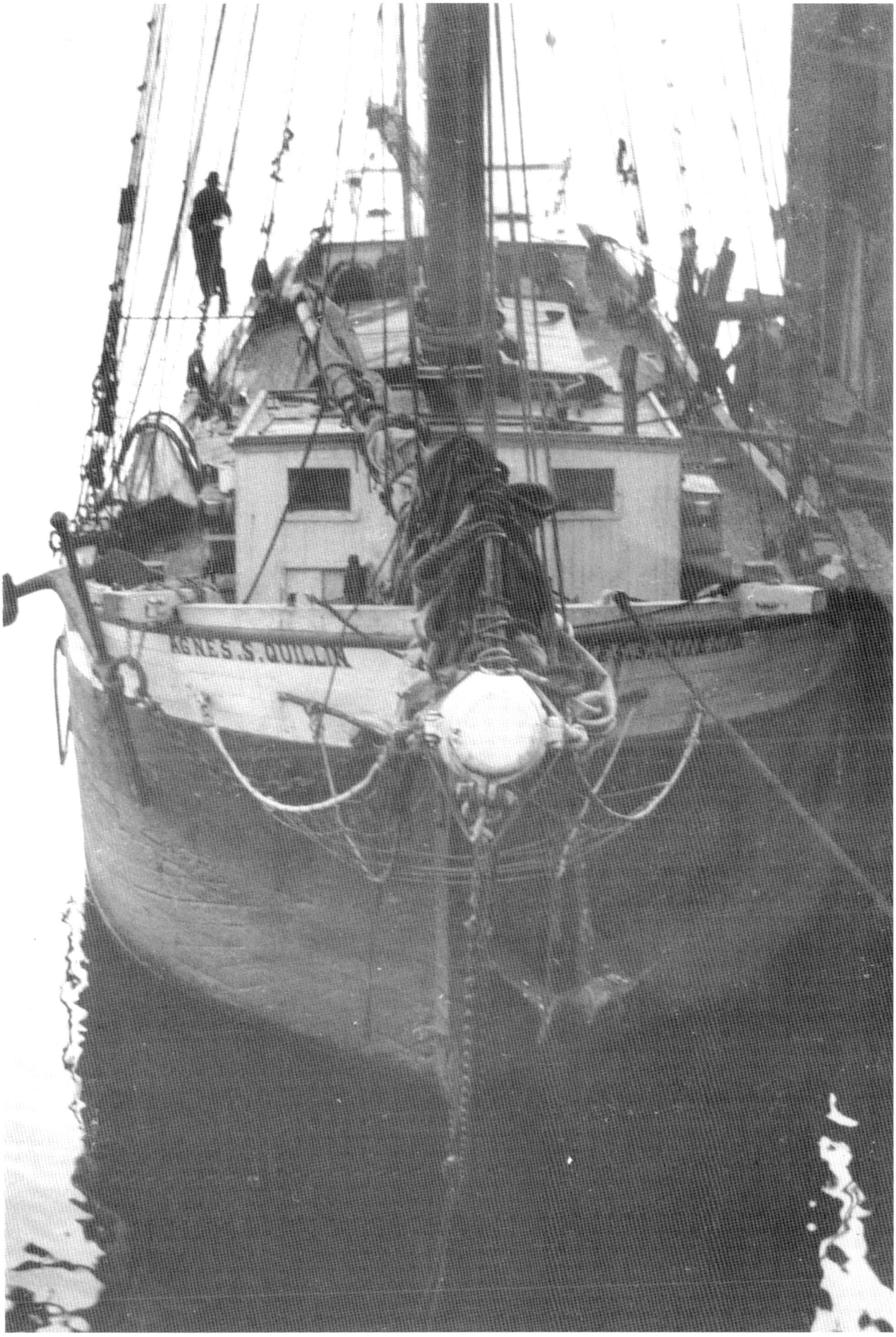
AGNES.S.QUILLIN

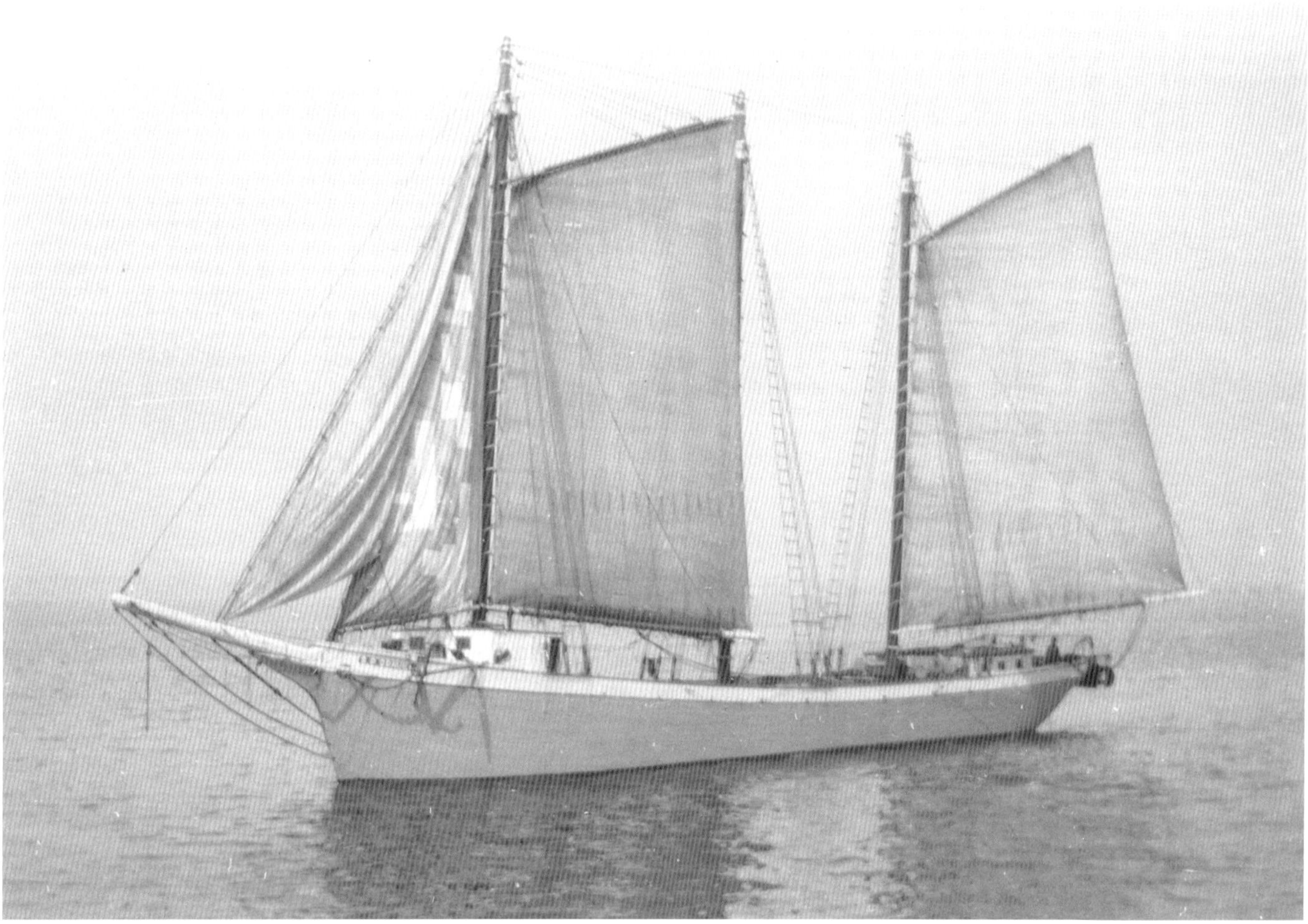

Ram *B.P. Gravenor* sails wing-and-wing before a light wind down Chesapeake Bay on August 11, 1935. Judging from her clean hull, she has just come out of a shipyard. Built as the *James H. Hargraves* in 1893 at Sharptown, Maryland, by J.M.C. Moore, this ram was one of several to have foretopmasts at the outset of their careers. Lack of sufficient headsails made rams difficult to steer, and it was felt that a foretopsail might ease that problem. That may have been true when the rams were going to sea, but there was no great need for a foretopmast in the Bay, and the *Gravenor's* was removed. In the earlier days of their careers the rams ranged along the East Coast. In June 1910, the *Hargraves* was at Providence, Rhode Island, loading oyster shells for Onset, Massachusetts, and at the latter port she was to take on a cargo of molding sand for Poughkeepsie, New York. On November 20, 1914, the *Hargraves* was sunk in the Bay after a collision with the fishing steamer *Fairfield.* She was raised, and sunk again within a year. On the latter occasion, rammed by the Chesapeake Line steamer *City of Richmond* off Point No Point, Maryland, during a fog, she was dismasted, stove in, and sunk. After she was raised, the cutter *Apache* took her in tow on November 22, 1915, to Baltimore for repairs. In August 1916, the vessel was renamed *B.P. Gravenor,* after a Sharptown sailmaker of that name, and was relatively free of trouble after that. In October 1932, with a cargo of slag loaded at Sparrows Point for the State Roads Commission, Rock Hall, she was driven ashore at Rock Point in a storm, but was refloated without damage. During the 1930s she was kept busy hauling lumber from North Carolina to Baltimore. Early during World War II she was sold and renamed *Mayfair.* While being taken south she was lost on the North Carolina coast on November 9, 1942, the last large commercial sailing vessel to be lost on the Outer Banks.

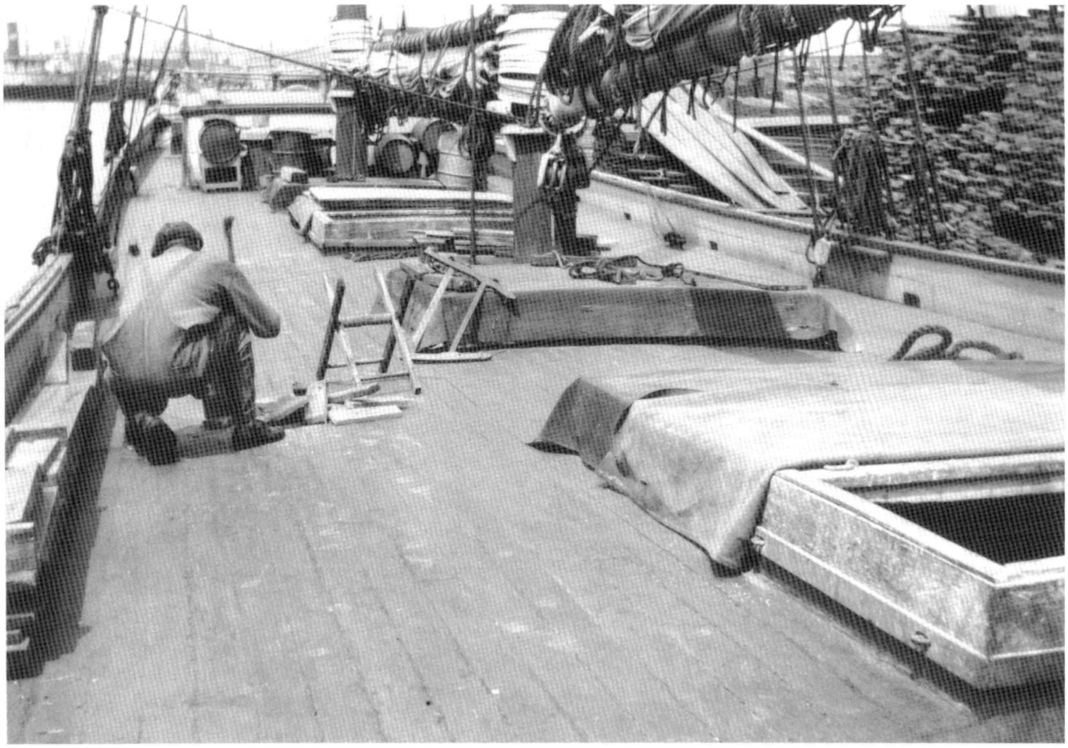

Looking aft on the *B.P. Gravenor.* A deckhand splits wood for the cookstove. This view presents an excellent comparison of the ram's beam with her length. With a length of 128 feet, her beam measured only 23.6 feet, just half a foot wider than the 71-foot bugeye *Emma A. Faulkner!* Yet with that narrow beam and knee-high waist, the rams did not hesitate to put to sea in pursuit of a livelihood even though they were not designed with that in mind.

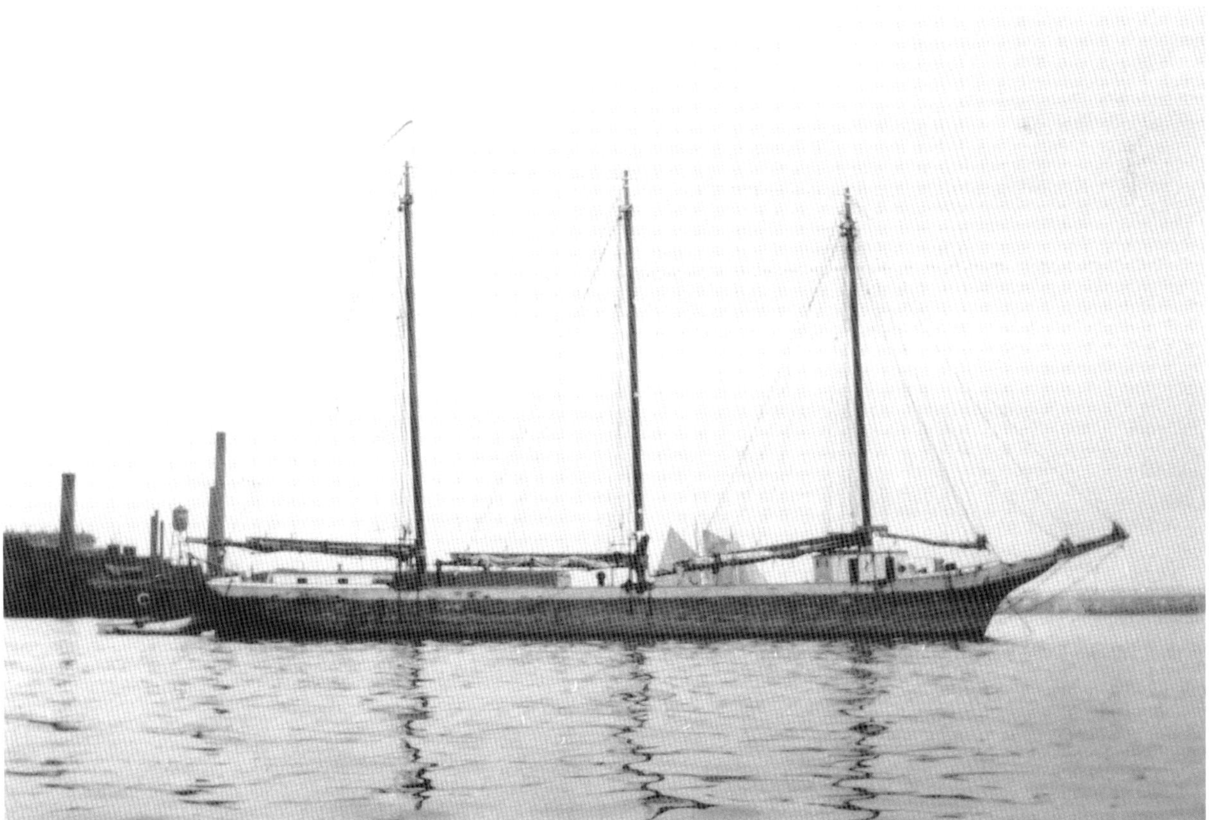

Clarence A. Holland anchored in Canton Hollow, Baltimore, October 29, 1933. This ram was built at Bethel, Delaware, in 1893, by J.M.C. Moore.

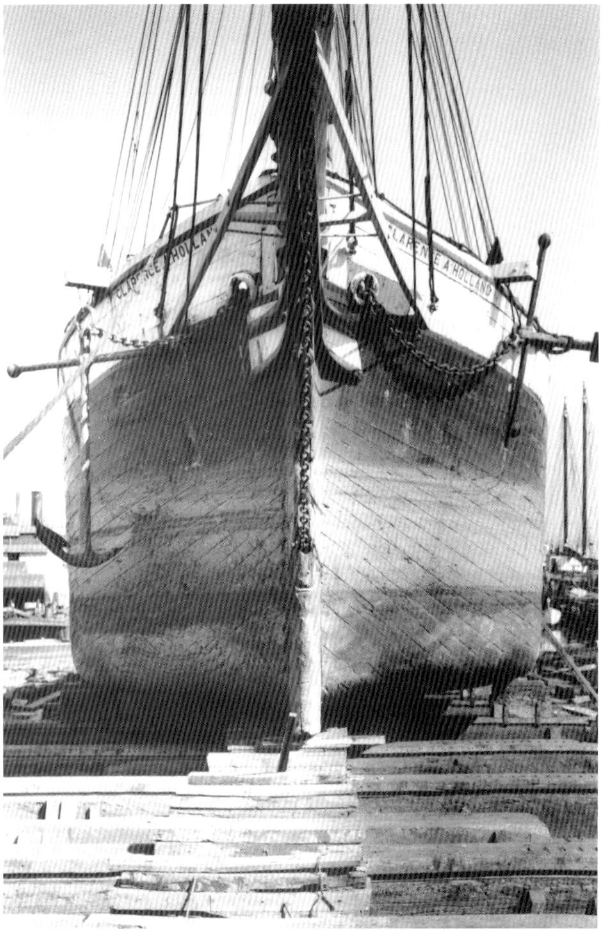

The bluff bows of the *Clarence A. Holland* are evident in this head-on view. The waterline marks on the hull are interesting. The lower mark, just below the bobstay chains, indicates the depth at which the ram rides when she is without cargo. The uppermost mark, where the dark hull meets the stained portion, shows her loaded depth, with intermediate markings caused by oily harbor waters as the cargo was discharged. Note the light chain from which the starboard anchor is suspended from the cathead.

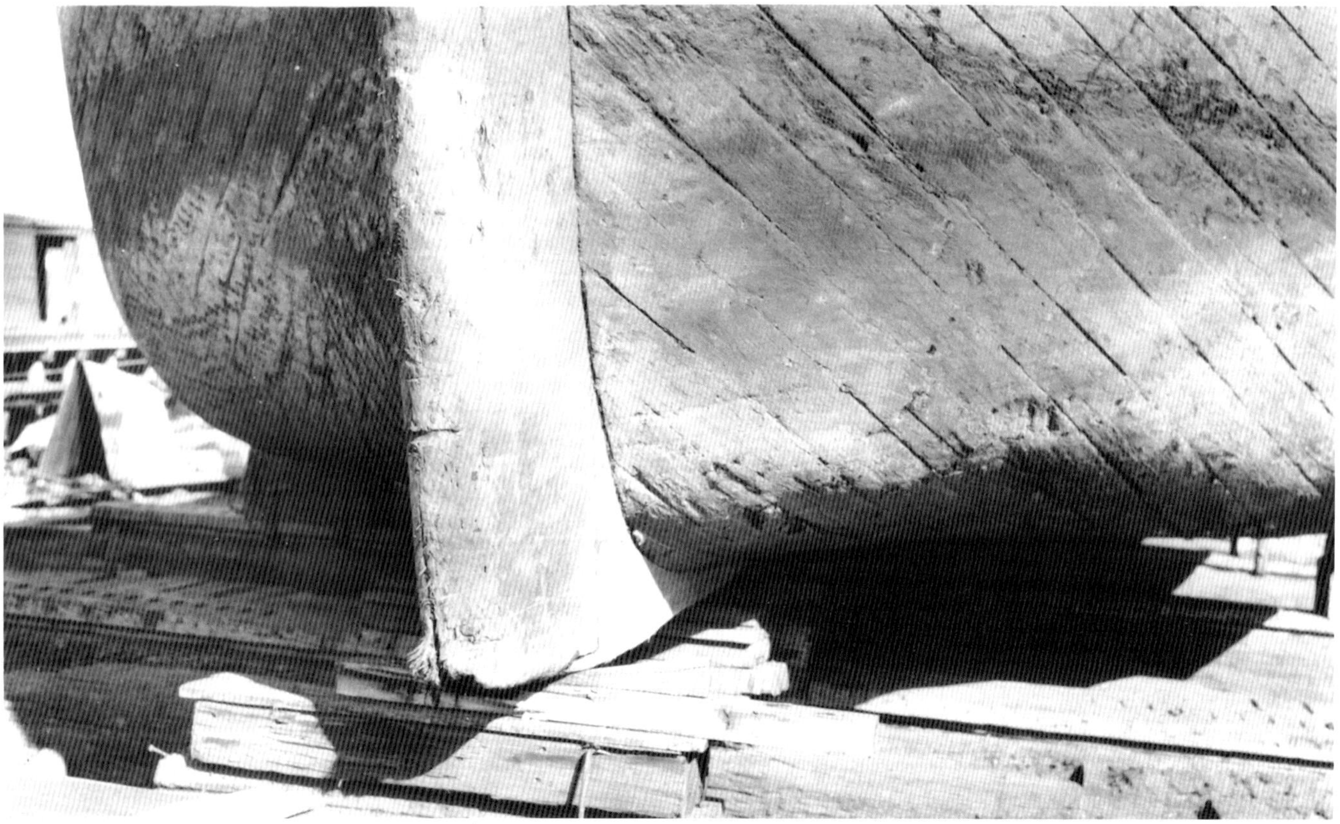

Close-up of the chunky forefoot of the *Clarence A. Holland.* Note the abrupt angle at which her planking was bent toward the stempiece. About 1940 this ram was abandoned at Elizabeth City, North Carolina. The timbers in this view now lie buried in the muddy bottom of the Pasquotank River.

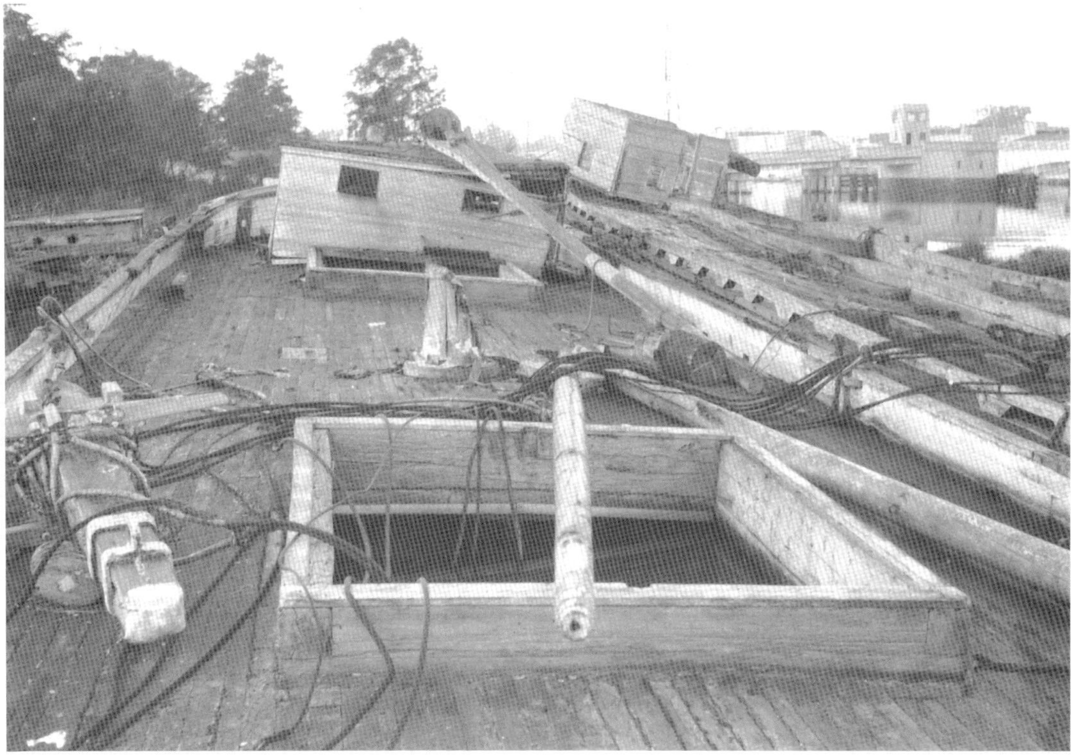

Ten years after her abandonment, the *Holland* appeared in this dilapidated state, October 14, 1950. Her booms, gaffs, mastheads, blocks, deadeyes, and other sundry gear litter the deck. The Pasquotank River Bridge at Elizabeth City is in the background.

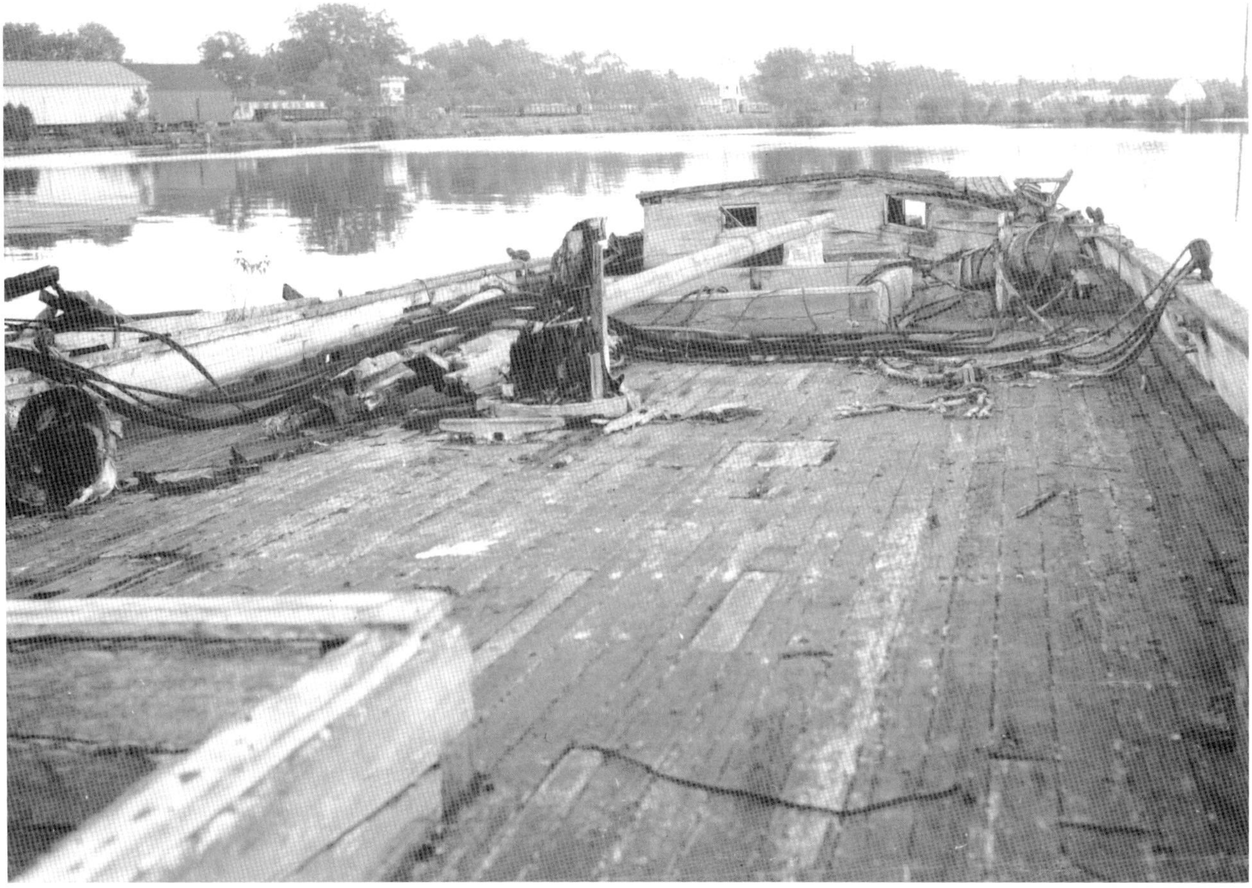

✪ Looking aft on the *Holland,* also on October 14, 1950. She presents a forlorn sight in her 57th year.

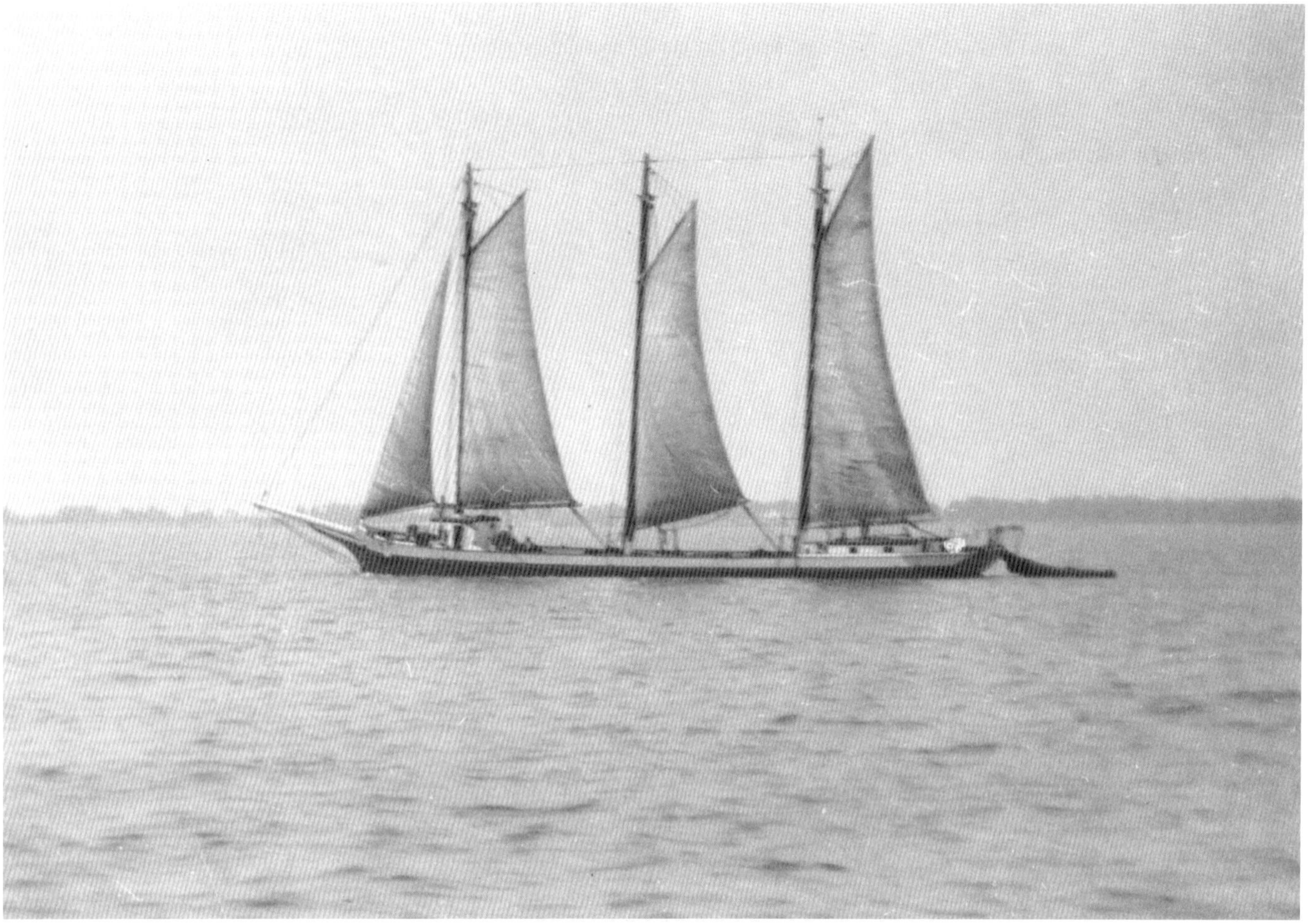

In July 1929, the ram *Corapeake* sails down Chesapeake Bay, the bow of her yawl boat lifted against her transom to assist her in the light wind. She is deep in the water and is probably loaded with slag, gravel, fertilizer, or cement. Nothing in this scene hints at the mishap-ridden existence she had experienced up to that time in her career, nor the tragedies that lay ahead. This ram was built as the *Ivy Blades* at Bethel, Delaware, in 1894. While under that name she suffered loss when, near midnight on April 14, 1907, she collided with the anchored steamer *Margaret* off Old Point Comfort, Virginia, causing considerable damage to the steamer. In 1909 the ram was renamed *Corapeake* after the North Carolina town of that name. The name change, however, failed to keep her out of trouble. On October 14, 1914, the *Corapeake* ran aground in Croatan Sound, North Carolina, while loaded with a cargo of lumber. The cutter *Pamlico* pulled her into deep water. Real trouble was to come her way in 1928. On April 17 of that year, at 8 a.m., the *Corapeake* left Norfolk for Baltimore. About noon, when two miles northwest of Thimble Shoal Light and heading northeast, the ram passed a beacon marking a wreck about 400 yards on the starboard side of the vessel. When in this position, the *Corapeake* suddenly fouled the wreck and began to take in water. The ram's pumps could not keep up with the leak, so word was sent by a nearby fisherman to go to Buckroe Beach and call for aid. About 2:30 p.m. the tug *John J. Chandler* arrived and took the stricken vessel in tow for Norfolk. When she arrived at Colonna's Shipyard at 6 p.m., she was completely filled. Repairs were made and the ram resumed her trade.

The Hampton Roads area spelled doom again for the *Corapeake.* On October 12, 1930, with 75 tons of fish scrap loaded at Baltimore for a Norfolk fertilizer plant, the ram was sailing down Hampton Roads channel. At 1:50 a.m. she struck a scow anchored without lights. The *Corapeake* sank. The next day salvage operations were commenced by Merritt-Chapman & Scott at a cost of $150 per day plus $75 per day for a diver. In five days the ram was refloated and taken to the Norfolk Shipbuilding and Drydock Company for a survey. This revealed 26 items of damage that required attention to restore the ship to the same condition as before the collision, at a cost of $14,250, certainly more than three times the worth of the old vessel at that time. In any event, she was temporarily repaired and her cargo trimmed for return delivery to Baltimore where it had been re-sold.

After a long period of idleness, tied up at her owner's pier in Baltimore, on July 2, 1933, she filled and sank during a severe northeast storm. Again she was raised, and after more weeks of languishing at her pier, she was finally put back into service. Then, in the early afternoon of November 6, 1934, the *Corapeake* was bound down Chesapeake Bay from Philadelphia to Baltimore. She was near Pooles Island, the wind at moderate to gale force from the northwest, and steering on a southwest course with only foresail and mainsail set. A tug with barge in tow was some distance astern of the ram but overtaking her on the sailing vessel's port side. It was reported that the tug passed the *Corapeake* on her port side without giving any signals and then proceeded to cross the bow of the ram and suddenly went aground just to the west of Pooles Island channel. At that time the tug was no more than 100 feet ahead of the ram. As soon as the tug went aground she blew distress signals, but the *Corapeake* was too close to avoid a collision. The port side of the ram swung across the starboard bow of the tug, causing considerable damage to the veteran sailer. Owners of the tug and the *Corapeake* brought libel suits against each other. The old ram was repaired once again, and she resumed her trade on the Bay. Still more trouble was in store for her.

On February 15, 1936, the *Corapeake* was in Petersburg, Virginia, and "jammed" a houseboat. The owner of the houseboat brought suit "by reason of alleged negligence of the captain in the handling of the *Corapeake*." He settled for $50 and the ram's owner received a document releasing him of any future claim as a result of the collision.

The *Corapeake* found herself in the North River, North Carolina, on June 15, 1936. With Captain S.W. Harrison on this passage were his wife and sons Jeff and John. At the mouth of the river a drum of gasoline became ignited onboard, setting the vessel on fire. The captain and his family attempted vainly to quell the flames, then lowered the yawl boat and came ashore. A Coast Guard cutter from Elizabeth City found the burning hulk and towed her out of the channel. Her total destruction ended a hard 42-year career, more than any single vessel should have had to endure.

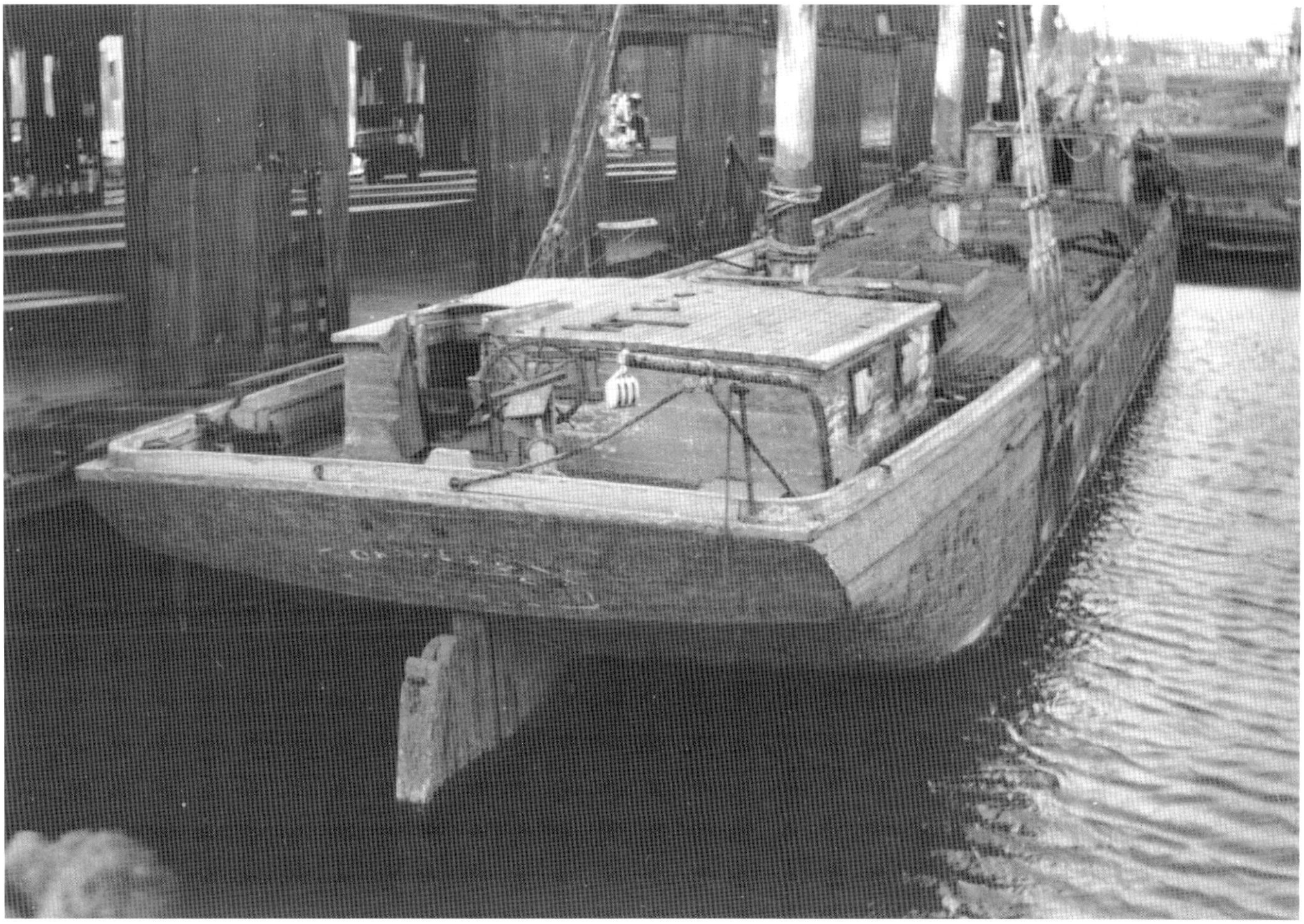

At Wathen's Andre Street pier, Baltimore, the tragedy-ridden *Corapeake*, on December 9, 1933, shows the effects of her sinking a few months earlier at the dock where she is shown. Her port davit is missing, booms, gaffs, and sails gone, steering gear askew, door to the after cabin missing, port shrouds slack, and hatch covers absent—just a few of the items needing attention before the ram can see service again.

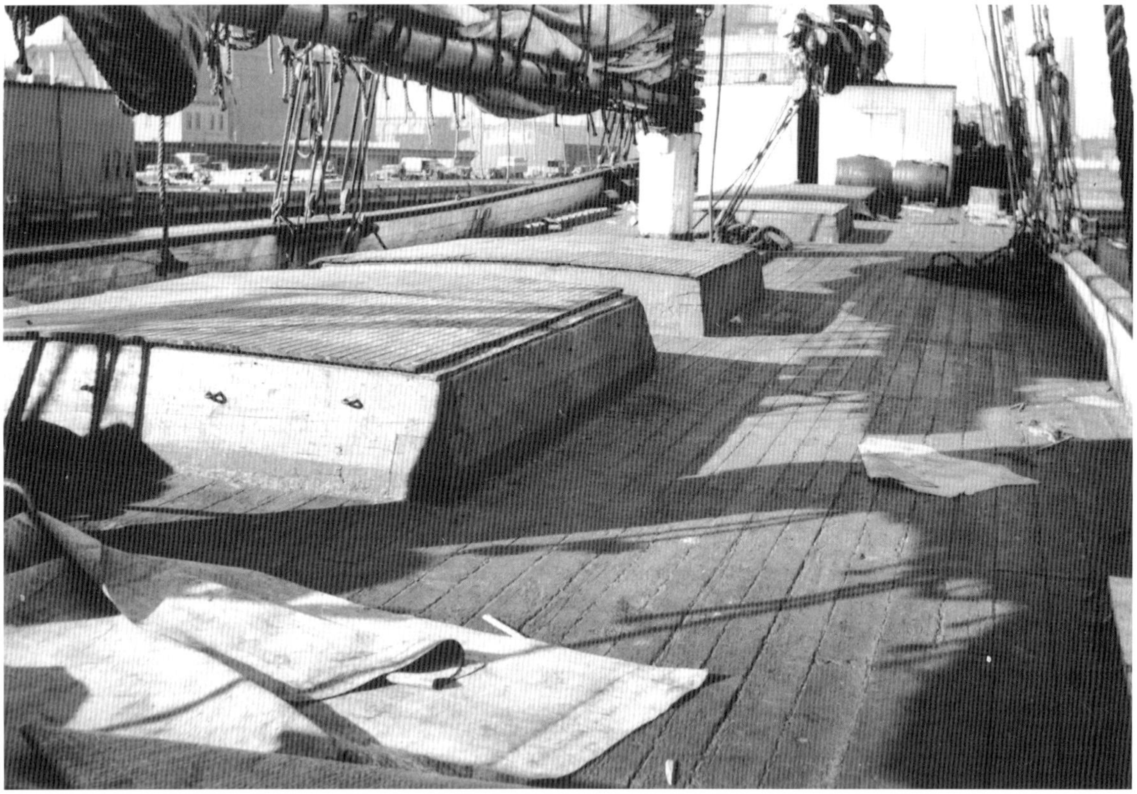

As evident in this view looking forward on the deck of the *Corapeake* on January 11, 1935, she was put back into service and has just discharged a cargo of lumber at Pier 6, Pratt Street, Baltimore. Considering all her tribulations, she makes a fine appearance on deck.

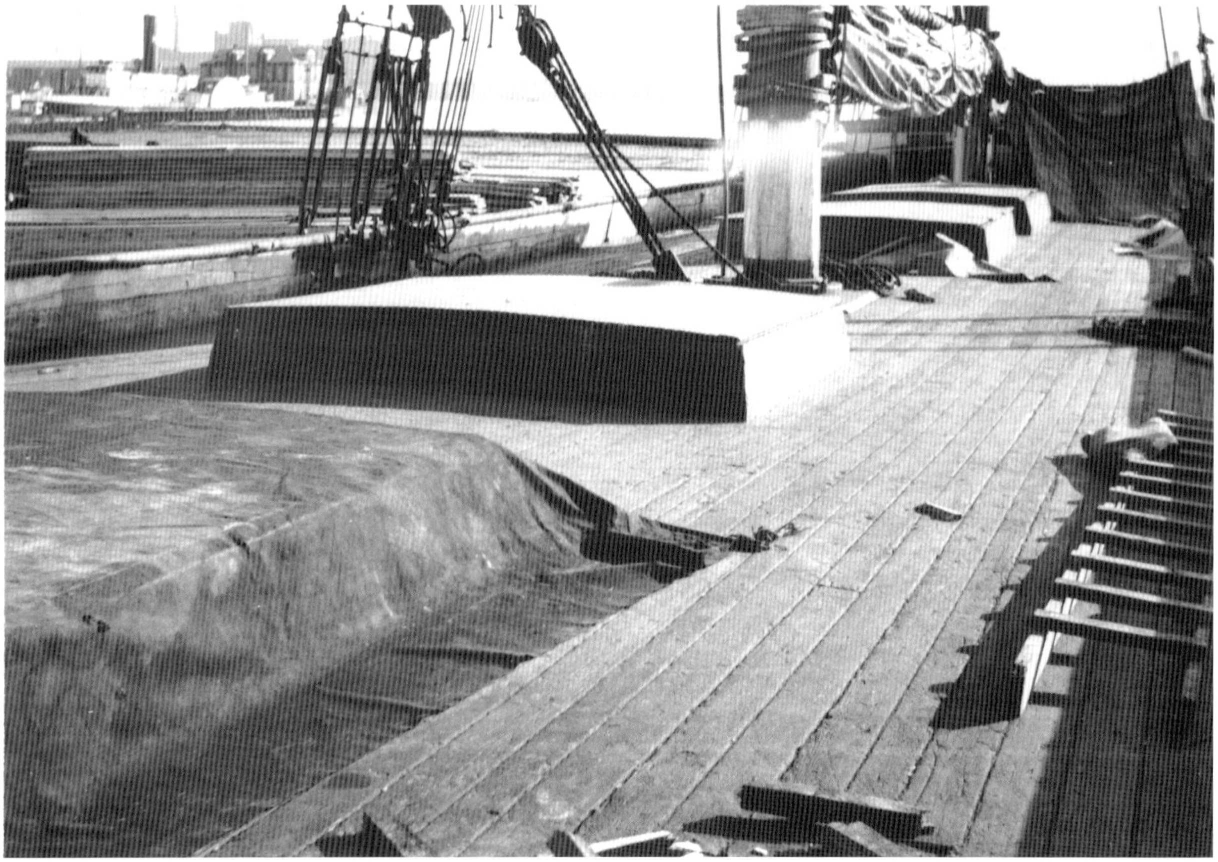

✪ Looking aft on the *Corapeake,* on the same day.

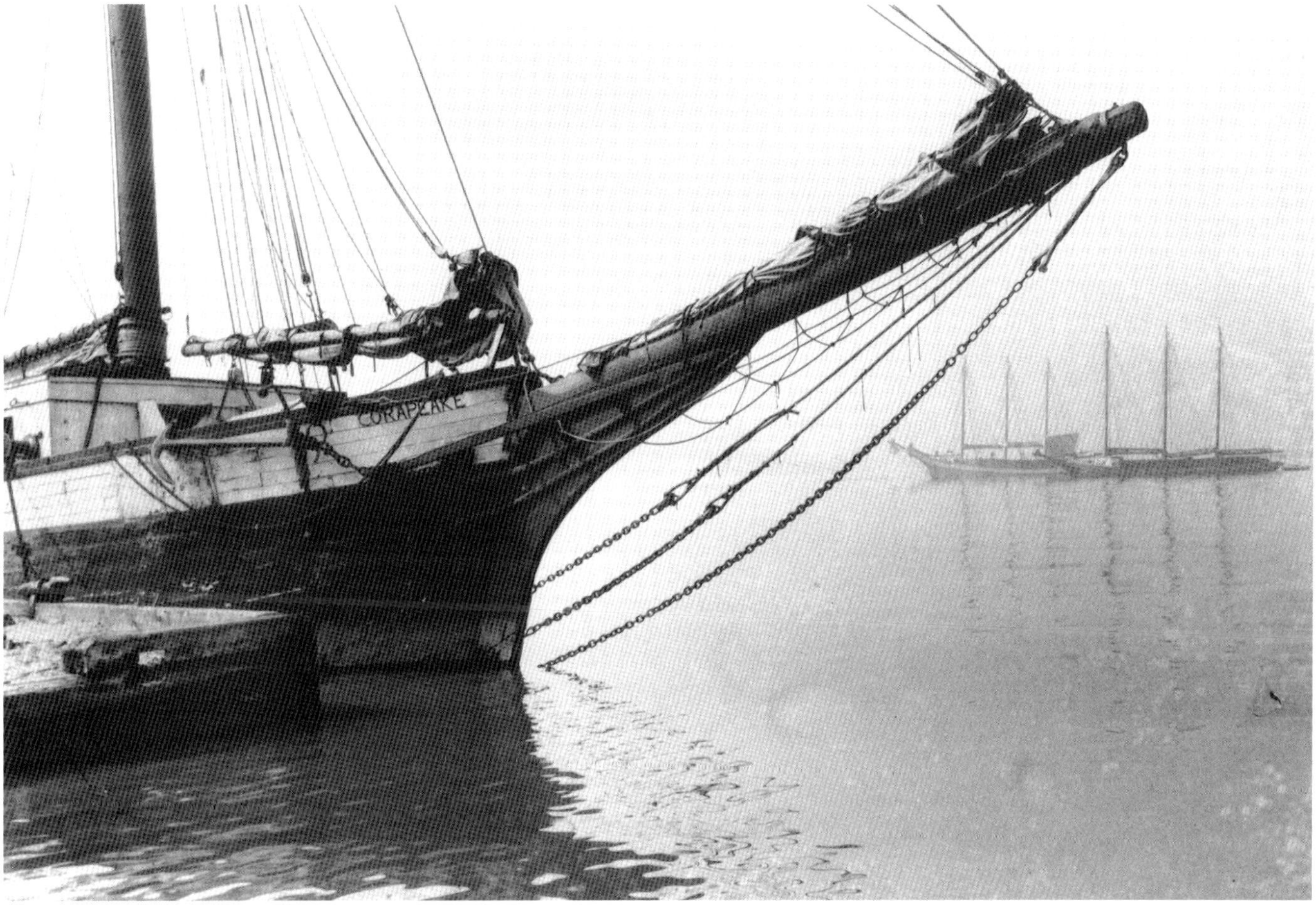

Ram *Corapeake* at McGrath's pier, Baltimore, March 15, 1936, presents a graceful clipper bow on an otherwise homely hull. The two lumber-laden rams anchored out in Canton Hollow are, at right, the *B.P. Gravenor*, and the *Edward R. Baird, Jr.* Three months to the day after this photograph was made, the *Corapeake* was destroyed by fire.

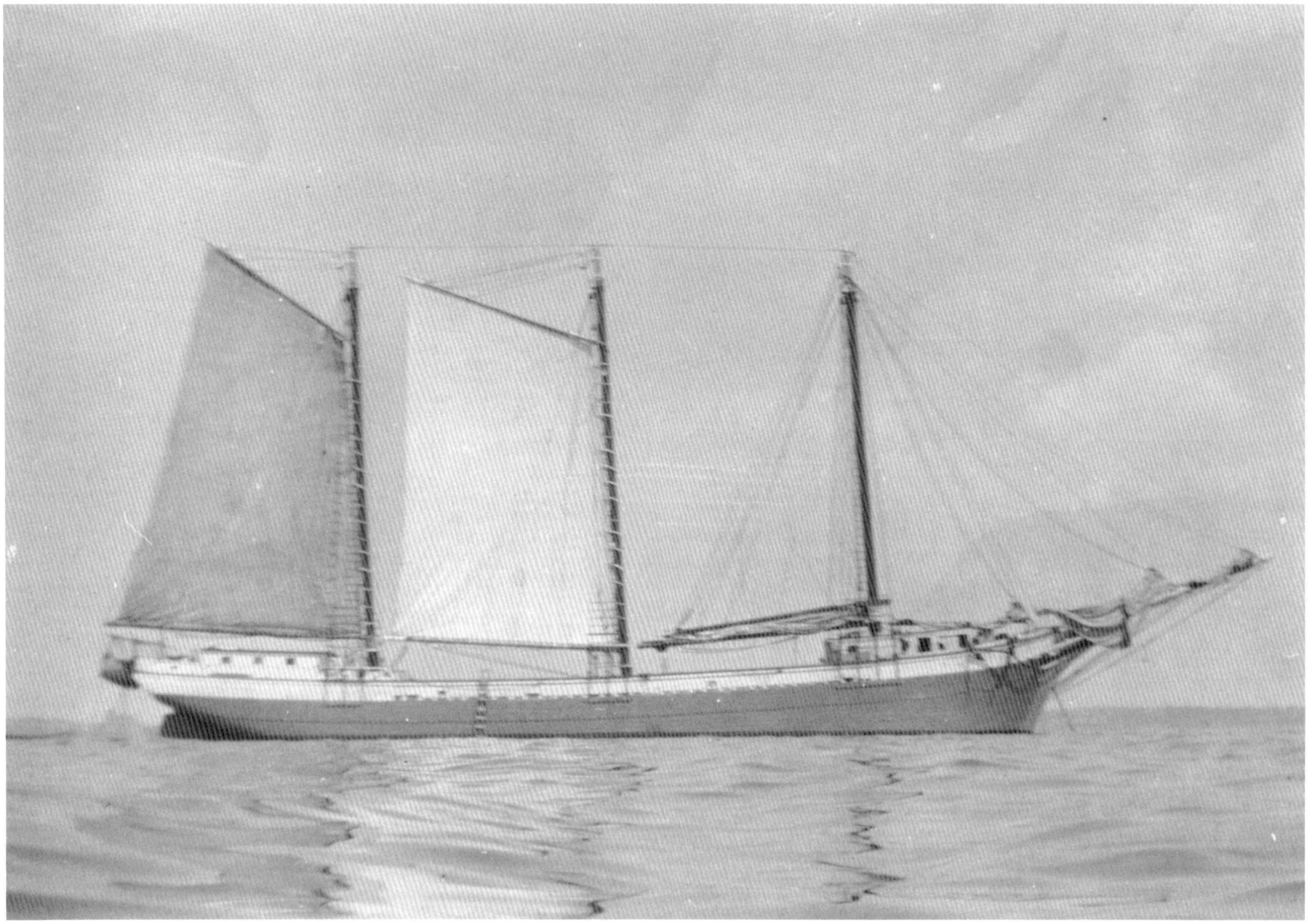

Of the 13 rams plying the Chesapeake in the early 1930s, one of those, the *Granville R. Bacon,* put into Norfolk only occasionally and cannot be considered a true Bay trader. She was the last ram built, in 1911, and the largest of the 13. The *Edward R. Baird, Jr.* had the most pleasing lines and the finest sheer. She was also the largest putting into Baltimore, measuring 132 feet. She had been built by G.K. Phillips and Company at Bethel, Delaware, in 1903. To bear out the fact that rams in their early days traded well beyond their usual routes in Chesapeake Bay, mention can be made of some of her ports of call along the East Coast. On September 26, 1907, she was bound from Sullivan, Maine, to New York with a cargo of stone. She missed stays and collided with the Pollock Rip Slue Lightship at the entrance to Nantucket Sound, Massachusetts. She was damaged and beached to prevent her from sinking; the next day a wrecking tug refloated her after temporary repairs had been made. On September 17, 1908, the *Baird* left Beaufort, North Carolina, for New York and had fine weather up to Scotland Lightship, where she ran into a thick fog and was compelled to anchor. When the fog lifted, her master sailed the ram up to Staten Island. In January 1914, she carried lumber for New York from Charleston, South Carolina, and then returned to that southern port with salt. In September 1914, she carried tankage from Barren Island to Philadelphia, sailing thence from Lewes, Delaware, to Charleston with fish scrap, and returned to New York with lumber. On March 12, 1916, she was off the North Carolina coast with a cargo of salt. She ran out of provisions and had to be resupplied by the Cape Lookout Life-Saving Station.

During World War I the *Baird* was shelled by a U-boat in the vicinity of Chesapeake Lightship after her crew had left in the yawl boat. The ram's lumber cargo kept her afloat and she was towed in to Norfolk for repairs to her port bow, holed by a shell. Later she was towed to New York to discharge her cargo. All this time she had been registered out of Wilmington, Delaware, but about 1924 she was purchased by C.C. Paul and Company, and Baltimore became her home port. On September 5, 1932, the *Baird* was anchored off the mouth of Curtis Bay, Baltimore, as shown here, awaiting a fair wind to take her down the Bay. Approaching her thirtieth year, she still made a fine appearance.

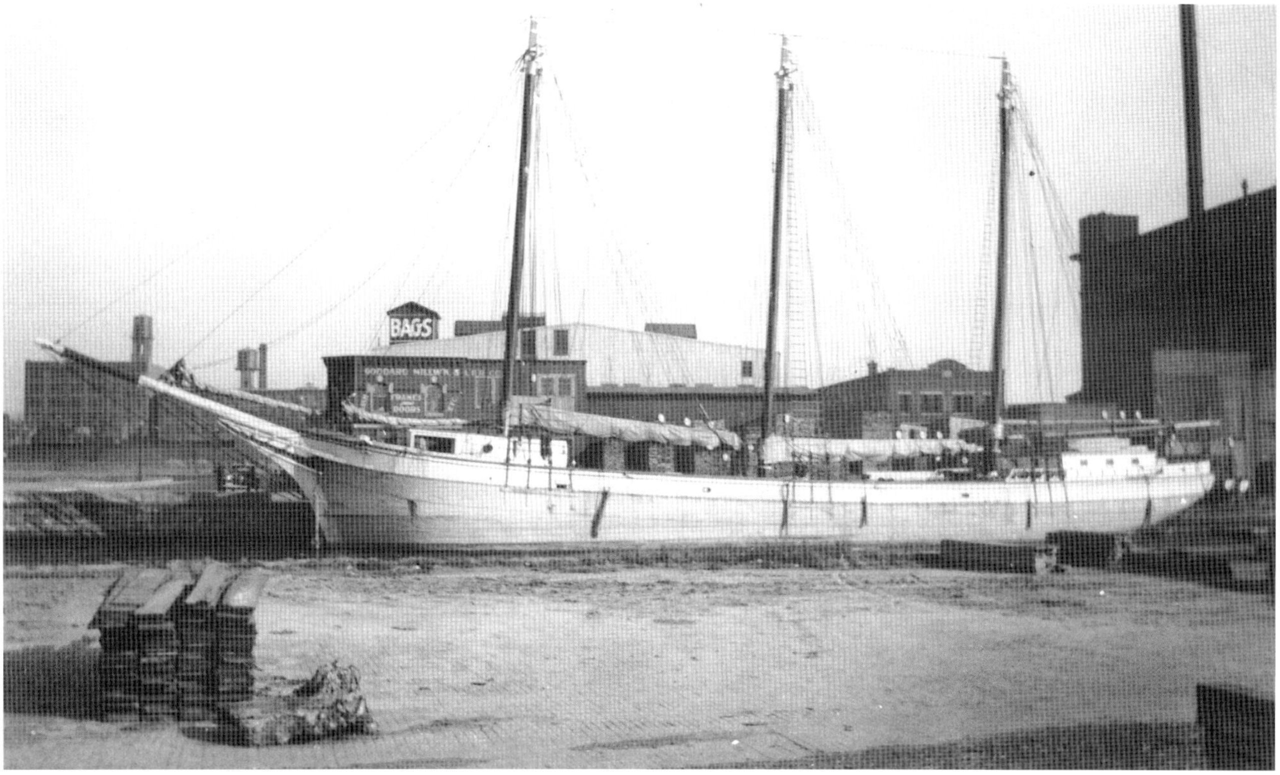

On November 25, 1933, the *Edward R. Baird, Jr.*, lies in Jones Falls, Baltimore. From the deck line down to the water her hull was painted a light gray, while her waist was white.

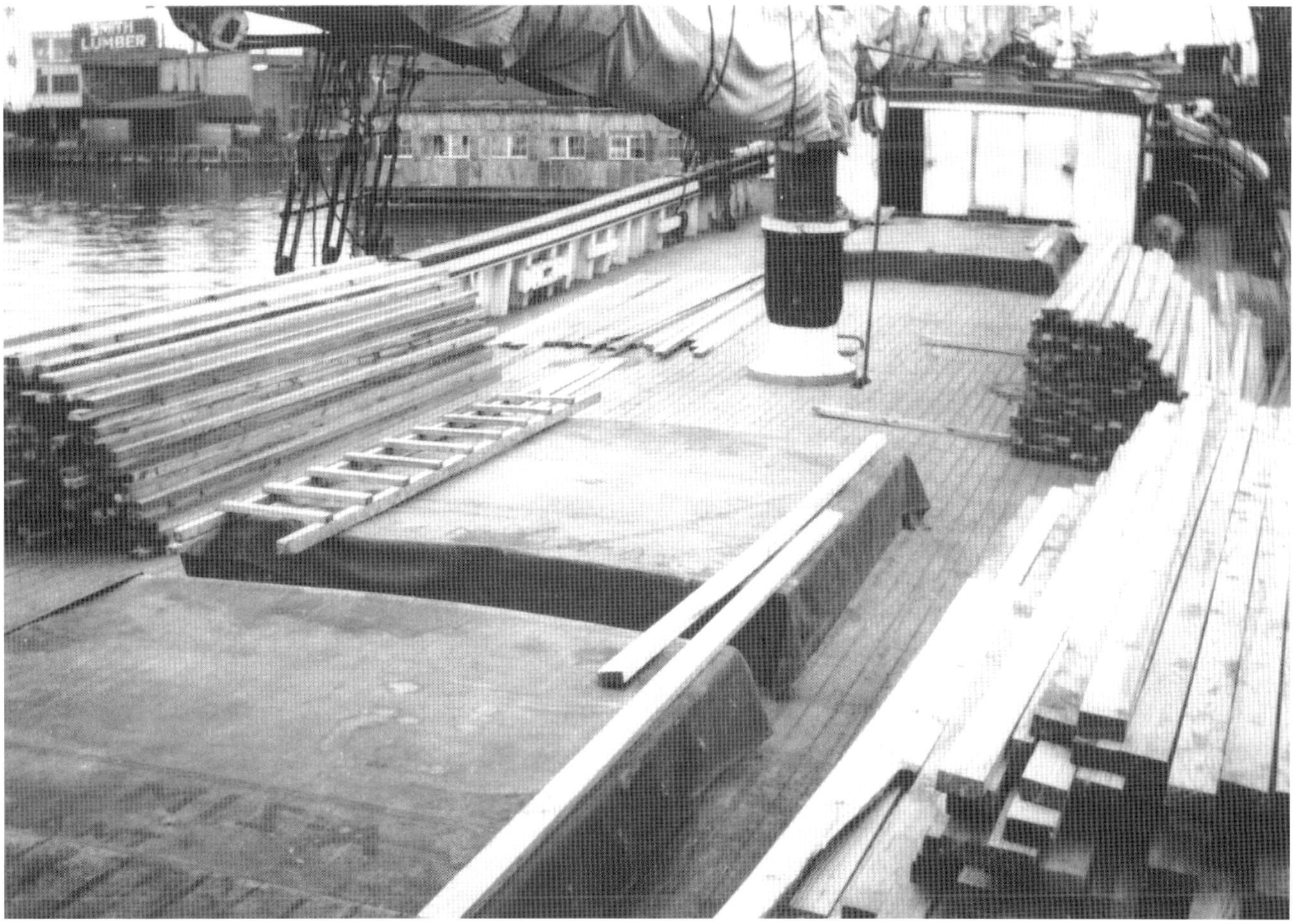

✪ Looking forward on the ram *Edward R. Baird, Jr.* Her centerboard rod is starboard of her mainmast.

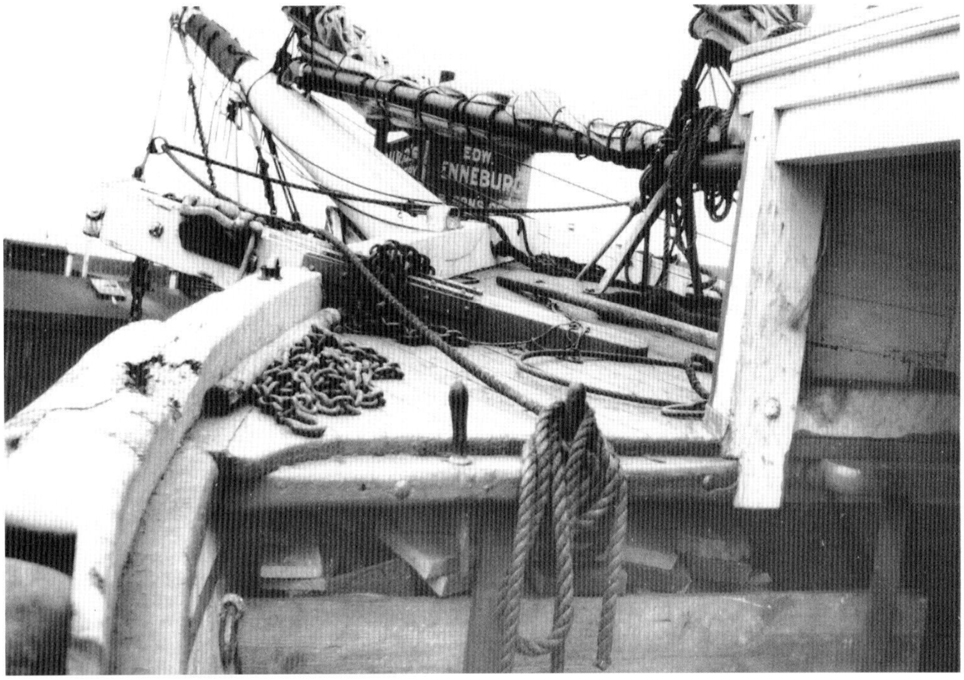

✪ The raised forecastle of the *Baird.* Her port cathead and her bowsprit dominate this view.

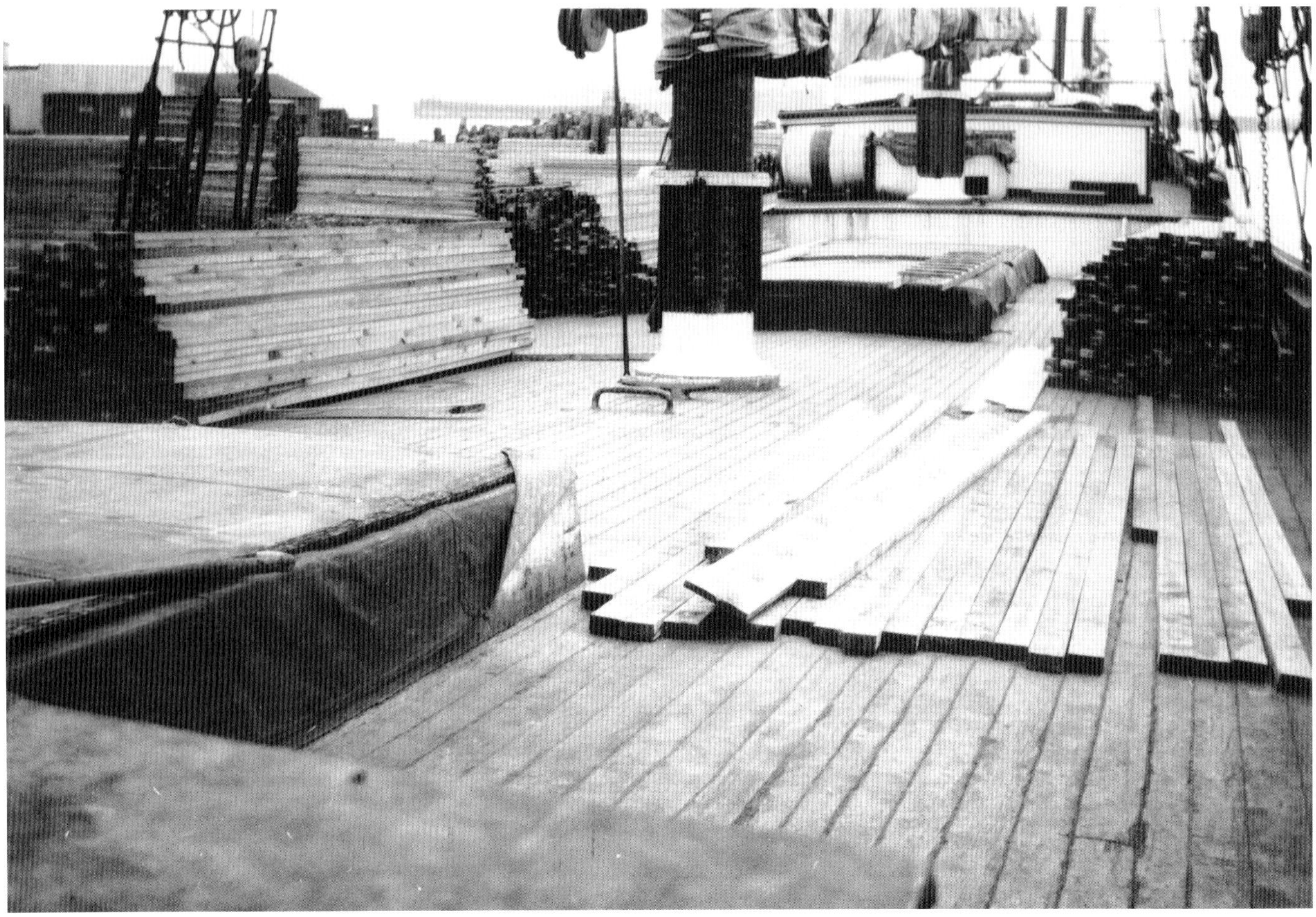

Looking aft on the deck of the *Baird.* The raised quarterdeck and the forecastle of this ram classified her as being of the "outside" type. The *Edwin and Maud* and the *Granville R. Bacon,* of this period, were similarly constructed. The *Edward R. Baird, Jr.* kept busy during the 1930s and early '40s and in the late 1940s was on a rather steady run between Baltimore and Snow Hill and Pocomoke City with fertilizer materials. In 1954 she was sold, and while being towed to Baltimore for a refit she developed a leak through open bow seams and sank in Tangier Sound on September 19, 1955. The hulk, the masts of which protruded more than half their length above the water, was abandoned and destroyed by the Coast Guard.

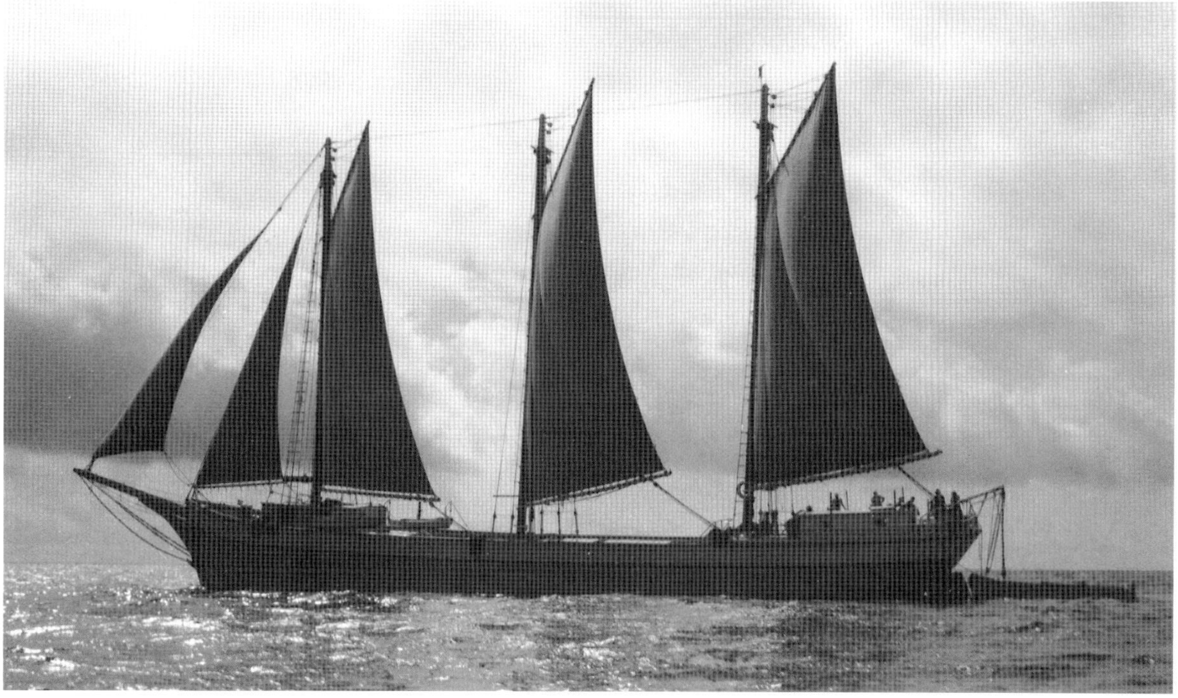

Her days of drudgery over, the ram *Edwin and Maud* sails across Chesapeake Bay from the Piankatank River to Tangier Island, July 10, 1950, in the role of a "dude cruiser." About 1946 she retired from her lumber trade, had staterooms built in her hold, and ports cut through her sides. Then she operated out of Annapolis on cruises of one and two weeks' duration to various parts of the Chesapeake country. This ram had been built in 1900 at Bethel, Delaware, by G.K. Phillips and Company.

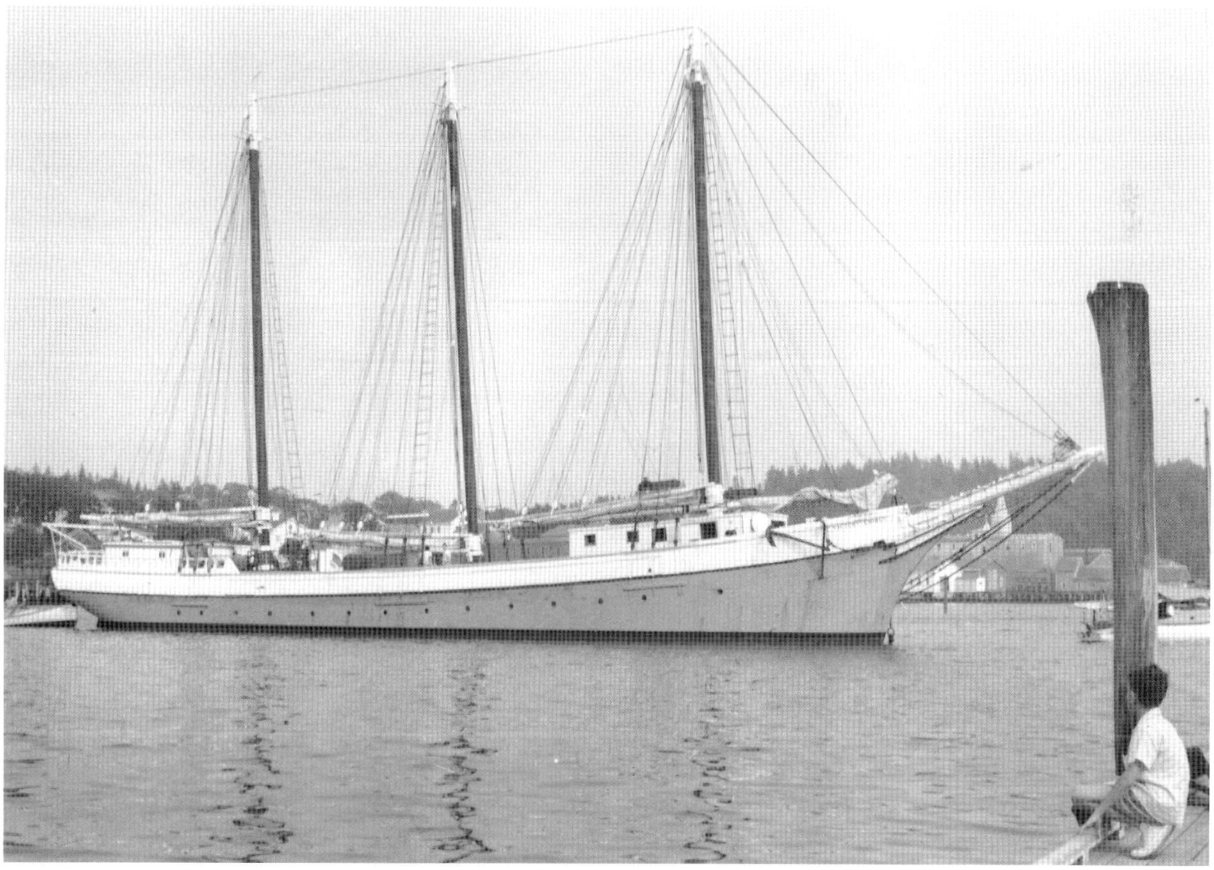

In 1954 the *Edwin and Maud* was sold to Maine owners, taken to Humphreys Shipyard, Weems, Virginia, for repairs, and renamed *Victory Chimes*. Registered out of Castine, she continued as a "dude cruiser" among the islands off the Maine coast and still plies in that trade there. Here she is seen at Boothbay Harbor, Maine, July 1955.

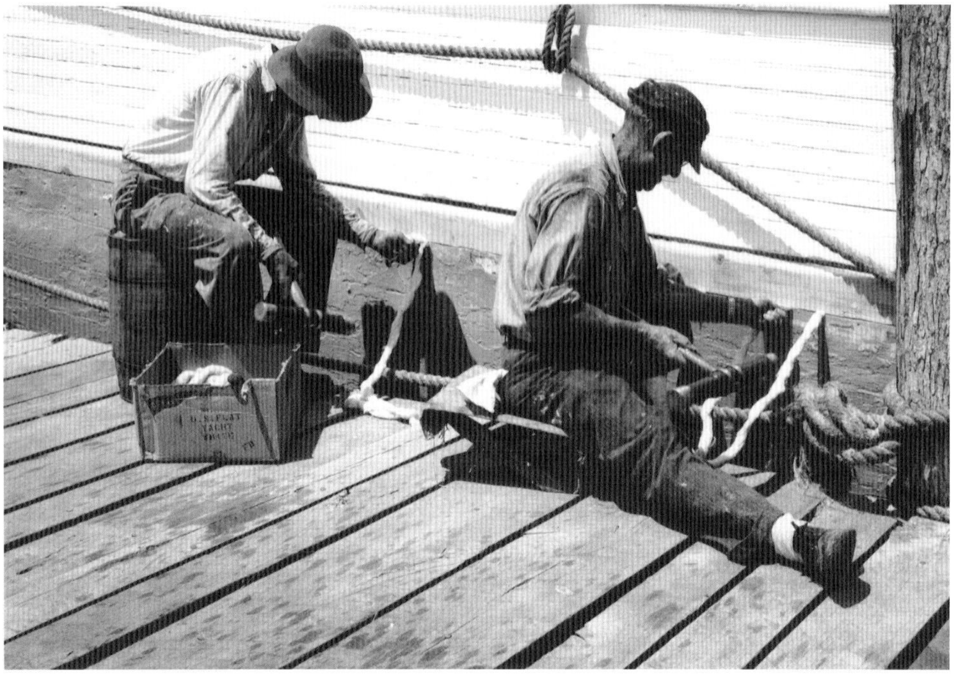

✪ Caulking above the waterline on the newly renamed *Victory Chimes,* probably at Weems, Virginia, on June 25, 1954.

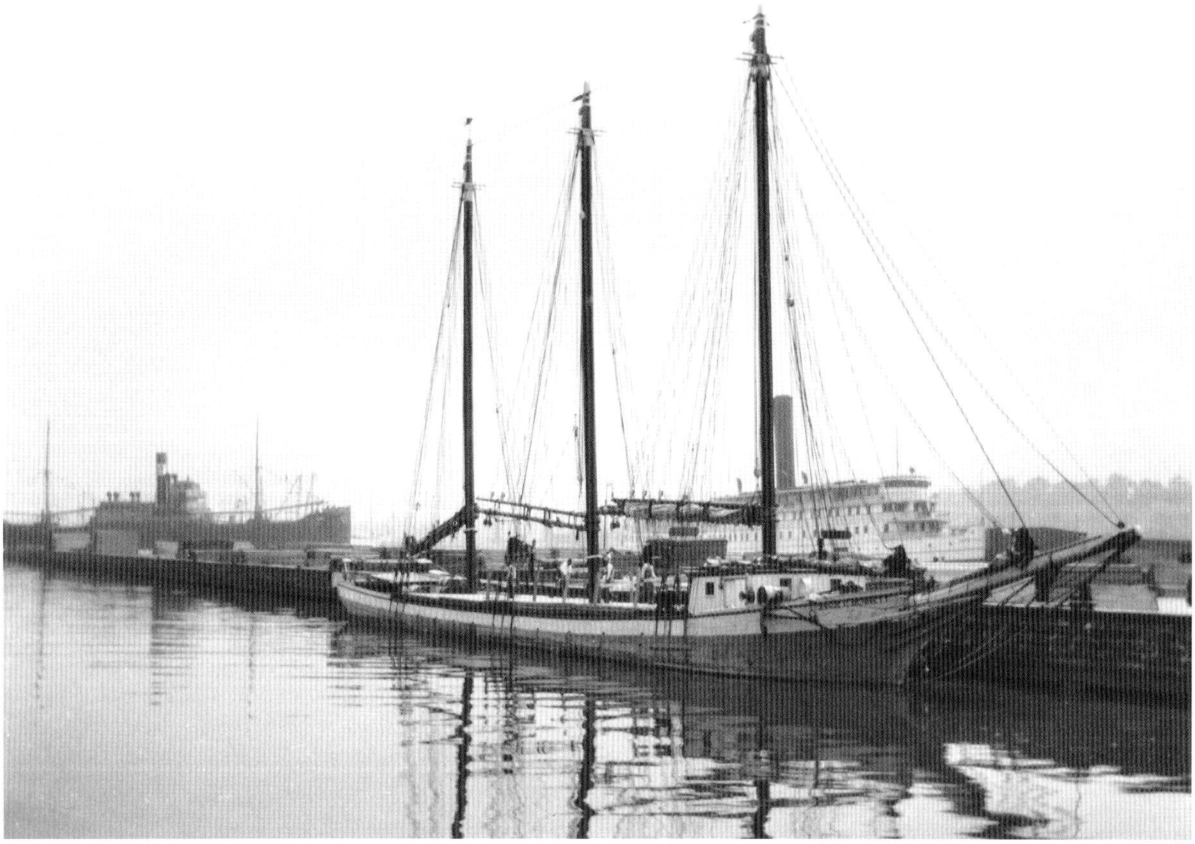

Ram *Grace G. Bennett* unloads her lumber at Pier 6, Pratt Street, Baltimore, May 11, 1934. Stevedores can be seen on her deck passing individual planks to men on the pier where the lumber is stacked in piles. At the end of that pier, at left, is the Bull Line steamer *Jean.* Behind the foremast of the *Bennett* is the Old Bay Line steamer *State of Virginia.* The *Bennett* was built at Bethel, Delaware, in 1893 by G.K. Phillips and Company. Since her name does not appear in the *Annual Reports* of the United States Life-Saving Service during her early career, it may be assumed that she was relatively free of serious mishaps. On September 3, 1913, however, while bound from Washington, North Carolina, to Baltimore, with a cargo of lumber, she foundered six and a half miles northwest of the Portsmouth Life-Saving Station, situated south of Ocracoke Inlet, North Carolina. That would have placed her in Pamlico Sound. Her master, Captain S.H. Larimore, described the storm that caused the sinking of his ram as "the worst I have ever experienced after years on the seas." The persons aboard included five women, and all were taken off the vessel. The *Bennett* was salvaged and put back into service again.

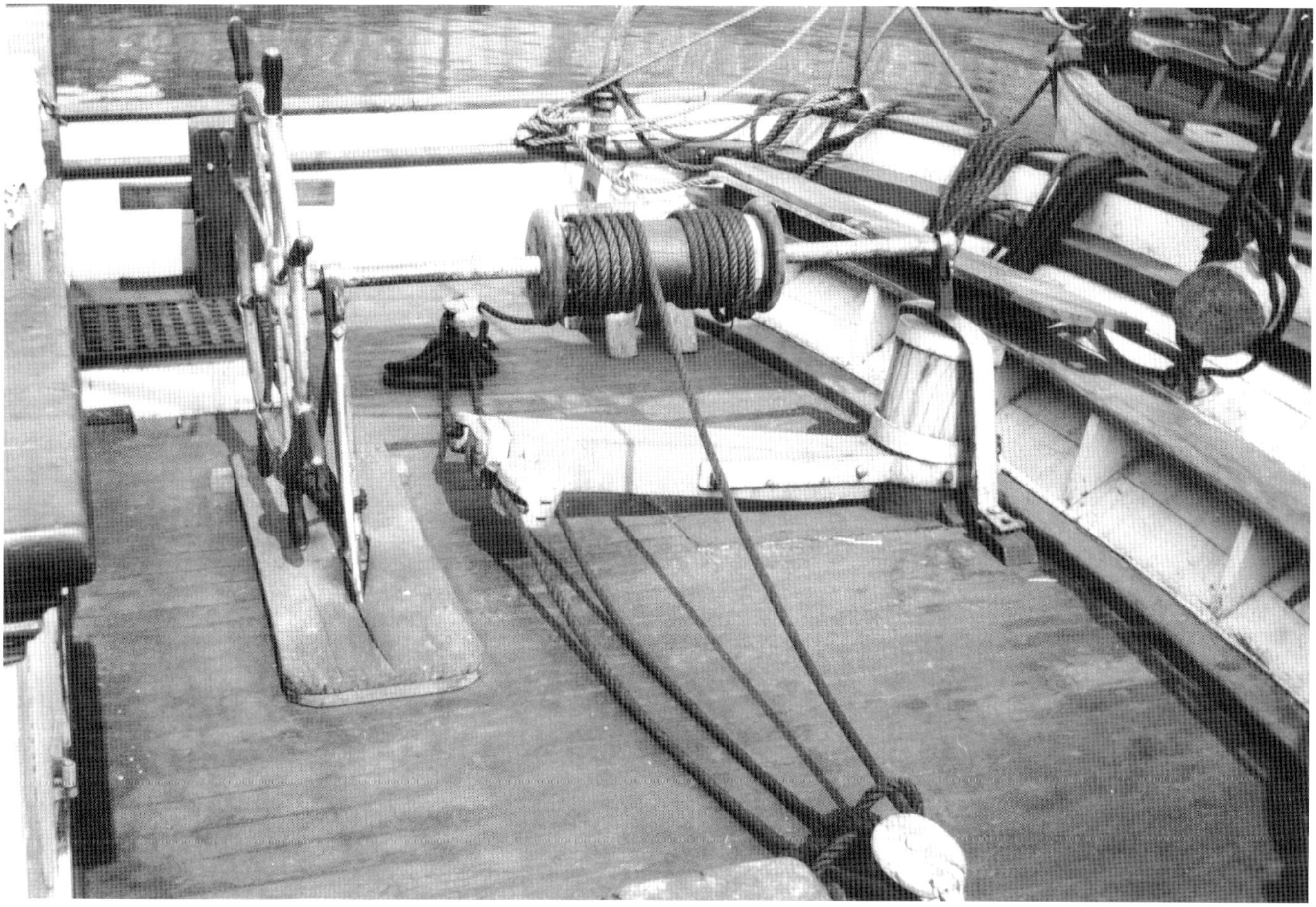

The primitive steering gear on the *Grace G. Bennett.* The metal steering wheel is affixed to a shaft that passes through an iron A-frame at its forward part to a yoke mounted over the rudder post. Midway on the shaft is a wooden drum around which are ropes extending to blocks attached to the tiller.

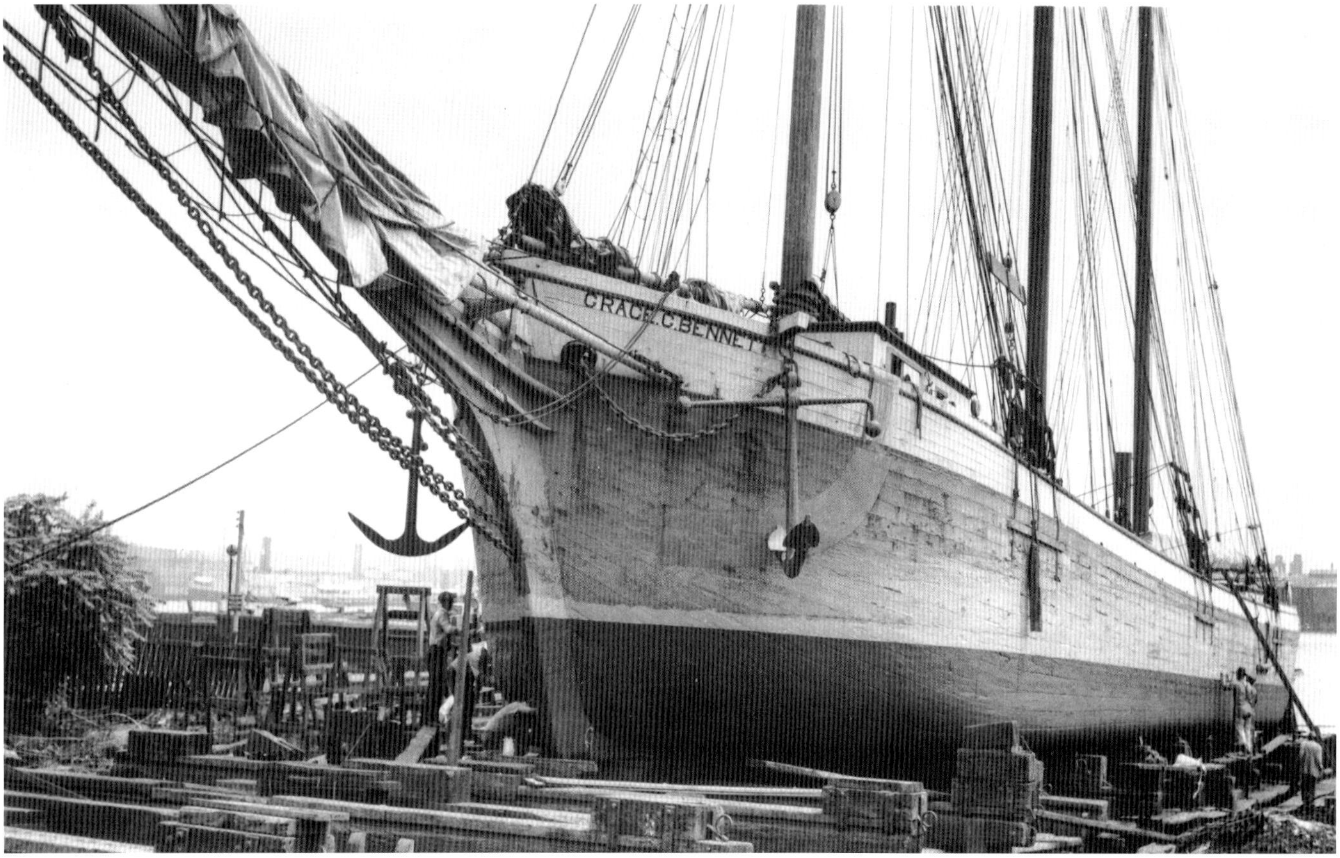

Grace G. Bennett has her bottom painted at the Redman and Vane Shipyard, Baltimore, in June 1936. The relatively shallow draft of the ram can be compared with the height of the man standing amidships on her port side. The steamers at left background are the icebreakers *F.C. Latrobe* and *Annapolis.*

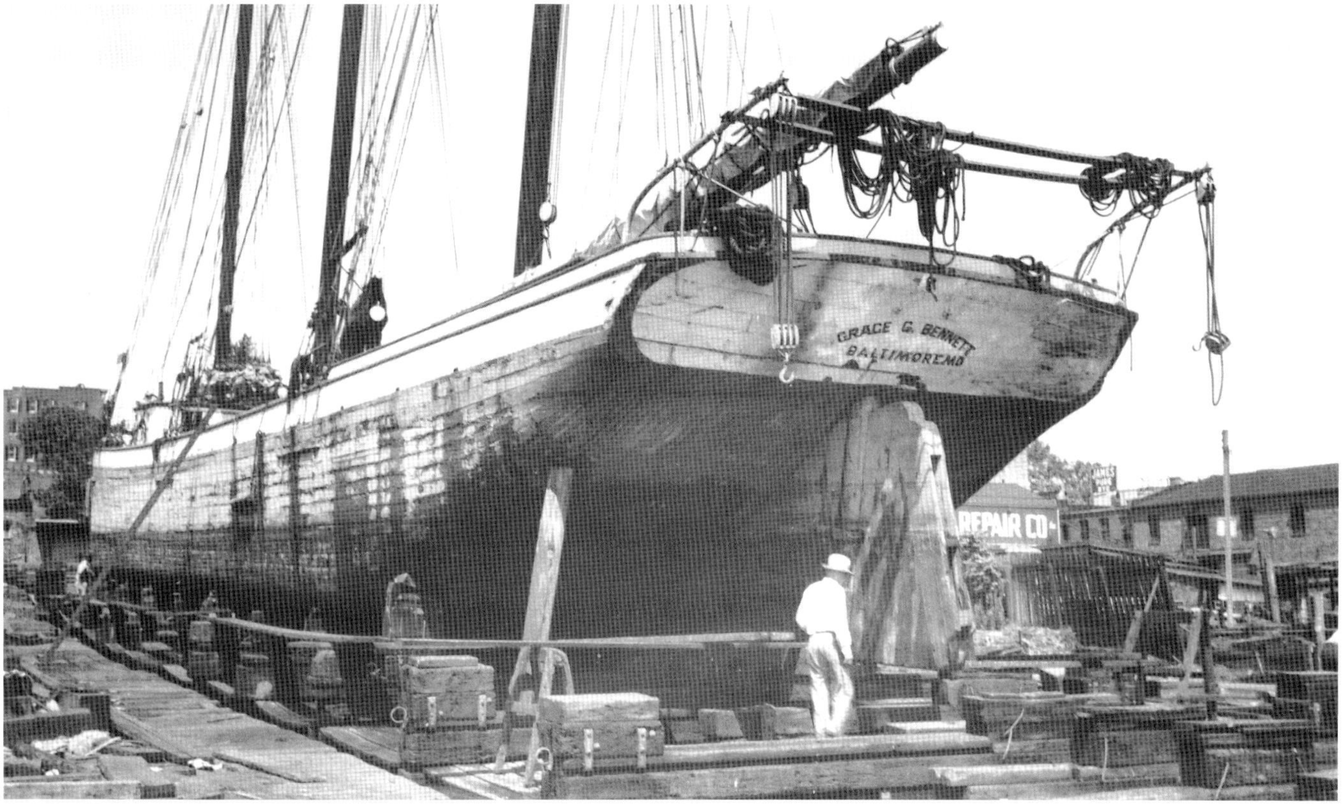

Stern of the *Grace G. Bennett,* revealing her bargelike run and skeg. A huge beam shores up the port quarter as her aging timbers need support when out of the water.

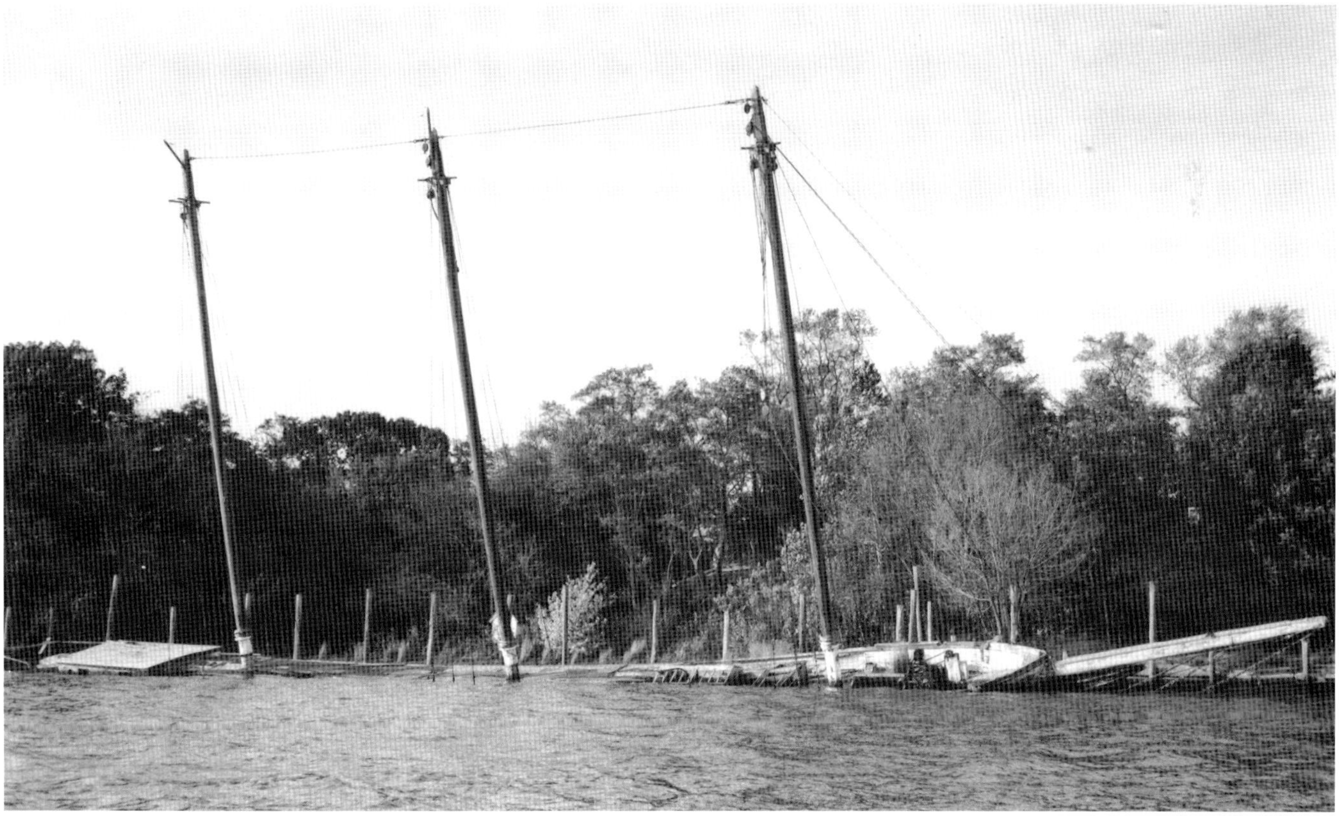

Noting the success of the rams *Levin J. Marvel* and *Edwin and Maud* as "dude cruisers," the new owners of the *Grace G. Bennett* fitted her out for that trade in the mid-1940s. A cabin was built above deck extending from abaft the foremast to forward of the mizzenmast, with nine rectangular ports on each side. While tied up out of season at a wharf at Crumpton, Maryland, she sank during hurricane Hazel on October 15, 1954. In 1960 she was dismantled. This view shows her at Crumpton in October 1959.

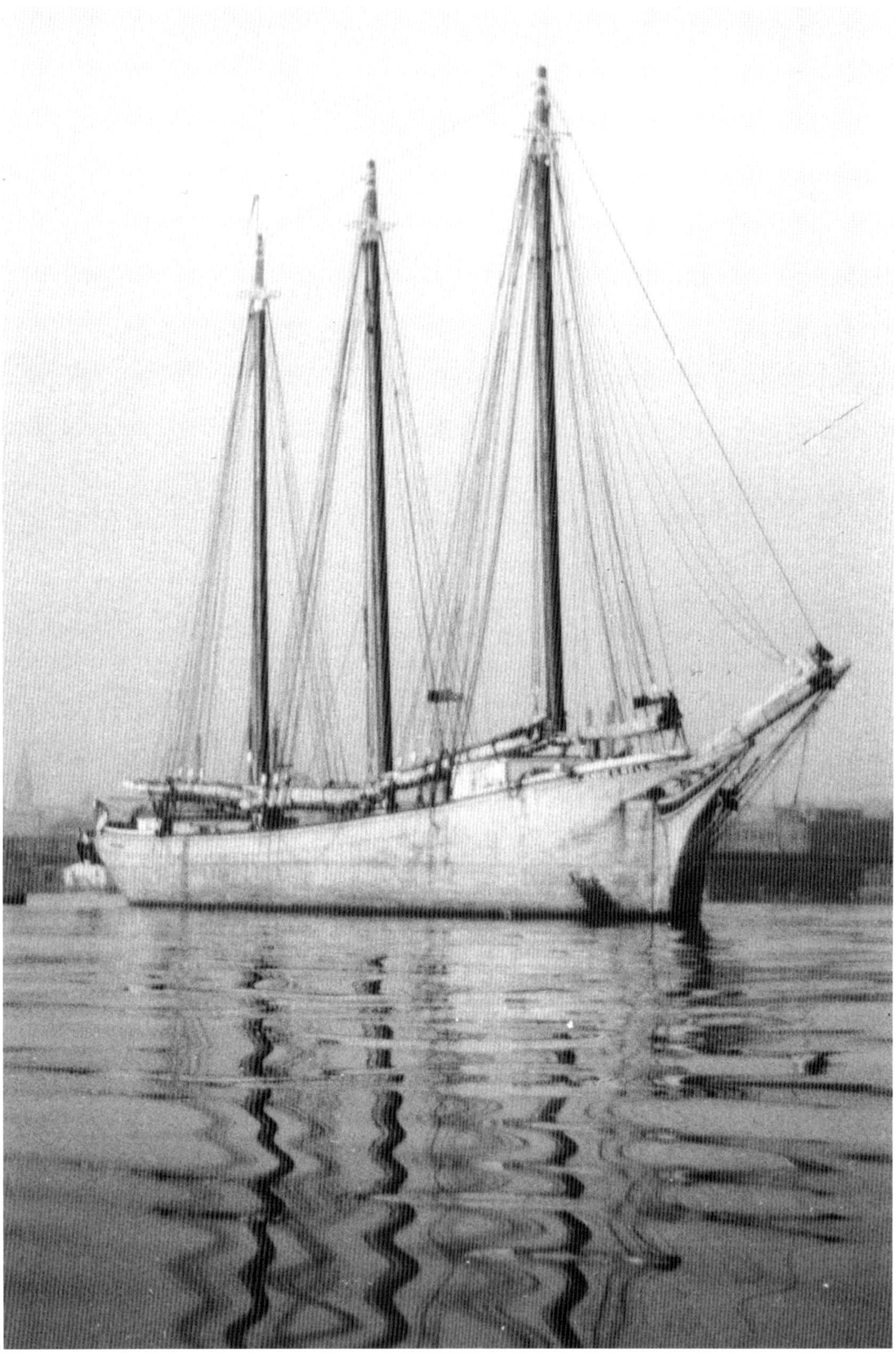

Rarely does a captain purposely put his ship ashore, but this strange fate befell the ram *Jennie D. Bell* at least twice in her career. On December 27, 1903, while at anchor in the channel a quarter-mile northeast of the Fire Island Life-Saving Station, New York, the *Bell* was endangered by large quantities of drift ice. Upon the request of her master, Captain William S. Bell, the lifesaving crew boarded the vessel and assisted in beaching her in a safe place. Owned in Seaford, Delaware, at the time, she was bound from Patchogue, New York, to New York City. The following March 31, surfmen assisted in floating the ram, which resumed her trade with little ill effect. Far from those times and places, the *Jennie D. Bell* rides peacefully at anchor in Canton Hollow, Baltimore, on Christmas Eve, 1933.

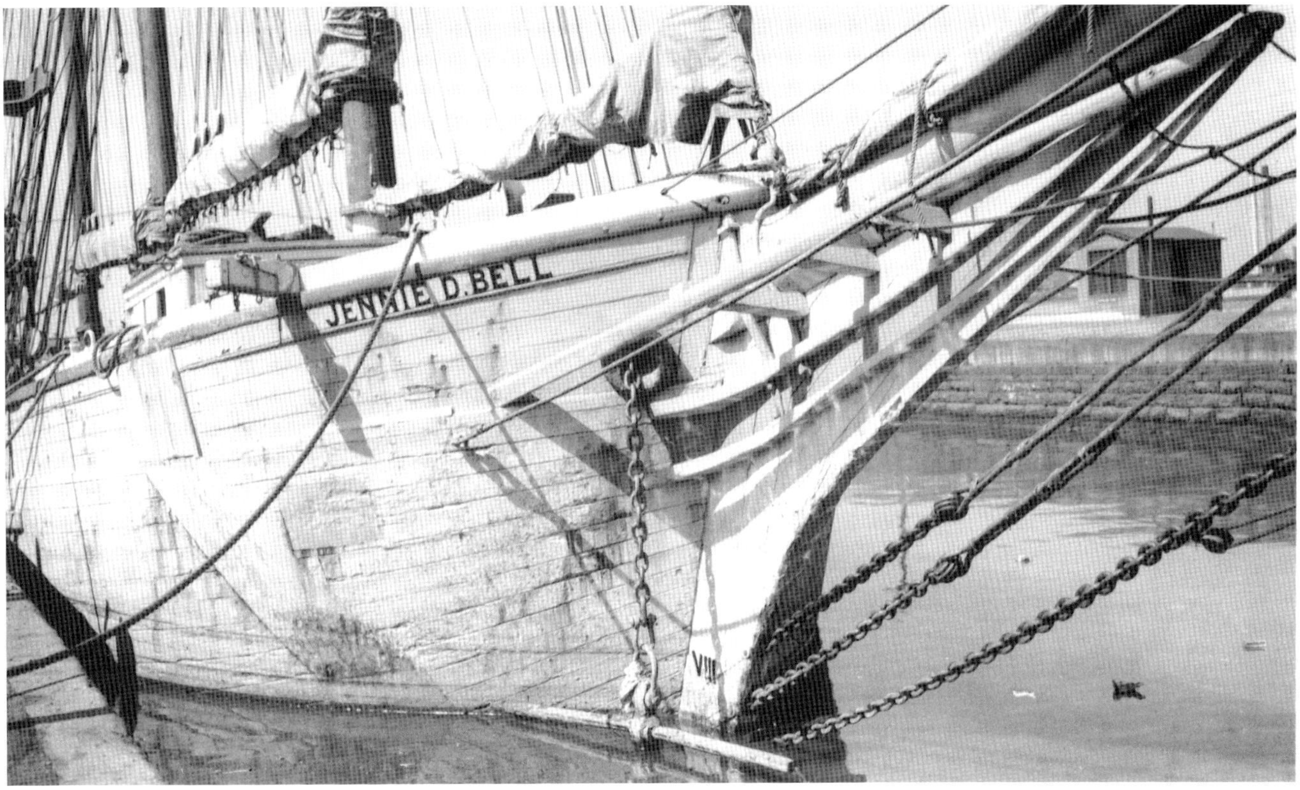

✪ Bow detail of the *Jennie D. Bell* at Montford Yard, Baltimore, on May 3, 1936.

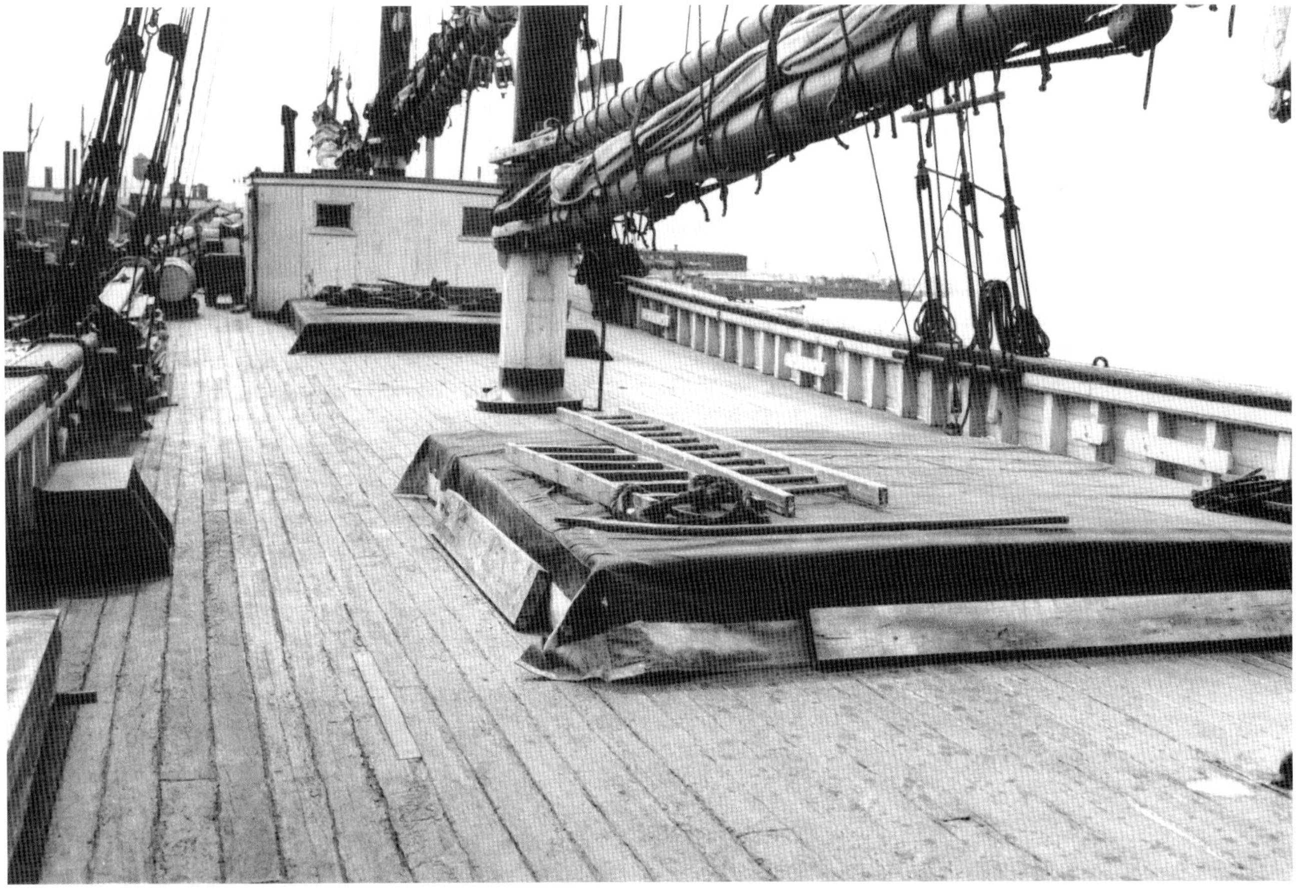

Looking forward along the deck of the *Jennie D. Bell* moored at Montford Yard, Baltimore, May 3, 1936. This ram was built in 1898 at Bethel, Delaware, by G.K. Phillips and Company and was destined to be the last vessel of its type on the Chesapeake and one of the last large sailing vessels to carry a cargo on those waters.

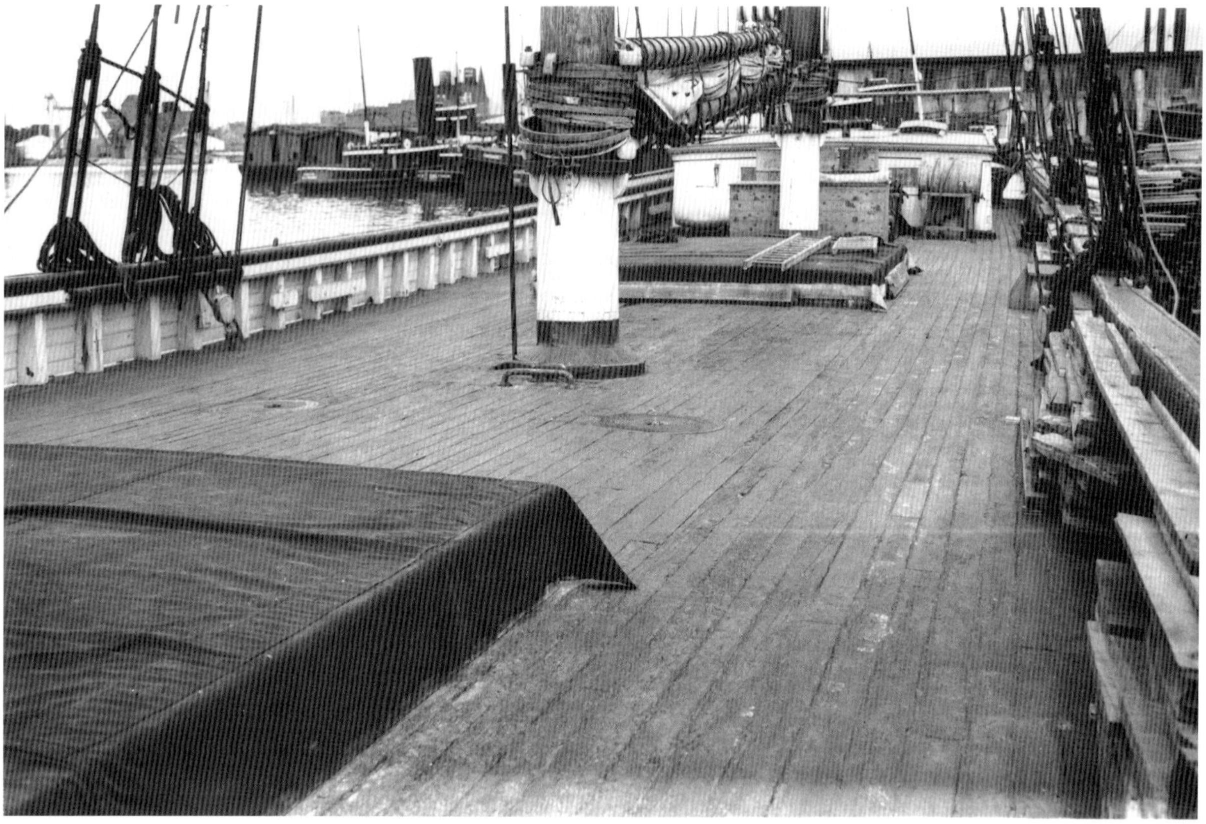

Looking aft on the deck of the *Jennie D. Bell.* Having delivered her cargo, her hold and deck swept clean, tarpaulins over her hatch covers, the *Bell* is ready to head down the Bay again for another cargo. Baltimore's insatiable appetite for lumber kept these old vessels employed steadily.

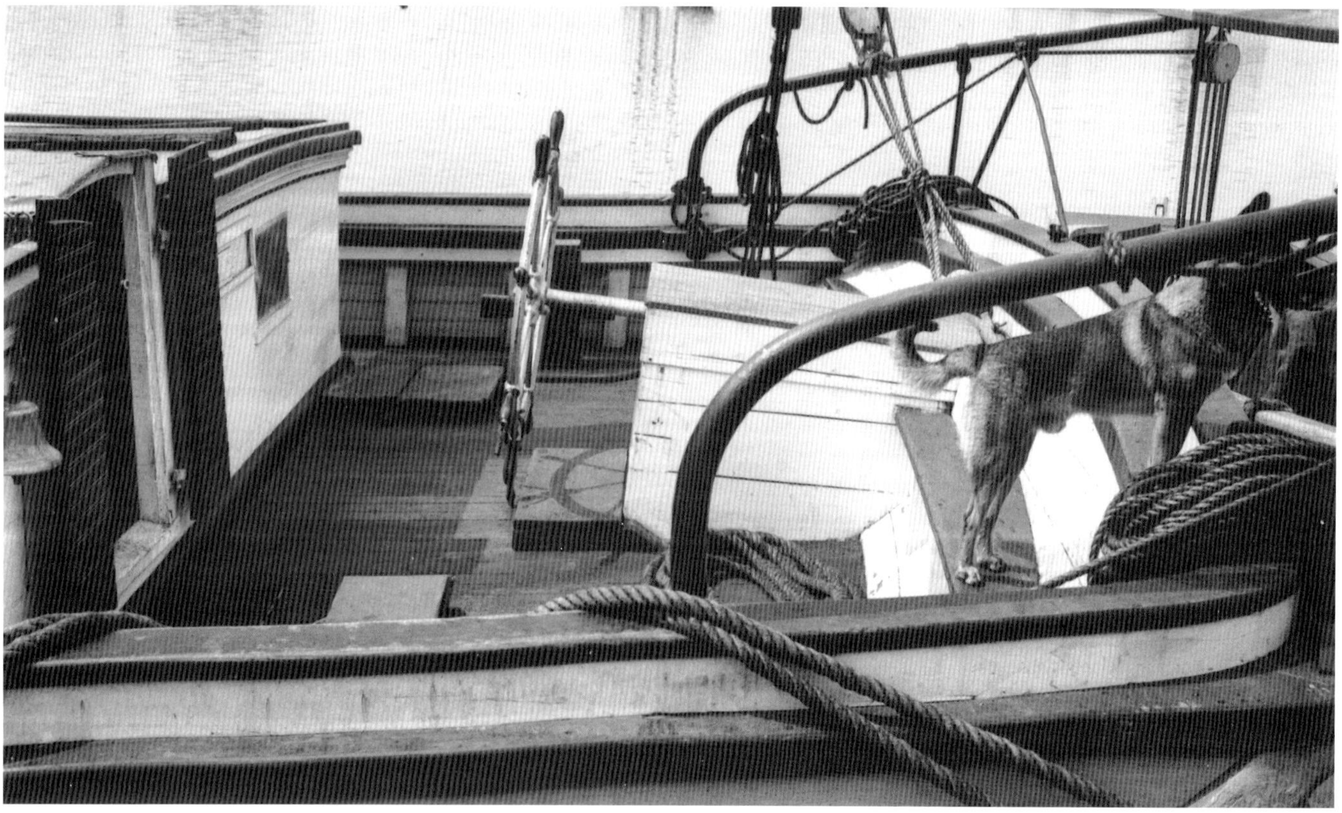

✪ The stern of the *Jennie D. Bell,* on the same day. Her mascot stands the watch. Captain Clarence Heath kept several dogs, and walking on deck could be treacherous.

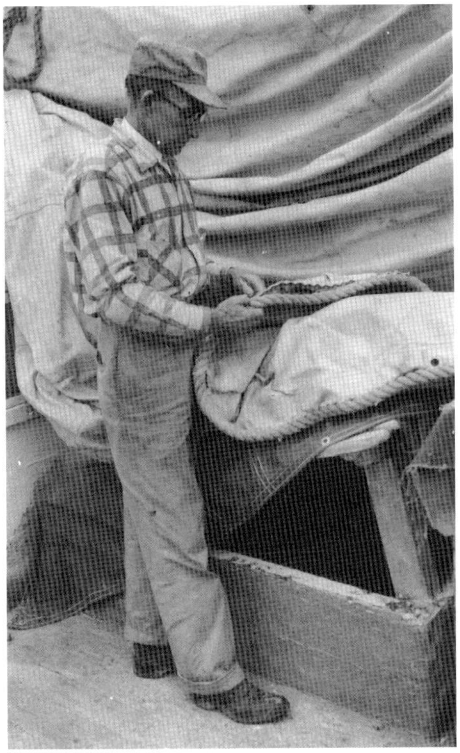

✪ Captain Clarence Heath mends sails on the ram *Jennie D. Bell* in October 1956.

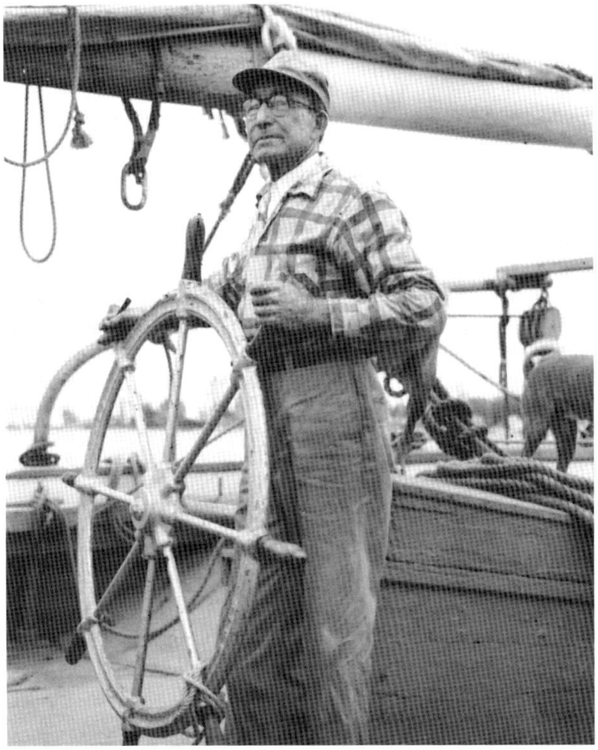

✪ Captain Heath at the wheel of the *Bell* in October 1956.

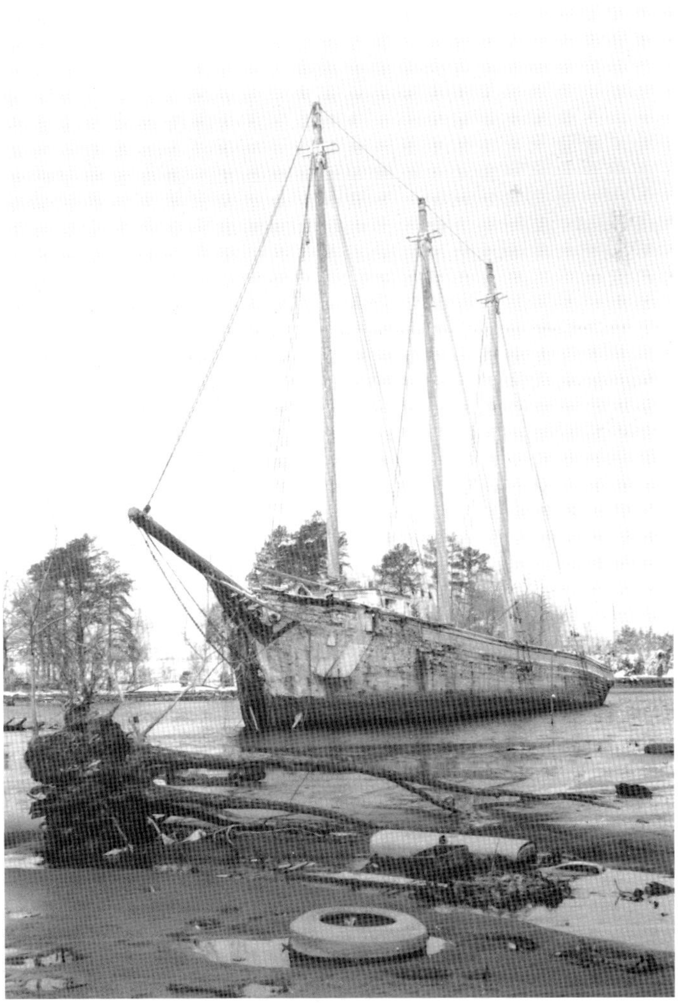

The last time the *Jennie D. Bell* was purposely run aground was on the flats of the southern shore of the Wicomico River at Salisbury, Maryland, in October 1961. That was to prevent her sinking at the Salisbury pier where she had laid since 1954, serving as the home for her owner and master, Captain Clarence Heath, and his wife. And that is where the old vessel died. She looked like this on February 29, 1963.

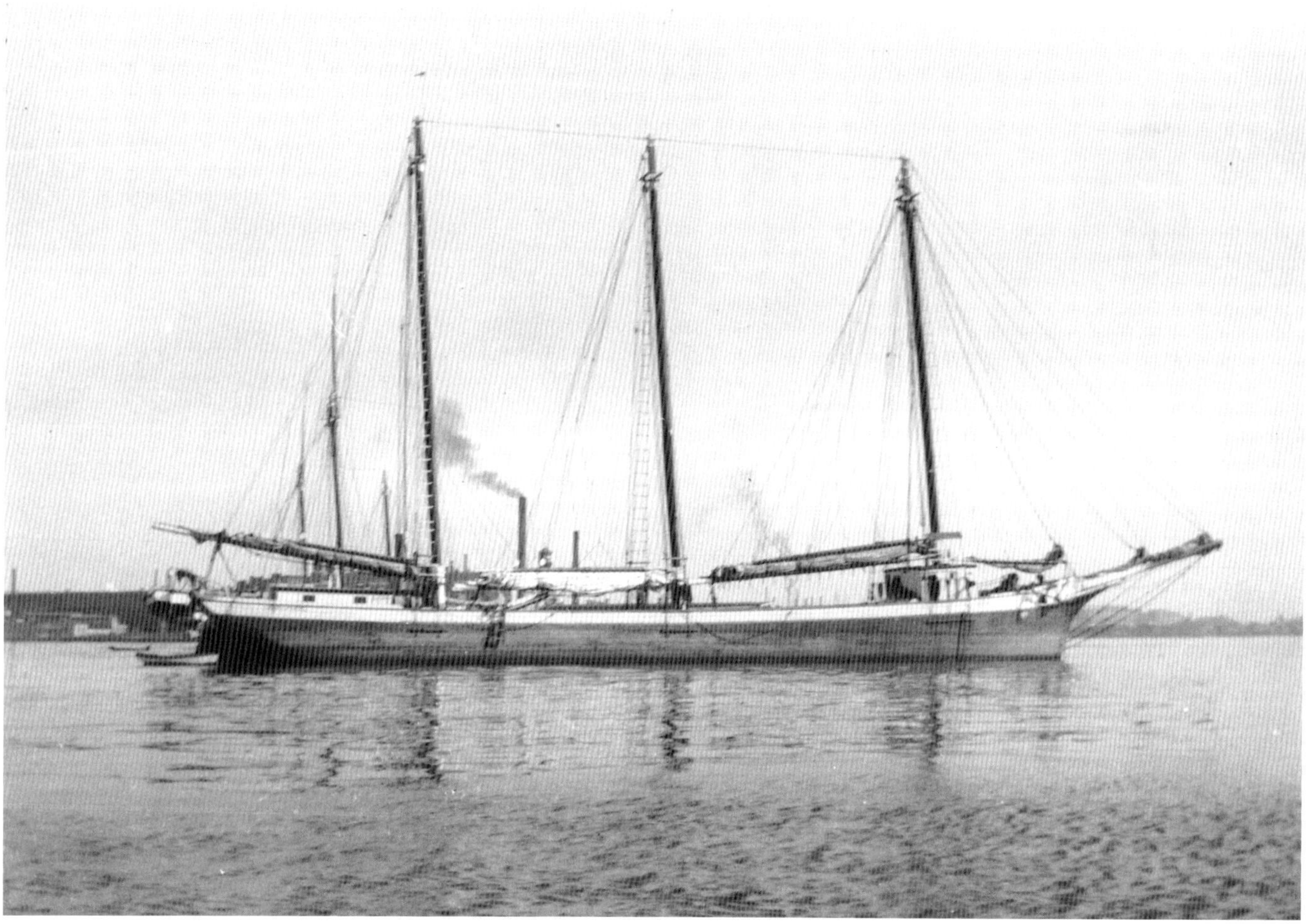

A light wind stirs the water in the foreground, and if it had been favorable perhaps the ram *Levin J. Marvel* would have been preparing to get under way down Baltimore harbor. But it was a headwind, as indicated by the direction of the smoke in the background, so she remains at anchor in Canton Hollow, October 29, 1933.

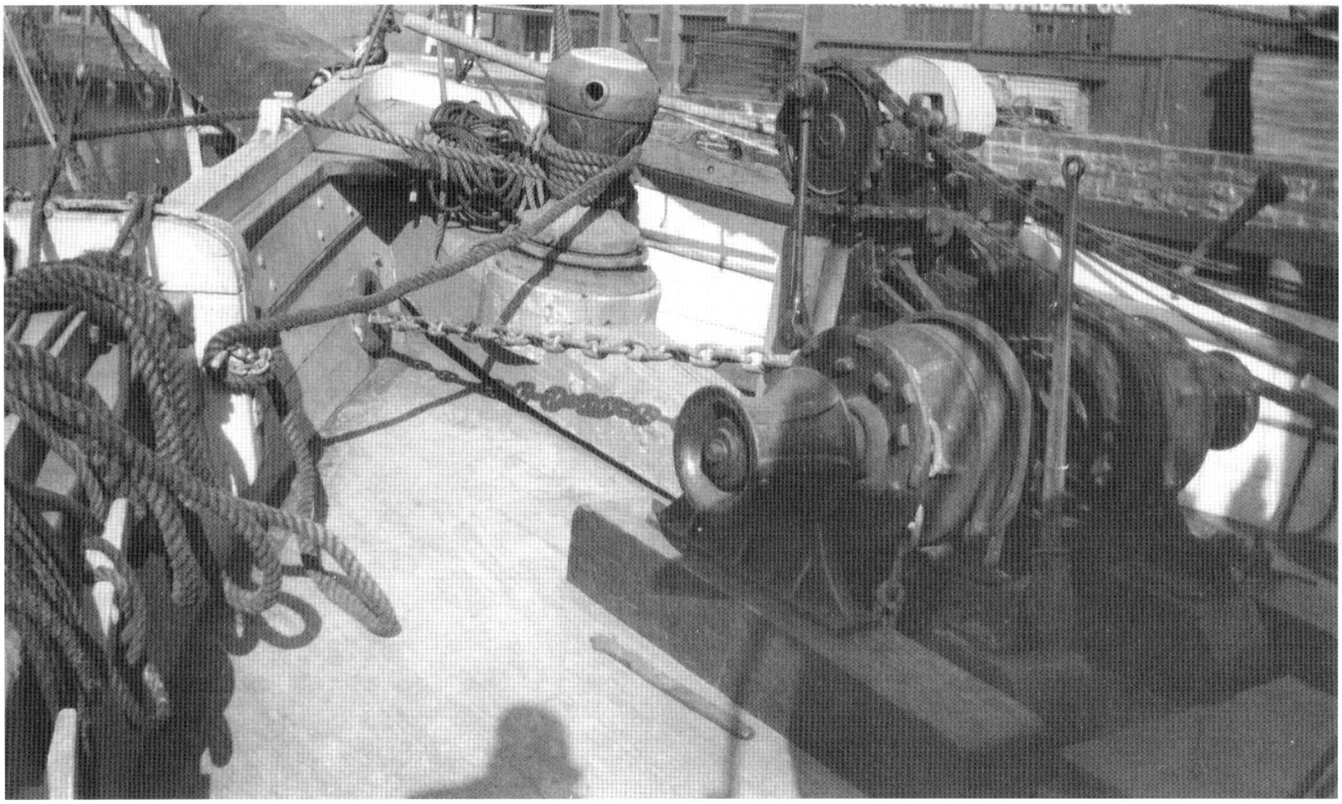

Foredeck of the ram *Levin J. Marvel* at Pier 6, Pratt Street, Baltimore, January 30, 1934. At right is the windlass activated by chain from the donkey engine in the forward house. The anchor chain passes over the barrel of the windlass and disappears through the deck into a chain locker below. The heavy rope mooring line passes through a chock on the port rail and then around the capstan mounted on the heel of the bowsprit.

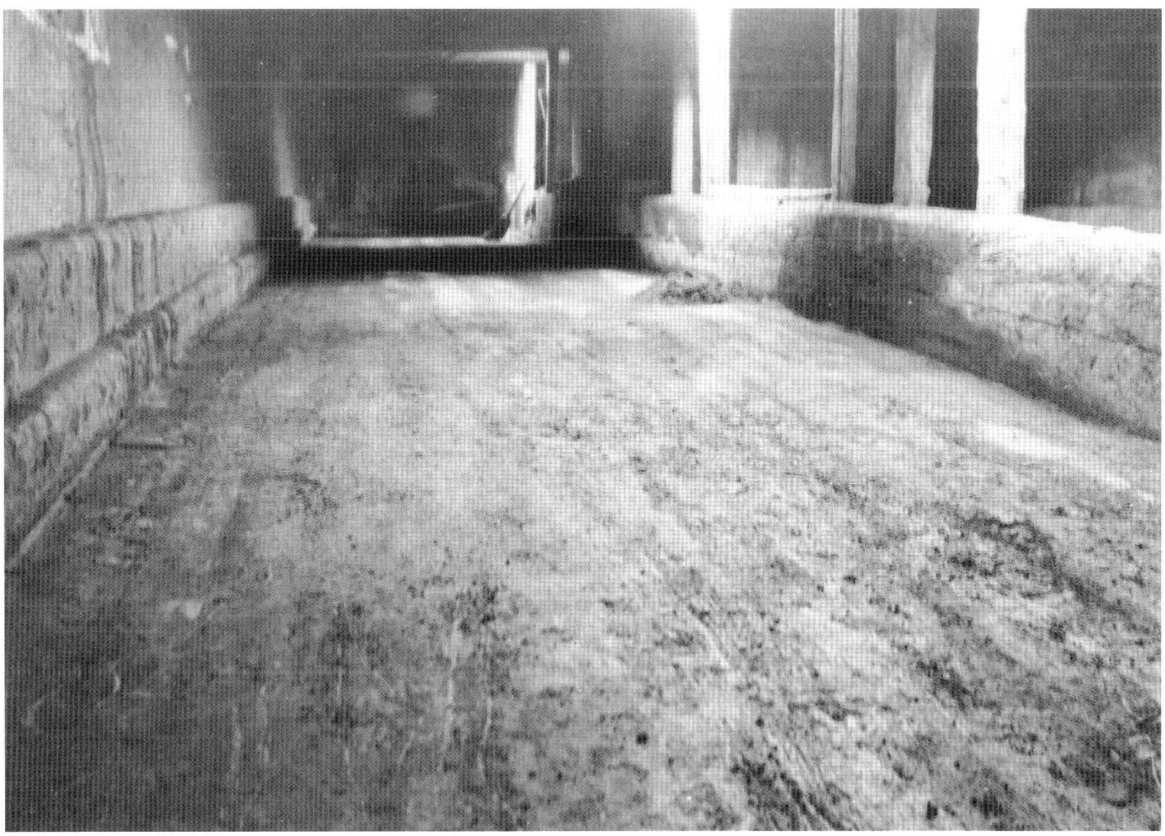

Looking forward in the hold of the *Levin J. Marvel,* which has been swept clean. The flat bottom and wall sides, characteristics of the ram, are most noticeable. At right is the keelson.

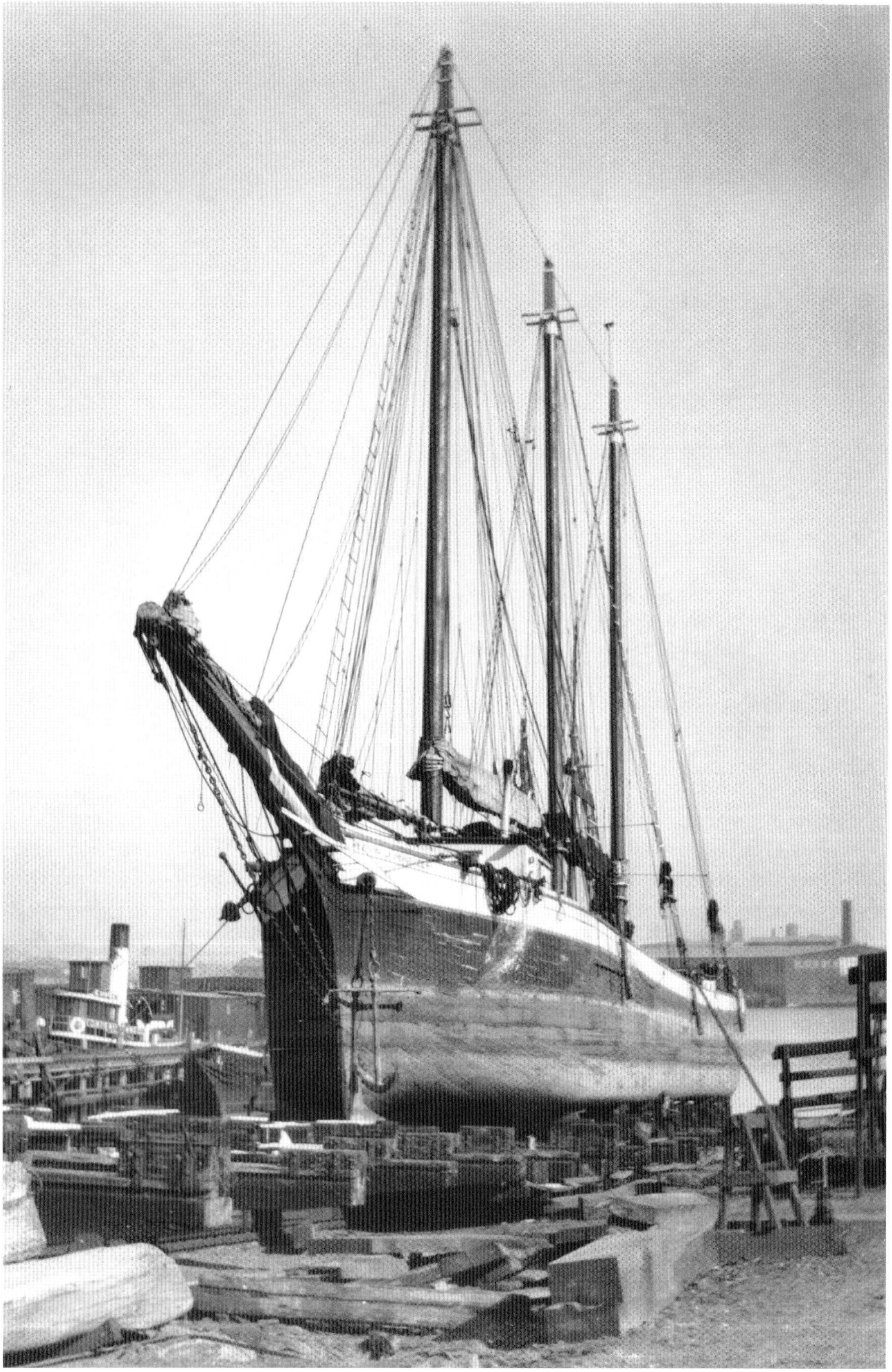

Levin J. Marvel needs little blocking up, other than some wedges under the turn of her bilge, as she sits on the marine railway at the Redman and Vane Shipyard, January 31, 1934. The tug at left is the *Canton* of the Curtis Bay Towing Company. The bow of the stripped schooner *Mary A. Fisher* can be seen across the pier from the tug.

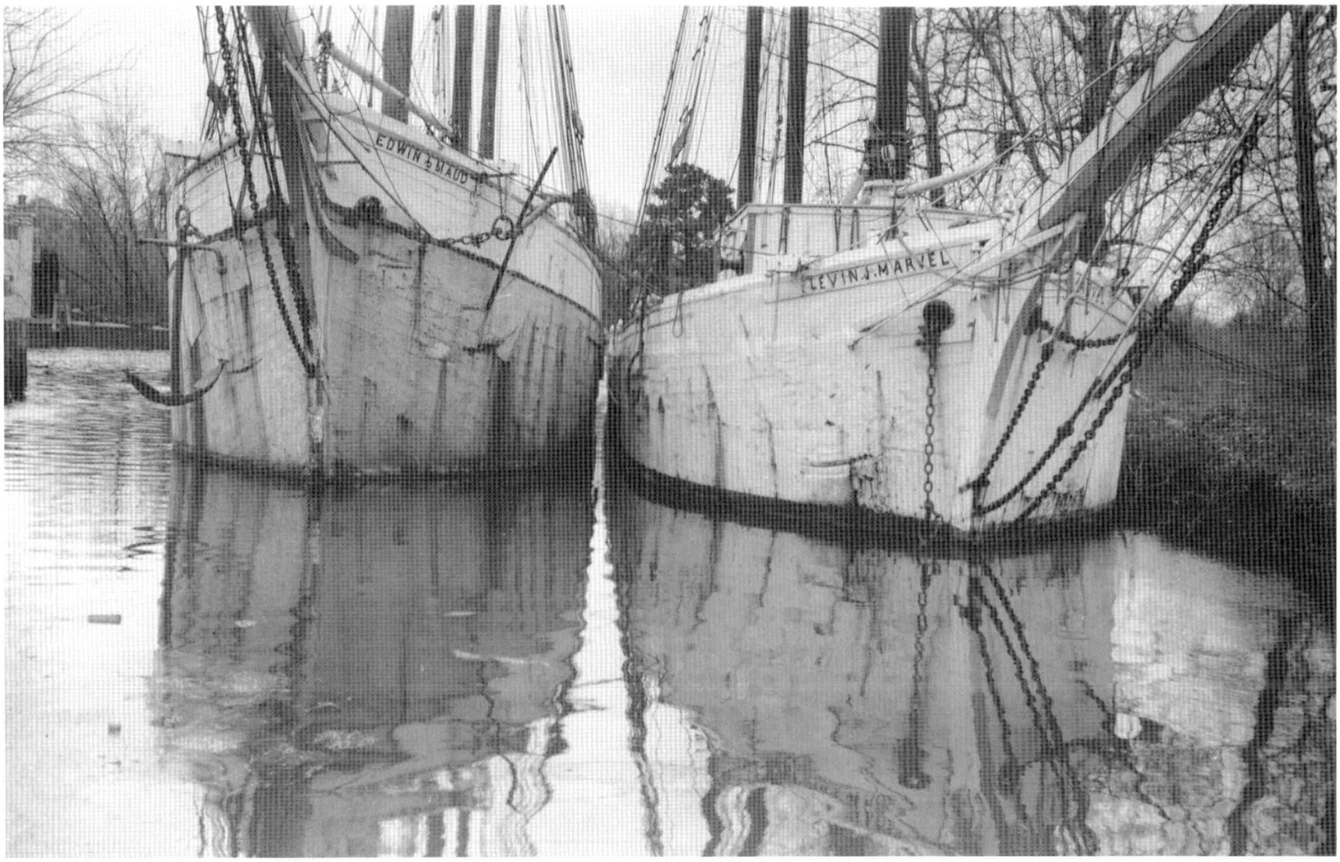

The "dude cruisers" *Edwin and Maud* and *Levin J. Marvel* tied up for the winter at Salisbury, Maryland, January 26, 1952. The *Marvel* was built in 1891 by J.M.C. Moore at Bethel, Delaware. For more than half a century she traded coastwise and on Chesapeake Bay. In 1945, with the *Edwin and Maud,* she was purchased by H.E. Knust and converted into a cruising ship. In 1954 he sold both vessels, the *Edwin and Maud* going to Maine as the *Victory Chimes* and the *Marvel* remaining on the Chesapeake—but not for long. While on a cruise, the *Marvel* encountered hurricane Connie on August 12, 1955, and while anchored off North Beach, Maryland, she was overwhelmed and lost with 14 of her passengers.

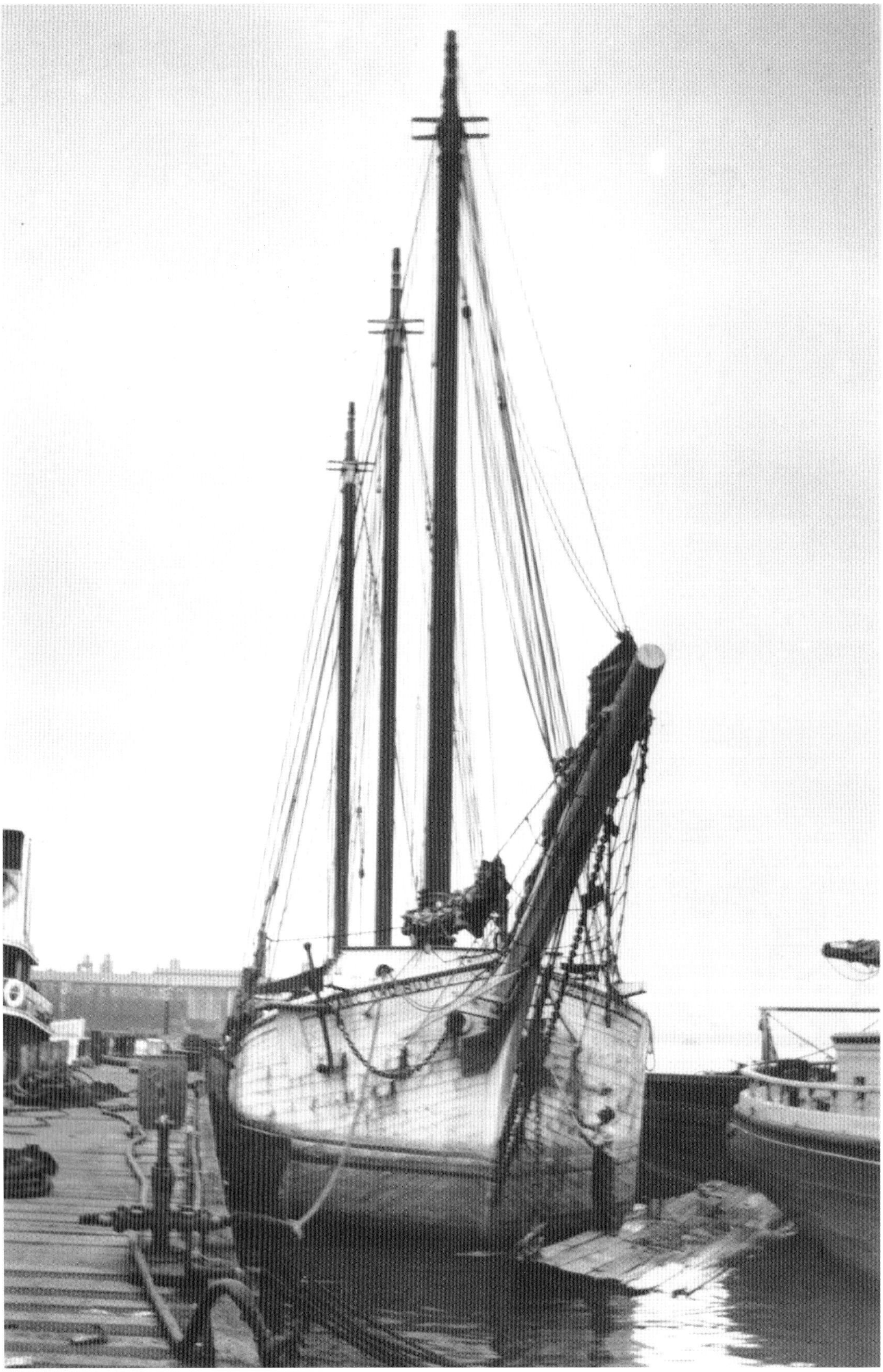

Ram *Mabel and Ruth* preparing for a haul-out at the Redman and Vane Shipyard, July 15, 1933. One of the unattractive features of the ram, the telephone-pole-like bowsprit, can be seen here. Few of the rams made any attempt at a tapered or more graceful spar. In the mid-1940s this ram was taken to Florida, had cabins built below and ports cut into her sides, and was renamed *City of St. Petersburg* to be used as a "dude cruiser" there. Her career in those waters was short-lived, for in 1948 she was abandoned. She had been built in 1896 at Bethel, Delaware, by J.M.C. Moore.

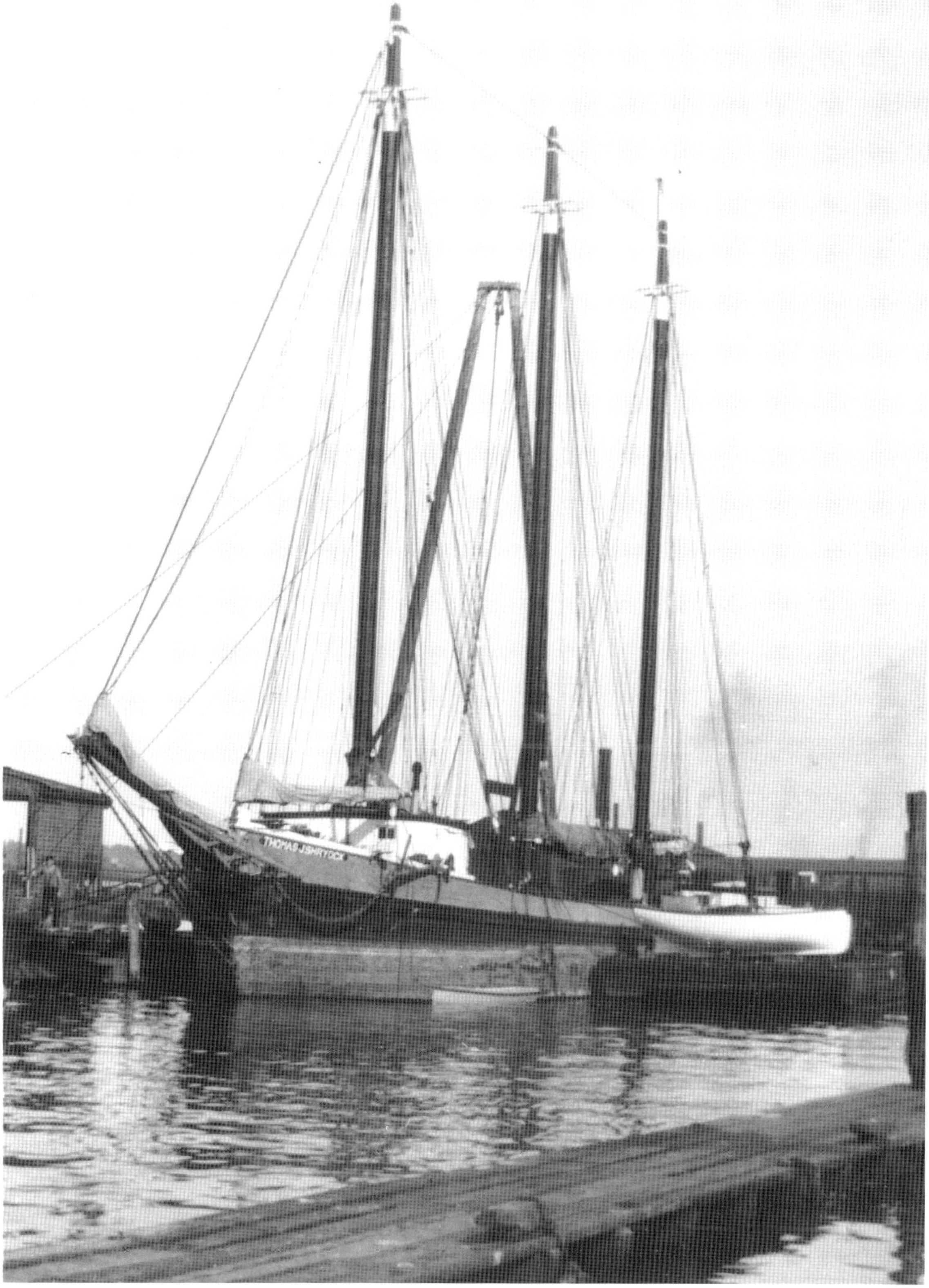

The ram *Thomas J. Shryock* was built in 1891 at Bethel, Delaware by G.K. Phillips and Company. On April 18, 1909, while bound from Baltimore to Norfolk with a cargo of coal, she foundered in Chesapeake Bay with the loss of one life. The vessel was raised and taken to a Bethel shipyard where she spent the entire summer being repaired and overhauled. In this view she is in a shipyard again—Redman and Vane's, Baltimore, on October 14, 1933—but only for a haul-out. Her shapely yawl boat had been placed on a scow for repairs and repainting at the same time and revealed a hull to be admired.

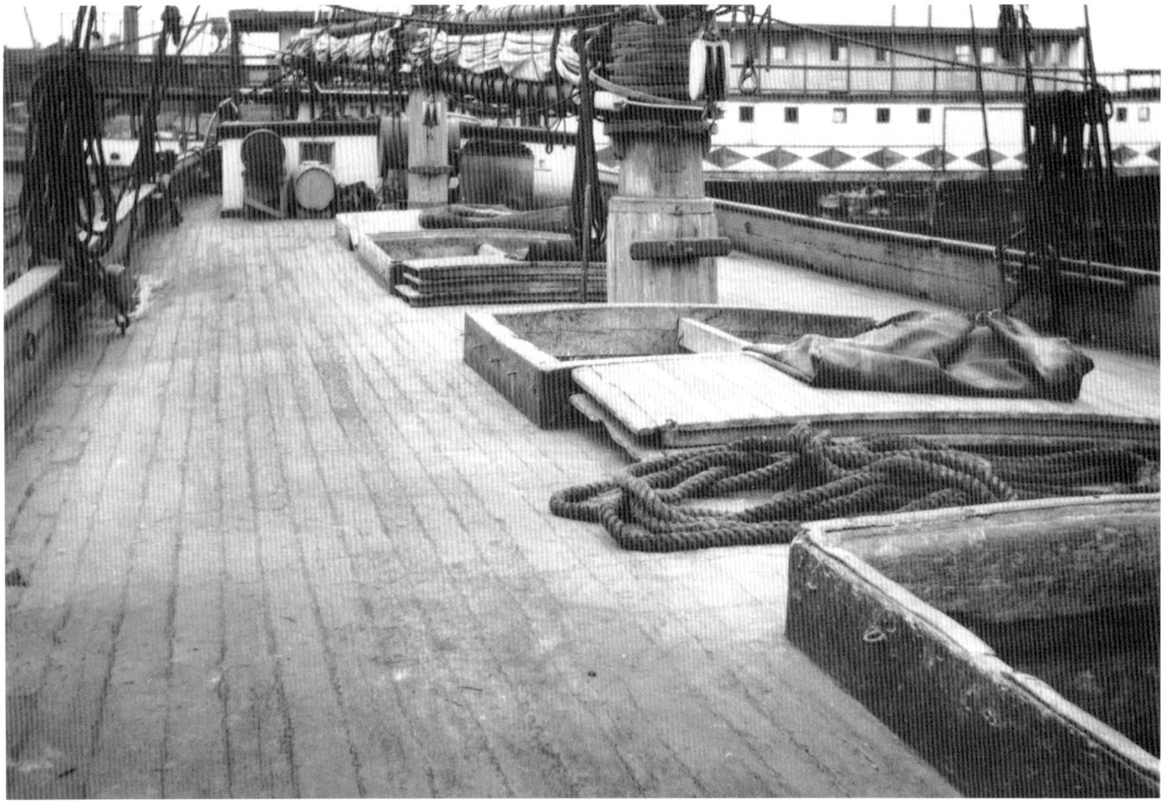

Looking aft on the *Thomas J. Shryock.* In the background is the concrete-hulled Community Cheer Center used as a refuge around Baltimore harbor during the Depression. The *Shryock* was wrecked on North River Bar, North Carolina, in the fall of 1940 and was later towed to Elizabeth City, North Carolina, and abandoned.

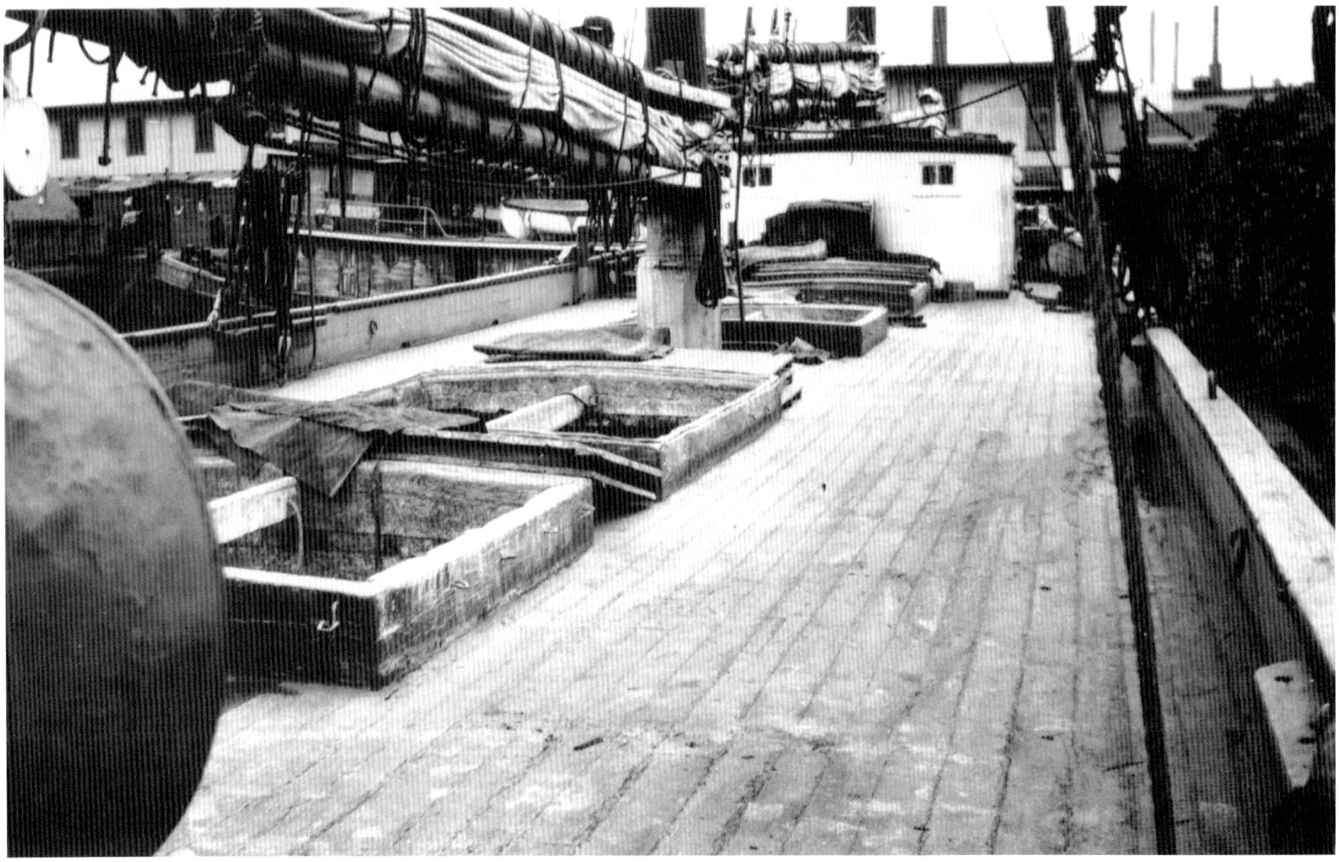

✪ Looking forward on the deck of the *Shryock,* at Baltimore, on October 14, 1933.

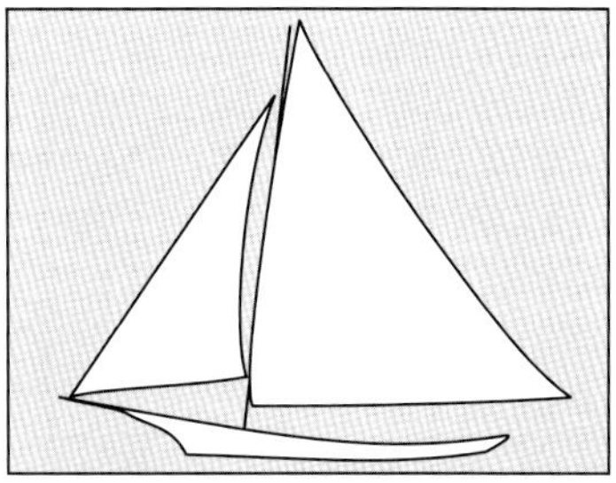

Skipjacks

The only sailing craft hulks found littering Chesapeake backwaters today are those of skipjacks that have served their time in the Bay's oyster dredging fleets, exhausted themselves structurally beyond repair, or were set aside during a dull season with the hope of being revived the next. That never occurred, and the only commercial sailing vessels active on the Bay, about 30 in number, are those skipjacks still licensed to dredge over Maryland's oyster beds. The roster grows smaller each year as the picturesque craft join their sisters on the mud banks.

Today, skipjacks are all the rage. They are paraded before the public as if they were the only commercial sailing craft to have ever sailed these waters. Two races are held for them each year, one off Deal Island on Labor Day, sponsored by the Deal Island-Chance Lions Club; the other off Sandy Point, above Annapolis, on the last weekend in October, just before the dredging season opens. The latter event is sponsored by the State of Maryland and the Windjammers, a sailing group centered around the Baltimore-Annapolis area. The Sandy Point event has been combined with Chesapeake Appreciation Day.

Skipjacks have visited Baltimore harbor as goodwill representatives of Maryland's sailing oystermen. In June 1972, a fleet of five spent a week in Washington, D.C., as part of the Folklife Festival saluting the State of Maryland. They make the headlines today for what was part of their everyday routine three and more decades ago. A skipjack, the *Joy Parks,* has been preserved at the Harry Lundeberg School, Piney Point, Maryland, and the city of Baltimore recently purchased the *Minnie V* as its very own to tie up on the waterfront near the frigate *Constellation.* No doubt the Chesapeake Bay Maritime Museum, St. Michaels, will eventually add one to its collection of floating exhibits. Steps are being taken to establish a skipjack museum at Deal Island, with plans to move a skipjack ashore and cover it with a housing for exhibit.

Perhaps the skipjack deserves some of this adulation, since it constitutes the last commercial sailing fleet in this country. It affords the observer an opportunity to see a true working vessel under sail. A great part of the population has never seen, and probably never will see, any other type of vessel that uses the wind to make a living.

The skipjack was the last of the commercial sailing types developed on the Chesapeake, at a time when oyster catches were diminishing and there was less need for large boats. Easier and cheaper to build and man, these craft filled a need just before the turn of the century and have been carrying on that tradition to the present day.

It is believed the skipjack developed from the earlier enlarged types of flat-bottomed and V-bottomed skiffs once so common and popular on the lower Eastern Shore of Maryland. It seems logical that this would have been the chain of development, but there were steps between the first of those craft and the skipjack as we know it today. The early boats were not flush-decked but, instead, had a "sunk-platform" where hand winders for bringing in the dredges were located. In the first decade of this century the flush deck was generally adopted due to the increased use of the power winder. The heavy winders, putting strain on masts and rigging, also resulted in the boats being fitted with shrouds.

How the term "skipjack" was derived or when it was first applied to the craft is not definitely known. In the area from Oxford to Deal Island the boats are known as bateaux, and the rig, skipjack. The typical skipjack with jib and mainsail is referred to as a two-sail bateau. The three-sail bateau has the rig of a bugeye and was once found in the dredging fleets.

Most of the skipjacks were built on the Eastern Shore, from Talbot County, Maryland, down to Accomack County, Virginia; few were built on the Western Shore of Maryland and Virginia. The last flurry of skipjack building occurred after World War II when several of the vessels came off the ways: the *Helen Virginia* at Crisfield in 1948; the *Caleb W. Jones* in 1953 at Reedville, Virginia, and the *City of Crisfield* and *Somerset* in 1949 at the same port; the *H.M. Krentz* at Harryhogan, Virginia, in 1955; and the *Rosie Parks* and *Martha Lewis* that same year at Wingate, Maryland. The last built was the *Lady Katie* in 1956, also at Wingate. It is unlikely that others will be built for commercial use. Those still in service will be repaired as long as feasible or until the state of Maryland changes its oyster laws.

Skipjacks never seemed to fit into any trade other than oyster dredging, except for seasonal use in transporting watermelons, hay, or tomatoes. Their cargo-carrying capacities were limited and they were unsuitable as strict powerboats. Some have been converted into yachts and though they appeared satisfactory for awhile after the conversion, they did not linger in the yachting fleets very long. Most of them require excessive rebuilding or altering and they look strange after a large house has been built on them to serve as a cabin.

In the 1960s some beverage firms chose skipjacks to help promote their bottled goods. Dredging craft were completely rebuilt, cabins erected amidships, and the craft followed the Chesapeake yachting circuit making themselves prominent. Sails or hulls were emblazoned with identifying marks so that viewers would be conscious of the products that supported the vessels' presence.

Most of the skipjacks retain an air of the past with the use of deadeyes and trailboards. On some the deadeyes have been replaced by turnbuckles. Trailboards are appreciated more now than they were years ago; the carvings are kept goldleafed and freshly painted, and some are even removed from the vessels when they are laid up after the dredging season. Old, battered boards are removed and replaced by new ones, carved usually by former dredger Dewey Webster, of Deal Island. A few skipjacks have even added quarterboards and stern decorations.

Innovations like depth finders, ship-to-shore phones, and propane gas stoves have appeared in the fleet. Part of the charm of the vessels years ago, on a still morning at a mooring, was the aroma of a wood fire from the stove in a skipjack's cabin as the cook prepared a substantial breakfast, a necessity for the men who still work as hard as their forebears, emptying dredges, culling, and shoveling oysters on deck, during the coldest part of the year, from November 1 through March 15.

Though the boats in the dredging fleet today are relatively few, conditions aren't all that bad judging from the prices the oystermen get per bushel. Sometimes it is difficult to imagine how hard times can get in that fishery but one has only to look back to 1938, for instance. Up to the end of November in that year's dredging season, only four dredging licenses had been issued. Maryland state law gave the dredgers a 30-day period of grace from November 1 to get licenses. This was to permit the dredgers to market their take and obtain money for the licenses, rather than call on them to advance money before they started dredging.

Reasons for a bad start that year were the abnormally warm weather and rock-bottom prices. Then there were more than a hundred skipjacks, bugeyes, schooners, and sloops in the fleet, and a third of them were anchored in Bodkin Creek to await the lifting of fog and a chance to sell the oysters in their holds. And 1938 was the year mussels plagued the oysters, impeding their growth and reducing the catch from 30 to 40 percent. Prices ranged from 30 to 40 cents per bushel but packers refused to buy the oysters because of the clusters of mussels. Oysters sold from 50 to 55 cents per bushel the previous year. At those prices the oystermen said they couldn't afford to buy licenses, let alone feed the crew and maintain the boat. The skipjacks and other dredging craft lost one day in two while cleaning off the mussels. It was reported that the demand for hatchets, used to hack mussels off oysters, exhausted the stocks of Annapolis stores.

Continuance of a bad season was reflected in a mishap that occurred within the dredging fleet in the Choptank River on February 3, 1939. As the boats were preparing to return to Cambridge after a day's dredging off Howell's Point, a sudden storm overwhelmed the craft. Three of the boats capsized, the skipjack *Annie Lee,* and bugeyes *Nora E. Lawson* and *Agnes,* and nine men were drowned. Among those lost was Captain William Bradford of the *Agnes,* aged 77, said to have been the oldest captain in the oyster dredging fleet. There have been dismastings, strandings, and fires onboard the skipjacks, but that was the last time there was such a large loss of life at one time on the dredgers.

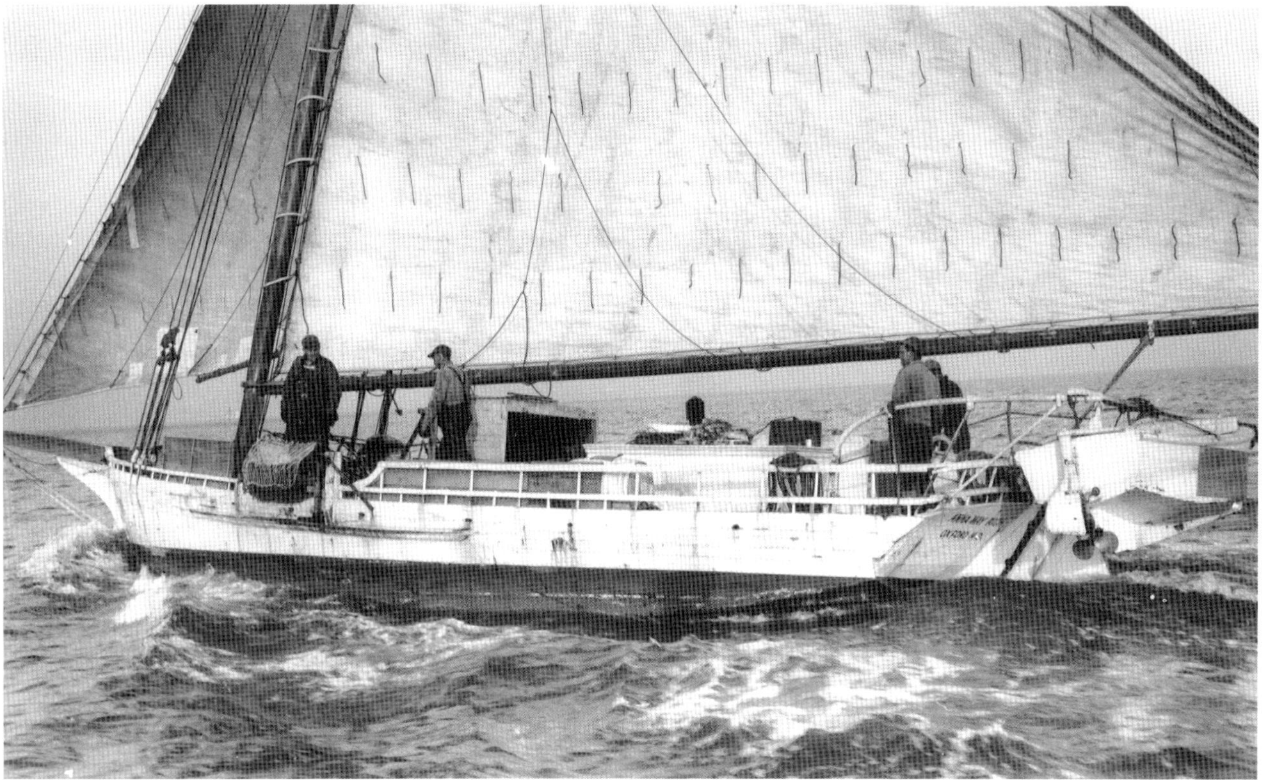

Skipjack *Anna May Rich* needs no assist from her yawl boat to get out to the oyster grounds. On the early morning of February 19, 1948, a rising wind sends her scudding across the Choptank River as she leaves Oxford astern. The port dredge is poised on the roller ready to be thrown over the side and pulled over the bottom by its steel cable when the skipjack arrives over the dredging area.

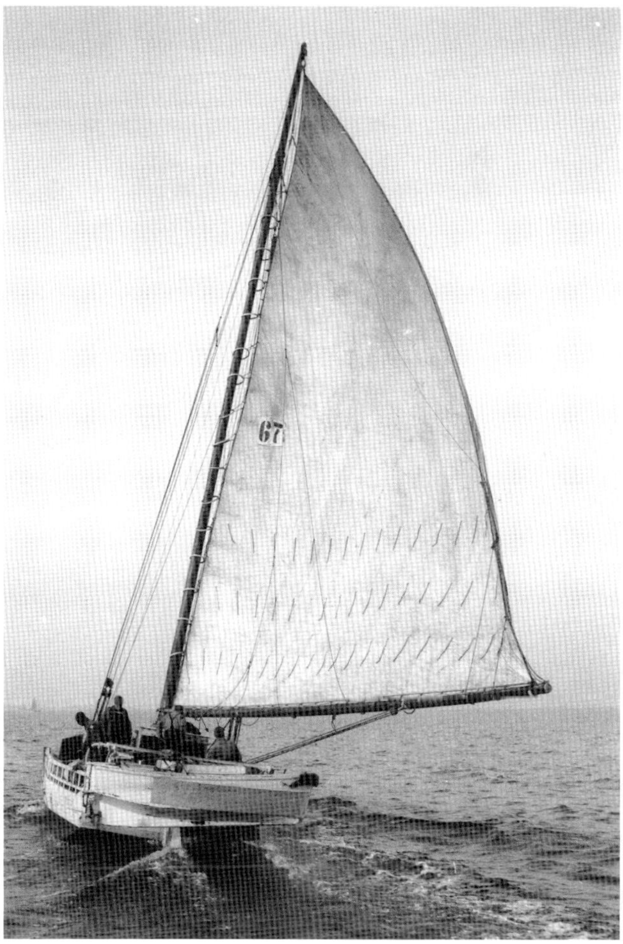

Sails on the horizon in left background indicate the Choptank dredging grounds the *Anna May Rich* steers for. Note the metal sheathing around the sides of the yawl boat for protection against the ice even then in Town Creek, Oxford. The *Rich* was built in 1908 at Oriole, Maryland. ✪ She was reported as "Abandoned" in the 1976 *MVUS.*

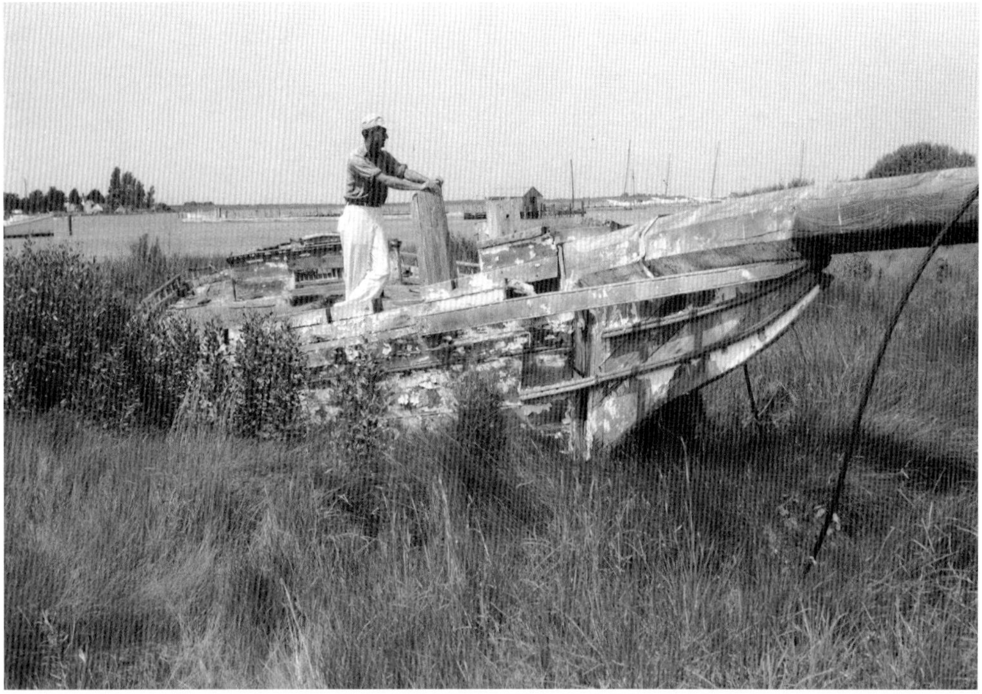

The *Catherine* was built at Pocomoke City, Maryland, in 1901 and did not quite make a half century on the Chesapeake's dredging grounds. She lies surrounded by marsh grasses at Ewell, Smith Island, Maryland, in 1951. Although recorded as "Abandoned" in *MVUS* for 1942, she held together rather well during the intervening years. ✪ The vessel was visited by the author and friend Fred Kaiser, shown on deck, during a cruise of the Chesaeake aboard Kaiser's two-masted schooner *Lady Churchill.*

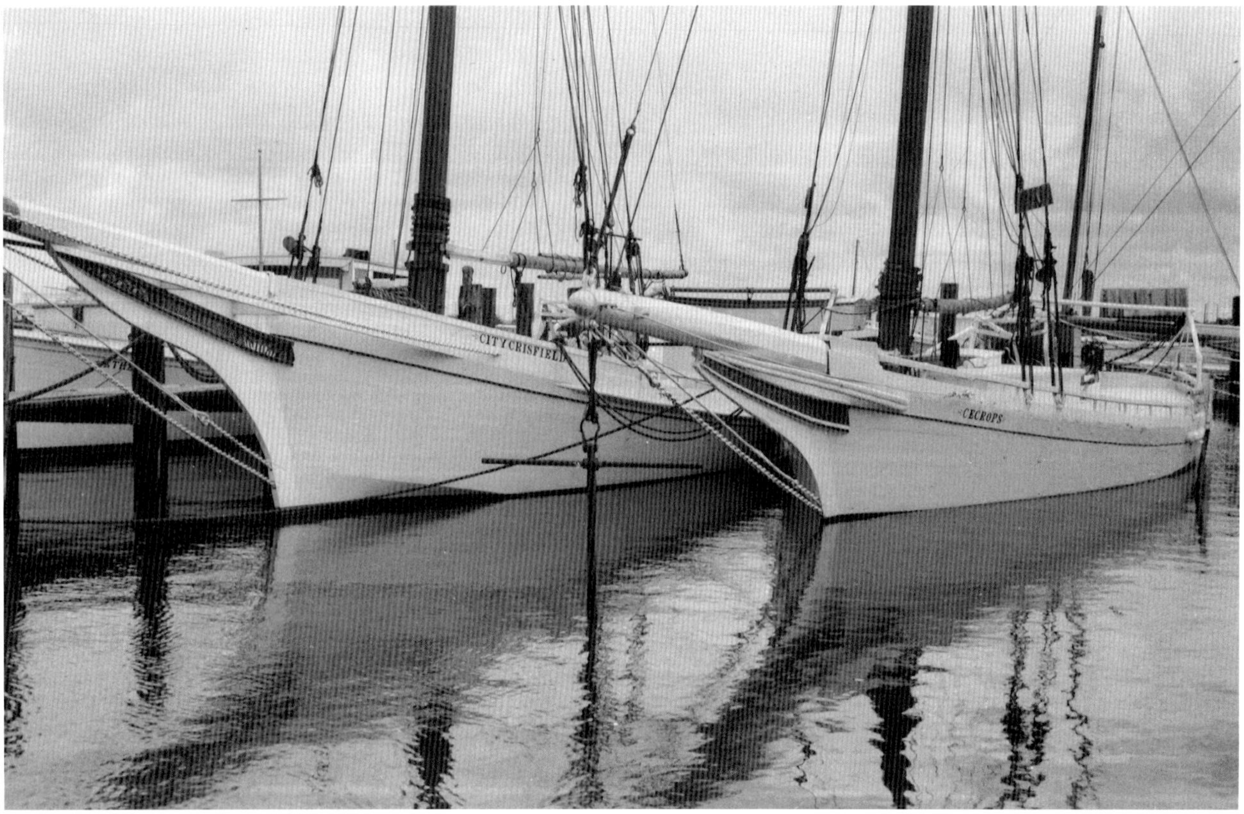

Just about a half century separates these two skipjacks, *City of Crisfield,* left, and the *Cecrops,* the former built at Reedville, Virginia, in 1949 and the latter at Hooper Island, Maryland, in 1900. For appearance, however, the *Cecrops* would win this author's vote. Note the awkward curve in the clipper bow of the *City of Crisfield* compared with the graceful curve in that of the *Cecrops.* ✪ The *City of Crisfield* was still active in 2005, but the *Cecrops* was reported as "Abandoned" in the 1978 *MVUS.*

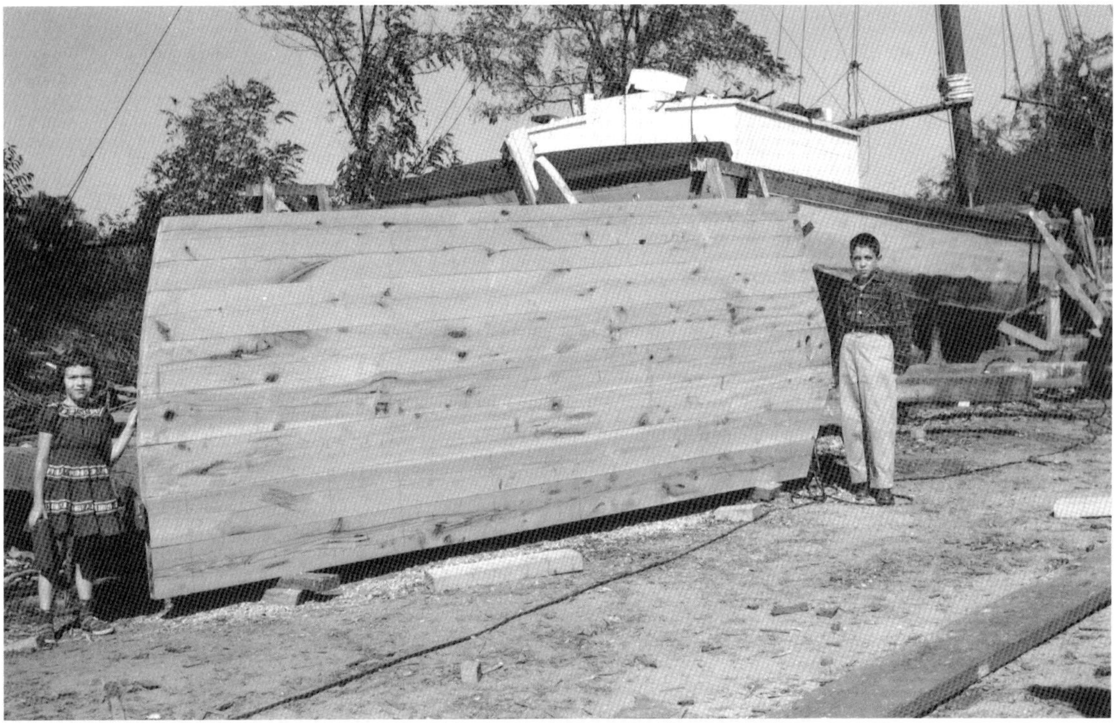

As the *Cecrops*, background, undergoes a rebuilding at the Krentz Shipyard, Harryhogan, Virginia, in the summer of 1954, her new oak centerboard is ready for painting and installation after the skipjack is waterborne again. Its size can be determined by comparing it with the author's son, Bruce, and daughter, Janet.

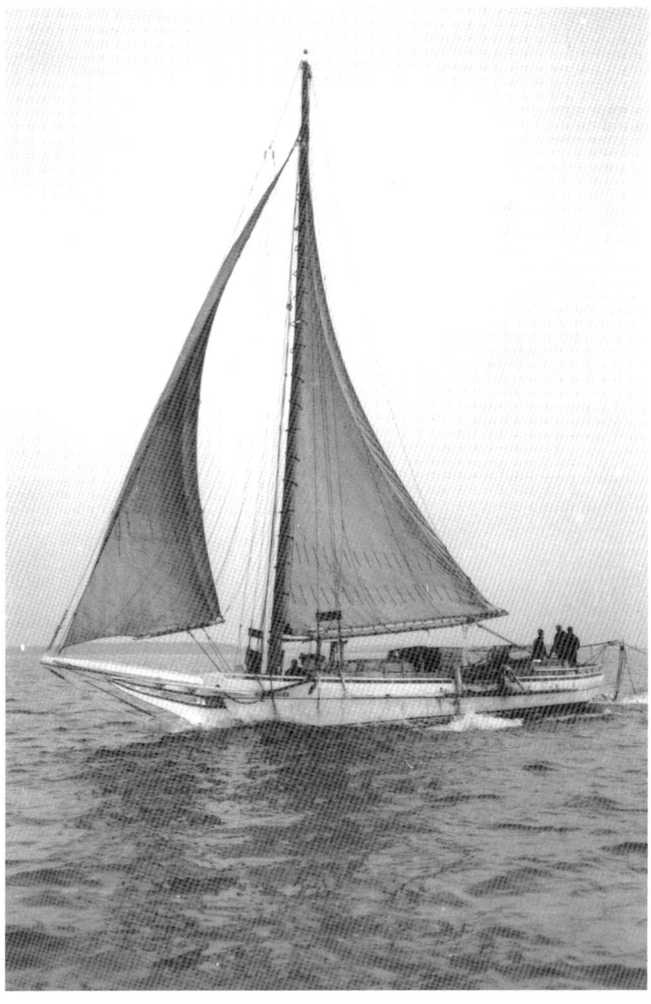

E.C. Collier sails and is pushed out of the Tred Avon River from Oxford for the Choptank oyster dredging grounds on February 21, 1948. ✪ Built at Deal Island in 1910, this veteran has been a land exhibit at the Chesapeake Bay Maritime Museum since 2001.

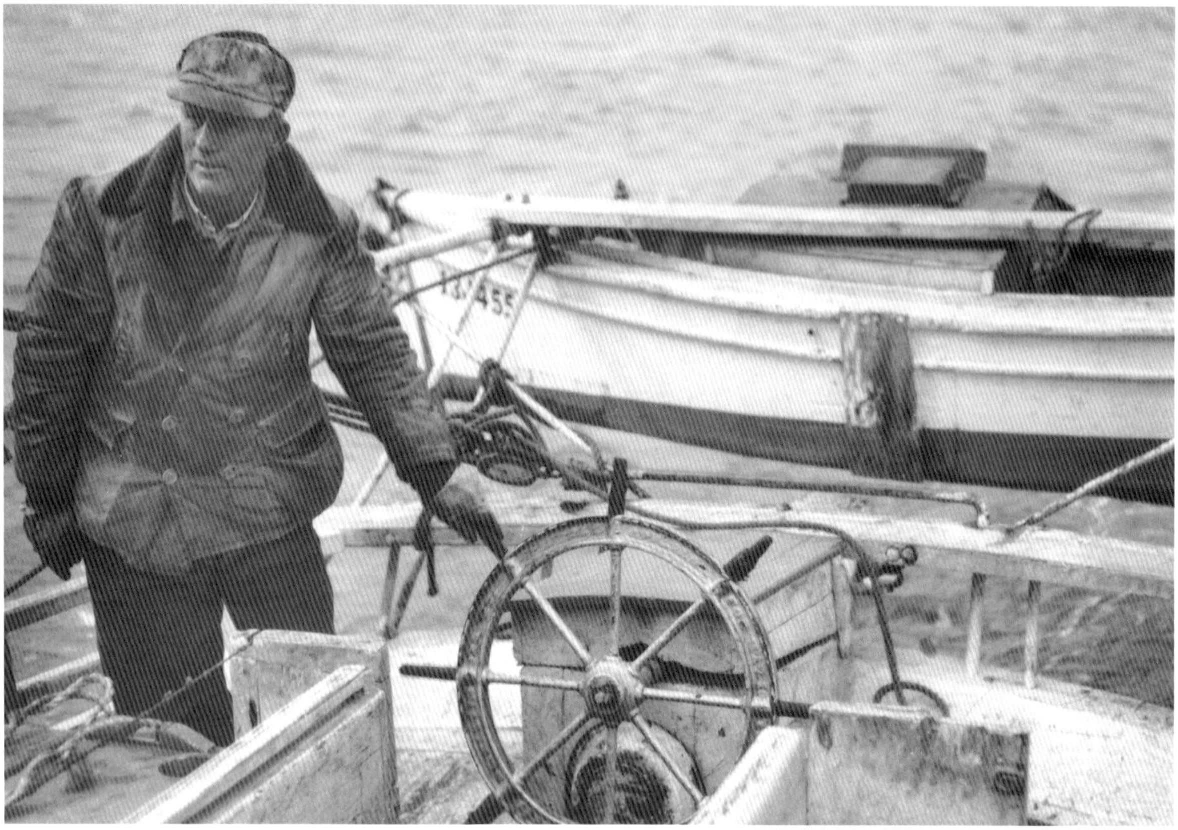

✪ Gordon Pope on the skipjack *E.C. Collier* in February 1948.

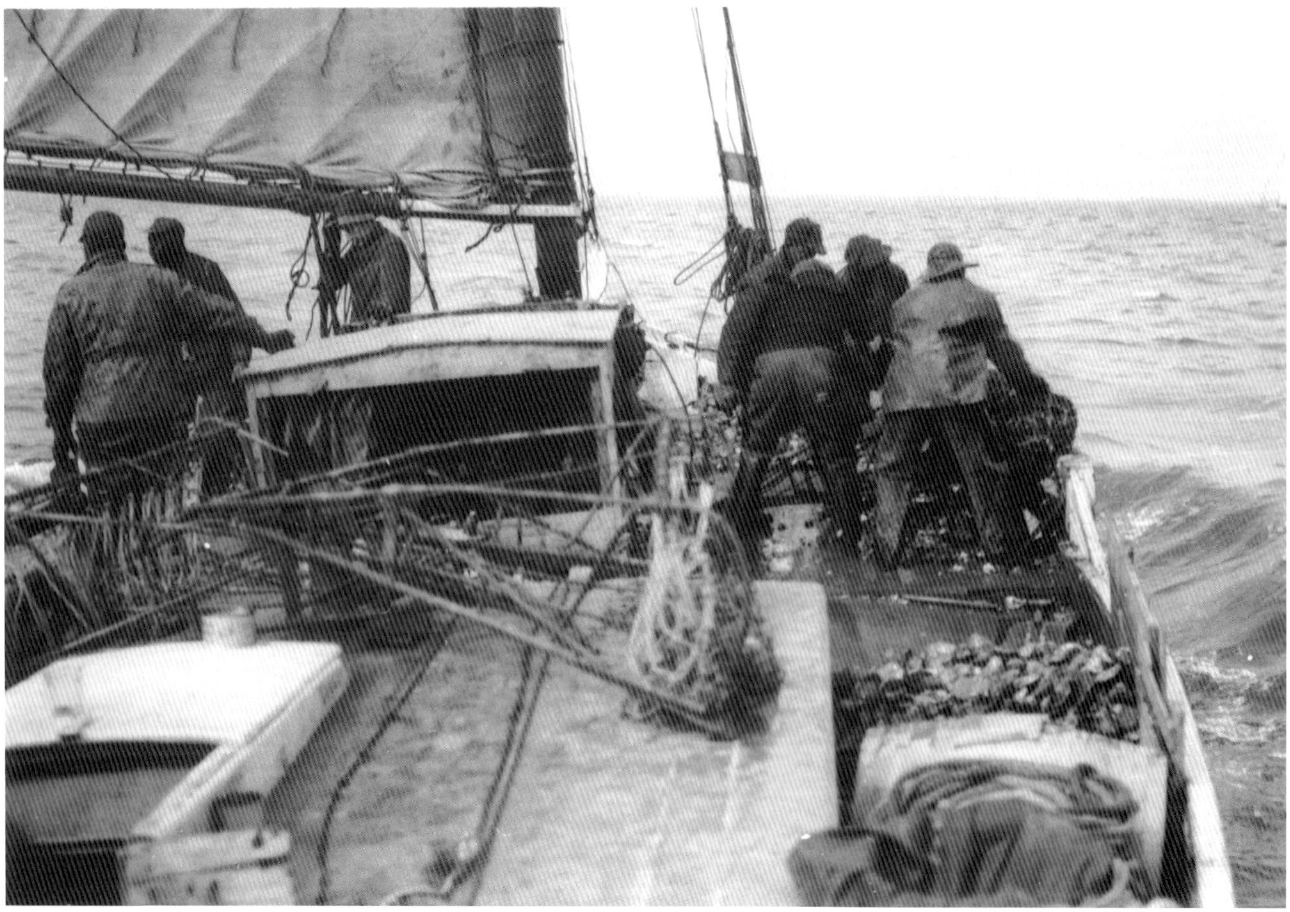

✪ The crew of the *Collier* hauls the dredge aboard.

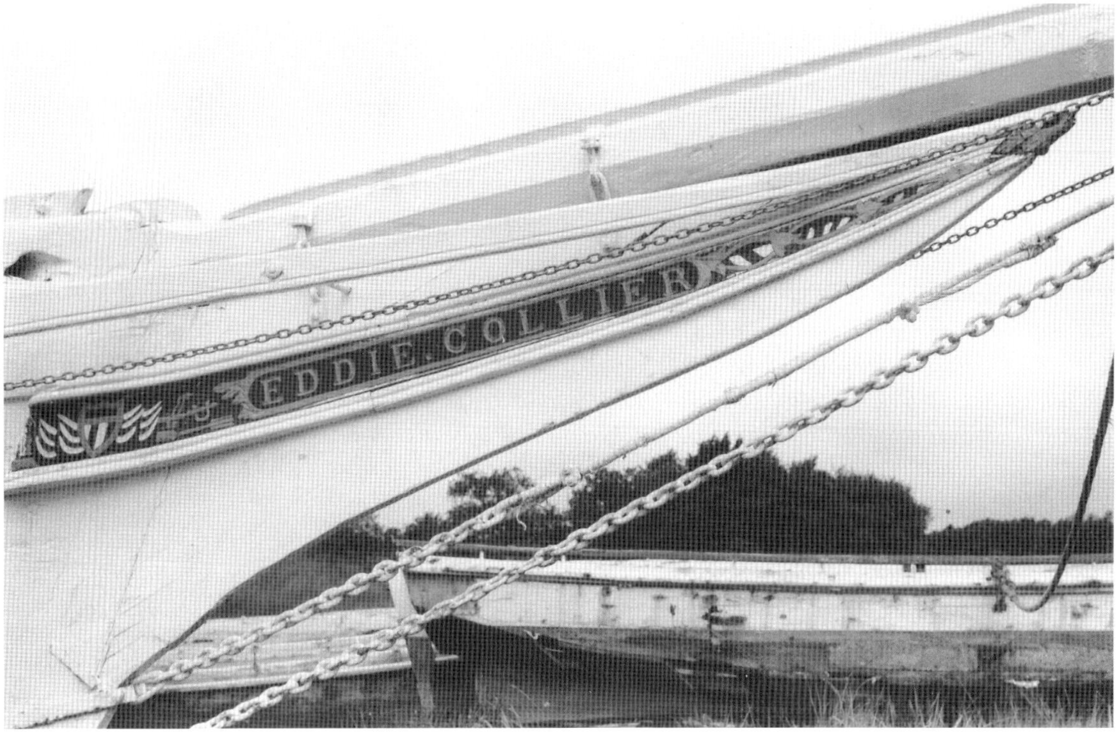

Eddie Collier, freshly painted and sitting on blocks in a marsh near Chance, Maryland, displays her trailboards and eagle figurehead. Built at Deal Island in 1902, that skipjack had been led by these carvings over Chesapeake waters for more than half a century when this scene was taken in 1960. But in the 1969 *MVUS* she appears "Out of Documentation."

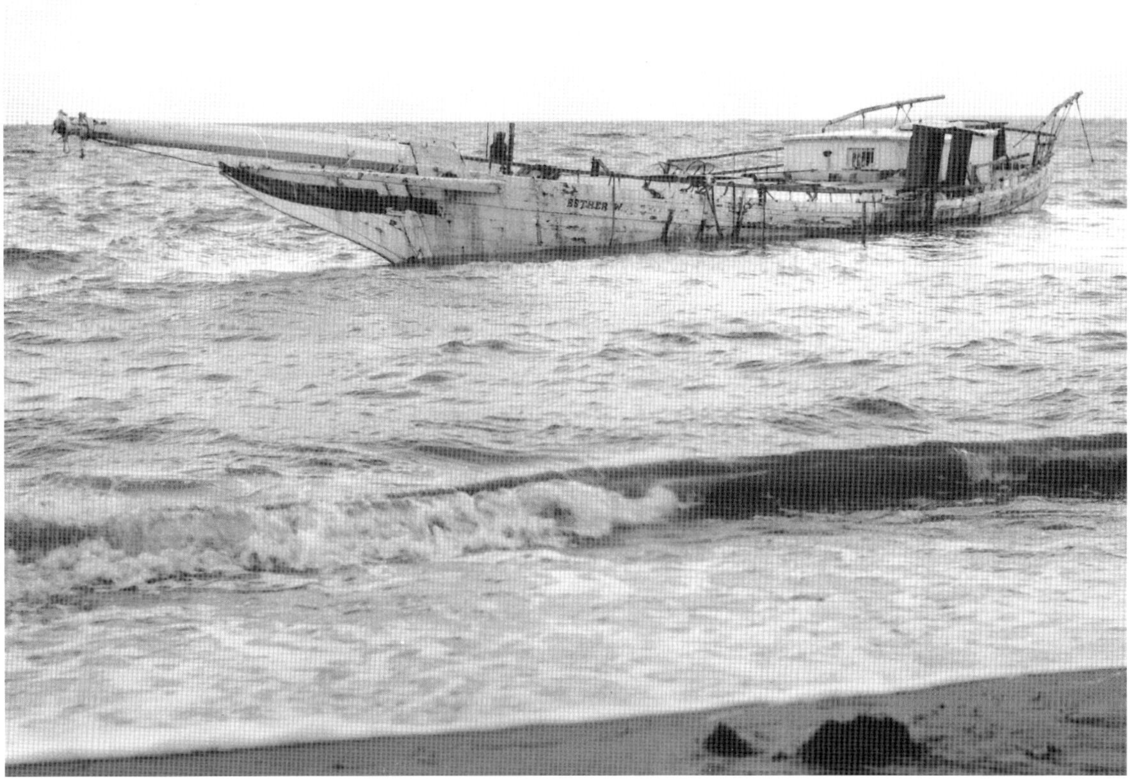

The *Esther W.* had outlived her usefulness as an oyster dredger but still found work to do if only as a breakwater on the Tangier Sound side of Deal Island, October 27, 1959. Stripped of everything useful to other skipjacks in the Deal Island fleet, *Esther W.* awaits battering by the strong northwest winds that sweep across the Sound in the wintertime. They soon reduced her to fragments and left the beach in lower foreground exposed to further erosion. She served well from the time she had been launched at Oriole, Maryland, in 1914.

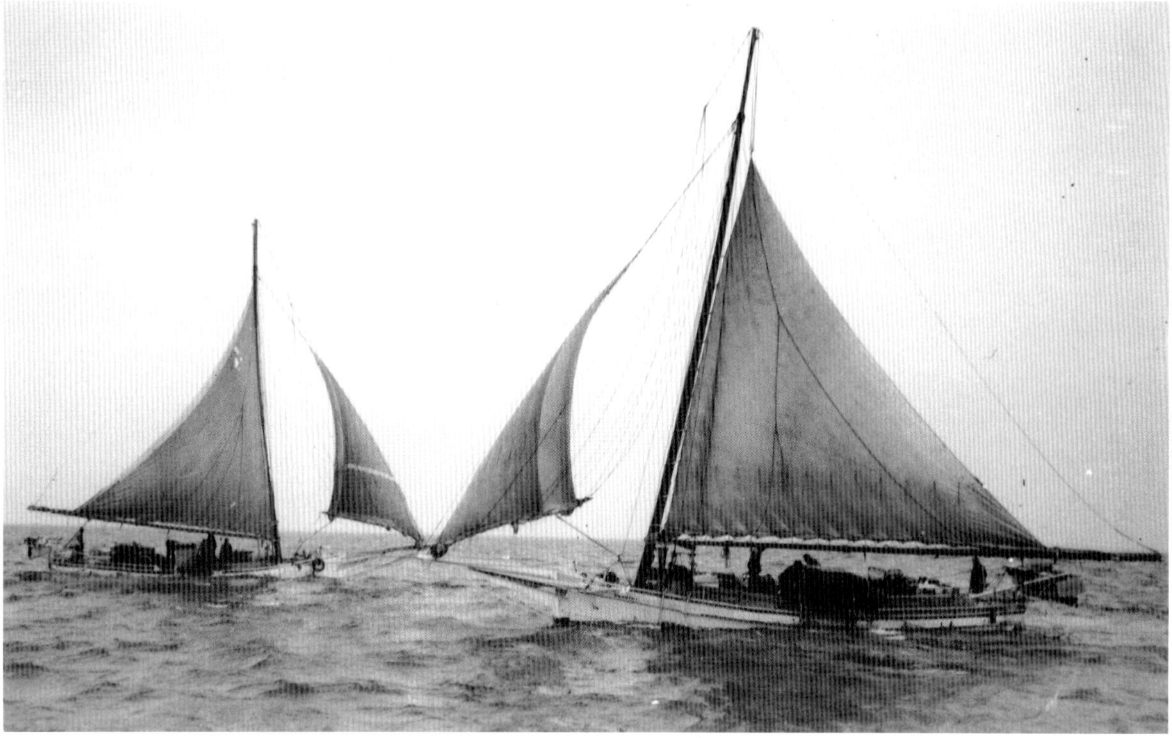

One not familiar with the principles of sailing would have difficulty in understanding how two sailing craft using the wind from the same direction could be sailing in opposite directions. A beam wind with proper trim of the sails permits these skipjacks to sail parallel courses as they dredge in the lower Choptank River on February 19, 1948. At left is the *Ethelyn Dryden,* with the *Dorothy.* The latter was built at Deal Island, Maryland, in 1905. ✪ She was reported as "Out of Documentation" in the 1980

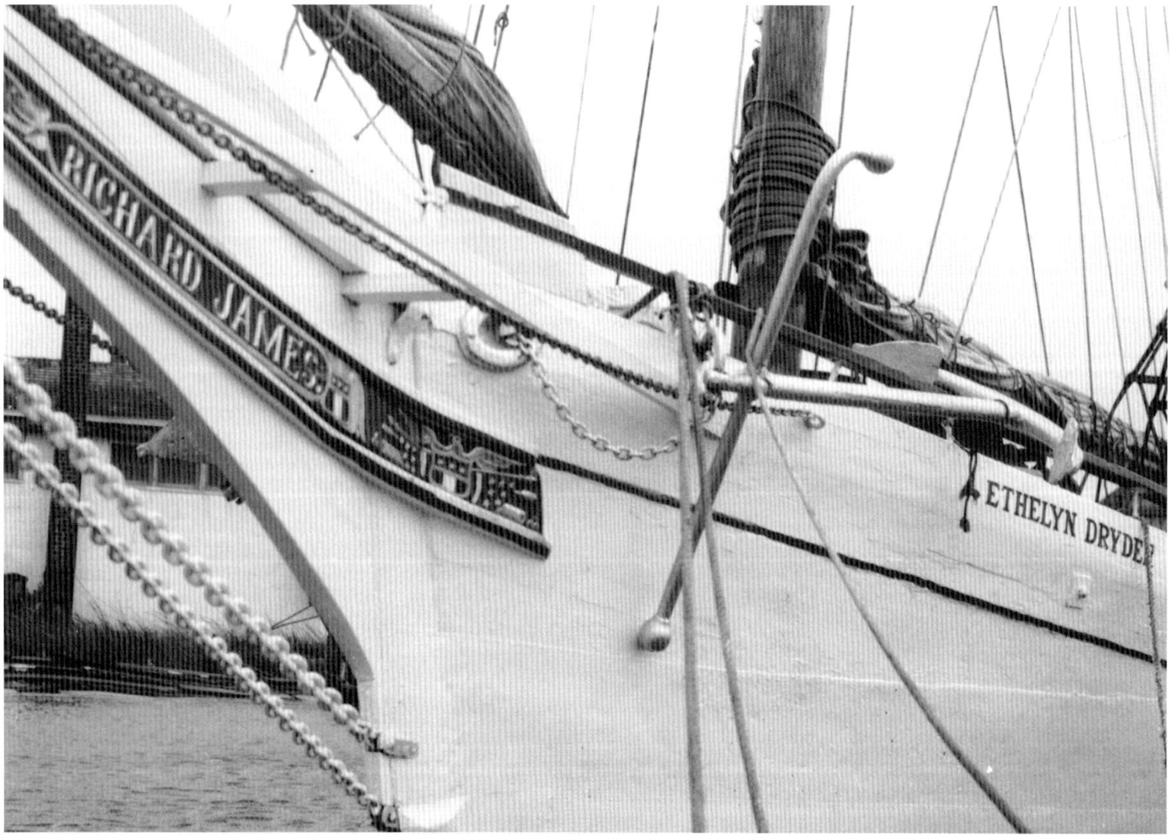

One rarely sees a skipjack bearing two names on her bows at the same time. Here in the early 1950s the *Ethelyn Dryden,* built in 1901 at Pocomoke City, Maryland, also carries trailboards with the name *Richard James.* But the latter name was not official. Her owner wanted to name the vessel after his two sons. The 1968 *MVUS* carries her as "Abandoned."

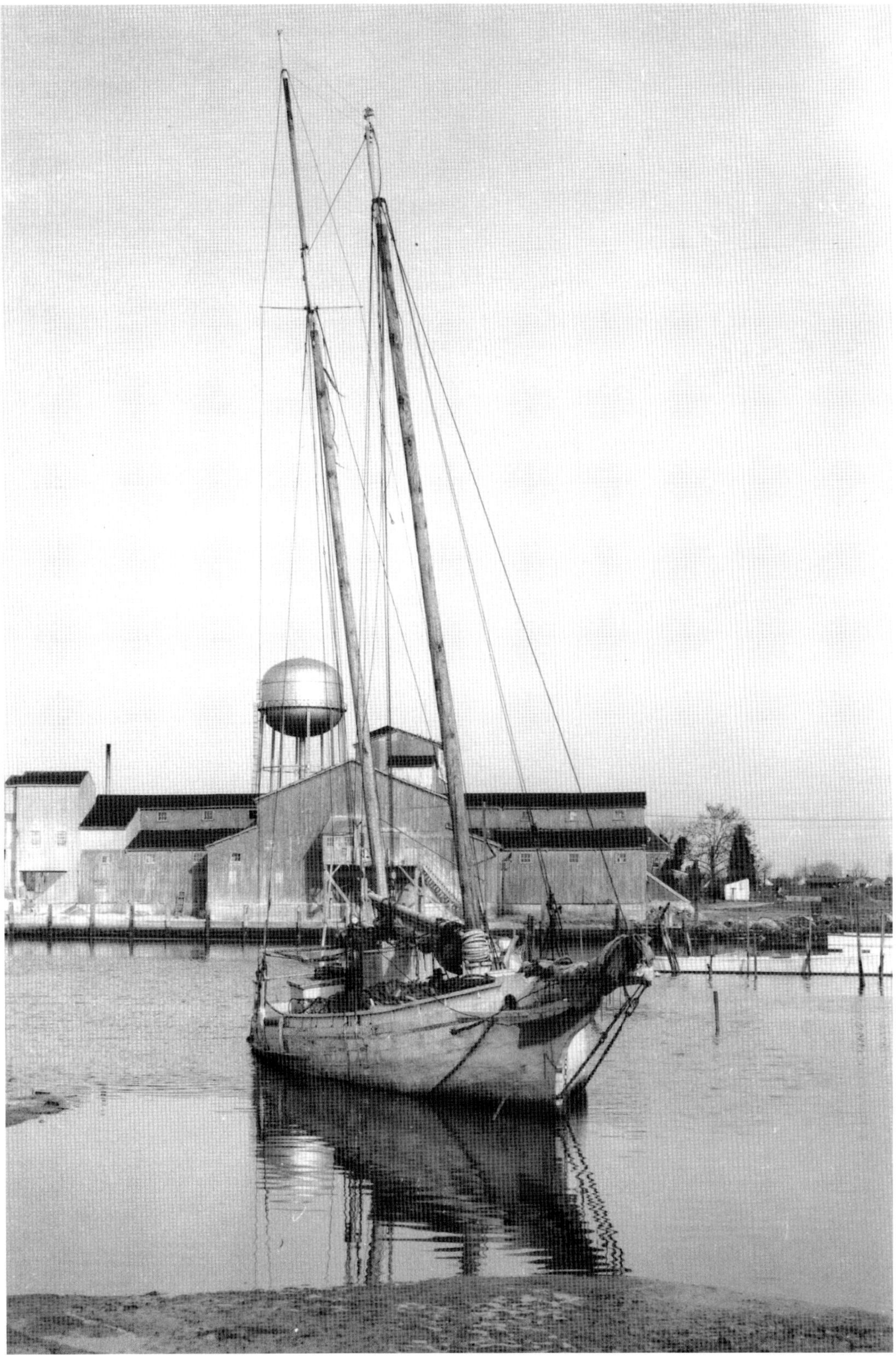

The 39-foot schooner-rigged deadrise hull *Eva* rests on the beach in the upper reaches of Cambridge Creek on January 30, 1949, after her captain died. She, too, died here, ending a career that started in Cambridge in 1890. At least, that is the date *MVUS* carries as her building date. According to H.I. Chapelle, "The owner of this bateau, Captain Charles A. Le Compte, claims this boat was the first 'bateau' or 'skipjack' built in Dorchester County. She was built prior to the date given in the Custom House Register, 1890, as the present owner (1944) bought her in 1888. She was twice rebuilt, once about 1890 when her rig was changed from three-sail bateau to schooner and again, about fifteen years later, when her stern was lengthened three feet to form a counter. She was built from scrap timber left over from the schooner *Rover,* which was built at Cambridge by Johnson in 1883."

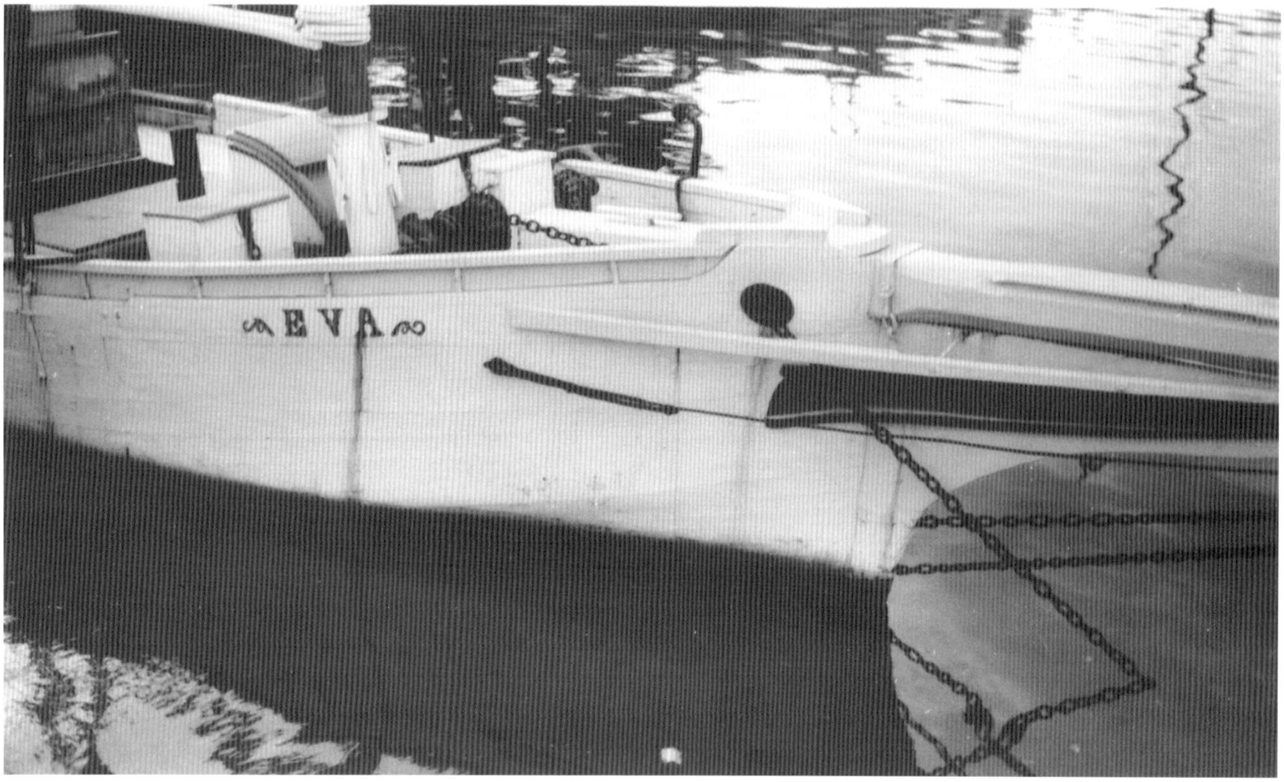

✪ The bow of the *Eva,* at anchor in Cambridge, in happier days.

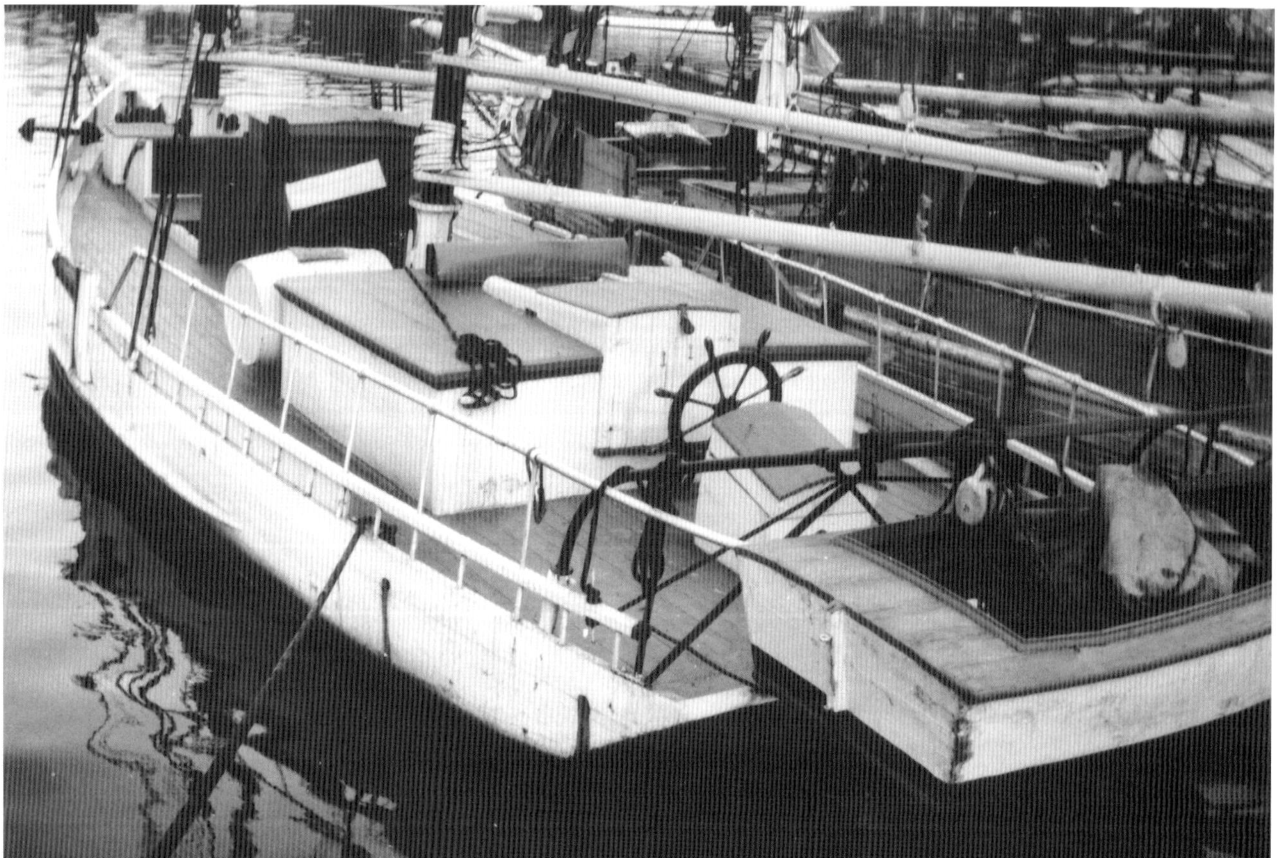

Looking down on the *Eva* at Cambridge, August 23, 1935. During the last half century, this vessel was the smallest schooner-rigged commercial vessel on the Chesapeake. H.I. Chapelle states: "This bateau was one of the very few built with frames, like a round bottom hull, and planked fore-and-aft. The *Eva* was a very able boat and sailed well."

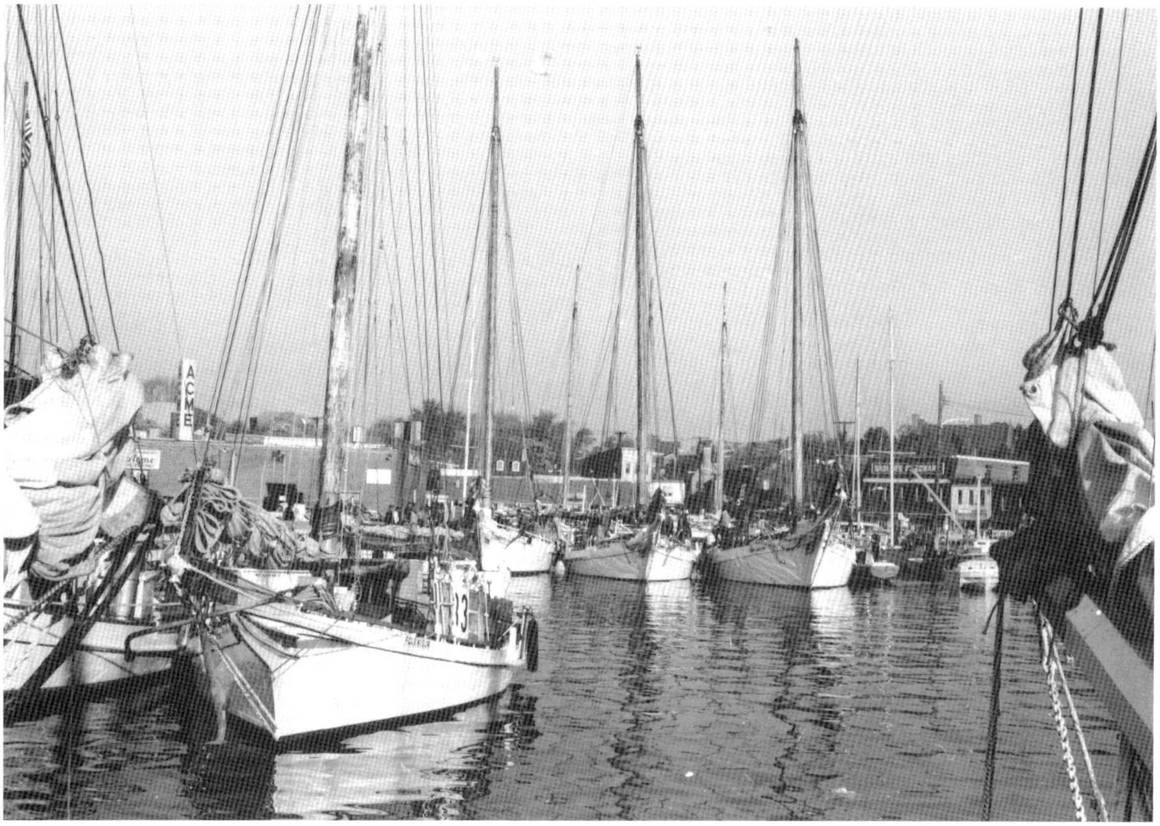

✪ The *F.C. Lewis, Jr.* rests at Annapolis on October 29, 1966. The *Thomas Clyde, Mamie A. Mister,* and *Robert L. Webster* (l-r) are in the background. Today she is blocked up on land as an exhibit in Denton, Maryland.

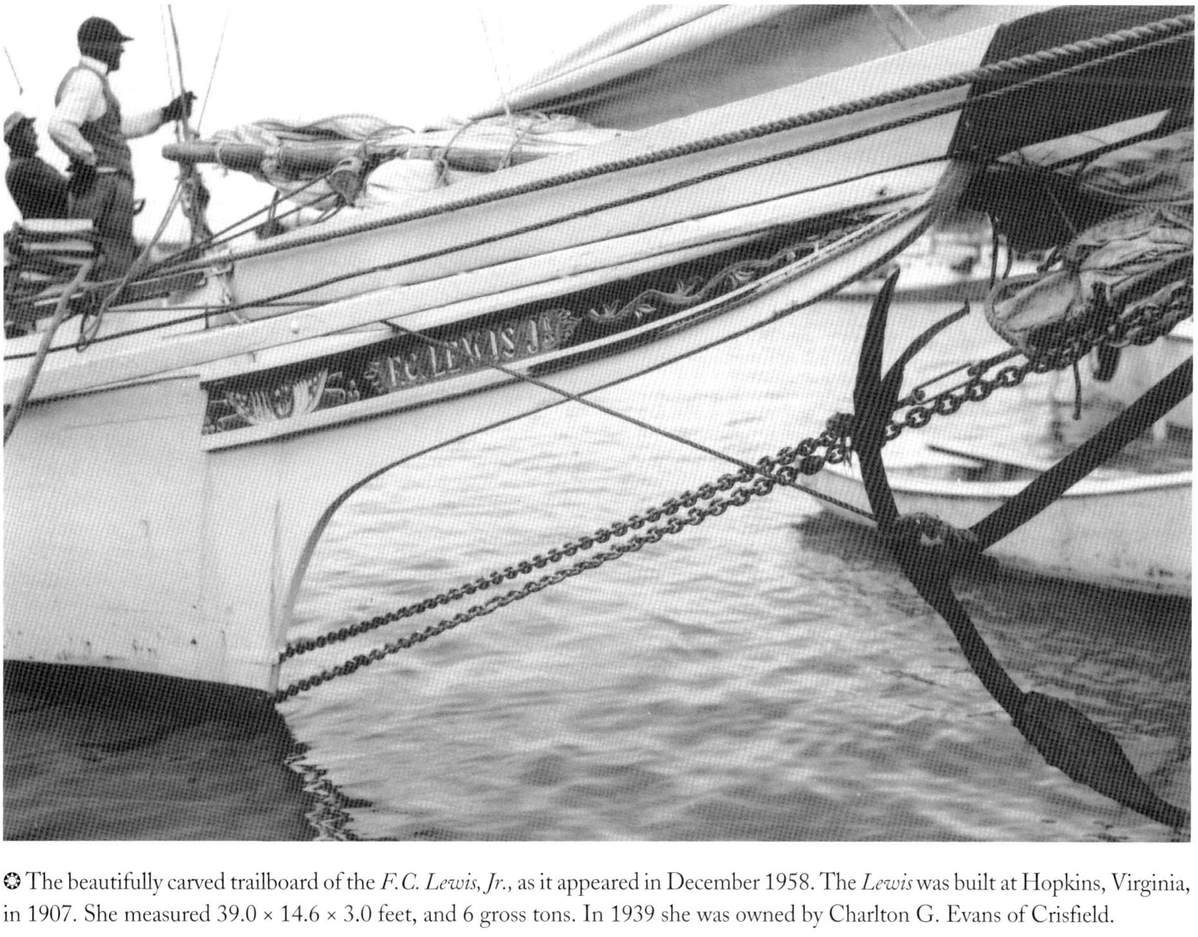

✪ The beautifully carved trailboard of the *F.C. Lewis, Jr.,* as it appeared in December 1958. The *Lewis* was built at Hopkins, Virginia, in 1907. She measured 39.0 × 14.6 × 3.0 feet, and 6 gross tons. In 1939 she was owned by Charlton G. Evans of Crisfield.

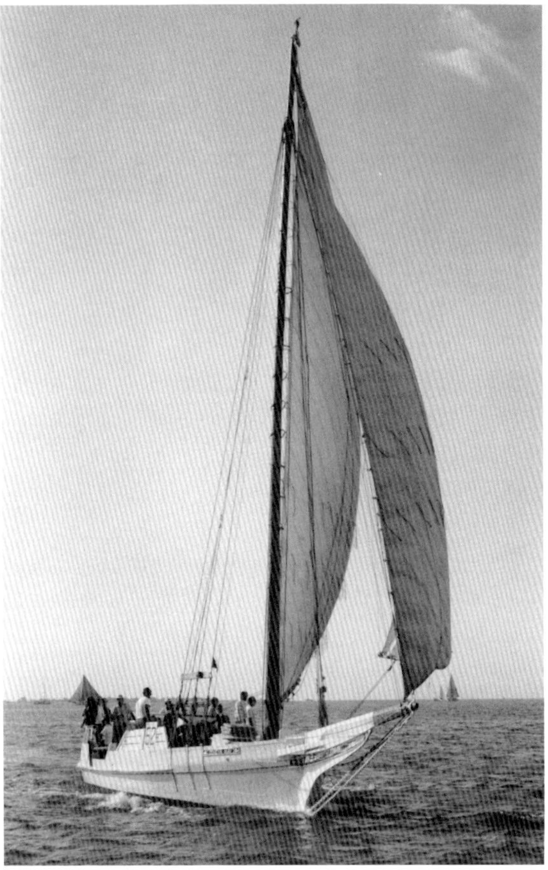

Geneva May maneuvers in Tangier Sound just prior to the start of the workboat races off Deal Island on Labor Day, 1970. ✪ Built at Wenona, Maryland, in 1908, this skipjack was listed as "Out of Documentation" in the 1980 *MVUS*.

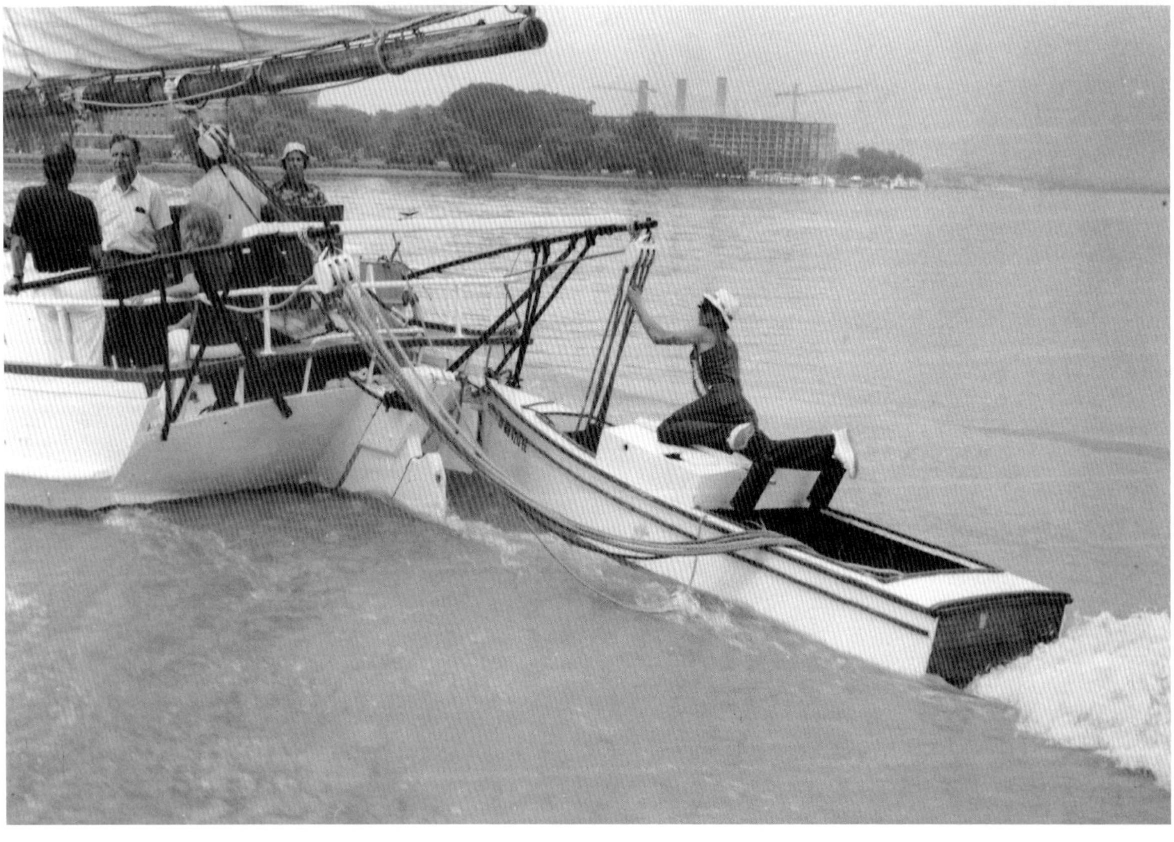

One of five skipjacks to participate in the Folklife Festival at Washington, D.C., in 1972, the *Geneva May* approaches that city on June 29 with the aid of her powerful yawl boat. At her wheel is the late Captain Paul Benton, Jr., and in the yawl boat, his son. Standing on deck in shirtsleeves is Ben H. Evans, who, with the Lions Club, revived the workboat races at Deal Island in 1960 and promotes watermen and the skipjacks at every opportunity.

❂ Working on the mast of the *Geneva May* on October 26, 1959.

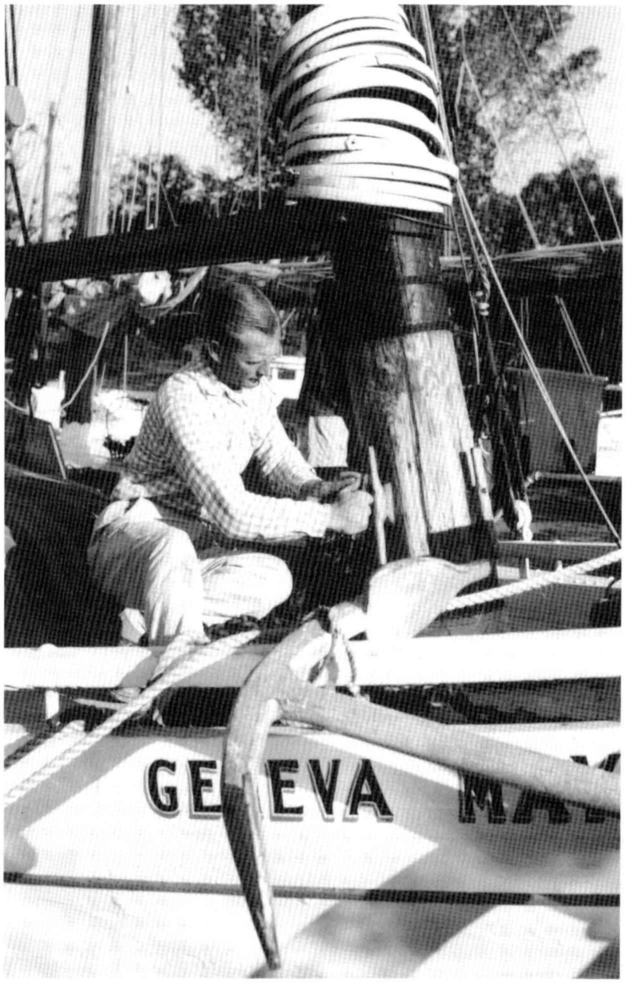

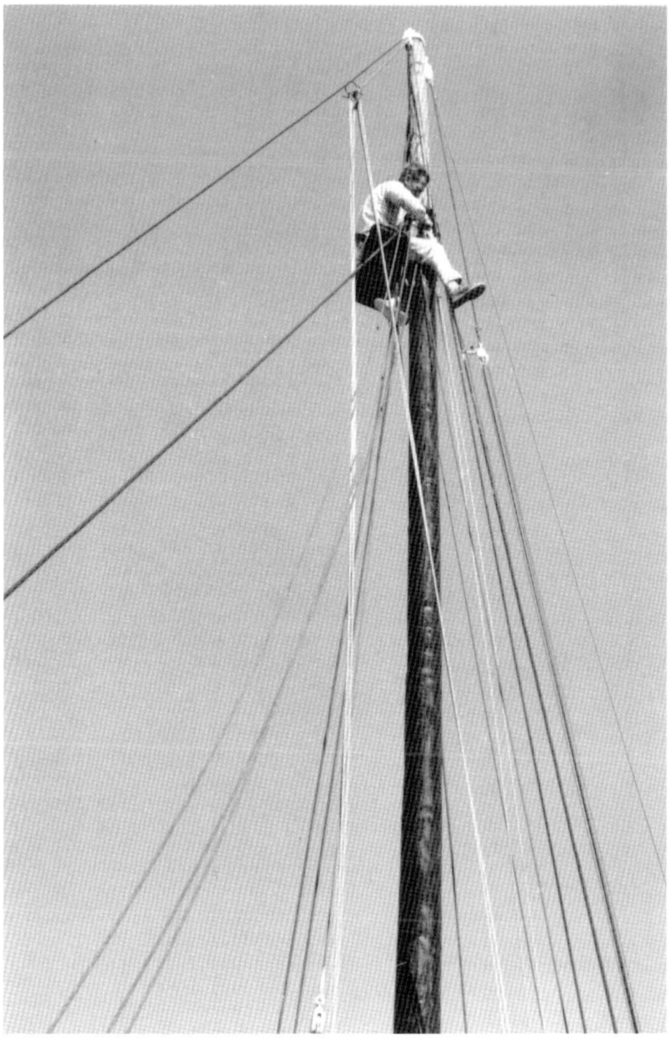

❂ Working aloft on the *Geneva May* on the same day.

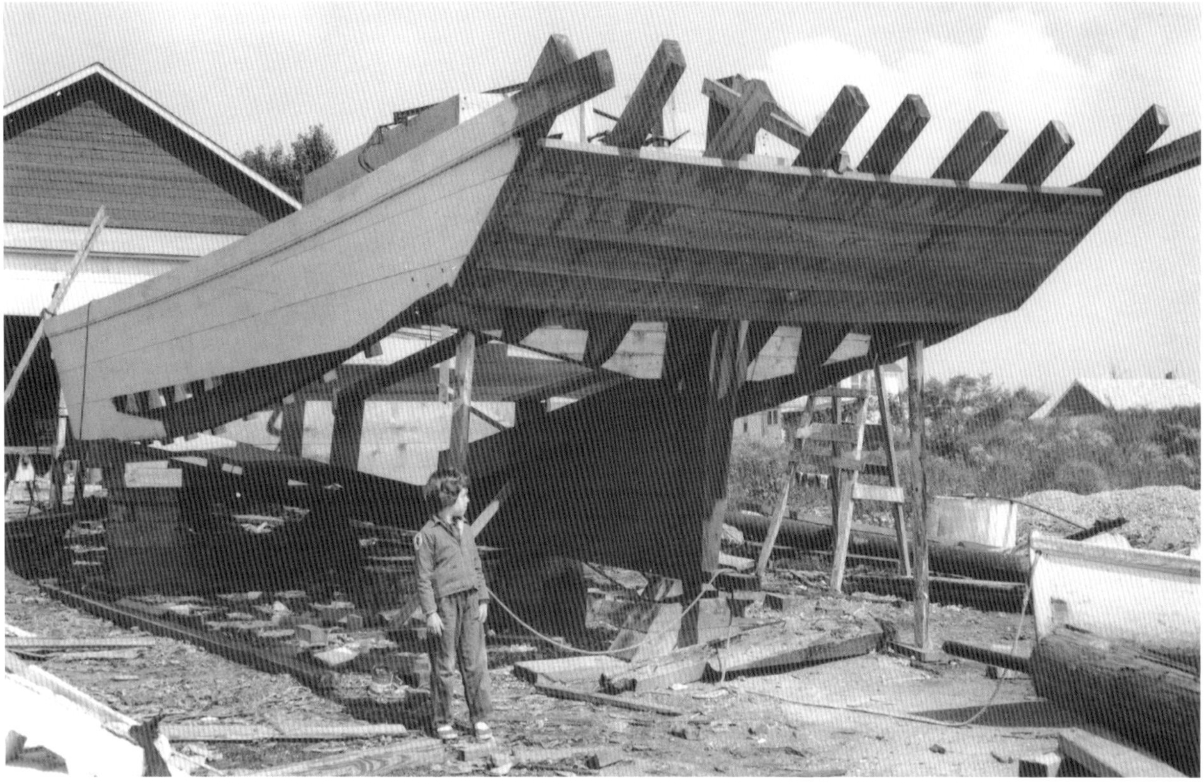

One of the last skipjacks built in the Chesapeake area for the Maryland oyster dredging fleet, and still working, was the *H.M. Krentz,* constructed at the Krentz Shipyard, Harryhogan, Virginia, in 1955. ✪ The vessel was still active in 2005. The author's daughter Janet is in the foreground.

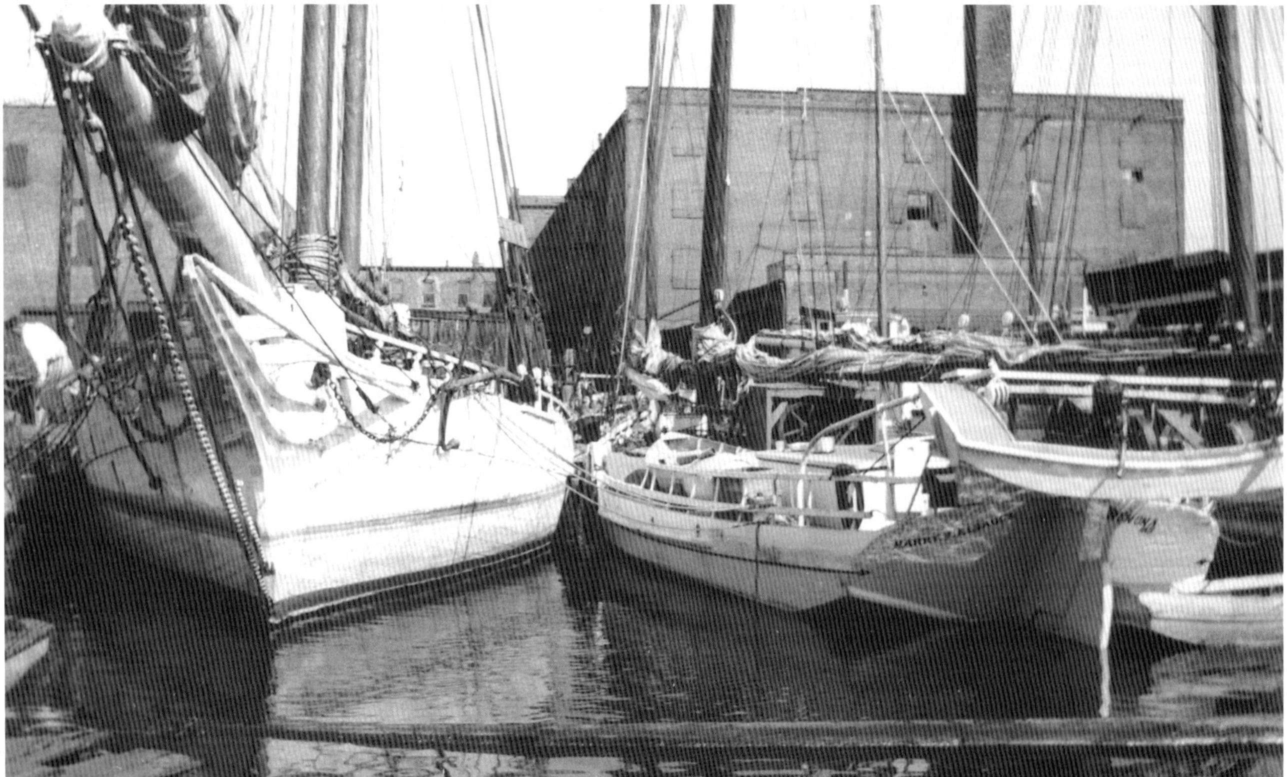

Harry F. Albaugh, built in 1910 at Champ, Maryland, at Webster's Wharf, Baltimore, October 29, 1933. To her port is the round-sterned bugeye *Catherine.* In 1952 the *Albaugh* was sold from Deal Island and while being taken to Norfolk on November 3 of that year, she was disabled in a squall off Smith Point, Virginia, and sank. Attempts to salvage her were futile.

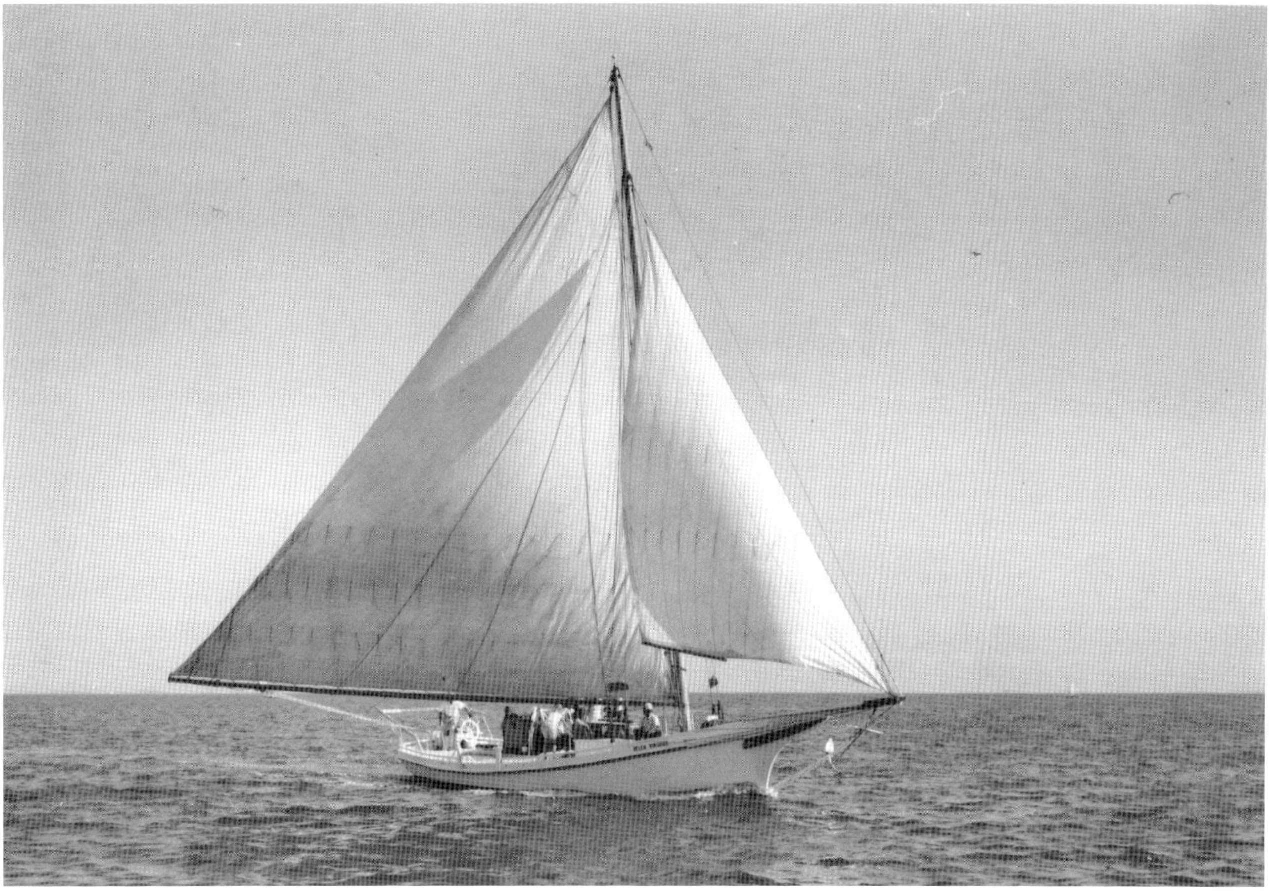

The *Helen Virginia* has proved to be a fast sailer in the workboat races and presents this picture during the Labor Day, 1970, event off Deal Island.

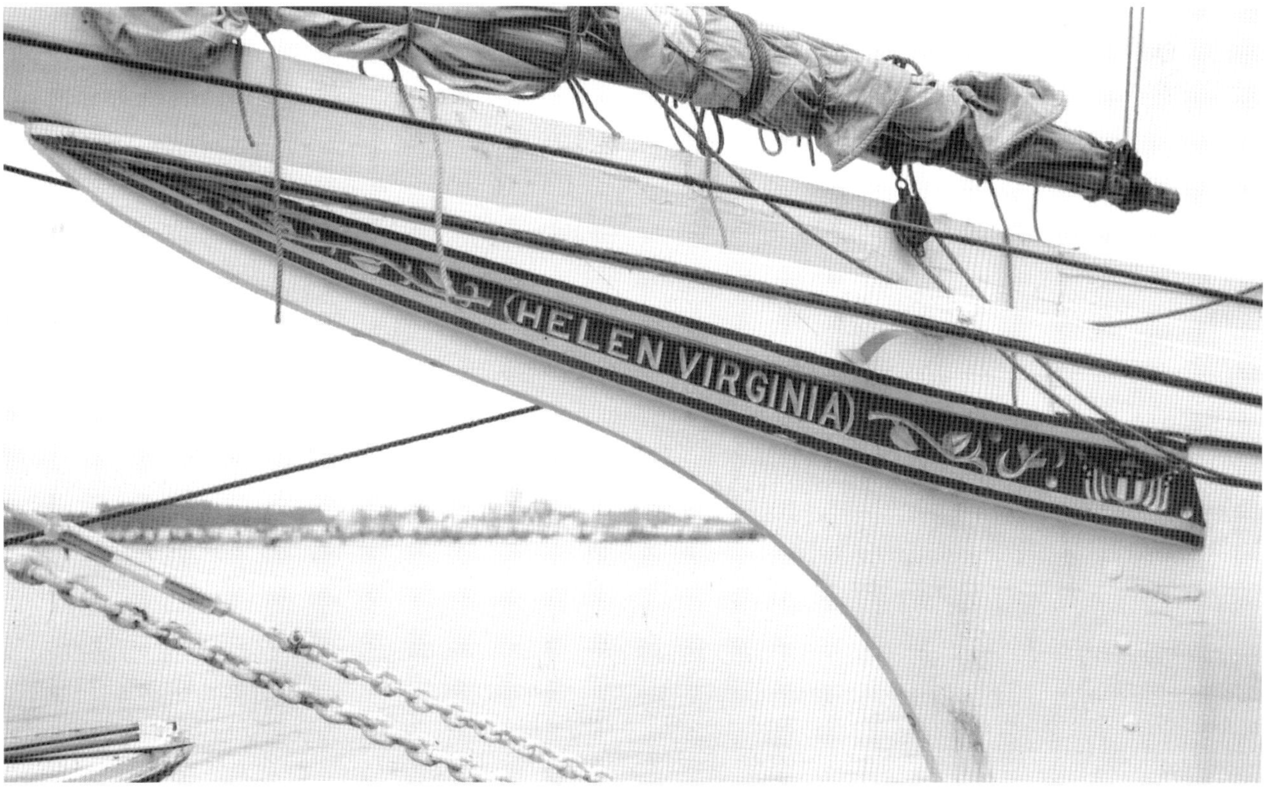

A comparatively new skipjack was the *Helen Virginia,* built at Crisfield, in 1948. Even on the later-built skipjacks the trailboard tradition is carried forth. ✪ This skipjack was still active in 2005.

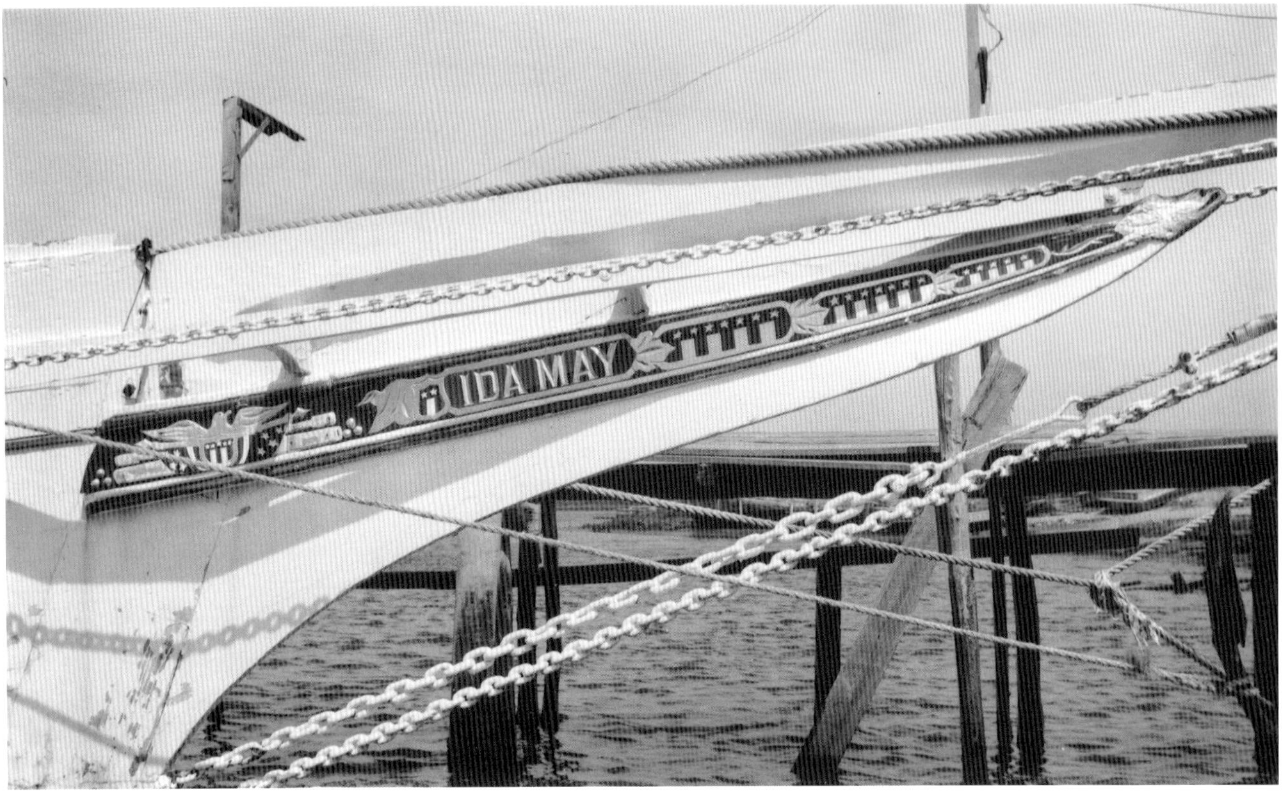

Built in 1906 at Deep Creek, Virginia, the skipjack *Ida May* shows off her trailboard and figurehead at Deal Island on April 25, 1970. She continues to sail out of Deal Island as an oyster dredger.

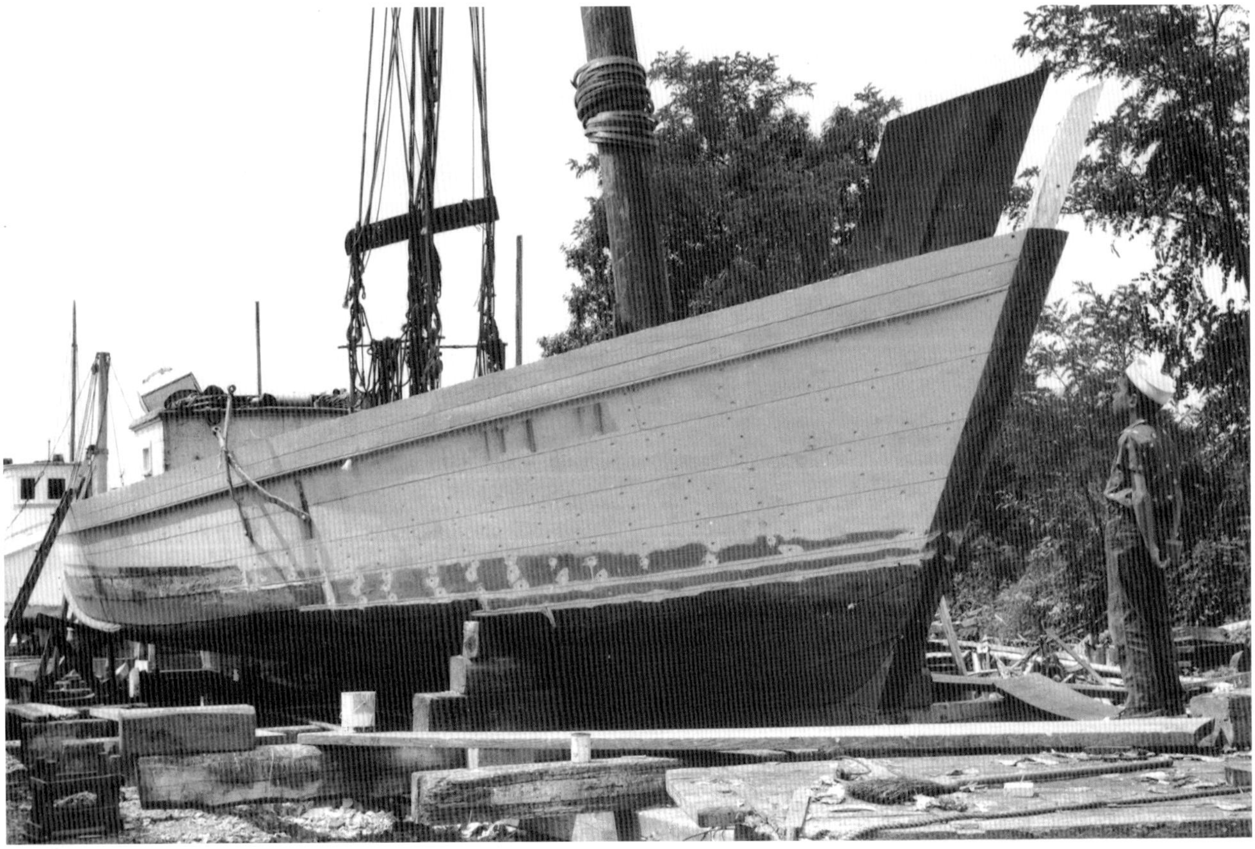

On August 8, 1954, the skipjack *Kathryn* is in the process of being rebuilt at the Krentz Shipyard. A great amount of her hull planking is replaced, new frames installed, and a complete deck with frames built. She will emerge from the ways almost as good as when she was built at Crisfield, Maryland, in 1901. ✪ The author's son Bruce is in the foreground.

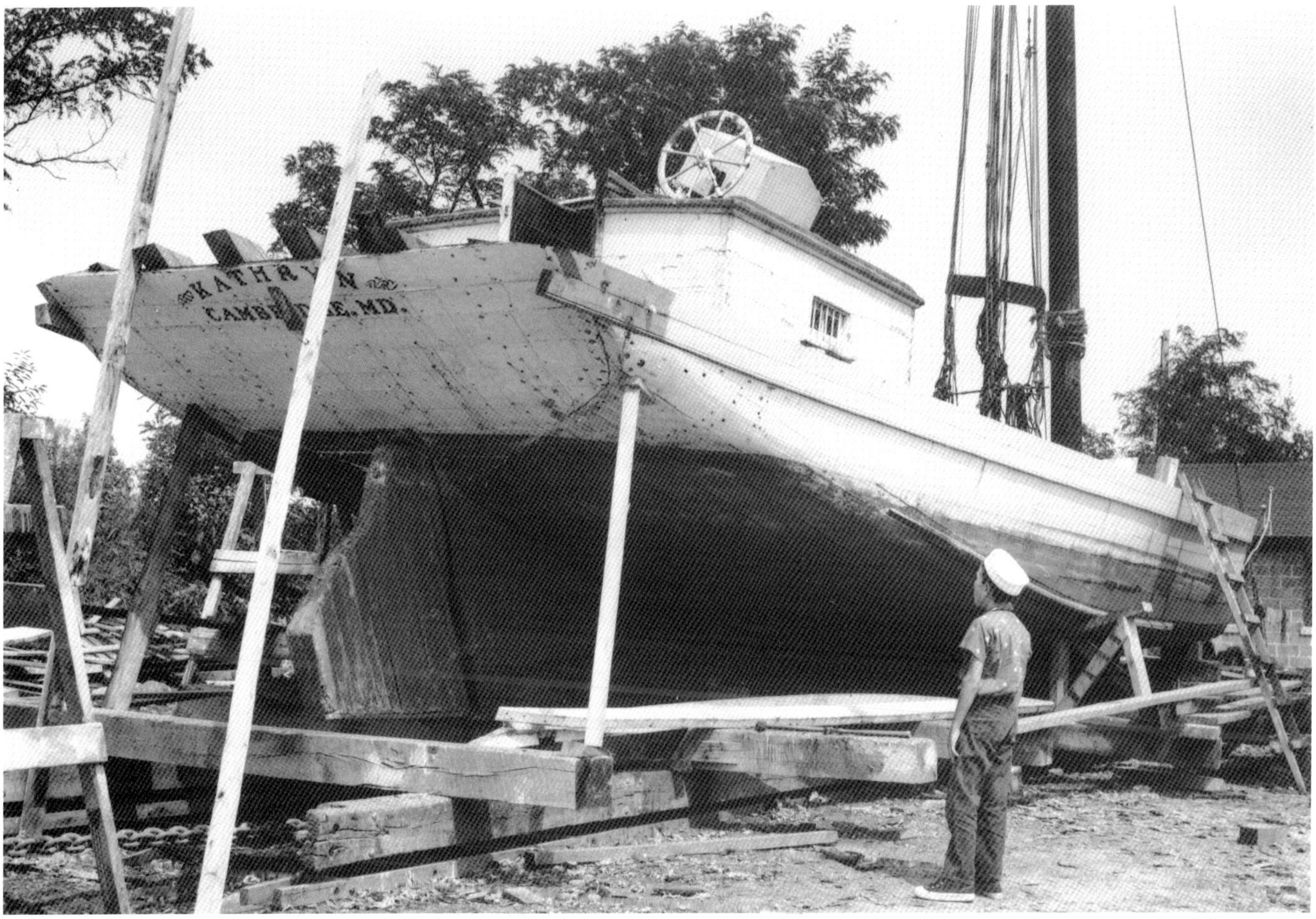

The steering wheel rests atop the cabin while her deck is being replaced. The *Kathryn* differed from the average skipjack in that her chine was rounded.

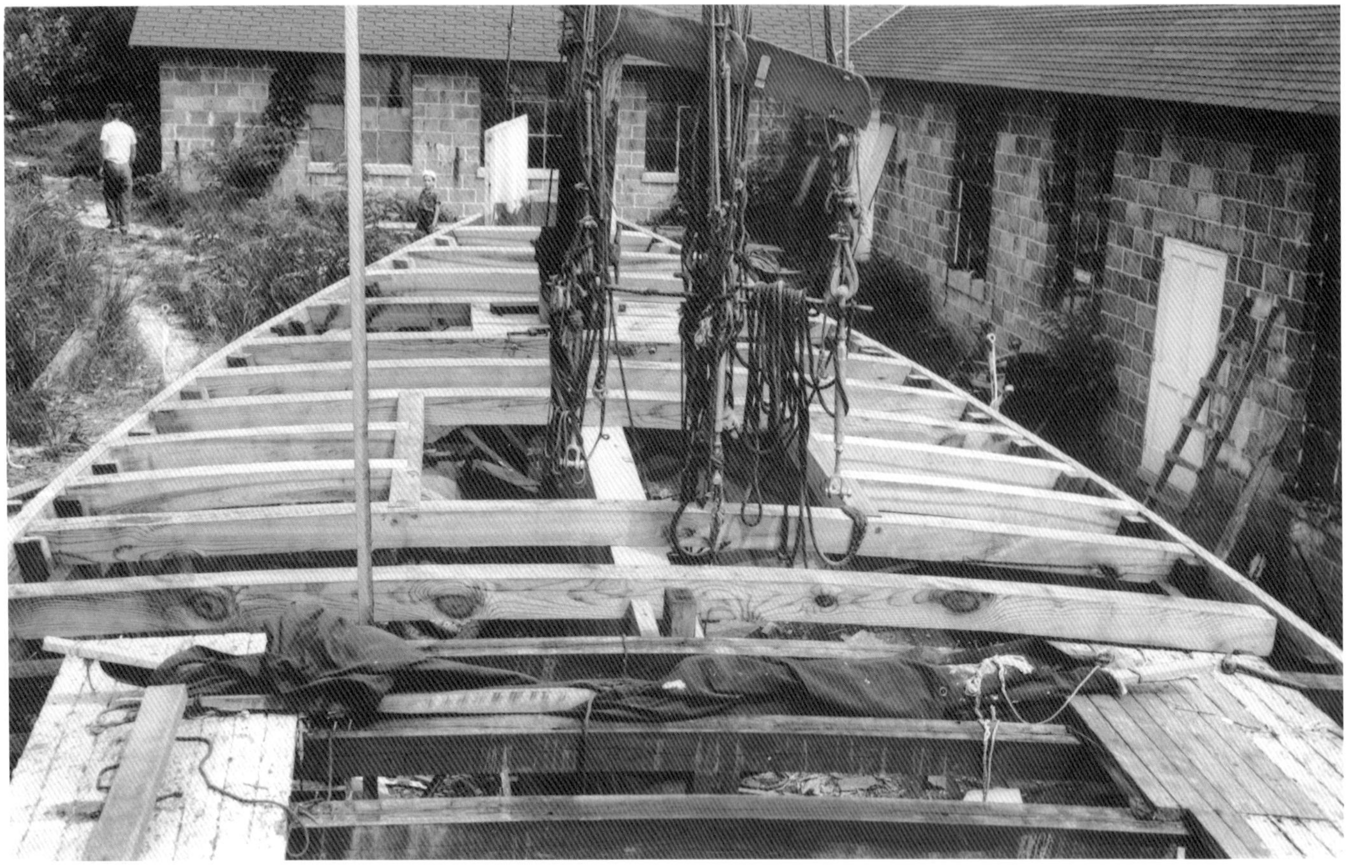

Looking forward on the *Kathryn* from the top of her after cabin. Shown is the arrangement of the deck beams. The shrouds with turnbuckles hang from the top of the mast.

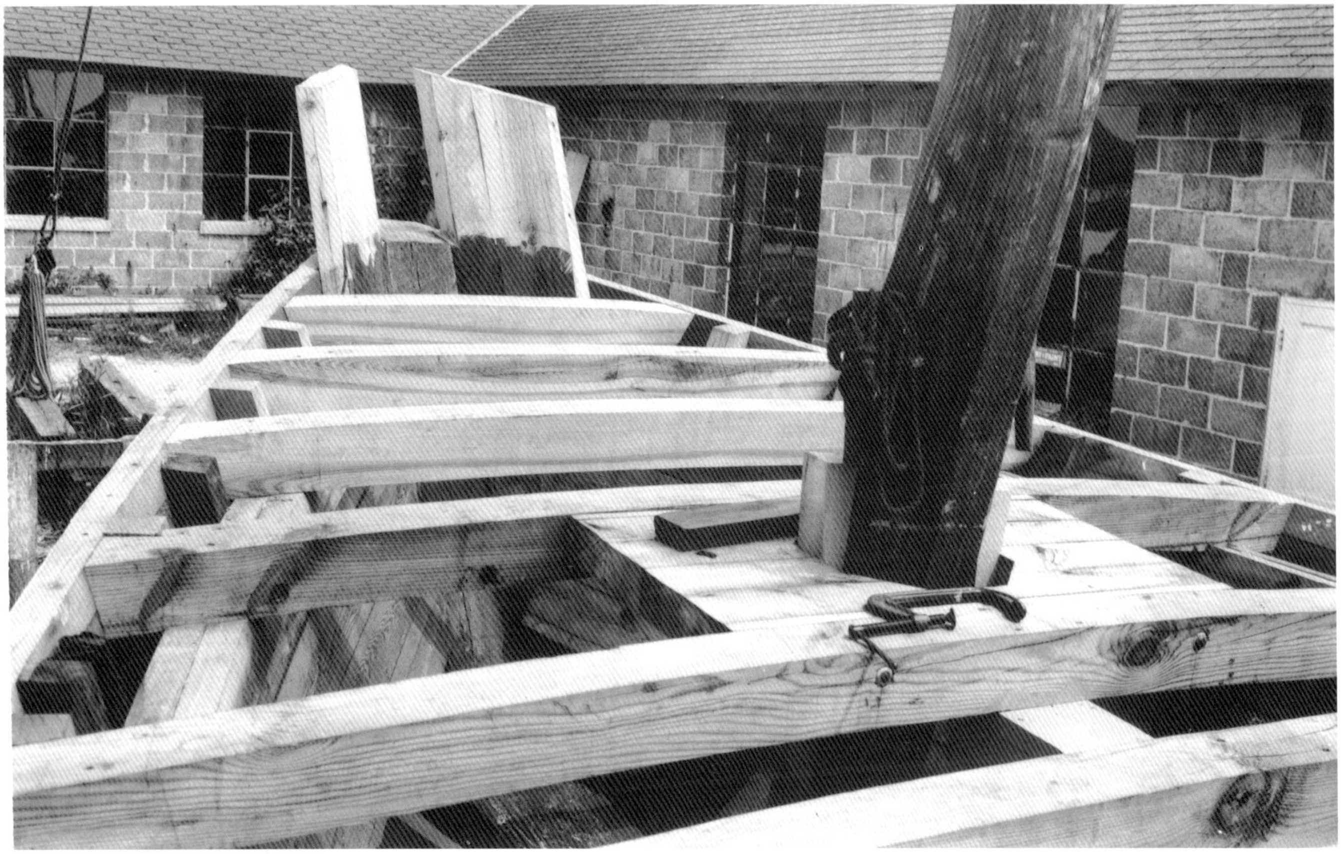

Beams of the *Kathryn's* foredeck and mast partners and carlings. The knightheads protrude above the deck line.

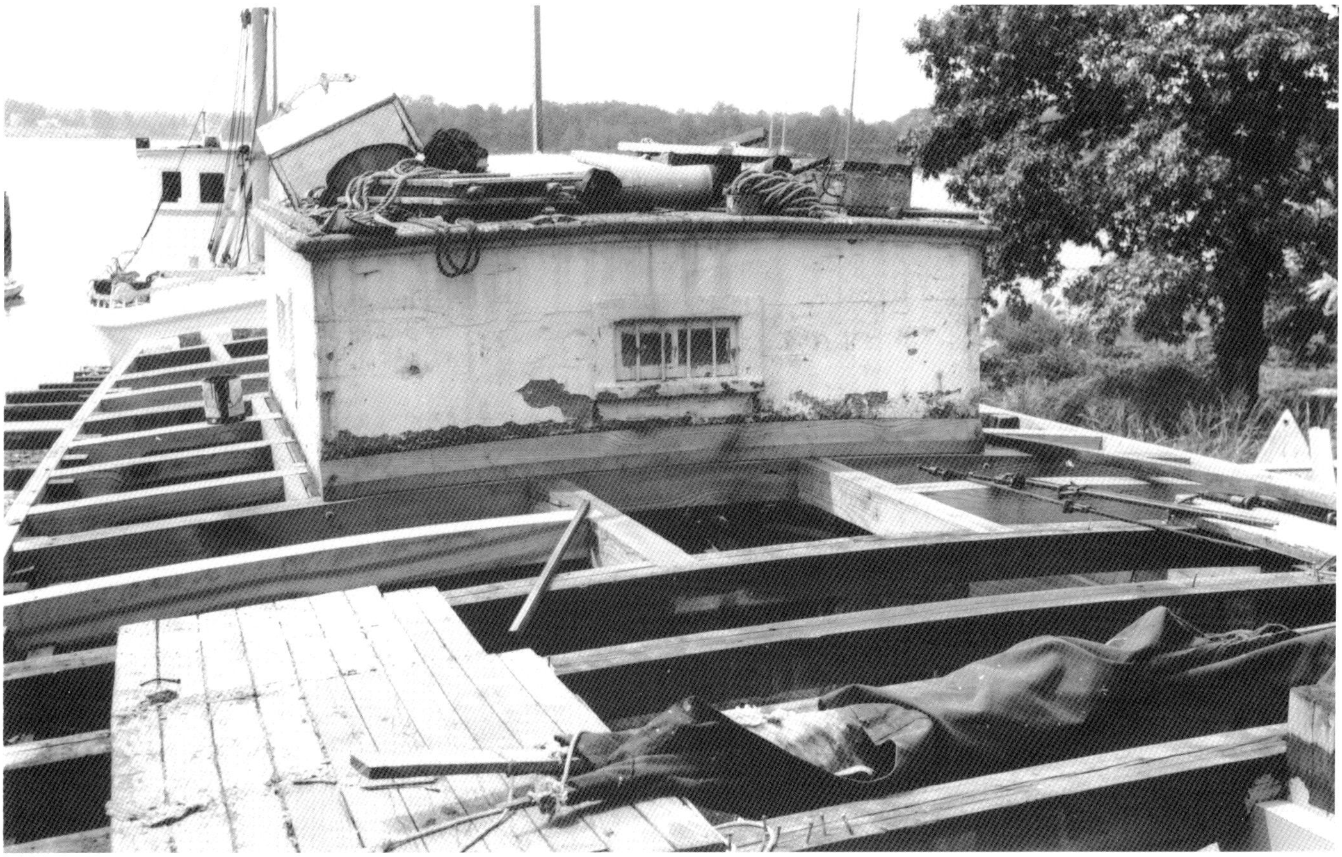

❂ The planking has been removed and deck beams have been replaced during the *Kathryn*'s rebuilding in 1954.

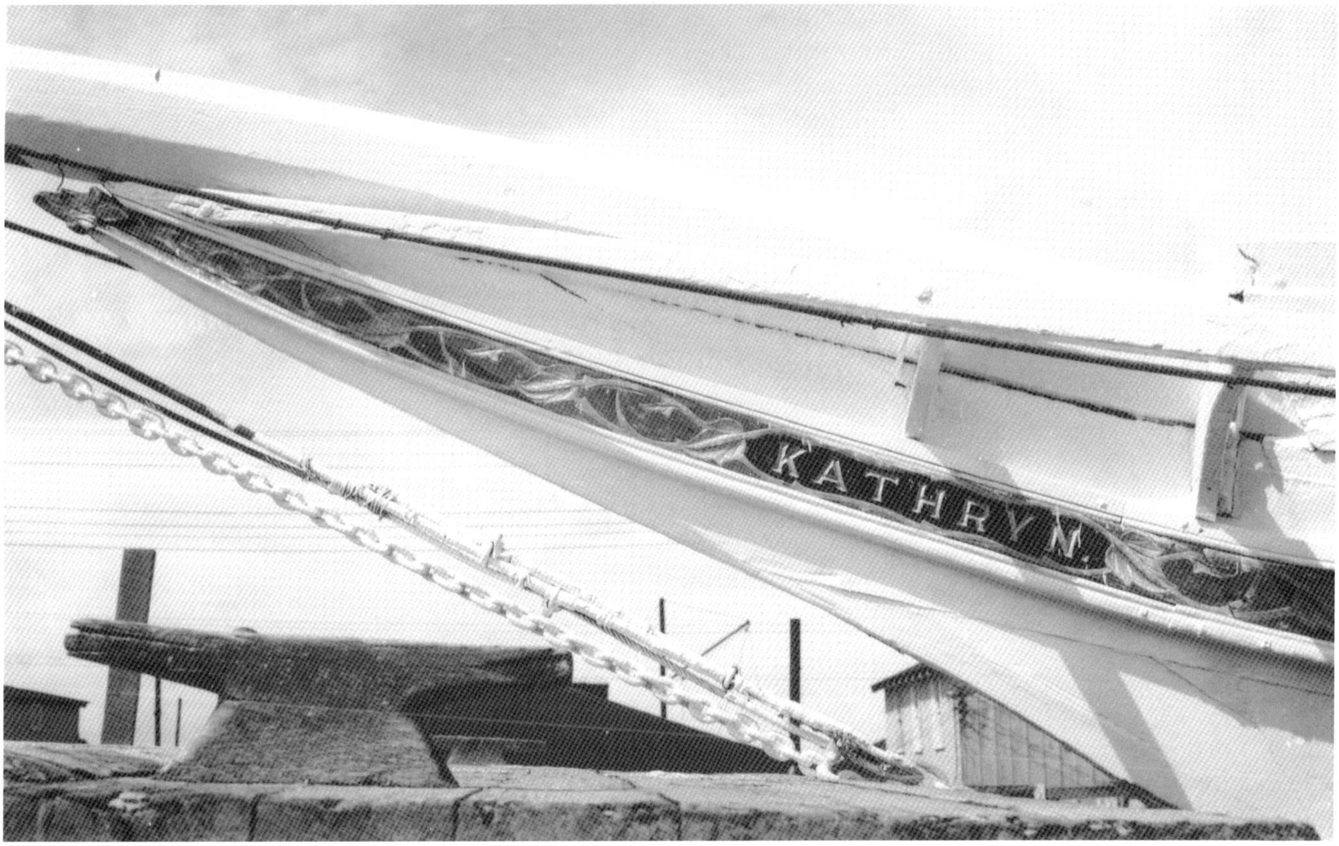

❂ Detail of the longhead and trailboard of the 49-year-old *Kathryn,* at Crisfield in 1950. In 2005 she was still active, dredging oysters out of Tilghman Island, Maryland.

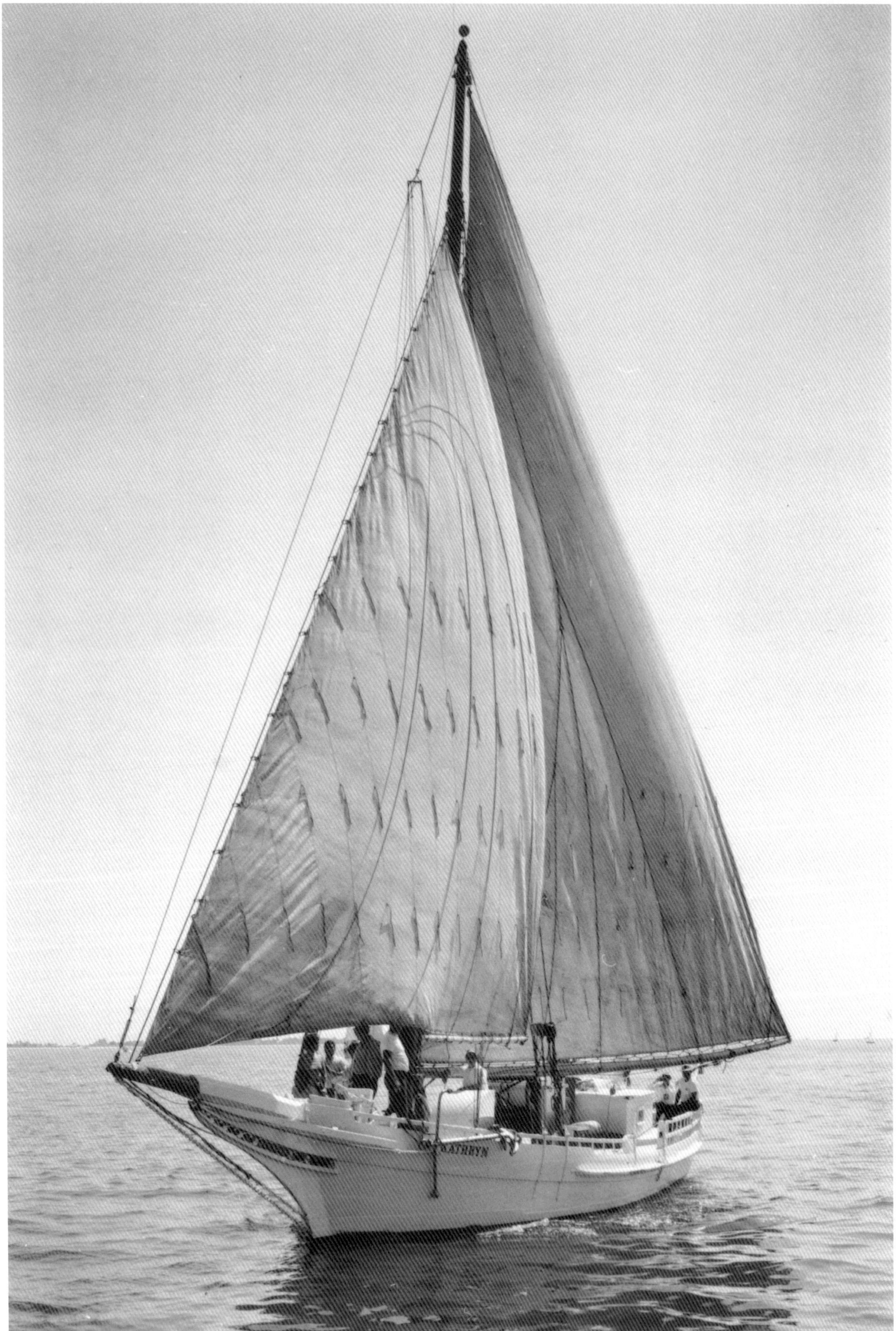

The *Kathryn* at the 1970 Labor Day workboat race at Deal Island. Sixteen years after her rebuilding, and statistically 69 years old at the time, she looks like a new boat. This 50-foot skipjack still works over the Maryland oyster beds.

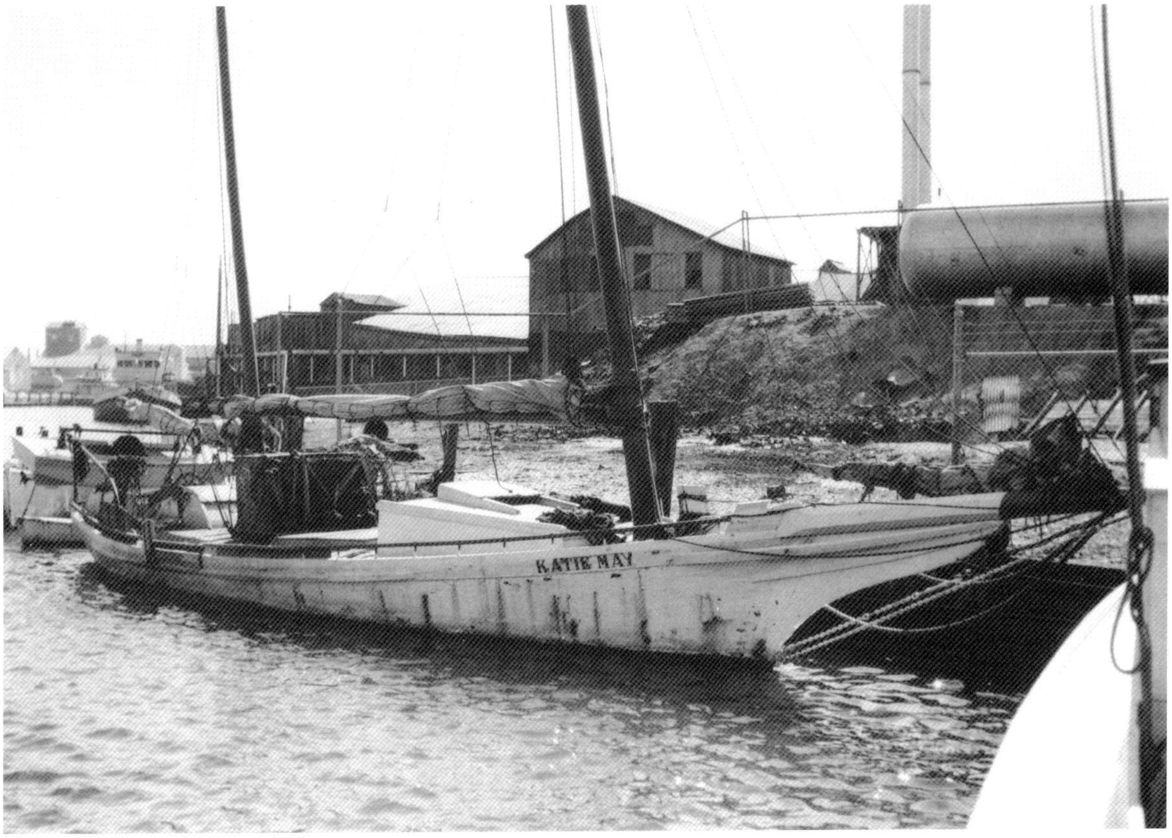

The skipjack with a bugeye rig, or three-sail bateau, is the *Katie May* at Cambridge in February 1948. She was built at Bishops Head, Maryland, where most craft of this type were constructed, in 1900. *Katie May* disappears from the 1970 *MVUS*.

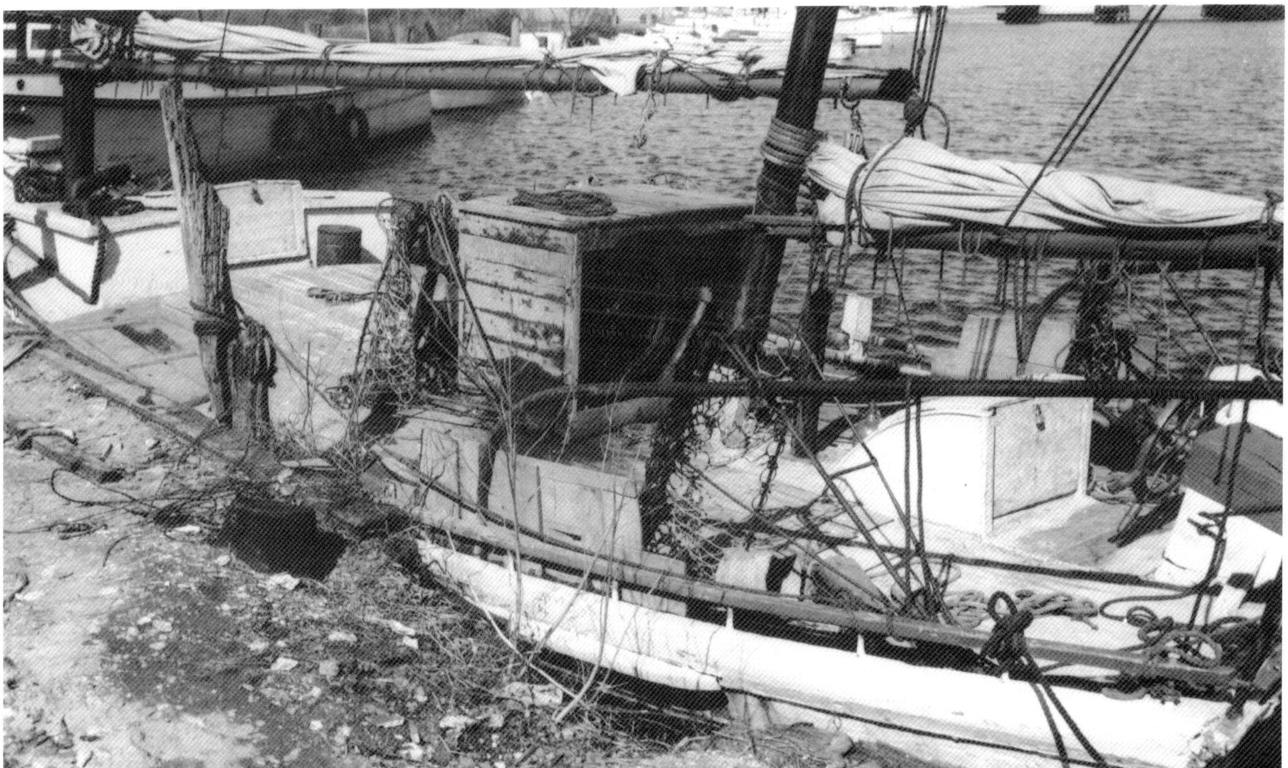

Looking forward on the *Katie May,* showing the arrangement of her forward house and the hatch leading to quarters under the after deck. The winder engine is just forward of the mainmast.

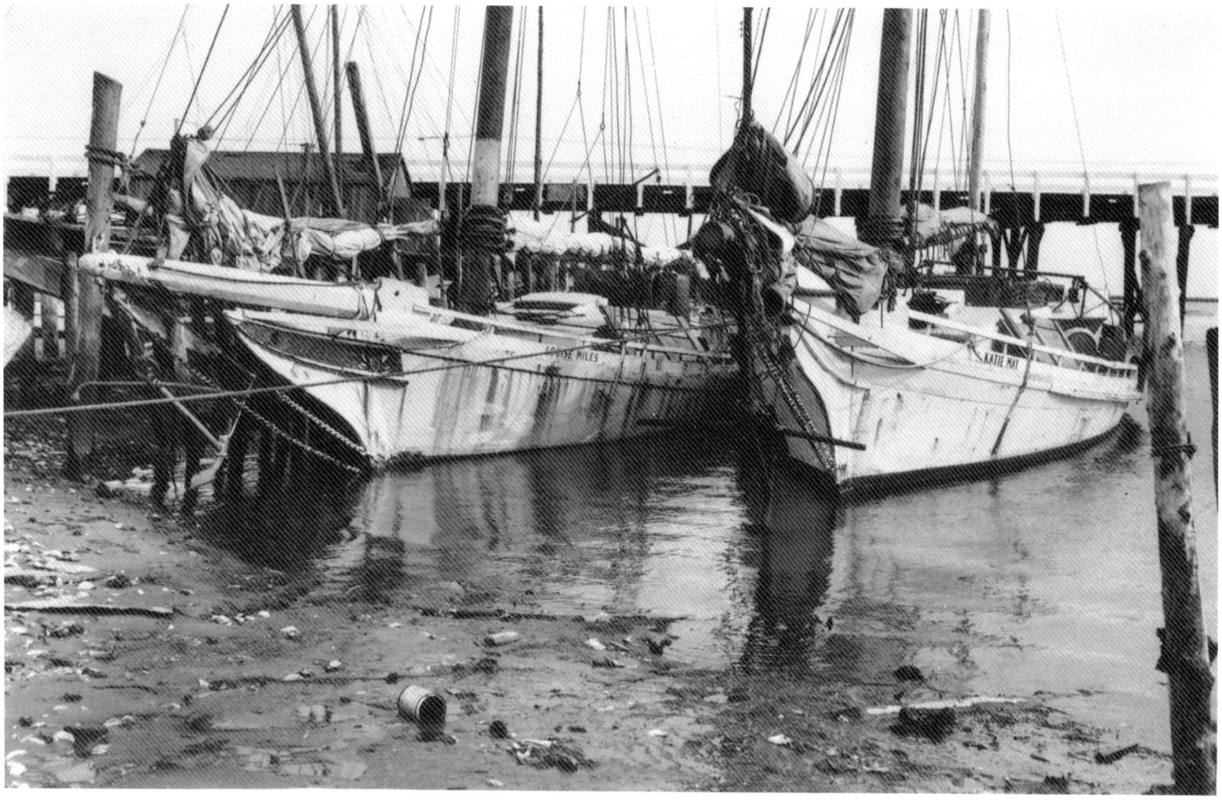

The skipjack *Louise Miles,* left, and *Katie May* at Chance, Maryland, in February 1950. The former was built at Crab Island, Maryland, in 1899. Not long after this view was taken she was abandoned in the marsh at Chance.

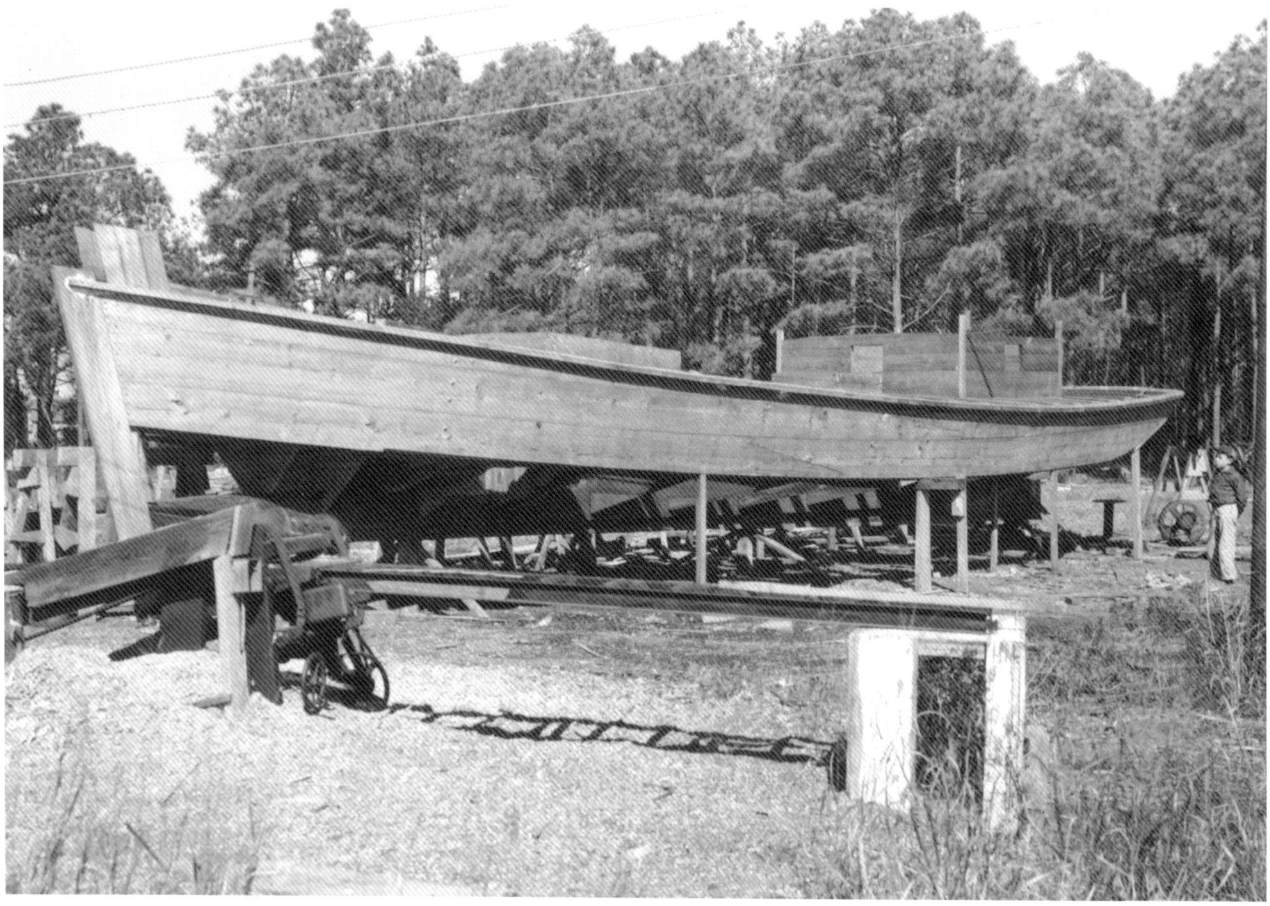

✪ The skipjack *Lady Katie* under construction on October 22, 1955. She was one of the last of her type to be built and was still afloat in 2005.

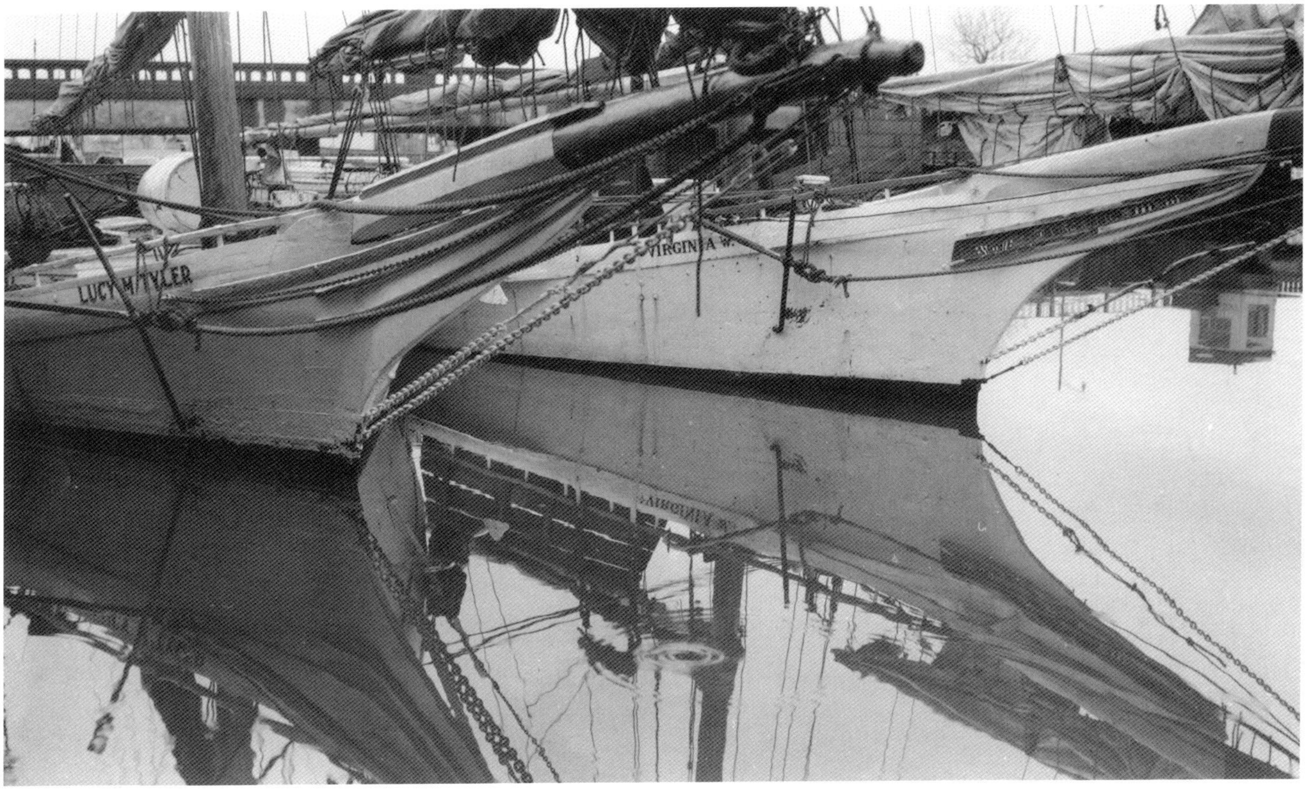

Reflections at Cambridge of the skipjacks *Lucy M. Tyler,* left, and *Virginia W.,* January 27, 1952. Both old-timers, the former was built at Hooper Island, Maryland, in 1906, and the latter at Guilford, Virginia, in 1904.

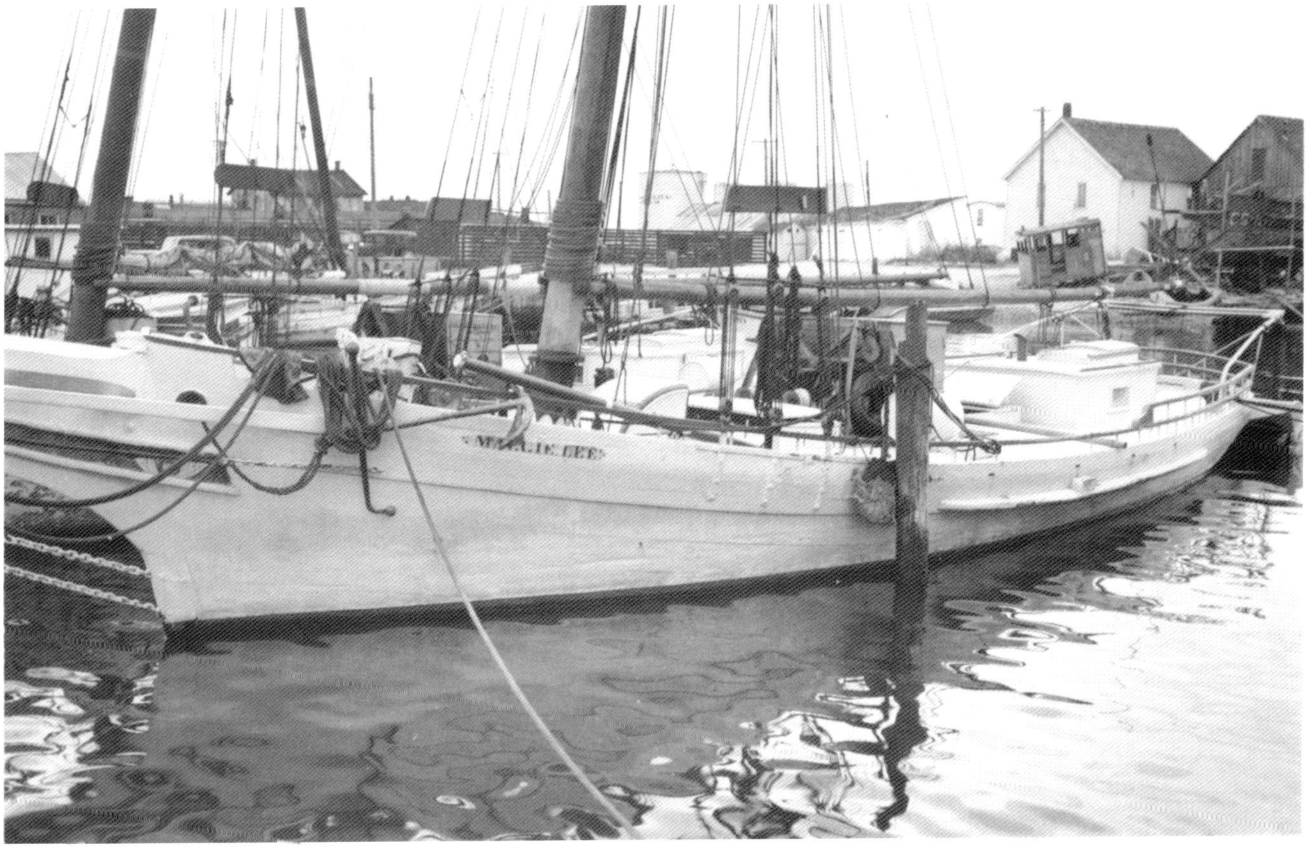

✪ The *Maggie Lee* rests from her labors at Crisfield, Maryland, on June 19, 1936. The *Ethelyn Dryden* is moored inboard of her. The two skipjacks were built at Pocomoke City, Maryland, in 1903 and 1901, respectively. *Maggie Lee* was abandoned on the shore of the Choptank at Denton early in the 21st century.

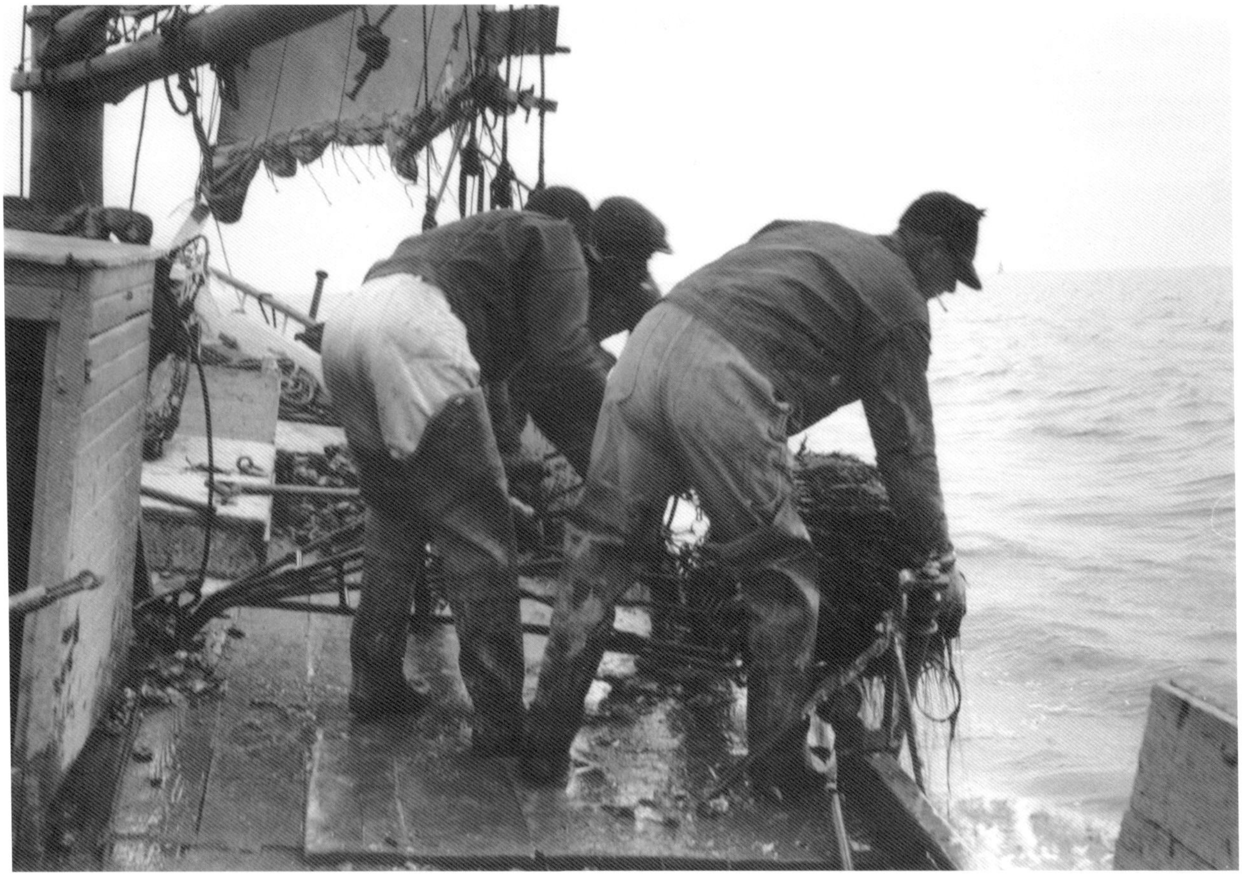

✪ Retrieving the dredge aboard the *Maggie Lee* in February 1948.

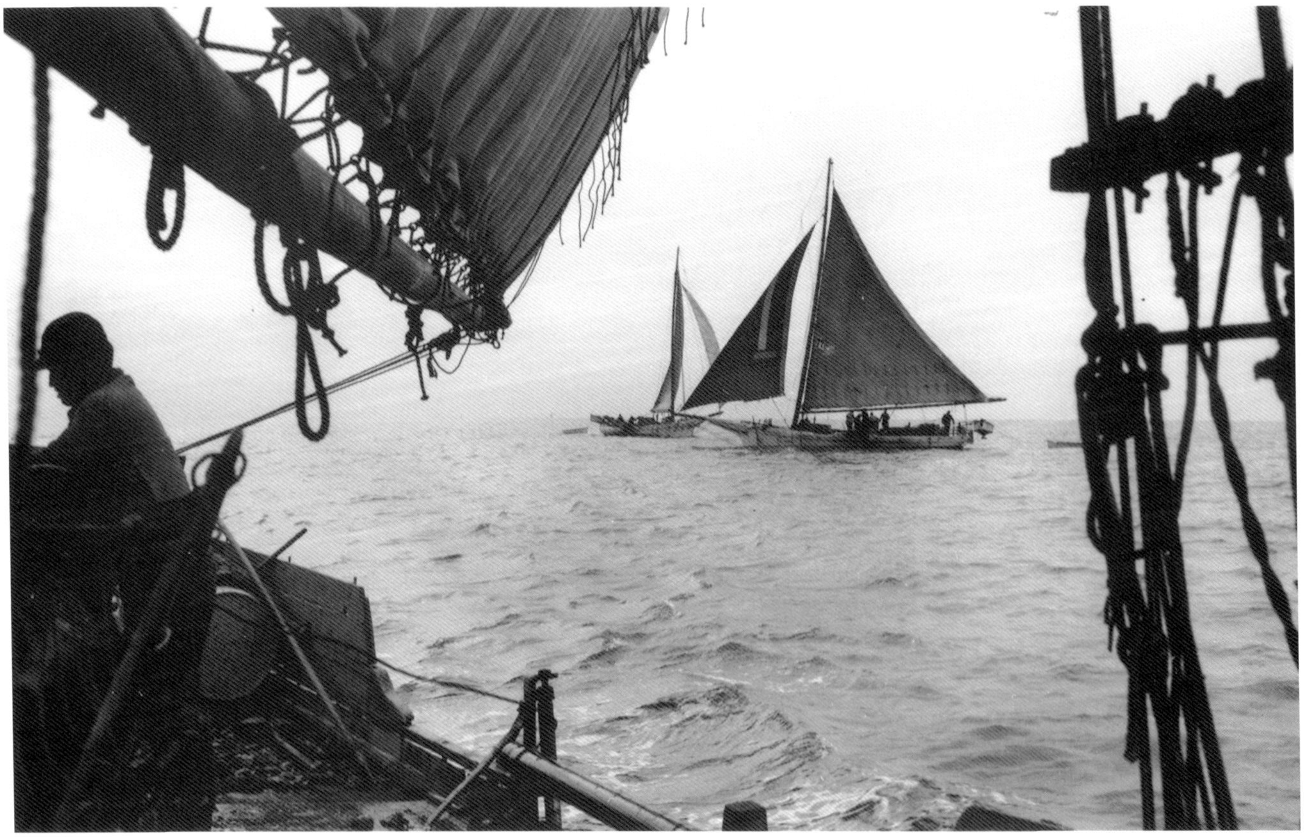

As the deckhand, at left, on the *Maggie Lee* pauses for a few moments while the dredge is making a lick over the bottom, two other skipjacks in the background drag their dredges over the bottom of Tangier Sound in February 1952.

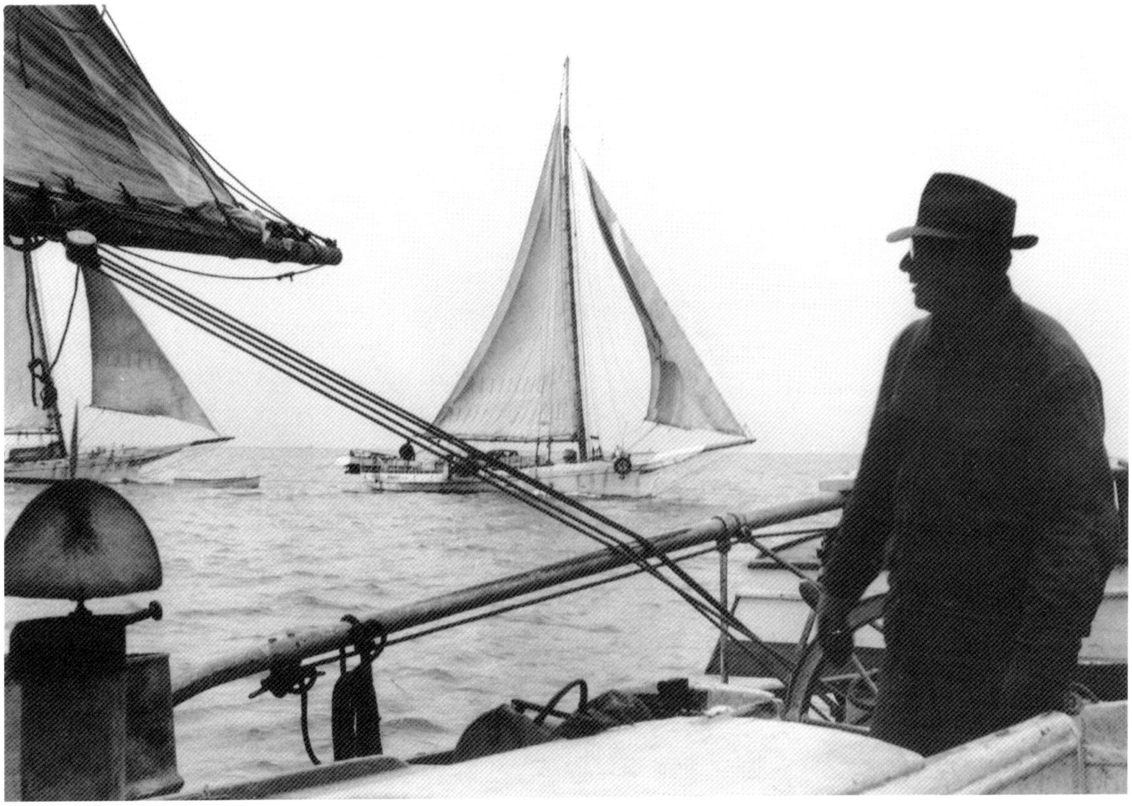

✪ The *Clifton Webster* as seen from the *Maggie Lee* in February 1952. R.H. Tawes is at the helm of the *Maggie Lee.*

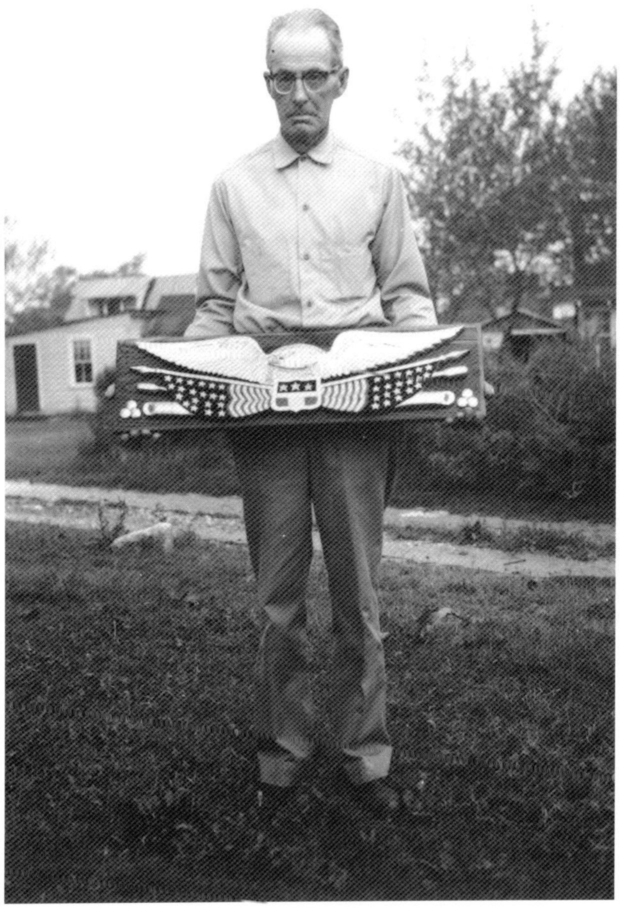

✪ Famed wood carver Dewey Webster of Deal Island, Maryland, carved for many Bay vessels, including the *Maggie Lee.*

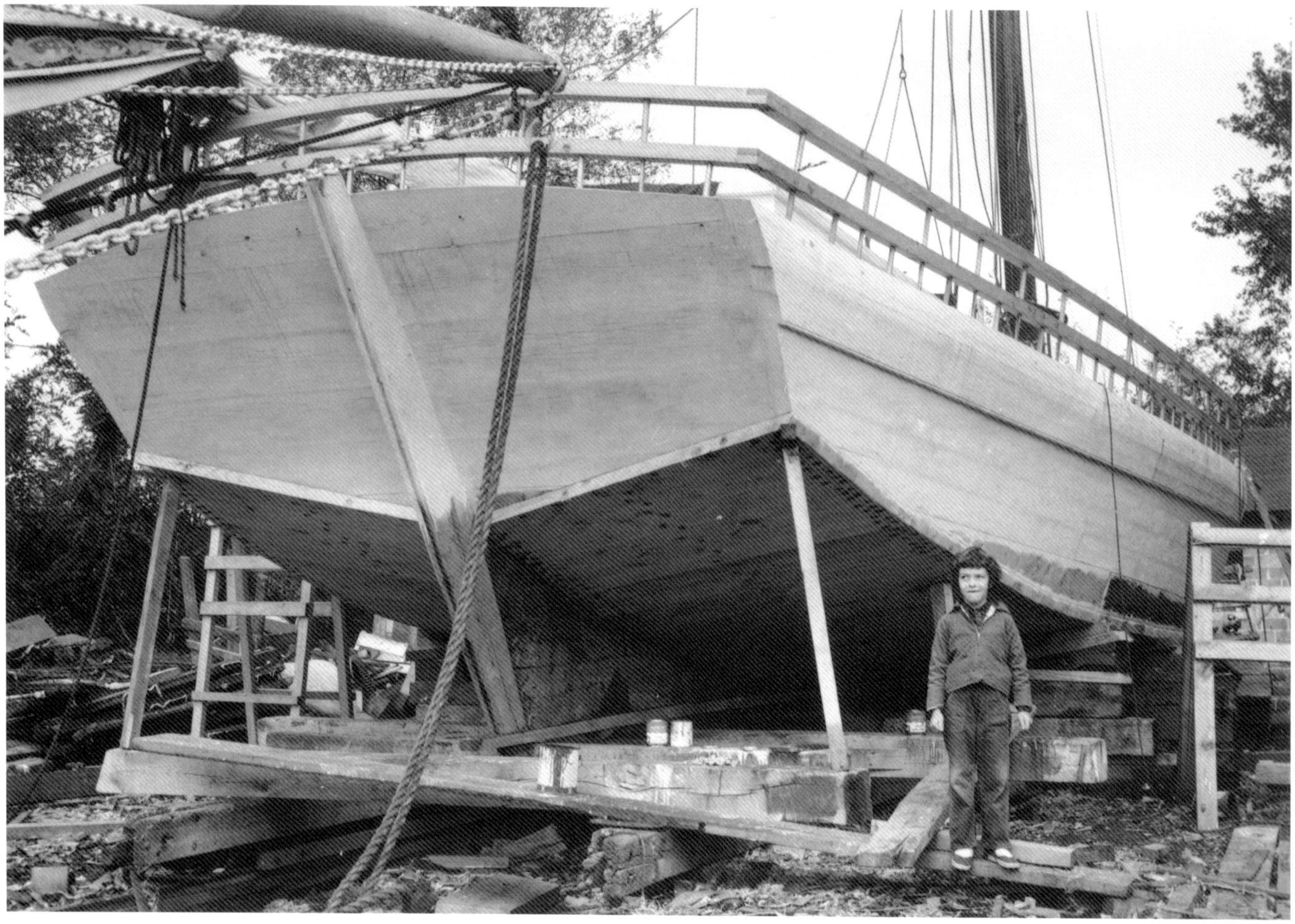

The *Mamie A. Mister,* originally built in 1910 at Champ, Maryland, is rebuilt at the Krentz Shipyard, October 1, 1955, virtually from the keel up. The only original wood visible in this view is the skeg and mast. ✪ The author's daughter Janet is in the foreground.

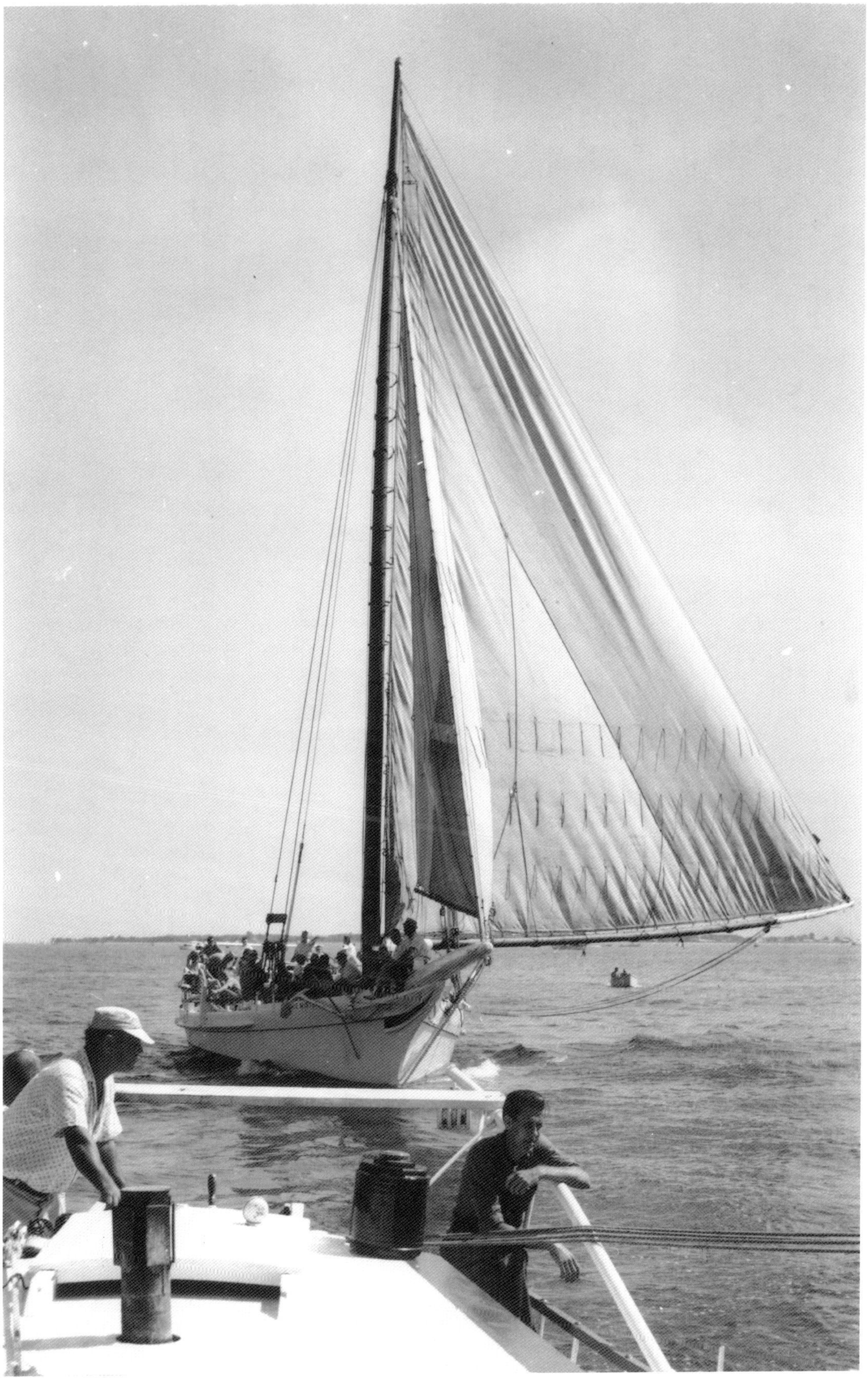

Mamie A. Mister, her deck crowded with visitors during the workboat race at Deal Island on Labor Day, 1968, tries to overtake the *F.C. Lewis, Jr.*, Captain Stanford White in lower left. Within a couple of years the *Mister* was sold away from the Bay and converted into a yacht. In the 1972 *MVUS* she still retained her original name but hailed out of New York as an auxiliary. ✪ She was reportedly still active in 2005.

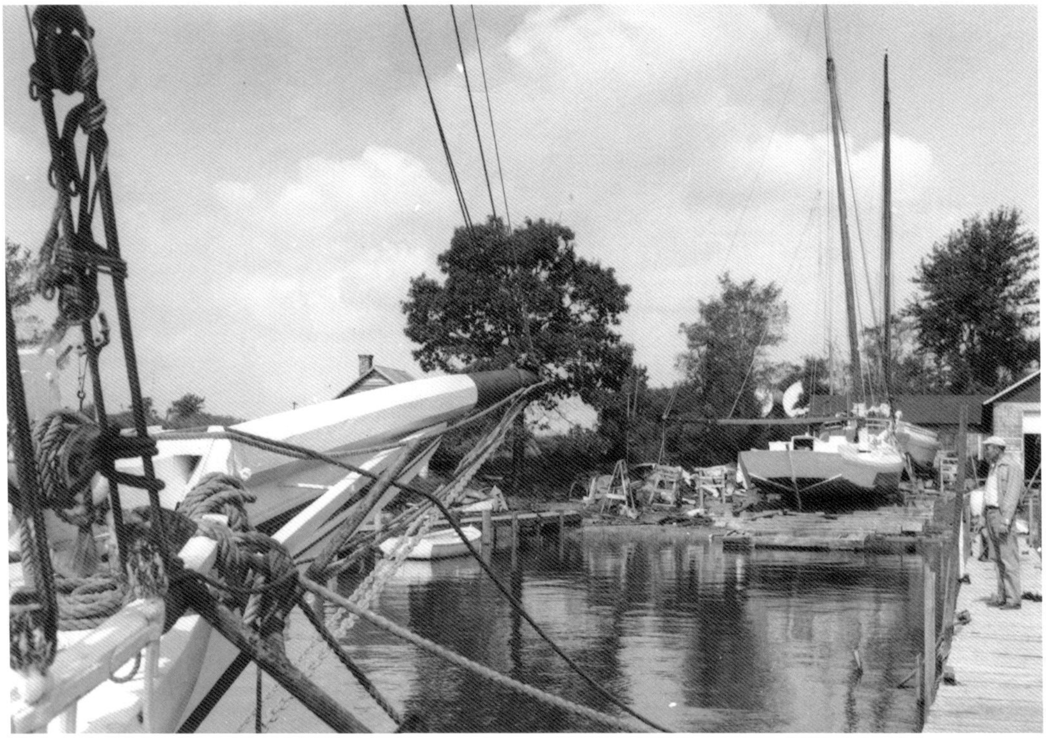

The H.M. Krentz Shipyard, Harryhogan, Virginia, builders of the skipjack *H.M. Krentz,* was a favorite among Maryland oyster dredgers when repairs or rebuilding of their skipjacks were required. Here, on October 1, 1955, two skipjacks, one the *Mamie A. Mister,* are hauled out on the ways as the *Elsworth,* at left, awaits her turn after the other boats are put overboard.

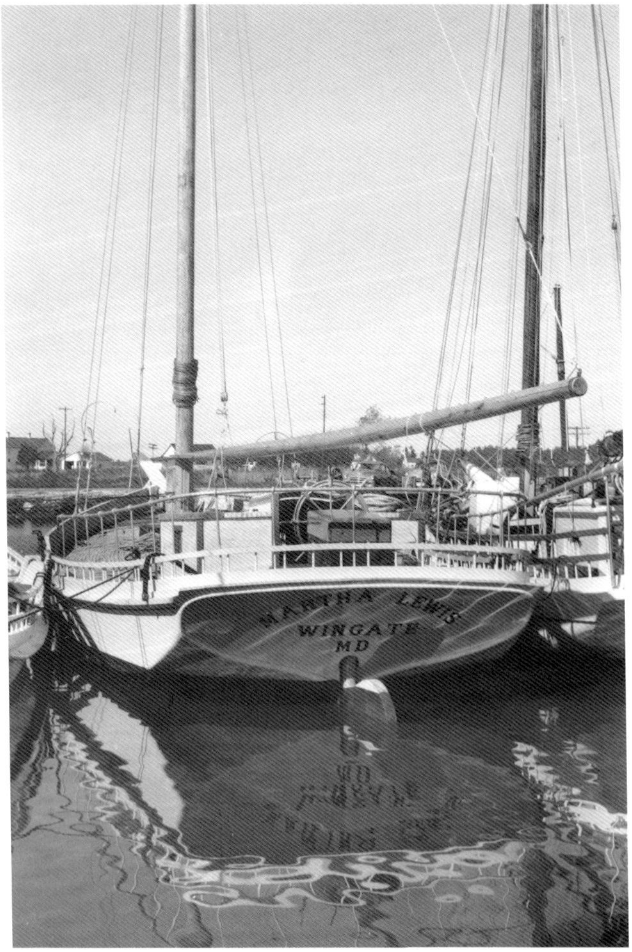

One of the newer skipjacks is the *Martha Lewis,* built in 1955 by Bronza Parks at Wingate, Maryland, and shown there on October 22 of that year. There is still some work to be done onboard her, including placement of the stern davits. To her starboard is the *Rosie Parks,* also built by Parks in 1955. ✪ The *Martha Lewis* was still active in 2005, but the *Rosie Parks* has been at the Chesapeake Bay Maritime Museum since 1975.

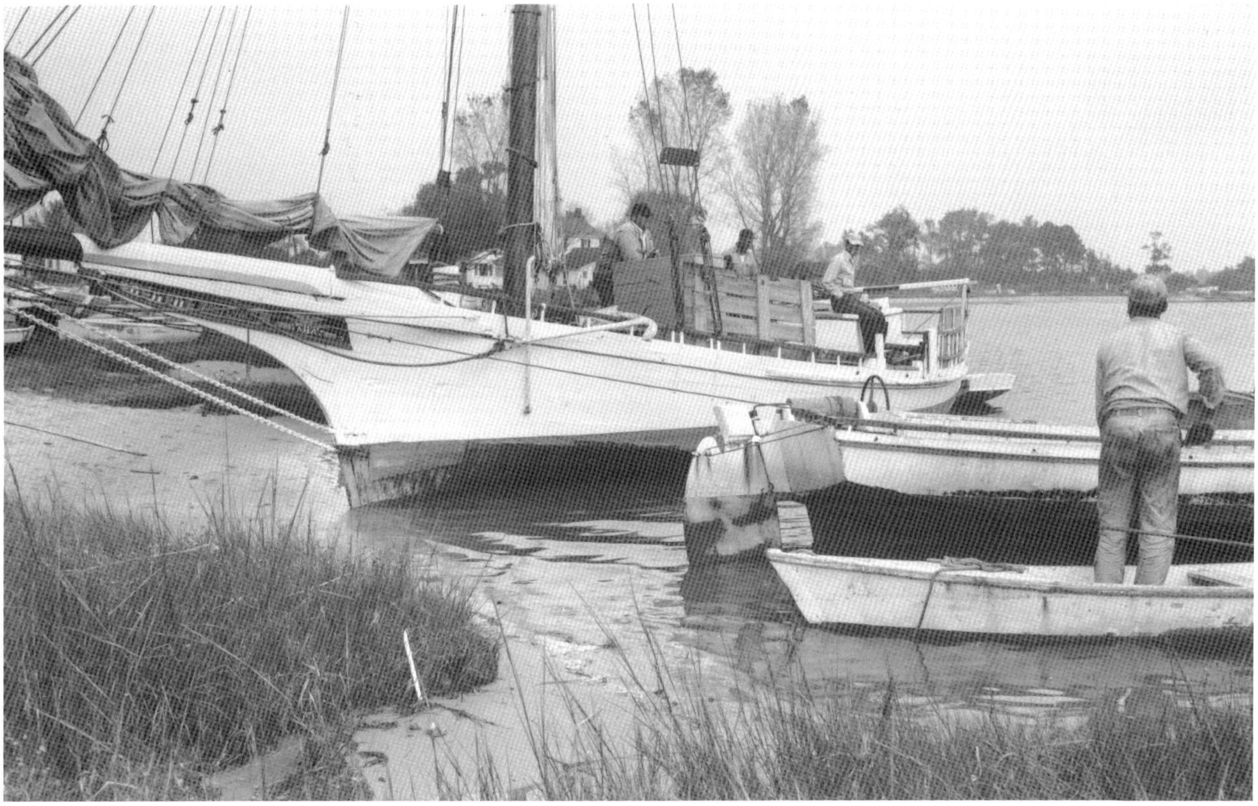

No casualty here. The skipjack *Mayflower* had been beached on soft sand at high tide to permit work on her hull, thereby eliminating shipyard charges. This was on October 27, 1953, just before the start of the dredging season. The powerboat, at right, has also been beached and her owner applies an old broom to her bottom to clean it off, a common procedure around the Bay as it was here at Wenona, Deal Island. *Mayflower* appeared in *MVUS* with date and place of building unknown. ✪ She was reported as "Abandoned" in the 1973 *MVUS*.

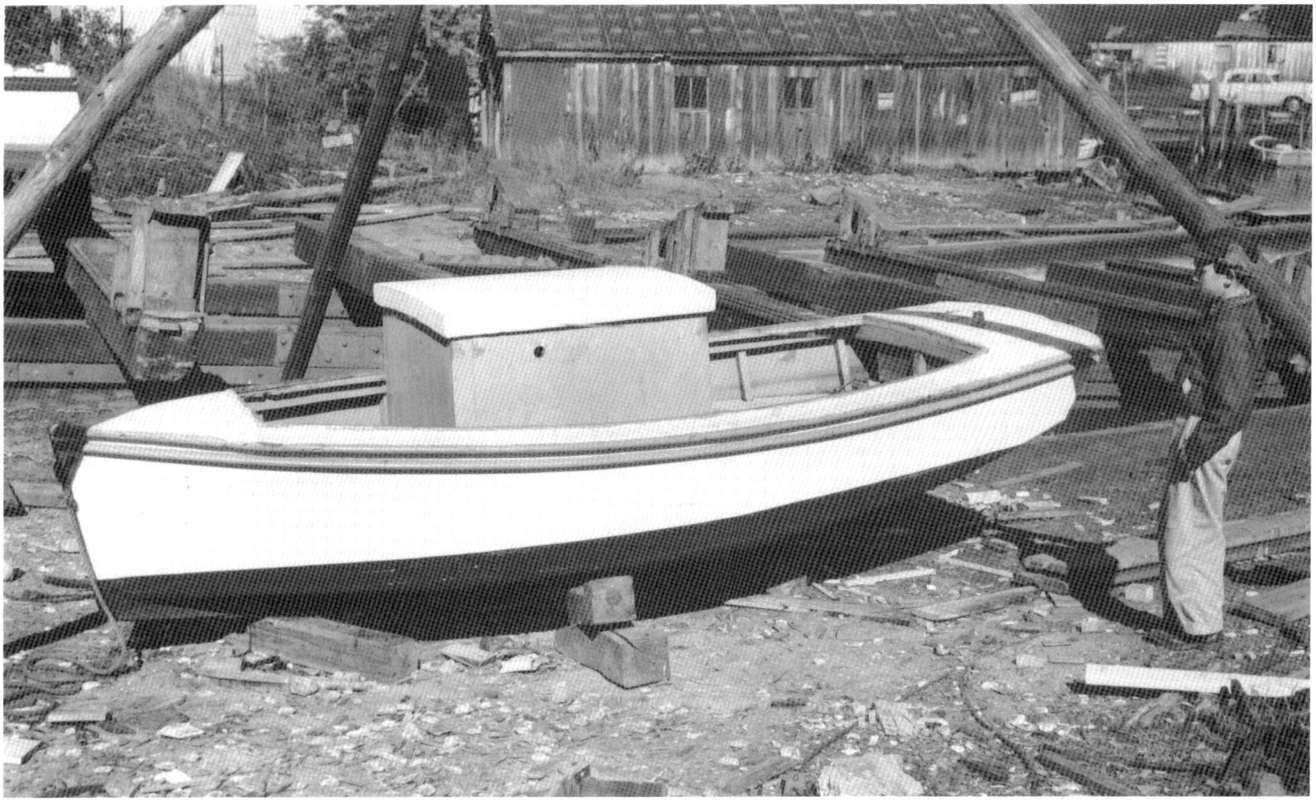

Compare this yawl boat belonging to the skipjack *Nellie L. Byrd* at Cambridge, October 22, 1955, to one of those used on the larger Bay sailing craft, such as the *Ruth Decker* 20 years earlier, on page 174. ✪ The author's son Bruce is in the foreground.

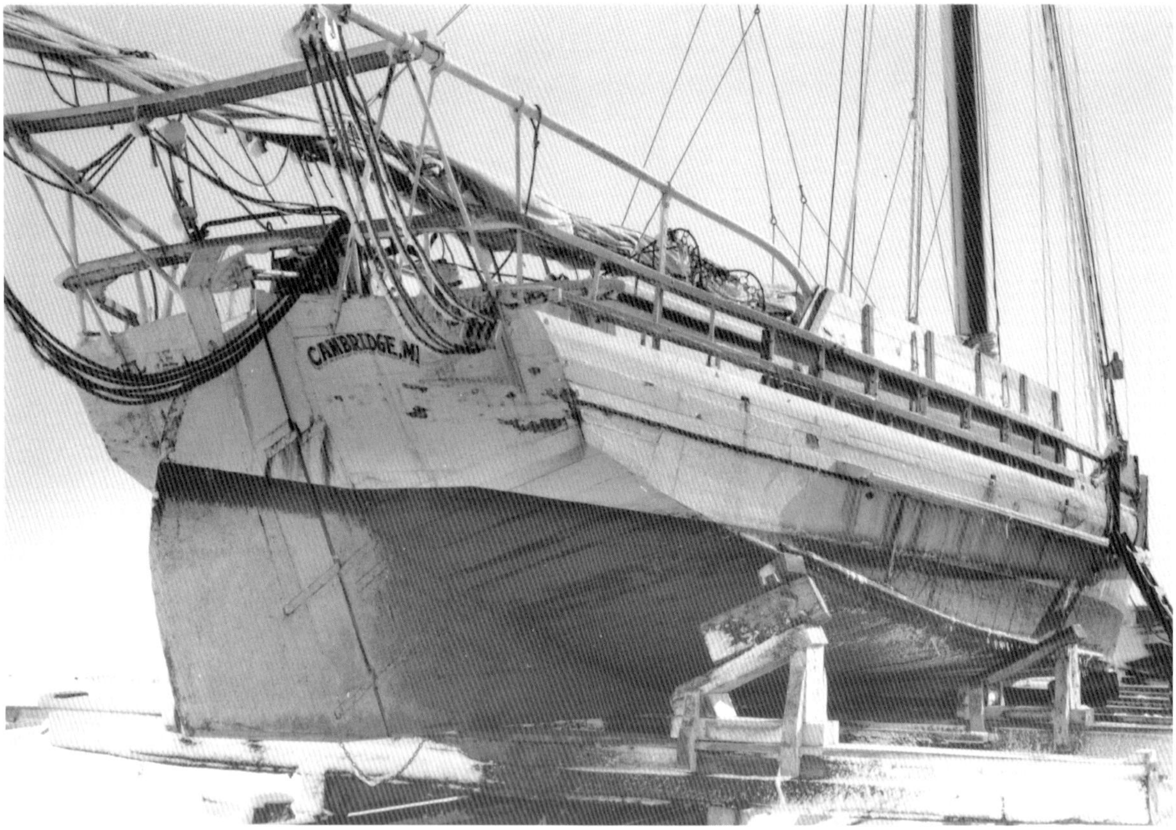

✪ The *Nellie L. Byrd* on the railway, probably at Cambridge on the same day as the previous photo. She was built at Oriole, Maryland, in 1911, and was still sailing in 2005.

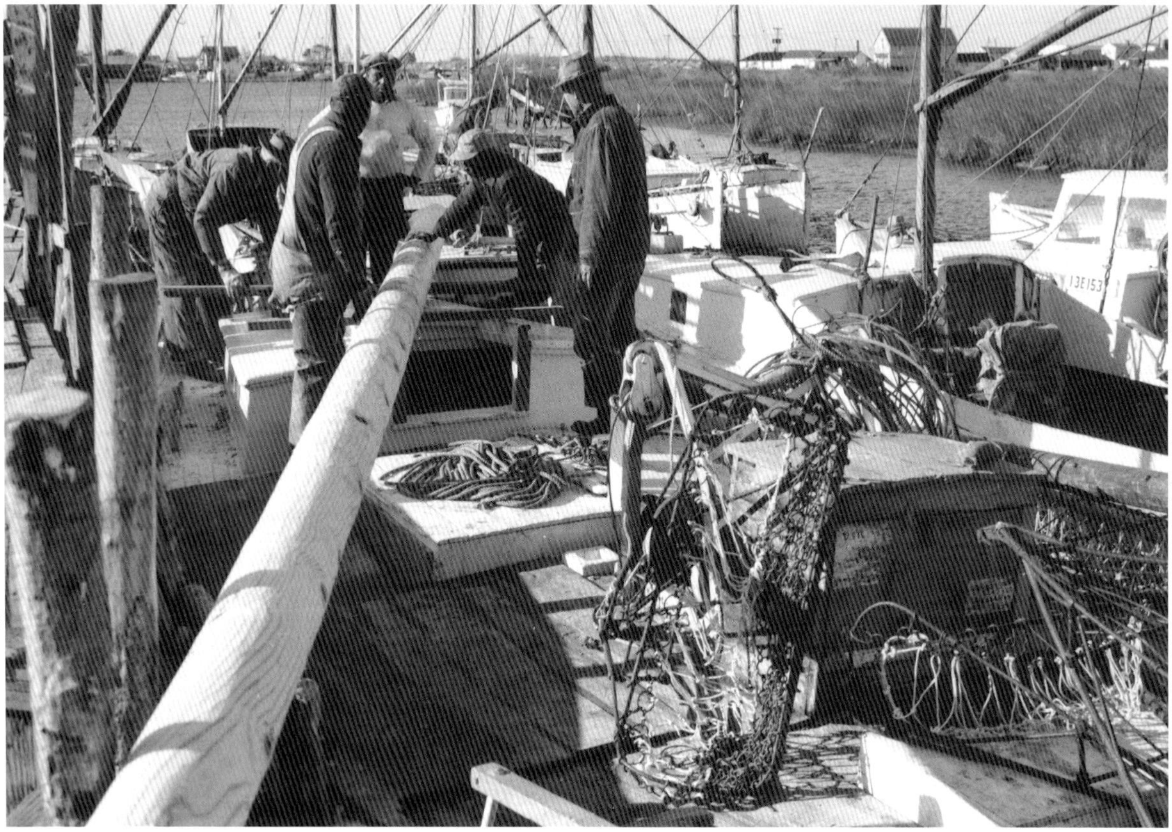

The crew of the skipjack *Noralee* steps a new mast at the outset of the 1957 dredging season, at Chance, Maryland. No imported tree is this, but a pine of proper dimensions from a forest on the Eastern Shore.

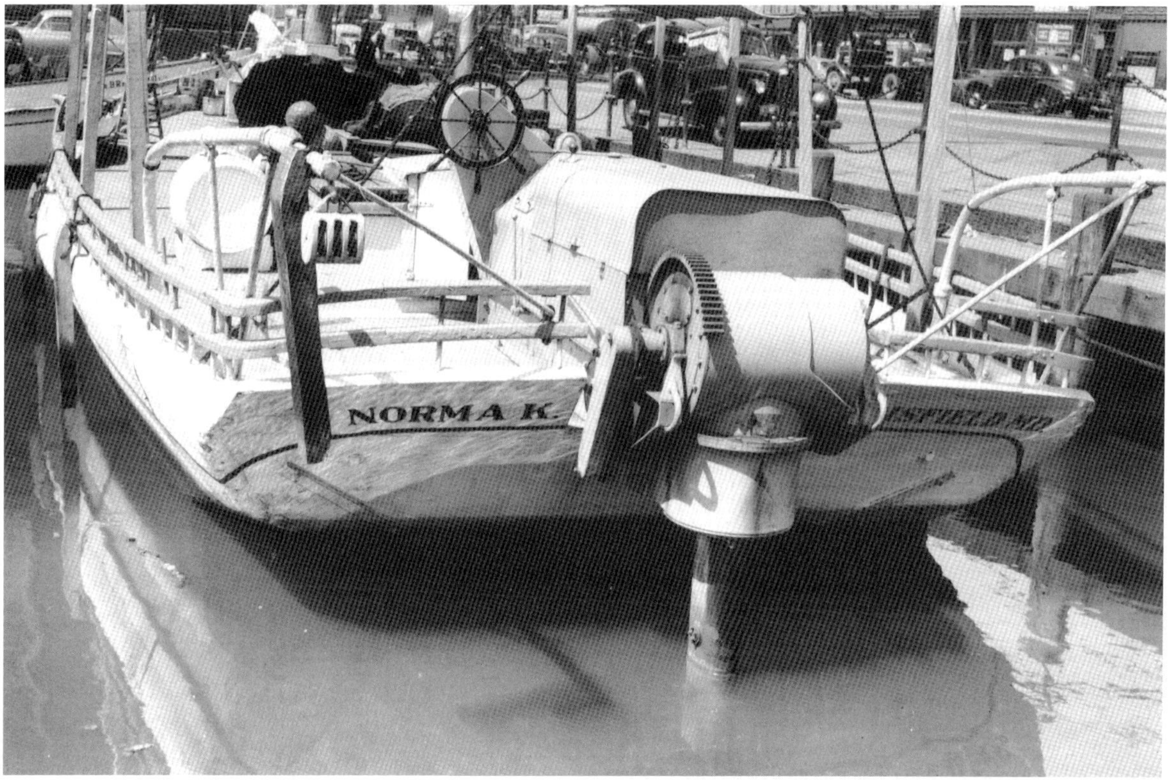

Someone was bound to try it—auxiliary power for a skipjack. The *Norma K.* is tied up at Pratt Street, Baltimore, after discharging a cargo of watermelons on August 30, 1950. Mounted on her stern is a surplus U.S. Navy "Rhino," a Chrysler auxiliary engine of about 100 horsepower normally used to propel barges. Although she still carried a mast, her sails were removed. The upright stakes along her railing support a canvas awning to shade her melons. *Norma K.* was built at Whitehaven, Maryland, in 1915. Not long after this picture was made she had reached the end of her usefulness and appeared in the 1956 *MVUS* as "Abandoned." Perhaps there was too much of a kick in the "Rhino" for her aging timbers. The author never saw this experiment tried on any other Bay sailing craft.

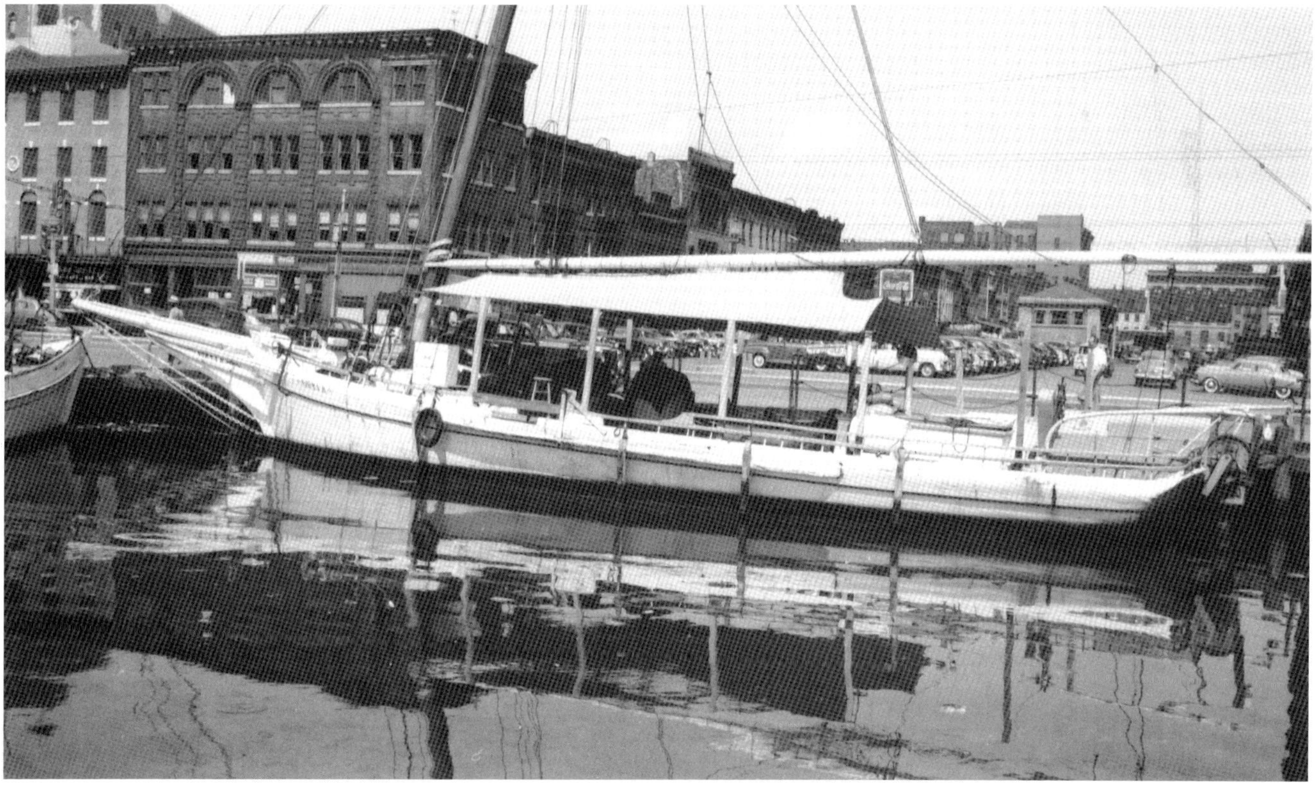

✪ A profile of the *Norma K.*, taken on the same day. A canvas shade has been rigged to protect her cargo and crew from the hot August sun.

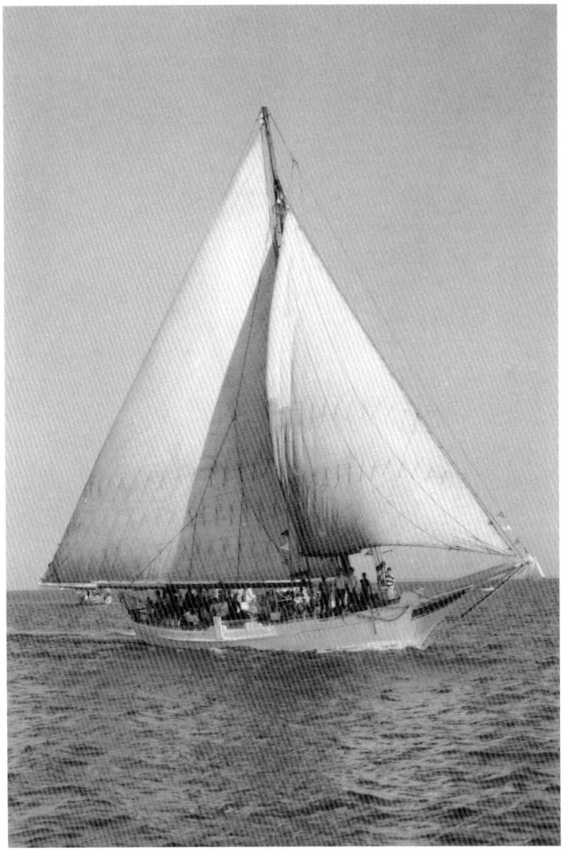

One of the largest skipjacks, the *Robert L. Webster,* sails down Tangier Sound during the 1970 Labor Day workboat races at Deal Island. During these races the skipjack takes aboard those observers who want to sail around the course; as many as 50 people have been counted on the *Webster's* deck. This 60-foot skipjack was built in 1915 at Oriole, Maryland. ✪ She was reportedly broken up in 1981.

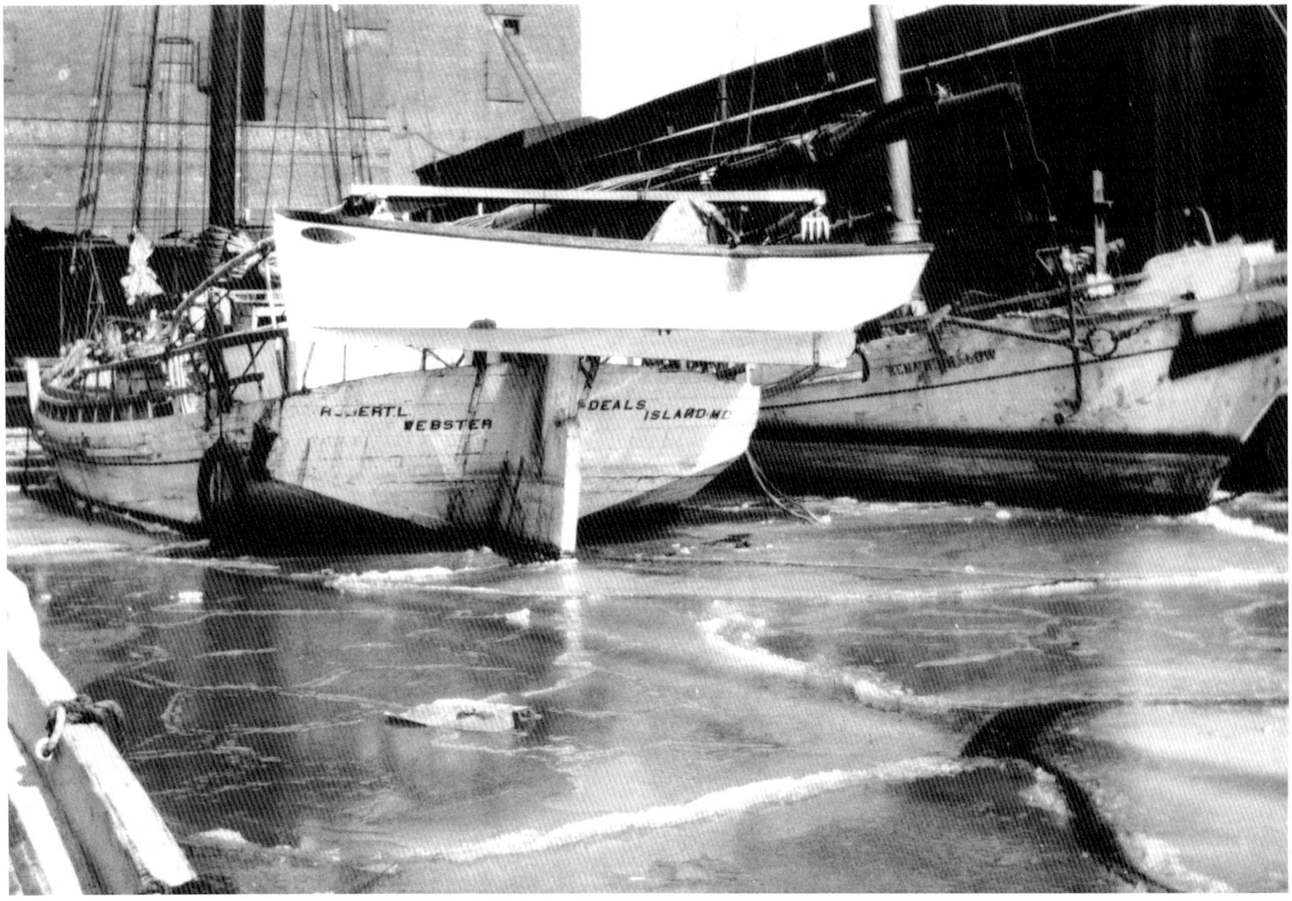

✪ Skipjacks *Robert L. Webster* and *Rena A. Callow* in the ice at Baltimore, sometime in the 1930s. The dirty harbor water has left its imprint on the hull of the *Callow.*

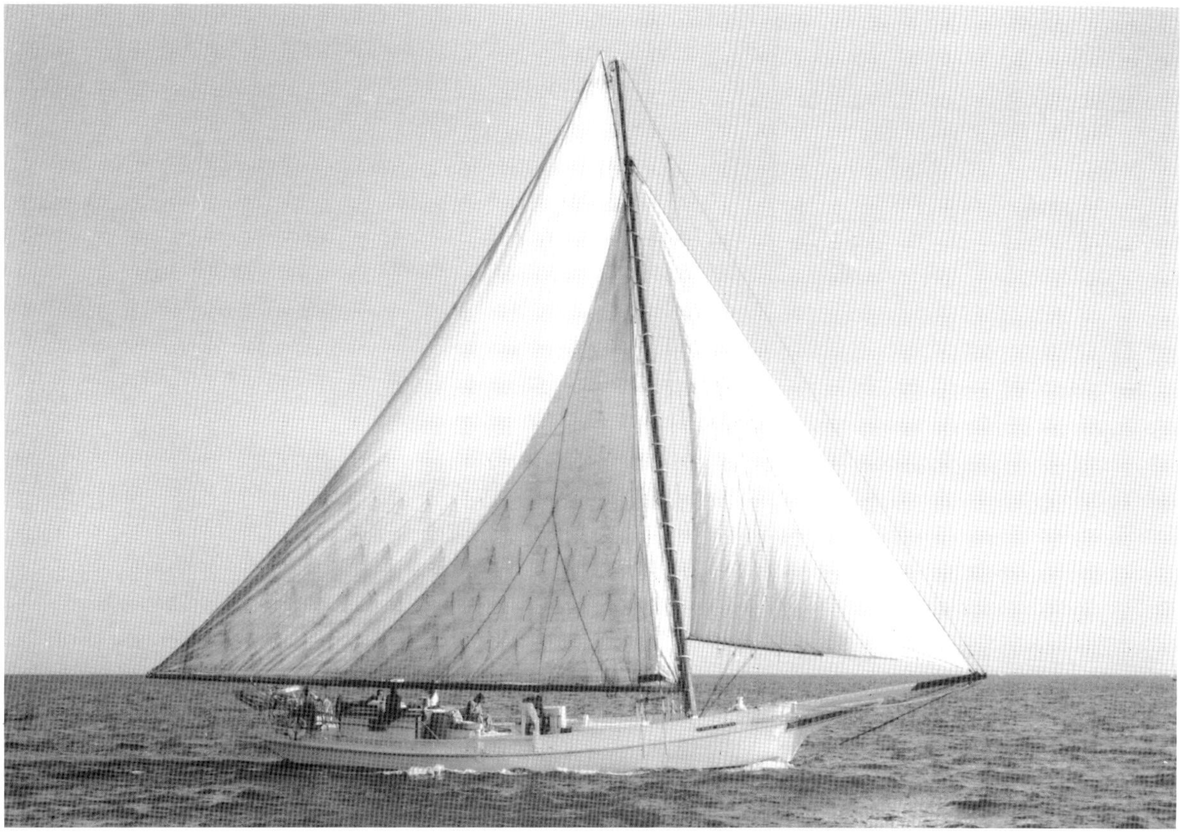

Rosie Parks was built for Captain Orville Parks by his brother Bronza Parks at Wingate, Maryland, in 1955, and named for their mother. Although a "drudgin'" vessel and working hard during the winter months, when lay-up season comes she is maintained like a yacht and looks that part as she participates in the 1970 Labor Day "Oyster Fleet Sailboat Race" sponsored by the Deal Island-Chance Lions Club. *Rosie Parks* has won more first places in that event than any other skipjack.

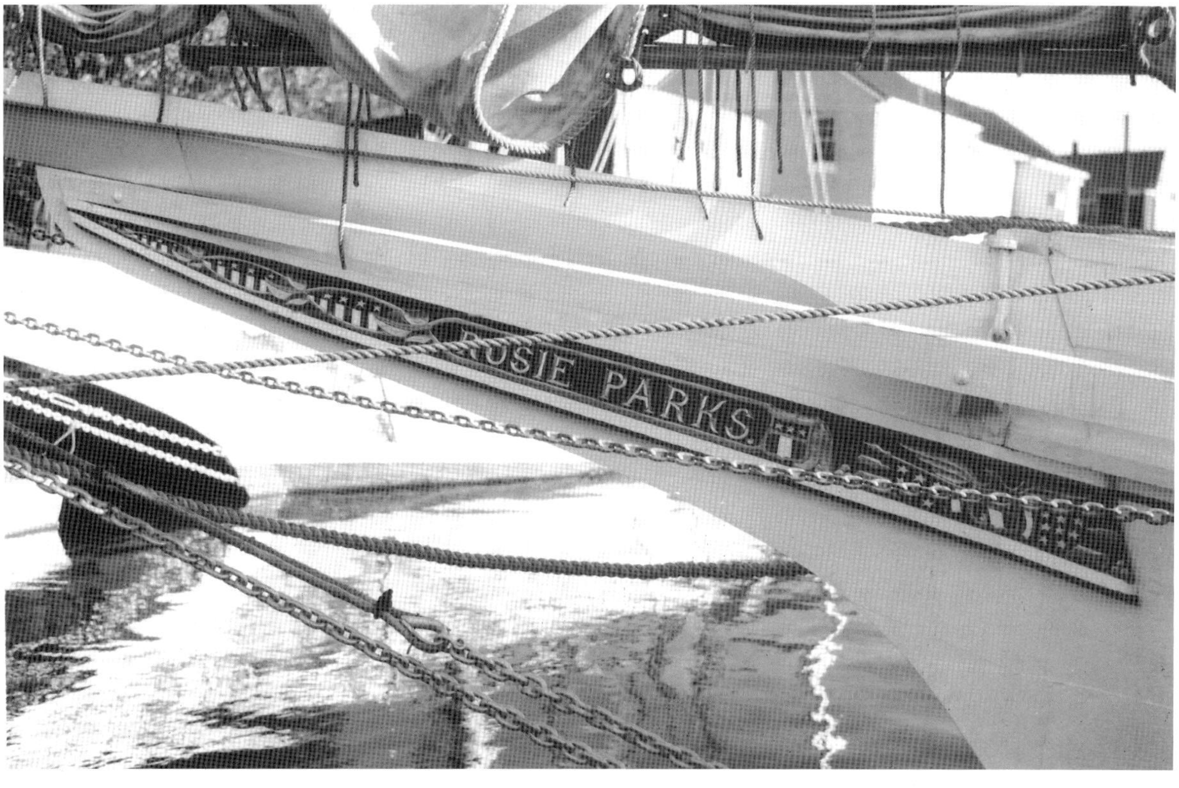

Trailboard of the *Rosie Parks,* carved by Captain Dewey Webster, carries on an old tradition on one of the newer skipjacks. This skipjack also has carved quarterboards and a stern decoration.

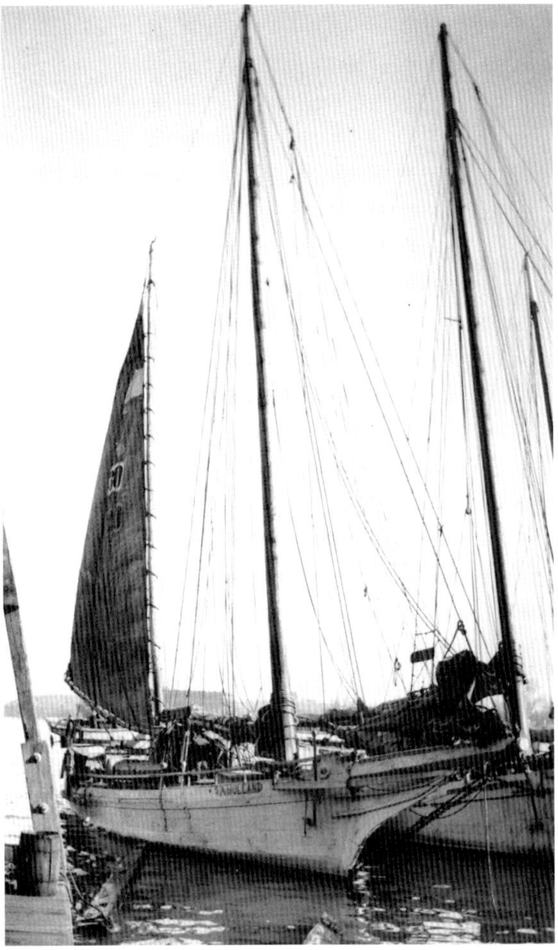

Three-sail bateau, or two-masted skipjack with bugeye rig, *S.A. Holland,* built 1893 at Marion, Maryland, at A. Smith & Sons Shipyard, Curtis Creek, Baltimore, December 23, 1933. To her port is the skipjack *Fannie L. Daugherty.* Tied up for the Christmas holidays, some of the vessels' crew takes the opportunity to go back to their Eastern Shore homes. The *Holland* carried on until listed as "Abandoned" in the 1941 *MVUS.*

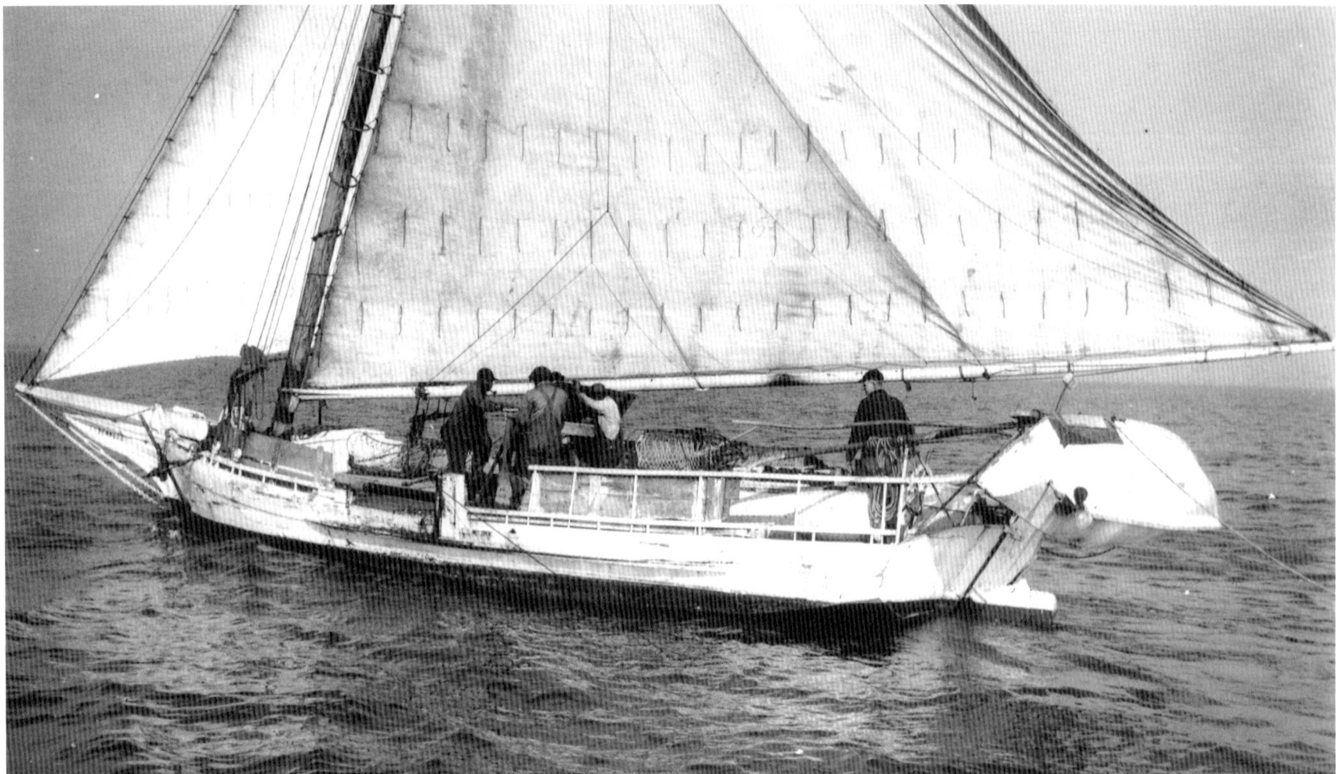

S.J. Bennett dredges oysters in the Choptank River, February 19, 1948. The cable attached to the dredge being pulled over the bottom can be seen extended over the port roller from the winder. The *Bennett* was built in 1903 at Inverness, Maryland, and last appeared in the 1965 issue of *MVUS.*

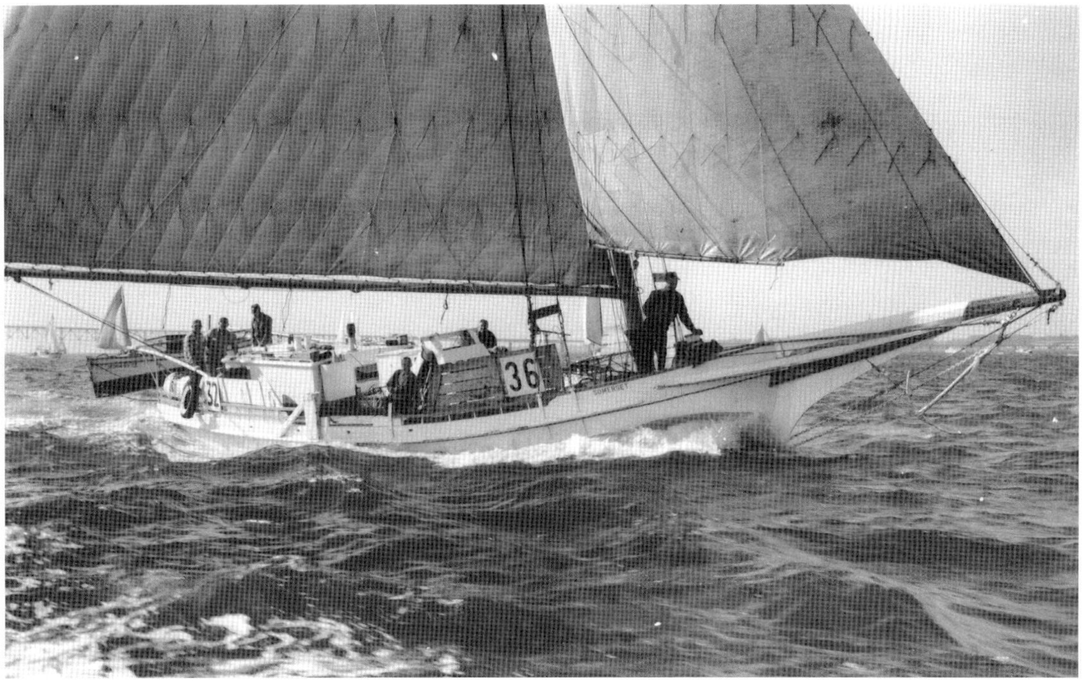

Skipjack *Somerset* races up Chesapeake Bay during the Chesapeake Appreciation Day workboat race held off Sandy Point, Maryland, October 29, 1966. One of the later additions to Maryland's dredging fleet, the *Somerset* was built in 1949 at Reedville, Virginia, and still plies her trade on the Chesapeake. The anchor can be seen suspended from the end of the bowsprit and attached to a line secured on deck. Most of the skipjacks carried two folding stock anchors, one on each bow with rope or chain cables led through the hawsepipes onto deck.

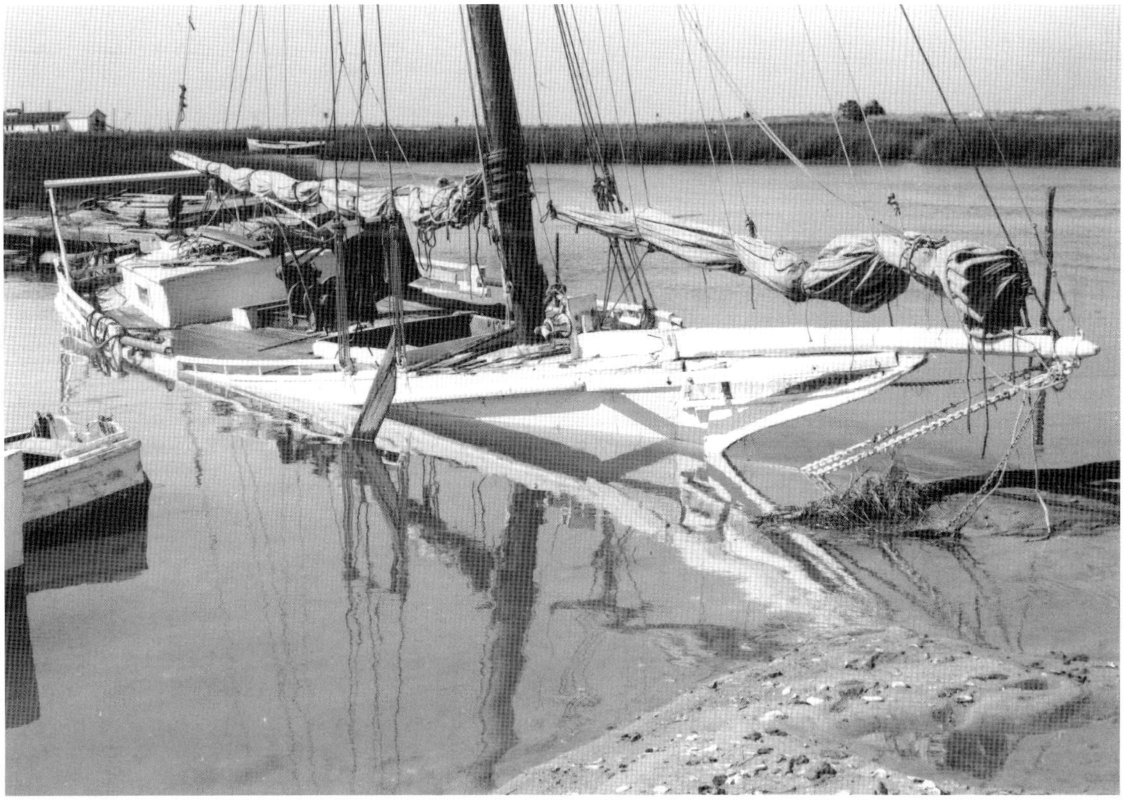

Hurricane Hazel, passing over the Eastern Shore, October 15, 1954, left considerable havoc in its wake. One of the skipjacks disabled at Deal Island-Chance, Maryland, was the *Upshur Q.* This disastrous storm came just as the dredging fleet was completing preparations to open the season. Some skipjacks were delayed for repairs, but there were few major losses. *Upshur Q.*, built at Justisville, Virginia, in 1905, was refloated to carry on as an oyster dredger. ✪ She was reported as "Dismantled" in the 1975 *MVUS*.

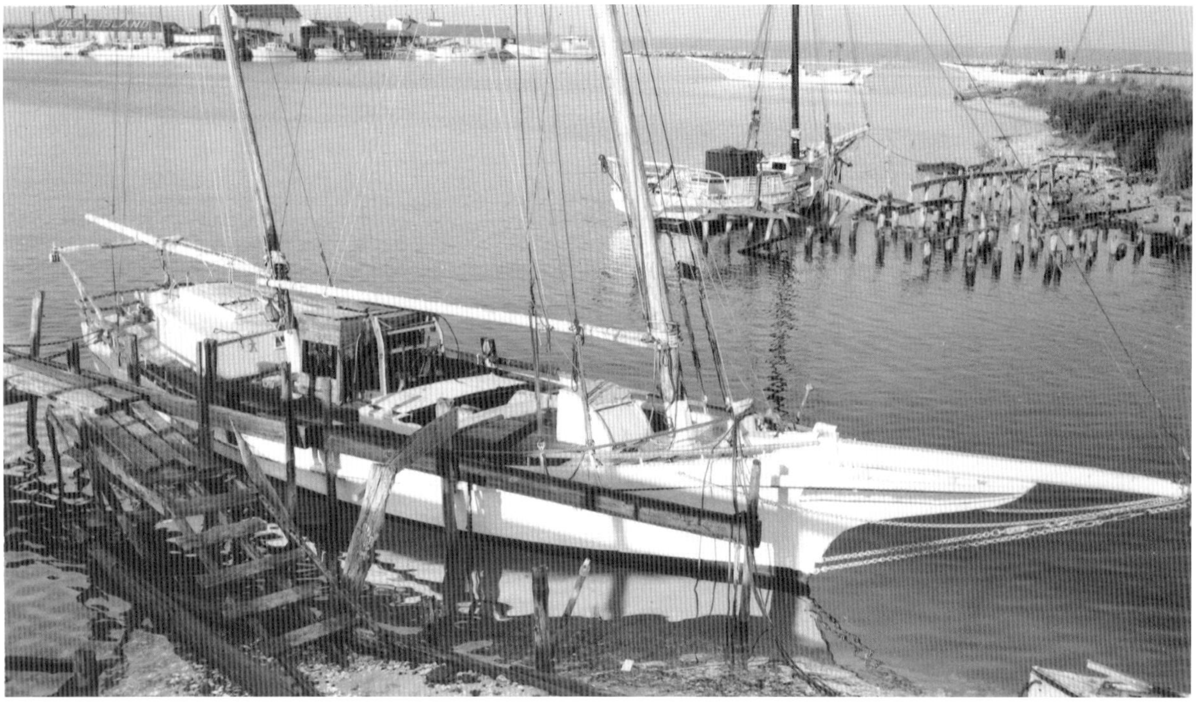

Three-sail bateau *Virgie G. Dean* at Chance, Maryland, October 27, 1952. Within a few days she would have had her sails bent and started another year over the oyster beds. In the left background are the oyster packing plants on Deal Island. The *Dean* was built in 1897 at Bishops Head, Maryland, and recorded as "Abandoned" in the 1972 *MVUS.*

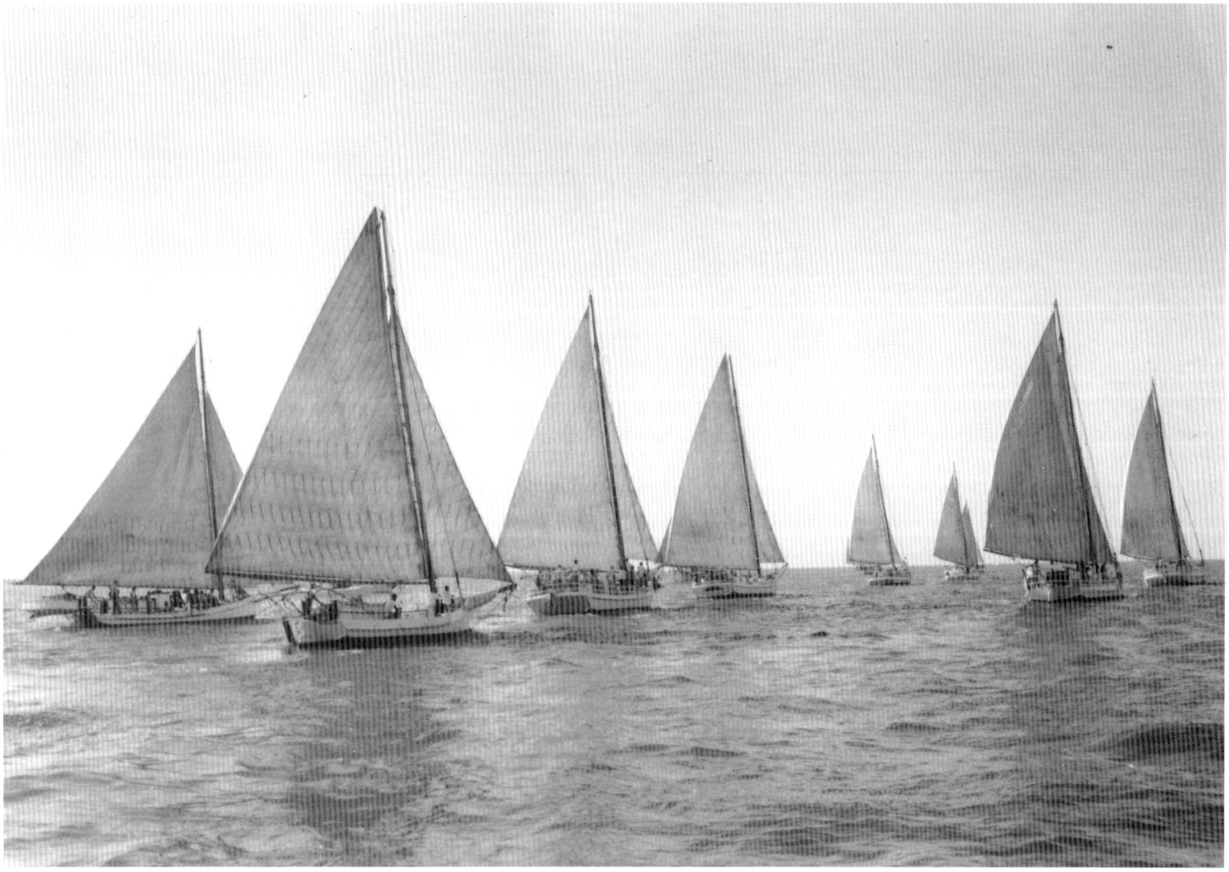

Twelve skipjacks get under way at the start of the Tenth Annual Oyster Fleet Sailboat Race on Tangier Sound, off Deal Island, Labor Day, 1969. The triangular course was laid down the Sound to Sound Point from Haines Point Light, then to Chain Shoal Bell Buoy and back to Haines Point. Starting at 9 a.m., the boats seldom get back to the finish line before 3 p.m. Shown is one of the better starts, with a favorable wind, during the decade of racing.

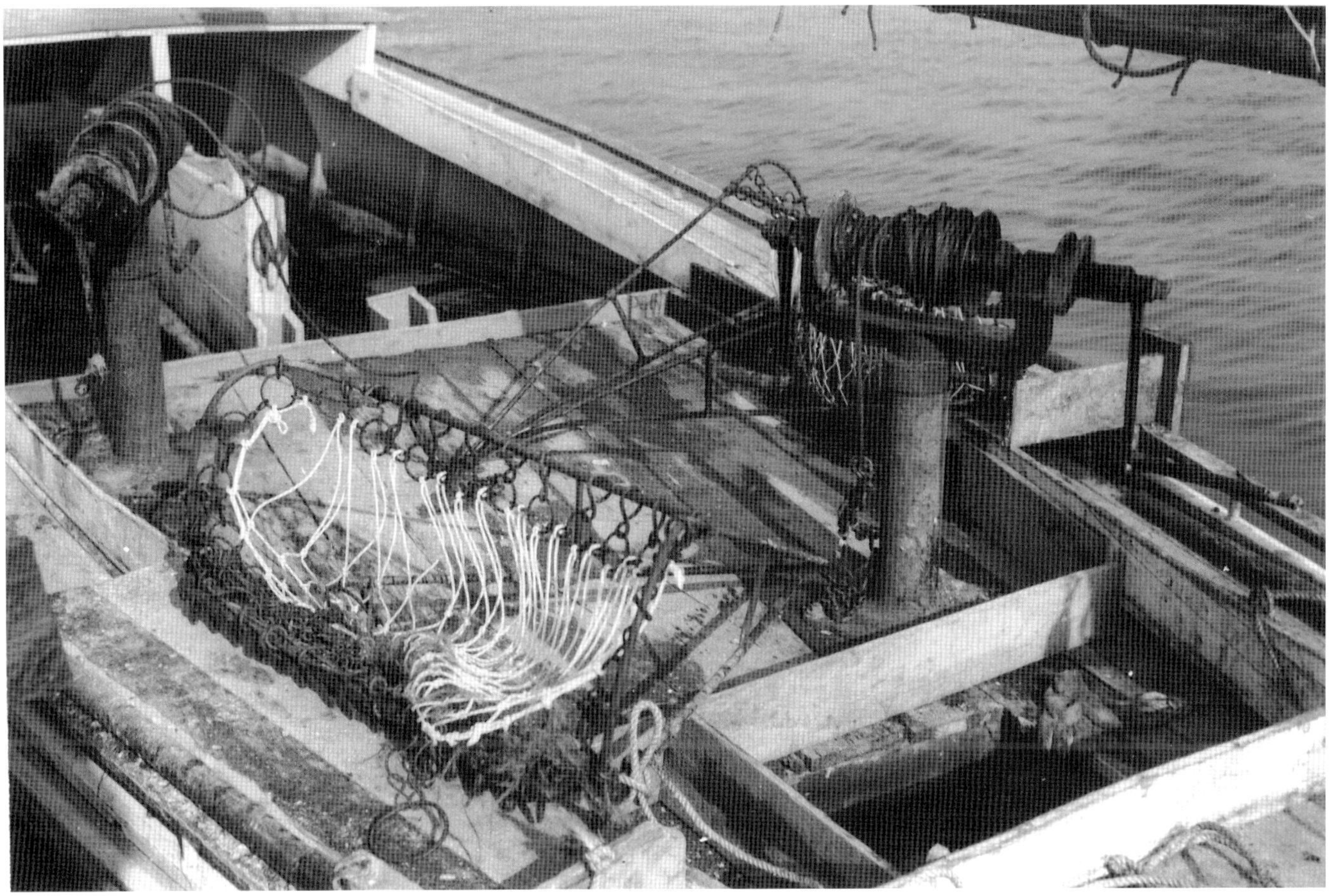

A set of hand winders still in use on an open-decked skipjack at Deal Island, November 1949. The dredge with rope bag is attached to a cable that is wound around the drum of the winder. When in use, the dredge is thrown over the side and the cable passes over the steel roller atop the rail. As the boat sails along, the dredge is pulled over the bottom and its steel teeth dig into the bottom and the oysters fall into the net. When full, it is pulled up to the deck by two men, one on each crank handle. Gas engine winders came into use early in the 20th century. The pungy *L.B. Platt,* in October 1906, was one of the first vessels on the Bay to have such an installation.

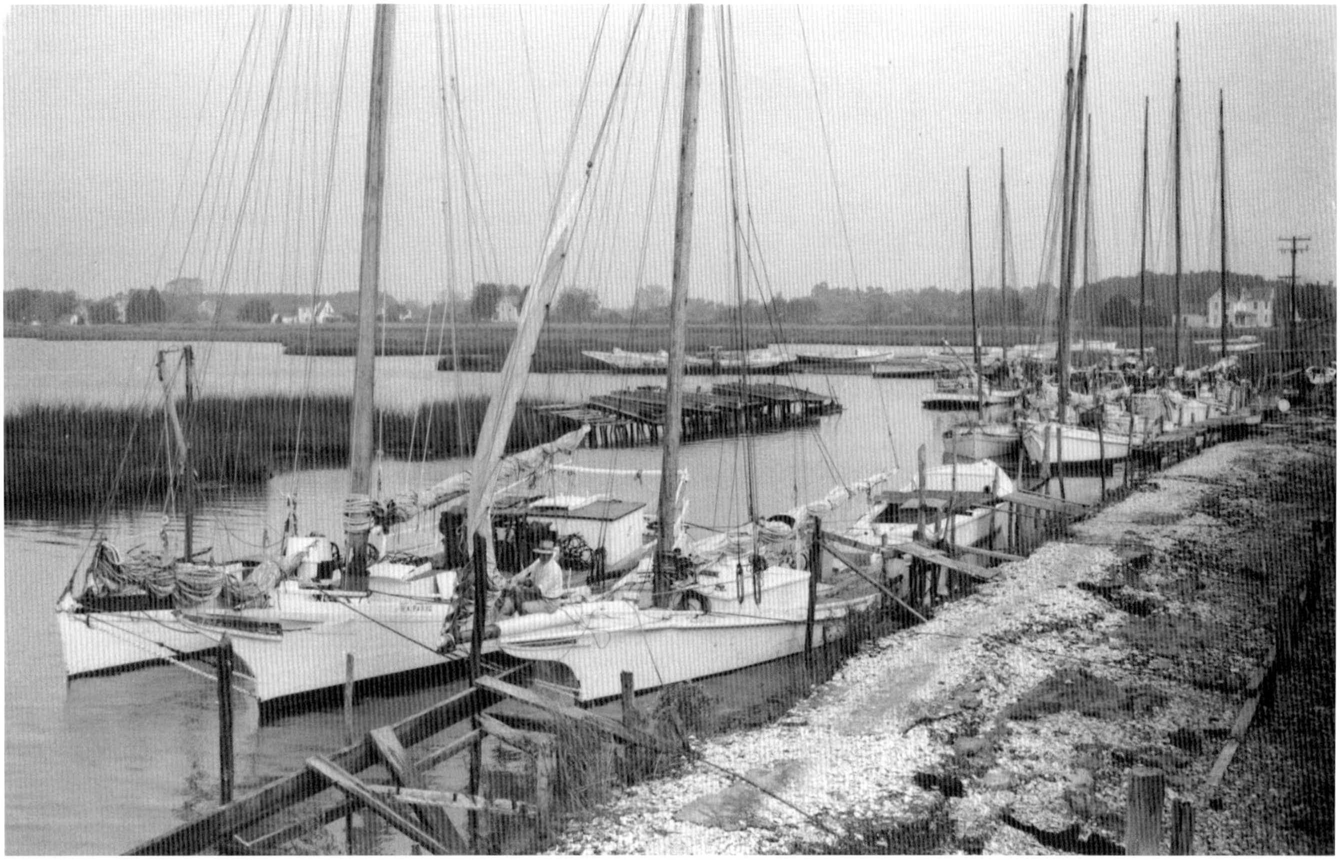

Some of the skipjacks of the Deal Island-Chance dredging fleet line up along the breakwater at the base of the bridge connecting the two settlements, October 27, 1953. Among the vessels are the *H.A. Parks* and *March Gale,* in foreground, and *Upshur Q.* and *Ida May,* near background.

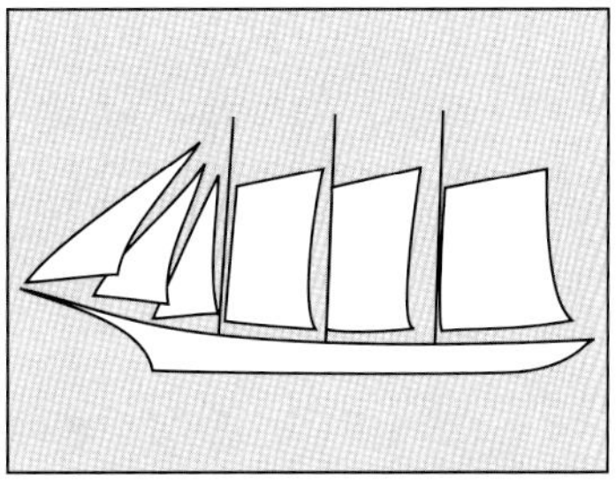

Three- and Four-masted Schooners

Windjammer Back, Sea Saga Relived; Doris Hamlin Shatters Record on Haiti Voyage; Calls Ended Voyage Worst and Longest; Nine From Here On Missing Schooner; Schooner Springs Leak On Way Here; Vessel Bound Here Long Overdue; Ship Takes 7 Days To Make 200 Miles.—All headlines from Baltimore newspapers in the mid-1930s, some exaggerated by reporters desiring to make their stories eye-catching, but all containing a breath of salt air linked to the past through the large sailing ship. These headlines might be called the death throes of the large schooner. The four-masters owned in Baltimore and clinging to trades out of the Chesapeake were kept active hauling coal, lumber, and logwood and were earning good freights for their owners. The hazards of World War II and not the decline in the trades they followed were to mark the end of the big schooner on the Chesapeake. Although the vessels were getting old in the late 1930s there were a few more years of good service remaining in their timbers if they had been left to their normal routine.

There were never great numbers of four-masted schooners owned in or operating out of Chesapeake ports. Only ten were ever built in the Chesapeake watershed states of Maryland, Virginia, and Delaware—and one rebuilt. The first four-masted schooner built here was the *William T. Hart*, at Alexandria, Virginia, in 1883. The first of two built in Baltimore was the *Van Lear Black*, in 1889; the other was *The Josephine*, in 1891. Then followed the *Sallie C. Marvil* at Sharptown, Maryland, in 1901; *Judge Pennewill* at Bethel, Delaware, in 1906; *Albert W. Robinson* at Sharptown, in 1907 (the last two were also referred to as four-masted rams); *Anna R. Heidritter*, rebuilt from the burned-out *Cohasset* at Sharptown, in 1910; *Charles M. Struven* at Pocomoke City, Maryland, in 1917; *Purnell T. White* at Sharptown, in 1917; *Alexander H. Erickson* at Seaford, Delaware, in 1918; and *Anandale* at Sharptown, in 1919. One might have expected more of these big schooners to have been built in the Bay area judging from the excellent shipbuilding timbers available in the tidewater forests. But large vessels built elsewhere, from Georgia to Nova Scotia, found their way to owners in Baltimore and other Bay ports. Ships, barks, and barkentines had thrived earlier in trades out of the Chesapeake but passed from the Bay fleets prior to the period covered in this volume.

Three-masters, schooners, and rams were numerous; many were built in the Bay country and several traded on the Chesapeake up into the 1950s. They rarely went to sea in the half century covered here (✪ 1925–1975) but usually confined their activities to the sheltered waters of the Bay in the same trades as the two-masted schooners, from northeastern North Carolina to Philadelphia. It was partly due to this sheltered existence that the majority of the vessels were permitted to carry on as long as they did. Many had rigorous early careers sailing coastwise and later thrived in the Chesapeake.

Around 1900 and shortly thereafter Chesapeake Bay was the destination of countless coasting schooners ranging from three to seven masts. Most of them were bound for Hampton Roads ports or Baltimore for coal, or for Washington, D.C. and Baltimore with ice. The coal trade out of Hampton Roads lasted until the early 1940s, as long as there were large schooners available to carry this cargo. Incoming cargoes to Baltimore in those latter days consisted of phosphate, lumber, and logwood. Logwood was a dark, heavy, gnarled tropical wood that came from various ports in Haiti. The last Chesapeake-owned schooners in that trade were the *G.A. Kohler, Doris Hamlin,* and *Albert F. Paul.* Schooners owned in other ports also brought in logwood and sheep and goat manure from Venezuela. They were the *Edward L. Swan, Herbert L. Rawding,* four-masted schooners, and the five master *Edna Hoyt.*

These big schooners, ranging from 174.5 to 224 feet in length, formed the last tie between the Chesapeake and deepwater sail. They were of wooden construction and built between 1917 and 1920, varying in appearance from the boxlike *Kohler* to the more graceful *Rawding.* They were engineless as far as propulsion was concerned but did carry gasoline or steam donkey engines for heavy lifting and pumping. Their crews were kept at a minimum—seven to nine men, including officers. The *Doris Hamlin* carried captain, mate or bos'n, four seamen, cook, and engineer. On some of her voyages, the *Hamlin's* master was the only white man onboard and the only one with any knowledge of navigation. She carried no refrigeration or wireless. The *Hamlin* was rare in having a Delco system for generating electricity for lights and radio receiver. She had steam heat in radiators originating from the donkey engine boiler but no running water. Water was heated on the galley stove or by exhausting live steam in a bucket of cold water until the proper temperature was reached.

The wages paid the *Hamlin's* crew were standard for sailing vessels of that period. In 1936 the captain was earning $125 per month plus his keep. In addition to their keep, the rest of the crew was paid as follows: cook $70; bos'n $55; engineer $50; able-bodied seamen $35; all per month. The schooner was the only home some of the men had. Upon arrival in port after a voyage the men would be paid off after the sails were removed if there was no charter in immediate sight. The cook or mate might live aboard to look after the ship; the captain had an apartment ashore.

Between trips the seamen might reside in a cheap "flophouse" near the waterfront. Daily they would frequent the store of Myer Saperstein who operated a hiring hall on the side. Schooner operators would contact Saperstein when crews were needed for their vessels whether they be the smaller Bay vessels or even oyster dredging craft, to the larger coasting schooner. Naturally, men with experience were preferred and there always seemed to have been a surplus of them along Pratt Street.

Rarely did one have the opportunity to see one of the four-masters under sail on the Chesapeake. Conditions had to be just right for them to sail between Baltimore and the Virginia capes. The waters were too confining to permit beat-and-tacking. But given a fair wind, they would carry all the sail they could to save the expense of a tow. I recall the *Doris Hamlin* sailing away from her pier at the J.S. Young and Company, Baltimore, on October 19, 1934. A strong northwest wind was blowing down the harbor and her master, Captain George H. Hopkins, carried just sufficient sail to keep way on her down the Patapsco River channel. As the river widened and the Bay was reached, the schooner had a good start on her voyage to Fort Liberte, Haiti. In August 1936, Captain Hopkins also eased his logwood-laden schooner up to the Young dock under sail.

If the big schooners needed a tow out of Baltimore, a tug belonging to Captain J. Herbert Mullen was usually hired. He had a fleet of old steam tugs as quaint as he was a character, the *Jesse Tyson, Easby,* and *Radnor.* In November 1933, I went with him on the 1873-built *Radnor* as he towed the *Doris Hamlin* downriver from Baltimore where she had room to proceed on her own. I had hoped to get some photographs of the schooner under sail after the tug had cast off the towline. But the wind came ahead so the vessel anchored off Bodkin Point to await a fair wind to send her on her way to Haiti.

I was aboard the *Hamlin* in October 1936, when she was towed away from Young's pier to an anchorage outside Baltimore harbor by her motor yawl boat and her owner's yacht *Melvin* in tandem. The next day the *Hamlin* was pushed by her yawl boat tied alongside a couple of miles farther down the Patapsco to anchorage off Stony Creek. The wind came around out of the northwest the following day and the schooner weighed anchor and sailed down the Chesapeake to anchor again 18 hours later off Thimble Shoal Light, entrance to Hampton Roads, Virginia. A tug would have charged $250 to tow the schooner that distance.

There was the inevitable, steady decline in large sailing craft (three- and four-masted schooners) in the 1920s and '30s as age, stress of the sea, and conversion into barges took their toll. None had been built since the World War I period. Some of those owned in Bay ports evaded me or had disappeared before my camera became active.

The ten-year period, 1924 through 1934, was a most disastrous one for the Chesapeake-owned four-masted schooners. Starting off this tragic decade was the loss of the *Dorothea L. Brinkmann* that stranded at Oregon Inlet, North Carolina, on March 24, 1924, while bound from Boston to Darien, Georgia. Then the *Jacob W. Hook* foundered off Cape Hatteras on November 11, 1928, with the loss of three men. On August 12, 1930, while bound to St. Georges, Bermuda, with a cargo of coal, the *Sally Wren* grounded on West Reef, off Bermuda, was freed, and towed into the harbor. She was discharged and then abandoned there in Mullet Bay.

In early 1931, the *Emerett,* originally the *Cumberland Queen,* left Jacksonville, Florida, with a cargo of lumber for Puerto Rico. She was sighted on March 4, dismasted, waterlogged, and abandoned. On the 21st, the Coast Guard cutter *Modoc* picked up the derelict and towed her into Norfolk. The hulk was later abandoned at Alexandria, Virginia, for use as a gravel storage barge and through the years she deteriorated there. Her bones were removed in 1972 to make way for a new marina.

On January 15, 1932, the *Charles D. Stanford* left La Vela de Coro, Venezuela, for Boston with a cargo of sheep and goat manure. A month later she put into Jacksonville and then left on the 17th. She was sighted on March 3 east of Cape Hatteras but not reported thereafter—lost with all hands. The next year, on August 23, the *G.A. Kohler* stranded at Gull Shoal, North Carolina, and became a total loss.

On February 9, 1934, the *Purnell T. White* capsized under tow south of the Virginia capes with the loss of four men. Her hulk was later taken to Baltimore and abandoned. This left only the *Doris Hamlin* and *Albert F. Paul* to carry on the tradition of deepwater sail out of the Chesapeake until the early years of the 1940s when they, too, were lost, after being sold by their Baltimore owners.

From the mid-1920s the following Bay-owned three-masted schooners, exclusive of rams, were gradually stricken from the active list: *Nantisco, Abel W. Parker, Lucie Wheatley, John L. Martino, Ida O. Robinson, William T. Parker, Robert, Virginia, Beulahland, William J. Stanford,* and *Edna Bright Hough.* The *Lydia Middleton, Josephine Wimsatt,* and *William Linthicum* were sold out of the Chesapeake but their careers thereafter were brief.

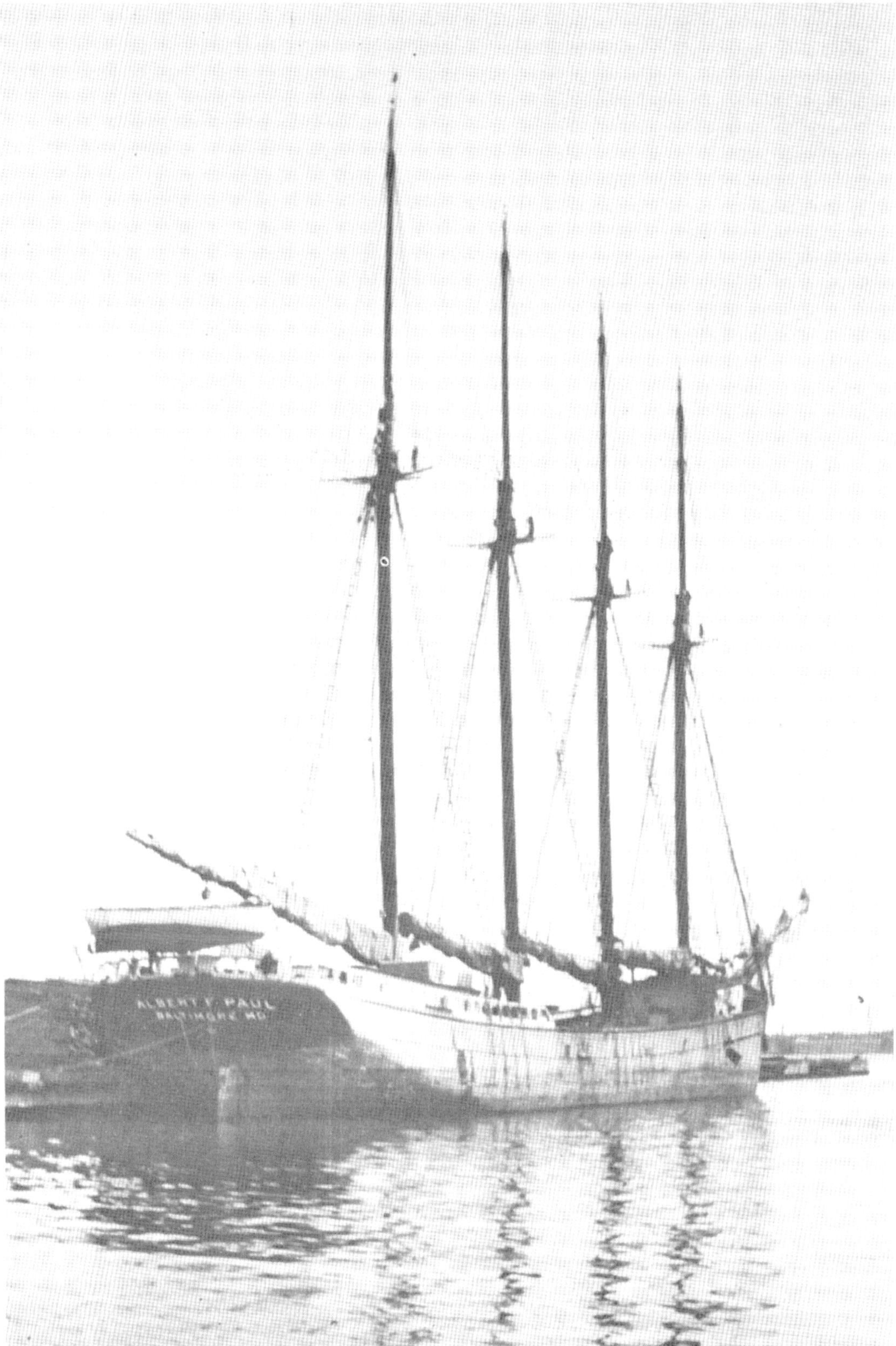

At Pier 5, Pratt Street, Baltimore, on May 10, 1931, lies the four-masted schooner *Albert F. Paul.* She had arrived on April 29 from Georgetown, South Carolina, with lumber and had discharged at Back Basin. Since these were Depression times, with no immediate charter for the big schooner she was tied up at Pier 5 for the summer. On September 1 she was towed down the Patapsco River to set sail for Georgetown again. The *Paul* was built by W.G. Abbott at Milford, Delaware, in 1917, for C.C. Paul and Company, Baltimore, as managing owners, and as shown here had just completed her 56th voyage since her maiden passage to Pernambuco, Brazil, in October 1917.

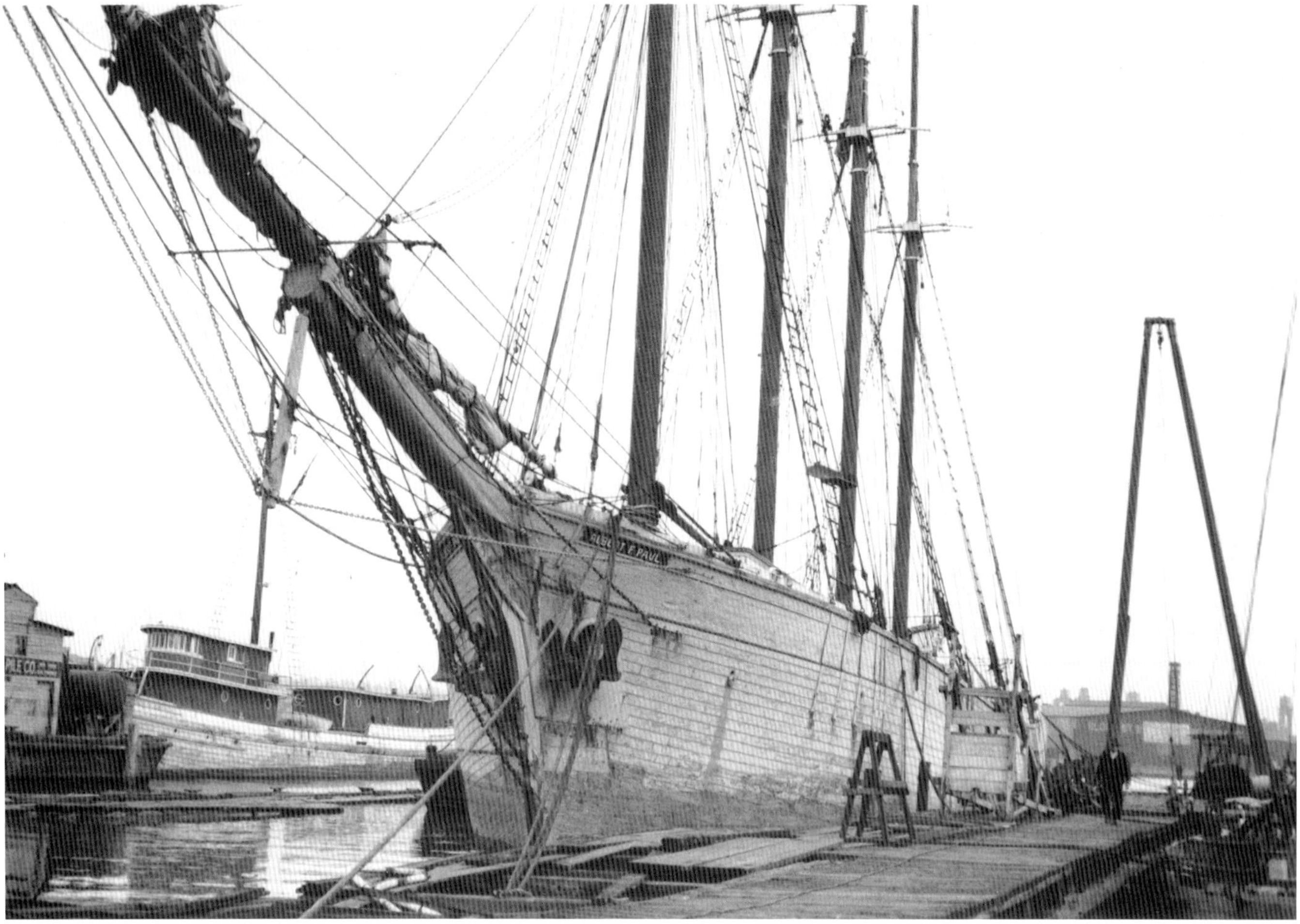

The *Albert F. Paul* had arrived at Baltimore with logwood from Cap Haitien, Haiti, on September 16, 1940, and after discharging was towed to the Redman and Vane Shipyard for slight repairs and is shown there on the 28th. She sailed on October 1 for Jacksonville, Florida. At the end of the pier are the shear legs with block and fall used for stepping and lifting the masts of the smaller Bay sailing craft. At left is the menhaden steamer *Alden S. Swan* at the dock of Booz Shipyard. She lacks a stack and was probably in the process of changing from steam to diesel power.

At this time, the *Paul* was nearing the end of her almost quarter century of sailing out of Baltimore, operating under the C.C. Paul flag the entire period. She had been one of the more successful coasting schooners, keeping active most of her career. In late 1941 the 174-foot schooner was sold to the Albert Shipping Company, Inc., of New York for $40,000. She had cost $76,000 to build and outfit 24 years earlier, so her Baltimore owners did well considering that they had profitable use of her during that period. Her new owners had her load coal in Hampton Roads for Bermuda and after discharging she sailed down to Turks Island, British West Indies, to load salt for Norfolk. In command was Captain William M. Martino, who had been her master in 1920. He had replaced Captain Robert O. Jones who had been in the *Paul* since 1924. About February 21, 1942, she departed from Turks Island. Weeks passed by without any word from her and she failed to be reported. In April the family of Captain Martino received word from the U.S. Government that he was missing while in the service of his country. Later the U.S. Navy publicly reported her missing. A recent German publication revealing the activities of German submarines during World War II is more specific. It indicates that the *Paul* was sunk by torpedo on March 15, 1942, by U-332 in the Atlantic at 26° N. Latitude, 72° W. Longitude.

Facing page: These massive bows of the wooden four-masted schooner *Doris Hamlin* had furrowed the waters of the North and South Atlantic, Mediterranean, and Gulf of Mexico before coming to the Chesapeake to be registered out of Baltimore. She had been purchased by W.B. Vane at a U.S. Marshal's sale in Portland, Maine, on January 14, 1931, while the schooner was laid up at Boothbay Harbor. Two months later she sailed for Norfolk to be reconditioned and then to load coal at Newport News for Martinique. The *Hamlin* arrived in her new home port on March 27, 1932, and then she was idled until June. On the 23rd of that month she sailed for Bermuda and West Indian ports with a group of college students on a cruise under the title of Caribbean Motion Picture Expedition. Due to desertions and lack of money, the cruise came to an end in Martinique when her master, Captain F.E. Garfield, had to wire her Baltimore owners for funds to return home. Back in the Chesapeake again, the *Hamlin* was tied up at the old Woodall Shipyard in Baltimore until November 1933. On the 15th of that month the schooner was towed to the Bethlehem Shipbuilding and Dry Dock Company and placed in a floating dry dock before sailing for Cap Haitien, Haiti, for logwood. She is shown on blocks in the dry dock as men scrub her bottom before painting.

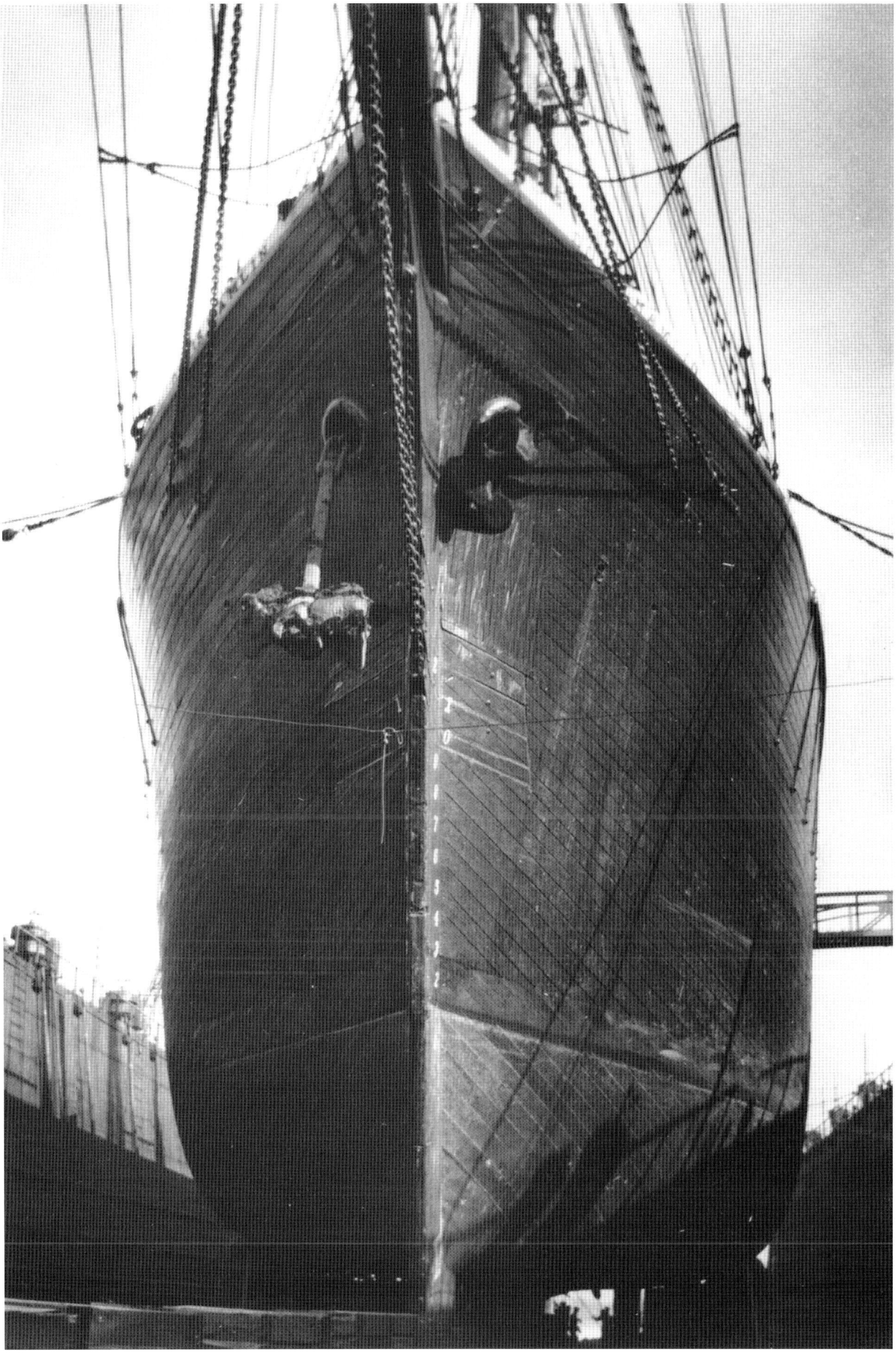

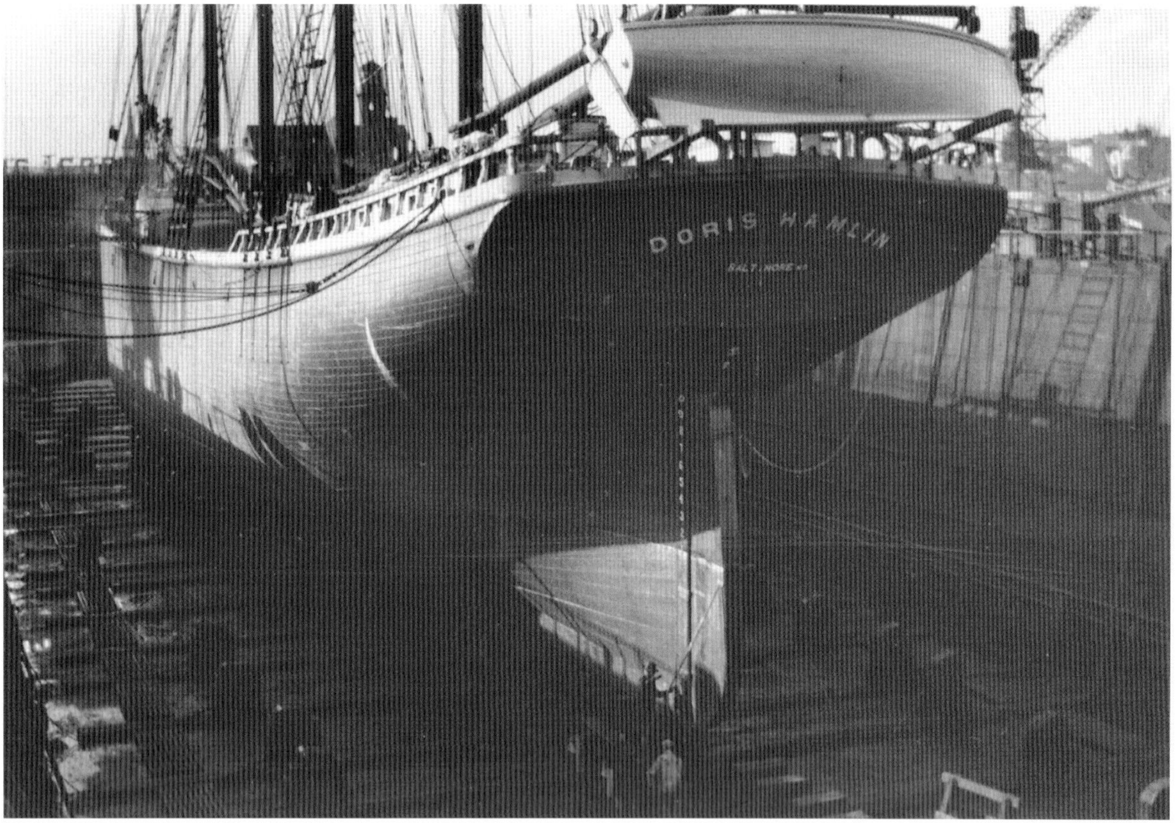

✪ The handsome stern of the *Hamlin* in dry dock, November 1933.

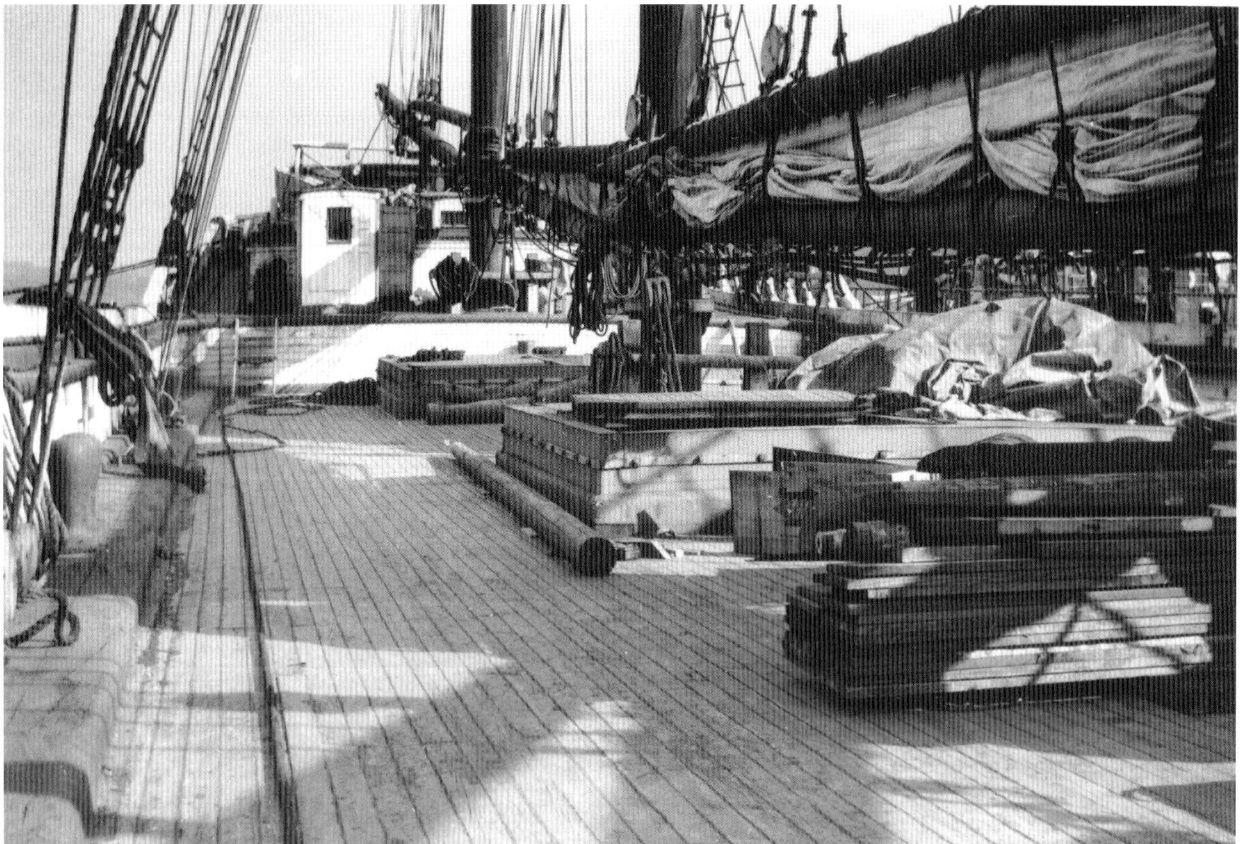

Looking aft on the deck of the *Doris Hamlin* while she was in dry dock. Her spanker lies atop her after cabin waiting to be bent. Before the ship gets to sea, the hatches will be covered and battened down and the dunnage secured on deck.

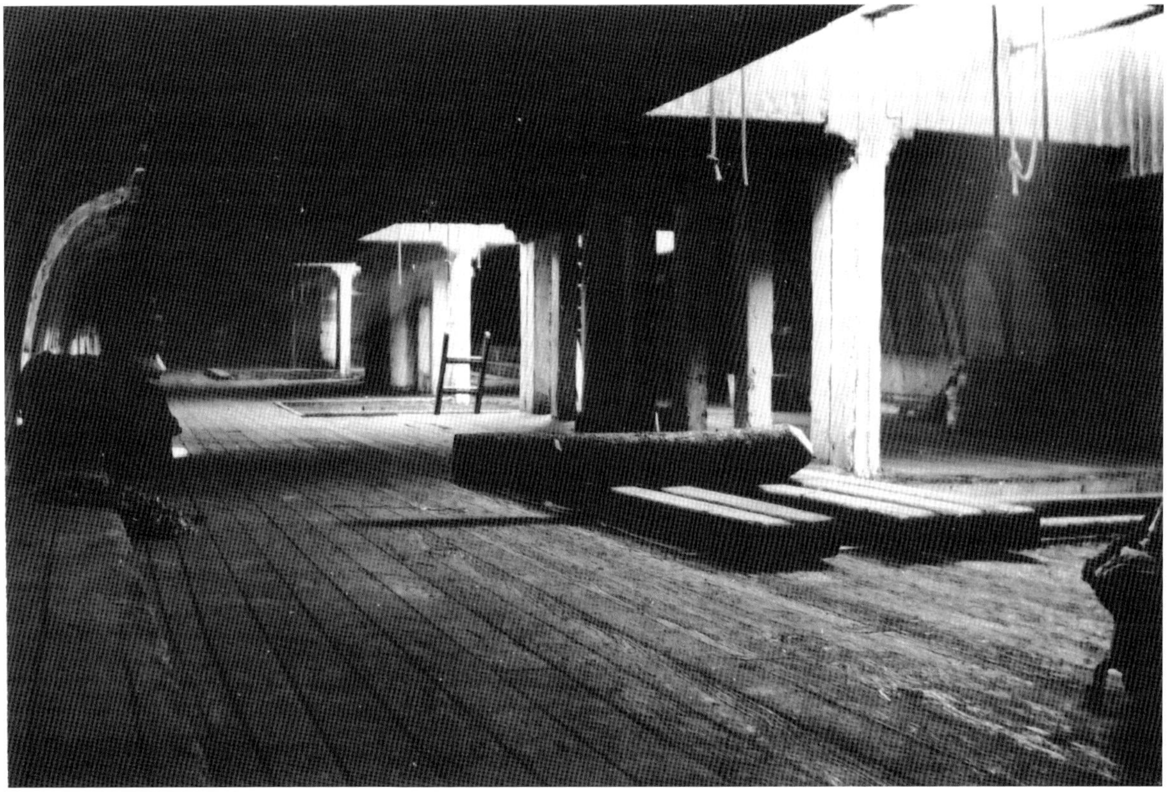

Looking aft in the 'tweendeck of the *Doris Hamlin* while in dry dock. The ladder in center leads to the lower hold. When she returned from Haiti this entire hold was filled with logwood and an additional supply was stowed on deck above the height of her railing.

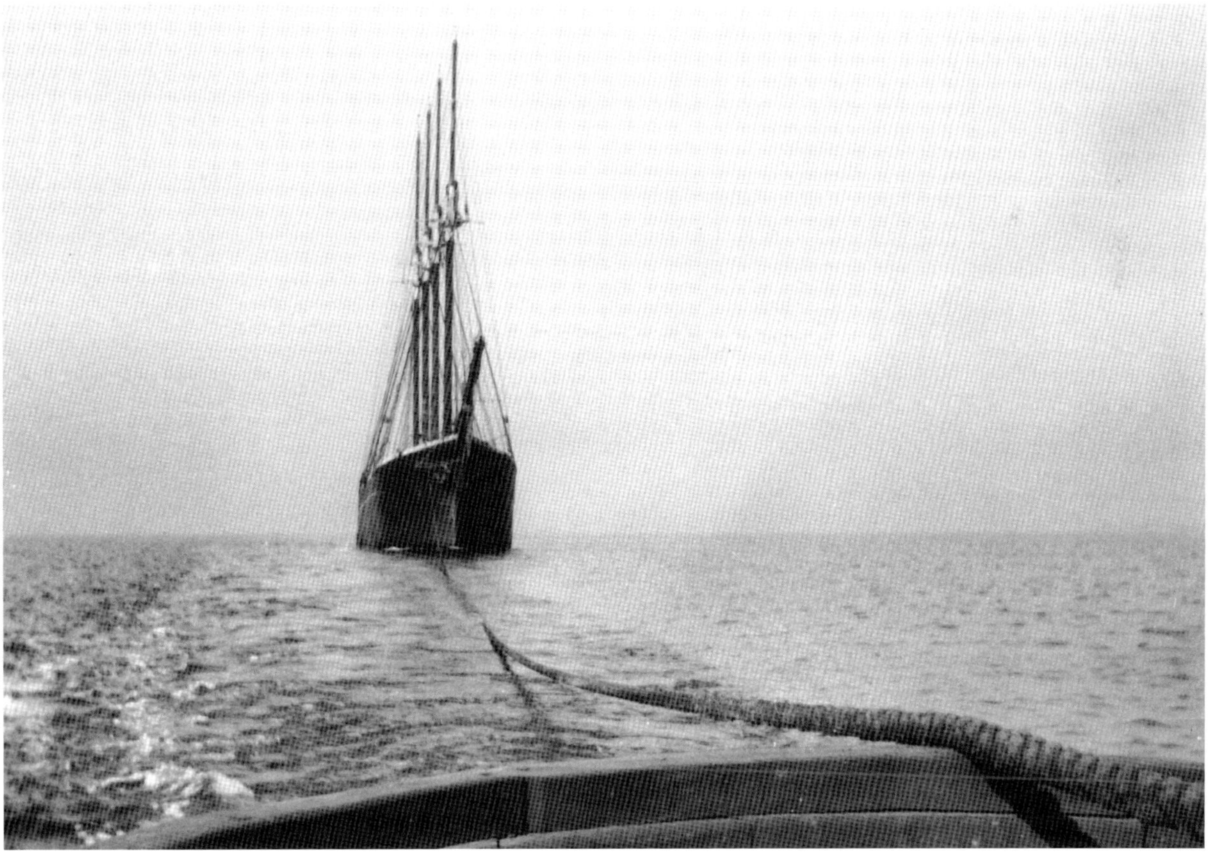

On November 17, 1933, the *Hamlin* is taken in tow by the tug *Radnor,* Captain Joseph H. Mullen, and is shown entering Chesapeake Bay at the mouth of the Patapsco River. If the wind had been fair, sails would have been set and the voyage would have begun.

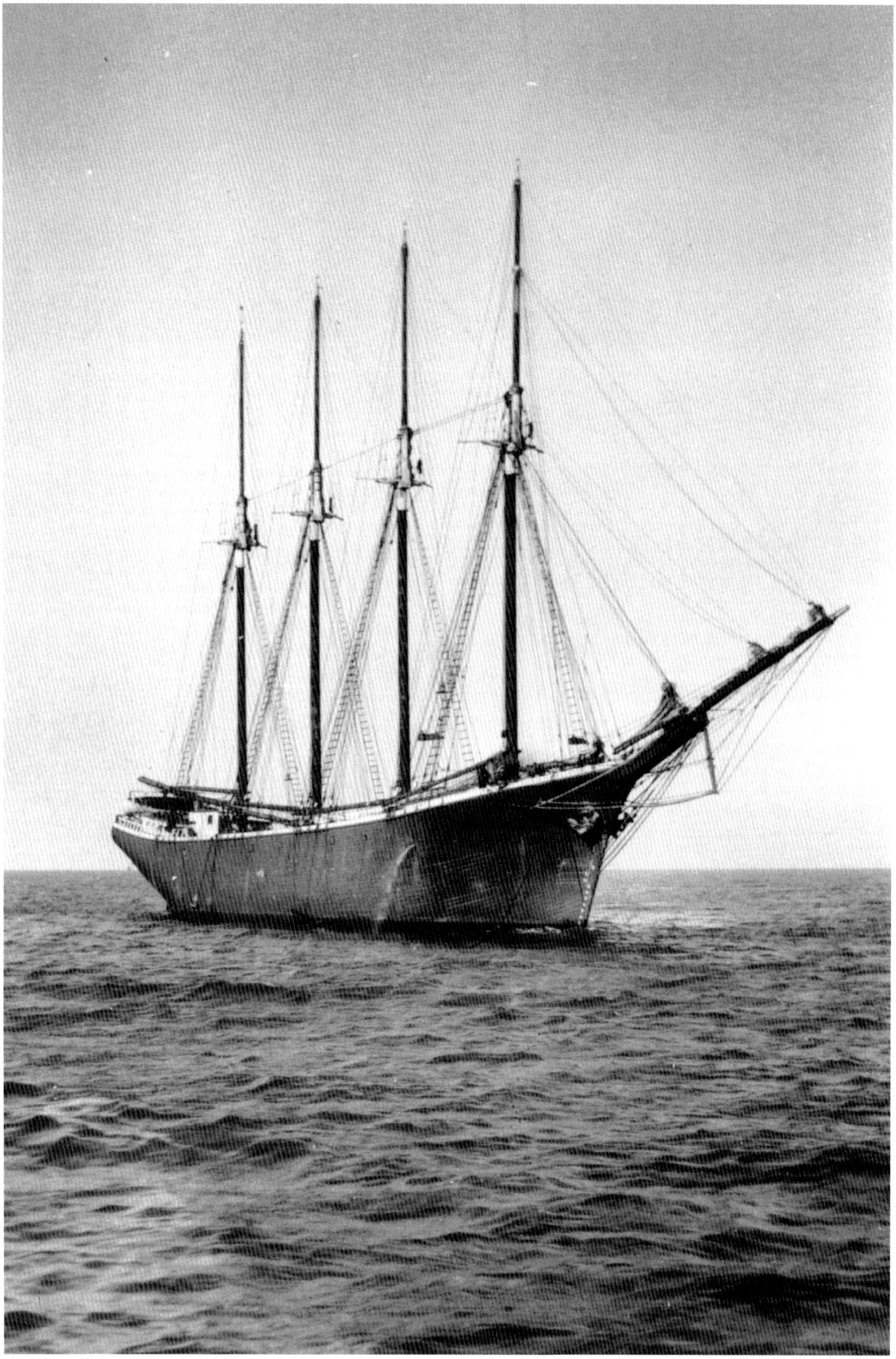

Due to a head wind, Captain George H. Hopkins of the *Doris Hamlin* decides to anchor south of Seven Foot Knoll Light to await a favorable slant down Chesapeake Bay. The tug has cast off the towline that is being taken aboard the schooner. As soon as headway is lost, the starboard anchor will be released. *Doris Hamlin* was a handsome schooner from this angle. Built by the Frye-Flynn Shipbuilding Company at Harrington, Maine, in 1919, the *Hamlin* was one of scores of similar large sailing vessels built during the World War I years, particularly in Maine. They were constructed to help relieve the shipping shortage caused by that conflict, but many were launched too late to be of any benefit or to reap any great profits.

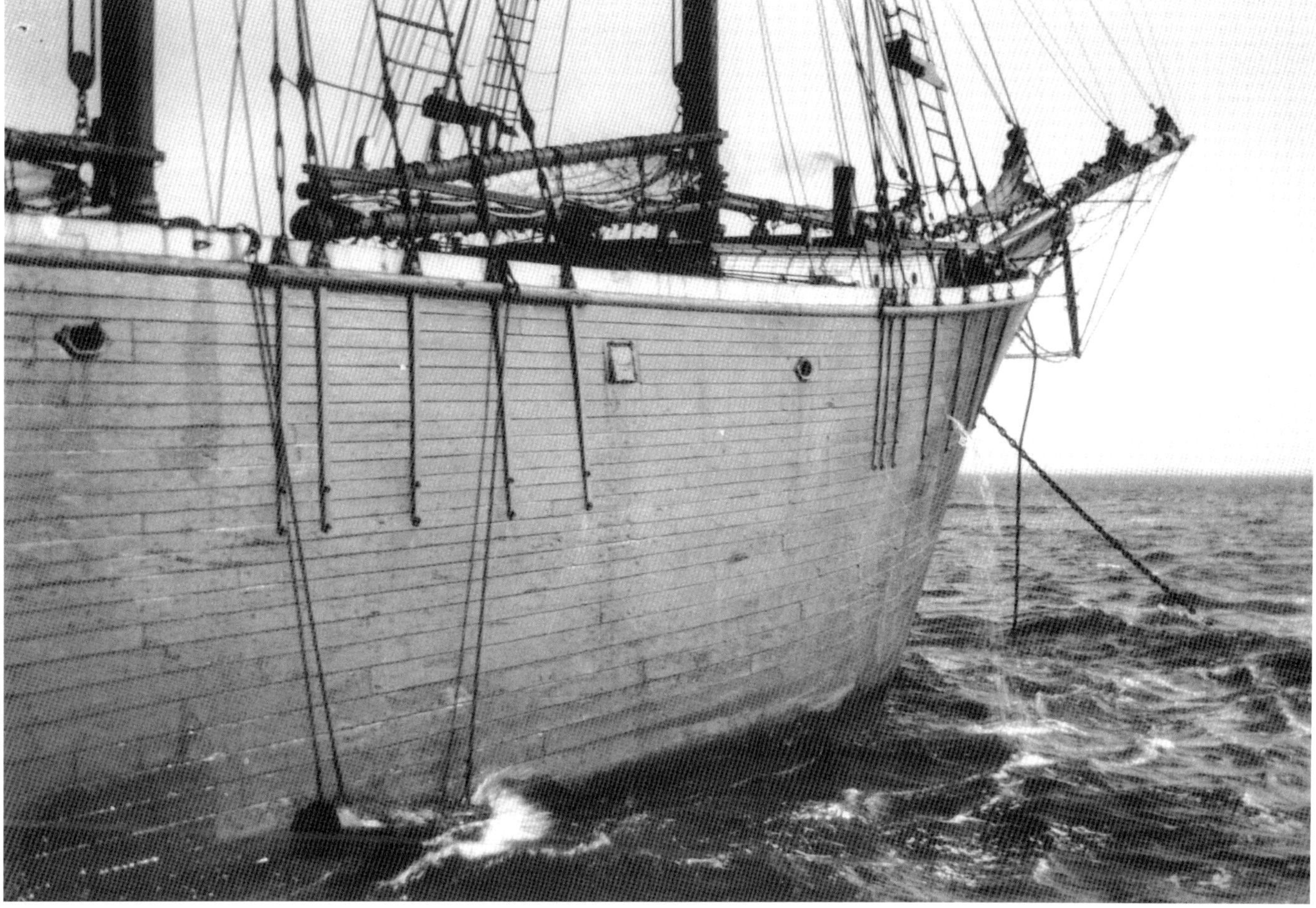

The *Hamlin* strains at her anchor as the towline is still being taken onboard. The tug is coming alongside to pick up any mail the crew has to send ashore. The schooner is only 14 years old at this time and the good condition of her topside planking is evident. Within a couple of days a fair wind favored the *Doris Hamlin,* sending her down the Chesapeake and out through the Virginia capes. She arrived in Haiti on December 4, loaded and departed for Baltimore on the 17th, and reached her destination on January 2, 1934. This schooner continued in this triangular trade, usually with coal to Bermuda, light down to Haiti, and back to Baltimore with logwood, with occasional voyages to Georgetown, South Carolina, New York, Martinique, or Guadeloupe until 1939. Sailing into the latter West Indian island, she grounded on a reef but was refloated and discharged her coal cargo. Upon her return to the Bay she was hauled out at Colonna's Shipyard, Berkley, for examination and then was towed to Baltimore, arriving there on March 10. She was tied up at the Redman and Vane Shipyard and sold on September 14 to the Penn-Lehigh Coal Company of New York. Her new owners had her towed back to Colonna's Shipyard for an overhaul and then loaded her with coal for Teneriffe, Canary Islands. In late December she left Norfolk in command of Captain Eric Isaksen. Off the Virginia coast a gale severely damaged her sails and she was towed back to Hampton Roads by a Coast Guard cutter. After repairs had been made, she set sail again and passed out of Cape Henry on January 1, 1940. Three days later she was back in Norfolk again after being buffeted by a storm 40 miles off Oregon Inlet, North Carolina, with more loss of sails. Word of her plight was relayed by the steamer *Winding Gulf* and the Coast Guard cutter *Sebago* brought the schooner back into port. Further repairs were made and for the third time the *Hamlin* left the Virginia capes astern and set a course for the Canaries. She was never reported after that, and was posted as missing with all hands.

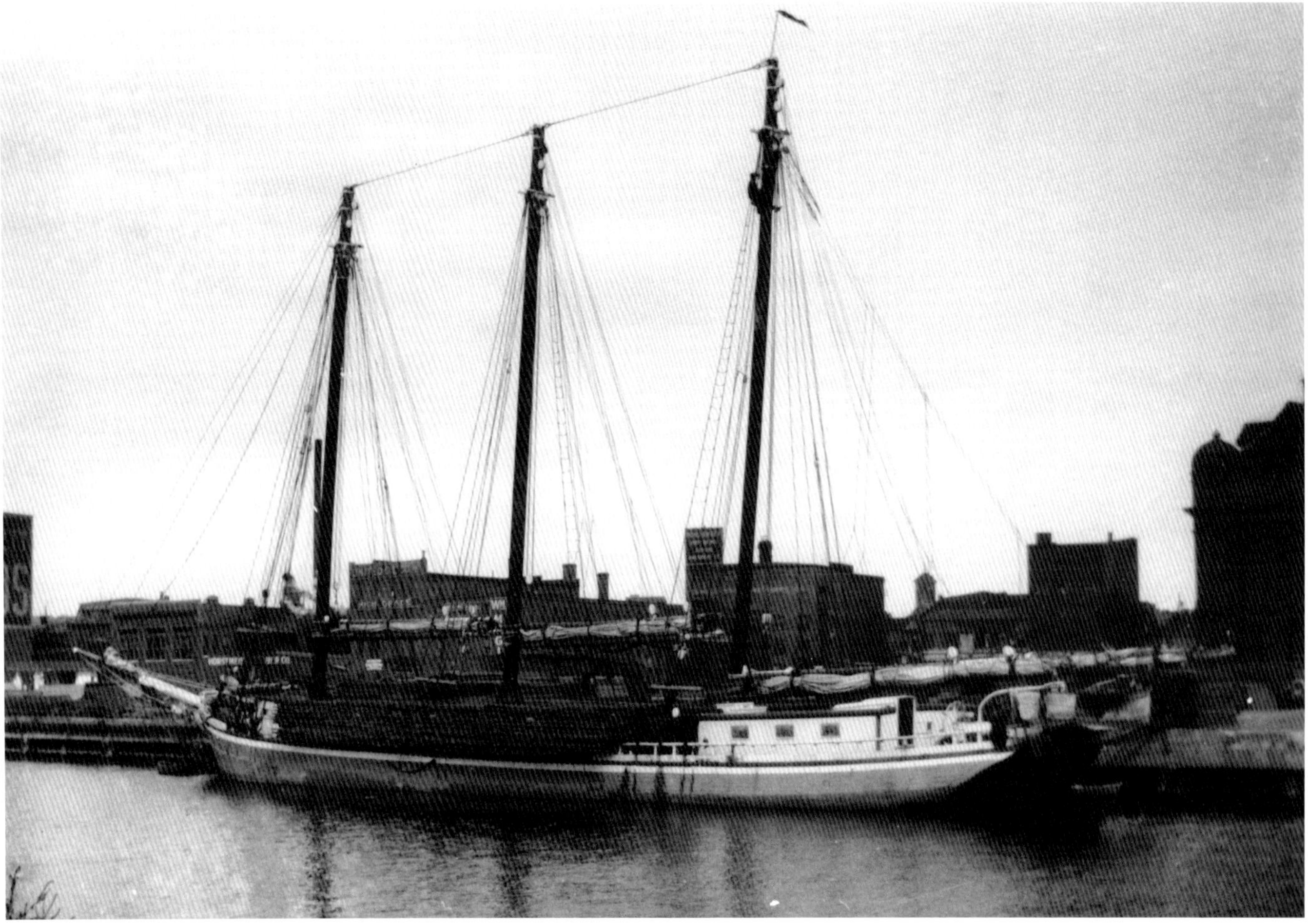

One of the strangest careers any Chesapeake sailing craft ever followed was that of the three-masted schooner *Venus.* She had been built by Sauerhoff at Cambridge, Maryland, in 1893, and was one of the larger vessels to engage in the pineapple trade between the Bahamas and Baltimore. In 1913 she left the Chesapeake when her Baltimore owner sold the schooner to Portuguese interests for use in the fishing trade. It was reported that she had been fitted out with a drying apparatus to cure her catch onboard. She was renamed *Santa Luiza* and hailed from Vianno do Castillo, a seaport of Minho, Portugal. Apparently she carried on in that business for little more than a decade. She last appears as the *Santa Luiza* in Lloyd's Register of 1926–27. Thereafter she appears under the American flag again. Renamed *Edna Bright Hough* in 1923, she reappeared on the Chesapeake. During that latter part of her career her topmasts were removed. The absence of those spars and her spike bowsprit caused her to be mistaken for a ram, but her rather shapely hull bore no resemblance to those bargelike craft. Here she is tied up at Pier 6, Pratt Street, Baltimore, on October 10, 1933, with a high deckload of lumber and bearing Moss Neck, a Virginia locality on the upper Rappahannock River, on her transom as her home port.

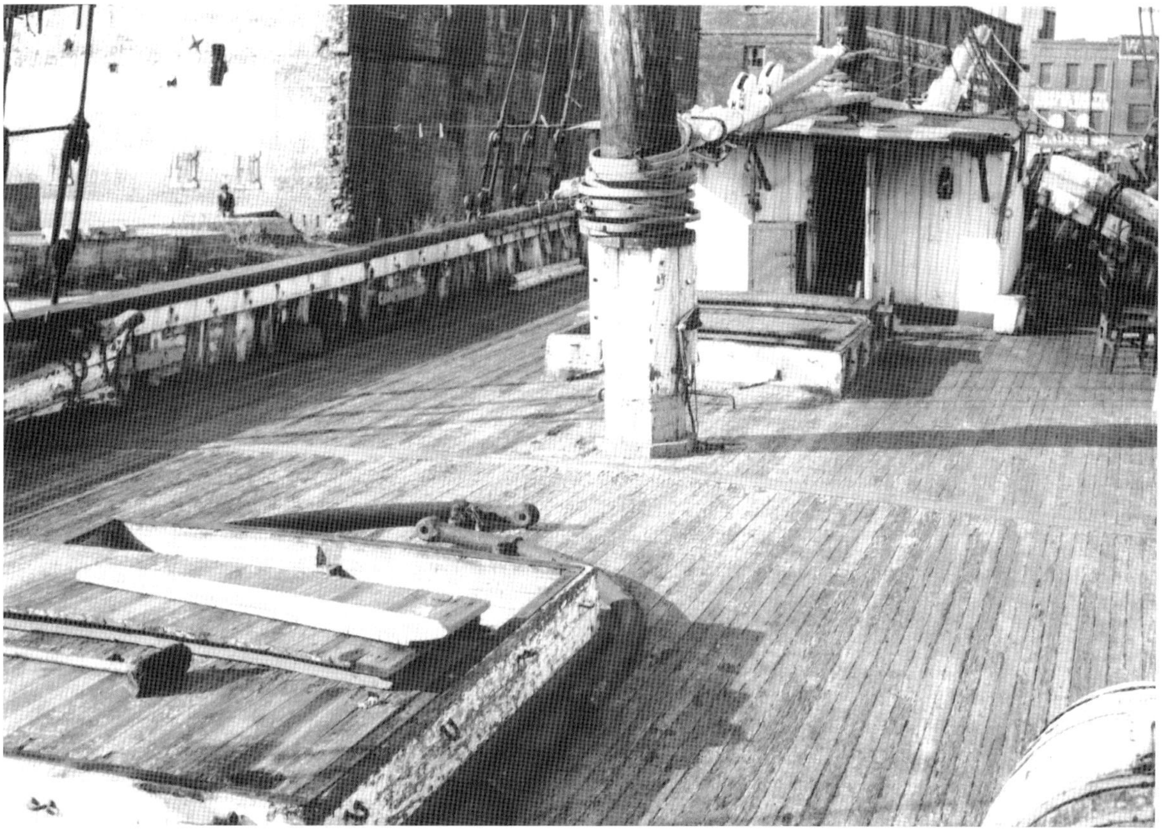

In 1935 the *Edna Bright Hough* was laid up at Norfolk. This view looking forward on deck on March 29, 1936, shows how quickly she started to deteriorate. A family was living aboard but no maintenance was carried out. Not long after this photograph was made, the schooner was moved away from the pier. She was listed as "Abandoned" in the 1938 *MVUS*.

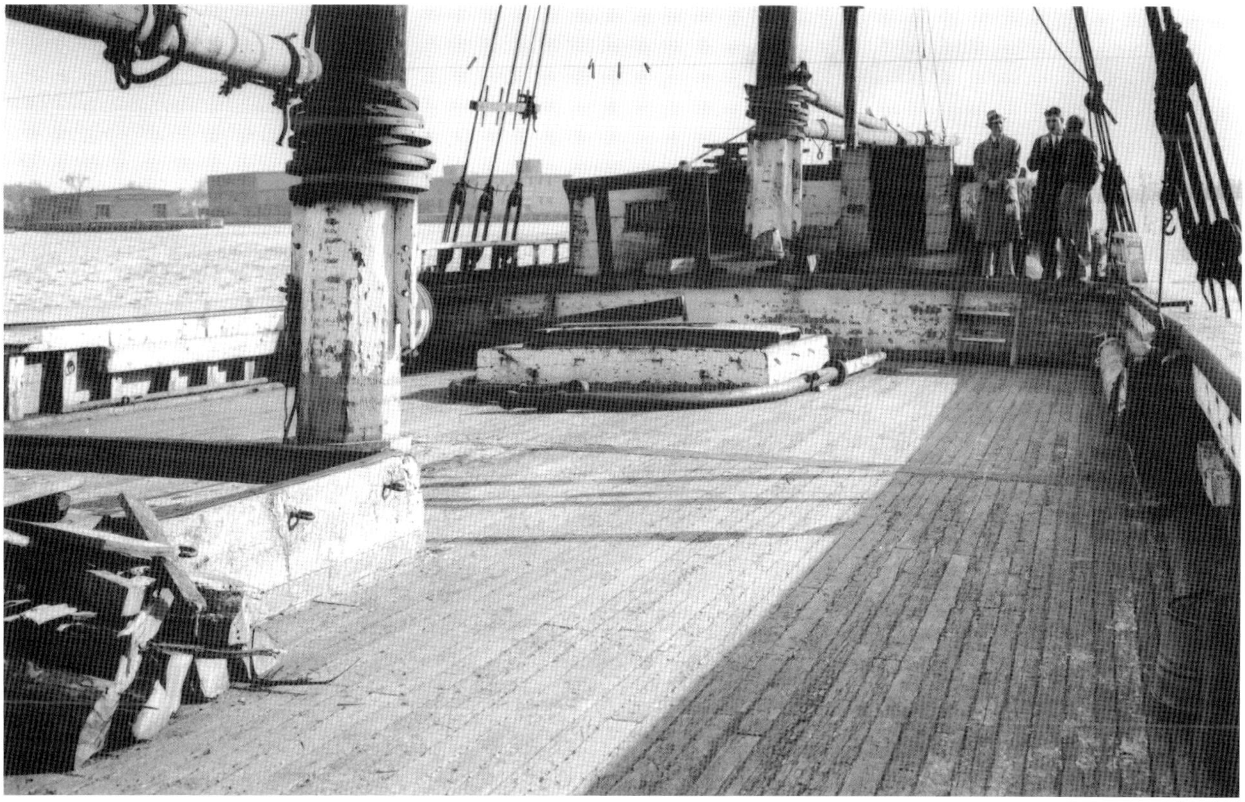

✪ Another main deck view of *Edna Bright Hough,* this one looking aft, also taken at Norfolk on March 29, 1936. Her lack of maintenance is obvious. The businessmen on her stern may have been discussing her fate.

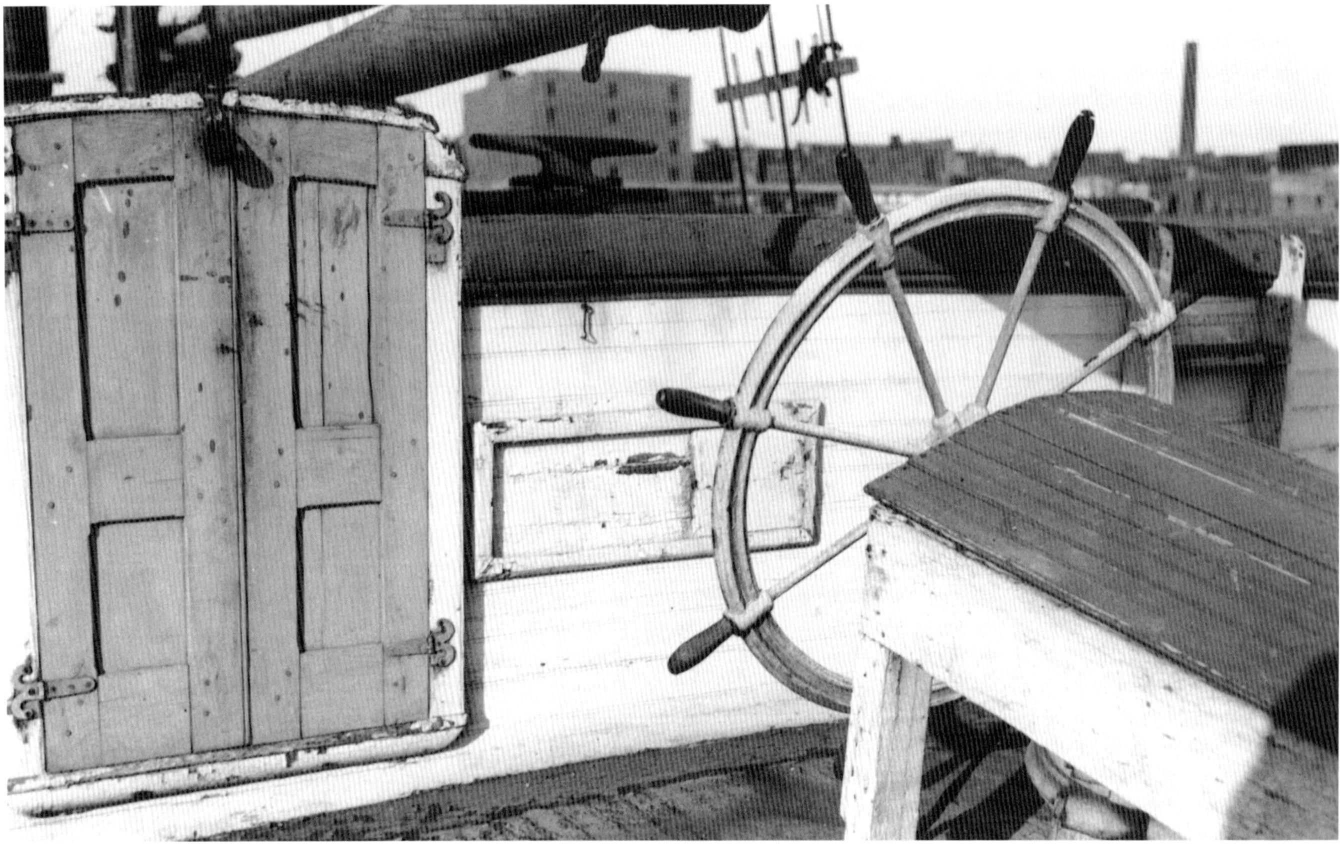

✪ The helm of *Edna Bright Hough* looking forward, as it appeared on March 29, 1936. Her compass was sheltered inside the small sliding door to starboard of her handsome companionway entrance. Her deteriorated condition is somewhat less evident in this view.

One of the largest commercial sailing vessels ever to be registered out of a Chesapeake Bay port was the four-masted schooner *G.A. Kohler,* shown moored in Jones Falls, Baltimore, July 27, 1933. Built as the *Charles S. Gawthrop* in 1919 by the American Car and Foundry Company, Wilmington, Delaware, she measured 212 feet in length and was of 1,462 gross tons. When originally built she had twin screws operated by 350-horsepower oil-fired Winton engines. These were later removed and she became a pure sailing vessel. In 1918 the Maryland Navigation Company was formed in Baltimore and their first vessel was the four-masted schooner *Jacob W. Hook.* She was sold to R.B. White in 1926 and foundered off Cape Hatteras on November 11, 1928. Maryland Navigation's next vessel was the four-masted schooner *Herdis,* originally a barkentine, which was later sold to Puerto Rican interests; her register was forfeited in 1936. Next were the four-masted schooners *Charles S. Gawthrop* and *W.H. Woodin,* sister ships that became the *Kohler* and *B.S. Taylor.* The *Taylor* was sold in 1928 and she eventually became a barge. The *Kohler* was sold to R.B. White in 1929. In the mid-1920s, when these schooners were tied up along Pratt Street, their long jibbooms would overhang that busy thoroughfare.

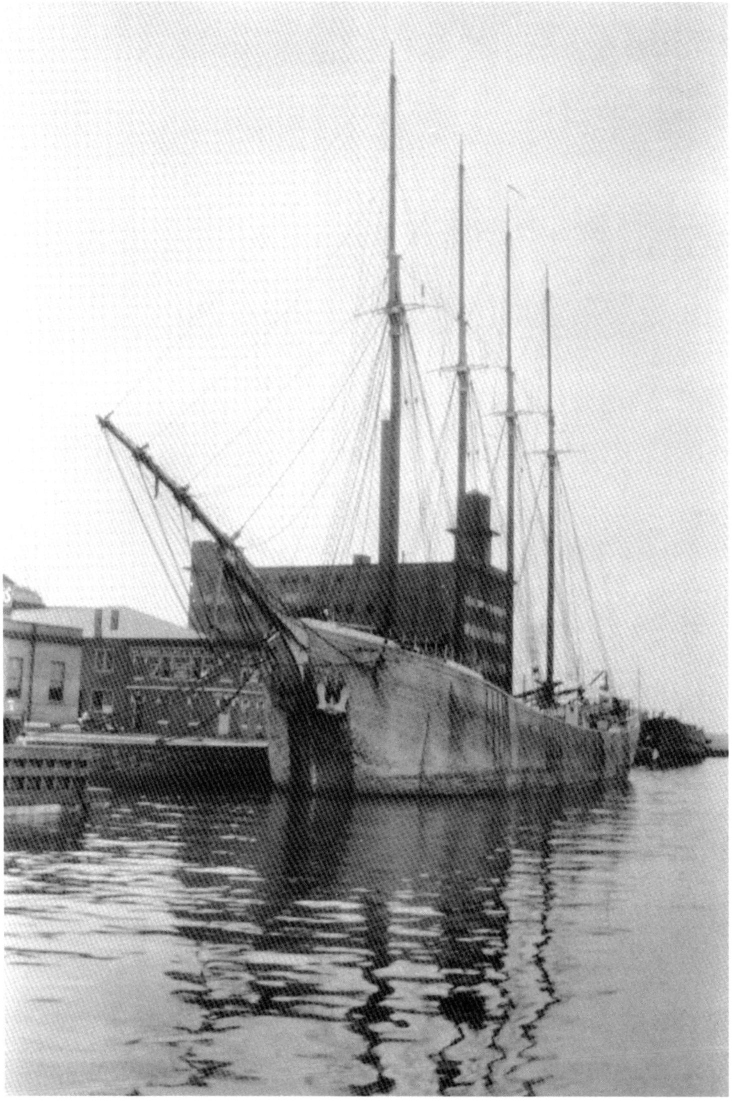

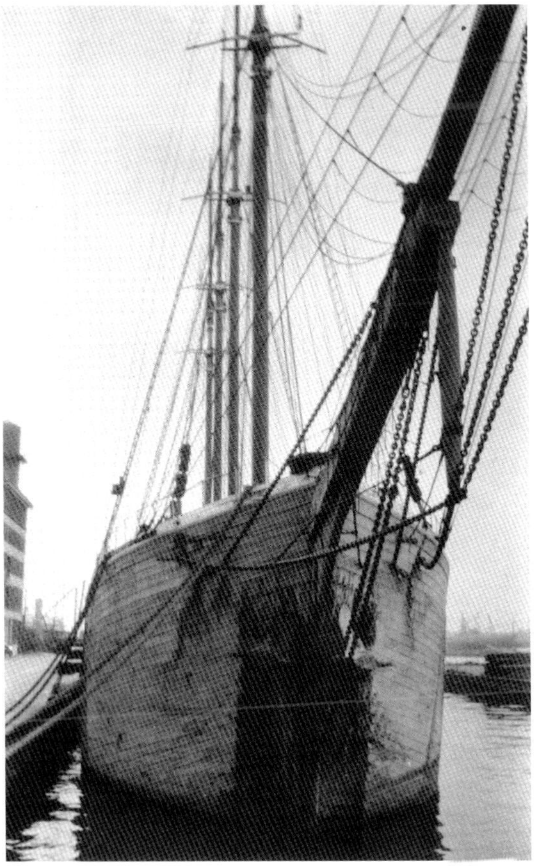

The *Kohler* was kept busy by R.B. White, who also owned the four-masted schooner *Purnell T. White* and rams *B.P. Gravenor* and *Clarence A. Holland.* In 1929, while in command of Captain W.M. Tappe, the *Kohler* drifted 12 days after losing rudder control but was towed to safety behind Cape Henry and later to Baltimore. Captain George H. Hopkins took over the *Kohler* in 1930 and remained in the big schooner for the balance of her career. The *Kohler* was a boxy-looking schooner, with bluff bows and little sheer, but she was considered a good sailer by Captain Hopkins.

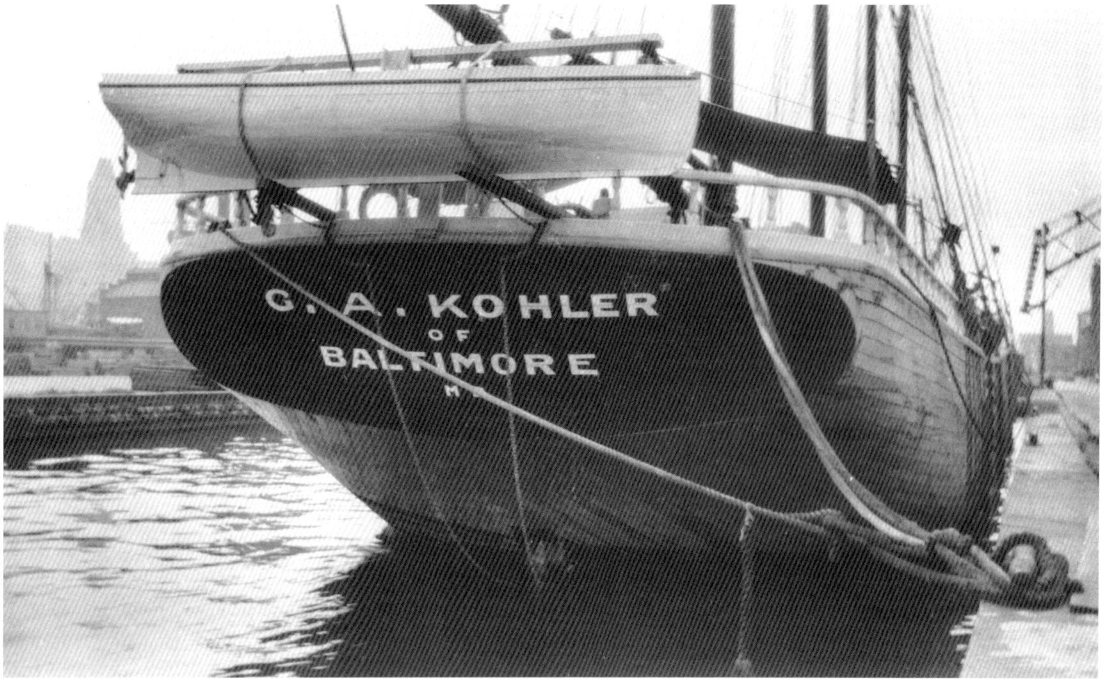

The *Kohler* was lying in the freshwater of Jones Falls to kill any toredos that might have been in her hull. She traded regularly to the West Indies and periodic immersion in freshwater was a deterrent against marine worms. After this "treatment" the schooner was moved across the harbor to the Key Highway yard of the Bethlehem Shipbuilding and Dry Dock Company to have her bottom coated with copper paint. Less than a month after this photograph was made on July 27, 1933, the *G.A. Kohler* was a total wreck on the North Carolina coast.

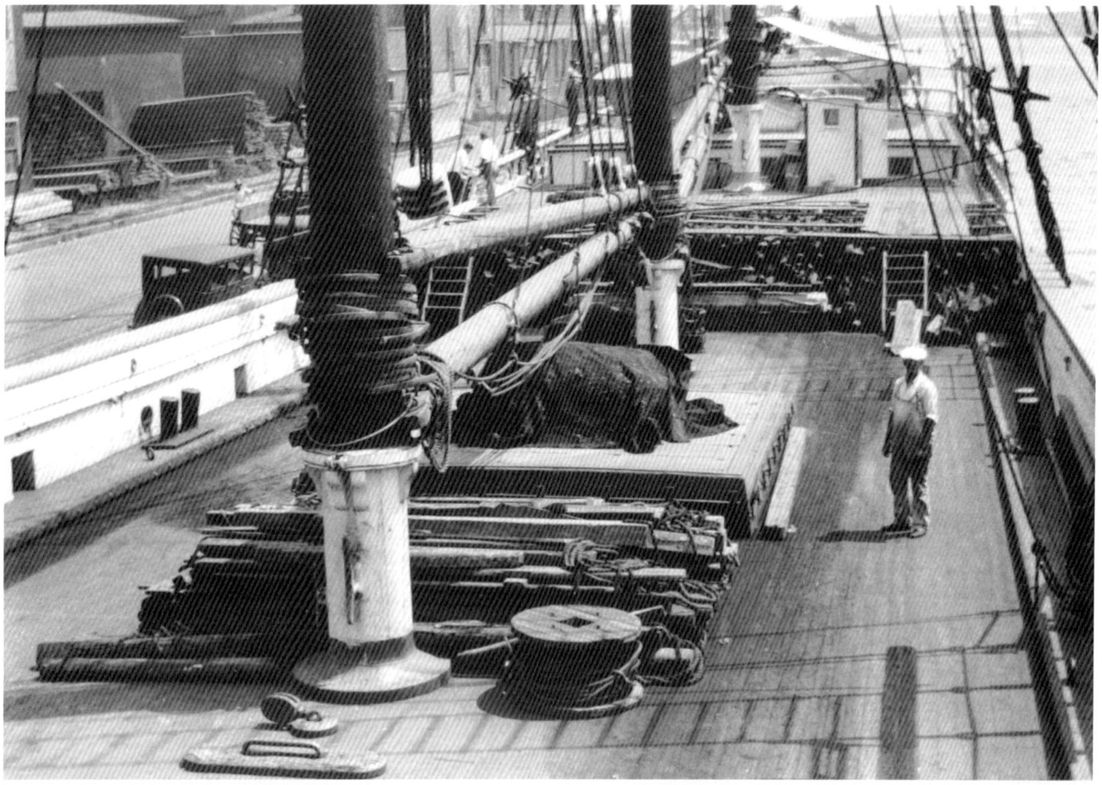

Looking aft on the deck of the *Kohler*, July 24, 1933. A portion of her deckload of logwood can still be seen forward of her quarterdeck. Her master, Captain George H. Hopkins, stands on the starboard quarterdeck talking to a tradesman whose horse and wagon wait in the street. The schooner's mate, Nels Sorensen, is standing on the starboard poop rail adjusting the awning over her cabin. The figure on the main deck was a member of the crew of the pungy *Amanda F. Lewis*, which was lying at a nearby pier and which the author had photographed earlier.

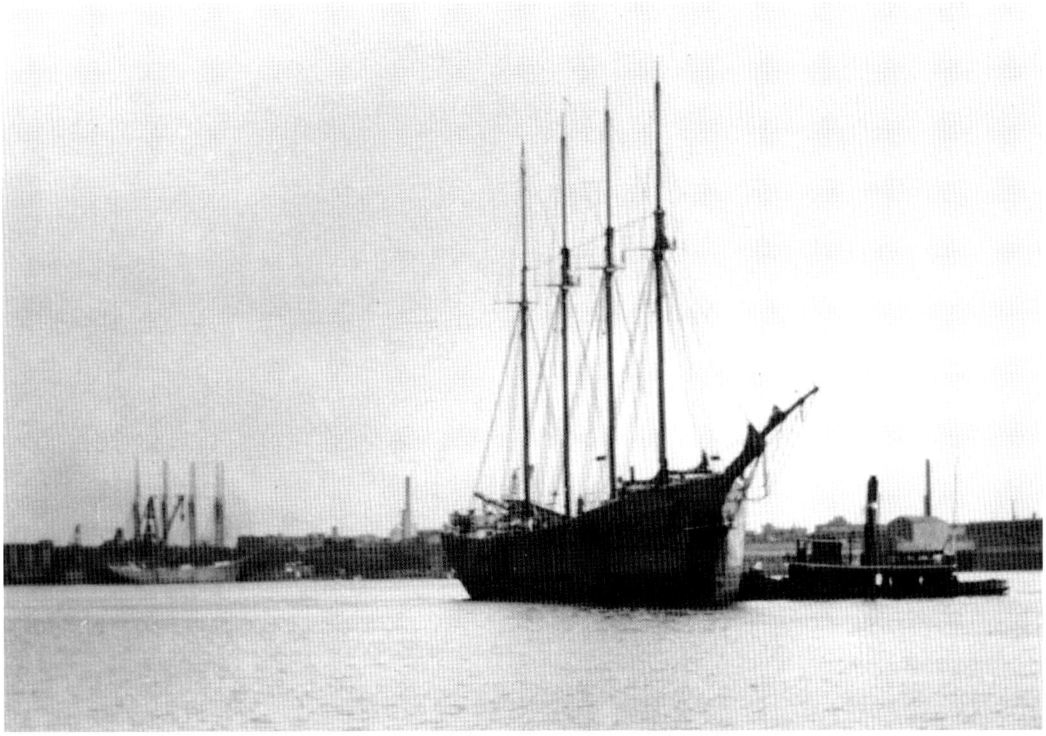

On August 19, 1933, the tug *Radnor* eases the *G.A. Kohler* away from the pier of the Redman and Vane Shipyard before taking her towline to move her down Baltimore harbor to Chesapeake Bay where she would set sail for Haiti. Before her was a terrific battle with a hurricane that caught her off the North Carolina coast on the early morning of the 22nd. Unable to make headway offshore, the vessel was anchored, but throughout the day she dragged until on the 23rd she struck bottom and was cast broadside onto the beach. Her crew was rescued by breeches buoy but the schooner had to be abandoned. This view was probably the last taken of the *G.A. Kohler* in her home port. In the left background are the four-masted schooners *Katherine May,* with hull showing, and *Doris Hamlin* tied up at the former Woodall Shipyard. The following November, the *Hamlin* took over the Haitian logwood trade that had kept the *Kohler* employed.

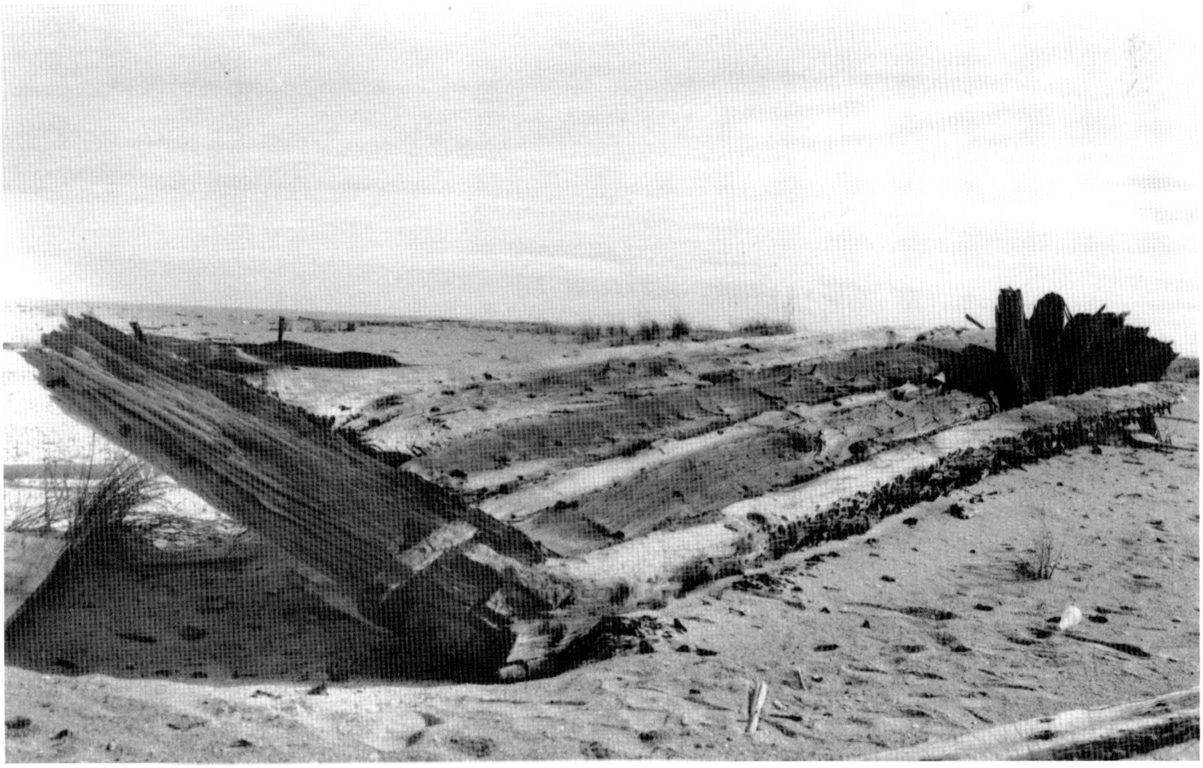

✪ Years later, her skeg was one of several portions of the *Kohler* still visible on the Outer Banks shore.

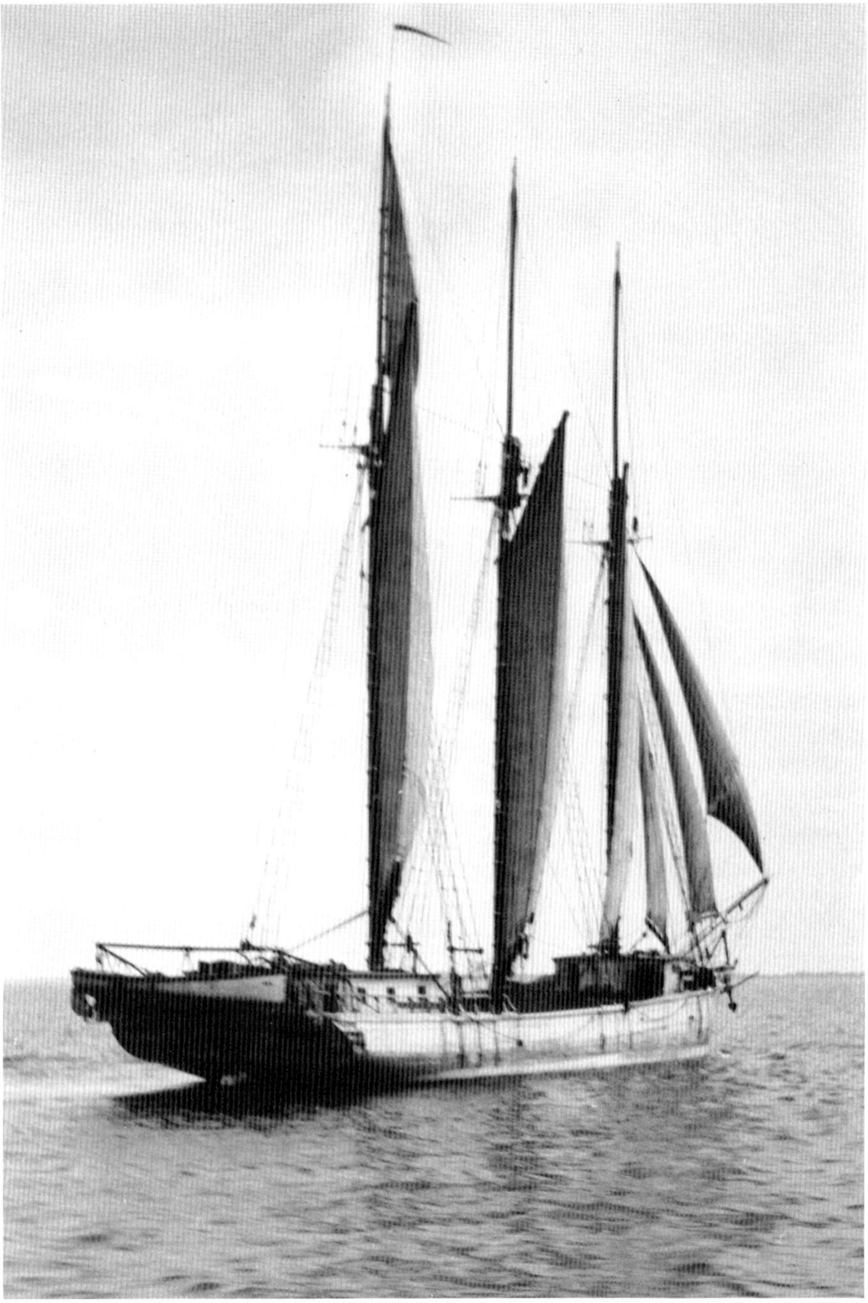

The shipbuilding firm of W.G. Abbott, Milford, Delaware, built some fine schooners during its lifetime, and one of those was the *May and Anna Beswick,* built in 1891. This three-master was only 95 feet in length, smaller than many of the two-masted schooners plying the Chesapeake, but she went coastwise in delivering her cargoes. On January 31, 1908, the *Beswick* stranded a mile east-northeast of the Portsmouth Life-Saving Station, North Carolina, while bound from New York City to New Bern, North Carolina. Apparently she was making for Ocracoke Inlet. The lifesaving crew in a surfboat went to her aid and ran out an anchor in an attempt to get the schooner into deep water, but the tide had fallen and their efforts were unsuccessful. During the entire month of February the *Beswick* remained grounded as the surfmen were daily engaged in carrying out anchors and hauling her toward deep water. Finally, at flood tide on February 27, the schooner was floated. The North Carolina shoals were to entrap her again, this time on December 1, 1915, when she stranded on Swan Island, in the Pamlico Sound area. There she was to remain for more than a week while the cutter *Pamlico* supplied coal for her galley, and other provisions. On the night of the llth a heavy northeaster drove the *Beswick* over the shoal and left her afloat. About 1921 the schooner was renamed *Josephine Wimsatt,* registered out of Washington, D.C., as she had been since 1911 as the *May and Anna Beswick.* In this view taken on July 14, 1933, the *Wimsatt* has a fair wind down Chesapeake Bay as she sails to a Virginia lumber port.

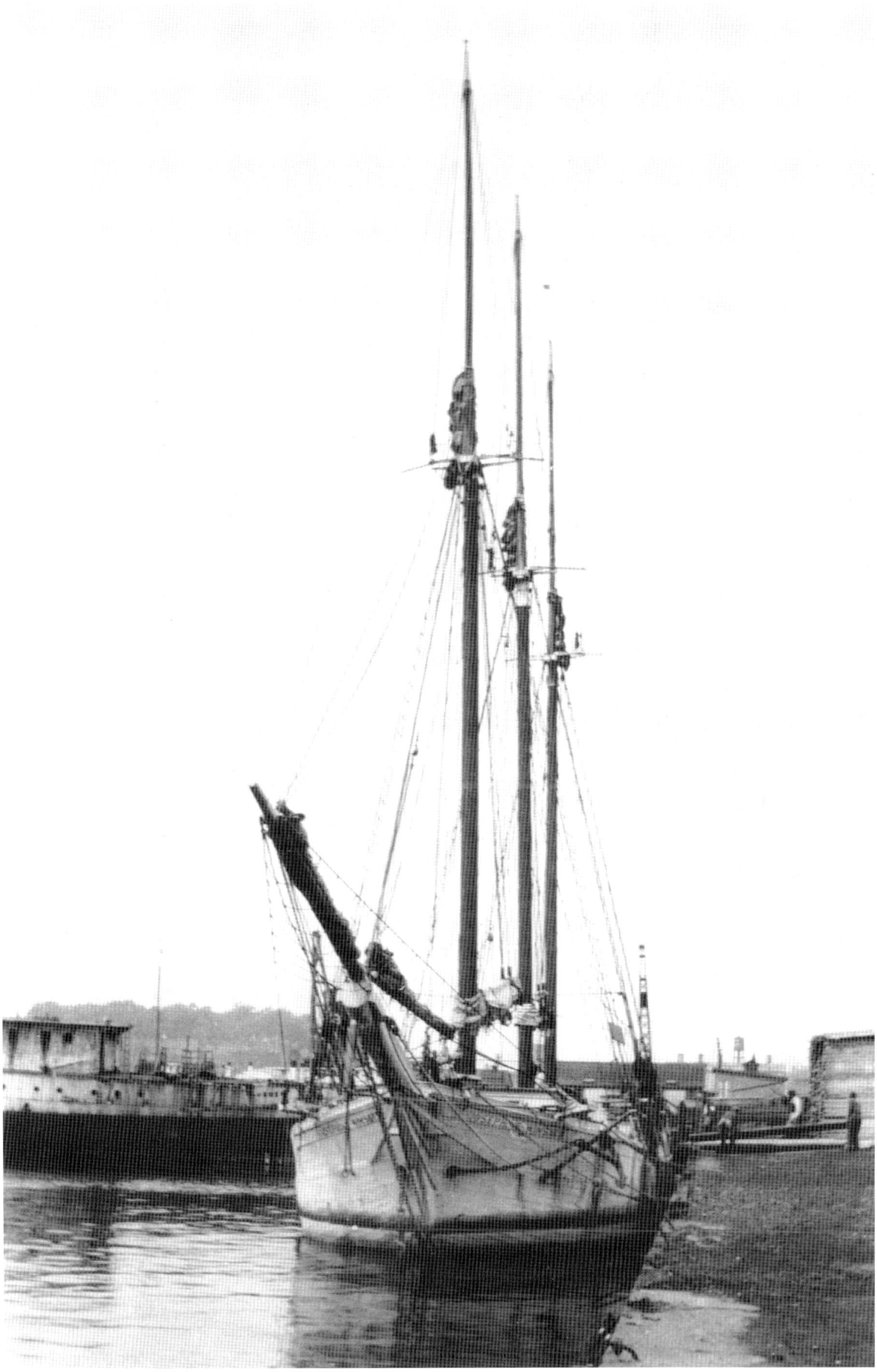

✪ The schooner *Josephine Wimsatt* in the 1930s.

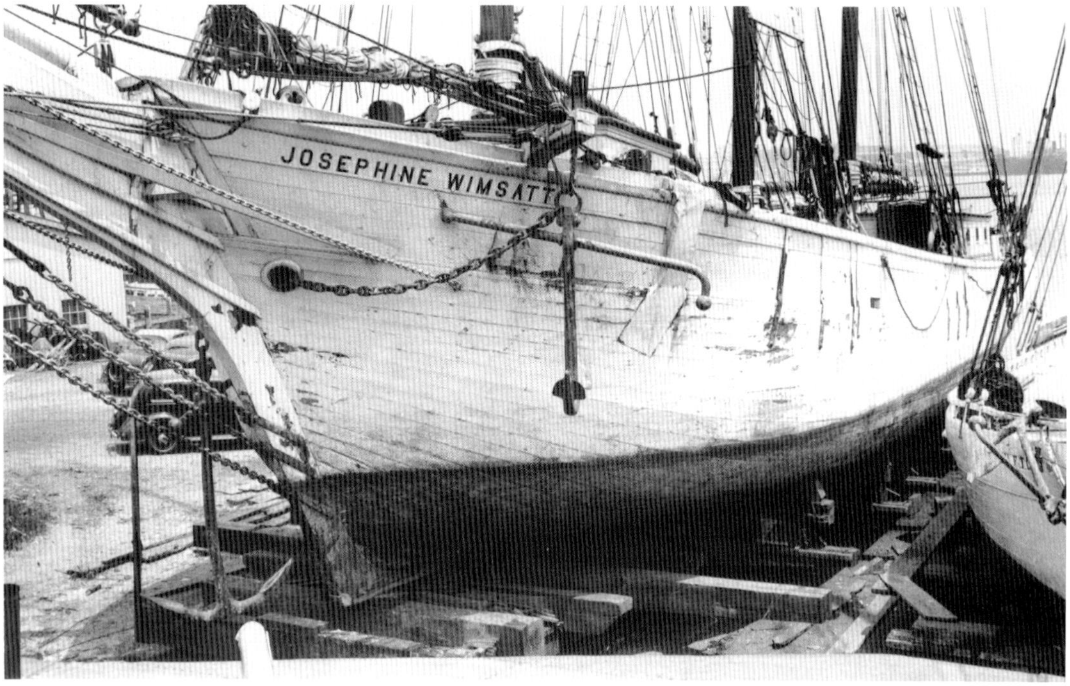

The *Wimsatt* pauses between her lumber-lugging trips for a bottom repainting at A. Smith & Sons Shipyard, Curtis Bay, Baltimore, on July 19, 1936. This little schooner was almost a small version of the larger sized coasting schooner. She had a raised quarterdeck and forecastlehead and her cabin joinery and molding were superior to that of the average Bay schooner. To the port of the *Wimsatt* is the bugeye *Edith F. Todd.* It is interesting to note that the total height of the bugeye's hull just about reaches to the *Wimsatt's* waterline, and the latter was of shallow draft.

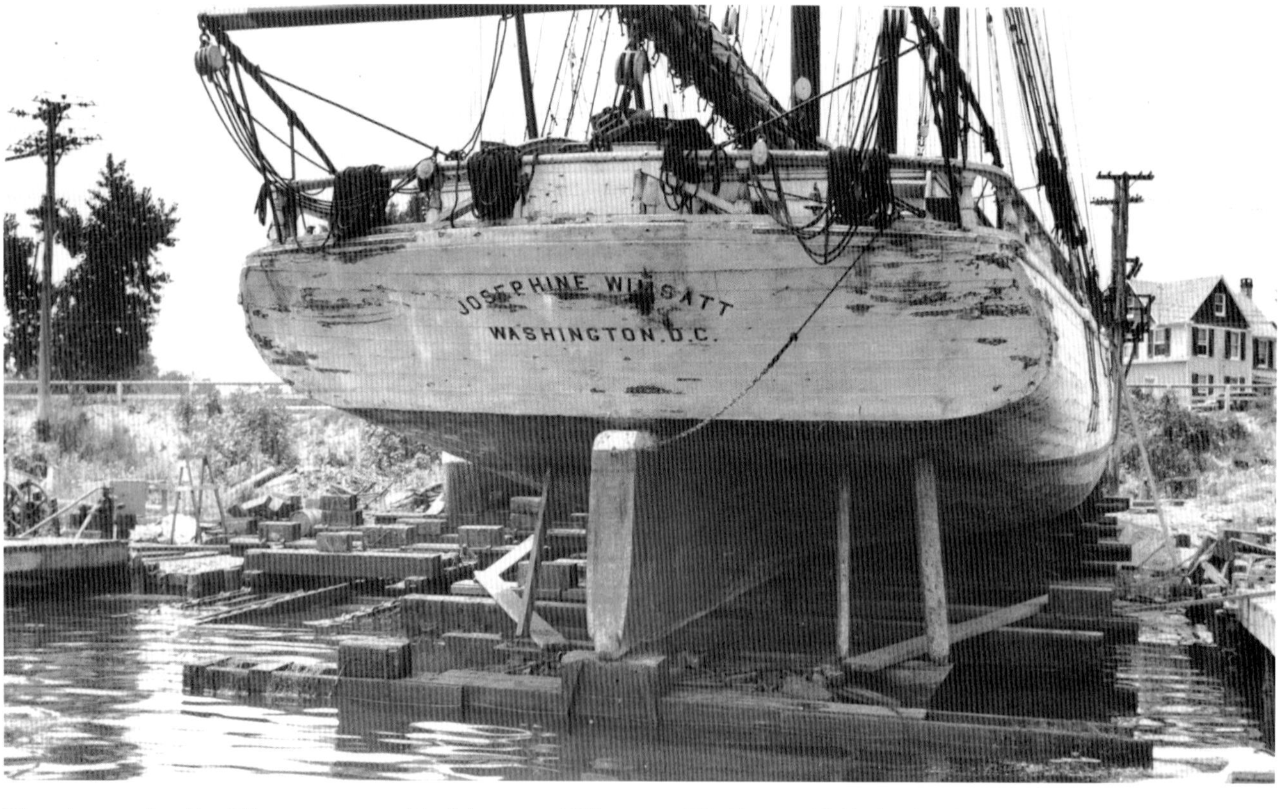

The schooner *Josephine Wimsatt* was owned by Johnson and Wimsatt, a Washington, D.C., lumber firm and was undoubtedly the last three-masted topmast schooner to sail the Potomac River, a waterway that saw countless such vessels before the turn of the century as they brought ice from Down East and took out coal from Alexandria, Virginia, and Georgetown, D.C. In 1939 the *Wimsatt* was sold to the British, renamed *Ruby W.*, and traded mostly between the Bahamas and Florida. Engines were installed and she became a twin-screw auxiliary. She was wrecked on August 26, 1945, forty miles south of Brownsville, Texas.

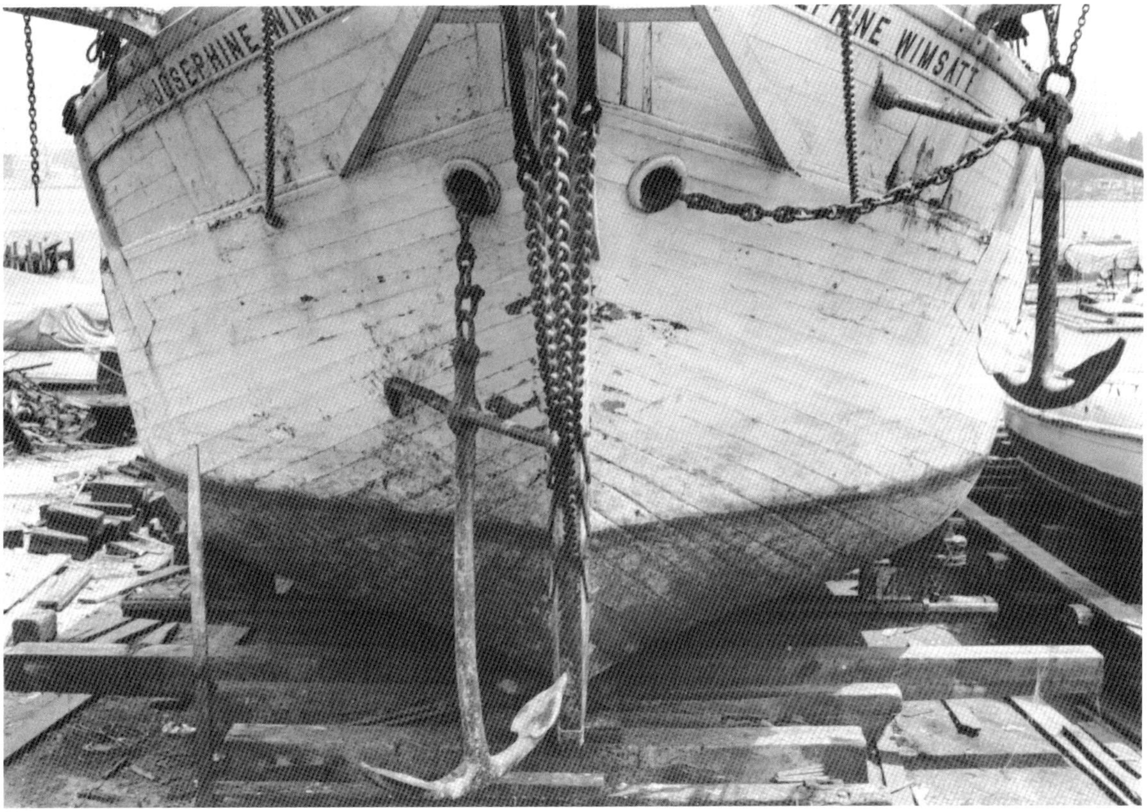

✪ The forefoot of the *Josephine Wimsatt* on the railway at A. Smith & Sons Shipyard, Curtis Bay, Baltimore, on July 19, 1936. Note the lumber ports in her bows.

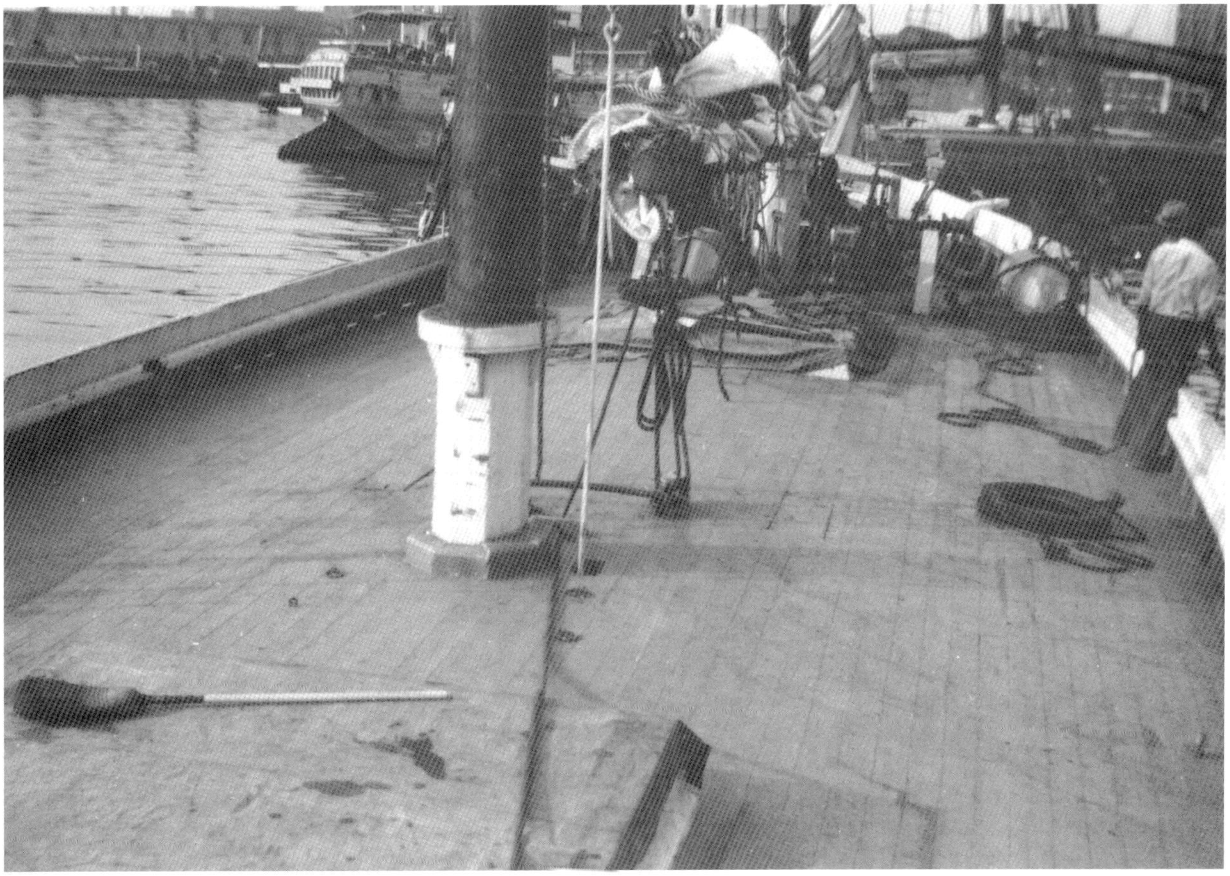

✪ Looking forward on deck on the *Wimsatt.*

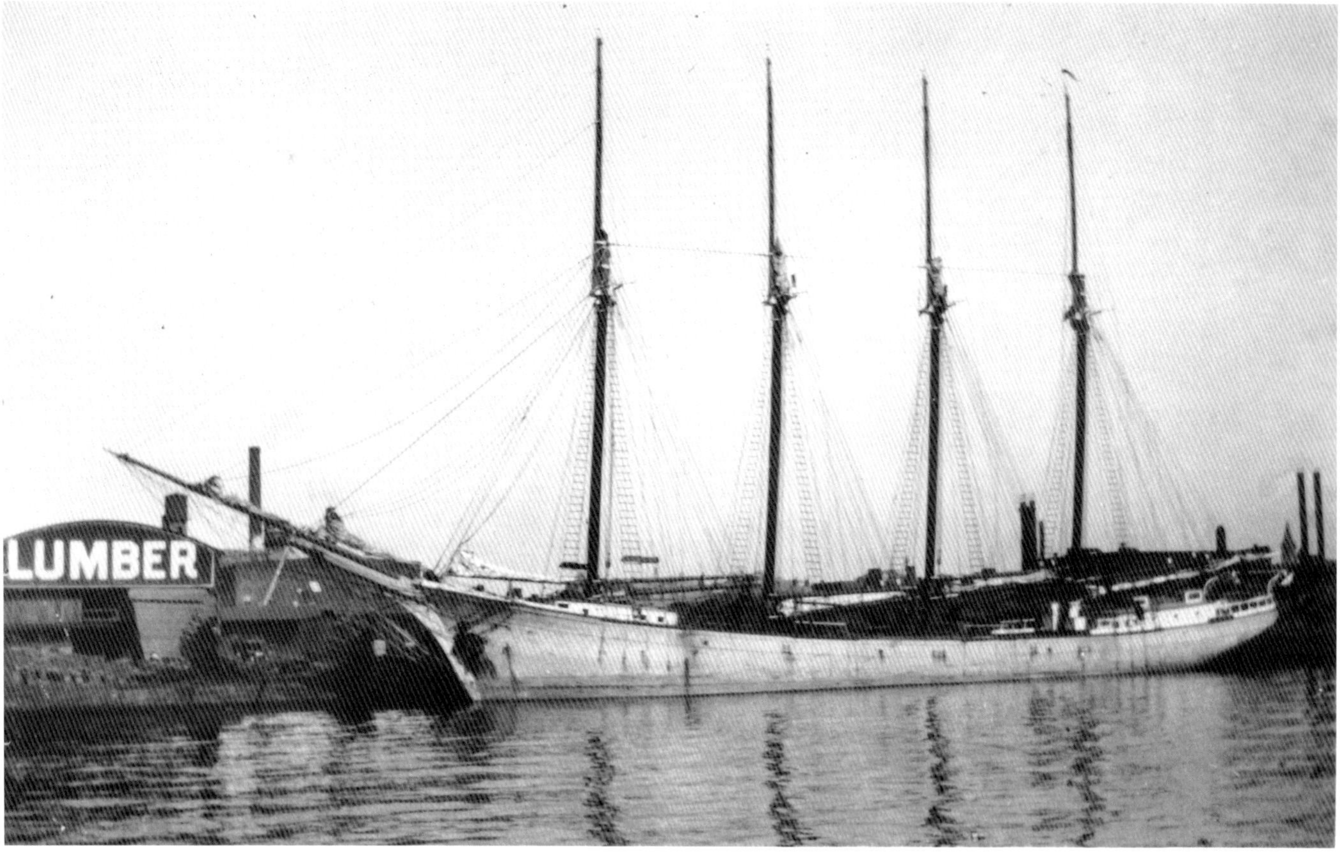

If there had been a title "Queen of Chesapeake Sail" it certainly would have been bestowed upon the four-masted schooner *Purnell T. White,* shown here discharging lumber from Georgetown, South Carolina, at the James Lumber Company, Back Basin, Baltimore, February 25, 1931. As long as this vessel plied the Chesapeake no other sailing craft there could match her in beauty and grace, even though she was a grime- and salt-streaked working ship. The author has seen virtually all of those commercial sailing craft that sailed the Bay during the past 50 years (✪ 1925–1975), has been aboard most of them, and sailed in some, and his vote for "Queen" would have been cast in favor of the *White.* Built in 1917 at Sharptown, Maryland, by Alonzo Conley, the *White* on her maiden voyage went to the Belgian Congo. Later voyages were to the Canary Islands, Brazil, France, Nova Scotia, England, West Indies, and Gulf of Mexico and East Coast ports. Starting the latter part of 1928, the schooner began a rather steady run between Baltimore and Georgetown, with an occasional delivery of lumber to New York. It was while on a passage from Georgetown to New York in January 1934 that the *Purnell T. White* came to the end of her career. She had left the South Carolina port on the 27th in command of Captain Charles Nicklas and encountered heavy weather during the following week. On February 7 she was only 200 miles east of Cape Fear, North Carolina, waterlogged and wallowing in heavy seas. The steamer *Maiden Creek* sighted her distress signals and radioed to the Norfolk Coast Guard base for assistance to be sent to the stricken vessel. The cutter *Mendota* rushed to the scene and took the *White* in tow for Chesapeake Bay. For two days the ships progressed slowly up the coast. On the evening of the 9th the schooner capsized in the trough of a heavy sea created by a rising storm. The *Mendota* cast off the towing hawser and swung around to approach the derelict and rescue her crew. But intense cold and rough seas were obstacles, and only three seamen were saved. The captain and three other crewmen were lost. The rescued men were hurried to Norfolk and the *White* was left to wallow in the shipping lanes, her cargo of lumber keeping her afloat. The next day the Coast Guard cutter *Ponchartrain* found the *White* and towed her into the Chesapeake, where a commercial tug took over the derelict to tow her to Colonna's Shipyard, Berkley. There the dismasted schooner was righted and pumped free of water. For more than a month the *White* lay at the shipyard as the debris of rigging was cleaned up. On March 23, in tow of the tug *Fannie May,* she reached Baltimore, where her cargo was discharged. Even though her hull was intact, re-rigging would have been too costly, so she was sold and moved to a ship graveyard in the middle branch of the Patapsco River. Plans to convert her into a barge never materialized. She filled and sat on the river's bottom, leaving her deck above water. As the adjacent shore was enlarged by landfill, the *White* was slowly engulfed until only her forward section remained exposed. That was set afire in order to remove her windlass for scrap. There she remained until 1957, when it was announced a pier was to be built at her grave site. But the hulk had to be removed. A clamshell bucket cleared the fill off her collapsed deck and out of her hold. Pumps were rigged and it was found that with slight caulking she could become waterborne. And after 23 years of total abandonment the battered hulk floated. She was towed farther down the Patapsco River and pushed aground again near Hawkins Point. In 1974 portions of her wooden hull were still jutting above the river's surface.

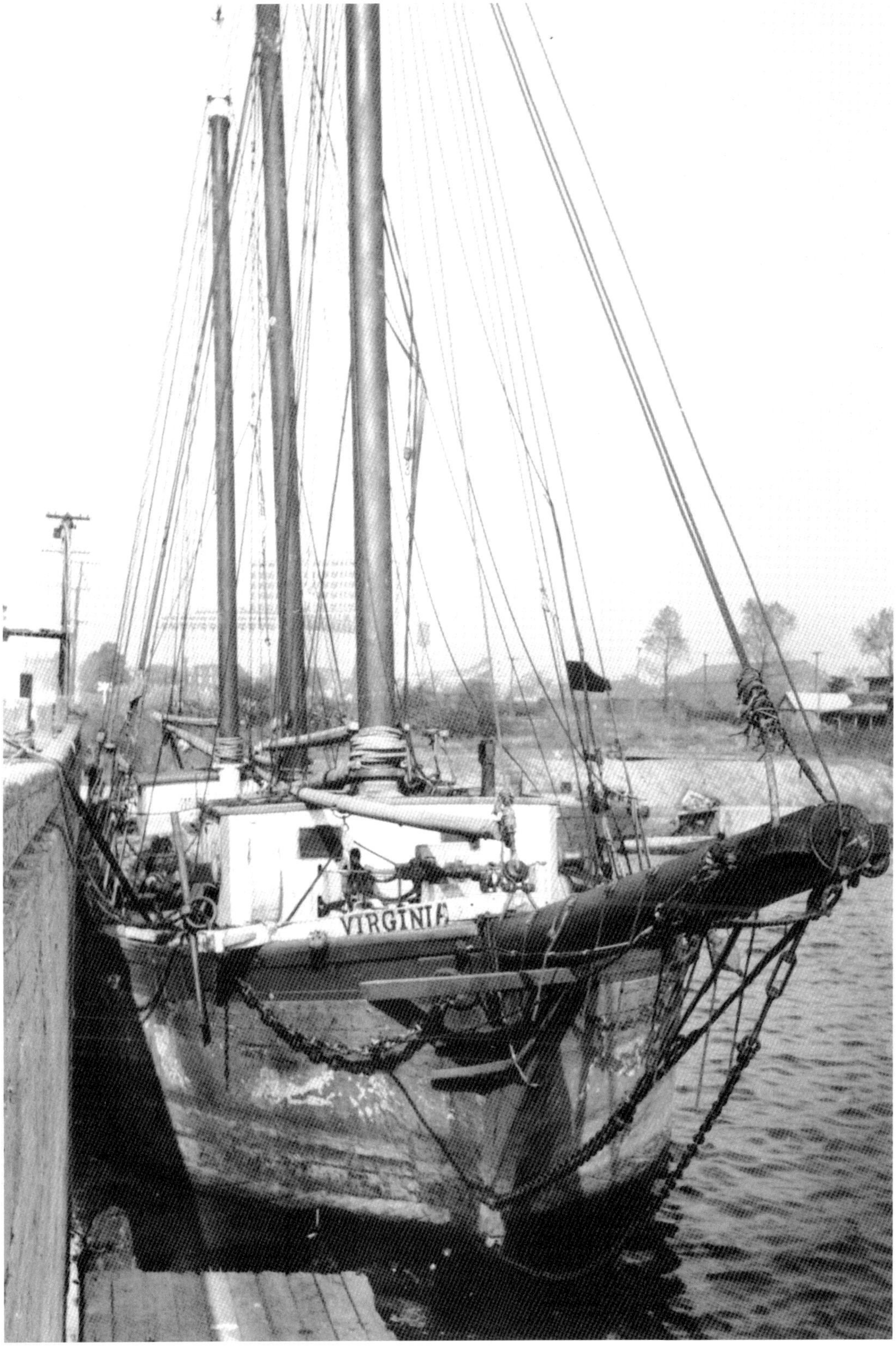

One of the ancient sailing craft around the Chesapeake up into the 1930s was the three-masted schooner *Virginia.* She started her career in 1857 as the two-masted schooner *James Diverty,* built at Dennisville, New Jersey. She was rebuilt at Sharptown, Maryland, in 1888 and emerged as the *John Q. Ferguson,* still a two-masted schooner with fore and main topmasts. Sometime after this she was transformed into a three-masted schooner, but her dimensions remained the same—105 feet in length. About 1919 she was renamed *Virginia* and was an attractive little schooner with jibboom but no topmasts. But by 1938 she had declined to the state shown in this photograph, featuring an awkward-looking bowsprit with slack stays and bobstay chains as she lay alongside a barge at the foot of Andre Street, Baltimore.

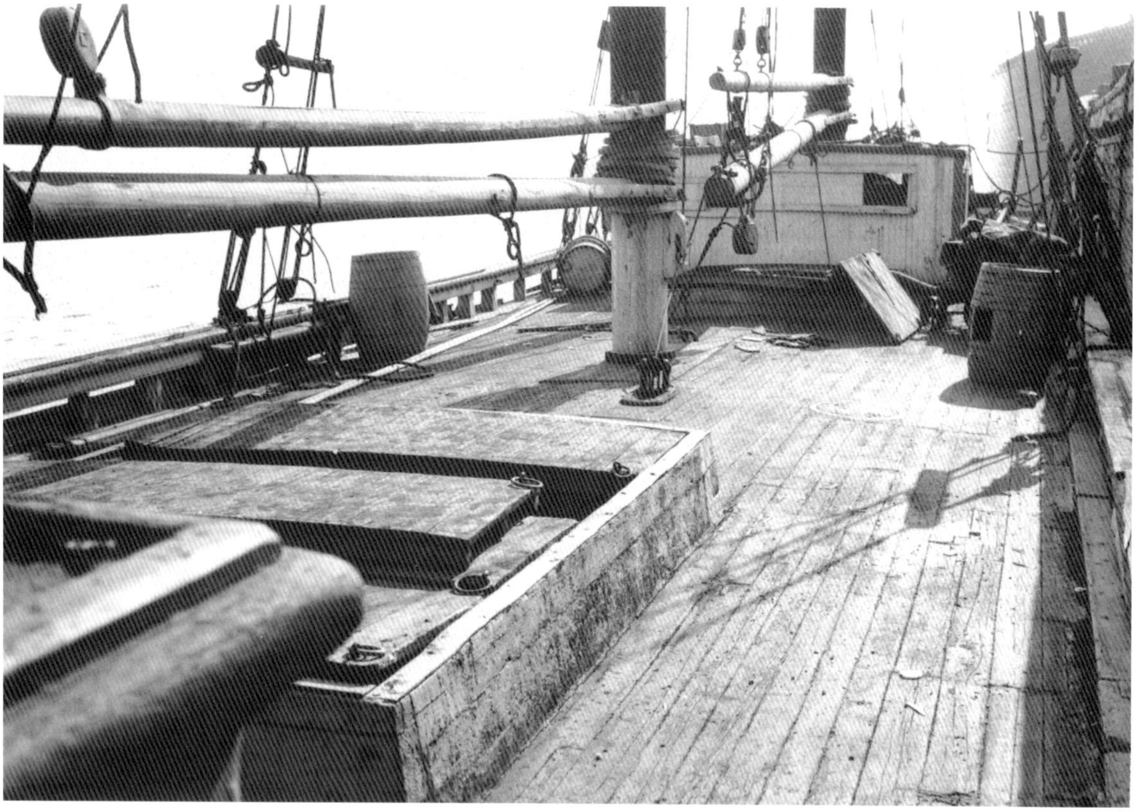

Looking forward on the deck of the *Virginia,* spars stripped of their sails and a look of abandonment about her.

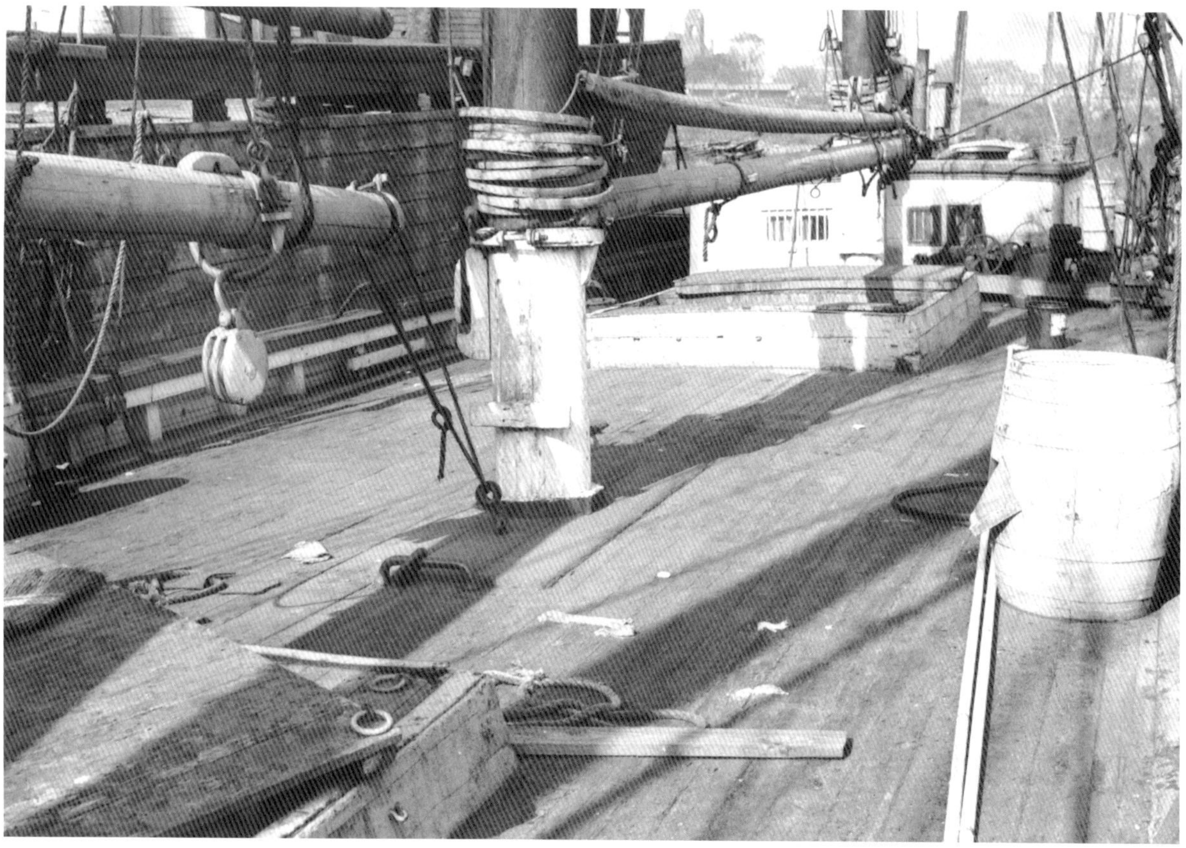

Looking aft along the deck of the *Virginia* at an after house that seems out of proportion. At right is an upended water barrel, and her deck is strewn with litter. The schooner never sailed again after this picture was made in December 1938. The next year *MVUS* recorded her as "Abandoned" as she was broken up for firewood not far from where she lay at that time.

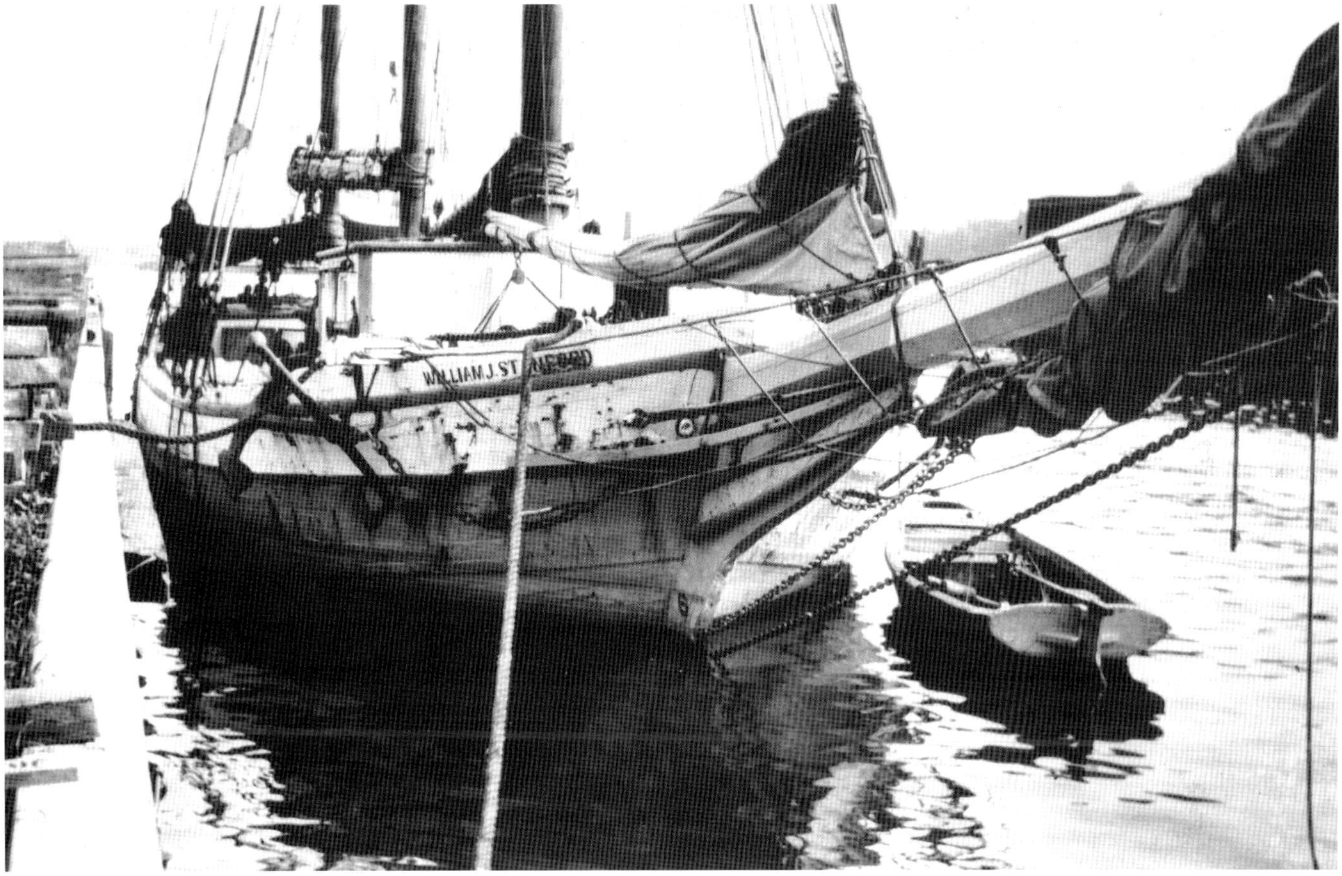

The smallest three-masted schooner on the Chesapeake in the last days of sail was the *William J. Stanford,* measuring only 89 feet in length. She had been built in 1868 as the two-masted schooner *John P. Connor* at Barker's Landing, Delaware, and then measuring 68 feet in length. In 1887 she was lengthened to 89 feet and was classified in the register as a tern, or three-masted schooner. About 1903 she was renamed *Priscilla* and registered out of Baltimore. About 1930 the schooner was purchased by Captain William J. Stanford of Colonial Beach, Virginia. He had her rebuilt at A. Smith & Sons Shipyard, Curtis Bay, Baltimore, that year and renamed the vessel after himself. Unfortunately, with the Depression's effect on Bay freighting Captain Stanford could not keep the schooner employed. Unable to pay the shipyard's bill, in the middle of 1931 he turned her over to the yard. Here the *Stanford* is tied up at Pier 6, Pratt Street, Baltimore, on July 10, 1934. Her yawl boat, at right, looks diminutive but had the power to push the schooner.

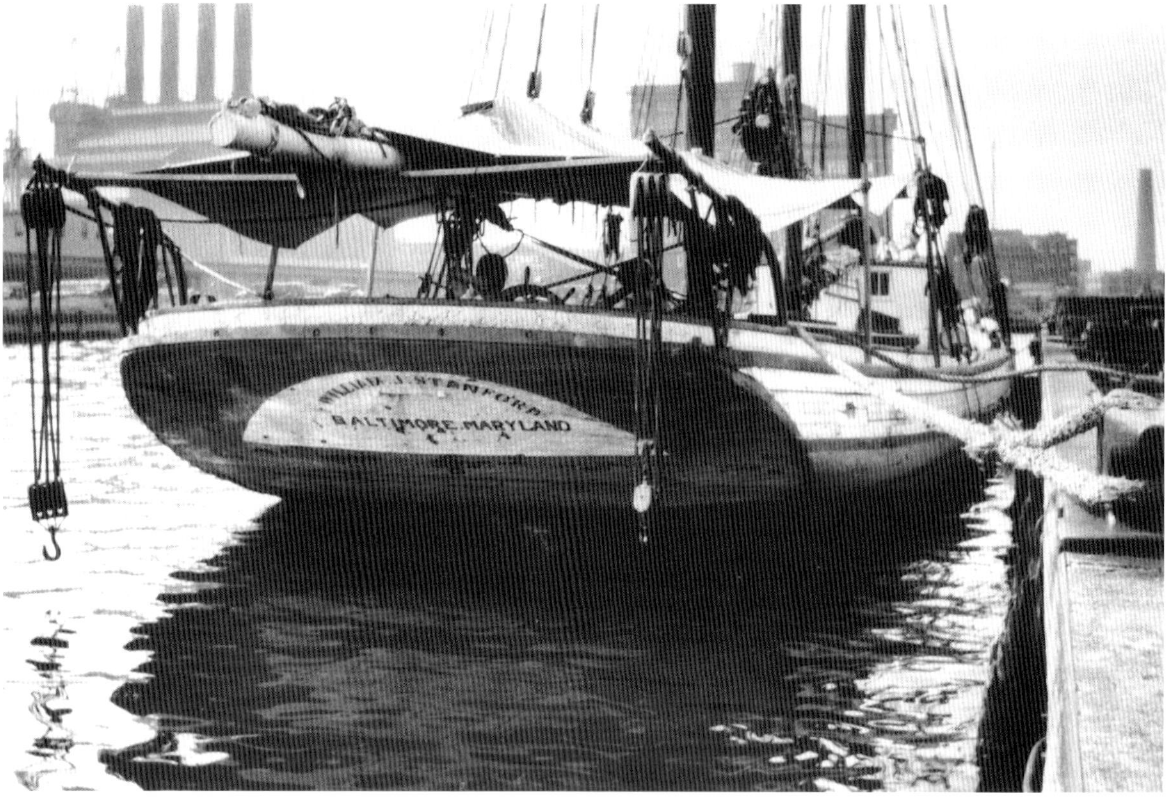

A tarpaulin stretched over the mizzen boom of the *Stanford* shades members of the captain's family as they look over a portion of the Baltimore skyline, a contrast to the rural area where the schooner loaded her lumber cargo. Familiar landmarks in the background are, at left, the generating station of the Consolidated Gas and Electric Company on Pratt Street; center, the Candler Building on the same street; and, at far right, the Shot Tower on Fayette Street.

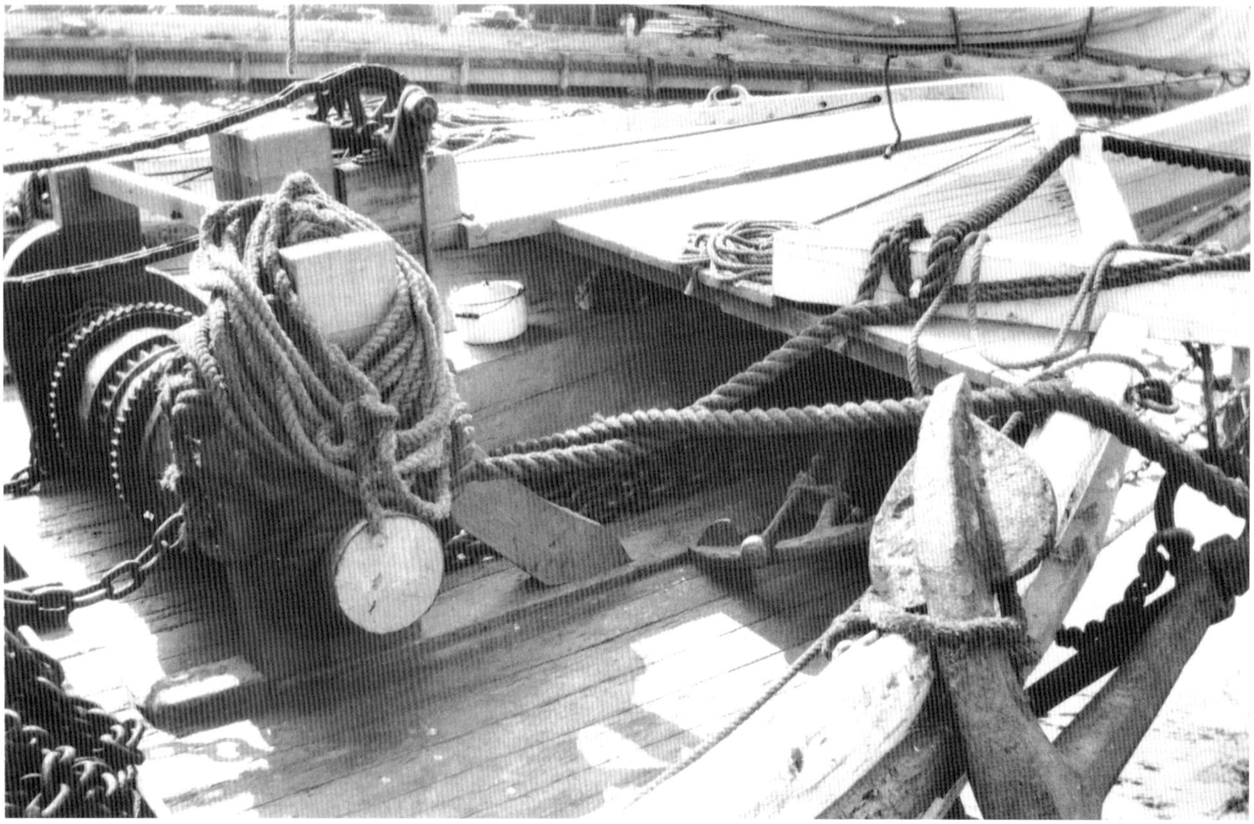

✪ The foredeck of the *William J. Stanford,* probably at Baltimore on July 10, 1934.

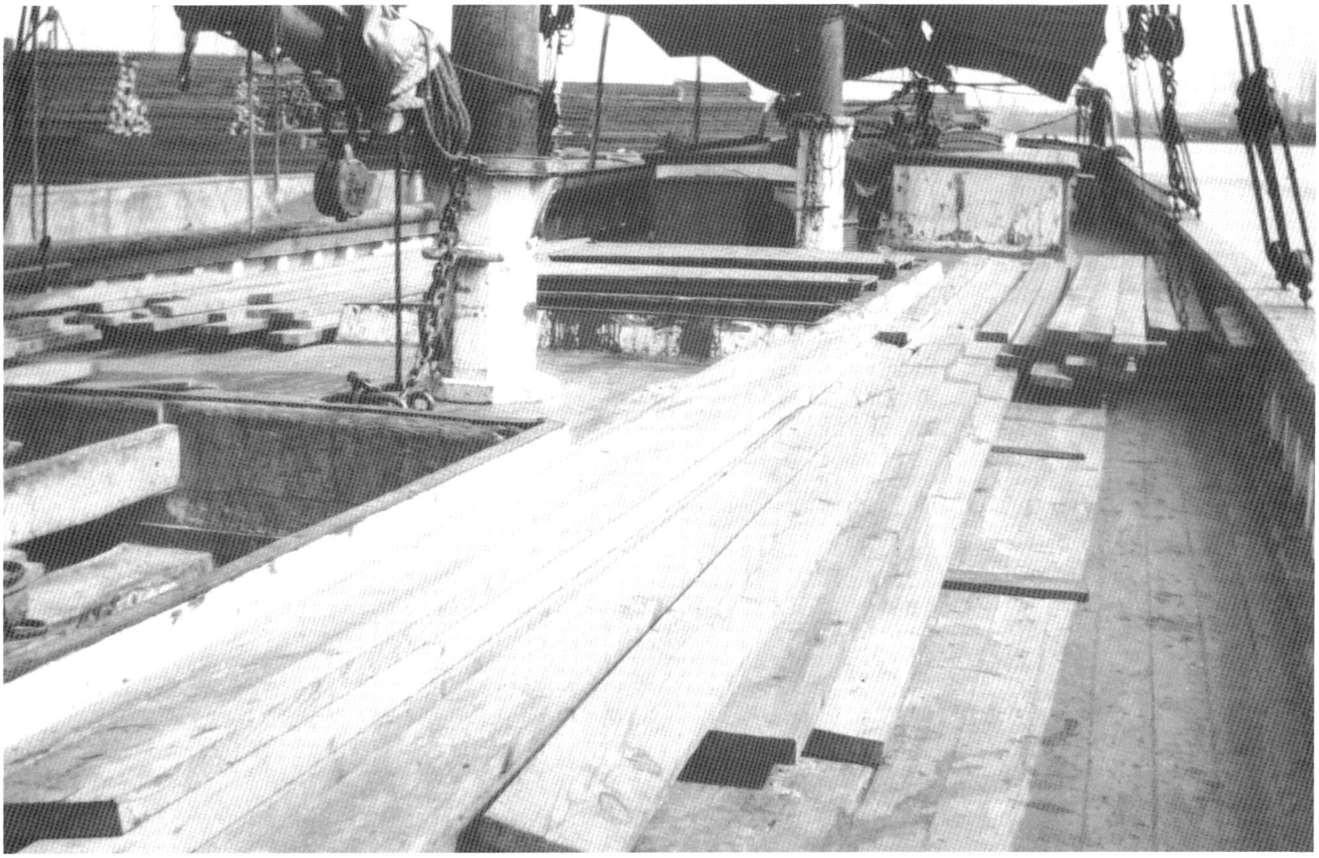

Looking aft on the deck of the *Stanford.*

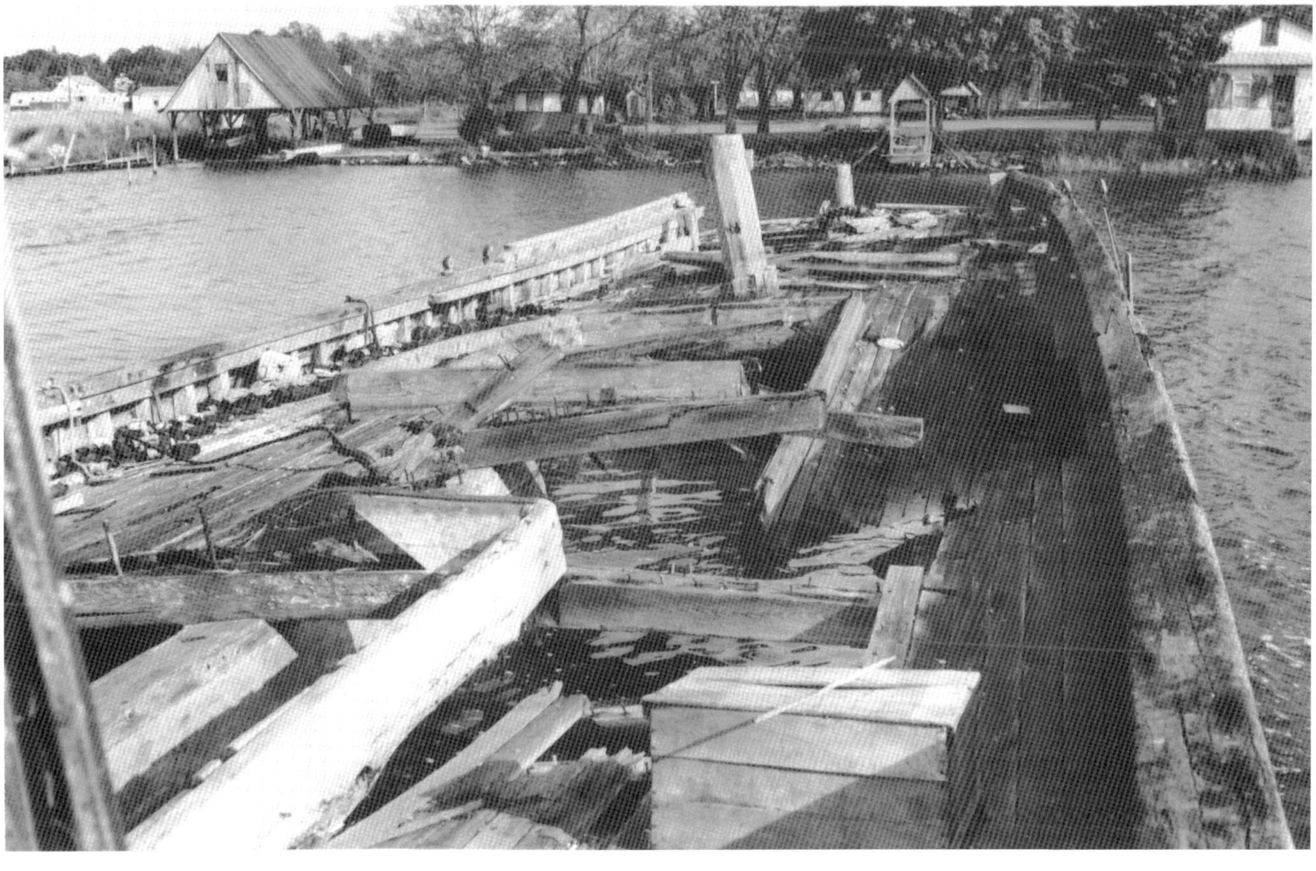

✪ Looking at the same scene on the ruined *Stanford,* at Rock Hall, Maryland, on October 17, 1969.

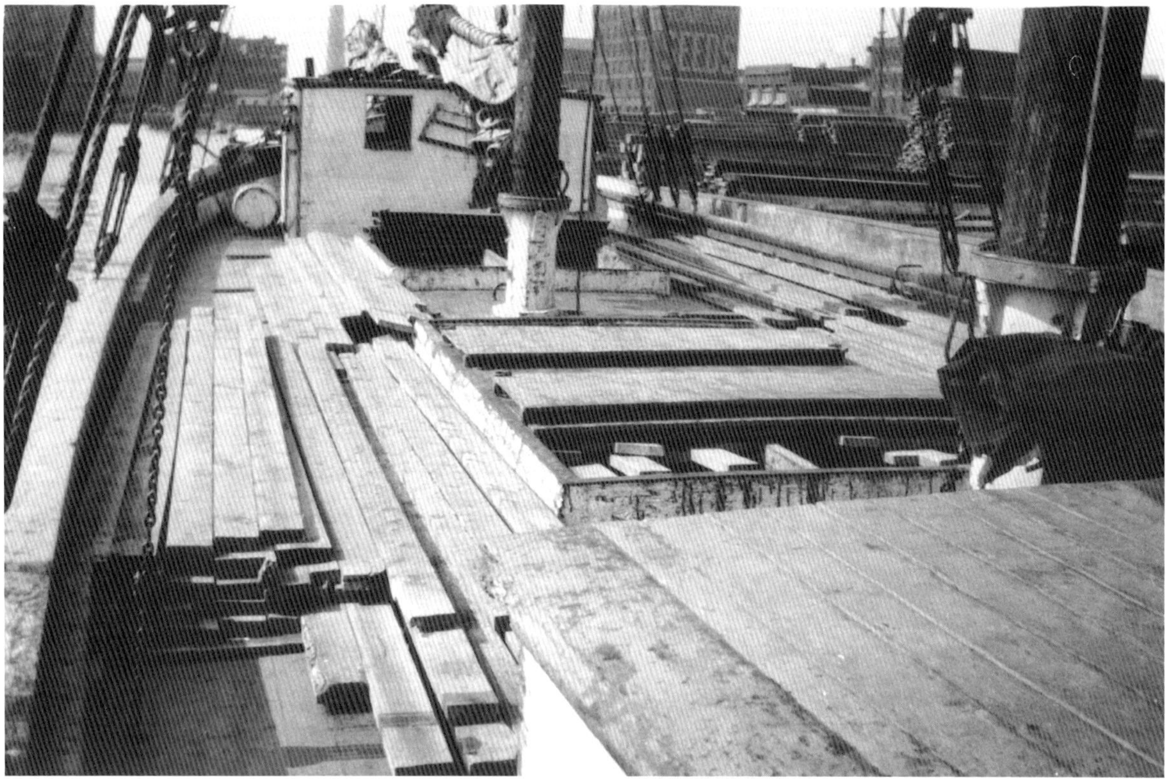

The hold of the *Stanford* still has a quantity of lumber stowed in it, as can be seen in the main hatch. The saw hanging on the after side of the forward house was almost a standard fixture on Bay sailing craft for sawing up firewood for the cookstove there.

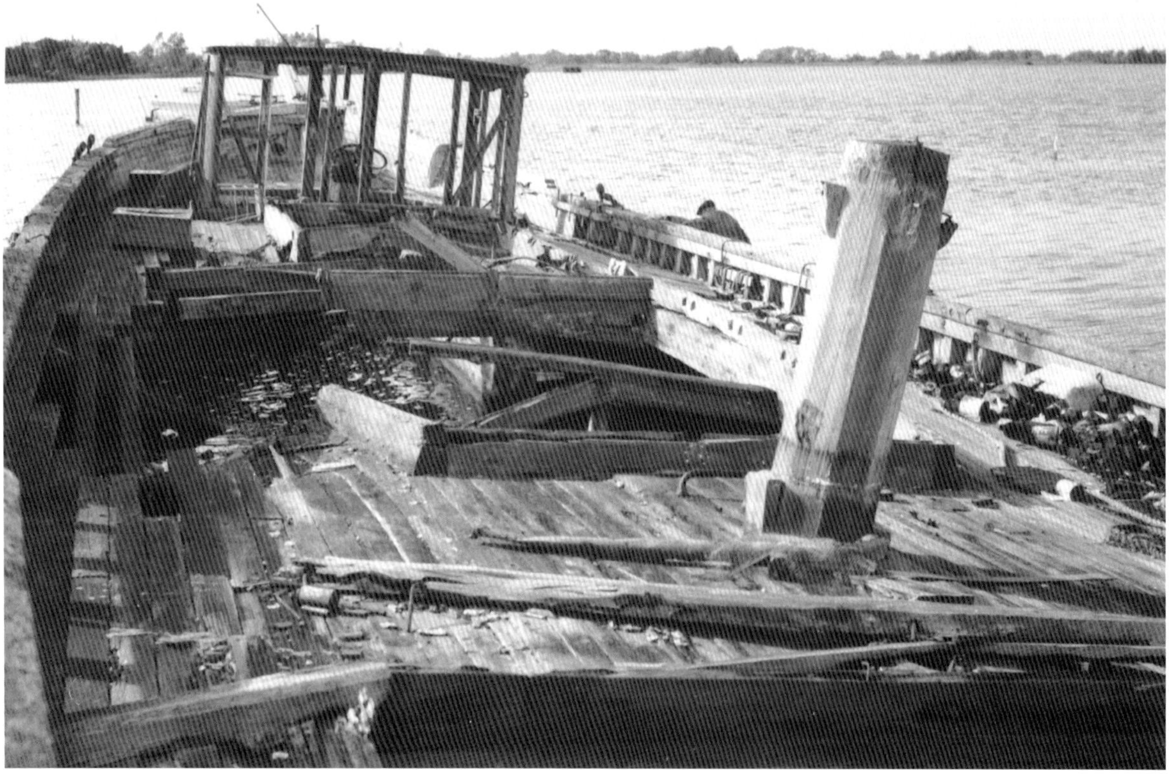

In October 1959, twenty-five years after the previous deck scene was taken, from an identical position, the total devastation of the *William J. Stanford* is evident. Rusting cans and other debris litter that portion of the deck that has not collapsed. The forward house has been stripped of its tongue-and-groove siding, leaving the upright supports and roof to shelter a long unused pump sitting on the stump of her foremast.

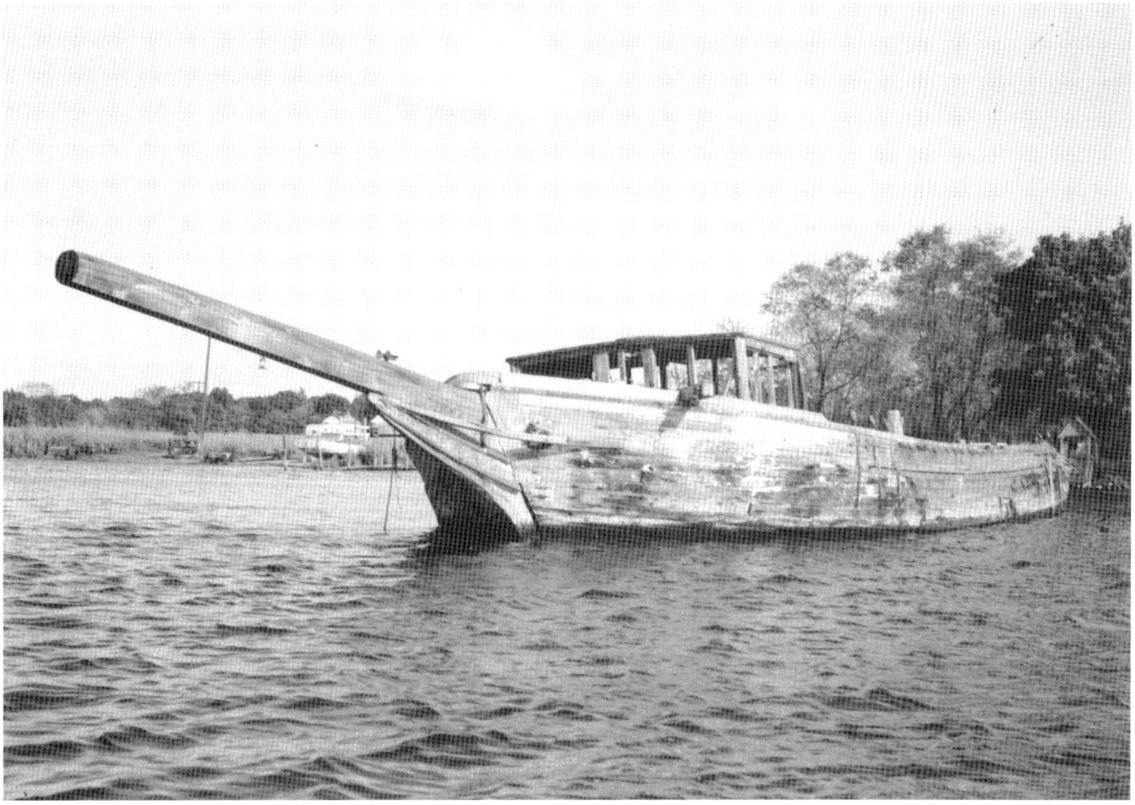

The A. Smith & Sons Shipyard operated the *Stanford* until 1937 and then sold her to Captain Levin P. Kelly, who had owned the schooner *Smith K. Martin.* In 1942, after Captain Kelly's death, the *Stanford* was sold to the Smith brothers of Galesville, Maryland. The 1948 *MVUS* lists her as "Abandoned" and her final resting place was in Swan Creek, near Rock Hall, Maryland. She appeared like this on October 25, 1959.

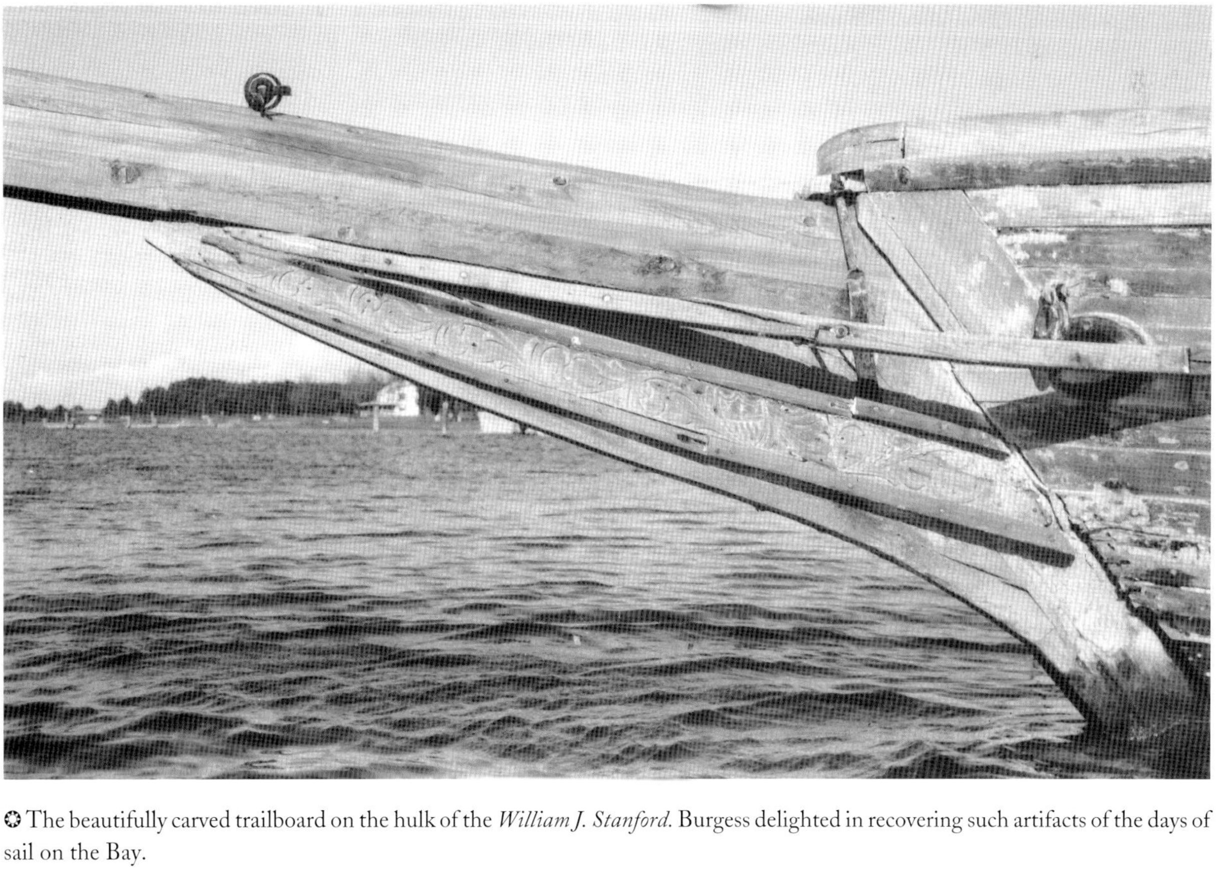

✪ The beautifully carved trailboard on the hulk of the *William J. Stanford.* Burgess delighted in recovering such artifacts of the days of sail on the Bay.

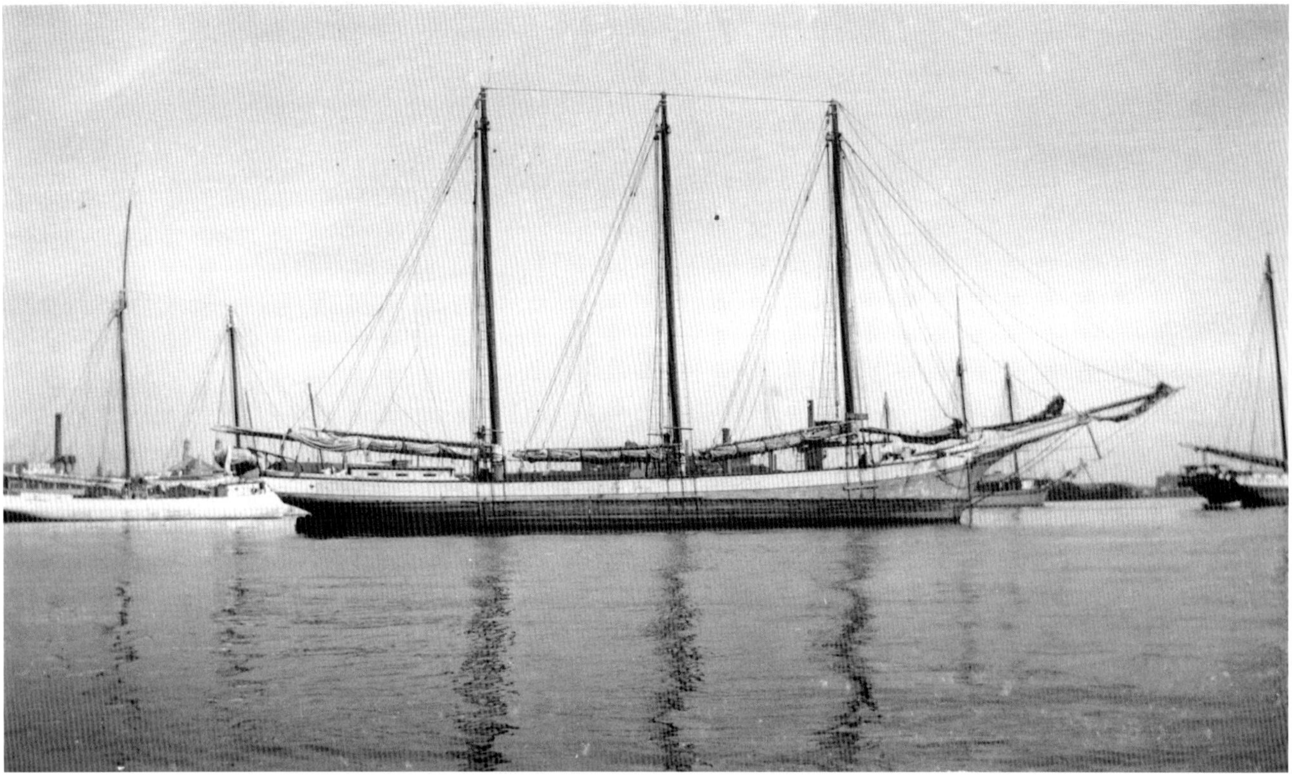

The schooner *William Linthicum* was built in 1894 at Church Creek, Maryland, and on October 29, 1933, while anchored in Canton Hollow, Baltimore, was still in fine condition although lacking her topmasts. Here the *Linthicum* shares the anchorage with, left to right, the schooner *Harvey H. Conway,* bugeye *Russell De Hughlett,* schooners *Minnie T. Phillips* and *Robert McClintock,* and ram *Levin J. Marvel.*

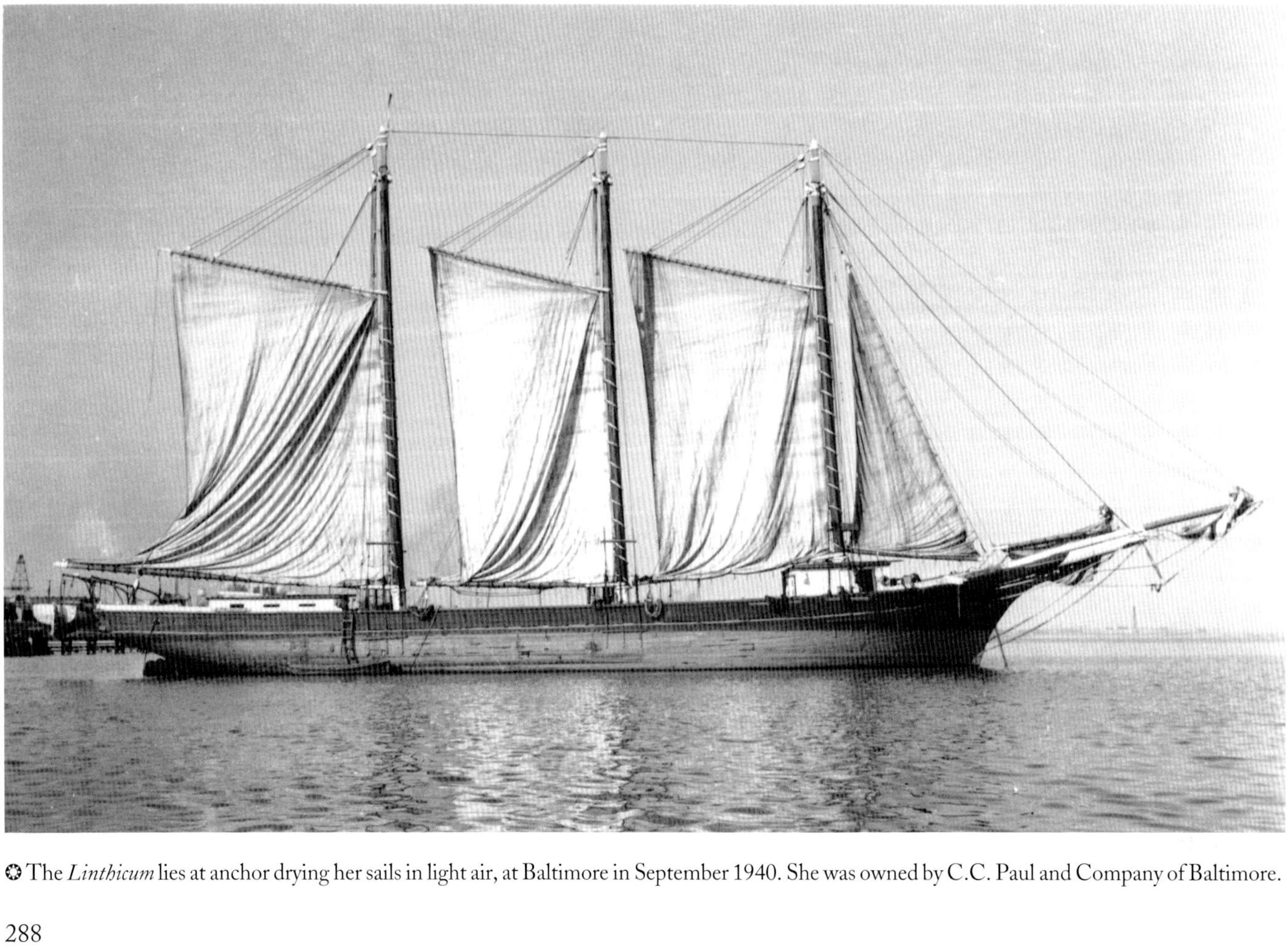

✪ The *Linthicum* lies at anchor drying her sails in light air, at Baltimore in September 1940. She was owned by C.C. Paul and Company of Baltimore.

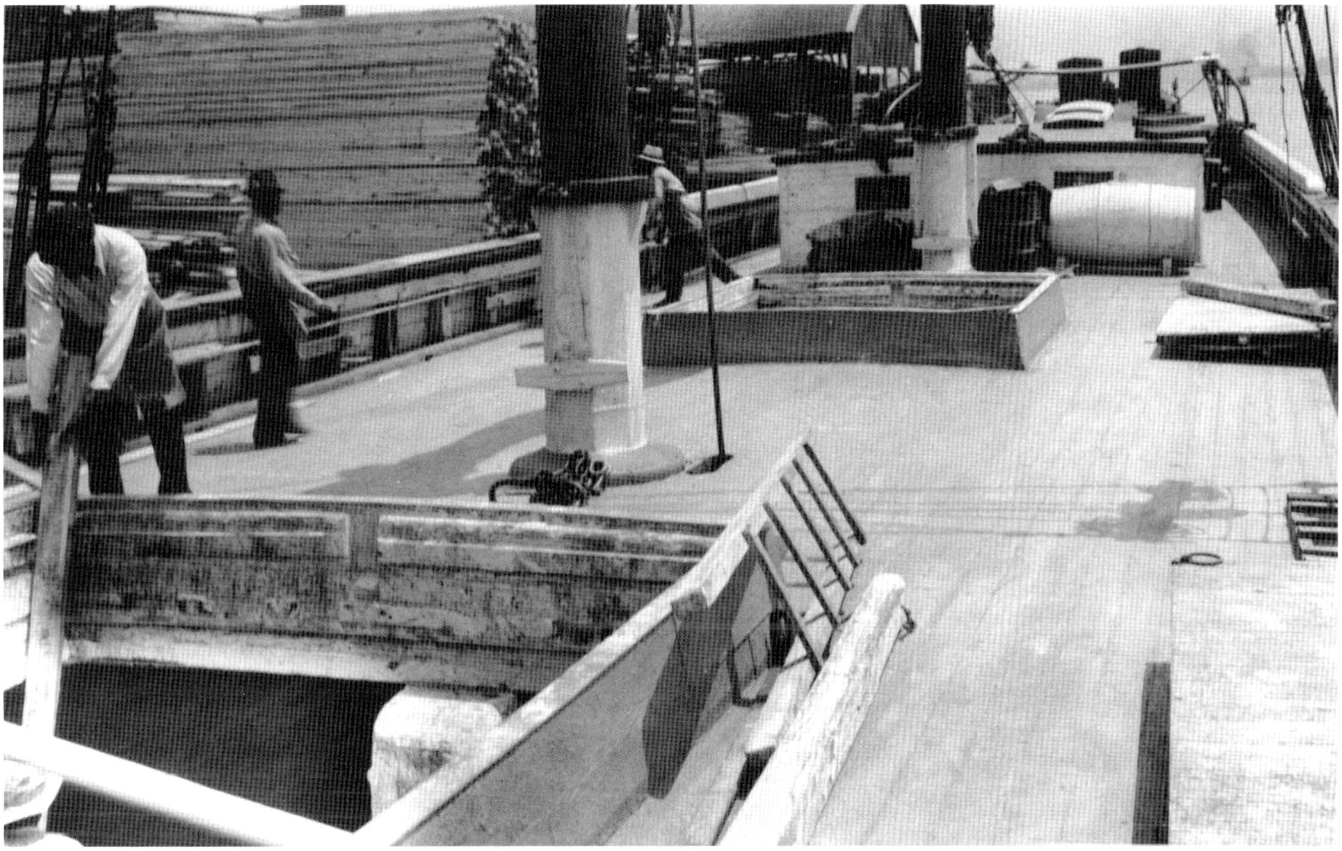

The *Linthicum* discharges lumber at the William D. Gill Lumber Company, Philpot Street, Baltimore, on June 26, 1936. Throughout this volume numerous deck scenes have appeared, but never has a hatch cover been shown placed upside down on deck. It was a tradition of the sea that doing so would bring the vessel bad luck. That she didn't need, for within a few years after this view she was to sail deep water again. In December 1940, it was announced that the *Linthicum* had been sold by C.C. Paul and Company to William E. Meyer and Company, Ltd., of St. Georges, Bermuda. After a refit at the Redman and Vane Shipyard, she took on a cargo of lumber and, in command of Captain William M. Martino, sailed from Baltimore on December 27. At Bermuda she was stripped of her masts and converted into a barge.

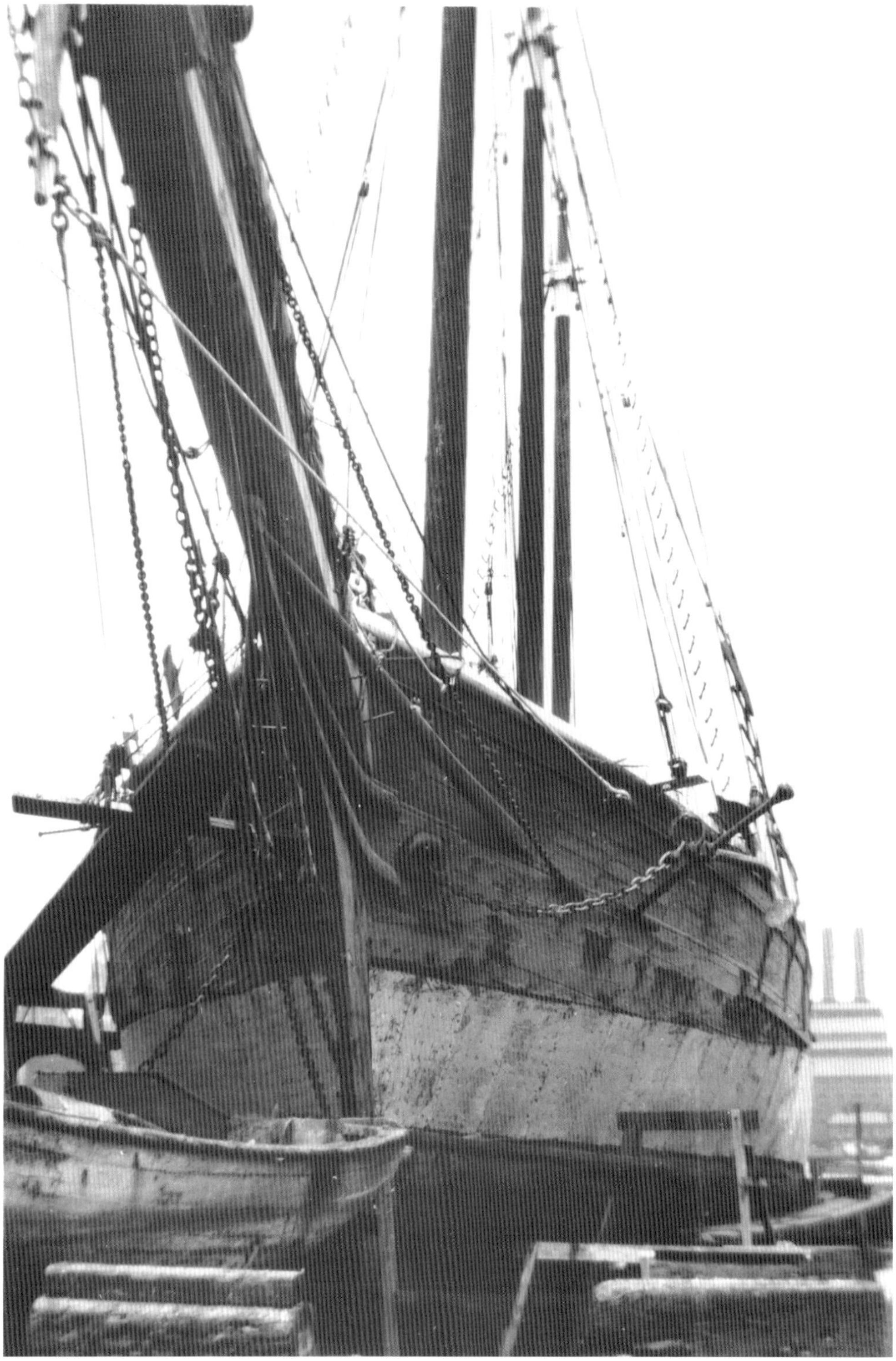

Not until 1917 did the schooner *William T. Parker* appear in the register as hailing from Baltimore. From the time of her building by C.C. Davidson at Milton, Delaware, in 1891, to her transfer to the Maryland port, she had been owned in Milton, with Wilmington as her home port. While sailing out of Delaware Bay she stranded on the point of Cape Henlopen, Delaware, on August 27, 1899, while bound from Bogue Inlet, North Carolina, to Patchogue, New York. She lost her deckload and, before arriving at a port of refuge after being released from her dangerous position, capsized twice. Then, on September 10, 1908, she was disabled in a gale off Ocracoke, North Carolina, but surfmen piloted her over the shoals to safety in Pamlico Sound. In 1915 she was reported to have been abandoned off the North Carolina coast and left to drift unmanned. She was eventually towed into port and put back into service. Not long after that incident, she came to the Chesapeake and spent most of her time in sheltered waters. On June 24, 1934, she was hauled out at the Chesapeake Marine Railway on Key Highway, Baltimore, where she was remetalled along the waterline and repainted.

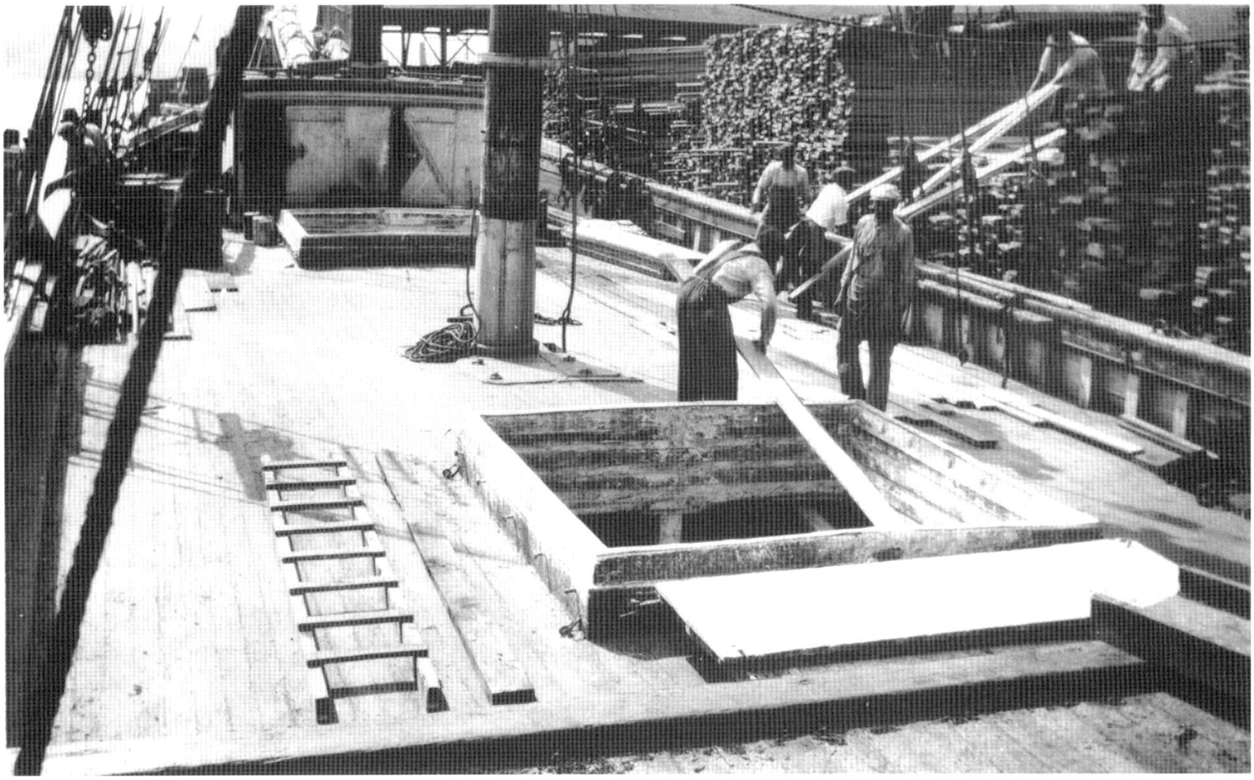

With each man handling two boards at a time, the cargo of the *William T. Parker* is slowly unloaded at Philpot Street, Baltimore, on May 19, 1934. This is how the billions of feet of lumber brought to the major Bay ports by sailing craft were discharged. By the 1940s trailer trucks, with lumber loaded onto pallets handled by forklifts, took over this trade to a great extent.

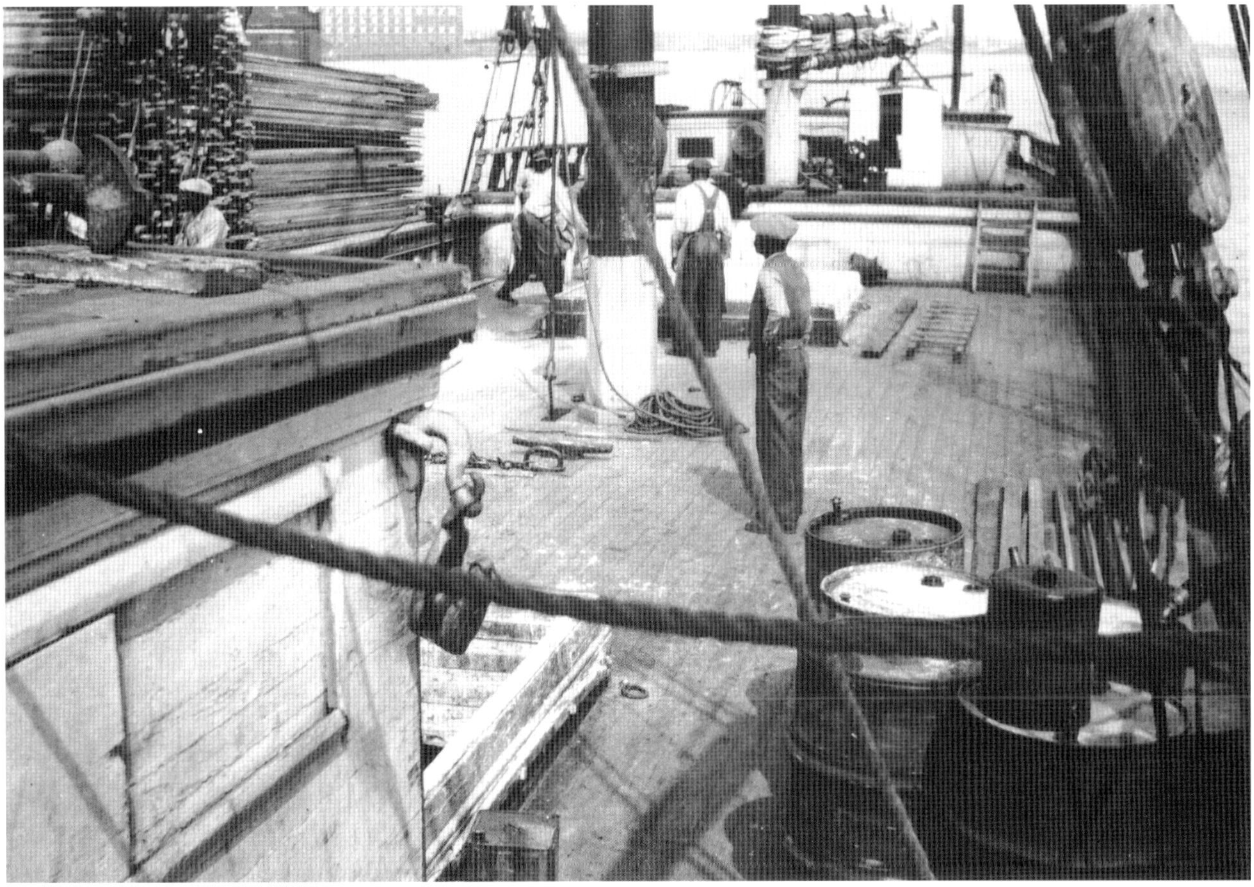

⊛ A deck view of the *Parker* unloading at Baltimore on May 19, 1934, looking aft.

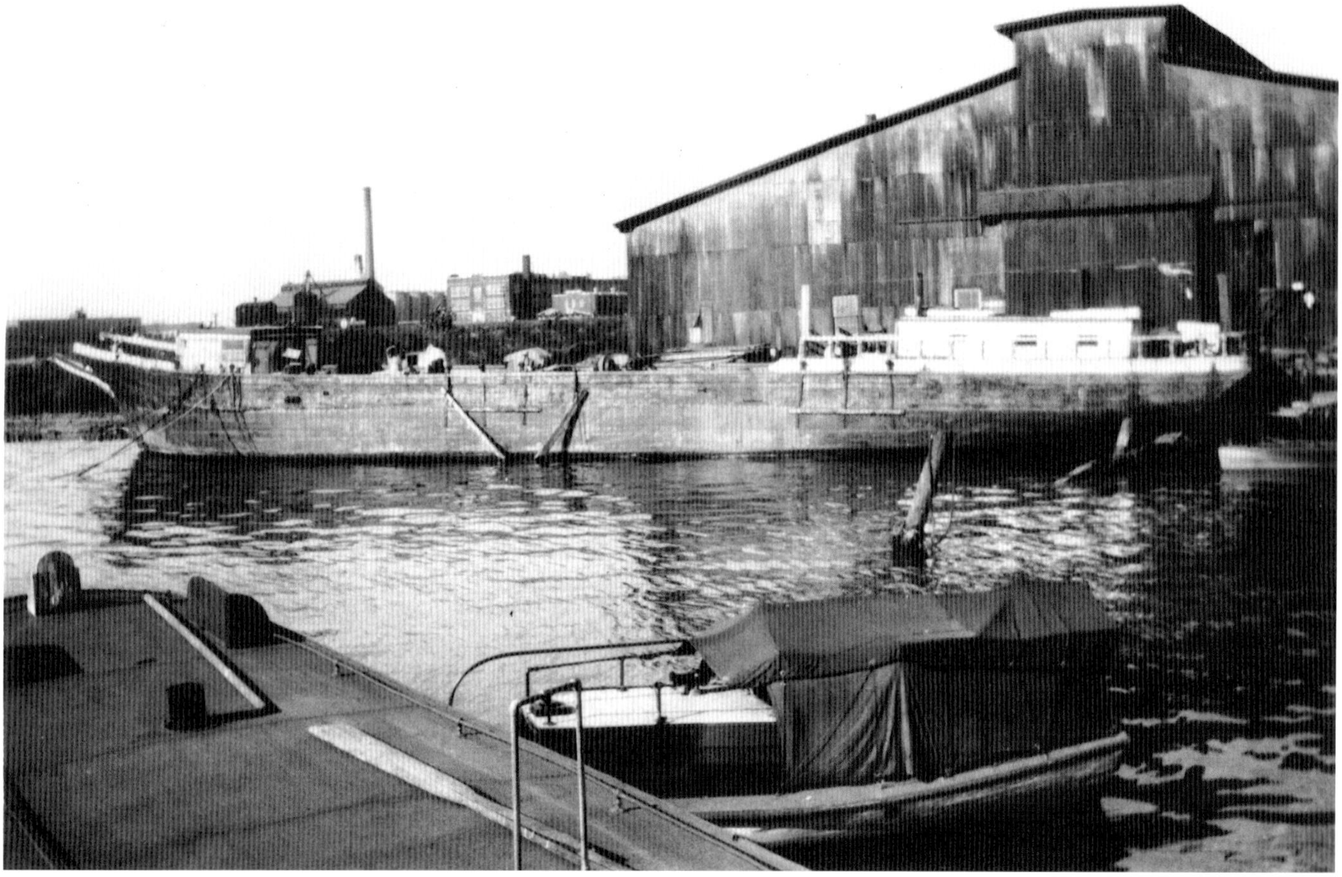

On the night of February 24, 1935, while bound up the Chesapeake with lumber from North Carolina for Baltimore, the *Parker* was rammed by the steamer *Commercial Bostonian* off Bloody Point. The schooner lost her jibboom and all her head gear and put into Annapolis before proceeding up the Bay. After cargo was discharged, her damage was surveyed and it was determined that repairs would be too costly, so she was dismantled. Her bare hull was moved to Locust Point, Baltimore, where a family moved aboard, reconditioned her main cabin for living quarters, and used her deck as a workshop.

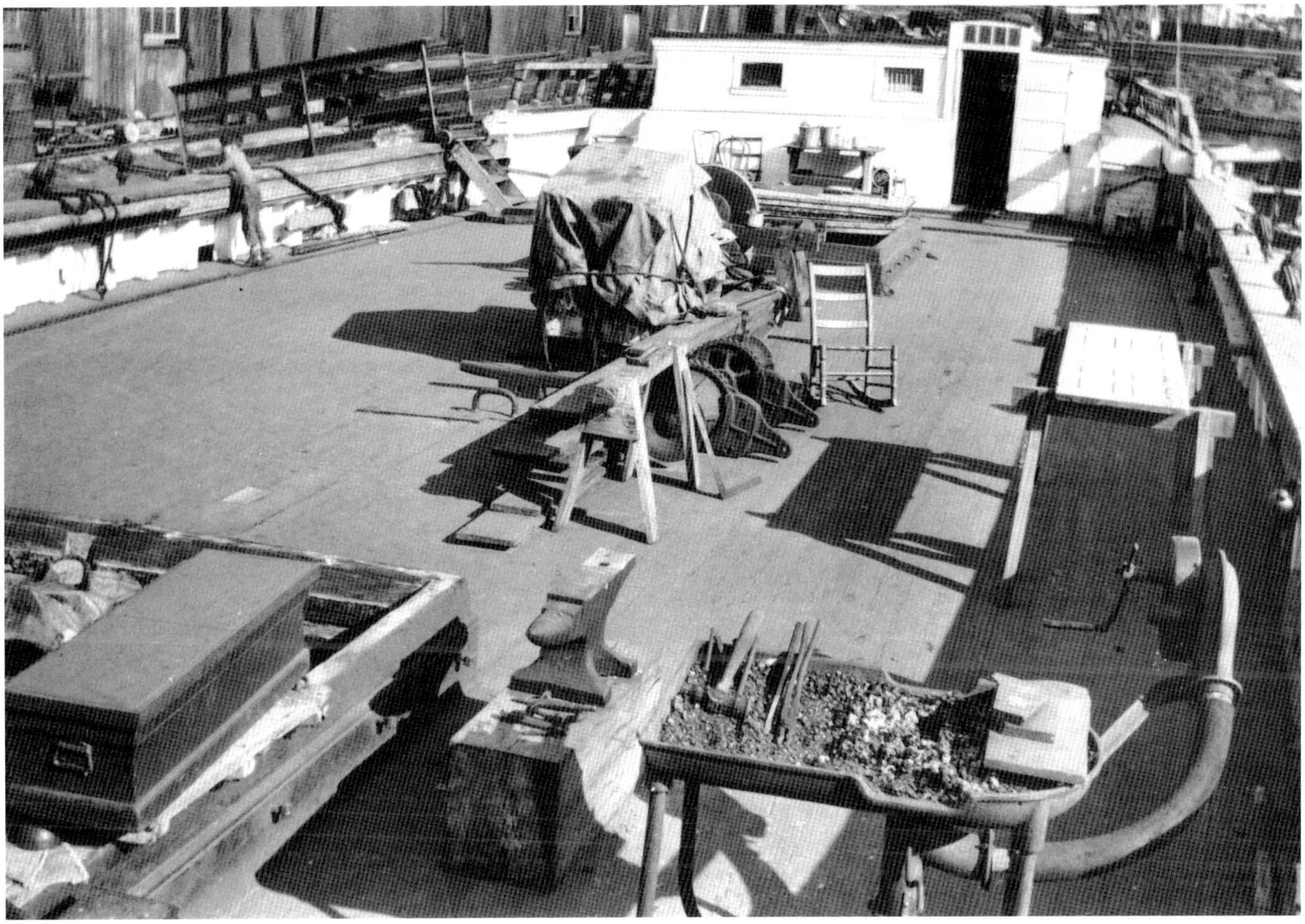

Eight months after her collision the *Parker* appeared as in this illustration, looking aft on deck. In the lower foreground are a blacksmith forge and anvil, with tool chest on her fore hatch. Amidships is the schooner's windlass, probably destined for the scrap heap, and abaft that is a large grindstone. Her quarterdeck has been cut back flush with the after cabin and a full-height door has replaced her companionway. The old hulk made a comfortable home at that site for seven years until she was moved to a ship graveyard in Curtis Creek.

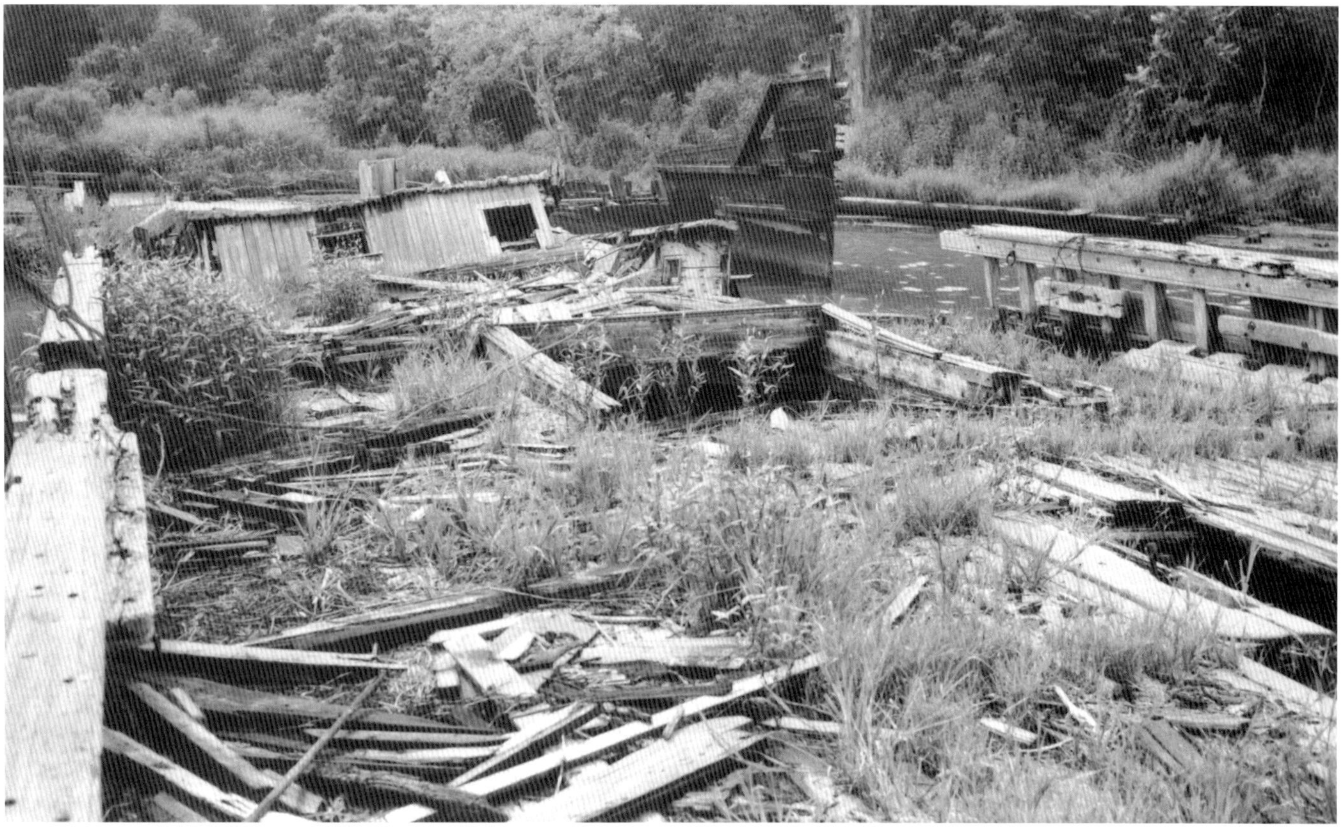

In the graveyard the *Parker* had her keel resting on the creek's bottom but another family moved aboard and lived there until she rotted away beneath them in the 1950s. On August 29, 1960, looking aft on the deck of the old hulk, this scene was presented, with grass and weeds sprouting from her deck seams.

Index

Abbreviations used in this index to indicate types of vessels:

bug.	Bugeye
C.G. ctr.	Coast Guard cutter
dsl. frtr.	diesel freighter
pb.	powerboat
pgy.	pungy
sch.	schooner
skip.	skipjack
slp.	sloop
stmbt.	steamboat
stmr.	steamer

More books on Sailing Craft and the Chesapeake Region from Tidewater Publishers

Annapolis: A Walk Through History
Elizabeth B. Anderson
ISBN 10: 0-87033-546-4. ISBN 13: 978-0-87033-546-4.

Annapolis Vignettes
Ginger Doyel
ISBN 10: 0-87033-571-5. ISBN 13: 978-0-87033-571-6.

Bodine's Chesapeake Bay Country
A. Aubrey Bodine
Edited by Jennifer B. Bodine
ISBN 10: 0-87033-562-6. ISBN 13: 978-0-87033-562-4.

Chesapeake Bay Buyboats
Larry S. Chowning
ISBN 10: 0-87033-553-7. ISBN 13: 978-0-87033-553-2.

Chesapeake Bay in the Civil War
Eric Mills
ISBN 10: 0-87033-479-4. ISBN 13: 978-0-87033-479-5.

Chesapeake Bay Log Canoes & Bugeyes
M. V. Brewington
ISBN 10: 0-87033-011-X. ISBN 13: 978-0-87033-011-7.

Chesapeake Bay Schooners
Quentin Snediker and Ann Jensen
ISBN 10: 0-87033-508-1. ISBN 13: 978-0-87033-508-2.

Chesapeake Bay Skipjacks
Pat Vojtech
ISBN 10: 0-87033-451-4. ISBN 13: 978-0-87033-451-1.

Chesapeake Rumrunners of the Roaring Twenties
Eric Mills
ISBN 10: 0-87033-518-9. ISBN 13: 978-0-87033-518-1.

Chesapeake Sails: A History of Yachting on the Bay
Richard Henderson
ISBN 10: 0-87033-514-6. ISBN 13: 978-0-87033-514-3.

Chesapeake Steamboats: Vanished Fleet
David C. Holly
ISBN 10: 0-87033-455-7. ISBN 13: 978-0-87033-455-9.

Dancing with the Tide: Watermen of the Chesapeake
Mick Blackistone
ISBN 10: 0-87033-532-4. ISBN 13: 978-0-87033-532-7.

Exploring the Chesapeake in Small Boats
John Page Williams, Jr.
ISBN 10: 0-87033-429-8. ISBN 13: 978-0-87033-429-0.

Fiberglass Repair: Polyester or Epoxy
David and Zora Aiken
ISBN 10: 0-87033-567-7. ISBN 13: 978-0-87033-567-9.

Fisherman's Knots & Nets
Rauol Graumont and Elmer Wenstrom
ISBN 10: 0-87033-024-1. ISBN 13: 978-0-87033-024-7.

Ghost Fleet of Mallows Bay: And Other Tales of the Lost Chesapeake
Donald G. Shomette
ISBN 10: 0-87033-480-8. ISBN 13: 978-0-87033-480-1.

Harvesting the Chesapeake: Tools and Traditions
Larry S. Chowning
ISBN 10: 0-87033-469-7. ISBN 13: 978-0-87033-469-6.

Lighting the Bay: Tales of Chesapeake Lighthouses
Pat Vojtech
ISBN 10: 0-87033-466-2. ISBN 978-0-87033-466-5.

Rivers of the Eastern Shore: Seventeen Maryland Rivers
Hulbert Footner
Illustrated by Aaron Sopher
ISBN 10: 0-87033-092-6. ISBN 13: 978-0-87033-092-6.

Rudow's Guide to Fishing the Chesapeake
Lenny Rudow
ISBN 10: 0-87033-568-5. ISBN 13: 978-0-87033-568-6.

Sequoia: *Presidential Yacht*
Giles M. Kelly, with photographs by Ann Stevens
ISBN 10: 0-87033-561-8. ISBN 13: 978-0-87033-561-7.

Shipwrecks on the Chesapeake:
Maritime Disasters on Chesapeake Bay and Its Tributaries, 1608–1978
Donald G. Shomette
ISBN 10: 0-87033-283-X. ISBN 13: 978-0-87033-283-8.

Tidewater Time Capsule: History Beneath the Patuxent
Donald G. Shomette
ISBN 10: 0-87033-463-8. ISBN 13: 978-0-87033-463-4.

Tidewater Triumph:
The Development and Worldwide Success of the Chesapeake Bay Pilot Schooner
Geoffrey M. Footner
ISBN 10: 0-87033-511-1. ISBN 13: 978-0-87033-511-2.

Twilight on the Bay: The Excursion Boat Empire of B. B. Wills
Brian J. Cudahy
ISBN 10: 0-87033-509-X. ISBN 13: 978-0-87033-509-9.

The Watermen of the Chesapeake Bay
John Hurt Whitehead III
ISBN 10: 0-87033-374-7. ISBN 13: 978-0-87033-374-3.